BASIC COLLEGE MATHEMATICS

D. FRANKLIN WRIGHT
CERRITOS COLLEGE

Editor: Cynthia Ellison
Development Director: Marcel Prevuznak
Production Editors: Harding Brumby, Phillip Bushkar, Mandy Glover
Bethany Loftis, Nina Miller, Kim Scott
Editorial Assistants: Kelly Epperson, D. Kanthi, B. Syam Prasad, Chris Yount
Layout: QSI (Pvt.) Ltd.: U. Nagesh, E. Jeevan Kumar
Art: Ayvin Samonte

HAWKES
LEARNING
SYSTEMS

A division of Quant Systems, Inc.

1023 Wappoo Rd. Suite A-6/Charleston, SC 29407
www.hawkeslearning.com

Library of Congress Control Number: 2006920801

Printed in the United States of America.

Student Softcover
978-1-932628-19-7
1-932628-19-3

Student Hardcover
978-1-932628-15-9
1-932628-15-0

Student Softcover Bundle
978-1-932628-20-3
1-932628-20-7

Student Hardcover Bundle
978-1-932628-16-6
1-932628-16-9

Contents

Chapter 3 Fractions 203

Chapter 4 Mixed Numbers 295

Chapter 5 Decimal Numbers 367

Chapter 6 Ratios and Proportions 487

Chapter 7 Percent 565

Chapter 8 Consumer Applications 651

Chapter 9 Introduction to Algebra 731

PREFACE

Purpose and Style

The purpose of *Basic College Mathematics*, Eighth Edition, is to provide students with a learning tool that will help them

1. develop basic arithmetic skills,
2. develop reasoning and problem-solving skills, and
3. achieve satisfaction in learning so that they will be encouraged to continue their education in mathematics.

The writing style gives carefully worded, thorough explanations that are direct, easy to understand, and mathematically accurate. The use of color, subheadings, and shaded boxes helps students understand and reference important topics.

The emphasis is on *why* basic operations and procedures work as they do, as well as on *how* to perform these operations and procedures. Point-by-point explanations are incorporated within the examples for better understanding, and directions are given in an easy-to-follow format. There are margin exercises that correspond to each example in almost every section. Practice problems (with answers) appear before the main exercises in every section to reinforce students' understanding of the concepts in that section and to provide the instructor with immediate classroom feedback.

The NCTM and AMATYC curriculum standards have been taken into consideration in the development of the topics throughout the text.

Enhancements to the New Edition

- The appearance of the Eighth Edition has been enhanced by a new, full color design.
- High quality images and drawings have been added throughout the text to illustrate mathematical concepts and create a relaxed atmosphere inviting to the reader.
- Real life application problems have been updated.
- The use of calculators is encouraged starting with Chapter 5.
- The new worktext edition gives students enough room to work out each problem.
- The objectives are clearly pointed out using arrows throughout each section.
- Completion examples and margin exercises help the student directly apply what is being taught in the section and examples.
- Geometric concepts such as recognizing geometric figures, and finding perimeter and area are integrated throughout the text.

Content

There is sufficient material for a three- or four-semester-hour course. The topics in Chapters 1 – 7 form the core material for a three-hour course. The topics in Chapters 8 and 9 provide additional flexibility in the course depending on students' background and the goals of the course.

CHAPTER 1, WHOLE NUMBERS, reviews the fundamental operations of addition, subtraction, multiplication, and division with whole numbers. Estimation is used to develop better understanding of whole number concepts, and word problems help to reinforce the need for these ideas and skills in common situations such as finding averages and making purchases. The last section of the chapter introduces the basic geometric concepts of plane geometry, polygons, length and perimeter, and area.

CHAPTER 2, EXPONENTS, PRIME NUMBERS, & LCM, introduces exponents, shows how to use the rules for order of operations, and defines prime numbers. The concepts of divisibility and factors are emphasized and related to finding prime factorizations, which are used to develop skills needed for finding the least common multiple (LCM) of a set of numbers. All of these ideas form the foundation of the development of fractions in Chapter 3.

CHAPTER 3, FRACTIONS, discusses the operations of multiplication, division, addition, and subtraction with fractions. A special effort is made to demonstrate the validity of the use of improper fractions, and exponents and the rules for order of operations are applied to fractions. Knowledge of prime numbers (and prime factorizations) underlies all of the discussions about fractions.

CHAPTER 4, MIXED NUMBERS, shows how mixed numbers are related to whole numbers and fractions. Topics include relating improper fractions and mixed numbers, the basic operations, the rules for order of operations, and denominate numbers.

CHAPTER 5, DECIMAL NUMBERS, covers the basic operations with decimal numbers, estimating and the use of calculators. To emphasize number concepts, there is a section on operating with decimal numbers, fractions, and mixed numbers in a single expression. Square roots and the Pythagorean theorem are covered toward the end of the chapter along with the geometric concepts of circles and volume.

CHAPTER 6, RATIOS AND PROPORTIONS, develops an understanding of ratios and introduces the idea of a variable. Techniques for solving equations are developed through finding the unknown term in a proportion. Angles and triangles are introduced at the end of this chapter.

CHAPTER 7, PERCENT, approaches percent as hundredths and uses this idea to discuss percent of profit and to find equivalent numbers in the form of percents, decimals, and fractions. The applications with percent are developed around proportions, the formula $R \times B = A$, and the skills of solving equations.

CHAPTER 8, CONSUMER APPLICATIONS, addresses the real life applications of simple interest, compound interest, balancing a checking account, buying and owning a car, buying and owning an home, and reading graphs. The use of calculators is encouraged and even necessary, such as in the case of using the formula for compound interest: $A = P(1 + r/n)^{nt}$. Calculating with this formula serves to emphasize the importance of following the rules for order of operations.

CHAPTER 9, INTRODUCTION TO ALGEBRA, provides a head start for those students planning to continue in their mathematics studies, either in a prealgebra course or a beginning course in algebra. Integers are introduced with the aid of number lines and are graphed on number lines. Topics included are absolute value, inequality symbols, order of operations, combining like terms, translating phrases, solving equations, and problem solving.

APPENDIX SECTIONS include five sections on various topics including the metric system, greatest common divisor, statistics, base two and base five, and ancient numeration systems. The first appendix on the metric system covers metric weight and volume and U. S. customary and metric equivalents. Greatest common divisor, statistics, and base two and base five are sections that provide further discussion on topics directly related to the text. The last appendix is an interesting look at ancient forms of different number systems.

Practice and Review

There are more than 4900 margin exercises, practice exercises, lesson exercises, and review items overall. Lesson exercises are carefully chosen and graded, proceeding from easy exercises to more difficult ones. Many sections contain a feature entitled Review Problems which generally has six to ten review exercises from previous chapters. Each chapter includes a Chapter Review, a Chapter Test, and a Cumulative Review (beginning with Chapter 2).

Many sections have exercises entitled Writing and Thinking about Mathematics, Check Your Number Sense, and Collaborative Learning. These exercises are an important part of the text and provide a chance for each student to improve communication skills, develop an understanding of general concepts, and communicate his or her ideas to the instructor. Written responses can be a great help to the instructor in identifying just what students do and do not understand. Many of these questions are designed for the student to investigate ideas other than those presented in the text, with responses that are to be based on each student's own experiences and perceptions. In most cases there is no right answer.

Answers to all margin exercises, many section exercises, all Chapter Review questions, all Chapter Test questions, and all Cumulative Review questions are provided in the back of the book. Answers to Practice Problems are given just below the problems themselves.

Features

What to Expect in Each Chapter

Each chapter opens with a list of sections included in the chapter and a "What to Expect" section which gives a brief description of the contents of the chapter.

1 Whole Numbers

Chapter 1 Whole Numbers

WHAT TO EXPECT IN CHAPTER 1

Chapter 1 provides a review of the basic operations (addition, subtraction, multiplication, and division) with whole numbers. Section 1.1 discusses number systems in general, with an emphasis on reading and writing whole numbers in the decimal system. Addition and subtraction with whole numbers are discussed in section 1.2 and 1.3 and the skills related to rounding and estimating with whole numbers are developed in Section 1.4. Sections 1.5 and 1.6 discuss multiplying and dividing with whole numbers while Section 1.7 (Problem Solving with Whole Numbers) outlines a four-step process for solving word problems, developed by George Pólya, a famous educator and mathematician from Stanford University.

Section 1.8 concludes the chapter with an introduction to geometric concepts, including plane geometry, polygons, length and perimeter, and area.

Chapter 1 Whole Numbers

Mathematics at Work!

We see whole numbers everywhere. People in various businesses, carpenters, teachers, auto mechanics, and airline pilots all deal with whole numbers on a daily basis. An example of using whole numbers in everyday life is deciding how many apples to buy at the grocery store.

As a student, you may be interested in information about colleges and the costs of attending different colleges. The following list contains this type of information about 15 well-known colleges and universities. Annual (yearly) tuition as well as room and board are given to the nearest hundred dollars for the year 2004. (Note: If you want this type of information for any college or university, write to the registrar of the school or check their website.)

			Tuition	
Institution	**Enrollment**	**Room/Board**	**Resident**	**Nonresident**
Univ. of Arizona	36,900	$7100	$4100	$13,100
Boston College	14,500	$10,900	$31,500	$31,500
Univ. of Colorado	29,200	$8000	$5400	$22,800
Florida A & M	13,100	$6000	$3200	$15,200
Purdue Univ.	38,700	$6800	$6500	$19,800
Johns Hopkins	18,200	$9900	$31,600	$31,600
MIT	10,300	$9500	$32,300	$32,300
Ohio State Univ.	51,000	$7300	$7400	$18,000
Princeton Univ.	6800	$8800	$31,500	$31,500
Rice Univ.	4900	$9000	$20,200	$20,200
Univ. of Tennessee	27,800	$6300	$4300	$9000
Univ. of Wisconsin	41,200	$6500	$6200	$21,100

For these universities and colleges:

a. Find the average of the yearly expenses for room and board.

b. Find the average difference between the tuition for the nonresident students and residents.

c. Determine which of these universities and colleges is the most expensive and which is the least expensive (including room and board and tuition) for nonresident students.

d. Determine which of these universities and colleges is the most expensive and which is the least expensive (including room and board and tuition) for resident students.

1.1 Reading and Writing Whole Numbers

The Decimal System

The abstract concept of numbers and the numeration systems used to represent numbers have been important, indeed indispensable, parts of intelligent human development. The Hindu-Arabic system of numeration that we use today was invented about 800 A.D. This system, called the **decimal system** (**deci** means **ten** in Latin), allows us to add, subtract, multiply, and divide faster and more easily than did any of the ancient number systems, such as Egyptian hieroglyphics and the Roman numeral system. The decimal system is more sophisticated than these earlier systems in that it uses a symbol for zero (0), the operation of multiplication, and the concept of place value.

Mathematics at Work!

Each chapter begins with a problem or example of mathematics as it applies to the chapter used in every day life. It is often referred to in an exercise later on in the chapter.

Objectives

Each section has a list of learning objectives that the student should keep in mind when reading through the section. Arrows that correspond to each objective are placed throughout the section so that the objectives are clearly identified.

The symbols used are called **digits**, and for the **whole numbers**, the value of a digit depends on its position to the left of a beginning point, called a **decimal point**. (See Figure 1.1.)

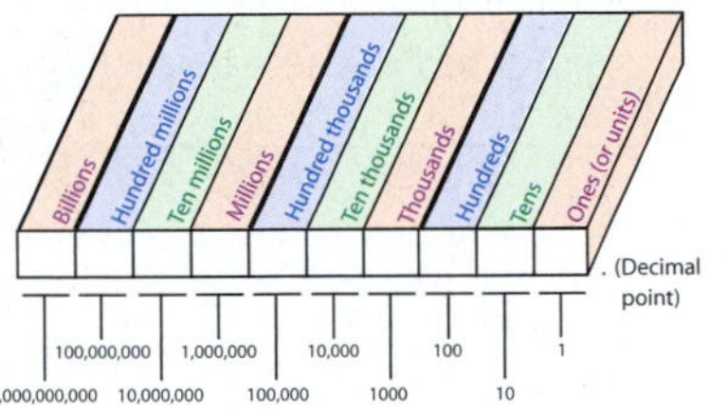

Figure 1.1

Objective 1

Objectives

1. Know the values of the places for whole numbers in the decimal system.
2. Understand the concept of a place value number system.
3. Know the powers of ten.
4. Read and write whole numbe[rs in] the decimal system.

Place Value System

The decimal system (or base ten system) is a **place value system** that depends on three things:

1. The **ten digits**: 0, 1, 2, 3, 4, 5, 6, 7, 8, 9;
2. The **placement** of each digit; and
3. The **value** of each place.

Objective 2

Each place to the left of the decimal point has a **power of ten** as its place value. The powers of ten are:

1 10 100 1000 10,000 100,000 1,000,000 and so on.

Objective 3

In **standard notation**, digits are written in places, and the value of each digit is to be multiplied by the value of the place. The value of the number is found by adding the results of multiplying the digits and place values. In **expanded notation**, the values represented by each digit in standard notation are written as a sum. The English word equivalents can then be easily rea[d] (or written) from these indicated sums. Examples 1 through 4 illustrate the[se ide]as.

Video and Software Component

Two icons appear at the beginning of each section underneath the objectives as a reminder of the *Hawkes Learning Systems: Basic Mathematics* software and video component that is available with the textbook. The videos include an instructor briefly covering material in the text. They may be viewed after installing the software and selecting a lesson.

Reading and Writing Whole Numbers

Objective 4

Whole numbers are the numbers used for counting and the number 0. They are the decimal numbers with digits written to the left of the decimal point and with only 0's (or no digits at all) written to the right of the decimal point. (Decimal numbers with digits to the right of the decimal point will be discussed in Chapter 5.) We use the capital letter **W** to represent the set of all whole numbers.

Whole Numbers

The **whole numbers** are the counting numbers and the number 0.

$\mathbf{W} = \{0, 1, 2, 3, 4, 5, 6, 7, 8, 9, 10, 11, 12, 13, 14, 15, \ldots\}$

The three dots in the list of whole numbers indicate that the pattern continues without end.

Figure 1.2 shows how large whole numbers are written and read with the digits in groups of three. Each group of digits is called a **period**. **Note that in writing whole numbers in standard form, the decimal point itself is optional.**

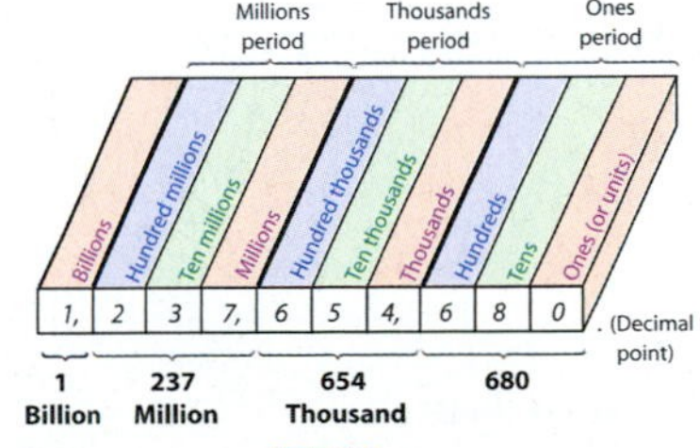

Figure 1.2

Reading and Writing Whole Numbers

You should note the following four things when reading or writing whole numbers:

1. Digits are read in **groups of three**. (See Figure 1.2.)
2. Commas are used to separate groups of three digits **if a number has more than four digits**.
3. The word **and** does not appear in English word equivalents. **And** is said only when reading a decimal point. (See Chapter 5.)
4. Hyphens (-) are used to write English words for the two-digit numbers from 21 to 99 except for those that end in 0.

Definitions, Key Concepts, Rules, Properties, and Procedures

In each section, definition boxes are purple and key concepts, rules, properties, and procedures are in yellow boxes. This makes them highly visible and easily found again.

Margin Exercises

Each example has its own margin exercise. The margin exercises give the student a chance to work out a problem similar to an example right after reading the example. They serve as a quick way for the student to apply the skills they've learned and test their understanding of the concepts.

Examples

Each section has multiple examples to help the student grasp the concepts covered. These examples also give a preview of what can be expected in the exercises.

8. What number should be added to 268 to get a sum of 654?

Example 8

What number should be ad

Solution

We know the sum and one
546 from 732.

$$\begin{array}{r} \overset{6}{\not{7}}\overset{12}{\not{3}}\overset{1}{2} \\ -546 \\ \hline 186 \end{array}$$ difference

The number to be added is 186.

Now work exercise 8 in the margin.

9. The cost of replacing Robert's used DVD player was going to be \$120 for parts (including tax) plus \$45 for labor. To buy a new DVD player, Robert was going to have to pay \$115 plus \$8 in tax and the dealer was going to pay him \$25 for his old DVD player. How much more would it cost to fix the old DVD player than it would to buy a new one?

Example 9

The cost of repairing Ed's used TV set was going to be \$250 for parts (including tax) plus \$85 for labor. To buy a new set, he was going to pay \$450 plus \$27 in tax and the dealer was going to pay him \$85 for his old set. How much more would Ed have to pay for a new set than to have his old set repaired?

Solution

Used Set	New Set	Difference
\$250 parts	\$450 cost	\$392
+ 85 labor	+ 27 tax	− 335
\$335	\$477	\$ 57
	− 85 trade-in	
	\$392 total	

Ed would pay \$57 more for the new set than for having his old set repaired. What would you do?

Now work exercise 9 in the margin.

Completion Examples

Most sections also have completion examples. Here, the student must complete the example by filling in the blanks. Completion examples are often similar to previous examples in the section.

10. Subtract using expanded notation.

$$\begin{array}{r} 577 \\ -236 \\ \hline \end{array}$$

Completion Example 10

Subtract 897 − 364 by using expanded notation.

Solution

897 = 800 + 90 + ___
−364 = 300 + ___ + ___
___ + ___ + ___ = ___

Now work exercise 10 in the margin.

30 Chapter 1 Whole Numbers

Name ________ Section ______ Date ______

Exercises 1.1

Write the following numbers in standard notation. See Examples 9 through 12.

1. Seventy-six
2. Five hundred eighty
3. Two thousand five
4. Seventy-eight thousand, nine hundred two
5. Thirty-three thousand, three hundred thirty-three
6. Five million, forty-five
7. Two hundred eighty-one million, three hundred thousand, five hundred one
8. Seven hundred fifty-eight million, three hundred fifty thousand, sixty
9. Eighty-two million, seven hundred thousand
10. Thirty-six

ANSWERS

1. ___
2. ___
3. ___
4. ___
5. ___
6. ___
7. ___
8. ___
9. ___
10. ___

Section Exercises

Each section concludes with a selection of exercises designed to allow the student to practice skills and master concepts. References to appropriate examples are clearly labeled for those who desire assistance.

59. ______

Writing and Thinking about Mathematics

59. Nothing was said in the text about a commutative property for subtraction. Do you think that this omission was intentional? Is there a commutative property of subtraction? Give several examples that justify your answer and discuss this idea in class.

60. ______

60. Nothing was said in the text about an associative property for subtraction. Do you think that this omission was intentional? Is there an associative property of subtraction? Give several examples that justify your answer and discuss this idea in class.

61. ______

61. You may have heard someone say that "you cannot subtract a larger number from a smaller number." Do you think that this statement is true? Does a subtraction problem such as 6 – 10 make sense to you? Can you think of any situation in which such a difference might seem reasonable?

36 **Chapter 1** Whole Numbers

Writing and Thinking about Mathematics

These exercises encourage students to express their ideas, interpretations, and understanding through writing. Many times, they challenge the student to analyze and combine new concepts as well and ideas that have been discussed in the section.

Name ______ Section ______ Date ______

ANSWERS

Collaborative Learning

Separate the class into teams of two to four students. Each team is to read and fill out the partial 1040A forms in Exercises 62 – 64. After these are completed, the team leader is to read the team's results and discuss any difficulties they had in reading and following directions while filling out the forms.

62. The Internal Revenue Service allows us to calculate our income tax returns with whole numbers. Calculate the taxable income on the portion of the Form 1040A shown here. Assume that you have 3 exemptions and you have not housed a person displaced by Hurricane Katrina.

62. [Respond in exercise.]

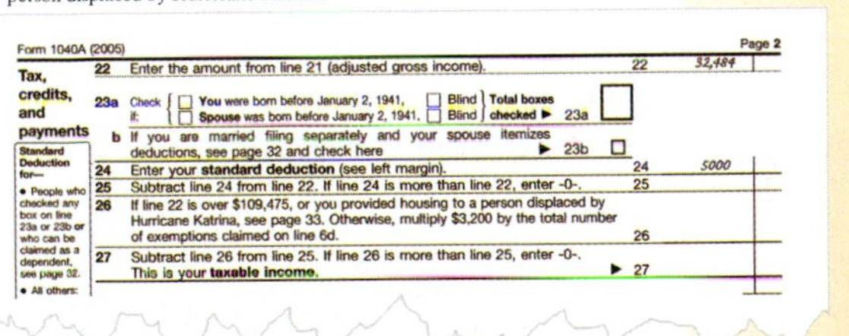

Form 1040A (2005) Page 2

Tax, credits, and payments

22 Enter the amount from line 21 (adjusted gross income). 22 32,484

23a Check if: You were born before January 2, 1941, Blind; Spouse was born before January 2, 1941, Blind. Total boxes checked ▶ 23a

b If you are married filing separately and your spouse itemizes deductions, see page 32 and check here ▶ 23b

Standard Deduction for— • People who checked any box on line 23a or 23b or who can be claimed as a dependent, see page 32. • All others:

24 Enter your **standard deduction** (see left margin). 24 5000

25 Subtract line 24 from line 22. If line 24 is more than line 22, enter -0-. 25

26 If line 22 is over $109,475, or you provided housing to a person displaced by Hurricane Katrina, see page 33. Otherwise, multiply $3,200 by the total number of exemptions claimed on line 6d. 26

27 Subtract line 26 from line 25. If line 26 is more than line 25, enter -0-. This is your **taxable income.** ▶ 27

Collaborative Learning

Collaborative learning exercises are designed to be completed in groups. Usually, the question involves a more in depth thought process that encourages interaction and discussion among students.

Check Your Number Sense

66. ______

66. Match each indicated quotient with the closest estimate of that product. Perform any calculations mentally.

	Quotient	Estimate
____	(a) $9\overline{)910}$	A. 6
____	(b) $34 \div 5$	B. 10
____	(c) $34\overline{)12{,}000}$	C. 100
____	(d) $18\overline{)3900}$	D. 200
____	(e) $216 \div 18$	E. 300

67. ______

67. Match each indicated quotient with the closest estimate of that product.

	Quotient	Estimate
____	(a) $3\overline{)870}$	A. 30
____	(b) $3\overline{)87{,}000}$	B. 300
____	(c) $3\overline{)87}$	C. 3000
____	(d) $3\overline{)8700}$	D. 30,000

Check Your Number Sense

These exercises are designed to help students develop confidence in their judgment, estimation, and mental calculation skills.

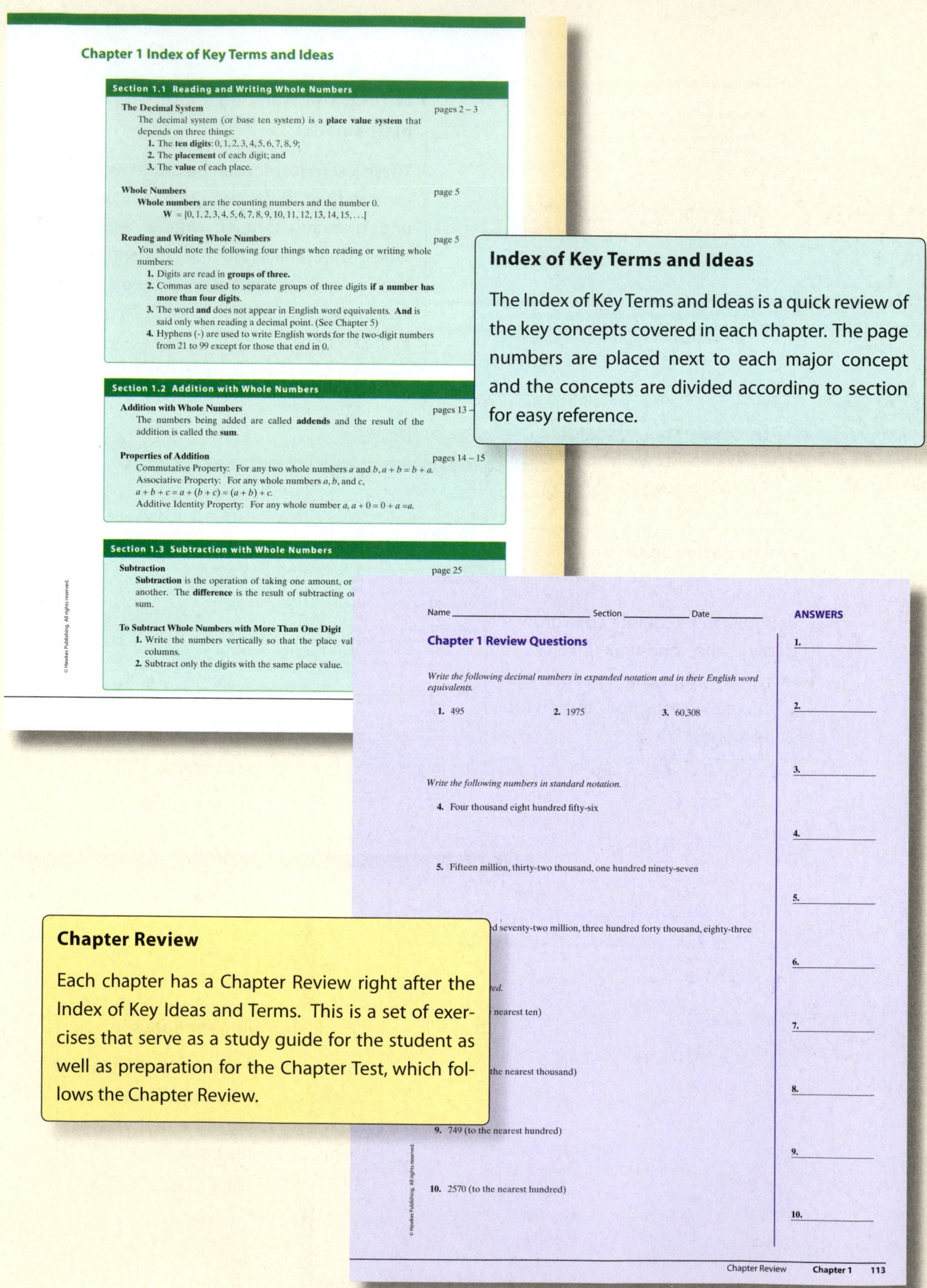

Chapter 1 Index of Key Terms and Ideas

Section 1.1 Reading and Writing Whole Numbers

The Decimal System — pages 2 – 3

The decimal system (or base ten system) is a **place value system** that depends on three things:

1. The **ten digits**: 0, 1, 2, 3, 4, 5, 6, 7, 8, 9;
2. The **placement** of each digit; and
3. The **value** of each place.

Whole Numbers — page 5

Whole numbers are the counting numbers and the number 0.

$\mathbf{W} = \{0, 1, 2, 3, 4, 5, 6, 7, 8, 9, 10, 11, 12, 13, 14, 15, \ldots\}$

Reading and Writing Whole Numbers — page 5

You should note the following four things when reading or writing whole numbers:

1. Digits are read in **groups of three.**
2. Commas are used to separate groups of three digits **if a number has more than four digits.**
3. The word **and** does not appear in English word equivalents. **And** is said only when reading a decimal point. (See Chapter 5)
4. Hyphens (-) are used to write English words for the two-digit numbers from 21 to 99 except for those that end in 0.

Section 1.2 Addition with Whole Numbers

Addition with Whole Numbers — pages 13 –

The numbers being added are called **addends** and the result of the addition is called the **sum**.

Properties of Addition — pages 14 – 15

Commutative Property: For any two whole numbers a and b, $a + b = b + a$.

Associative Property: For any whole numbers a, b, and c, $a + b + c = a + (b + c) = (a + b) + c$.

Additive Identity Property: For any whole number a, $a + 0 = 0 + a = a$.

Section 1.3 Subtraction with Whole Numbers

Subtraction — page 25

Subtraction is the operation of taking one amount, or another. The **difference** is the result of subtracting o sum.

To Subtract Whole Numbers with More Than One Digit

1. Write the numbers vertically so that the place val columns.
2. Subtract only the digits with the same place value.

Name ____________ Section ________ Date ________

ANSWERS

Chapter 1 Review Questions

Write the following decimal numbers in expanded notation and in their English word equivalents.

1. 495 **2.** 1975 **3.** 60,308

Write the following numbers in standard notation.

4. Four thousand eight hundred fifty-six

5. Fifteen million, thirty-two thousand, one hundred ninety-seven

…d seventy-two million, three hundred forty thousand, eighty-three

…ed.

… nearest ten)

…the nearest thousand)

9. 749 (to the nearest hundred)

10. 2570 (to the nearest hundred)

1. ______ 2. ______ 3. ______ 4. ______ 5. ______ 6. ______ 7. ______ 8. ______ 9. ______ 10. ______

Chapter Review **Chapter 1** **113**

Index of Key Terms and Ideas

The Index of Key Terms and Ideas is a quick review of the key concepts covered in each chapter. The page numbers are placed next to each major concept and the concepts are divided according to section for easy reference.

Chapter Review

Each chapter has a Chapter Review right after the Index of Key Ideas and Terms. This is a set of exercises that serve as a study guide for the student as well as preparation for the Chapter Test, which follows the Chapter Review.

Chapter Test and Cumulative Review

Each chapter ends with a Chapter Test and, starting with chapter 2, a Cumulative Review. These exercises give the student one more chance to test their level of mastery before moving on to the next chapter.

Name ______ Section ______ Date ______ **ANSWERS**

Chapter 1 Test

1. Write 8952 in expanded notation and in its English word equivalent.
2. The number 1 is called the multiplicative ________.
3. Give an example that illustrates the commutative property of multiplication.
4. Round 997 to the nearest thousand.
5. Round 135,721 to the nearest ten-thousand.

First estimate each sum; then find the sum.

6. 9586 + 345 + 2078
7. 37 + 486 + 493 + ...
8. 1,480,900 + 2,576,850

Subtract.

9. 850 − 362
10. ...

Name ______ Section ______ Date ______ **ANSWERS**

Cumulative Review: Chapters 1 – 2

1. Write the number 50,732 in **a.** expanded notation and **b.** its English word equivalent.

In Exercises 2 – 5, name the property illustrated.

2. $45 \cdot 2 = 2 \cdot 45$
3. $7 \cdot (3 \cdot 4) = (7 \cdot 3) \cdot 4$
4. $19 + 0 = 19$
5. $25 + 3 = 3 + 25$
6. Round 41,624 to the nearest thousand.

*In Exercises 7 – 10, first **a.** estimate each answer and then **b.** find the answer by performing the indicated operation.*

7. 83 + 947 + 1035
8. 6003 − 759
9. 74 × 86
10. $17\overline{)2210}$
11. Multiply mentally: 90×300.

Use the Rules for Order of Operations to evaluate the expressions in Exercises 12 and 13.

12. $15 + 2(8 - 2^3) - 3 \cdot 2^2$
13. $75 \div 5 \cdot 3 + 4(7 - 2^2)$
14. Without actually dividing, determine which of the numbers 2, 3, 5, 6, 9, and 10 will divide into 10,840. Give a brief reason for each decision.

Cumulative Review **Chapters 1 - 2** 201

Acknowledgements

I would like to thank Editor Cynthia Ellison and Developmental Editor Marcel Prevuznak for their hard work and invaluable assistance in the development and production of this text.

Many thanks go to the following reviewers who offered their constructive and critical comments:

Kinley Alston, *Trident Technical College, SC*
Carol Ellerbusch, *Iowa Lakes Community College, IA*
Charyl Link, *Kansas City Kansas Community College, KS*
Lee McCarty, *Hinds Community College, MS*
Barbara Odegard, *College of the Siskiyous, CA*
Elaine Richards, *Eastern Michigan University, MI*
Harriette Roadman, *New River Community College, VA*
Jennifer Smiley, *Walla Walla Community College, WA*
Karen Stewart, *College of the Mainland, TX*
Nan Strebeck, *Navarro College, TX*
Jay Villanueva, *Florida Memorial University, FL*
Shelley Walker, *Eastern Michigan University, MI*
Amy Young, *Navarro College, TX*

Finally, special thanks go to James Hawkes and Greg Hill for their faith in this edition and their willingness to commit so many resources to guarantee a top-quality product for students and teachers.

D. Franklin Wright

Hawkes Learning Systems: Basic Mathematics

About the Software:

This multimedia courseware allows students to become better problem-solvers by creating a mastery level of learning in the classroom. The software includes an "instruct," "practice," "tutor," and "certify" mode in each lesson, allowing students to learn through step-by-step interactions with the software. This automated homework system's tutorial and assessment modes extend instructional influence beyond the classroom. Intelligence is what makes the tutorials so unique. By offering intelligent tutoring and mastery level testing to measure what has been learned, the software extends the instructor's ability to influence students to solve problems. This courseware can be ordered either separately or bundled together with this text.

Minimum Requirements:

In order to run ***HLS: Basic Mathematics***, you will need:

Intel® Pentium® 800 MHz or faster processor
Windows® 98SE or later
128 MB RAM
150 MB hard drive space
256 color display (800x600, 16-bit color recommended)
Internet Explorer 6.0 or later
CD-ROM drive

Getting Started:

Before you can run ***HLS: Basic Mathematics***, you will need an access code. This 30 character code is your personal access code. To obtain an access code, go to http://www.hawkeslearning.com and follow the links to the access code request page (unless directed otherwise by your instructor.)

Installation:

1. Place CD #1 in the CD-ROM drive. (CD#1 is the only CD needed for the installation.)
2. Double-click on the **My Computer** icon.
3. Double-click on the CD-ROM drive (has a picture of a CD-ROM disk).
4. Double-click on Setup.exe.
5. Follow the on-screen instructions.
6. You will be prompted for a Course ID.

If you have internet access, select "Yes, the Course ID is:" and enter the Hawkes Course ID provided by your instructor in the box provided. If you do not have internet access, select the option that says "No, I will not be accessing an online progress report from this computer."

Starting the Courseware:

After you have installed ***HLS: Basic Mathematics*** on your computer, to run the courseware select Start/Programs/Hawkes Learning Systems/Basic Mathematics.

You will be prompted to enter your access code with a message box similar to the following:

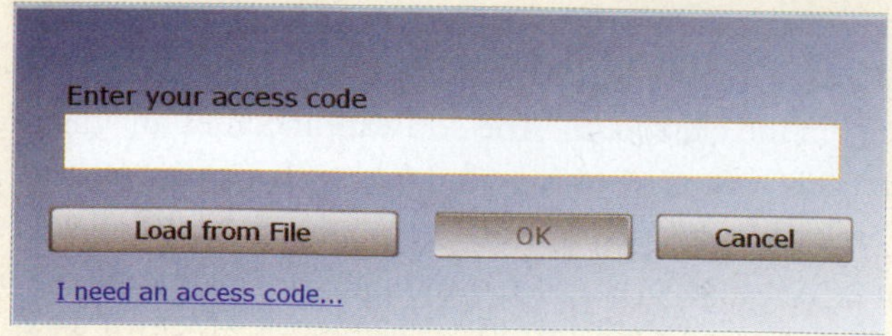

Type your entire access code in the box. No spaces or dashes are necessary; they will be supplied automatically. When you are finished, press OK.

If you typed in your access code correctly, you will be prompted to save the code to a disk. If you choose to save your code to a disk, typing in the authorization code each time you run ***HLS: Basic Mathematics*** will not be necessary. Instead, select the

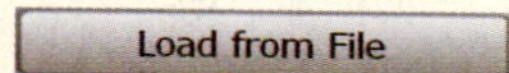

button when prompted to enter your access code and choose the path to your saved access code.

Now that you have entered your authorization code and saved it to a disk, you are ready to run a lesson. From the table of contents screen, choose the appropriate chapter and then choose the lesson you wish to run.

Features:

Each lesson in ***HLS: Basic Mathematics*** has four modes: Instruct, Practice, Tutor, and Certify.

Instruct: Instruct provides an expository on the material covered in the lesson in a multimedia environment. This same instruct mode can be accessed via the tutor mode.

Practice: Practice allows you to hone your problem-solving skills. It provides an unlimited number of randomly generated problems. Practice also provides access to the Tutor mode by selecting the Tutor button located by the Submit button.

Tutor: Tutor mode is broken up into several parts: Instruct, Explain Error, Step by Step, and Solution.

1. **Instruct**, which can also be selected directly from Practice mode, contains a multimedia lecture of the material covered in a lesson.
2. **Explain Error** is active whenever a problem is incorrectly answered. It will attempt to explain the error that caused you to incorrectly answer the problem.
3. **Step by Step** is an interactive "step through" of the problem. It breaks each problem into several steps, explains to you each step in solving the problem, and asks you a question about the step. After you answer the last step correctly, you have solved the problem.
4. **Solution** will provide you with a detailed "worked-out" solution to the problem.

Throughout the Tutor, you will see words or phrases colored green with a dashed underline. These are called **Hot Words**. Clicking on a Hot Word will provide you with more information on these word(s) or phrases.

Certify: Certify is the testing mode. You are given a finite number of problems and a certain number of strikes (problems you can get wrong). If you correctly answer the required number of questions, you will receive a certification code and a certificate. Write down your certification code and/or print out your certificate. The certification code will be used by your instructor to update your records. The Tutor mode is not available in Certify.

Support:

If you have questions about ***HLS: Basic Mathematics*** or are having technical difficulties, we can be contacted as follows:

Phone: (843) 571-2825
Email: techsupport@hawkeslearning.com
Web: www.hawkeslearning.com

Support hours are 8:30 am to 5:30 pm, Eastern Time, Monday through Friday..

To The Student

The goal of this text and of your instructor is for you to succeed in basic math. Certainly, you should make this your goal as well. What follows is a brief discussion about developing good work habits and using the features of this text to your best advantage. For you to achieve the greatest return on your investment of time and energy you should practice the following three rules of learning.

1. **Reserve a block of time for study everyday.**
2. **Study what you don't know.**
3. **Don't be afraid to make mistakes.**

How to Use This Book

The following ten-step guide will not only make using this book a more worthwhile and efficient task, but it will also help you benefit more from classroom lectures or the assistance that you receive in a math lab.

1. Before you begin a chapter, read the discussion of What to Expect in This Chapter. It will give you some insight on the direction that the chapter will take, the topics that you will be learning about, and how those topics are interrelated.
2. Try to look over the assigned section(s) before attending class or lab. In this way, new ideas may not sound so foreign when you hear them mentioned again. This will also help you see where you need to ask questions about material that seems difficult to you.
3. Read examples carefully. They have been chosen and written to show you all of the problem-solving steps that you need to be familiar with. You might even try to solve example problems on your own before studying the solutions that are given.
4. Work the completion examples whenever they appear. If you cannot complete the steps in one of these special examples, go back to review the standard examples that precede it, then try again. When you are satisfied with your solution to a completion example, compare your work to the answers given at the end of the section. Your ability to work completion examples successfully means that you are learning the problem-solving methods being developed in that section.
5. Work margin exercises when asked to do so throughout the lessons. They play an important role in reinforcing the ideas of each lesson and preparing you to work the section exercises. If the margin exercises seem too difficult for you, review the discussion and examples that precede them, then try again. Check your answers against those given in the Answer Key at the back of the book.
6. Work the section exercises faithfully as they are assigned. Problem-solving practice is the single most important element in achieving success in any math class, and there is no good substitute for actually doing this work yourself. Demonstrating that you can think indepen-

dently through each step of each type of problem will also give you confidence in your ability to answer questions on quizzes and exams. Check the Answer Key periodically while working section exercises to be sure that your have the right ideas and are proceeding in the right manner.

7. Use the Writing and Thinking About Mathematics questions as an opportunity to explore the way that you think about math. A big part of learning and understanding mathematics is being able to talk about mathematical ideas and communicate the thinking that you do when you approach new concepts and problems. These questions can help you analyze your own approach to mathematics and, in class or group discussions, learn from ideas expressed by your fellow students.)
8. Use the Chapter Index of Key Ideas and Terms as a recap when you begin to prepare for a Chapter Test. It will reference all the major ideas that you should be familiar with from that chapter and indicate where you can turn if review is needed. You can also use the Chapter Index as a final checklist once you feel you have completed your review and are prepared for the Chapter Test.
9. Chapter Tests are provided so that you can practice for the tests that are actually given in class or lab. To simulate a test situation, block out a one-hour, uninterrupted period in a quiet place where your only focus is on accurately completing the Chapter Test. Use the Answer Key at the back of the book as a self-check only after you have completed all of the questions on the test.
10. Cumulative Reviews will help you retain the skills that you acquired in studying earlier chapters. They appear after every chapter beginning with Chapter 2. Approach them in much the same manner as you would the Chapter Tests in order to keep all of your skills sharp throughout the entire course.

How to Prepare for an Exam

Gaining skill and Confidence

The stress that many students feel while trying to succeed in mathematics is what you have probably heard called "math anxiety." It is a real-life phenomenon, and many students experience such a high level of anxiety during mathematics exams in particular that they simply cannot perform to the best of their abilities. It is possible to overcome this stress simply by building your confidence in your ability to do mathematics and by minimizing your fears of making mistakes.

No matter how much it may seem that in mathematics you must either be right or wrong, with no middle ground, you should realize that you can be learning just as much from the times that you make mistakes as you can from the times that your work is correct. Success will come. Don't think that making mistakes at first means that you'll never be any good at mathematics. Learning mathematics requires lots of practice. Most importantly, it requires a true confidence in yourself and in the fact that with practice and persistence the mistakes will become fewer, the successes will become greater, and you will be able to say, "I *can* do this."

Showing What You Know

If you have attended class or lab regularly, taken good notes, read your textbook, kept up with homework exercises, and asked for help when it was needed, then you have already made significant progress in preparing for an exam and conquering any anxiety. Here are a few other suggestions to maximize your preparedness and minimize your stress.

1. Give yourself enough time to review. You will generally have several days advance notice before an exam. Set aside a block of time each day with the goal of reviewing a manageable portion of the material that the test will cover. Don't cram!
2. Work lots of problems to refresh your memory and sharpen your skills. Go back to redo selected exercises from all of your homework assignments.
3. Reread your text and your notes, and use the Chapter Index of Key Ideas and Terms and the Chapter Text to recap major ideas and do a self-evaluated test simulation.
4. Be sure that you are well-rested so that you can be alert and focused during the exam.
5. Don't study up to the last minute. Give yourself some time to wind down before the exam. This will help you to organize your thoughts and feel more calm as the test begins.

1

Whole Numbers

Chapter 1 Whole Numbers

WHAT TO EXPECT IN CHAPTER 1

Chapter 1 provides a review of the basic operations (addition, subtraction, multiplication, and division) with whole numbers. Section 1.1 discusses number systems in general, with an emphasis on reading and writing whole numbers in the decimal system. Addition and subtraction with whole numbers are discussed in Sections 1.2 and 1.3 and the skills related to rounding and estimating with whole numbers are developed in Section 1.4. Sections 1.5 and 1.6 discuss multiplying and dividing with whole numbers while Section 1.7 (Problem Solving with Whole Numbers) outlines a four-step process for solving word problems, developed by George Pólya, a famous educator and mathematician from Stanford University.

Section 1.8 concludes the chapter with an introduction to geometric concepts, including plane geometry, polygons, length and perimeter, and area.

Chapter 1 Whole Numbers

Mathematics at Work!

We see whole numbers everywhere. People in various businesses, carpenters, teachers, auto mechanics, and airline pilots all deal with whole numbers on a daily basis. An example of using whole numbers in everyday life is deciding how many apples to buy at the grocery store.

As a student, you may be interested in information about colleges and the costs of attending different colleges. The following list contains this type of information about 15 well-known colleges and universities. Annual (yearly) tuition as well as room and board are given to the nearest hundred dollars for the year 2004. (Note: If you want this type of information for any college or university, write to the registrar of the school or check their website.)

			Tuition	
Institution	**Enrollment**	**Room/Board**	**Resident**	**Nonresident**
Univ. of Arizona	36,900	$7100	$4100	$13,100
Boston College	14,500	$10,900	$31,500	$31,500
Univ. of Colorado	29,200	$8000	$5400	$22,800
Florida A & M	13,100	$6000	$3200	$15,200
Purdue Univ.	38,700	$6800	$6500	$19,800
Johns Hopkins	18,200	$9900	$31,600	$31,600
MIT	10,300	$9500	$32,300	$32,300
Ohio State Univ.	51,000	$7300	$7400	$18,000
Princeton Univ.	6800	$8800	$31,500	$31,500
Rice Univ.	4900	$9000	$20,200	$20,200
Univ. of Tennessee	27,800	$6300	$4300	$9000
Univ. of Wisconsin	41,200	$6500	$6200	$21,100

For these universities and colleges:

a. Find the average of the yearly expenses for room and board.

b. Find the average difference between the tuition for the nonresident students and residents.

c. Determine which of these universities and colleges is the most expensive and which is the least expensive (including room and board and tuition) for nonresident students.

d. Determine which of these universities and colleges is the most expensive and which is the least expensive (including room and board and tuition) for resident students.

1.1 Reading and Writing Whole Numbers

The Decimal System

The abstract concept of numbers and the numeration systems used to represent numbers have been important, indeed indispensable, parts of intelligent human development. The Hindu-Arabic system of numeration that we use today was invented about 800 A.D. This system, called the **decimal system** (**deci** means **ten** in Latin), allows us to add, subtract, multiply, and divide faster and more easily than did any of the ancient number systems, such as Egyptian hieroglyphics and the Roman numeral system. The decimal system is more sophisticated than these earlier systems in that it uses a symbol for zero (0), the operation of multiplication, and the concept of place value.

The symbols used are called **digits**, and for the **whole numbers**, the value of a digit depends on its position to the left of a beginning point, called a **decimal point**. (See Figure 1.1.)

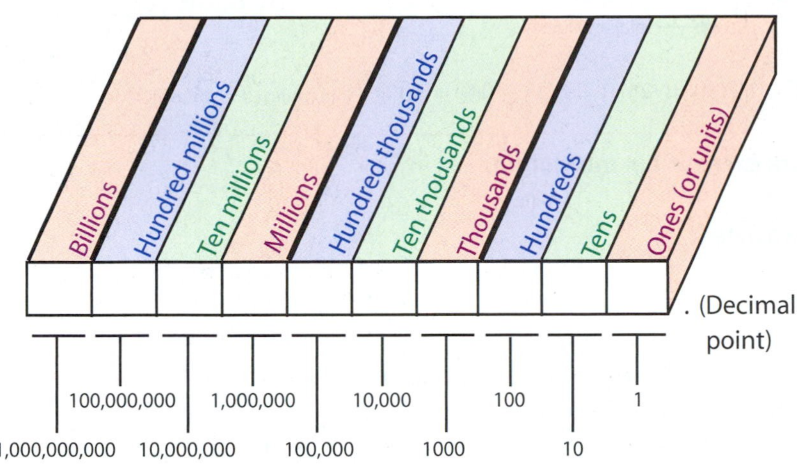

Figure 1.1

Objective 1

Objectives

1. Know the values of the places for whole numbers in the decimal system.
2. Understand the concept of a place value number system.
3. Know the powers of ten.
4. Read and write whole numbers in the decimal system.

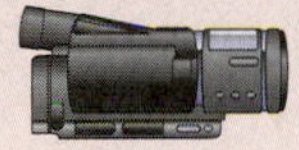

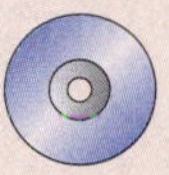

Objective 2

Place Value System

The decimal system (or base ten system) is a **place value system** that depends on three things:

1. The **ten digits**: 0, 1, 2, 3, 4, 5, 6, 7, 8, 9;
2. The **placement** of each digit; and
3. The **value** of each place.

Each place to the left of the decimal point has a **power of ten** as its place value. The powers of ten are:

1 10 100 1000 10,000 100,000 1,000,000 and so on.

Objective 3

In **standard notation**, digits are written in places, and the value of each digit is to be multiplied by the value of the place. The value of the number is found by adding the results of multiplying the digits and place values. In **expanded notation**, the values represented by each digit in standard notation are written as a sum. The English word equivalents can then be easily read (or written) from these indicated sums. Examples 1 through 4 illustrate these ideas.

1. Write 463 in expanded notation.

2. Write 7302 in expanded notation.

3. Write 29,524 in expanded notation.

4. Write 808,491 in expanded notation.

Example 1

954 (standard notation)

9	5	4	← digits (standard notation)
100	10	1	← place values

954 = 9(100) + 5(10) + 4(1) = 900 + 50 + 4 (expanded notation)

Now work exercise 1 in the margin.

Example 2

6507 (standard notation)

6	5	0	7	← digits (standard notation)
1000	100	10	1	← place values

6507 = 6(1000) + 5(100) + 0(10) + 7(1)

= 6000 + 500 + 0 + 7 (expanded notation) [**Note :** The 0 is optional.]

Now work exercise 2 in the margin.

Completion Example 3

Complete the expanded notation form of 32,081.

Solution

32,081 = 30,000 + _______ + _______ + _______ + 1

NOTE: Completion Examples are answered at the end of each section.

Now work exercise 3 in the margin.

Completion Example 4

Complete the expanded notation form of 497,500.

Solution

497,500 = 400,000 + _______ + _______ + _______ + 0 + 0

Now work exercise 4 in the margin.

Reading and Writing Whole Numbers

Objective 4

Whole numbers are the numbers used for counting and the number 0. They are the decimal numbers with digits written to the left of the decimal point and with only 0's (or no digits at all) written to the right of the decimal point. (Decimal numbers with digits to the right of the decimal point will be discussed in Chapter 5.) We use the capital letter **W** to represent the set of all whole numbers.

> **Whole Numbers**
>
> The **whole numbers** are the counting numbers and the number 0.
>
> $$\mathbf{W} = \{0, 1, 2, 3, 4, 5, 6, 7, 8, 9, 10, 11, 12, 13, 14, 15, \ldots\}$$

The three dots in the list of whole numbers indicate that the pattern continues without end.

Figure 1.2 shows how large whole numbers are written and read with the digits in groups of three. Each group of digits is called a **period**. **Note that in writing whole numbers in standard form, the decimal point itself is optional.**

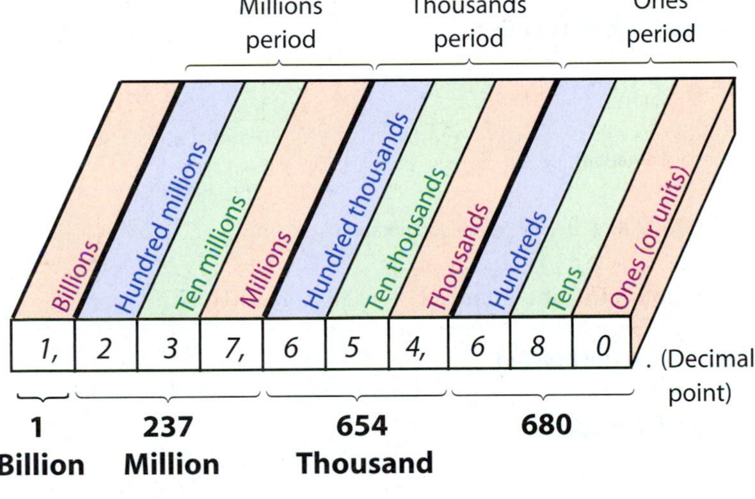

Figure 1.2

> **Reading and Writing Whole Numbers**
>
> You should note the following four things when reading or writing whole numbers:
>
> 1. Digits are read in **groups of three**. (See Figure 1.2.)
> 2. Commas are used to separate groups of three digits **if a number has more than four digits**.
> 3. The word **and** does not appear in English word equivalents. **And** is said only when reading a decimal point. (See Chapter 5.)
> 4. Hyphens (-) are used to write English words for the two-digit numbers from 21 to 99 except for those that end in 0.

We can write whole numbers in standard form, using digits and the place value system, in expanded notation, and lastly in their English word equivalents. For example,

268 (standard notation)

200 + 60 + 8 (expanded notation)

two hundred sixty-eight (english word equivalent)

Examples 5 through 8 illustrate whole numbers written in both expanded notation and English word equivalents.

5. Write 567 in expanded notation and in its English word equivalent.

Example 5

653 (standard notation)

600 + 50 + 3 (expanded notation)

six hundred fifty-three (English word equivalent)

Now work exercise 5 in the margin.

6. Write 25,400 in expanded notation and in its English word equivalent.

Example 6

75,900 (standard notation)

70,000 + 5000 + 900 + 0 + 0 (expanded notation)

seventy-five thousand, nine hundred (English word equivalent)

Now work exercise 6 in the margin.

7. Write 6791 in expanded notation and in its English word equivalent.

Completion Example 7

8407

8407 = 8000 + _______ + _______ + _______

eight thousand ________________________

Now work exercise 7 in the margin.

8. Write 23,507 in expanded notation and in its English word equivalent.

Completion Example 8

15,352

15,352 = 10,000 + _______ + _______ + _______ + _______

fifteen thousand, ________________________________

Now work exercise 8 in the margin.

Example 9

Write four hundred eighty thousand, five hundred thirty-three in standard notation.

Solution

480,533

Now work exercise 9 in the margin.

Example 10

Write five hundred seventy-two thousand, six hundred in standard notation.

Solution

572,600

Now work exercise 10 in the margin.

Completion Example 11

Write two million, eight hundred thousand, thirty-five in standard notation.

Solution

2,800, __________

Now work exercise 11 in the margin.

Completion Example 12

Write five million, three hundred fifty thousand in standard notation.

Solution

5, ______, ______

Now work exercise 12 in the margin.

9. Write twenty-eight thousand, six hundred forty-two in standard notation.

10. Write three hundred sixty-three thousand, seventy-five in standard notation.

11. Write six million, three hundred thousand, five hundred in standard notation.

12. Write four million, eight hundred seventy-five thousand in standard notation.

Completion Example Answers

3. 32,081 = 30,000 + **2000** + **0** + **80** + 1

4. 497,500 = 400,000 + **90,000** + **7000** + **500** + 0 + 0

7. 8407 = 8000 + **400** + **0** + **7**
eight thousand **four hundred seven**

8. 15,352 = 10,000 + **5000** + **300** + **50** + **2**
fifteen thousand, **three hundred fifty-two**

11. 2,800,**035**

12. 5,**350,000**

Practice Problems

1. Write the following numbers in expanded notation and in their English word equivalents.

a. 512 **b.** 6394

2. Write one hundred eighty thousand, five hundred forty-three in standard notation.

Answers to Practice Problems:

1. a. 500 + 10 + 2; five hundred twelve

b. 6000 + 300 + 90 + 4; six thousand three hundred ninety-four

2. 180,543

Name ______________________ Section ____________ Date __________

Exercises 1.1

Write the following numbers in standard notation. See Examples 9 through 12.

1. Seventy-six

2. Five hundred eighty

3. Two thousand five

4. Seventy-eight thousand, nine hundred two

5. Thirty-three thousand, three hundred thirty-three

6. Five million, forty-five

7. Two hundred eighty-one million, three hundred thousand, five hundred one

8. Seven hundred fifty-eight million, three hundred fifty thousand, sixty

9. Eighty-two million, seven hundred thousand

10. Thirty-six

11. Seven hundred fifty-seven

12. List the ten digits used to write whole numbers.

13. For whole numbers, the value of a digit depends on its position to the ________ of the beginning point, called a ____________________.

14. The powers of ten are ________________.

ANSWERS

1. ____________

2. ____________

3. ____________

4. ____________

5. ____________

6. ____________

7. ____________

8. ____________

9. ____________

10. ____________

11. ____________

12. ____________

13. ____________

14. ____________

15. ______________

16. ______________

17. ______________

18. ______________

19. ______________

20. ______________

21. ______________

22. ______________

23. ______________

24. ______________

25. ______________

26. ______________

27. ______________

28. ______________

29. ______________

30. ______________

31. ______________

32. ______________

33. ______________

34. ______________

35. ______________

36. ______________

15. The word ______________ does not appear in English word equivalents for whole numbers.

16. Use a comma to separate groups of three digits if a number has more than ______ digits.

17. Hyphens are used to write English words for numbers from ______ to ______ that do not end in ______.

Write the following whole numbers in expanded notation. See Examples 1 through 4.

18. 37 **19.** 84 **20.** 56 **21.** 821

22. 1892 **23.** 2059 **24.** 25,658 **25.** 32,341

Write the following whole numbers in their English word equivalents. See Examples 5 through 8.

26. 83 **27.** 122 **28.** 10,500

29. 683,100 **30.** 592,300 **31.** 16,302,590

32. 71,500,000

Write the following whole numbers in standard notation. See Examples 9 through 12.

33. Seventy-five **34.** Ninety-eight

35. One hundred forty-two **36.** Five hundred seventy-three

Name ______________________ Section ____________ Date __________

ANSWERS

37. Three thousand eight hundred thirty-four

38. Ten thousand, eleven

39. Four hundred thousand, seven hundred thirty-six

40. Five hundred thirty-seven thousand, eighty-two

41. Sixty-three million, two hundred fifty-one thousand, sixty-five

42. Name the position of each nonzero digit in the following number: 2,403,189,500.

Write the English word equivalent for the number(s) in each sentence. See Examples 5 through 8.

43. Population. The population of Los Angeles is 3,845,541.

44. Astronomy. The distance from the earth to the sun is about 93,000,000 miles, or 149,730,000 kilometers.

45. Geography. The country of Chile averages about 110 miles in width and is about 2650 miles long.

46. Geography. The Republic of Venezuela covers an area of 352,143 square miles, or about 912,050 square kilometers.

47. Architecture. One of the world's tallest buildings is the Taipei Financial Center in Taipei, Taiwan. The building has 101 stories and reaches 1761 feet tall.

48. Oceans. The Pacific Ocean has an area of approximately 63,800,000 square miles and has an average depth of 14,040 feet.

37. ______________

38. ______________

39. ______________

40. ______________

41. ______________

42. [Respond below exercise.]

43. [Respond below exercise.]

44. [Respond below exercise.]

45. [Respond below exercise.]

46. [Respond below exercise.]

47. [Respond below exercise.]

48. [Respond below exercise.]

49. a. ______

b. ______

c. ______

50. a. ______

b. ______

c. ______

51. ______

52. [Respond below exercise.] ______

49. Given the number 284,065, which digit indicates the number of
a. tens?
b. ten thousands?
c. hundreds?

50. Given the number 13,476,582, which digit indicates the number of
a. thousands?
b. millions?
c. ten millions?

Writing and Thinking about Mathematics

51. Large numbers. A *googol* is the name of a very large power of ten. Look up the meaning of this term in a dictionary or in a book that discusses the history of mathematics.
(**Note:** Ten to the power of a googol is called a *googolplex*.)

52. Population. According to the 2000 Census, Pennsylvania is the sixth most populous state in the United States with 12,281,054 people. Name the value of each nonzero digit in this number. In an almanac or other source, find the populations of the five most populous states.

1.2 Addition with Whole Numbers

Addition with Whole Numbers

Objective ①

Addition with whole numbers is indicated either by writing the numbers horizontally with a plus sign (+) between them or by writing the numbers vertically in columns with instructions to add.

$$6 + 23 + 17 \quad \text{or} \quad \text{Add} \quad \begin{array}{r} 6 \\ 23 \\ \underline{17} \end{array} \quad \text{or} \quad \begin{array}{r} 6 \\ 23 \\ \underline{+17} \end{array}$$

The numbers being added are called **addends**, and the result of the addition is called the **sum**.

$$\text{Add} \quad \begin{array}{rl} 6 & \text{addend} \\ 23 & \text{addend} \\ \underline{17} & \text{addend} \\ 46 & \text{sum} \end{array}$$

Be sure to keep the digits aligned (in column form) so you will be adding units to units, tens to tens, and so on. Neatness is a necessity in mathematics.

To be able to add with speed and accuracy, you must **memorize** the basic addition facts, which are given in Table 1.1. Practice all the combinations so you can give the answers immediately.

Objectives

① Be able to add whole numbers in both a vertical format and a horizontal format.

② Know the basic facts of addition with whole numbers.

③ Know the following properties of addition: commutative, associative, additive identity (0).

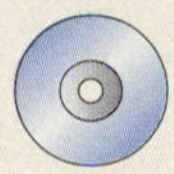

Objective ②

Table 1.1 Basic Addition Facts

+	0	1	2	3	4	5	6	7	8	9
0	**0**	1	2	3	4	5	6	7	8	9
1	1	**2**	3	4	5	6	7	8	9	10
2	2	3	**4**	5	6	7	8	9	10	11
3	3	4	5	**6**	7	8	9	10	11	12
4	4	5	6	7	**8**	9	10	11	12	13
5	5	6	7	8	9	**10**	11	12	13	14
6	6	7	8	9	10	11	**12**	13	14	15
7	7	8	9	10	11	12	13	**14**	15	16
8	8	9	10	11	12	13	14	15	**16**	17
9	9	10	11	12	13	14	15	16	17	**18**

$3 + 8 = 8 + 3$

$7 + 9 = 9 + 7$

As an exercise to help find out which combinations you need special practice with, write, on a piece of paper in mixed order, all one hundred possible combinations to be added. Perform the operations as quickly as possible. Then, using the table, check to find the ones you missed. Study these frequently until you are confident that you know them as well as all the others.

Your adding speed can be increased if you learn to look for combinations of digits that total 10.

1. Using the method of adding combinations of digits that total 10, find the following sum:

52 + 13 + 45.

Example 1

To add the following numbers, we note the combinations that total 10 and find the sums quickly.

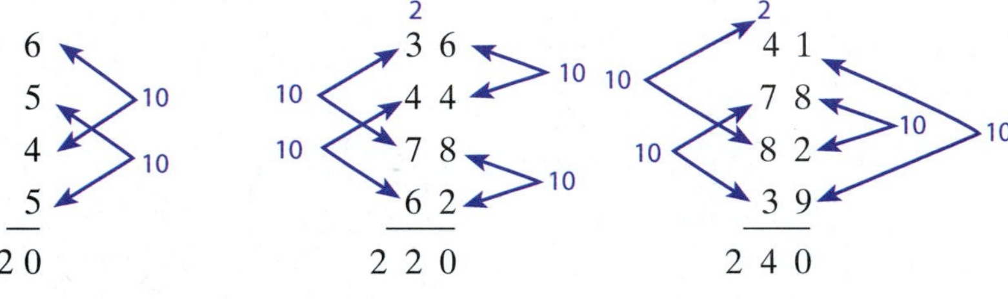

Now work exercise 1 in the margin.

When two numbers are added, the order of the numbers does not matter. That is, 4 + 9 = 13 and 9 + 4 = 13. Also,

$$\begin{array}{r} 5 \\ +\ 7 \\ \hline 12 \end{array} \quad \text{and} \quad \begin{array}{r} 7 \\ +\ 5 \\ \hline 12 \end{array}$$

By looking at Table 1.1 again, we can see that reversing the order of any two addends will not change their sum. This fact is called the **commutative property of addition**. We can state this property using letters to represent whole numbers.

Objective 3

Commutative Property of Addition

For any two whole numbers a and b, $a + b = b + a$.

To add three numbers, such as 6 + 3 + 5, we can add only two at a time, then add the third. Which two should we add first? The answer is that the sum is the same either way:

$$6 + 3 + 5 = (6 + 3) + 5 = 9 + 5 = 14$$

and

$$6 + 3 + 5 = 6 + (3 + 5) = 6 + 8 = 14$$

we can write

$$8 + 4 + 7 = (8 + 4) + 7 = 8 + (4 + 7) = 19$$

These examples illustrate the **associative property of addition**.

Associative Property of Addition

For any whole numbers a, b, and c,

$$a + b + c = (a + b) + c = a + (b + c).$$

Another property of addition with whole numbers is addition with 0. Whenever we add 0 to a number, the result is the original number:

$$8 + 0 = 8, \quad 0 + 13 = 13, \quad \text{and} \quad 38 + 0 = 38.$$

Zero (0) is called the **additive identity** or **the identity element for addition**.

Additive Identity Property

For any whole number a, $a + 0 = 0 + a = a$.

Example 2

a. $(15 + 20) + 6 = 15 + (20 + 6)$ — associative property

b. $8 + 3 = 3 + 8$ — commutative property

c. $17 + 0 = 17$ — additive identity property

Now work exercise 2 in the margin.

Note that Example **2a.** shows a change in the **grouping** (association) of the numbers, while Example **2b.** shows a change in the **order** of two numbers.

Example 3

Another illustration of the associative property:

$$\begin{array}{r} 8 \\ 2 \\ +\ 3 \\ \hline \end{array} \quad \rightarrow \quad \begin{array}{r} 10 \\ +\ 3 \\ \hline 13 \end{array} \qquad\qquad \begin{array}{r} 8 \\ 2 \\ +\ 3 \\ \hline \end{array} \quad \rightarrow \quad \begin{array}{r} 8 \\ +\ 5 \\ \hline 13 \end{array}$$

Now work exercise 3 in the margin.

Adding large numbers or several numbers can be done by writing the numbers vertically so that the place values are **lined up** in columns. Then only the digits with the same place value are added.

In the following exercises, find both sums and tell which property of addition is illustrated.

2. $\begin{array}{r} 8 \\ +5 \\ \hline \end{array} \qquad \begin{array}{r} 5 \\ +8 \\ \hline \end{array}$

3. $(4 + 7) + 9 = 4 + (7 + 9)$

Find each sum.

4.
$$\begin{array}{r} 4361 \\ +2528 \\ \hline \end{array}$$

5.
$$\begin{array}{r} 3590 \\ +4207 \\ \hline \end{array}$$

Example 4

If no sum of digits is more than 9:

Add 5623 + 3172.

$$\begin{array}{r} 5623 \\ +3172 \\ \hline 5 \end{array}$$ ← addend, ← addend, Add ones.

$$\begin{array}{r} 5623 \\ +3172 \\ \hline 95 \end{array}$$ Add tens.

$$\begin{array}{r} 5623 \\ +3172 \\ \hline 795 \end{array}$$ Add hundreds.

$$\begin{array}{r} 5623 \\ +3172 \\ \hline 8795 \end{array}$$ Add thousands.

$$\begin{array}{r} 5623 \\ +3172 \\ \hline 8795 \end{array}$$ ← Do not write all the steps. Write only the addends and the sum. ← sum

Now work exercise 4 in the margin.

Completion Example 5

If no sum of digits is more than 9:

Add 4375 + 5423.

$$\begin{array}{r} 4375 \\ +5423 \\ \hline \_ \end{array}$$ ← addend, ← addend, Add ones.

$$\begin{array}{r} 4375 \\ +5423 \\ \hline \_\_ \end{array}$$ Add tens.

$$\begin{array}{r} 4375 \\ +5423 \\ \hline \_\_\_ \end{array}$$ Add hundreds.

$$\begin{array}{r} 4375 \\ +5423 \\ \hline \_\_\_\_ \end{array}$$ Add thousands. ← sum

Now work exercise 5 in the margin.

Example 6

If the sum of the digits in one column is more than 9:

Add 217 + 389 + 634 + 536.

a. Write the ones digit in that column.
b. **Carry** the other digit to the column to the left.

$$\begin{array}{r} {}^{2} \\ 217 \\ 389 \\ 634 \\ +\,536 \\ \hline 6 \end{array}$$

Add 7 + 9 + ④ + ⑥ = 7 + 9 + 10
= 26.

10

Carry the 2.

$$\begin{array}{r} {}^{1\,2} \\ 217 \\ 389 \\ 634 \\ +\,536 \\ \hline 76 \end{array}$$

10

Add ② + 1 + ⑧ + 3 + 3 = 10 + 1 + 3 + 3
= 17

$$\begin{array}{r} {}^{1\,2} \\ 217 \\ 389 \\ 634 \\ +\,536 \\ \hline 1776 \end{array}$$

10

Add ① + 2 + ③ + ⑥ + 5 = 10 + 2 + 5
= 17

Now work exercise 6 in the margin.

Find the following sums.

6.

$$\begin{array}{r} 252 \\ 127 \\ 395 \\ +\,443 \\ \hline \end{array}$$

Completion Example 7

If the sum of the digits in one column is more than 9:

Add 328 + 604 + 517 + 192.

a. Write the ones digit in that column.
b. **Carry** the other digit to the column to the left.

$$\begin{array}{r} 328 \\ 604 \\ 517 \\ +\,192 \\ \hline \_ \end{array}$$

$$\begin{array}{r} 328 \\ 604 \\ 517 \\ +\,192 \\ \hline \_\_ \end{array}$$

7.

$$\begin{array}{r} 348 \\ 851 \\ 279 \\ +\,136 \\ \hline \end{array}$$

Continued on next page...

$$\begin{array}{r} 328 \\ 604 \\ 517 \\ +192 \\ \hline \_\_\_\_ \end{array}$$

Now work exercise 7 in the margin.

8. Diedra bought a couch for $315, a coffee table for $199, and a washer and dryer for $863. What total amount did she spend?

Example 8

Stan bought a television set for $859, a stereo for $697, and a computer (with printer) for $1285. What total amount did he spend?

Solution

The total amount spent is the sum:

$$\begin{array}{r} \$\ 859 \\ 697 \\ +\ 1285 \\ \hline \$\ 2841 \end{array}$$

Now work exercise 8 in the margin.

Completion Example Answers

5.
$$\begin{array}{r} 4375 \\ +5423 \\ \hline \mathbf{8} \end{array} \quad \begin{array}{r} 4375 \\ +5423 \\ \hline \mathbf{98} \end{array} \quad \begin{array}{r} 4375 \\ +5423 \\ \hline \mathbf{798} \end{array} \quad \begin{array}{r} 4375 \\ +5423 \\ \hline \mathbf{9798} \end{array}$$

7.
$$\begin{array}{r} 3\overset{2}{2}8 \\ 604 \\ 517 \\ +192 \\ \hline \mathbf{1} \end{array} \quad \begin{array}{r} \overset{1}{3}\overset{2}{2}8 \\ 604 \\ 517 \\ +192 \\ \hline \mathbf{41} \end{array} \quad \begin{array}{r} \overset{1}{3}\overset{2}{2}8 \\ 604 \\ 517 \\ +\ 192 \\ \hline \mathbf{1641} \end{array}$$

Practice Problems

Which property of addition is illustrated?

1. $12 + 0 = 12$

2. $15 + 3 = 3 + 15$

Find the following sums.

3. $\begin{array}{r} 57 \\ +\ 98 \\ \hline \end{array}$

4. $\begin{array}{r} 36 \\ 78 \\ +\ 89 \\ \hline \end{array}$

Answers to Practice Problems:

1. additive identity property

2. commutative property

3. 155

4. 203

Name ______________________ Section ____________ Date __________

Exercises 1.2

Do the following exercises mentally and write only the answers. See Examples 1 through 3.

1. $9 + (3 + 7)$

2. $(4 + 5) + 6$

3. $(2 + 3) + 8$

4. $(3 + 6) + (4 + 3)$

5. $8 + (3 + 4) + 6$

6. $9 + (8 + 3)$

7. $3 + 2 + 5$

8. $7 + (6 + 3)$

9. $(2 + 3) + 7$

10. $8 + 9 + 5$

11. $9 + 6 + 3$

12. $8 + 8 + 7$

13. $4 + 4 + 4$

14. $9 + 1 + 5 + 6$

15. $5 + 3 + 7 + 2$

16. $4 + 9 + 7 + 5$

17. $5 + 4 + 6 + 4 + 8$

18. $6 + 2 + 9 + 1$

19. $6 + 3 + 7 + 5 + 3$

20. $8 + 8 + 7 + 6 + 3$

Show that the following statements are true by performing the addition. State which property of addition is being illustrated. See Examples 2 and 3.

21. $9 + 3 = 3 + 9$

22. $8 + (5 + 2) = 8 + (2 + 5)$

23. $4 + (5 + 3) = (4 + 5) + 3$

24. $4 + 8 = 8 + 4$

25. $2 + (1 + 6) = (2 + 1) + 6$

26. $(8 + 7) + 3 = 8 + (7 + 3)$

ANSWERS

1. ________
2. ________
3. ________
4. ________
5. ________
6. ________
7. ________
8. ________
9. ________
10. ________
11. ________
12. ________
13. ________
14. ________
15. ________
16. ________
17. ________
18. ________
19. ________
20. ________
21. ________
22. ________
23. ________
24. ________
25. ________
26. ________

27. ______

28. ______

29. ______

30. ______

31. ______

32. ______

33. ______

34. ______

35. ______

36. ______

37. ______

38. ______

39. ______

40. ______

41. ______

42. ______

43. ______

44. ______

45. ______

27. $9 + 0 = 9$

28. $(2 + 3) + 4 = (3 + 2) + 4$

29. $7 + (6 + 0) = 7 + 6$

30. $8 + 20 + 1 = 8 + 21$

Add. See Examples 4 through 8.

31. 65 + 43

32. 24 + 78

33. 73 + 68 + 98

34. 165 + 276 + 394

35. 876 + 279 + 143

36. 268 + 93

37. 981 + 146

38. 2112 + 147 + 904 + 1005

39. 114 + 5402 + 710 + 643

40. 1403 + 7010 + 622 + 29

41. 213,116 + 116,018

42. 21,442 + 32,462

43. 438,966 + 572,486 + 227,462

44. 123,456 + 456,123 + 1,879,282

45. Travel. Mr. Juarez kept the mileage records indicated in the table shown here. How many miles did he drive during the six months?

Month	Mileage
Jan.	546
Feb.	378
Mar.	496
Apr.	357
May	503
June	482

Name ______________ Section ______________ Date ______________

ANSWERS

46. ______________

47. ______________

48. ______________

49. ______________

50. ______________

46. Profits. The Modern Products Corp. showed profits as indicated in the table for the years 2003 – 2006. What were the company's total profits for the years 2003 – 2006?

Year	Profits
2003	$ 1,078,416
2004	$ 1,270,842
2005	$ 2,000,593
2006	$ 1,963,472

47. Educational expenses. During four years of college, Fred's expenses were $12,035; $13,300; $14,000; and $15,750. What were his total expenses for four years of schooling?

48. Budget. The Magley family has the following monthly budget: $815 mortgage; $69 electric; $47 water; and $122 phone bills (including cell phones). What is the family's budget for each month for these expenses?

49. Enrollment. The following numbers of students at South Junior College are enrolled in mathematics courses: 303 in arithmetic; 476 in algebra; 293 in trigonometry; 257 in college algebra; and 189 in calculus. Find the total number of students taking mathematics.

50. Manufacturing. In one year, Stream Line Appliance Co. made 4217 gas stoves; 3947 electric stoves; and 9576 toasters. What was the total number of appliances Stream Line Appliance Co. produced that year?

51. [Respond in exercise.]

52. [Respond in exercise.]

Complete each of the following addition tables. You may want to make up your own tables for practicing difficult combinations.

51.

+	5	8	7	9
3				
6				
5				
2				

52.

+	6	7	1	4	9
3					
5					
4					
8					
2					

Name ______________________ Section ____________ Date __________

ANSWERS

46. ______________

47. ______________

48. ______________

49. ______________

50. ______________

46. Profits. The Modern Products Corp. showed profits as indicated in the table for the years 2003 – 2006. What were the company's total profits for the years 2003 – 2006?

Year	Profits
2003	$ 1,078,416
2004	$ 1,270,842
2005	$ 2,000,593
2006	$ 1,963,472

47. Educational expenses. During four years of college, Fred's expenses were $12,035; $13,300; $14,000; and $15,750. What were his total expenses for four years of schooling?

48. Budget. The Magley family has the following monthly budget: $815 mortgage; $69 electric; $47 water; and $122 phone bills (including cell phones). What is the family's budget for each month for these expenses?

49. Enrollment. The following numbers of students at South Junior College are enrolled in mathematics courses: 303 in arithmetic; 476 in algebra; 293 in trigonometry; 257 in college algebra; and 189 in calculus. Find the total number of students taking mathematics.

50. Manufacturing. In one year, Stream Line Appliance Co. made 4217 gas stoves; 3947 electric stoves; and 9576 toasters. What was the total number of appliances Stream Line Appliance Co. produced that year?

51. [Respond in exercise.]

52. [Respond in exercise.]

Complete each of the following addition tables. You may want to make up your own tables for practicing difficult combinations.

51.

+	5	8	7	9
3				
6				
5				
2				

52.

+	6	7	1	4	9
3					
5					
4					
8					
2					

Name ______________________ Section ______________ Date ____________

ANSWERS

46. **Profits.** The Modern Products Corp. showed profits as indicated in the table for the years 2003 – 2006. What were the company's total profits for the years 2003 – 2006?

Year	Profits
2003	$ 1,078,416
2004	$ 1,270,842
2005	$ 2,000,593
2006	$ 1,963,472

46. ____________

47. ____________

47. **Educational expenses.** During four years of college, Fred's expenses were $12,035; $13,300; $14,000; and $15,750. What were his total expenses for four years of schooling?

48. ____________

48. **Budget.** The Magley family has the following monthly budget: $815 mortgage; $69 electric; $47 water; and $122 phone bills (including cell phones). What is the family's budget for each month for these expenses?

49. ____________

49. **Enrollment.** The following numbers of students at South Junior College are enrolled in mathematics courses: 303 in arithmetic; 476 in algebra; 293 in trigonometry; 257 in college algebra; and 189 in calculus. Find the total number of students taking mathematics.

50. **Manufacturing.** In one year, Stream Line Appliance Co. made 4217 gas stoves; 3947 electric stoves; and 9576 toasters. What was the total number of appliances Stream Line Appliance Co. produced that year?

50. ____________

51. [Respond in exercise.]

52. [Respond in exercise.]

Complete each of the following addition tables. You may want to make up your own tables for practicing difficult combinations.

51.

+	5	8	7	9
3				
6				
5				
2				

52.

+	6	7	1	4	9
3					
5					
4					
8					
2					

1.3 Subtraction with Whole Numbers

Subtraction with Whole Numbers

A long-distance runner who is 25 years old had run 7 miles of his usual 9-mile workout when a storm forced him to quit for the day. How many miles short was he of his usual daily training?

Objectives

① Be able to subtract whole numbers without borrowing.

② Be able to subtract whole numbers with borrowing.

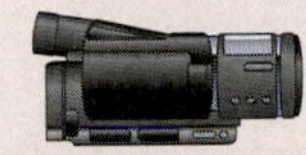

What *thinking* did you do to answer the question? You may have thought something like this: "Well, I don't need to know how old the runner is to answer the question, so his age is just extra information. Since he had already run 7 miles, I need to know what to add to 7 miles to get 9 miles. Since 7 + 2 = 9, he was 2 miles short of his usual workout."

In this problem, the sum of two addends was given, and only one of the addends was given. The other addend was the unknown quantity.

7	+	□	=	9
addend		missing addend		sum

As you may know, this kind of addition problem is called **subtraction** and can be written:

9	–	7	=	□	(read: "9 minus 7 equals blank.")
sum		addend		missing addend (or difference)	

Or

$$\begin{array}{r} 9 \\ -\ 7 \\ \hline \square \end{array}$$

sum
addend
difference or missing addend

We generally think of subtraction as the reverse of addition, or the process of "taking away" a number from the total to find the difference.

Subtraction

Subtraction is the operation of taking one amount, or number, away from another. (This is the opposite of adding the two amounts.)

The **difference** is the result of subtracting one addend (also called the **subtrahend**) from the sum (also called the **minuend**).

1. Perform the following subtraction:
$15-4$

Example 1

$$10 - 7 = \boxed{?}$$

sum ↑ 10, addend ↑ 7, missing addend (or difference) ↑ ?

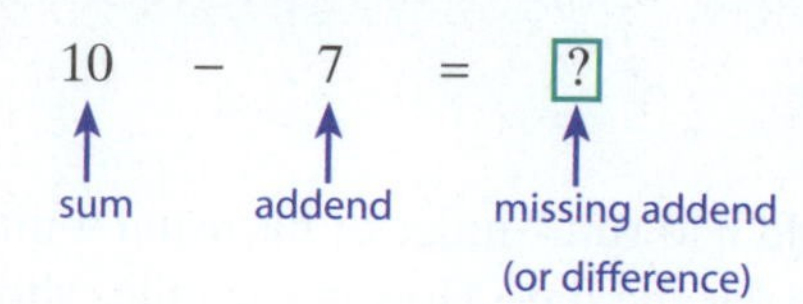

Think: What number added to 7 will give 10?
Since $7 + 3 = 10$, we have $10 - 7 = 3$; or

$$\begin{array}{r} 10 \\ \underline{-7} \\ 3 \end{array}$$

10 ⟵ minuend
−7 ⟵ subtrahend
3 ⟵ difference or missing addend

Now work exercise 1 in the margin.

Objective 1

To Subtract Whole Numbers with More Than One Digit:

1. Write the numbers vertically so that the place values are **lined up** in columns.
2. Subtract only the digits with the same place value.

2. Subtract.

$$\begin{array}{r} 654 \\ \underline{-421} \end{array}$$

Example 2

Subtract.

$496-342$

Solution

Step 1:

$$\begin{array}{r} 4\ 9\ 6 \\ \underline{-3\ 4\ 2} \\ 4 \end{array}$$

Subtract ones. $6-2=4$

Step 2:

$$\begin{array}{r} 4\ 9\ 6 \\ \underline{-3\ 4\ 2} \\ 5\ 4 \end{array}$$

Subtract tens. $9-4=5$

Step 3:

$$\begin{array}{r} 4\ 9\ 6 \\ \underline{-3\ 4\ 2} \\ 1\ 5\ 4 \end{array}$$

Subtract hundreds. $4-3=1$

154 ⟵ difference

Or, using expanded notation, we get the same result:

$$\begin{aligned} 496 &= \quad 400+90+6 \\ \underline{-342} &= \underline{-[300+40+2]} \\ & \quad\ 100+50+4=154 \end{aligned}$$

⟵ difference

Now work exercise 2 in the margin.

Borrowing

1. **Borrowing** is necessary when a digit is smaller than the digit being subtracted.

2. The process starts from the rightmost digit. **Borrow** from the digit to the left.

Objective 2

Example 3

Find the difference using expanded notation.

65 – 28

Solution

$$\begin{aligned} 65 &= \quad 60+5 \\ -28 &= -[20+8] \end{aligned}$$

Starting from the right, 5 is smaller than 8. We cannot subtract 8 from 5, so we **borrow** 10 from 60.

Borrow 10 from 60. ↓ — 10 borrowed from 60, plus 5. ↓

$$\begin{aligned} 65 &= \quad 60+5 &&= \quad 50+10+5 &&= \quad 50+15 \\ -28 &= -[20+8] &&= -[20+0+8] &&= -[20+\ 8] \\ & && && \quad 30+\ 7 = 37 \end{aligned}$$

Now subtract.

Now work exercise 3 in the margin.

3. Find the difference using expanded notation.

$$\begin{array}{r} 91 \\ -\ 43 \\ \hline \end{array}$$

Example 4

Use expanded notation and borrowing to find the difference.

536 – 258

Solution

$$\begin{aligned} 536 &= \quad 500+30+6 \\ -258 &= -[200+50+8] \end{aligned}$$

Since 6 is smaller than 8, borrow 10 from 30. — Since 20 is smaller than 50, borrow 100 from 500.

$$\begin{aligned} 536 &= \quad 500+30+6 &&= \quad 500+20+16 &&= \quad 400+120+16 \\ -258 &= -[200+50+8] &&= -[200+50+\ 8] &&= -[200+\ 50+\ 8] \\ & && && \quad 200+\ 70+\ 8 \\ & && && = \quad 278 \end{aligned}$$

Now work exercise 4 in the margin.

4. Find the difference using expanded notation and borrowing.

$$\begin{array}{r} 722 \\ -359 \\ \hline \end{array}$$

A common practice, with which you are probably familiar, is to indicate the borrowing by crossing out digits and writing new digits instead of using expanded notation. The expanded notation technique has been shown so that you can have a "picture" in your mind to help you understand the procedure of subtraction with borrowing.

5. Subtract.

$$\begin{array}{r} 867 \\ -\,328 \\ \hline \end{array}$$

Example 5

Subtract.

$$\begin{array}{r} 65 \\ -28 \\ \hline \end{array}$$

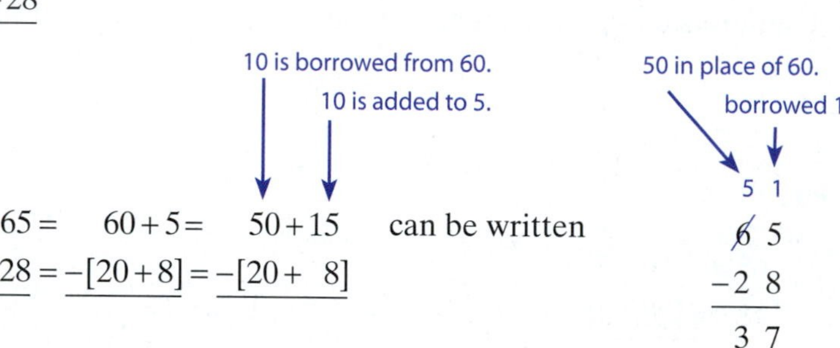

$$\begin{array}{r} 65 = \\ -28 = \\ \hline \end{array} \begin{array}{r} 60+5= \\ -[20+8]= \\ \hline \end{array} \begin{array}{r} 50+15 \\ -[20+\;\;8] \\ \hline \end{array}$$ can be written $$\begin{array}{r} \scriptstyle 5\;\;1 \\ \not{6}\;5 \\ -2\;8 \\ \hline 3\;7 \end{array}$$

Now work exercise 5 in the margin.

6. Subtract.

$$\begin{array}{r} 426 \\ -\,388 \\ \hline \end{array}$$

Example 6

Subtract.

$$\begin{array}{r} 536 \\ -258 \\ \hline \end{array}$$

Solution

Step 1:

Since 6 is smaller than 8, borrow 10 from 30 and add this 10 to 6 to get 16. This leaves 20, so cross out 3 and write 2.

$$\begin{array}{r} \scriptstyle 2\;1\; \\ 5\,\not{3}\,6 \\ -2\;5\;8 \\ \hline \end{array}$$

Step 2:

Since 2 is smaller than 5, borrow 100 from 500. This leaves 400, so cross out 5 and write 4.

$$\begin{array}{r} \scriptstyle 4\;12\;1\; \\ \not{5}\,\not{3}\,6 \\ -2\;5\;8 \\ \hline \end{array}$$

Step 3:

Now subtract.

$$\begin{array}{r} \scriptstyle 4\;12\;1\; \\ \not{5}\,\not{3}\,6 \\ -\,2\;5\;8 \\ \hline 2\;7\;8 \end{array}$$

Now work exercise 6 in the margin.

Example 7

Subtract.

$$\begin{array}{r} 8000 \\ -\ \ 657 \\ \hline \end{array}$$

Solution

Step 1:

Trying to borrow from 0 each time, we end up borrowing 1000 from 8000. Cross out 8 and write 7.

$$\begin{array}{r} {\scriptstyle 7\,1} \\ \not{8}000 \\ -\ \ 657 \\ \hline \end{array}$$

Step 2:

Borrow 100 from 1000. Cross out 10 and write 9.

$$\begin{array}{r} {\scriptstyle 9} \\ {\scriptstyle 7\,\not{1}1} \\ \not{8}\not{0}00 \\ -\ \ 657 \\ \hline \end{array}$$

Step 3:

Borrow 10 from 100. Cross out 10 and write 9.

$$\begin{array}{r} {\scriptstyle 9\,9} \\ {\scriptstyle 7\,\not{1}\,\not{1}1} \\ \not{8}\not{0}\not{0}0 \\ -\ \ 657 \\ \hline \end{array}$$

Step 4:

Now subtract.

$$\begin{array}{r} {\scriptstyle 9\,9} \\ {\scriptstyle 7\,\not{1}\,\not{1}1} \\ \not{8}\not{0}\not{0}0 \\ -\ \ 657 \\ \hline 7343 \end{array}$$

Note: Subtraction can be "checked" by addition. The sum of the difference and the subtrahend must be equal to the minuend. Thus, in Example 7, check:

$$\begin{array}{r} 7343 \\ +\ \ 657 \\ \hline 8000 \end{array}$$

Now work exercise 7 in the margin.

Find the following differences without using expanded notation.

7. $\begin{array}{r} 7000 \\ -\ 423 \\ \hline \end{array}$

8. What number should be added to 268 to get a sum of 654?

Example 8

What number should be added to 546 to get a sum of 732?

Solution

We know the sum and one addend. To find the missing addend, subtract 546 from 732.

$$\begin{array}{r} {\scriptstyle 6\ 12\ 1} \\ \not{7}\,\not{3}\,2 \\ -5\,4\,6 \\ \hline 1\,8\,6 \end{array} \quad \text{difference}$$

The number to be added is 186.

Now work exercise 8 in the margin.

9. The cost of replacing Robert's used DVD player was going to be $120 for parts (including tax) plus $45 for labor. To buy a new DVD player, Robert was going to have to pay $115 plus $8 in tax and the dealer was going to pay him $25 for his old DVD player. How much more would it cost to fix the old DVD player than to buy a new one?

Example 9

The cost of repairing Ed's used TV set was going to be $250 for parts (including tax) plus $85 for labor. To buy a new set, he was going to pay $450 plus $27 in tax and the dealer was going to pay him $85 for his old set. How much more would Ed have to pay for a new set than to have his old set repaired?

Solution

Used Set	New Set	Difference
$250 parts	$450 cost	$392
+ 85 labor	+ 27 tax	− 335
$335	$477	$ 57
	− 85 trade-in	
	$392 total	

Ed would pay $57 more for the new set than for having his old set repaired. What would you do?

Now work exercise 9 in the margin.

10. Subtract using expanded notation.

$$\begin{array}{r} 5\,7\,7 \\ -\,2\,3\,6 \\ \hline \end{array}$$

Completion Example 10

Subtract 897 – 364 by using expanded notation.

Solution

$$\begin{array}{rcl} 8\,9\,7 & = & 8\,0\,0 + 9\,0 + \_\_\_\_ \\ -3\,6\,4 & = & 3\,0\,0 + \_\_\_\_ + \_\_\_\_ \\ \hline & & \_\_\_\_ + \_\_\_\_ + \_\_\_\_ = \_\_\_\_ \end{array}$$

Now work exercise 10 in the margin.

Completion Example 11

Two painters bid on painting the same house. The first painter bid $2738 and the second painter bid $2950. What was the difference between the two bids?

Solution

This is a subtraction problem because we are asked for a difference.

$$\begin{array}{r} \$\,2950 \\ -\ 2738 \\ \hline \underline{\qquad} \end{array} \quad \leftarrow \text{difference in bids}$$

Now work exercise 11 in the margin.

11. Two neighboring towns have populations of 15,239 and 12,683. What is the difference in their populations?

Completion Example 12

After selling their house for $132,000, the owner paid the realtor $7920, back taxes of $450, and $350 in other fees. If they also paid off the bank loan of $57,000, how much cash did the owners receive from the sale?

Solution

In this problem, experience tells us that we must add and subtract even though there are no specific directions to do so. We add the expenses and then subtract this sum from the selling price to find the cash that the owners received.

$$\begin{array}{r} \$\ 7920 \\ 450 \\ 350 \\ +57{,}000 \\ \hline \underline{\qquad} \end{array} \leftarrow \text{total expenses} \qquad \begin{array}{rl} \$132{,}000 & \text{selling price} \\ - & \\ \underline{\qquad} & \text{total expenses} \\ \underline{\qquad} & \text{cash to owners} \end{array}$$

Now work exercise 12 in the margin.

12. You have $50,000 invested in stocks, but decide to sell off $31,897 of your holdings. What is the value of your remaining stock?

Completion Example Answers

10.

$$\begin{array}{r} 897 = 800 + 90 + \mathbf{7} \\ \underline{-364} = \underline{300 + \mathbf{60} + \mathbf{4}} \\ \mathbf{500 + 30 + 3 = 533} \end{array}$$

11.

$$\begin{array}{r} \$2950 \\ \underline{-\ 2738} \\ \mathbf{\$\ 212} \end{array}$$ ← difference in bids

12.

$$\begin{array}{r} \$\ 7{,}920 \\ 450 \\ 350 \\ \underline{+57{,}000} \\ \mathbf{\$65{,}720} \end{array}$$ ← total expenses

$$\begin{array}{rl} \$132{,}000 & \text{selling price} \\ \underline{-\ \mathbf{65{,}720}} & \text{total expenses} \\ \$\ \mathbf{66{,}280} & \text{cash to owners} \end{array}$$

Practice Problems

Find the following differences.

1. $\begin{array}{r} 83 \\ \underline{-\ 54} \end{array}$

2. $\begin{array}{r} 600 \\ \underline{-\ 368} \end{array}$

3. $\begin{array}{r} 7856 \\ \underline{-\ 6397} \end{array}$

Answers to Practice Problems:

1. 29 **2.** 232 **3.** 1459

Name ______________________ Section ____________ Date __________

ANSWERS

1. ____
2. ____
3. ____
4. ____
5. ____
6. ____
7. ____
8. ____
9. ____
10. ____
11. ____
12. ____
13. ____
14. ____
15. ____
16. ____
17. ____
18. ____
19. ____
20. ____
21. ____
22. ____
23. ____
24. ____
25. ____
26. ____
27. ____
28. ____

Exercises 1.3

Subtract. Do as many problems as you can mentally. See Examples 1 through 8.

1. $8 - 5$ **2.** $19 - 6$ **3.** $14 - 14$ **4.** $17 - 9$

5. $20 - 11$ **6.** $17 - 0$ **7.** $17 - 8$ **8.** $16 - 16$

9. $11 - 6$ **10.** $13 - 7$ **11.** $17 - 17$ **12.** $42 - 31$

13. $89 - 76$ **14.** $53 - 33$ **15.** $47 - 27$ **16.** $96 - 27$

17. $23 - 18$ **18.** $74 - 29$ **19.** $61 - 48$ **20.** $52 - 27$

21. $126 - 32$ **22.** $174 - 48$ **23.** $347 - 129$ **24.** $256 - 118$

25. $692 - 217$ **26.** $543 - 167$ **27.** $900 - 307$ **28.** $603 - 208$

29. ______

30. ______

31. ______

32. ______

33. ______

34. ______

35. ______

36. ______

37. ______

38. ______

39. ______

40. ______

41. ______

42. ______

43. ______

44. ______

45. ______

46. ______

47. ______

48. ______

49. ______

29. $\begin{array}{r} 474 \\ -\ 286 \\ \hline \end{array}$ **30.** $\begin{array}{r} 657 \\ -\ 179 \\ \hline \end{array}$ **31.** $\begin{array}{r} 7843 \\ -\ 6274 \\ \hline \end{array}$ **32.** $\begin{array}{r} 6793 \\ -\ 5827 \\ \hline \end{array}$

33. $\begin{array}{r} 4376 \\ -\ 2808 \\ \hline \end{array}$ **34.** $\begin{array}{r} 3275 \\ -\ 1744 \\ \hline \end{array}$ **35.** $\begin{array}{r} 3546 \\ -\ 3546 \\ \hline \end{array}$ **36.** $\begin{array}{r} 4900 \\ -\ 3476 \\ \hline \end{array}$

37. $\begin{array}{r} 5070 \\ -\ 4376 \\ \hline \end{array}$ **38.** $\begin{array}{r} 8007 \\ -\ 2136 \\ \hline \end{array}$ **39.** $\begin{array}{r} 4065 \\ -\ 1548 \\ \hline \end{array}$ **40.** $\begin{array}{r} 7602 \\ -\ 2985 \\ \hline \end{array}$

41. $\begin{array}{r} 7,085,076 \\ -\ 4,278,432 \\ \hline \end{array}$ **42.** $\begin{array}{r} 6,543,222 \\ -\ 2,742,663 \\ \hline \end{array}$ **43.** $\begin{array}{r} 4,000,000 \\ -\ 2,993,042 \\ \hline \end{array}$

44. $\begin{array}{r} 8,000,000 \\ -\ 647,561 \\ \hline \end{array}$ **45.** $\begin{array}{r} 6,000,000 \\ -\ 328,989 \\ \hline \end{array}$

46. What number should be added to 978 to get a sum of 1200?

47. What number should be added to 860 to get a sum of 1000?

48. If the sum of two numbers is 537 and one of the numbers is 139, what is the other number?

49. Schooling. A man is 36 years old, and his wife is 34 years old. Together, they have attended 28 years of school, including college. If the man attended 12 years of school, how many years did his wife attend?

Name ____________________ Section ____________ Date __________

ANSWERS

50. ____________

50. Basketball. Team A has twelve players and won its first three games by the following scores: 84 to 73, 97 to 78, and 101 to 63. Team B has ten players and won its first three games by 76 to 75, 83 to 70, and 94 to 84. What is the difference between the total of the differences of Team A's scores and those of its opponents and the total of the differences of Team B's scores and those of its opponents?

51. ____________

51. Construction. The Kingston Construction Co. made a bid of $7,043,272 to build a stretch of freeway, but the Beach City Construction Co. made a lower bid of $6,792,868. How much lower was the Beach City bid?

52. ____________

52. Landscaping. Two landscaping companies made bids on the landscaping of a new apartment complex. Company A bid $550,000 for materials and plants and $225,000 for labor. Company B bid $600,000 for materials and plants and $182,000 for labor. Which company had the lower total bid? How much lower was it?

Company A   Company B

green landscaping   LAWNS 'R' US

53. ____________

53. Checking. In June, Ms. White opened a checking account and deposited $1342, $238, $57, and $486. She also wrote checks for $132, $76, $42, $480, $90, and $327. What was her balance at the end of June?

54. ____________

54. Manufacturing. A manufacturing company had assets of $5,027,479, which included $1,500,000 in real estate. The liabilities were $4,792,023. By how much was the company "in the black"?

55. Real estate. A couple sold their house for $135,000. They paid the realtor $8100, and other expenses of the sale came to $800. If they owed the bank $87,000 for the mortgage, what were their net proceeds from the sale?

55. ____________

56. ____________

56. Home loans. A woman bought a condominium for a price of $150,000. She also had to pay other expenses of $750. If the local savings and loan association agreed to loan her $105,500 as a first trust deed on the house, how much cash did she need to make the purchase?

57. ____________

57. Automobile purchase. Junior bought a red car with a sticker price of $10,000, but the salesman added $1200 for taxes, license, and extras. The bank agreed to give Junior a loan of $7500 on the car. How much cash did Junior need to buy the car?

58. ____________

58. Automobile prices. In pricing a four-door car, Pat found she would have to pay a base price of $9500 plus $570 in taxes and $250 for license fees. For a two-door of the same make, she would pay a base price of $8700 plus $522 in taxes and $230 for license fees. Including all expenses, how much cheaper was the two-door model?

Writing and Thinking about Mathematics

59. ____________

59. Nothing was said in the text about a commutative property for subtraction. Do you think that this omission was intentional? Is there a commutative property of subtraction? Give several examples that justify your answer and discuss this idea in class.

60. ____________

60. Nothing was said in the text about an associative property for subtraction. Do you think that this omission was intentional? Is there an associative property of subtraction? Give several examples that justify your answer and discuss this idea in class.

61. ____________

61. You may have heard someone say that "you cannot subtract a larger number from a smaller number." Do you think that this statement is true? Does a subtraction problem such as 6 – 10 make sense to you? Can you think of any situation in which such a difference might seem reasonable?

Name ______________________ Section ____________ Date __________

ANSWERS

Collaborative Learning

Separate the class into teams of two to four students. Each team is to read and fill out the partial 1040A forms in Exercises 62 – 64. After these are completed, the team leader is to read the team's results and discuss any difficulties they had in reading and following directions while filling out the forms.

62. The Internal Revenue Service allows us to calculate our income tax returns with whole numbers. Calculate the taxable income on the portion of the Form 1040A shown here. Assume that you have 3 exemptions and you have not housed a person displaced by Hurricane Katrina.

62. [Respond in exercise.]

Form 1040A (2005) Page **2**

Tax, credits, and payments	Line	Description	Line	Amount
	22	Enter the amount from line 21 (adjusted gross income).	22	32,484
	23a	Check if: { ☐ **You** were born before January 2, 1941, ☐ Blind } **Total boxes** / { ☐ **Spouse** was born before January 2, 1941, ☐ Blind } **checked** ▶	23a	☐
Standard Deduction for—	**b**	If you are married filing separately and your spouse itemizes deductions, see page 32 and check here ▶	23b	☐
• People who checked any box on line 23a or 23b **or** who can be claimed as a dependent, see page 32.	**24**	Enter your **standard deduction** (see left margin).	24	5000
• All others:	**25**	Subtract line 24 from line 22. If line 24 is more than line 22, enter -0-.	25	
	26	If line 22 is over $109,475, or you provided housing to a person displaced by Hurricane Katrina, see page 33. Otherwise, multiply $3,200 by the total number of exemptions claimed on line 6d.	26	
	27	Subtract line 26 from line 25. If line 26 is more than line 25, enter -0-. This is your **taxable income.** ▶	27	

63. Calculate the taxable income on the portion of the Form 1040A shown here. Assume that you have 4 exemptions and you have not housed a person displaced by Hurricane Katrina.

63. [Respond in exercise.]

Form 1040A (2005) Page **2**

Tax, credits, and payments	Line	Description	Line	Amount
	22	Enter the amount from line 21 (adjusted gross income).	22	56,382
	23a	Check if: { ☐ **You** were born before January 2, 1941, ☐ Blind } **Total boxes** / { ☐ **Spouse** was born before January 2, 1941, ☐ Blind } **checked** ▶	23a	☐
Standard Deduction for—	**b**	If you are married filing separately and your spouse itemizes deductions, see page 32 and check here ▶	23b	☐
• People who checked any box on line 23a or 23b **or** who can be claimed as a dependent, see page 32.	**24**	Enter your **standard deduction** (see left margin).	24	10,000
• All others:	**25**	Subtract line 24 from line 22. If line 24 is more than line 22, enter -0-.	25	
	26	If line 22 is over $109,475, or you provided housing to a person displaced by Hurricane Katrina, see page 33. Otherwise, multiply $3,200 by the total number of exemptions claimed on line 6d.	26	
	27	Subtract line 26 from line 25. If line 26 is more than line 25, enter -0-. This is your **taxable income.** ▶	27	

64. [Respond in exercise.]

64. Calculate the taxable income on the portion of the Form 1040A shown here. Assume that you have 2 exemptions and that you have not housed a person displaced by Hurricane Katrina.

Form 1040A (2005) Page **2**

Tax, credits, and payments	**22**	Enter the amount from line 21 (adjusted gross income).	22	81,695
	23a	Check if: ☐ **You** were born before January 2, 1941, ☐ Blind; ☐ **Spouse** was born before January 2, 1941, ☐ Blind. **Total boxes checked** ► 23a ☐		
Standard Deduction for— • People who checked any box on line 23a or 23b **or** who can be claimed as a dependent, see page 32. • All others:	**b**	If you are married filing separately and your spouse itemizes deductions, see page 32 and check here ► 23b ☐		
	24	Enter your **standard deduction** (see left margin).	24	10,000
	25	Subtract line 24 from line 22. If line 24 is more than line 22, enter -0-.	25	
	26	If line 22 is over $109,475, or you provided housing to a person displaced by Hurricane Katrina, see page 33. Otherwise, multiply $3,200 by the total number of exemptions claimed on line 6d.	26	
	27	Subtract line 26 from line 25. If line 26 is more than line 25, enter -0-. This is your **taxable income.** ►	27	

1.4 Rounding and Estimating with Whole Numbers

Rounding Whole Numbers

Objectives

① Know the rule for rounding whole numbers to any given place.

② Use the rule of rounding to estimate sums and differences.

③ Develop a sense of the expected size of a sum or difference so that major errors can be easily spotted.

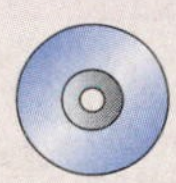

To **round** a given number means to find another number close to the given number. The desired place of accuracy must be stated. For example, if you were asked to round 872, you would not know what to do unless you were told the position or place of accuracy desired. The number lines in Figure 1.3 help to illustrate the problem.

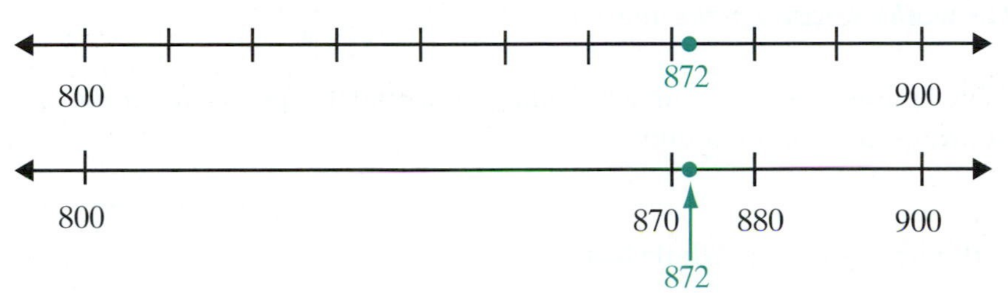

Figure 1.3

We can see that 872 is closer to 900 than to 800. So, **to the nearest hundred**, 872 rounds to 900. Also, 872 is closer to 870 than to 880. So, **to the nearest ten**, 872 rounds to 870.

Inaccuracy in numbers is common and rounded answers can be quite acceptable in many situations. For example, what do you think is the distance across the United States from the east coast to the west coast? Most people will use the rounded answer of 3000 miles. This answer is appropriate and acceptable.

Number lines are used as visual aids in rounding in the following examples.

Example 1

Round 47 to the nearest ten.

Solution

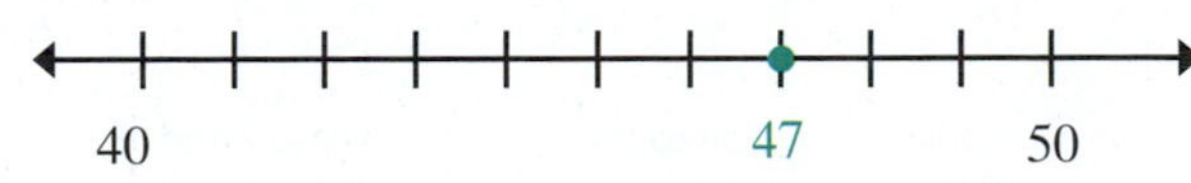

47 is closer to 50 than to 40. So 47 rounds to 50 (to the nearest ten).

Now work exercise 1 in the margin.

1. Use the number line to round 23 to the nearest ten.

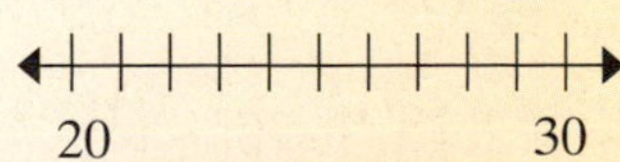

2. Use the number line to round 782 to the nearest hundred.

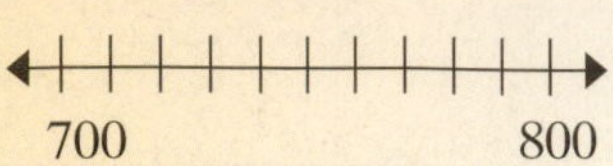

3. Round 2345 to the nearest thousand.

Example 2

Round 238 to the nearest hundred.

Solution

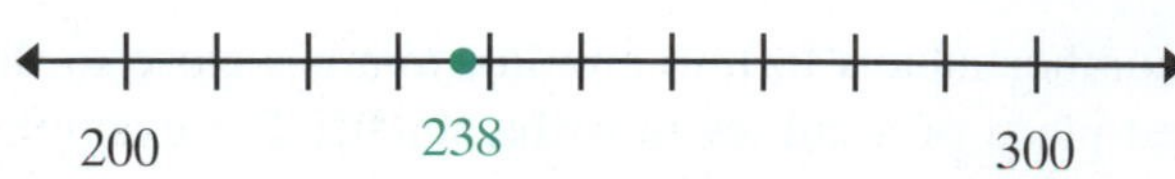

238 is closer to 200 than to 300. So 238 rounds off to 200 (to the nearest hundred).

Now work exercise 2 in the margin.

Using figures as an aid to understanding is fine but, for practical purposes, the following rule is more useful.

Objective 1

Rounding Rule for Whole Numbers

1. Look at the single digit just to the right of the digit that is in the place of desired accuracy.
2. If this digit is 5 or greater, make the digit in the desired place of accuracy one larger and replace all digits to the right with zeros. All digits to the left remain unchanged unless a 9 is made one larger.
3. If this digit is less than 5, leave the digit in the place of desired accuracy as it is and replace all digits to the right with zeros. All digits to the left remain unchanged.

The Rounding Rule is used in the following examples.

Example 3

Round 5749 to the nearest hundred.

Solution

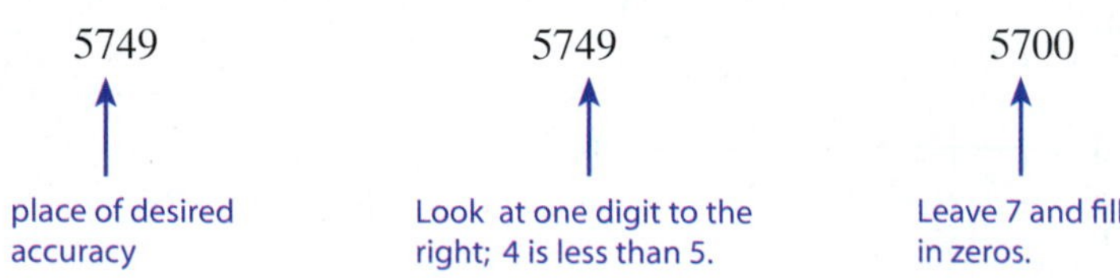

So 5749 rounds off to 5700 (to the nearest hundred).

Now work exercise 3 in the margin.

Example 4

Round 6500 to the nearest thousand.

Solution

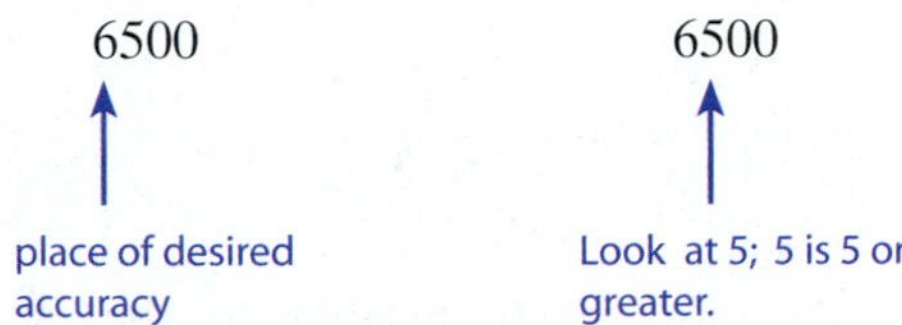

7000

Increase 6 to 7 (one larger) and fill in zeros.

So 6500 rounds off to 7000 (to the nearest thousand).

Now work exercise 4 in the margin.

4. Round 5300 to the nearest thousand.

Example 5

Round 983 to the nearest hundred.

Solution

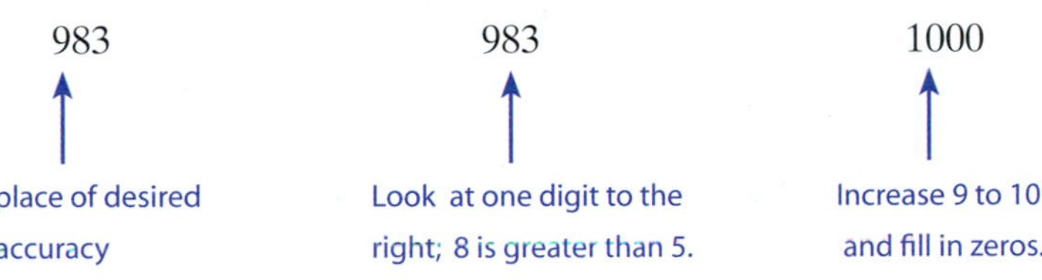

1000

Increase 9 to 10 and fill in zeros.

So 983 rounds off to 1000 (to the nearest hundred).

Now work exercise 5 in the margin.

5. Round 252 to the nearest hundred.

Estimating Answers

One very good use for rounded numbers is to **estimate** an answer before any calculations with the given numbers are made. Thus, **answers that are not at all reasonable**, possibly due to some large calculation error, **can be spotted**. Usually, we simply repeat the calculations and the error is found. There are also times, particularly in working word problems, when the wrong operation has been performed. For example, someone may have added when he should have subtracted.

To **estimate an answer** means to use rounded numbers in a calculation to get some idea of what the size of the actual answer should be. This is a form of checking your work before you do it.

Objective 2

To Estimate an Answer:

1. Round each number to the place of the leftmost digit. (Single-digit numbers are already considered to be rounded.)

2. Perform the indicated operation with these rounded numbers.

6. First estimate the sum, then find the sum.

$$\begin{array}{r} 885 \\ 467 \\ +\ 624 \\ \hline \end{array}$$

Example 6

Estimate the sum; then find the sum.

$$\begin{array}{r} 468 \\ 936 \\ +687 \\ \hline \end{array}$$

Solution

a. Estimate the sum first by rounding each number to the nearest hundred (in this case), and then by adding. In actual practice, many of these steps can be done mentally.

$$\begin{array}{rcr} 468 & \longrightarrow & 500 \\ 936 & \longrightarrow & 900 \\ +687 & \longrightarrow & +\ 700 \\ \hline & & 2100 \end{array} \quad \text{estimated sum}$$

Objective 3

b. Now we find the sum with the knowledge that the answer should be close to 2100.

$$\begin{array}{r} 468 \\ 936 \\ +\ 687 \\ \hline 2091 \end{array} \longrightarrow \text{This sum is quite close to 2100.}$$

Now work exercise 6 in the margin.

7. First estimate the difference, then find the difference.

$$\begin{array}{r} 780 \\ -\ 575 \\ \hline \end{array}$$

Example 7

Estimate the difference; then find the difference.

$$\begin{array}{r} 758 \\ -289 \\ \hline \end{array}$$

Solution

a. Estimate the difference first by rounding each number to the place of the leftmost digit (in this case, the nearest hundred) and then subtracting.

$$\begin{array}{rcr} 758 & \longrightarrow & 800 \\ -289 & \longrightarrow & -300 \\ \hline & & 500 \end{array} \longleftarrow \text{estimated difference}$$

b. Now we find the difference with the knowledge that the answer should be close to 500.

$$\begin{array}{r} {}^{6}\not{7}\,{}^{4\,1}\not{5}\,8 \\ -289 \\ \hline 469 \end{array} \longleftarrow \text{This difference is close to 500.}$$

Now work exercise 7 in the margin.

Completion Example 8

Add:

$$\begin{array}{r} 5483 \\ 232 \\ +\ 657 \\ \hline \end{array}$$

Solution

a. First, estimate the sum by rounding each number to the place of its leftmost digit and adding these rounded numbers.

$$\begin{array}{rcr} 5483 & \longrightarrow & 5000 \\ 232 & \longrightarrow & 200 \\ +\ 657 & \longrightarrow & +\ \underline{\qquad} \\ \hline & & \underline{\qquad} \end{array}$$

← estimated sum

b. Find the sum and compare your answer with the estimated sum. They should be "close."

$$\begin{array}{r} 5483 \\ 232 \\ +\ 657 \\ \hline \underline{\qquad} \end{array}$$

← actual sum

Now work exercise 8 in the margin.

8. Add by first estimating the sum and then finding the actual sum:

$$\begin{array}{r} 7824 \\ 133 \\ +\ 904 \\ \hline \end{array}$$

Completion Example Answers

8. a.

$$\begin{array}{r} 5483 \\ 232 \\ +\ 657 \\ \hline \end{array} \quad \begin{array}{c} \longrightarrow \\ \longrightarrow \\ \longrightarrow \\ \\ \end{array} \quad \begin{array}{r} 5000 \\ 200 \\ \mathbf{700} \\ \hline \mathbf{5900} \end{array} \longleftarrow \text{estimated sum}$$

b.

$$\begin{array}{r} 5483 \\ 232 \\ +\ 657 \\ \hline \mathbf{6372} \end{array} \longleftarrow \text{actual sum}$$

Practice Problems

Round as indicated.

1. 55 (nearest ten)

2. 43,610 (nearest thousand)

3. Estimate the sum; then find the sum.

$$\begin{array}{r} 147 \\ 172 \\ +\ 54 \\ \hline \end{array}$$

4. Estimate the difference; then find the difference.

$$\begin{array}{r} 854 \\ -\ 367 \\ \hline \end{array}$$

Answers to Practice Problems:

1. 60 **2.** 44,000 **3.** 350; 373 **4.** 500; 487

Name ______________________ Section ____________ Date __________

Exercises 1.4

Round as indicated. See Examples 1 through 5.

To the nearest ten:

1. 763 **2.** 31 **3.** 82 **4.** 503

5. 296 **6.** 722 **7.** 987 **8.** 347

To the nearest hundred:

9. 4163 **10.** 4475 **11.** 495 **12.** 572

13. 637 **14.** 3789 **15.** 76,523 **16.** 7007

To the nearest thousand:

17. 6912 **18.** 5500 **19.** 7500 **20.** 7499

21. 13,499 **22.** 13,501 **23.** 62,265 **24.** 47,800

To the nearest ten thousand:

25. 78,419 **26.** 125,000 **27.** 256,000 **28.** 62,200

29. 118,200 **30.** 312,500 **31.** 184,900 **32.** 615,000

33. 87 to the nearest hundred **34.** 46 to the nearest hundred

35. 532 to the nearest thousand

ANSWERS

1. ______
2. ______
3. ______
4. ______
5. ______
6. ______
7. ______
8. ______
9. ______
10. ______
11. ______
12. ______
13. ______
14. ______
15. ______
16. ______
17. ______
18. ______
19. ______
20. ______
21. ______
22. ______
23. ______
24. ______
25. ______
26. ______
27. ______
28. ______
29. ______
30. ______
31. ______
32. ______
33. ______
34. ______
35. ______

36. ____________

37. ____________

38. ____________

39. ____________

40. ____________

41. ____________

42. ____________

43. ____________

44. ____________

45. ____________

46. ____________

47. ____________

48. ____________

49. ____________

50. ____________

51. ____________

First estimate the answers by using rounded numbers; then find the following sums and differences. See Examples 6 through 8.

36. $\begin{array}{r} 83 \\ 62 \\ +\ 78 \\ \hline \end{array}$ **37.** $\begin{array}{r} 96 \\ 46 \\ +\ 25 \\ \hline \end{array}$ **38.** $\begin{array}{r} 146 \\ 259 \\ +\ 384 \\ \hline \end{array}$ **39.** $\begin{array}{r} 475 \\ 126 \\ +\ 572 \\ \hline \end{array}$

40. $\begin{array}{r} 600 \\ 542 \\ +\ 483 \\ \hline \end{array}$ **41.** $\begin{array}{r} 851 \\ 736 \\ +\ 294 \\ \hline \end{array}$ **42.** $\begin{array}{r} 5742 \\ 6271 \\ 8156 \\ +\ 972 \\ \hline \end{array}$ **43.** $\begin{array}{r} 483 \\ 1681 \\ 3054 \\ +\ 4006 \\ \hline \end{array}$

44. $\begin{array}{r} 22,506 \\ 38,700 \\ +\ 10,465 \\ \hline \end{array}$ **45.** $\begin{array}{r} 8742 \\ -\ 3275 \\ \hline \end{array}$ **46.** $\begin{array}{r} 6421 \\ -\ 1652 \\ \hline \end{array}$ **47.** $\begin{array}{r} 10,531 \\ -\ 4600 \\ \hline \end{array}$

48. $\begin{array}{r} 275,600 \\ -\ 94,300 \\ \hline \end{array}$ **49.** $\begin{array}{r} 63,504 \\ -\ 42,700 \\ \hline \end{array}$ **50.** $\begin{array}{r} 74,305 \\ -\ 33,082 \\ \hline \end{array}$

51. Population. If the population of the People's Republic of China is 1,292,270,000 and the population of the United States is 293,655,000, about how many more people live in China than live in the United States?

Name ______________________ Section ____________ Date __________

ANSWERS

52. Enrollment. In 2004, the University of California campuses had the following undergraduate enrollments:

UC Berkeley	22,880	UC Riverside	15,174
UC Davis	23,264	UC San Diego	20,339
UC Irvine	19,994	UC Santa Barbara	18,132
UC Los Angeles	24,946	UC Santa Cruz	13,694

Approximately how many undergraduate students were enrolled in the University of California system in 2004?

52. ____________

53. Book purchases. If you looked at the book list at school and found that the prices of the textbooks that you need to buy this semester are \$65, \$76, \$52, and \$25, about how much cash should you take with you, assuming you want to pay in cash: \$100, \$150, \$200, \$250, or \$300? (Don't forget that you will need to pay some sales tax.)

53. ____________

1.5 Multiplication with Whole Numbers

Basic Multiplication and Powers of Ten

Objectives

① Know the basic multiplication facts (the products of the numbers from 0 to 9).

② Develop the skill of mentally multiplying numbers involving powers of ten.

③ Be able to multiply whole numbers by using partial products.

④ Be able to multiply whole numbers by using the standard short method.

⑤ Estimate products.

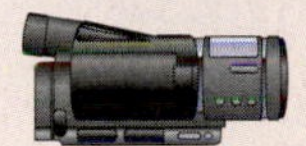
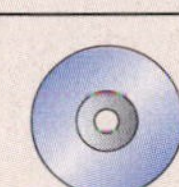

Multiplication is a process that shortens repeated addition with the same number. For example, we can write

$$8+8+8+8+8 = 5 \cdot 8 = 40.$$

The repeated addend (8) and the number of times it is used (5) are both called **factors** of 40, and the result (40) is called the **product** of the two factors.

$$8+8+8+8+8 = 5 \cdot 8 = 40$$

factor factor product

Terms Related to Multiplication with Whole Numbers

Multiplication: Repeated addition

Product: Result of multiplication

Factors: Numbers that are multiplied (Each whole number multiplied is a **factor** of the product.)

Several notations can be used to indicate multiplication. In this text, we will use the raised dot (as with $5 \cdot 8$) and parentheses much of the time. The cross sign (×) will also be used; however, we must be careful not to confuse it with the letter x used in some of the later chapters and in algebra.

Symbols for Multiplication

Symbol		Example
·	raised dot	$5 \cdot 8$
( )	numbers inside or next to parentheses	5(8) or (5)8 or (5)(8)
×	cross sign	5×8 or $\begin{array}{r} 8 \\ \underline{\times 5} \end{array}$

To change a multiplication problem to a repeated addition problem every time we multiply two numbers would be ridiculous. For example, $48 \cdot 137$ would mean using 137 as the addend 48 times. The first step in learning the multiplication process is to **memorize** the basic multiplication facts shown in Table 1.2. The factors in the table are only the digits 0 through 9. Using other factors involves the place value concept, as we shall see.

Objective 1

Table 1.2 Basic Multiplication Facts

·	**0**	**1**	**2**	**3**	**4**	**5**	**6**	**7**	**8**	**9**
0	**0**	0	0	0	0	0	0	0	0	0
1	0	**1**	2	3	4	5	6	7	8	9
2	0	2	**4**	6	8	10	12	14	16	18
3	0	3	6	**9**	12	15	18	21	24	27
4	0	4	8	12	**16**	20	24	28	32	36
5	0	5	10	15	20	**25**	30	35	40	45
6	0	6	12	18	24	30	**36**	42	48	54
7	0	7	14	21	28	35	42	**49**	56	63
8	0	8	16	24	32	40	48	56	**64**	72
9	0	9	18	27	36	45	54	63	72	**81**

If you have difficulty with **any** of the basic facts in the table, write all the possible combinations in a mixed-up order on a sheet of paper. Write the products down as quickly as you can and then find the ones you missed. Practice these in your spare time until you are sure you know them.

The table indicates three properties of multiplication. Since the table is a mirror image of itself on either side of the main diagonal (the numbers 0, 1, 4, 9, 16, 25, 36, 49, 64, 81), multiplication is **commutative**. Multiplication by 0 gives 0, and this result is called the **zero factor law**. Multiplication by 1 gives the number being multiplied, and thus 1 is called the **multiplicative identity** or the **identity element for multiplication**.

Find each of the indicated products and tell which property of multiplication is illustrated.

Commutative Property of Multiplication

For any two whole numbers a and b, $a \cdot b = b \cdot a$.

1. $7 \cdot 8 = 8 \cdot 7$

Example 1

$5 \cdot 3 = 15$ and $3 \cdot 5 = 15$. Thus, $5 \cdot 3 = 3 \cdot 5$.

Now work exercise 1 in the margin.

2. $\begin{array}{r} 5 \\ \times 1 \\ \hline \end{array} \quad \begin{array}{r} 1 \\ \times 5 \\ \hline \end{array}$

Example 2

$$\begin{array}{r} 9 \\ \times 8 \\ \hline 72 \end{array} \qquad \begin{array}{r} 8 \\ \times 9 \\ \hline 72 \end{array}$$

Thus, $8 \times 9 = 9 \times 8$.

Now work exercise 2 in the margin.

Zero Factor Law

For any whole number a, $a \cdot 0 = 0$.

Multiplicative Identity Property

There is a unique whole number 1 with the property that, for any whole number a, $a \cdot 1 = a$.

Example 3

Zero Factor Law		**Multiplicative Identity Property**	
a.	$0 \cdot 7 = 0$	**e.**	$1 \cdot 7 = 7$
b.	$9 \cdot 0 = 0$	**f.**	$9 \cdot 1 = 9$
c.	$83 \cdot 0 = 0$	**g.**	$83 \cdot 1 = 83$
d.	$0 \cdot 654 = 0$	**h.**	$1 \cdot 654 = 654$

Now work exercise 3 in the margin.

The number 1 is called the **multiplicative identity** or **identity element for multiplication**.

If three or more numbers are to be multiplied, we have to decide which two to **group** or **associate** together first. The **associative property of multiplication** indicates that we can multiply different pairs of numbers and arrive at the same product.

Associative Property of Multiplication

If a, b, and c are whole numbers, then $a \cdot b \cdot c = a(b \cdot c) = (a \cdot b)c$.

Example 4

$2 \cdot 3 \cdot 7$
$= 6 \cdot 7$
$= 42$

$2 \cdot 3 \cdot 7$
$= 2 \cdot 21$
$= 42$

Thus, $(2 \cdot 3)7 = 2(3 \cdot 7)$

Now work exercise 4 in the margin.

Example 5

$9 \cdot 2 \cdot 5 = (9 \cdot 2) \cdot 5 = 18 \cdot 5 = 90$

$9 \cdot 2 \cdot 5 = 9 \cdot (2 \cdot 5) = 9 \cdot 10 = 90$

Now work exercise 5 in the margin.

Find each of the indicated products and tell which property of multiplication is illustrated.

3. $0 \cdot 7$

4. $(7 \cdot 4)2 = 7(4 \cdot 2)$

5. $1(6 \cdot 3) = (1 \cdot 6)3$

Powers of 10

A **power** of any number is 1, that number, or that number multiplied by itself one or more times. We will discuss this idea in more detail in Chapter 2. Some of the powers of 10 are: 1, 10, 100, 1000, 10,000, and 100,000, and so on.

$$\begin{aligned} &1 \\ &10 \\ 10 \cdot 10 &= 100 \\ 10 \cdot 10 \cdot 10 &= 1000 \\ 10 \cdot 10 \cdot 10 \cdot 10 &= 10{,}000 \\ 10 \cdot 10 \cdot 10 \cdot 10 \cdot 10 &= 100{,}000 \end{aligned}$$

and so on.

Multiplication by powers of 10 is useful in explaining multiplication with whole numbers in general. Such multiplication should be done mentally and quickly. The following examples illustrate an important pattern:

$$\begin{aligned} 6 \cdot 1 &= 6 \\ 6 \cdot 10 &= 60 \\ 6 \cdot 100 &= 600 \\ 6 \cdot 1000 &= 6000 \end{aligned}$$

If one of two whole number factors is 1000, the product will be the other factor with three zeros (000) written to the right of it. Two zeros (00) are written to the right of the other factor when multiplying by 100, and one zero (0) is written when multiplying by 10. Will multiplication by one million (1,000,000) result in writing six zeros to the right of the other factor? The answer is yes.

To multiply a whole number

by 10, write 0 to the right;
by 100, write 00 to the right;
by 1000, write 000 to the right;
by 10,000, write 0000 to the right;
and so on.

Many products can be found mentally by using the properties of multiplication and the techniques of multiplying by powers of 10. The processes are written out in the following examples, but they can easily be done mentally with practice.

Objective 2

To Multiply Two or More Whole Numbers that End with 0's:

1. Count the ending 0's.
2. Multiply the numbers without the ending 0's.
3. Write the product from Step 2 with the counted number of 0's at the end.

Example 6

a. $6 \cdot 90 = 6(9 \cdot 10) = (6 \cdot 9)10 = 54 \cdot 10 = 540$
b. $3 \cdot 400 = 3(4 \cdot 100) = (3 \cdot 4)100 = 12 \cdot 100 = 1200$
c. $2 \cdot 300 = 2(3 \cdot 100) = (2 \cdot 3)100 = 6 \cdot 100 = 600$
d. $6 \cdot 700 = 6(7 \cdot 100) = (6 \cdot 7)100 = 42 \cdot 100 = 4200$
e. $40 \cdot 30 = (4 \cdot 10)(3 \cdot 10) = (4 \cdot 3)(10 \cdot 10) = 12 \cdot 100 = 1200$
f. $50 \cdot 700 = (5 \cdot 10)(7 \cdot 100) = (5 \cdot 7)(10 \cdot 100) = 35 \cdot 1000 = 35{,}000$

Now work exercise 6 in the margin.

Example 7

$200 \cdot 800 = (2 \cdot 100)(8 \cdot 100) = (2 \cdot 8)(100 \cdot 100) = 16 \cdot 10{,}000 = 160{,}000$

Now work exercise 7 in the margin.

Example 8

$$7000 \cdot 9000 = (7 \cdot 1000)(9 \cdot 1000) = (7 \cdot 9)(1000 \cdot 1000) = 63 \cdot 1{,}000{,}000 = 63{,}000{,}000$$

Now work exercise 8 in the margin.

A whole number that ends with 0's has a power of 10 as a factor. Thus, we can easily find at least two factors of a whole number that ends with one or more 0's.

Example 9

$70 = 7 \cdot 10$

factor factor

Now work exercise 9 in the margin.

Example 10

$500 = 5 \cdot 100$

factor factor

Now work exercise 10 in the margin.

Example 11

$46{,}000 = 46 \cdot 1000$

factor factor

Now work exercise 11 in the margin.

Find the following products using the properties of multiplication and the techniques of multiplying by powers of 10.

6. $8 \cdot 1000$

7. $400 \cdot 100$

8. $3000 \cdot 5000$

Find two factors of the given whole number. At least one of the numbers must be a power of 10.

9. 50

10. 800

11. 68,000

Multiplication with Whole Numbers

We can use expanded notation and our skills with multiplication by powers of 10 to help in understanding the technique for multiplying two whole numbers. We also need the **distributive property of multiplication over addition** illustrated in the following discussion.

To multiply $3(70 + 2)$, we can add first, then multiply:

$$3(70+2) = 3(72) = 216$$

Objective 3

But we can also multiply first, then add, in the following manner:

$$\begin{aligned} 3(70+2) &= 3\cdot 70 + 3\cdot 2 \\ &= 210 + 6 \quad \text{(210 and 6 are called \textbf{partial products}.)} \\ &= 216 \end{aligned}$$

Or, arranging the problem vertically,

$$\begin{array}{r} 70+2 \\ 3 \\ \hline 210+6=216 \end{array}$$

partial products (210, 6) — product of 72 and 3 (216)

We say that 3 is **distributed** over the sum of 70 and 2.

Find the product by using partial products.

12. $\begin{array}{r} 256 \\ \times\ 7 \\ \hline \end{array}$

Distributive Property of Multiplication Over Addition

For any whole numbers a, b, and c, $a(b+c) = a\cdot b + a\cdot c$.

Example 12

Multiply $6\cdot 39$.

Solution

$$\begin{array}{r} 39 \\ \times\ 6 \\ \hline \end{array} \qquad \begin{array}{r} 30+9 \\ 6 \\ \hline 180+54=234 \end{array} \qquad \text{or} \qquad \begin{array}{r} 3\,9 \\ 6 \\ \hline 5\,4 \\ 1\,8\,0 \\ \hline 2\,3\,4 \end{array}$$

180 + 54: partial products; 234: product

54 ← $6\cdot 9$, 180 ← $6\cdot 30$ (partial product); 234 ← product

The expanded notation is shown so that you can see that 6 is multiplied by 30 and not just by 3. However, the vertical notation is generally used because it is easier to add the partial products in this form.

Now work exercise 12 in the margin.

Example 13

Multiply 37 · 42.

Solution

$$\begin{array}{r} 42 \\ \times\ 37 \\ \hline \end{array} \qquad \begin{array}{r} 40+\ \ 2 \\ 30+\ \ 7 \\ \hline 280+14 \\ 1200+\ 60\qquad \\ \hline 1200+340+14=1554 \end{array} \qquad \text{or} \qquad \begin{array}{rl} 42 & \text{factor} \\ \underline{37} & \text{factor} \\ 14 & (7\cdot 2=14) \\ 280 & (7\cdot 40=280) \\ 60 & (30\cdot 2=60) \\ \underline{1200} & (30\cdot 40=1200) \\ 1554 & \text{product} \end{array}$$

Now work exercise 13 in the margin.

13. $\begin{array}{r} 83 \\ \times\ 49 \\ \hline \end{array}$

Example 14

Multiply 26 · 276.

Solution

$$\begin{array}{r} 276 \\ \times\ 26 \\ \hline \end{array} \qquad \begin{array}{r} 200+\ \ 70+\ \ 6 \\ 20+\ \ 6 \\ \hline 1200+420+36 \\ 4000+1400+120\qquad \\ \hline 4000+2600+540+36=7176 \end{array} \qquad \text{or} \qquad \begin{array}{rl} 276 & \text{factor} \\ \underline{26} & \text{factor} \\ 36 & (6\cdot 6=36) \\ 420 & (6\cdot 70=420) \\ 1200 & (6\cdot 200=1200) \\ 120 & (20\cdot 6=120) \\ 1400 & (20\cdot 70=1400) \\ \underline{4000} & (20\cdot 200=4000) \\ 7176 & \text{product} \end{array}$$

Now work exercise 14 in the margin.

14. $\begin{array}{r} 372 \\ \times\ 64 \\ \hline \end{array}$

Objective 4

There is a shorter and faster method of multiplication used most of the time. In this method:

a. digits are carried to be added to the next partial product.

b. sums of some partial products are found mentally.

Find each of the following products by using the Short Method.

15. $\begin{array}{r} 42 \\ \times \quad 6 \\ \hline \end{array}$

16. $\begin{array}{r} 37 \\ \times 85 \\ \hline \end{array}$

Example 15

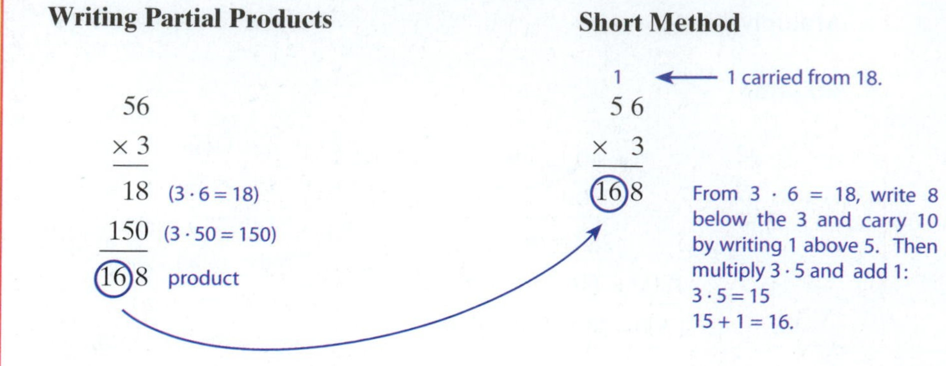

Now work exercise 15 in the margin.

Example 16

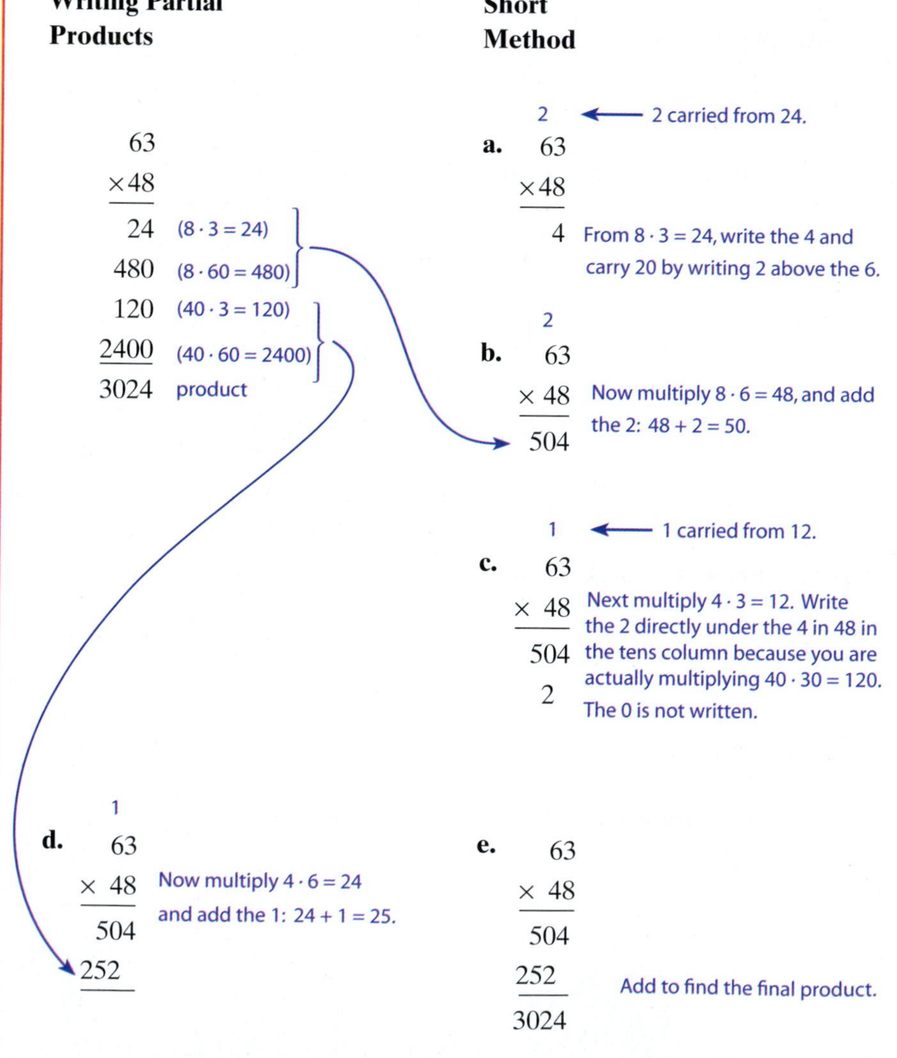

Steps **a.**, **b.**, **c.**, and **d.** are shown for clarification of the Short Method. You will write only step **e.** as shown in Example 17.

Now work exercise 16 in the margin.

Example 17

$$\begin{array}{r} {\scriptstyle 4} \\ {\scriptstyle 1} \\ 87 \\ \times\ 26 \\ \hline 522 \\ 174 \\ \hline 2262 \end{array}$$

522 ← 6 · 7 = 42. Write 2, carry 4. 6 · 8 = 48. Add 4: 48 + 4 = 52.

174 ← 2 · 7 = 14. Write 4, carry 1. 2 · 8 = 16. Add 1: 16 + 1 = 17.

2262 ← product

Now work exercise 17 in the margin.

17. $\begin{array}{r} 73 \\ \times\ 24 \\ \hline \end{array}$

Completion Example 18

$$\begin{array}{r} {\scriptstyle 3} \\ {\scriptstyle 1} \\ 45 \\ \times\ 27 \\ \hline 315 \\ 90 \\ \hline 1215 \end{array}$$

(31)5 — 31 comes from 7 · 4 = 28 and 28 + 3 = 31

(9)0 — 9 comes from ______________

Now work exercise 18 in the margin.

18. $\begin{array}{r} 62 \\ \times\ 81 \\ \hline \end{array}$

Suppose you want to multiply 2000 by 423. Would you write

$$\begin{array}{r} 2000 \\ \times\ 423 \\ \hline 6000 \\ 4000 \\ 8000 \\ \hline 846{,}000 \end{array} \qquad \text{or} \qquad \begin{array}{r} 423 \\ \times\ 2000 \\ \hline 000 \\ 000 \\ 000 \\ 846 \\ \hline 846{,}000 \end{array}$$

Both answers are correct; neither technique is wrong. However, writing all the 0's is a waste of time, so knowledge about powers of ten is helpful. We can write

$$\begin{array}{r} 423 \\ \times\ 2\ 000 \\ \hline 846{,}000 \end{array}$$

We know 2000 = 2 · 1000, so we are simply multiplying 423 · 2 and then multiplying the result by 1000.

19. Multiply $354 \cdot 4000$.

Example 19

Multiply $596 \cdot 3000$.

Solution

$$\begin{array}{r} 596 \\ \times \quad 3\,000 \\ \hline 1,788,000 \end{array}$$

Now work exercise 19 in the margin.

20. Multiply $797 \cdot 11,000$.

Example 20

Multiply $265 \cdot 15,000$.

Solution

$$\begin{array}{l} \quad\;\; 265 \\ \times \quad 15,000 \\ \hline 1\;\, 325\;000 \\ 2\;65 \\ \hline 3,975,000 \end{array}$$

Now work exercise 20 in the margin.

Objective 5

Estimating Products with Whole Numbers

Estimating the product of two numbers before actually performing the multiplication can be of help in detecting errors. Just as with addition and subtraction, estimations are done with rounded numbers.

To Estimate a Product:

1. Round each number to the place of the **leftmost** digit.
2. Multiply the rounded numbers by using your knowledge of multiplication by powers of 10.

First estimate, and then find the following product.

21. $18 \cdot 74$

Example 21

Find the product $62 \cdot 38$, but first estimate the product by multiplying the numbers in rounded form.

a.

$$\begin{array}{rcrl} 62 & \longrightarrow & 60 & \text{rounded value of 62} \\ \times\, 38 & \longrightarrow & \times\, 40 & \text{rounded value of 38} \\ \hline & & 2400 & \text{estimated product} \end{array}$$

b.
$$\begin{array}{r} \overset{1}{62} \\ \times\ 38 \\ \hline 496 \\ 186\ \ \\ \hline 2356 \end{array}$$
← The actual product is close to 2400.

Now work exercise 21 in the margin.

Example 22

Use rounded values to estimate the product 267 · 93; then find the product.

22. Use rounded values to estimate the product 551 · 35; then find the product.

Solution

a. Estimation:
$$\begin{array}{r} 267 \\ \times\ 93 \\ \hline \end{array} \begin{array}{c} \longrightarrow \\ \longrightarrow \end{array} \begin{array}{r} 300 \\ \times\ \ \ 90 \\ \hline 27{,}000 \end{array}$$

b. The product should be near 27,000.
$$\begin{array}{r} 267 \\ \times\ 93 \\ \hline 801 \\ 24\ 03\ \ \\ \hline 24{,}831 \end{array}$$

In this case, the estimated value of 27,000 is considerably larger than 24,831. However, we only expect an estimation to indicate that our answer is reasonable. For example, if we had arrived at an answer of 248,310, then we would know that a mistake had been made because this number is not reasonably close to 27,000.

Now work exercise 22 in the margin.

Completion Example Answers

18.
$$\begin{array}{r} \overset{\overset{3}{1}}{4}5 \\ \times\ \ 27 \\ \hline 315 \\ 90\ \ \\ \hline 1215 \end{array}$$
31 comes from 7 · 4 = 28 and 28 + 3 = 31
9 comes from **2 · 4 = 8 and 8 + 1 = 9**

Practice Problems

Which property of multiplication is illustrated?

1. $5 \cdot (3 \cdot 7) = (5 \cdot 3) \cdot 7$ **2.** $32 \cdot 0 = 0$ **3.** $6 \cdot 8 = 8 \cdot 6$

Estimate and then find the following products.

4. $\begin{array}{r} 18 \\ \times\ 24 \\ \hline \end{array}$ **5.** $\begin{array}{r} 354 \\ \times\ 50,000 \\ \hline \end{array}$ **6.** $\begin{array}{r} 129 \\ \times\ 39 \\ \hline \end{array}$

Answers to Practice Problems:

1. associative property **2.** zero factor law
3. commutative property **4.** 400; 432
5. 20,000,000; 17,700,000 **6.** 4000; 5031

Name ______________ Section ______________ Date ______________

Exercises 1.5

Do the following problems mentally and write only the answers. See Examples 1 and 2.

1. $8 \cdot 9$ **2.** $7 \cdot 6$ **3.** $6(4)$ **4.** $5(0)$

5. $1(7)$ **6.** $9 \cdot 5$ **7.** $0(3)$ **8.** $(7)(7)$

Find each product and tell what property of multiplication is illustrated. See Examples 1 through 5.

9. $4 \cdot 7 = 7 \cdot 4$ **10.** $2(1 \cdot 6) = (2 \cdot 1)6$

11. $5 \cdot 1$ **12.** $8 \cdot 0$

Use the technique of multiplying by powers of ten to find the following products mentally. See Examples 6 through 11.

13. $25 \cdot 10$ **14.** $47 \cdot 1000$ **15.** $20 \cdot 200$ **16.** $500 \cdot 700$

17. $200 \cdot 80$ **18.** $40 \cdot 2000$ **19.** $30 \cdot 30$ **20.** $300 \cdot 600$

21. $4000 \cdot 4000$ **22.** $900 \cdot 3000$ **23.** $10 \cdot 500$ **24.** $120 \cdot 300$

25. $\begin{array}{r} 6000 \\ \times \quad 6 \\ \hline \end{array}$ **26.** $\begin{array}{r} 90 \\ \times \ 90 \\ \hline \end{array}$ **27.** $\begin{array}{r} 300 \\ \times \ 500 \\ \hline \end{array}$ **28.** $\begin{array}{r} 700 \\ \times \quad 80 \\ \hline \end{array}$

ANSWERS

1. ______
2. ______
3. ______
4. ______
5. ______
6. ______
7. ______
8. ______
9. ______
10. ______
11. ______
12. ______
13. ______
14. ______
15. ______
16. ______
17. ______
18. ______
19. ______
20. ______
21. ______
22. ______
23. ______
24. ______
25. ______
26. ______
27. ______
28. ______

29. ______

30. ______

31. ______

32. ______

33. ______

34. ______

35. ______

36. ______

37. ______

38. ______

39. ______

40. ______

41. ______

42. ______

43. ______

44. ______

45. ______

46. ______

47. ______

48. ______

49. ______

50. ______

51. ______

52. ______

53. ______

54. ______

55. ______

56. ______

In the following problems, first estimate each product and then find the product. (Remember that single-digit numbers are already considered to be rounded.) See Examples 12 through 22.

29. 56×4

30. 27×6

31. 48×9

32. 65×5

33. 84×3

34. 95×8

35. 42×56

36. 25×33

37. 48×20

38. 93×30

39. 83×85

40. 96×62

41. 17×32

42. 28×91

43. 20×44

44. 16×26

45. 25×15

46. 93×47

47. 24×86

48. 72×65

49. 12×13

50. 81×36

51. 126×41

52. 232×76

53. 114×25

54. 72×106

55. 207×143

56. 420×104

Name ______________________ Section ____________ Date ____________

ANSWERS

57. ____________

58. ____________

59. ____________

60. ____________

61. ____________

62. ____________

63. ____________

64. ____________

65. ____________

57. $\begin{array}{r} 200 \\ \times\ 49 \\ \hline \end{array}$

58. $\begin{array}{r} 849 \\ \times\ 205 \\ \hline \end{array}$

59. $\begin{array}{r} 673 \\ \times\ 186 \\ \hline \end{array}$

60. $\begin{array}{r} 192 \\ \times\ 467 \\ \hline \end{array}$

61. Find the sum of eighty-four and one hundred forty-seven. Find the difference between ninety-six and thirty-eight. Then find the product of the sum and the difference.

62. Find the sum of thirty-four and fifty-five. Find the sum of one hundred seventeen and two hundred twenty. Find the product of the two sums.

63. **Salary.** If your beginning salary is $2300 per month and you are to get a raise once a year of $230 per month, by how much will your pay increase from year to year? How much money will you make over a five-year period?

64. **Rent.** If you and a roommate rent an apartment with two bedrooms for $740 per month and you know that the rent will increase $15 per month every 12 months, what will you and your roommate pay in rent during the first three years of living in this apartment? By how much does the total rent increase each year?

65. **Automobile purchases.** Your company bought 18 new cars, each with air conditioning and antilock brakes, at a price of $15,800 per car. How much did your company pay for these cars?

66. ______

66. Fundraising. The Math Club members decided to attend the national meeting of the NCTM (National Council of Teachers of Mathematics) and had a book sale to raise money for the event. Registration fees were $85 per student and the club had 35 members. How much money did the club need to raise for registration fees?

The menu prices (to the nearest dollar) for certain items at a local fast food restaurant are shown in the table. Use this table to answer Exercises 67 and 68.

67. ______

Item	Price ($)
Hamburger	4
Turkey burger	3
French fries	2
Onion rings	3
Soda	1
Milkshake	2
Cookie	1

68. ______

67. Fast Food. Sally and her two friends ordered 2 hamburgers, 1 turkey burger, 3 orders of French fries, and 3 milkshakes. What was the total cost of their order?

68. Fast Food. George is hungry and hasn't decided just what to order. He is trying to decide between

(a) 2 hamburgers, 1 order of onion rings, 1 soda, and 2 cookies; or

(b) 1 hamburger, 2 orders of french fries, 1 milkshake, and 1 cookie.

Which of the two orders is more expensive? By how much?

69. ______

69. Commercials. Network television has 12 minutes of commercial time in each hour. How many minutes of commercial time does a network have in a one-day programming schedule of 20 hours? In one week?

Name ______________________ Section ____________ Date __________

ANSWERS

70. Bird importing. According to the U.S. Fish and Wildlife Service, migratory birds are imported at a value of about $19. Suppose that about 800,000 live birds are imported each year. What is the total value of these imported birds?

70. ______________

71. Travel. According to the National Household Travel Survey, U.S. citizens drove a particular vehicle almost 13,500 miles per person in 2001. If the estimated 2001 census population is 284,796,887, how many miles did U.S. citizens drive altogether?

71. ______________

Check Your Number Sense

72. You want to pay cash for your textbooks at the bookstore because you know that the cash line is shorter and moves faster than the other lines. If you are going to buy four books and their prices are between $35 and $45 each, about how much cash do you need to take with you: $100, $200, $300, or $400? (Don't forget sales tax will be involved.)

72. ______________

73. Match each indicated product with the closest estimate of that product. Perform any calculations mentally.

	Product	Estimate
_____	(a) $16 \cdot 18$	A. 210
_____	(b) $6(78)$	B. 300
_____	(c) $11 \cdot 32$	C. 400
_____	(d) $(8)(69)$	D. 480
_____	(e) $25(7)$	E. 560

73. ______________

74. ____________

75. ____________

76. ____________

74. Match each indicated product with the closest estimate of that product. Perform any calculations mentally.

	Product	Estimate
_____	(a) 37(500)	A. 200
_____	(b) 37(50)	B. 2000
_____	(c) 37(5)	C. 20,000
_____	(d) 37(5000)	D. 200,000

Writing and Thinking about Mathematics

75. Write, in your own words, the meaning of the zero factor law.

76. Explain, in your own words, why 1 is called the multiplicative identity.

1.6 Division with Whole Numbers

Division with Whole Numbers

We know that $6 \cdot 10 = 60$ and that 6 and 10 are **factors** of 60. They are also called **divisors**. The process of division can be thought of as the reverse of multiplication. In division, we want to find how many times one number is contained in another. How many sixes are in 60? There are 10 sixes in 60, and we say that 60 **divided by** 6 is 10 (or $60 \div 6 = 10$).

The relationship between multiplication and division can be seen from the following table format:

Division						Multiplication		
Dividend		Divisor		Quotient		Factors		Product
21	÷	7	=	3	since	7 · 3	=	21
24	÷	6	=	4	since	6 · 4	=	24
36	÷	4	=	9	since	4 · 9	=	36
12	÷	2	=	6	since	2 · 6	=	12

This table indicates that the number being divided is called the **dividend**; the number doing the dividing is called the **divisor**; and the result of division is called the **quotient**.

Division does not always involve factors (or exact divisors). Suppose we want to divide 23 by 4. By using repeated subtraction, we can find how many 4's are in 23. We continuously subtract 4 until the number left (called the **remainder**) is less than 4.

$$\begin{array}{r}23\\-\ 4\\\hline 19\end{array}\quad\begin{array}{r}19\\-\ 4\\\hline 15\end{array}\quad\begin{array}{r}15\\-\ 4\\\hline 11\end{array}\quad\begin{array}{r}11\\-\ 4\\\hline 7\end{array}\quad\begin{array}{r}7\\-\ 4\\\hline 3\end{array}\ \leftarrow \text{remainder}$$

(subtraction 5 times)

Another form used to indicate division as repeated subtraction is shown in Examples 1 and 2.

Objectives

① Know the terms related to division: dividend, divisor, quotient, and remainder.

② Be able to divide with whole numbers.

③ Understand that when the remainder is 0, both the divisor and quotient are factors of the dividend.

④ Estimate quotients.

⑤ Know that division by 0 is not possible.

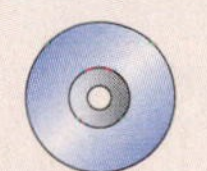

Objective ①

Example 1

How many 7's are in 185?

Solution

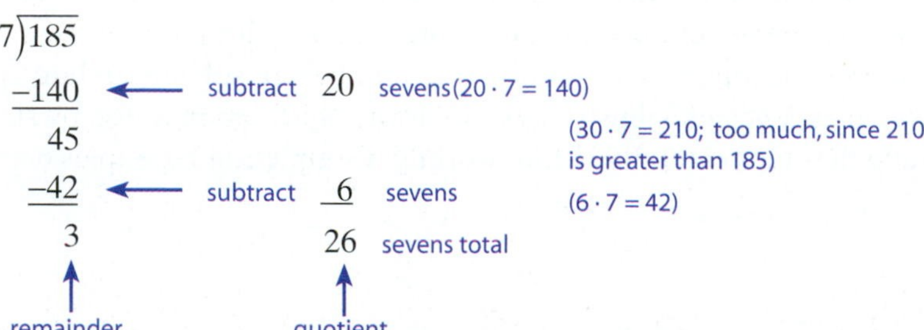

$$\begin{array}{r}7\overline{)185}\\-140\\\hline 45\\-42\\\hline 3\end{array}$$

subtract 20 sevens ($20 \cdot 7 = 140$)
($30 \cdot 7 = 210$; too much, since 210 is greater than 185)
subtract 6 sevens ($6 \cdot 7 = 42$)
26 sevens total

3 ← remainder; 26 ← quotient

1. How many 6's are in 97?

Divide by using repeated subtraction, then check your work.

2. $9\overline{)849}$

Check:

Division is checked by multiplying the quotient and the divisor and then adding the remainder. The result should be the dividend.

$$\begin{array}{rl} 26 & \text{quotient} \\ \times\ 7 & \text{divisor} \\ \hline 182 & \end{array} \qquad \begin{array}{rl} 182 & \\ +\ 3 & \text{remainder} \\ \hline 185 & \text{dividend} \end{array}$$

Now work exercise 1 in the margin.

Example 2

Find 275 ÷ 6 using repeated subtraction, and check your work.

Solution

$$\begin{array}{rlrll} 6\overline{)275} & & & & \\ -180 & \text{subtract} & 30 & \text{sixes} & (30 \cdot 6 = 180) \\ \hline 95 & & & & \\ -60 & \text{subtract} & 10 & \text{sixes} & (10 \cdot 6 = 60) \\ \hline 35 & & & & \\ -18 & \text{subtract} & 3 & \text{sixes} & (3 \cdot 6 = 18) \\ \hline 17 & & & & \\ -12 & \text{subtract} & \underline{2} & \text{sixes} & (2 \cdot 6 = 12) \\ \hline 5 & & 45 & \text{sixes total} & \\ \uparrow & & \uparrow & & \\ \text{remainder} & & \text{quotient} & & \end{array}$$

Now work exercise 2 in the margin.

> **Note:** You can subtract any number of sixes less than the quotient. But this will not lead to a good explanation of the shorter division algorithm. In general, at each step you should subtract the largest number of thousands, hundreds, tens, or units of the divisor that you can.

Objective 2

The repeated subtraction technique for division provides a basis for understanding the **division algorithm**, a much shorter method of division with which we are generally familiar. (**Note**: An algorithm is a process or pattern of steps to be followed in working with numbers or solving a related problem.) Examples 3, 4, and 5 illustrate the division algorithm in a step by step format. Study the steps carefully before working Completion Examples 6 and 7.

Example 3

Find $2076 \div 8$.

Solution

Step 1:

```
     2        ← Write 2 in hundreds position.
 8)2076
  -1600        200 eights (200 · 8 = 1600)
    476
```

Step 2:

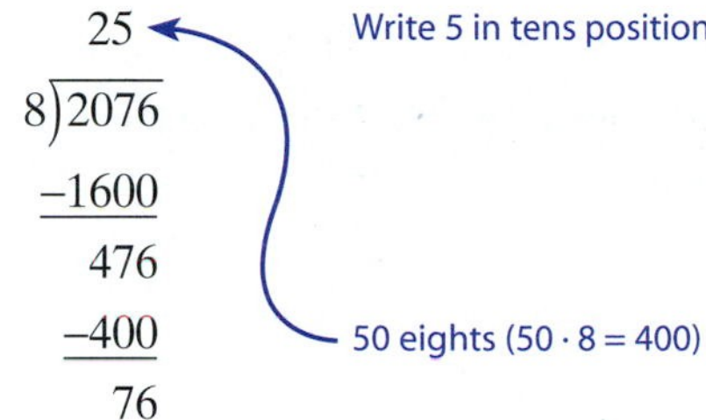

Step 3:

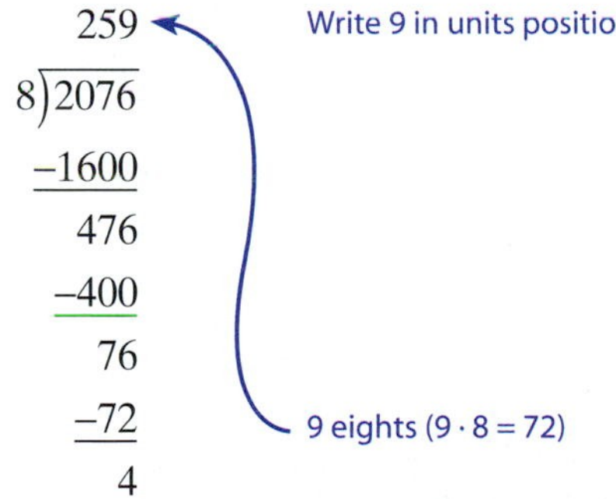

Summary

The process can be shortened by not writing all the 0's and "bringing down" only one digit at a time.

```
    259 R4
 8)2076
   16
   47     ← "Bring down" the 7 only; then divide 8 into 47.
   40
    76
    72
     4
```

Now work exercise 3 in the margin.

Divide by using the division algorithm, then check your answer.

3. $327 \div 7$

Divide by using the division algorithm, then check your answer.

4. $16\overline{)324}$

Example 4

Find 746 ÷ 32.

Solution

Step 1:

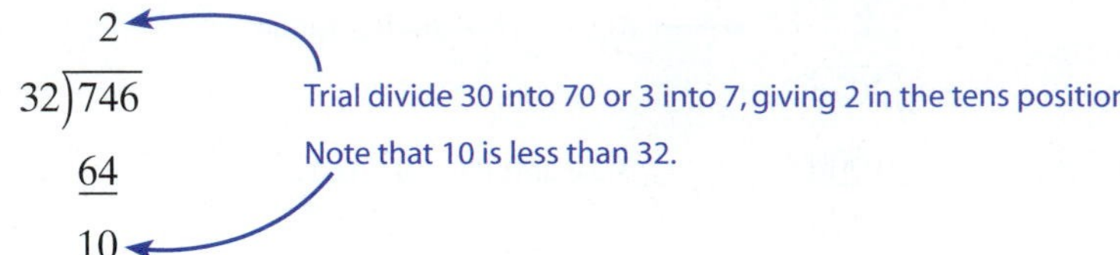

Step 2:

```
    23  R10
32)746
   64
   106
    96
    10
```

Trial divide 30 into 100 or 3 into 10, giving 3 in the units position.

Check:

```
   23         736
 × 32       +  10
   46         746
  69
  736
```

Now work exercise 4 in the margin.

Example 5

Find 9325 ÷ 45.

Solution

Step 1:

```
    2
45)9325
   90
    3
```

Trial divide 40 into 90 or 4 into 9, giving 2 in the hundreds position.

Step 2:

```
    20
45)9325
   90
    32
     0
```

45 divides into 32 zero times. So, write 0 in the tens column and multiply $0 \cdot 45 = 0$.

Step 3:

```
    208
45)9325
   90
    32
     0
    325
    360
```

Trial divide 45 into 325 or 4 into 32. But the trial quotient 8 is too large since $8 \cdot 45 = 360$ is larger than 325.

Step 4:

```
    207   R 10
45)9325
   90
    32
     0
    325
    315
     10
```

Now the trial quotient is 7. Since $7 \cdot 45 = 315$ and 315 is smaller than 325, 7 is the desired number.

Check:

```
   207        9315
 ×  45      +   10
  1035        9325
  828
  9315
```

Now work exercise 5 in the margin.

5. $31\overline{)9571}$

Find the quotient and remainder.

6. $12\overline{)1869}$

7. $39\overline{)8370}$

Common Error:

In Step 2 of Example 5, we wrote 0 in the quotient because 45 did not divide into 32. Many students fail to write the 0. This error obviously changes the value of the quotient.

Wrong Solution

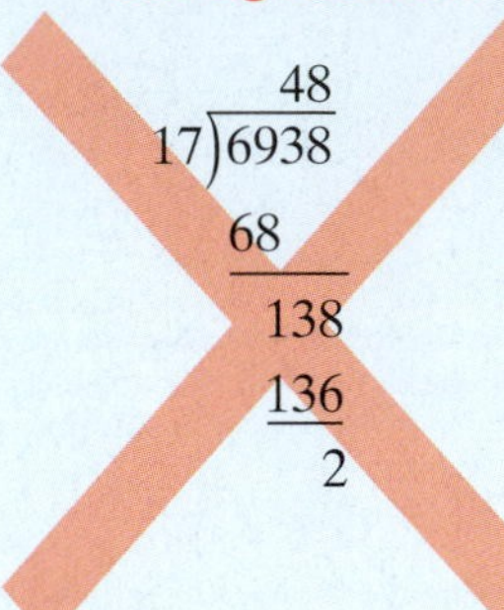

Correct Solution

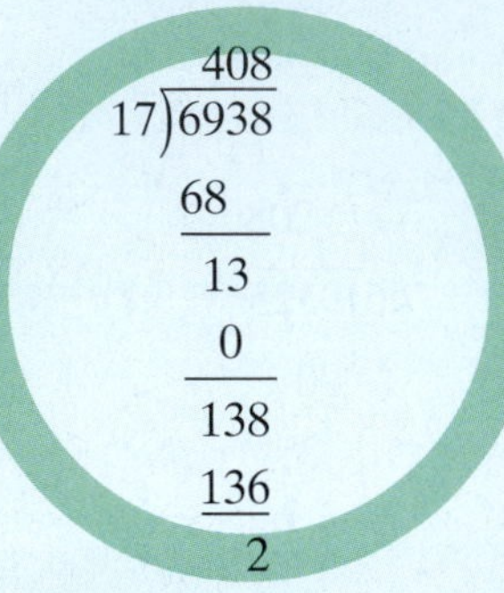

Writing the 0 in the quotient gives the correct quotient of 408, which is considerably different from 48.

Completion Example 6

Find the quotient and remainder. Use the gray lines to keep the numbers aligned.

```
   8
4)3 3 4
  3 2
  ---
    1
```

remainder

Now work exercise 6 in the margin.

Completion Example 7

Find the quotient and remainder.

```
    2
12)2 4 5 1
   2 4
   ---
     5
```

remainder

Now work exercise 7 in the margin.

Step 2:

$$\begin{array}{r} 20 \\ 45\overline{)9325} \\ \underline{90} \\ 32 \\ \underline{0} \end{array}$$

45 divides into 32 zero times. So, write 0 in the tens column and multiply $0 \cdot 45 = 0$.

Step 3:

$$\begin{array}{r} 208 \\ 45\overline{)9325} \\ \underline{90} \\ 32 \\ \underline{0} \\ 325 \\ \underline{360} \end{array}$$

Trial divide 45 into 325 or 4 into 32. But the trial quotient 8 is too large since $8 \cdot 45 = 360$ is larger than 325.

Step 4:

$$\begin{array}{r} 207 \quad \text{R } 10 \\ 45\overline{)9325} \\ \underline{90} \\ 32 \\ \underline{0} \\ 325 \\ \underline{315} \\ 10 \end{array}$$

Now the trial quotient is 7. Since $7 \cdot 45 = 315$ and 315 is smaller than 325, 7 is the desired number.

Check:

$$\begin{array}{r} 207 \\ \times \quad 45 \\ \hline 1035 \\ \underline{828} \\ 9315 \end{array} \qquad \begin{array}{r} 9315 \\ + \quad 10 \\ \hline 9325 \end{array}$$

Now work exercise 5 in the margin.

5. $31\overline{)9571}$

Common Error:

In Step 2 of Example 5, we wrote 0 in the quotient because 45 did not divide into 32. Many students fail to write the 0. This error obviously changes the value of the quotient.

Wrong Solution

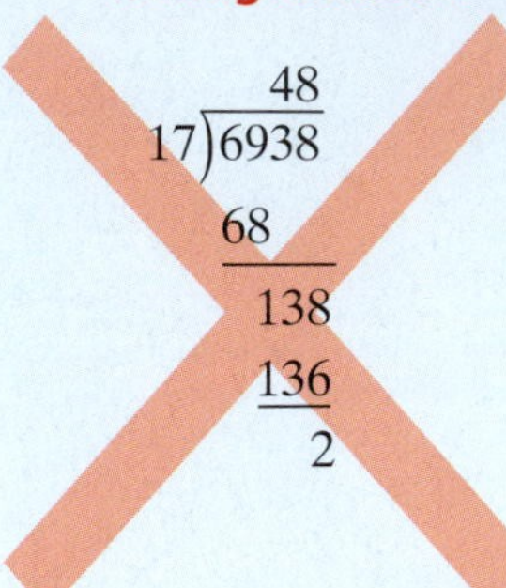

Correct Solution

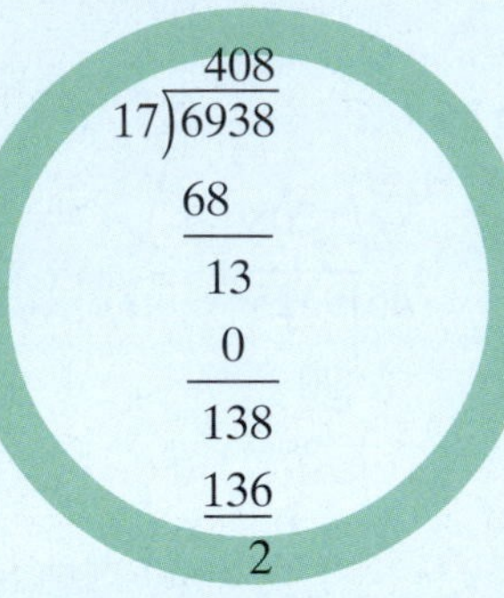

Writing the 0 in the quotient gives the correct quotient of 408, which is considerably different from 48.

Find the quotient and remainder.

6. 12)1869

Completion Example 6

Find the quotient and remainder. Use the gray lines to keep the numbers aligned.

```
     8
4)3 3 4
  3 2
  ---
    1
```

remainder

Now work exercise 6 in the margin.

7. 39)8370

Completion Example 7

Find the quotient and remainder.

```
      2
12)2 4 5 1
   2 4
   ---
       5
```

remainder

Now work exercise 7 in the margin.

If the remainder is 0, then the following statements are true:

1. Both the divisor and quotient are **factors** of the dividend.

2. We say that both factors **divide exactly** into the dividend.

3. Both factors are called **divisors** of the dividend.

Objective 3

Example 8

Show that 17 and 36 are factors (or divisors) of 612.

Solution

$$\begin{array}{r} 36 \\ 17\overline{)612} \\ \underline{51} \\ 102 \\ \underline{102} \\ 0 \end{array}$$

Since 0 is the remainder, both 17 and 36 are factors (or divisors) of 612. Just to double-check, we find 612 ÷ 36.

$$\begin{array}{r} 17 \\ 36\overline{)612} \\ \underline{36} \\ 252 \\ \underline{252} \\ 0 \end{array}$$

Yes, both 17 and 36 divide exactly into 612.

Now work exercise 8 in the margin.

8. Show that 21 and 14 are divisors of 294.

Example 9

A plumber purchased 17 special pipe fittings. What was the price of one fitting if the bill was $544 before taxes?

Solution

We need to know how many times 17 goes into 544.

$$\begin{array}{r} 32 \\ 17\overline{)544} \\ \underline{51} \\ 34 \\ \underline{34} \\ 0 \end{array}$$

The price for one fitting was $32.

Now work exercise 9 in the margin.

9. If seven identical bicycles cost $805 total, what is the price of each bike?

Objective 4

Estimating Quotients with Whole Numbers

By rounding both the divisor and the dividend, we can estimate the quotient. Estimation can help identify unreasonable answers when the actual value is calculated.

To Estimate a Quotient:

1. Round both the divisor and dividend to the place of the last digit on the left.
2. Divide with the rounded numbers. (This process is very similar to the trial dividing step in the division algorithm.)

10. Estimate the quotient $882 \div 76$ by using rounded values, then find the quotient.

Example 10

Estimate the quotient $325 \div 42$ by using rounded values; then find the quotient.

Solution

a. Estimation: $325 \div 42 \longrightarrow 300 \div 40$

$$\begin{array}{r} 7 \\ 40\overline{)300} \\ \underline{280} \\ 20 \end{array}$$

7 estimated quotient

b. The quotient should be near 7.

$$\begin{array}{r} 7 \\ 42\overline{)325} \\ \underline{294} \\ 31 \end{array}$$

7 quotient; 31 remainder

In this case, the quotient is the same as the estimated value. The true remainder is different.

Now work exercise 10 in the margin.

Example 11

Estimate the quotient 48,062 ÷ 26; then find the quotient.

Solution

a. Estimation: 48,062 ÷ 26 ⟶ 50,000 ÷ 30

```
     1 666    approximate quotient
30)50,000
   30
   20 0
   18 0
    2 00
    1 80
      200
      180
       20
```

b. The quotient should be near 1600 or 1700.

```
     1 848
26)48,062
   26
   220
   208
    126
    104
     222
     208
      14     remainder
```

Now work exercise 11 in the margin.

An Adjustment to the Process of Estimating Answers

The rule for rounding off to the leftmost digit for estimating an answer is flexible. We sometimes adjust the estimating process; there are times when using two digits gives simpler calculations or more accurate estimates. In Example 11, for instance, 3 can be seen to divide into 48, so we use two-digit rounding to make the estimate. (See the discussion on the next page.) Thus, estimating answers does involve some basic understanding and intuitive judgment, and there is no one best way to estimate answers. However, "adjustment" to the leftmost digit is more applicable to division than it is to addition, subtraction, or multiplication.

11. Estimate the quotient 40,284 ÷ 59; then find the quotient.

As an example of another technique for estimating answers, we could have proceeded as follows in Example 11, part **a.**:

Estimation: $48{,}062 \div 26 \longrightarrow 48{,}000 \div 30$

Use 48,000 instead of 50,000 because 3 divides evenly into 48. Thus, the estimate could be

$$\begin{array}{r} 1\,600 \\ 30\overline{)48{,}000} \\ \underline{30} \\ 180 \\ \underline{180} \\ 00 \\ \underline{0} \\ 00 \\ \underline{0} \\ 0 \end{array}$$

approximate quotient

12. Find the quotient and remainder, then estimate the quotient and find the difference in the two values.

$36\overline{)1340}$

Completion Example 12

Find the quotient and remainder.

$$\begin{array}{r} 3\,0 \\ 21\overline{)6\,4\,6\,1} \\ \underline{6\,3} \\ 1\,6 \end{array}$$

remainder

Now estimate the quotient. Do this by mentally dividing rounded numbers:

← estimate

$20\overline{)6000}$

Is your estimate close to the actual quotient? __________

What is the difference? __________

Now work exercise 12 in the margin.

For completeness, we close this section with two rules about division involving 0. These rules will be discussed in detail in Chapter 3.

Division with 0

1. If a is any nonzero whole number, then $0 \div a = 0$.
   For example, $0 \div 6 = 0$ because $6 \cdot 0 = 0$,
   just as $12 \div 6 = 2$ because $6 \cdot 2 = 12$.
2. If a is any whole number, then $a \div 0$ is **undefined**.
   For example, $8 \div 0$ is undefined because 0 times any number is 0 and not 8.

Objective 5

Completion Example Answers

6.
$$\begin{array}{r} 8\mathbf{3} \\ 4\overline{)334} \\ \underline{32} \\ 14 \\ \underline{\mathbf{12}} \\ \mathbf{2} \end{array} \text{ remainder}$$

7.
$$\begin{array}{r} 2\mathbf{04} \\ 12\overline{)2451} \\ \underline{24} \\ 5 \\ \underline{\mathbf{0}} \\ \mathbf{51} \\ \underline{\mathbf{48}} \\ \mathbf{3} \end{array} \text{ remainder}$$

12.
$$\begin{array}{r} 30\mathbf{7} \\ 21\overline{)6461} \\ \underline{63} \\ 16 \\ \underline{\mathbf{0}} \\ \mathbf{161} \\ \underline{\mathbf{147}} \\ \mathbf{14} \end{array} \text{ remainder}$$

Now estimate the quotient by mentally dividing rounded numbers:

$$\begin{array}{r} \mathbf{300} \\ 20\overline{)6000} \end{array} \longleftarrow \text{ estimate}$$

Is your estimate close to the actual quotient? **Yes**
What is the difference? **7**

Practice Problems

Find the quotient and remainder for each of the following problems.

1. $800 \div 9$ **2.** $27\overline{)356}$ **3.** $73\overline{)19,637}$

Answers to Practice Problems:

1. 88 R 8 **2.** 13 R 5 **3.** 269 R 0

Name ______________________ Section ____________ Date __________

Exercises 1.6

Find the quotient and remainder, if necessary, for each of the following problems by using the method of repeated subtraction. See Examples 1 and 2.

1. $240 \div 6$ **2.** $189 \div 9$ **3.** $210 \div 7$ **4.** $140 \div 14$

5. $168 \div 8$ **6.** $70 \div 5$ **7.** $132 \div 11$ **8.** $120 \div 4$

9. $75 \div 15$ **10.** $51 \div 3$ **11.** $52 \div 8$ **12.** $44 \div 6$

13. $600 \div 25$ **14.** $413 \div 20$ **15.** $161 \div 15$ **16.** $182 \div 13$

17. $150 \div 13$ **18.** $500 \div 14$ **19.** $205 \div 5$ **20.** $321 \div 7$

Divide and check using the division algorithm. See Examples 3 through 7.

21. $6\overline{)32}$ **22.** $7\overline{)17}$ **23.** $4\overline{)25}$ **24.** $5\overline{)35}$

ANSWERS

1. ________
2. ________
3. ________
4. ________
5. ________
6. ________
7. ________
8. ________
9. ________
10. ________
11. ________
12. ________
13. ________
14. ________
15. ________
16. ________
17. ________
18. ________
19. ________
20. ________
21. ________
22. ________
23. ________
24. ________

25. ______
26. ______
27. ______
28. ______
29. ______
30. ______
31. ______
32. ______
33. ______
34. ______
35. ______
36. ______
37. ______
38. ______
39. ______
40. ______
41. ______
42. ______
43. ______
44. ______
45. ______
46. ______
47. ______
48. ______
49. ______
50. ______
51. ______
52. ______

25. $8\overline{)48}$ **26.** $6\overline{)72}$ **27.** $9\overline{)81}$ **28.** $2\overline{)76}$

29. $3\overline{)98}$ **30.** $14\overline{)52}$ **31.** $12\overline{)108}$ **32.** $11\overline{)424}$

33. $16\overline{)128}$ **34.** $20\overline{)305}$ **35.** $18\overline{)206}$ **36.** $30\overline{)847}$

37. $10\overline{)423}$ **38.** $15\overline{)750}$ **39.** $13\overline{)260}$ **40.** $17\overline{)340}$

41. $12\overline{)360}$ **42.** $19\overline{)7603}$ **43.** $16\overline{)4813}$ **44.** $11\overline{)4406}$

45. $13\overline{)3917}$ **46.** $73\overline{)148}$ **47.** $68\overline{)207}$ **48.** $49\overline{)993}$

49. $50\overline{)3065}$ **50.** $40\overline{)2163}$ **51.** $68\overline{)210}$ **52.** $116\overline{)232}$

Name ______________________ Section ____________ Date __________

ANSWERS

53. $213\overline{)4760}$ **54.** $716\overline{)3056}$ **55.** $630\overline{)4768}$ **56.** $414\overline{)83276}$

57. $502\overline{)98762}$ **58.** $317\overline{)70365}$ **59.** $471\overline{)50612}$ **60.** $215\overline{)64930}$

61. Income. Suppose your income for one year was $30,576 and your income was the same each month. Estimate your monthly income. What was your exact monthly income?

62. School supplies. A high school bought 3075 new textbooks for a total price of $116,850. What was the approximate price of each text? What was the exact price of each text?

63. Show that 28 and 36 are both factors of 1008 by using division.

64. Show that 45 and 702 are both factors of 31,590 by using division.

53. ____________

54. ____________

55. ____________

56. ____________

57. ____________

58. ____________

59. ____________

60. ____________

61. ____________

62. ____________

63. ____________

64. ____________

65. ______

65. Population density. The United States has a population of about 293,655,000 people and has a land area of about 3,600,000 square miles. About how many people are there for each square mile? (Do you think that the population per square mile is the same for New York City or for Chicago or for the state of New Mexico as it is for the entire United States? Raw numbers do not always give a true-to-life picture when taken out of context. In many applications, some judgment and numerical skills must be used as well.)

Check Your Number Sense

66. ______

66. Match each indicated quotient with the closest estimate of that product. Perform any calculations mentally.

	Quotient	Estimate
____	(a) $9\overline{)910}$	A. 6
____	(b) $34 \div 5$	B. 10
____	(c) $34\overline{)12,000}$	C. 100
____	(d) $18\overline{)3900}$	D. 200
____	(e) $216 \div 18$	E. 300

67. ______

67. Match each indicated quotient with the closest estimate of that product.

	Quotient	Estimate
____	(a) $3\overline{)870}$	A. 30
____	(b) $3\overline{)87,000}$	B. 300
____	(c) $3\overline{)87}$	C. 3000
____	(d) $3\overline{)8700}$	D. 30,000

1.7 Problem Solving with Whole Numbers

Strategy for Solving Word Problems

In this section, problem solving can involve various combinations of the four operations of addition, subtraction, multiplication, and division. Decisions on which operations to use are generally based on experience and practice as well as certain key words.

Objective ①

> **Note:** If you are not exactly sure of what operations to use, at least try something. Even by making errors in technique or judgment you are learning what does not work. If you do nothing, then you learn nothing. Do not be embarrassed by mistakes.

Objectives

① Realize that problem solving takes time and that the learning process also involves learning from making mistakes.

② Know and understand the Basic Strategy for Solving Word Problems.

③ Be able to apply this strategy to problems involving Consumer Items, Checking Accounts, and Averages.

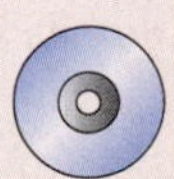

The problems discussed here will come under one of the following headings: Consumer Items, Checking Accounts, and Average. The steps in the basic strategy listed here will help give an organized approach to problem solving regardless of the type of problem. These steps were developed by George Pólya, a famous educator and mathematician from Stanford University.

Objective ②

Basic Strategy for Solving Word Problems

1. Read each problem carefully until you understand the problem and know what is being asked.
2. Draw any type of figure or diagram that might be helpful and decide what operations are needed.
3. Perform these operations.
4. Mentally check to see if your answer is reasonable and see if you can think of another more efficient or more interesting way to do the same problem.

Consumer Items

Objective ③

Example 1

Bill bought a car for \$9000. The salesman added \$540 for taxes and \$150 for license fees. If Bill made a down payment of \$3500 and financed the rest with his credit union, how much did he finance?

Continued on next page...

1. John bought $10 worth of meat and $7 worth of potatoes at the grocery store. If his total bill was $38, how much of the total was not spent on meat or potatoes?

2. Susan deposits checks in the amounts of $52, $37 and $65, then writes a check for $20. If her beginning balance was $280, what is the balance after these transactions?

Solution

Find the total cost and subtract the down payment.

$$\begin{array}{rl} \$9000 & \\ 540 & \\ +\ 150 & \\ \hline \$9690 & \text{total cost} \end{array} \qquad \begin{array}{rl} \$9690 & \text{total cost} \\ -\ 3500 & \text{down payment} \\ \hline \$6190 & \text{to be financed} \end{array}$$

Bill financed $6190.

(Mental checking is difficult here. But you might think that if he paid about $10,000 and put down $3500, an answer around $6500 is reasonable.)

Now work exercise 1 in the margin.

Checking Accounts

Example 2

In July, Ms. Smith opened a checking account and deposited $1528. She wrote checks for $132, $425, $196, and $350. What was her balance at the end of July?

Solution

To find the balance, find the sum of the checks and subtract the sum from $1528.

$$\begin{array}{rl} \$\ \ 132 & \\ 425 & \\ 196 & \\ +\ 350 & \\ \hline \$1103 & \text{total checks} \end{array} \qquad \begin{array}{rl} \$1528 & \\ -\ 1103 & \\ \hline \$\ \ 425 & \text{balance} \end{array}$$

The balance was $425 at the end of July.

(Mentally round each check to the nearest hundred to see if the answer is reasonable.)

Now work exercise 2 in the margin.

Average

A topic closely related to addition and division is **average**. Your grade in this course may be based on the average of your exam scores. Newspapers and magazines have information about the Dow-Jones stock averages, the average income of American families, the average life expectancy of dogs and cats, and so on. The average of a set of numbers is a sort of "middle number" of the set.

The average of a set of numbers is found by dividing the sum of the values by the number of values in the set. The average is also called the **arithmetic average,** or **mean.**

The average of a set of whole numbers is not always a whole number. However, in this section, the problems are set up so that the averages are whole numbers. Other cases involving fractions and decimals will be discussed later in the chapters on fractions and decimals (Chapters 3 and 5). A useful way to look at numbers is to use a bar graph. In Example 3, a bar graph is used to display a company's profits over 6 months. Bar graphs will be discussed further in Chapter 8.

Example 3

Chancellor Sporting Goods recorded its sales for tennis rackets for six months. They found their profits to be: January, \$5380; February, \$7590; March, \$6410; April, \$4530; May, \$5840; June, \$6250. Below is a bar graph of their findings. **a.** In what month did they have the most sales of tennis rackets? **b.** What is the average sales per month over the six months?

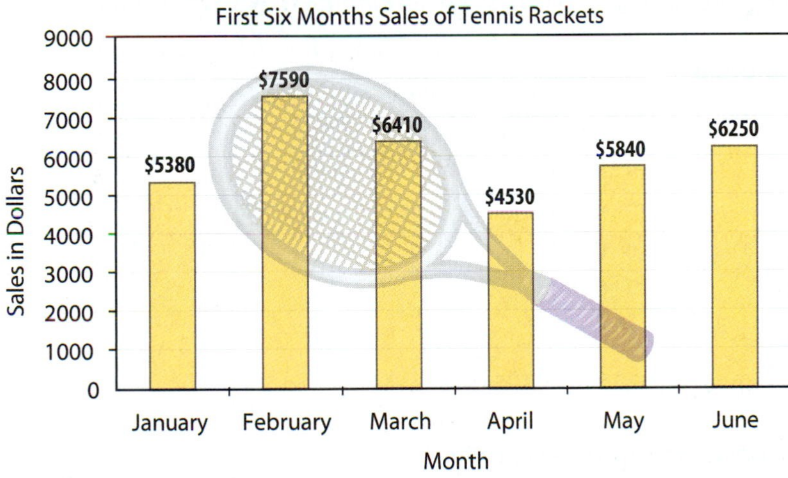

Solution

a. From the bar graph, we can see that the month with the most sales is February with \$7590.

b. The average can be found by finding the sum of the numbers and dividing the sum by the number of numbers.

$$\begin{array}{r} \$\ 5380 \\ 7590 \\ 6410 \\ 4530 \\ 5840 \\ +\ 6250 \\ \hline \$\ 36{,}000 \end{array} \qquad \begin{array}{r} \$6\ 000 \\ 6\overline{)\ 36{,}000} \\ \underline{36{,}000} \\ 0 \end{array} \quad \text{(average)}$$

The sum, 36,000, is divided by 6 because six numbers are being added. The average sales per month over six months is \$6,000.

Now work exercise 3 in the margin.

3. A baseball player hits 3[illegible], 48, 31, and 35 home r[illegible]s in four consecutive [illegible]ea-sons. What is his a[illegible]erage number of home [illegible]ns per year?

4. Find the average of the three numbers 93, 105, and 129.

Example 4

Find the average of the three numbers 32, 47, and 23.

Solution

$$\begin{array}{r} 32 \\ 47 \\ +\ 23 \\ \hline 102 \end{array} \qquad \begin{array}{r} 34 \\ 3\overline{)102} \\ \underline{9} \\ 12 \\ \underline{12} \\ 0 \end{array} \quad \text{(average)}$$

The sum, 102, is divided by 3 because three numbers are being added.

Now work exercise 4 in the margin.

5. Five cities have populations of 180,000, 190,000, 200,000, 210,000 and 220,000. What is the average population of the cities?

Example 5

Suppose five people had the following incomes for one year: $23,000; $24,000; $25,000; $26,000; $27,000. Find their average income.

Solution

$$\begin{array}{r} \$\ \ 23{,}000 \\ 24{,}000 \\ 25{,}000 \\ 26{,}000 \\ +\ 27{,}000 \\ \hline \$125{,}000 \end{array} \qquad \begin{array}{r} \$\ 25{,}000 \\ 5\overline{)125{,}000} \\ \underline{125{,}000} \\ 0 \end{array}$$

The average income is $25,000.

Now work exercise 5 in the margin.

6. Five cities have populations of 5000, 10,000, 15,000, 20,000 and 950,000. What is the average population of the cities?

Example 6

Suppose five people had the following incomes for one year: $8000; $8000; $8000; $8000; $93,000. Find their average income.

Solution

$$\begin{array}{r} \$\ \ 8000 \\ 8000 \\ 8000 \\ 8000 \\ +\ 93{,}000 \\ \hline \$125{,}000 \end{array} \qquad \begin{array}{r} \$\ 25{,}000 \\ 5\overline{)125{,}000} \\ \underline{125{,}000} \\ 0 \end{array}$$

The average income is $25,000.

Now work exercise 6 in the margin.

The average of a set of numbers can be very useful, but it can also be misleading. Judging the importance of an average is up to you, the reader of the information.

In Example 5, the average of $25,000 serves well as a "middle score" or "representative" of all incomes. However, in Example 6, not one of the incomes is even close to $25,000. The one large income completely destroys the "representativeness" of the average. Thus it is useful to see the numbers or at least know something about them before attaching too much importance to an average.

Example 7

On an English exam, two students score 95 points, five students score 86 points, one student scores 82 points, one student scores 78 points, and six students score 75 points. What is the mean score of the class?

$$\begin{array}{r}95\\ \times\ 2\\ \hline 190\end{array}\qquad \begin{array}{r}86\\ \times\ 5\\ \hline 430\end{array}\qquad \begin{array}{r}82\\ \times\ 1\\ \hline 82\end{array}\qquad \begin{array}{r}78\\ \times\ 1\\ \hline 78\end{array}\qquad \begin{array}{r}75\\ \times\ 6\\ \hline 450\end{array}$$

We multiply rather than write down all 15 scores. However, when the five products are added together, we divide the sum by 15 because the sum represents 15 scores.

$$\begin{array}{r}190\\ 430\\ 82\\ 78\\ +\ 450\\ \hline 1230\end{array}\qquad \begin{array}{r}82\\ 15\overline{)1230}\\ \underline{120}\\ 30\\ \underline{30}\end{array}\ \text{mean score}$$

The class mean is 82 points.

Now work exercise 7 in the margin.

7. In a recent high school basketball game, two of the starters scored 18 points each, two scored 10 points, and one scored 4 points. What was the average number of points scored among the starters?

Practice Problems

1. David bought a new stereo which was originally priced at $120. He had a coupon which gave him $10 off the price of the stereo. He also bought three CDs at $15 apiece. How much did David pay overall for the stereo and CDs?

2. Mr. Morris opened a checking account and deposited $4000 in it. He wrote two checks for $175 each and one for $300. What was his balance after writing these checks?

3. The Lee family spent $238 on groceries in June, $207 in July and $218 in August. What was their average amount spent on groceries per month?

Answers to Practice Problems:

1. $155 **2.** $3350 **3.** $221

Name ______________________ Section __________ Date __________

ANSWERS

Exercises 1.7

Consumer Items

1. ______________

1. To purchase a new refrigerator for $1200 including tax, Mr. Kline paid $240 down and the remainder in six equal monthly payments. What were his monthly payments?

2. ______________

2. Miguel decided to go shopping for school clothes before college started in the fall. How much did he spend if he bought four pairs of pants for $21 each, five shirts for $18 each, three pairs of socks for $4 a pair, and two pairs of shoes for $38 a pair?

3. ______________

3. To purchase a new dining room set for $1200, Mrs. Steel had to pay an additional $72 in sales tax. If she made a deposit of $486, how much did she still owe?

4. ______________

4. Alan wants to buy a new car. He could buy a red one for $8500 plus $510 in sales tax and $135 in fees, or he could buy a blue one for $8700 plus $522 in sales tax and $140 in fees. If the manufacturer is giving a $250 rebate on the blue model, which car would be cheaper for Alan? How much cheaper would it be?

5. ______________

5. Lynn decided to take up surfing. She bought a new surfboard for $675, a wet suit for $130, a beach towel for $12, and a new swimsuit for $57. How much money did she spend? (Sales tax was included in the prices.)

6. ______________

6. Pat needed art supplies for a new course at the local community college. She bought a portfolio for $32, a zinc plate for $44, etching ink for $12, and three sheets of rag paper for a total of $6. She received a student discount of $9. How much did she spend on art supplies?

Checking Accounts

7. ______________

7. If you opened a checking account with $875, then wrote checks for $20, $35, $115, $8, and $212, what would be your balance?

8. ______________

8. Your friend had a checking account balance of $1250 and wrote checks for $375, $52, $83, and $246. What was her new balance?

9. ______________

9. On August 1, Matt had a balance of $250 in his checking account. During August, he made deposits of $200, $350, and $236. He wrote checks for $487, $25, $33, and $175. What was his balance on September 1?

10. ______________

10. Melissa deposited $500, $2470, $800, $3562, and $2875 in her checking account over a five-month period. She wrote checks totaling $6742. If her beginning balance was $1400, what was her balance at the end of the five months?

Name ______________ Section ______________ Date ______________

ANSWERS

Average

11. Below is a bar graph showing the number of hybrid cars sold in the U.S. from the year 2000 to 2004.
    a. In what year was the least number of hybrid cars sold?
    b. What was the average number of hybrid cars sold over the five years?

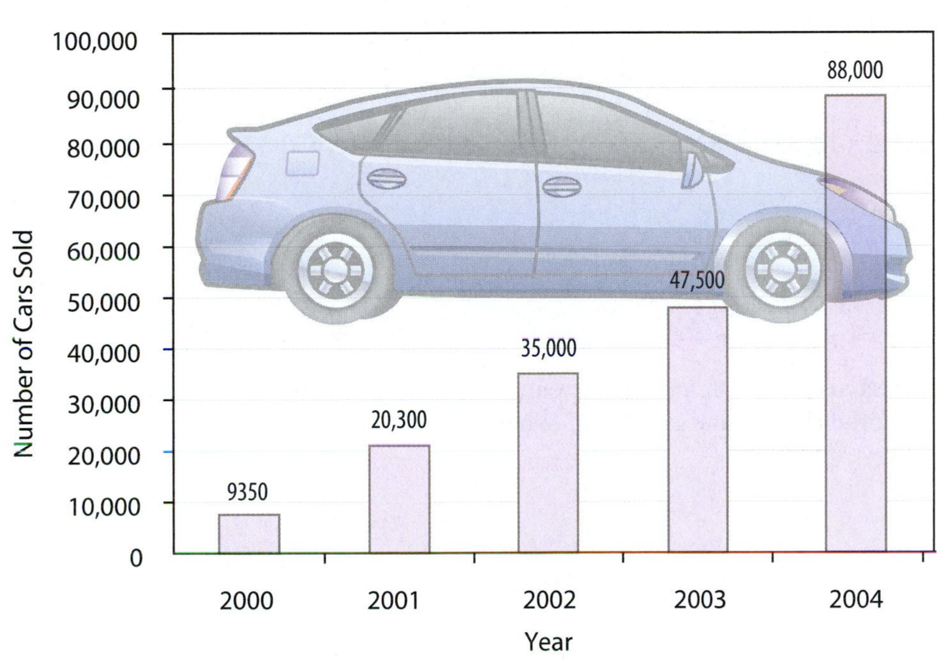

12. If Rina's cell phone bills for the past five months have been \$56, \$63, \$52, \$85, and \$49, what was her average cell phone bill for the past 5 months?

13. Bill wanted to compare car insurance rates to find out the average amount he should pay for car insurance. If he looked at four different companies and the monthly rates were \$164, \$107, \$131, and \$98, what is the average monthy rate of car insurance?

14. During a sports trivia game, one team scored 35 points, three teams scored 23 points, two teams scored 18 points, and two teams scored 14 points. What was the mean score of the teams?

11. a. ______________

b. ______________

12. ______________

13. ______________

14. ______________

15. ______

16. ______

17. ______

18. ______

19. ______

20. ______

21. ______

22. ______

23. ______

In Exercises 15 – 20, find the average (or mean) of each set of numbers. See Examples 3 through 6.

15. 102, 113, 97, 100

16. 56, 64, 38, 58

17. 6, 8, 7, 4, 4, 5, 6, 8

18. 5, 4, 5, 6, 5, 8, 9, 6

19. 512, 618, 332, 478

20. 436, 520, 630, 422

21. Test scores. On a history exam, two students scored 95, six scored 90, three scored 80, and one scored 50. What was the class average?

22. Sales. A salesman sold items from his sales list for $972, $834, $1005, $1050, and $799. What was the average price per item?

23. Stocks. Ms. Lee bought 150 shares of stock in Microsoft at $24 per share. Two months later, she bought another 100 shares at $29 per share. What average price per share did she pay? If she sold all 250 shares at $28 per share, what was her profit?

Name ______________________ Section ____________ Date __________

ANSWERS

24. ____________

25. a. ____________

b. ____________

26. ____________

27. ____________

24. Income. Three families, each with two children, had incomes of $56,000. Two families, each with four children, had incomes of $62,000. Four families, each with two children, had incomes of $45,000. One family had no children and an income of $37,000. What was the mean income per family?

25. Checking. During July, Mr. Rodriguez made deposits in his checking account of $400 and $750 and wrote checks totaling $625. During August, his deposits were $632, $322, and $798, and his checks totaled $978. In September, his deposits were $520, $436, $200, and $376, and his checks totaled $836.

a. What was the average monthly difference between his deposits and his withdrawals?

b. What was his bank balance at the end of September if he had a balance of $500 on July 1?

26. Flight time. In one month (30 days), an airline pilot spent the following number of hours in preparation for and flying each of 12 flights: 6, 8, 9, 6, 7, 7, 7, 5, 6, 6, 6, and 11 hours. What was the average (mean) amount of time the pilot spent per flight?

27. Population. The 10 largest cities in South Carolina have the following approximate populations:

Cities	Population	Cities	Population
Columbia	116,331	Greenville	56,291
Charleston	104,883	Sumter	39,671
North Charleston	84,271	Spartanburg	38,599
Rock Hill	57,902	Hilton Head	34,371
Mount Pleasant	56,350	Summerville	34,241

What is the average population of these cities?

28. ______

29. a. ______

b. ______

c. ______

d. ______

28. The five longest rivers in the world are the:

River	Distance (miles)
Nile (Africa)	4180
Amazon (South America)	3900
Mississippi-Missouri-Red Rock (North America)	3880
Yangtze (China)	3600
Ob (Russia)	3460

What is the mean length of these rivers?

29. The following list is repeated from **Mathematics at Work!** at the beginning of this chapter. Tuition, room, and board are given as annual figures (rounded to the nearest ten and subject to change) for the year 2004.

			Tuition	
Institution	**Enrollment**	**Room/Board**	**Resident**	**Nonresident**
Univ. of Arizona	36,900	$7100	$4100	$13,100
Boston College	14,500	$10,900	$31,500	$31,500
Univ. of Colorado	29,200	$8000	$5400	$22,800
Florida A & M	13,100	$6000	$3200	$15,200
Purdue Univ.	38,700	$6800	$6500	$19,800
Johns Hopkins	18,200	$9900	$31,600	$31,600
MIT	10,300	$9500	$32,300	$32,300
Ohio State Univ.	51,000	$7300	$7400	$18,000
Princeton Univ.	6800	$8800	$31,500	$31,500
Rice Univ.	4900	$9000	$20,200	$20,200
Univ. of Tennessee	27,800	$6300	$4300	$9000
Univ. of Wisconsin	41,200	$6500	$6200	$21,100

For these universities and colleges, find (to the nearest whole number):

a. The average enrollment,

b. The average tuition for residents,

c. The average tuition for nonresidents, and

d. The average cost of room and board.

Name ______________________ Section ____________ Date __________

ANSWERS

Writing and Thinking about Mathematics

30. If the product of 607 and 93 is divided by 3, what is the quotient? What is the remainder? Is 3 a factor of the product? Explain briefly.

30. ____________

31. If the difference between 347 and 196 is multiplied by 15, what is the product? Is 5 a factor of this product? Explain briefly.

31. ____________

1.8 Geometry: Perimeter and Area

Introduction to Geometry

Plane geometry is the study of the properties of figures in a plane. Geometry is an important part of the development of mathematics because basic mathematical concepts are geometric as well as arithmetic and algebraic.

The three most basic ideas in plane geometry are the **point**, **line**, and **plane**. These concepts are so fundamental that they have historically been used without definitions. As we will see, these undefined terms provide the foundation for the study of geometry and the definitions of other geometric terms such as **line segment**, **polygon**, **triangle** (3-sided figure), **quadrilateral** (4-sided figure), and so on.

Objective 1

Objectives

1. Recognize the terms **point**, **line**, and **plane** and know that they are undefined terms.
2. Know what a **polygon** is.
3. Recognize and be able to define the geometric figures **triangles**, **rectangles**, and **squares**.
4. Understand the concepts of **length** and **perimeter**.
5. Understand the concept of **area**.

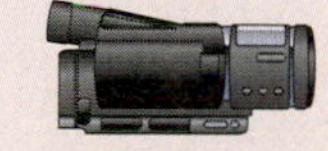
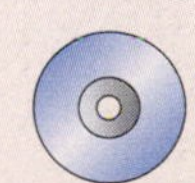

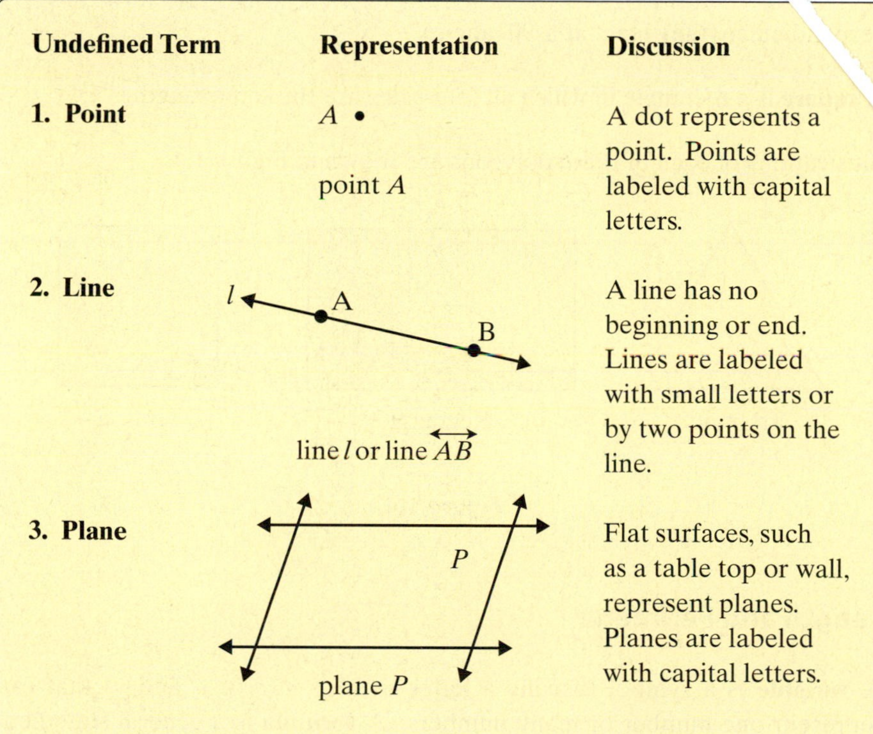

Undefined Term	Representation	Discussion
1. Point	A • point A	A dot represents a point. Points are labeled with capital letters.
2. Line	(line l through points A and B) line l or line $\overleftrightarrow{AB}$	A line has no beginning or end. Lines are labeled with small letters or by two points on the line.
3. Plane	(plane P) plane P	Flat surfaces, such as a table top or wall, represent planes. Planes are labeled with capital letters.

In a more formal approach to plane geometry, these undefined terms are studied in detail along with definitions, certain assumptions known as axioms, other statements called theorems, and a formal system of logic. This approach to the study of geometry is credited to Euclid (about 300 B. C.), which is why the plane geometry courses generally given in high school are known as courses in Euclidean geometry.

Polygons

A **line segment** consists of two points on a line and all the points on the line between these two points.

Objective ②

Polygon

A **polygon** is a closed plane figure, with three or more sides, in which each side is a line segment.

(Each point where two sides meet is called a **vertex**.)

(**Note**: A **closed figure** begins and ends at the same point.)

Three commonly studied types of polygons are triangles, rectangles, and squares.

Objective ③

A **triangle** is a polygon with three sides.

A **rectangle** is a polygon with four sides in which adjacent sides are perpendicular (they meet at a $90°$ angle).

A **square** is a rectangle in which all four sides are the same length.

Illustrations of each of these polygons are shown in Figure 1.4.

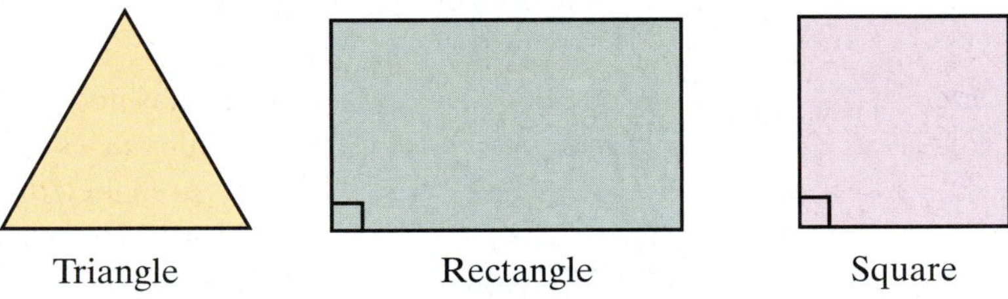

Figure 1.4

Length and Perimeter

A **variable** is a symbol (usually a letter such as *a, b, s, x,* and *y*) that can represent one number or many numbers. A **formula** is a general statement (usually an equation) that relates two or more variables. The formulas in this section are used to represent perimeter and area of polygons. Some of the units of length from the metric system are millimeter (mm), centimeter (cm), meter (m), and kilometer (km). From the U.S. customary system, some units of length are inch (in.), foot (ft), yard (yd), and mile (mi). The formulas do not depend on any particular measurement system.

Be sure to label all answers with the correct units of measure.

Objective ④

Perimeter

The **perimeter**, P, of a polygon is the sum of the lengths of its sides.

The formulas for the perimeters of triangles, rectangles and squares are shown in Figure 1.5. In the triangle, a, b, and c represent the lengths of the three sides. In the rectangle, l represents length and w represents width. (Usually, the width is the shorter of the two.) In the square, s is the length of one side, and all four sides have the same length.

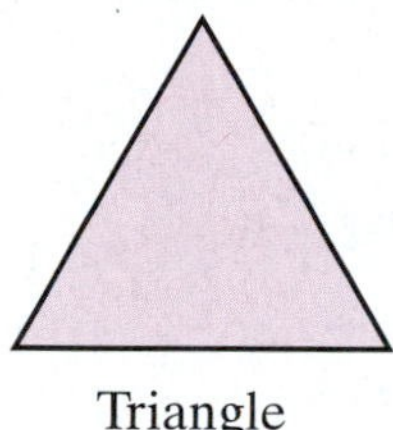

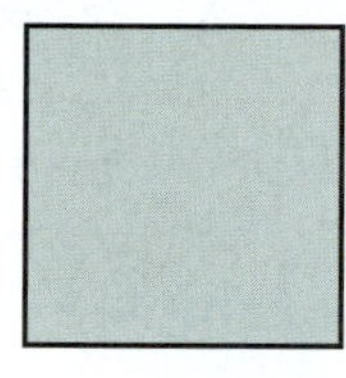

Triangle	Rectangle	Square
$P = a + b + c$	$P = 2l + 2w$	$P = 4s$

Figure 1.5

Example 1

Find the perimeter of the square shown.

Solution

Because the figure is a square, each side is 16 in. long. Thus the perimeter can be found as follows:

$$P = 16 + 16 + 16 + 16 = 64 \text{ in.}$$

Or, by using the formula, we have

$$P = 4 \cdot 16 = 64 \text{ in.}$$

Now work exercise 1 in the margin.

1. Find the perimeter of a square with one side labeled 8 ft.

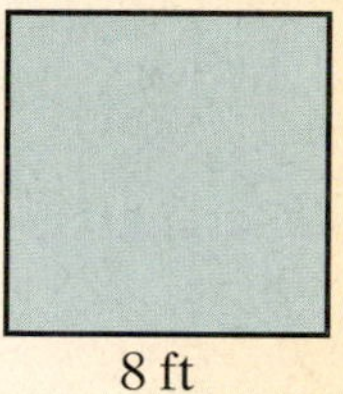

Example 2

Find the perimeter of the polygon shown.
(**Note:** A 5-sided polygon is called a **pentagon**.)

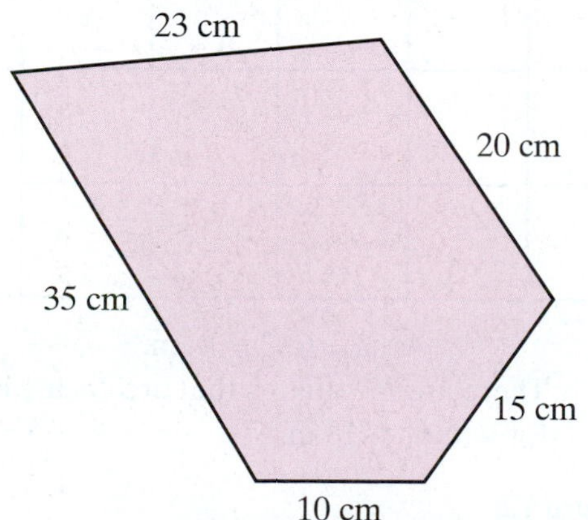

2. Find the perimeter of the figure shown below.

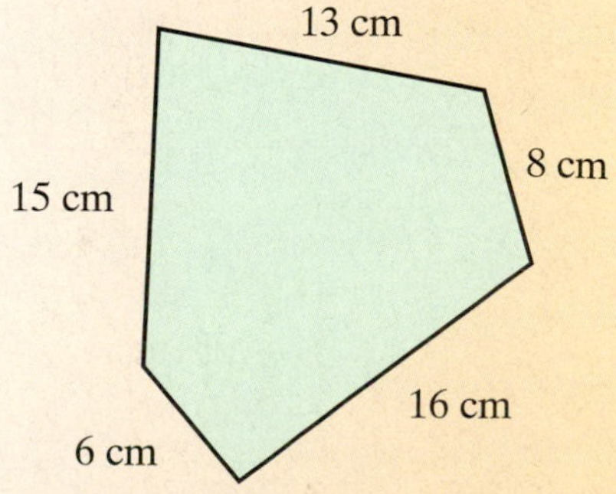

Solution

The perimeter of any polygon is found by adding the lengths of the sides. Thus, for this pentagon we have

$$P = 10 + 15 + 20 + 23 + 35 = 103 \text{ cm}$$

Now work exercise 2 in the margin.

3. Find the perimeter of the following rectangle.

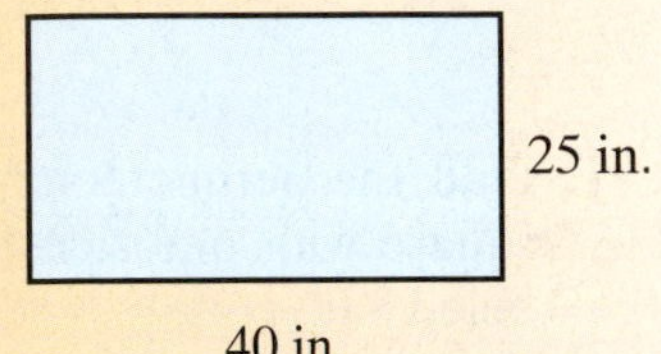

Example 3

Find the perimeter of the rectangle shown.

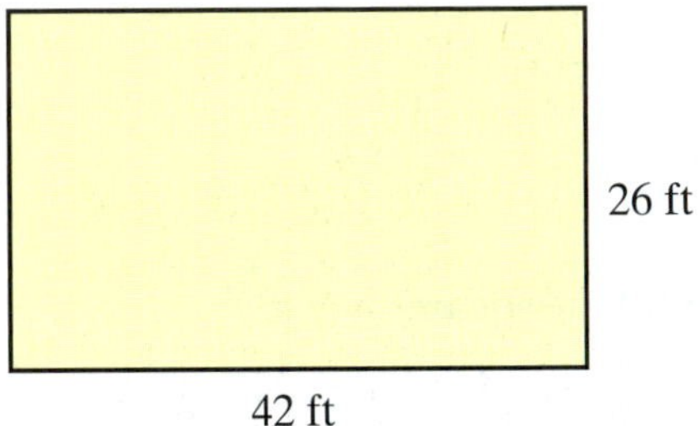

Solution

The perimeter can be found by adding the lengths of the sides. Thus,

$$P = 42 + 26 + 42 + 26 = 136 \text{ ft}$$

Or, by using the formula we have,

$$P = 2 \cdot 42 + 2 \cdot 26 = 84 + 52 = 136 \text{ ft}$$

Now work exercise 3 in the margin.

Area

Objective 5

Area is a measure of the interior of (or surface enclosed by) a figure in a plane and is measured in **square units**. The concept of area is illustrated in Figure 1.6. [**Note:** in.^{2} is read "inches squared" or "square inches."]

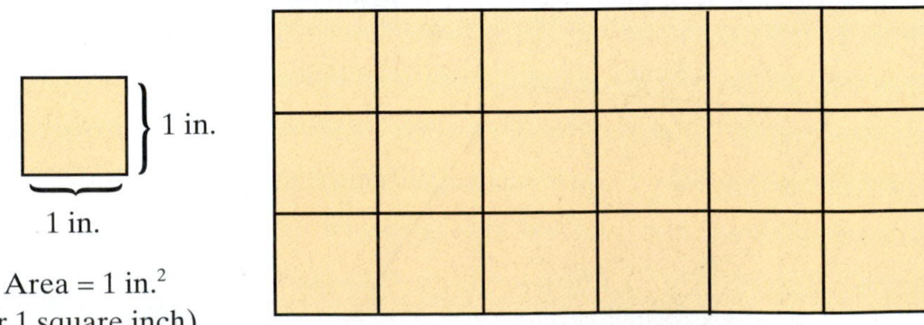

Figure 1.6

In the metric system, some units of area are square meters (m^2), square centimeters (cm^2), square millimeters (mm^2), and square kilometers (km^2). In the U.S. customary system, some units of area are square feet (ft^2), square inches ($in.^2$), square yards (yd^2), and square miles (mi^2). Figure 1.7 shows three polygons and the corresponding formulas for finding the areas. In the triangle b and h represent the base and height, respectively. The base and height are perpendicular to each other.

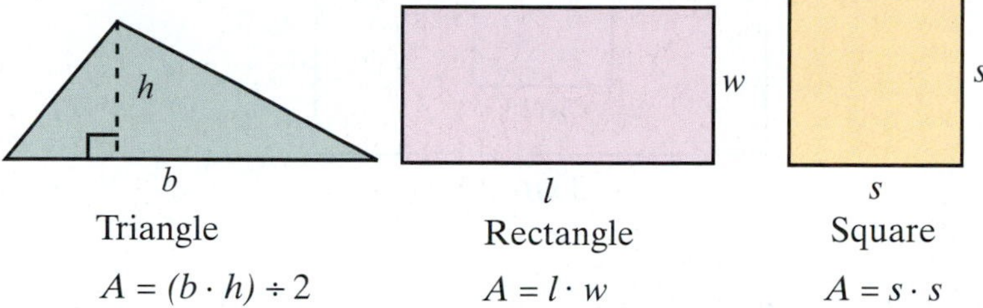

Figure 1.7

Find the area of each figure.

Example 4

Find the area of the triangle shown.

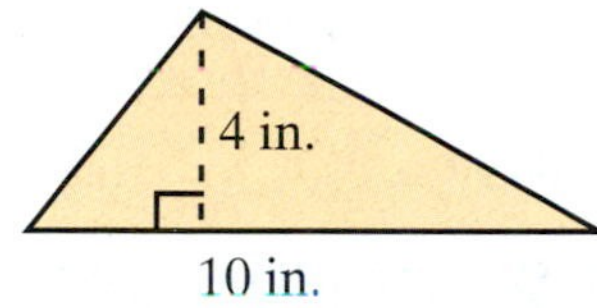

Solution

To find the area of the triangle, multiply the base times the height and divide by 2.

$$A = (10 \cdot 4) \div 2 = 40 \div 2 = 20 \text{ in.}^2$$

Now work exercise 4 in the margin.

4.

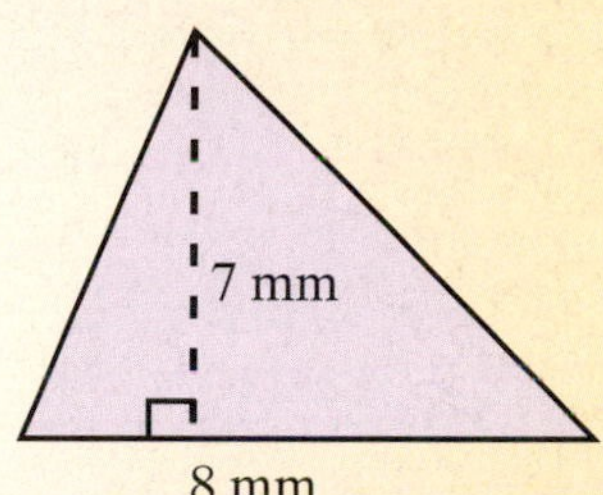

Example 5

Find the area of the rectangle shown.

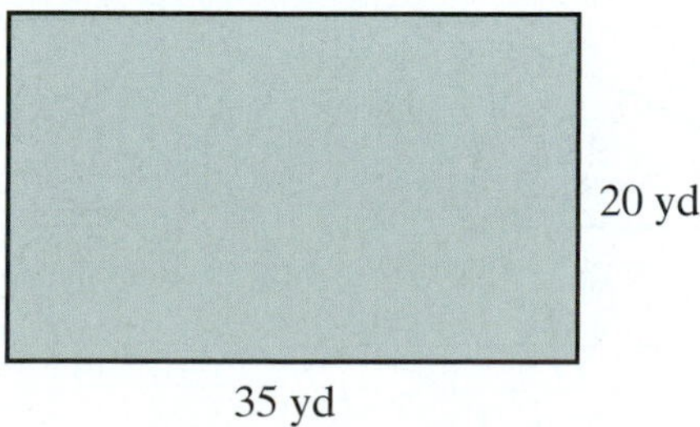

Solution

To find the area of the rectangle, multiply the length times the width.

$$A = 35 \cdot 20 = 700 \text{ yd}^2$$

Now work exercise 5 in the margin.

5.

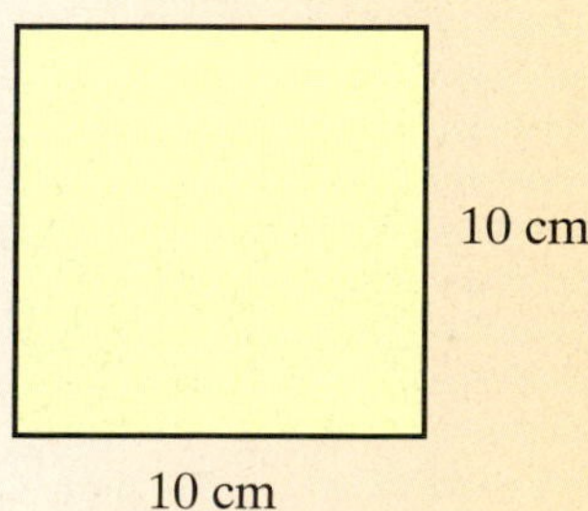

6. A rectangle is inside a rectangle as shown. Find the area of the shaded region.

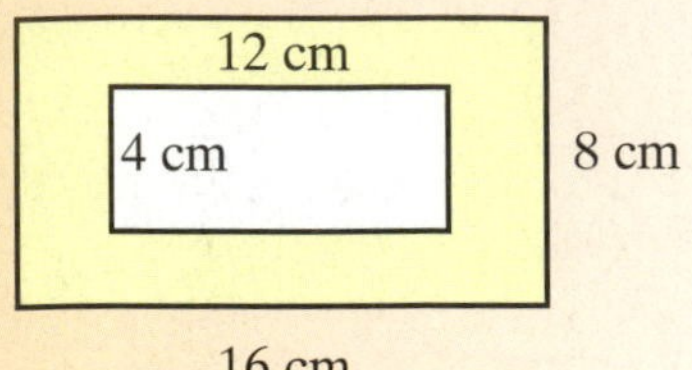

Example 6

A square is inside a rectangle as shown. Find the area of the shaded region.

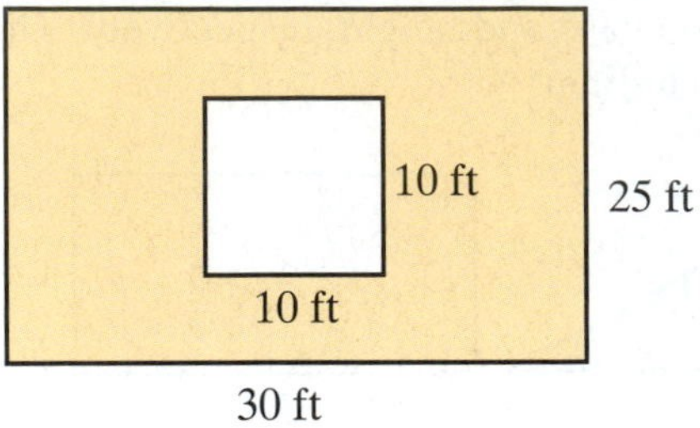

To find the area of the shaded region:

First, find the area of the rectangle. Second, find the area of the square.

$A = 30 \cdot 25 = 750 \text{ ft}^2$ $\qquad$ $A = 10 \cdot 10 = 100 \text{ ft}^2$

Third, find the difference of the two areas.

Area of shaded region = $750 - 100 = 650 \text{ ft}^2$

Now work exercise 6 in the margin.

Practice Problems

1. Find the perimeter of the following figure.

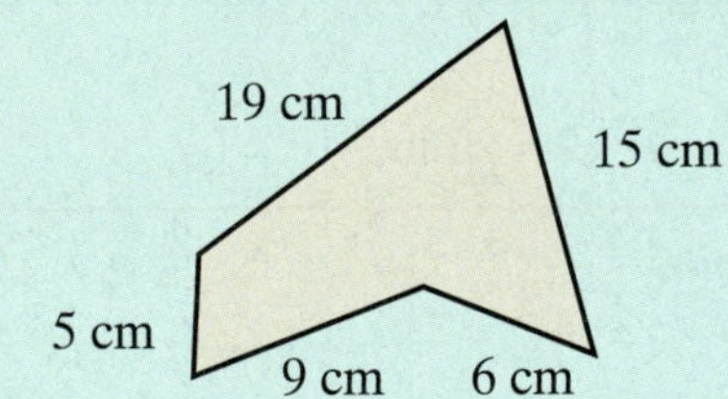

2. Find the area of the following figures.

a.

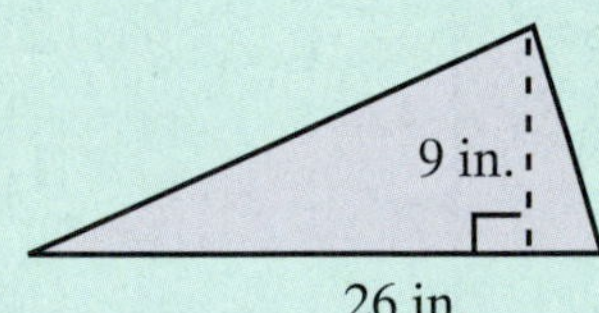

b.

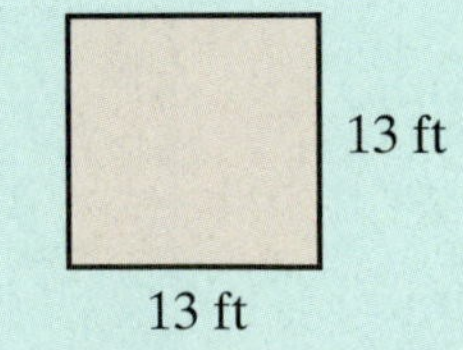

Answers to Practice Problems:

1. 54 cm
2. a. 117 in.^{2} **b.** 169 ft^2

Name ______________ Section ____________ Date __________

Exercises 1.8

1. Name the three undefined terms used in geometry.

2. Write the definition of a polygon.

3. True or False:
   a. Every square is a rectangle.
   b. Every rectangle is a square.

Find the perimeter of each figure in Exercises 4 – 15. See Examples 1 through 3.

4. 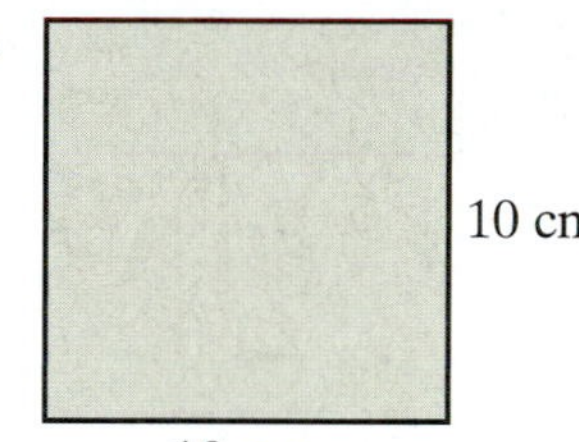

5. 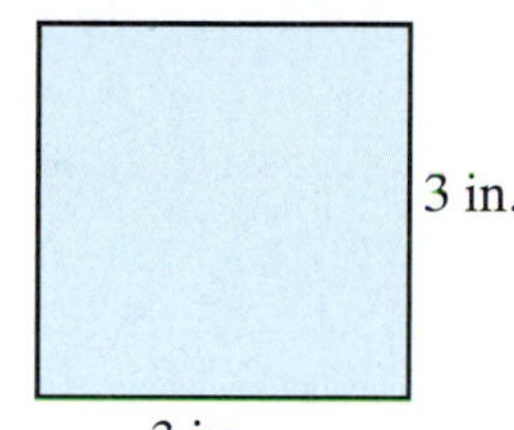

6. 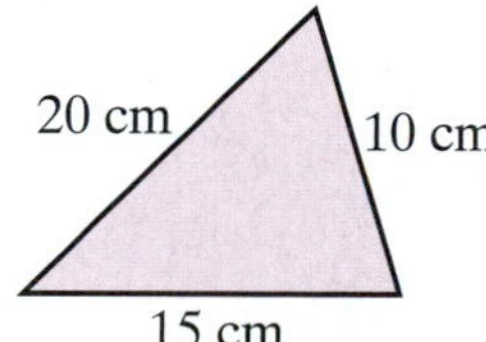

7. 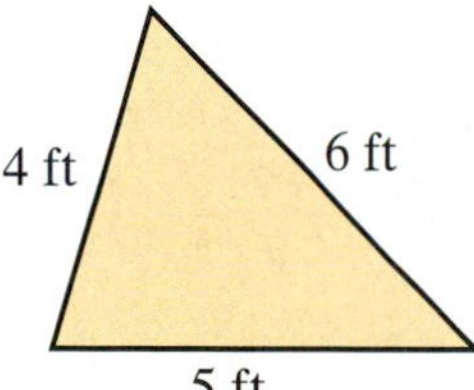

8. 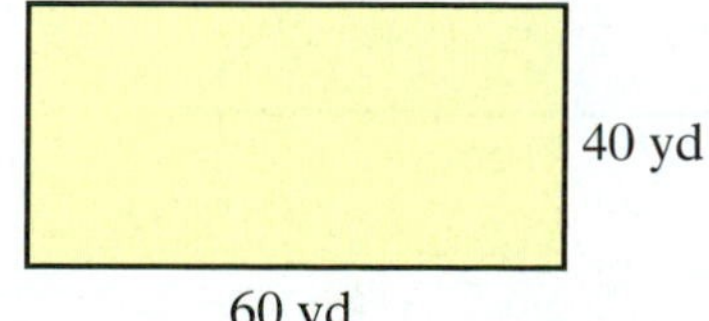

9. 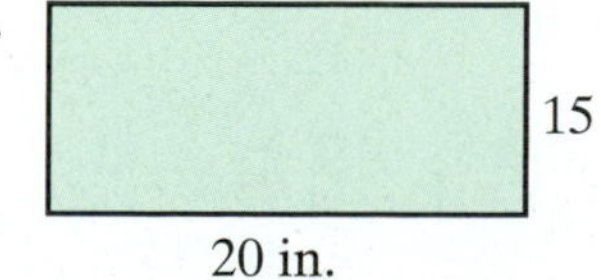

ANSWERS

1. ______
2. ______
3. a. ______
b. ______
4. ______
5. ______
6. ______
7. ______
8. ______
9. ______

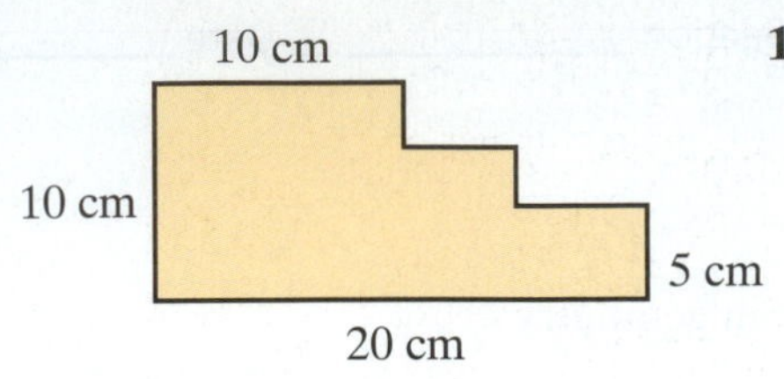

10. ______

11. ______

12. ______

13. ______

14. ______

15. ______

16. ______

17. ______

10.

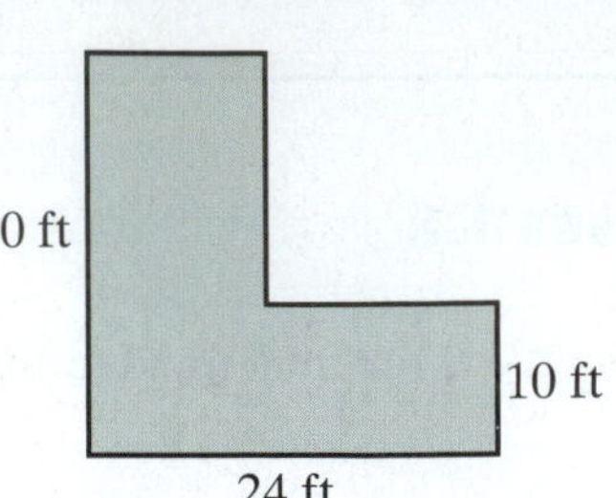

11.

12.

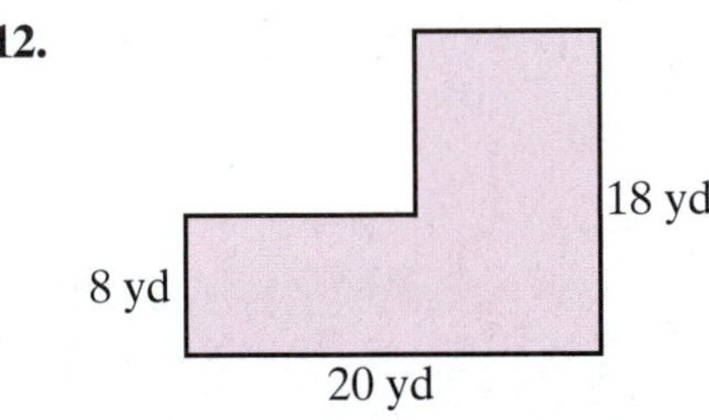

13.

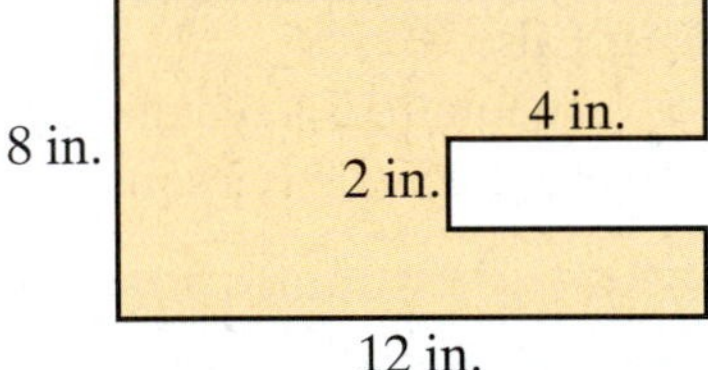

14.

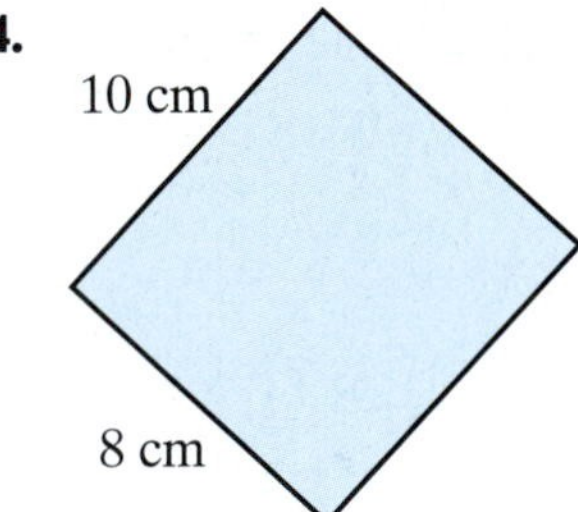

15.

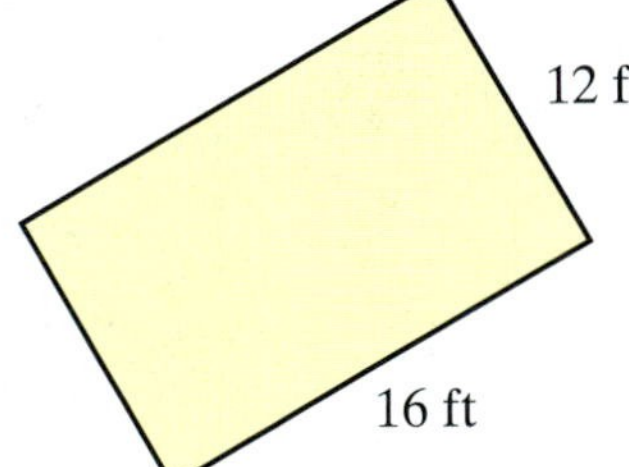

Find the area of each figure in Exercises 16 – 25. See Examples 4 and 5.

16.

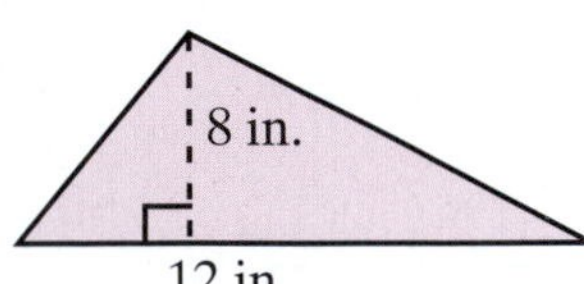

17.

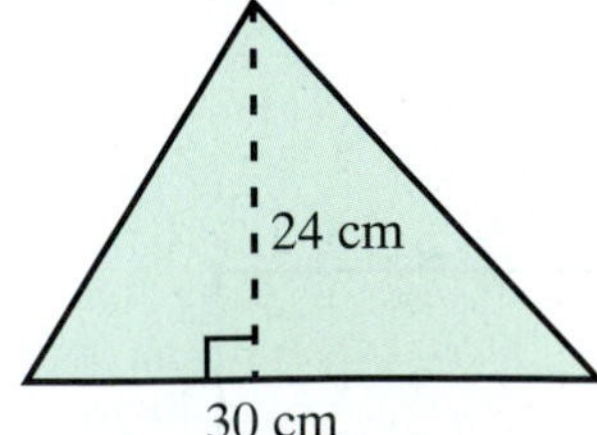

Name ______________ Section ______________ Date ______________

ANSWERS

18.

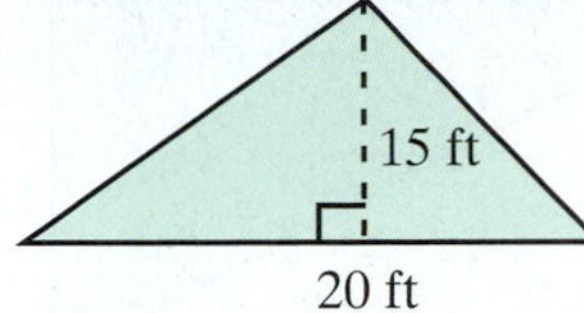

19.

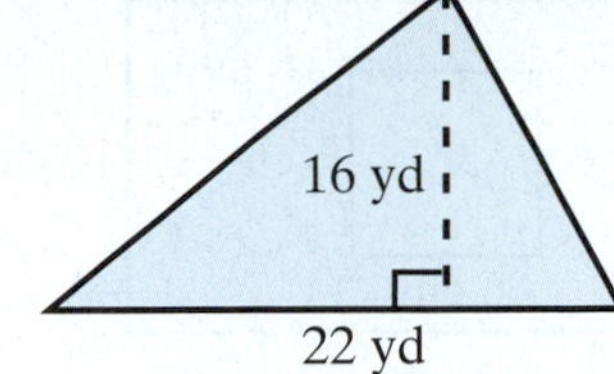

20.

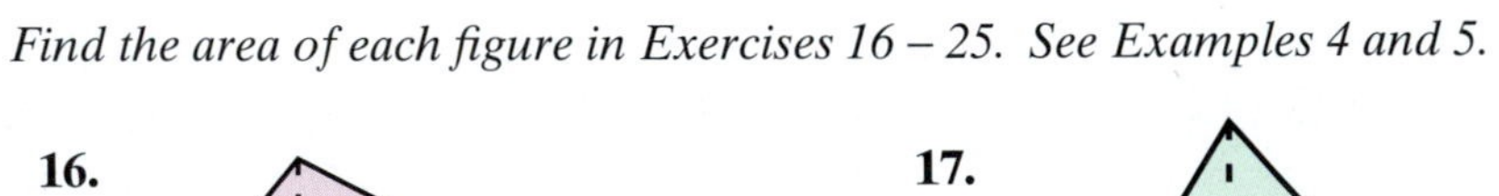

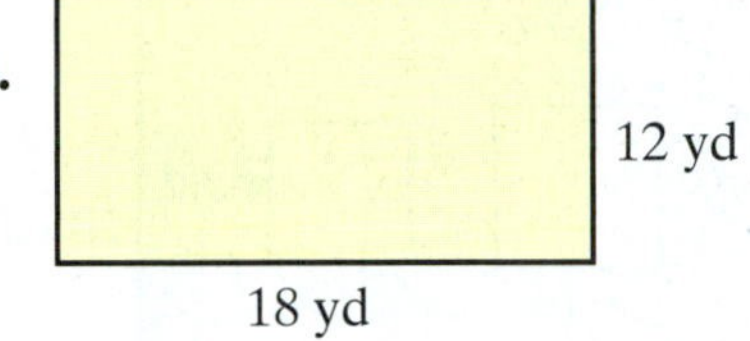

21.

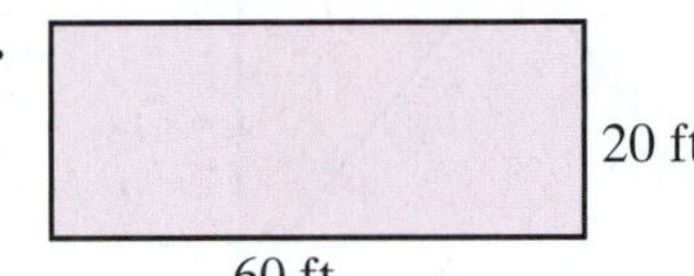

22.

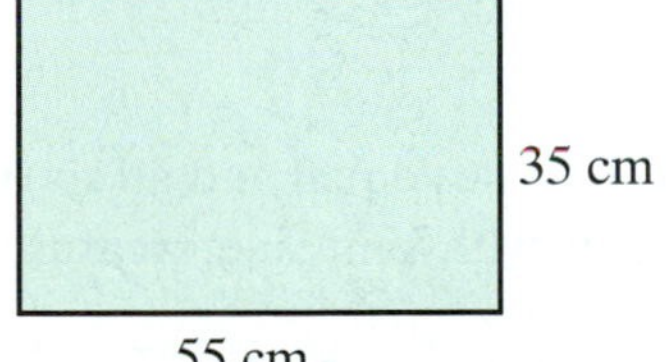

23.

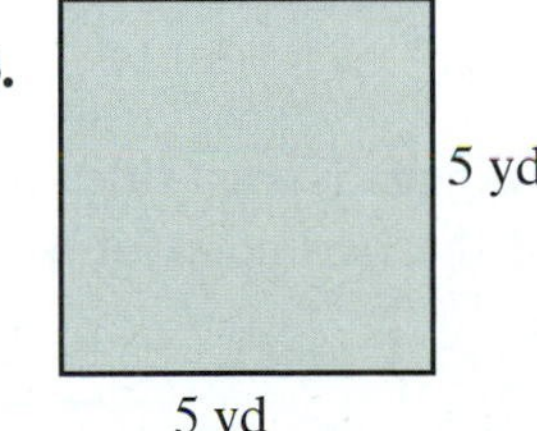

24.

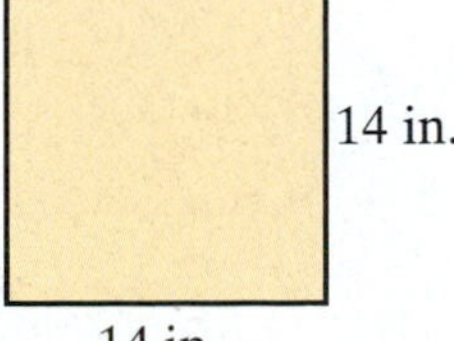

25.

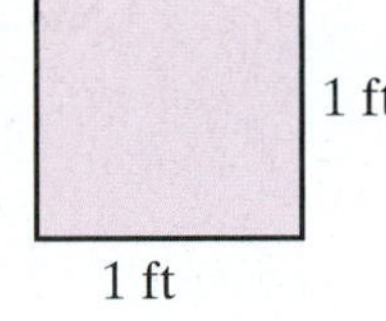

18. ______________

19. ______________

20. ______________

21. ______________

22. ______________

23. ______________

24. ______________

25. ______________

26. ______

27. ______

28. ______

29. ______

30. ______

In Exercises 26 – 30, find the area of the shaded part of each figure. See Example 6.

26.

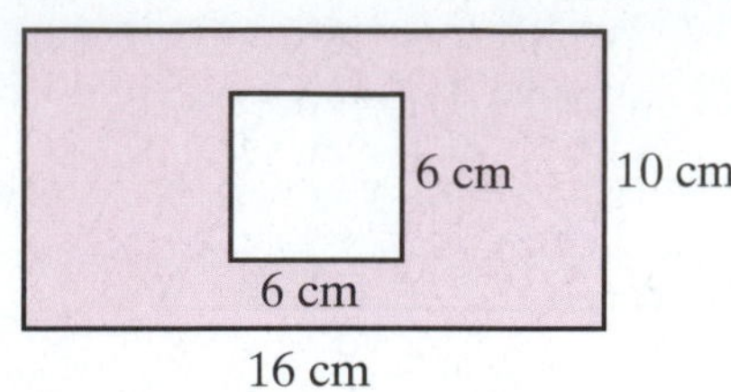

27.

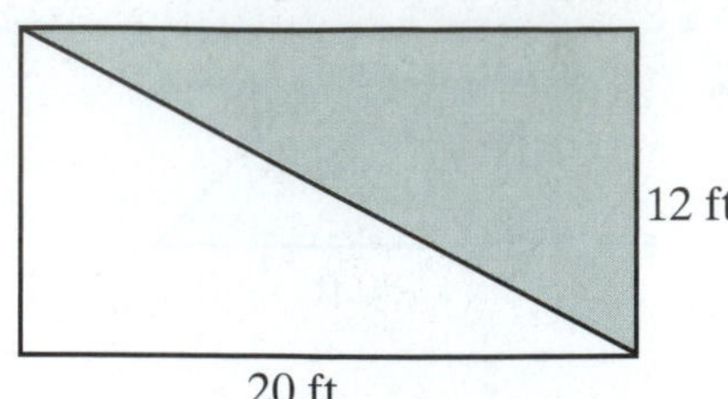

28.

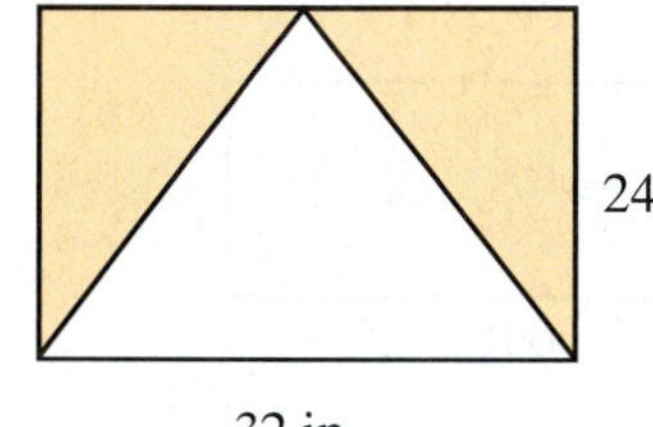

29.

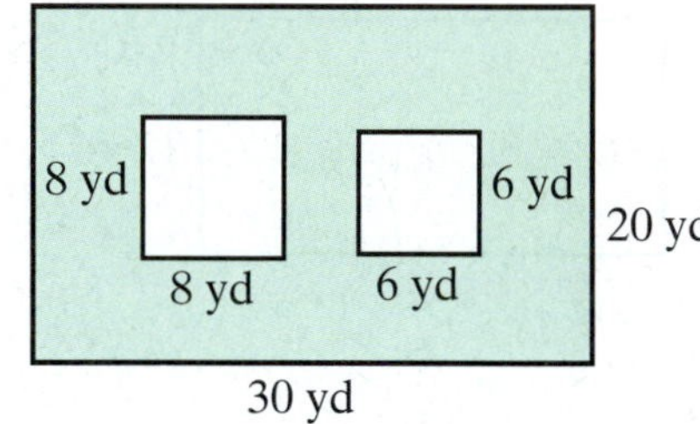

30. A **regular hexagon** is a six-sided figure with all six sides equal and all six angles equal. Find the perimeter of a regular hexagon with one side measuring 19 centimeters.

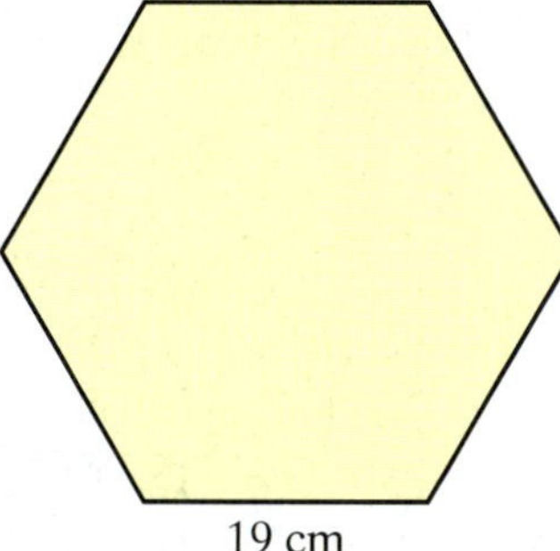

Name ______________________ Section ____________ Date ___________

31. An **isosceles triangle** (two sides equal) is placed on top of a square to form the figure shown below. If each of the two equal sides of the triangle is 18 inches and the square is 28 inches on a side, what is the perimeter of the figure?

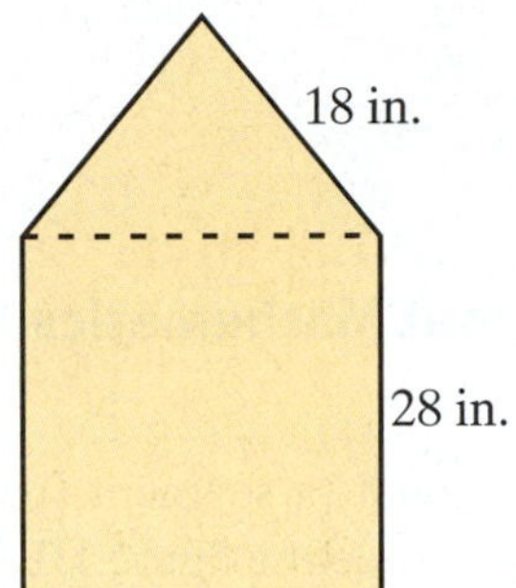

32. A **right triangle** is a triangle with one angle of 90°, which means that two sides are perpendicular. The base and the height are the two perpendicular sides. Find the perimeter and the area for each of the following right triangles.

a.

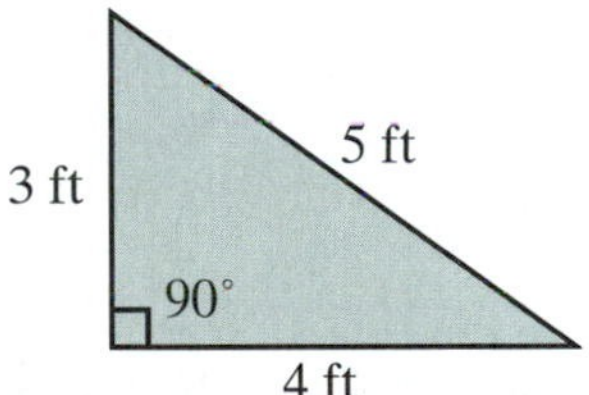

b.

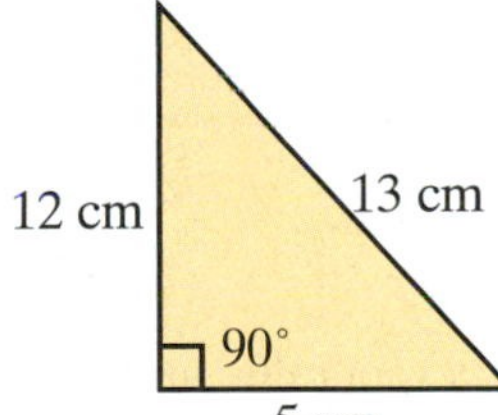

c.

12 in.
20 in.
90°
16 in.

ANSWERS

31. ____________

32. a. ____________

b. ____________

c. ____________

33. ____________

33. A **regular octagon** is an eight-sided figure with all eight sides equal and all eight angles equal. Find the perimeter of a regular octagon if one side measures 12 inches.
(**Note**: Where do you see regular octagons on a regular basis?)

Writing and Thinking about Mathematics

34. Draw a rectangle and a diagonal (a segment from one vertex to an opposite vertex) forming two triangles. (See the figure.) Discuss the fact that the area of the triangle can be found by using the formula $A = (b \cdot h) \div 2$.

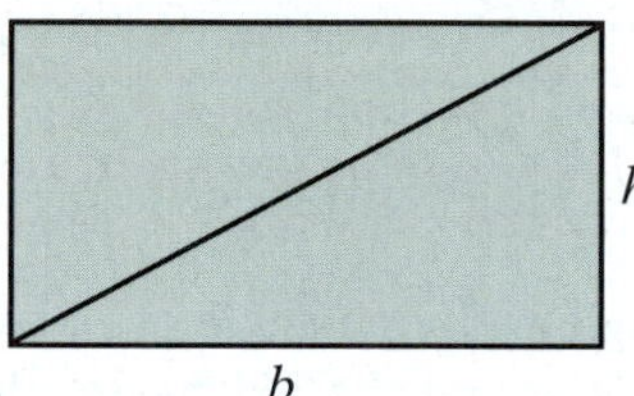

34. ____________

35. Draw a rectangle and choose any point on one side of the rectangle. Draw line segments to the vertices on the opposite side (forming a triangle). Now cut out the two triangles on each end. Place these triangles over the remaining triangle to show that the total of the two areas is equal to the area of the remaining triangle. Do this three different times choosing a different point each time. (See the figure.) What fact does this illustrate about the area of a triangle?

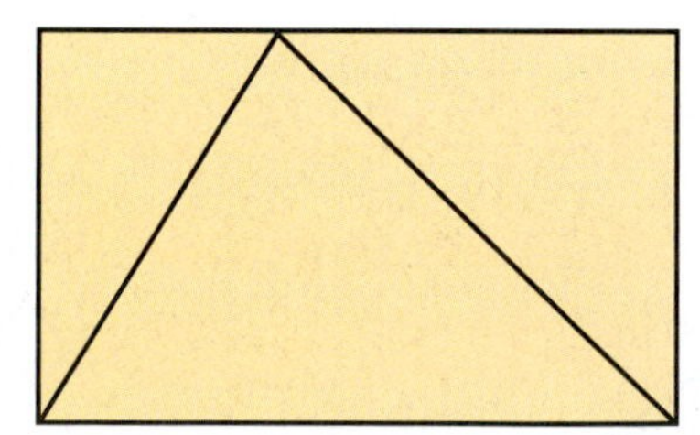

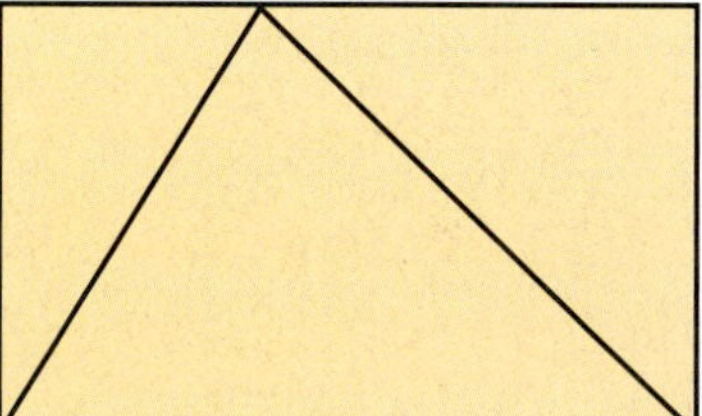

35. ____________

Chapter 1 Index of Key Terms and Ideas

Section 1.1 Reading and Writing Whole Numbers

The Place Value System pages 2 – 3

The decimal system (or base ten system) is a **place value system** that depends on three things:

1. The **ten digits**: 0, 1, 2, 3, 4, 5, 6, 7, 8, 9;
2. The **placement** of each digit; and
3. The **value** of each place.

Whole Numbers page 5

Whole numbers are the counting numbers and the number 0.

$\mathbf{W} = \{0, 1, 2, 3, 4, 5, 6, 7, 8, 9, 10, 11, 12, 13, 14, 15, \ldots\}$

Reading and Writing Whole Numbers page 5

You should note the following four things when reading or writing whole numbers:

1. Digits are read in **groups of three.**
2. Commas are used to separate groups of three digits **if a number has more than four digits**.
3. The word **and** does not appear in English word equivalents. **And** is said only when reading a decimal point. (See Chapter 5)
4. Hyphens (-) are used to write English words for the two-digit numbers from 21 to 99 except for those that end in 0.

Section 1.2 Addition with Whole Numbers

Addition with Whole Numbers pages 13 – 14

The numbers being added are called **addends**, and the result of the addition is called the **sum**.

Properties of Addition pages 14 – 15

Commutative Property: For any two whole numbers a and b, $a + b = b + a$.

Associative Property: For any whole numbers a, b, and c,
$a + b + c = a + (b + c) = (a + b) + c$.

Additive Identity Property: For any whole number a, $a + 0 = 0 + a = a$.

Section 1.3 Subtraction with Whole Numbers

Subtraction page 25

Subtraction is the operation of taking one amount, or number, away from another. The **difference** is the result of subtracting one addend from the sum.

To Subtract Whole Numbers with More Than One Digit page 26

1. Write the numbers vertically so that the place values are **lined up** in columns.
2. Subtract only the digits with the same place value.

Continued on next page...

Section 1.3 Subtraction with Whole Numbers (continued)

Borrowing page 27

1. **Borrowing** is necessary when a digit is smaller than the digit being subtracted.
2. The process starts from the rightmost digit. **Borrow** from the digit to the left.

Section 1.4 Rounding and Estimating with Whole Numbers

Rounding Whole Numbers page 39

To **round** a given number means to find another number close to the given number.

Rounding Rule for Whole Numbers page 40

1. Look at the single digit just to the right of the digit that is in the place of desired accuracy.
2. If this digit is 5 or greater, make the digit in the desired place of accuracy one larger and replace all digits to the right with zeros. All digits to the left remain unchanged unless a 9 is made one larger.
3. If this digit is less than 5, leave the digit in the place of desired accuracy as it is and replace all digits to the right with zeros. All digits to the left remain unchanged.

To Estimate an Answer page 41

1. Round each number to the place of the leftmost digit.
2. Perform the indicated operation with these rounded numbers.

Section 1.5 Multiplication with Whole Numbers

Multiplication with Whole Numbers pages 49 – 50

Multiplication is a process that shortens repeated addition with the same number. The result of multiplying two numbers is called the **product**, and the two numbers are called **factors** of the product.

Properties of Multiplication pages 50 – 54

Commutative Property: $a \cdot b = b \cdot a$

Associative Property: $a \cdot b \cdot c = a(b \cdot c) = (a \cdot b)c$

Multiplicative Identity Property: $a \cdot 1 = a$

Zero Factor Law: $a \cdot 0 = a$

Distributive Property of Multiplication over Addition:

$$a(b + c) = a \cdot b + a \cdot c$$

To Multiply Two or More Whole Numbers that End with 0's page 52

1. Count the ending 0's.
2. Multiply the numbers without the ending 0's.
3. Write the product from Step 2 with the counted number of 0's at the end.

Continued on next page...

Section 1.5 Multiplication with Whole Numbers (continued)

To Estimate a Product page 58

1. Round each number to the place of the **leftmost** digit.
2. Multiply the rounded numbers by using your knowledge of multiplication by powers of 10.

Section 1.6 Division with Whole Numbers

Division with Whole Numbers pages 67 – 68

The process of division can be thought of as the reverse of multiplication. The number dividing is called the **divisor**, the number being divided is called the **dividend**, the result is called the **quotient**, and the number left over is called the **remainder** (it must be less than the divisor.)

If the remainder is 0, then the following statements are true. page 73

1. Both the divisor and quotient are **factors** of the dividend.
2. We say that both factors **divide exactly** into the dividend.
3. Both factors are called **divisors** of the dividend.

To Estimate the Quotient page 74

1. Round both the divisor and dividend to the place of the last digit on the left.
2. Divide with the rounded numbers. (This process is very similar to the trial dividing step in the division algorithm.)

Division with 0 page 77

1. If a is any nonzero whole number, then $0 \div a = 0$.
2. If a is any whole number, then $a \div 0$ is **undefined**.

Section 1.7 Problem Solving with Whole Numbers

Basic Strategy for Solving Word Problems page 83

1. Read each problem carefully until you understand the problem and know what is being asked.
2. Draw any type of figure or diagram that might be helpful and decide what operations are needed.
3. Perform these operations.
4. Mentally check to see if your answer is reasonable and see if you can think of another more efficient or more interesting way to do the same problem.

Average page 84 – 85

An **average** (or **mean**) of a set of numbers is the number found by adding the values in the set and then dividing this sum by the number of values in the set.

Section 1.8 Geometry: Perimeter and Area

Plane Geometry page 97

The fundamental concepts of **plane geometry** are the **point**, **line**, and **plane**, which are generally used without definitions.

Polygon pages 97 – 98

A **polygon** is a closed plane figure, with three or more sides, in which each side is a line segment. (Each point where two sides meet is called a **vertex**.) (**Note:** A **closed figure** begins and ends at the same point.)

Geometric Figures pages 98

A **triangle** is a polygon with three sides.
A **rectangle** is a polygon with four sides in which adjacent sides are perpendicular (they meet at a 90° angle).
A **square** is a rectangle in which all four sides are the same length.
Illustrations of each of these polygons are shown below.

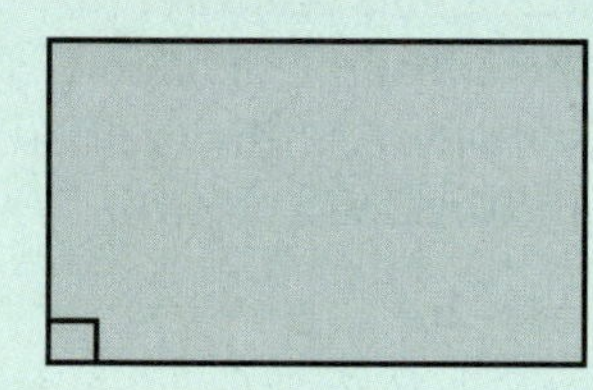

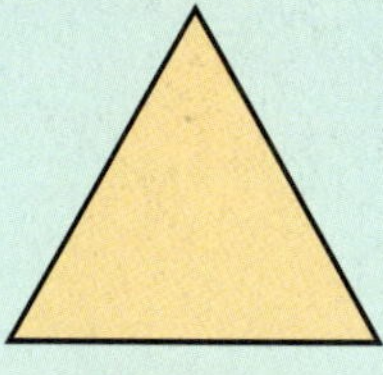

Triangle

Rectangle

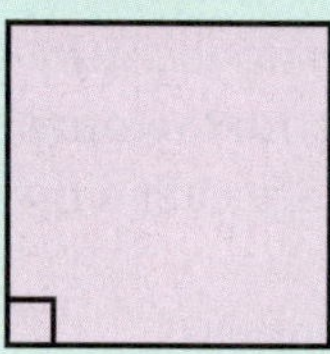

Square

Perimeter pages 98 – 99

The **perimeter**, P, of a polygon is the sum of the lengths of its sides.

Area pages 100 – 101

Area is a measure of the interior of (or surface enclosed by) a figure in a plane.

Name ______________________ Section ____________ Date __________

ANSWERS

Chapter 1 Review Questions

Write the following decimal numbers in expanded notation and in their English word equivalents.

1. 495 **2.** 1975 **3.** 60,308

Write the following numbers in standard notation.

4. Four thousand eight hundred fifty-six

5. Fifteen million, thirty-two thousand, one hundred ninety-seven

6. Six hundred seventy-two million, three hundred forty thousand, eighty-three

Round as indicated.

7. 625 (to the nearest ten)

8. 14,620 (to the nearest thousand)

9. 749 (to the nearest hundred)

10. 2570 (to the nearest hundred)

1. __________

2. __________

3. __________

4. __________

5. __________

6. __________

7. __________

8. __________

9. __________

10. __________

11.

12.

13.

14.

15.

16.

17.

18.

19.

20.

21.

22.

23.

24.

25.

In Exercises 11 – 14, state which property of addition or multiplication is illustrated.

11. $17 + 32 = 32 + 17$

12. $3(22 \cdot 5) = (3 \cdot 22)5$

13. $28 + (6 + 12) = (28 + 6) + 12$

14. $72 \cdot 89 = 89 \cdot 72$

First, estimate each sum; then find the sum.

15.
$$\begin{array}{r} 8445 \\ 267 \\ 1351 \\ +\ 478 \\ \hline \end{array}$$

16.
$$\begin{array}{r} 39 \\ 487 \\ 966 \\ +\ 182 \\ \hline \end{array}$$

Subtract.

17.
$$\begin{array}{r} 647 \\ -\ 139 \\ \hline \end{array}$$

18.
$$\begin{array}{r} 7036 \\ -\ 4652 \\ \hline \end{array}$$

19.
$$\begin{array}{r} 5000 \\ -\ 2898 \\ \hline \end{array}$$

Multiply.

20. $0 \cdot 36$

21. $70 \cdot 80$

22. $90 \cdot 4000$

First estimate each product; then find the product.

23.
$$\begin{array}{r} 98 \\ \times\ 52 \\ \hline \end{array}$$

24.
$$\begin{array}{r} 8975 \\ \times\ 436 \\ \hline \end{array}$$

25.
$$\begin{array}{r} 4837 \\ \times\ 5000 \\ \hline \end{array}$$

Name ______________________ Section ____________ Date __________

ANSWERS

First estimate each quotient; then find the quotient.

26. $7\overline{)2044}$ **27.** $38\overline{)23,028}$ **28.** $529\overline{)71,496}$

29. If the product of 17 and 51 is added to the product of 16 and 12, what is the sum?

30. Find the average of 33, 42, 25, and 40.

31. If the quotient of 546 and 6 is subtracted from 100, what is the difference?

32. Two years ago, Ms. Miller bought five shares of stock at $353 per share. One year ago, she bought another ten shares at $290 per share. Yesterday, she sold all her shares at $410 per share. What was her total profit? What was her average profit per share?

33. On a history exam, two students scored 98 points, five students scored 87 points, one student scored 81 points, and six students scored 75 points. What was the average score in the class?

26. ____________

27. ____________

28. ____________

29. ____________

30. ____________

31. ____________

32. ____________

33. ____________

34. ______

34. What number should be added to seven hundred forty-three to get a sum of eight hundred thirteen?

35. ______

35. If you buy a car for $10,000 plus taxes and license fees totaling $1200 and make a down payment of $3500, how much will you finance?

36. ______

36. Fill in the missing numbers in the chart according to the directions.

Given Number	Add 100	Double	Subtract 200
3	103	206	?
20	120	?	?
15	?	?	?
?	?	?	16

37. ______

37. Find the perimeter of the polygon shown.

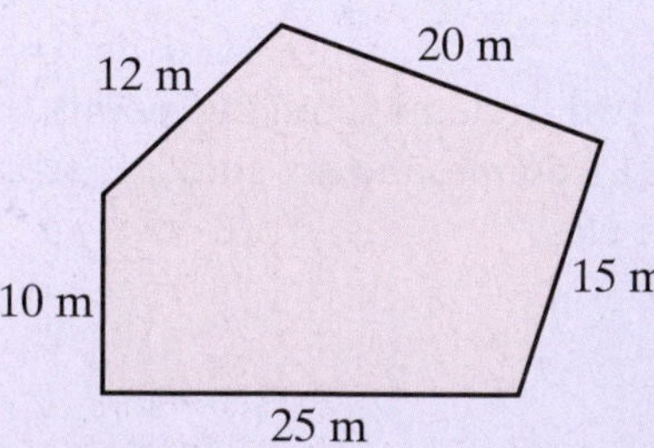

Name ______________________ Section ______________ Date ___________

ANSWERS

38. Find the perimeter of the rectangle shown.

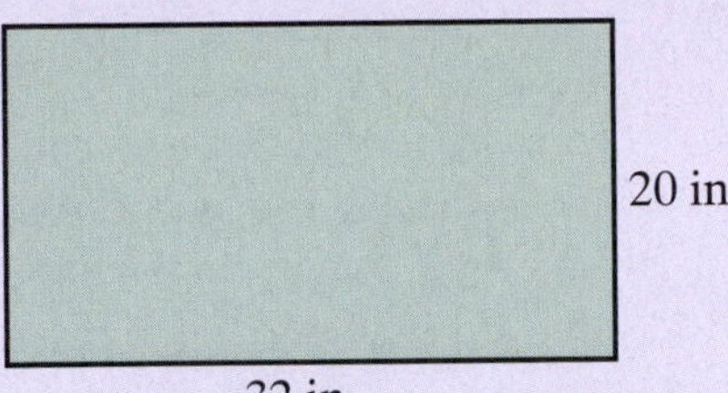

38. ______________

39. a. ______________

39. Find **a.** the perimeter and **b.** the area of the triangle shown.

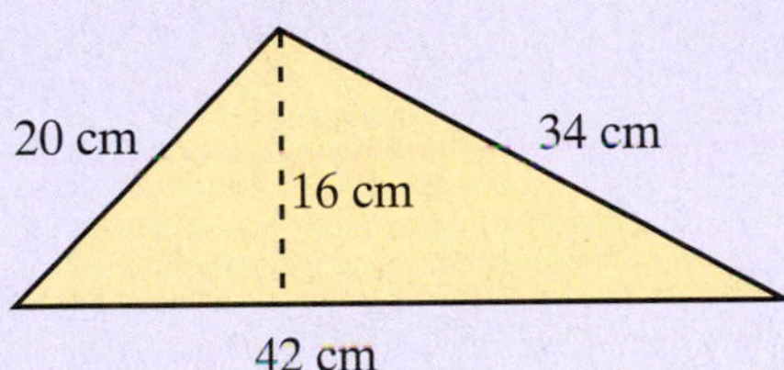

b. ______________

40. Find the area of a rectangle with width 55 yards and length 120 yards.

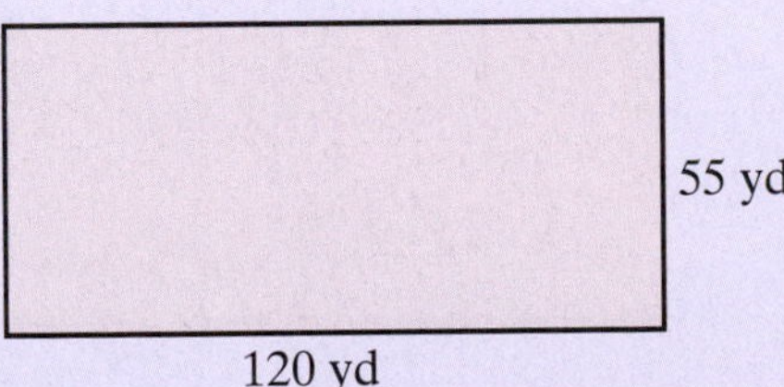

40. ______________

Name ______________________ Section ____________ Date ____________

ANSWERS

Chapter 1 Test

1. Write 8952 in expanded notation and in its English word equivalent.

2. The number 1 is called the multiplicative __________.

3. Give an example that illustrates the commutative property of multiplication.

4. Round 997 to the nearest thousand.

5. Round 135,721 to the nearest ten-thousand.

First estimate each sum; then find the sum.

6.
$$\begin{array}{r} 9586 \\ 345 \\ +\ 2078 \\ \hline \end{array}$$

7.
$$\begin{array}{r} 37 \\ 486 \\ 493 \\ 162 \\ +\ 557 \\ \hline \end{array}$$

8.
$$\begin{array}{r} 1,480,900 \\ +\ 2,576,850 \\ \hline \end{array}$$

Subtract.

9.
$$\begin{array}{r} 850 \\ -\ 362 \\ \hline \end{array}$$

10.
$$\begin{array}{r} 5097 \\ -\ 3868 \\ \hline \end{array}$$

11.
$$\begin{array}{r} 6000 \\ -\ 293 \\ \hline \end{array}$$

1. ____________
2. ____________
3. ____________
4. ____________
5. ____________
6. ____________
7. ____________
8. ____________
9. ____________
10. ____________
11. ____________

12. ______

13. ______

14. ______

15. ______

16. ______

17. ______

18. ______

19. ______

20. a. ______

b. ______

c. ______

21. ______

First estimate each product; then find the product.

12. $\begin{array}{r} 34 \\ \times\ 76 \\ \hline \end{array}$

13. $\begin{array}{r} 2593 \\ \times\ \ 85 \\ \hline \end{array}$

14. $\begin{array}{r} 793 \\ \times\ 266 \\ \hline \end{array}$

Divide.

15. $25\overline{)10,075}$

16. $462\overline{)79,852}$

17. $603\overline{)1,209,015}$

18. Find the average of 82, 96, 49, and 69.

19. If the quotient of 51 and 17 is subtracted from the product of 19 and 3, what is the difference?

20. Robert and his brother were saving money to buy a new TV for their parents. **a.** If Robert saved $23 a week and his brother saved $28 a week, how much did they save in six weeks? **b.** What was their average weekly savings? **c.** If the TV they wanted to buy cost $530 including tax, how much did they still need after the six weeks?

21. You open a checking account with a deposit of $2500. What is your balance if you write checks for $520, $35, $70, and $230 and make further deposits of $200 and $180?

Name ______________________ Section ____________ Date __________

ANSWERS

22. Find the perimeter of the polygon shown.

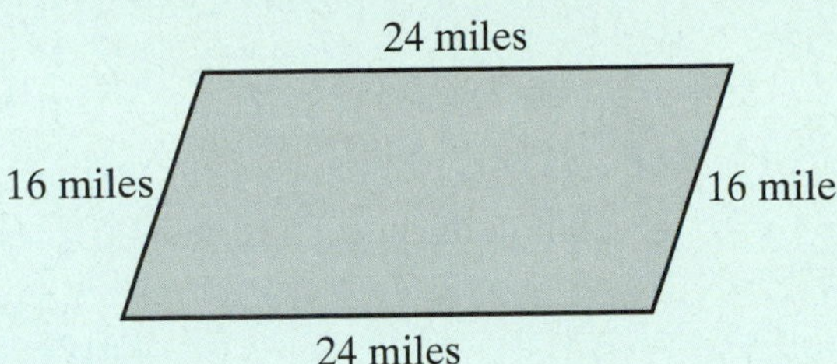

23. Find **a.** the perimeter and **b.** the area of a rectangle that is 34 inches wide and 42 inches long.

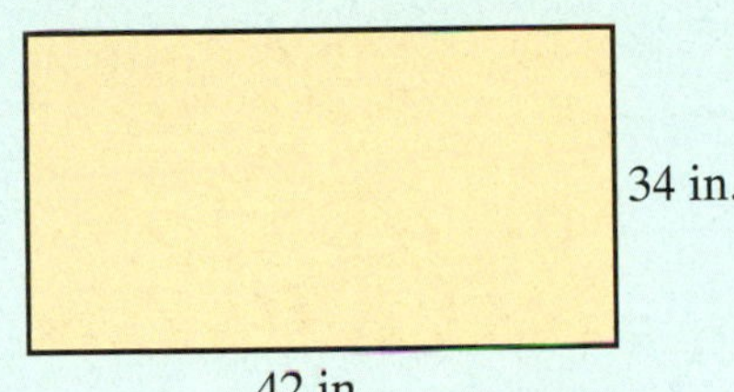

24. Find **a.** the perimeter and **b.** the area of the right triangle shown.

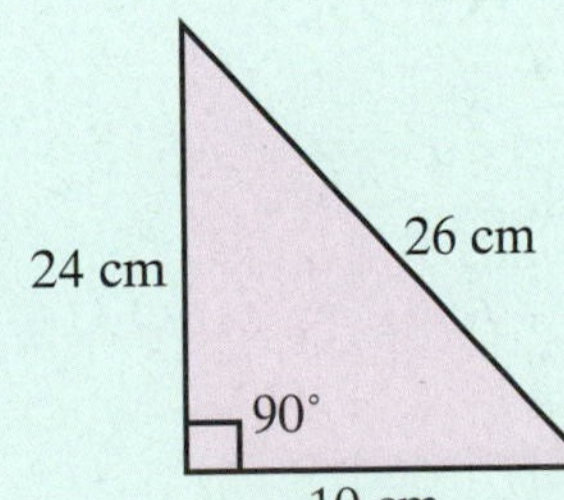

22. ____________

23. a. ____________

b. ____________

24. a. ____________

b. ____________

2 Exponents, Prime Numbers, & LCM

WHAT TO EXPECT IN CHAPTER 2

The topics in Chapter 2 form the foundation for all the work in Chapter 3 (Fractions) and Chapter 4 (Mixed Numbers). Exponents are introduced in Section 2.1, and they are used throughout Chapter 2 and in appropriate places in the rest of the text to simplify expressions and represent powers. The rules for order of operations, discussed in Section 2.2, are necessary so that everyone will arrive at the same answer when evaluating an expression involving more than one operation.

Section 2.3 provides a few tests for divisibility by 2, 3, 5, 6, 9, and 10 that indicate divisibility without actually requiring you to divide by any of these numbers. The tests are valuable tools for working with prime factorizations and simplifying fractions. The ideas discussed in the first three sections are carried over and used throughout the remainder of the chapter with prime numbers, prime factorizations, and the least common multiple (LCM) in Sections 2.4, 2.5, and 2.6. The concepts of prime numbers and prime factorizations are an integral part of the development of fractions and mixed numbers in Chapters 3 and 4.

Chapter 2 Exponents, Prime Numbers, & LCM

Mathematics at Work!

The concepts of multiples and least common multiples are an important part of this chapter. These ideas are used throughout our work with fractions and mixed numbers in Chapters 3 and 4. The following problem is an example of working with the least common multiple (or LCM). (See Section 2.6, Example 10 on page 186).

Suppose three weather satellites A, B, and C take different amounts of time to orbit the earth: satellite A takes 24 hours for a complete orbit, B takes 18 hours, and C takes 12 hours. If they are directly above each other now (as shown in part (a) of the figure), in how many hours will they again be directly above each other in the position shown in part (a)?

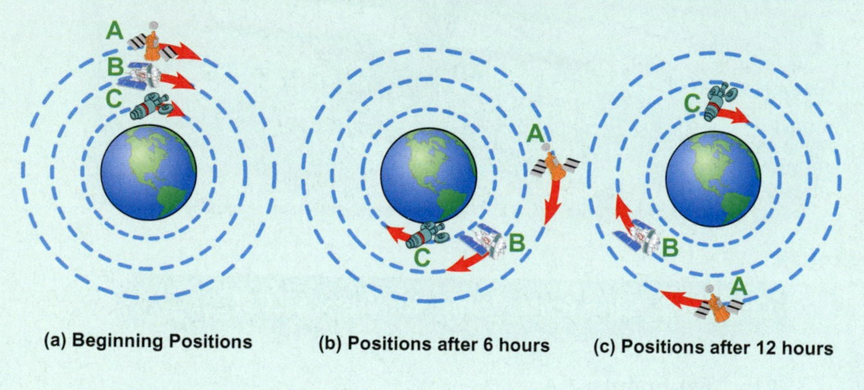

(a) Beginning Positions (b) Positions after 6 hours (c) Positions after 12 hours

2.1 Exponents

Exponents

We know that repeated addition of the same number can be shortened by using multiplication:

$$5 + 5 + 5 = 3 \cdot 5 = 15$$

factors product

The result is called the **product**, and the whole numbers that are multiplied are called **factors** of the product.

In a similar manner, repeated multiplication by the same whole number (called the **base**) can be shortened by using **exponents**. (Exponents are written slightly to the right of, and above, the base.)

Thus if 5 is used as a factor three times, we can write

$$\underset{\text{base}}{5}^{\overset{\text{exponent}}{3}} = \underbrace{5 \cdot 5 \cdot 5}_{\text{factors}} = \underbrace{125}_{\text{product}}$$

(The base 5 is used as a factor three times.)

Note that the base and exponent are not multiplied. Only the base is multiplied by itself. That is, $5^3 \neq 3 \cdot 5$. (The symbol $\neq$ is read "is not equal to.")

Exponents

An **exponent** is a number that tells how many times its base is to be used as a factor.

(An exponent indicates repeated multiplication of the base times itself.)

Objectives

① Know that an exponent is used to indicate repeated multiplication.

② Know the basic property: For any whole number a, $a = a^1$.

③ Know the basic property: For any nonzero whole number a, $a^0 = 1$.

④ Recognize the squares of the whole numbers from 1 to 20.

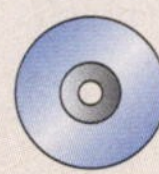

Example 1

Objective ①

Repeated Multiplication	Using Exponents
a. $7 \cdot 7 = 49$	$7^2 = 49$ **Note:** 7^2 is read "7 to the second power" or "7 squared."
b. $3 \cdot 3 = 9$	$3^2 = 9$ **Note:** 3^2 is read "3 to the second power" or "3 squared."
c. $2 \cdot 2 \cdot 2 = 8$	$2^3 = 8$ **Note:** 2^3 is read "2 to the third power" or "2 cubed."
d. $10 \cdot 10 \cdot 10 = 1000$	$10^3 = 1000$ **Note:** 10^3 is read "10 to the third power" or "10 cubed."
e. $2 \cdot 2 \cdot 2 \cdot 2 \cdot 2 = 32$	$2^5 = 32$ **Note:** 2^5 is read "2 to the fifth power."

Now work exercise 1 in the margin.

1. Rewrite the following expressions using an exponent and then evaluate.

a. $5 \cdot 5$

b. $2 \cdot 2 \cdot 2 \cdot 2$

c. $3 \cdot 3 \cdot 3$

d. $4 \cdot 4 \cdot 4$

e. $9 \cdot 9$

As illustrated in Example 1, we say that the base is "**squared**" when the exponent is 2 and "**cubed**" when the exponent is 3. With other exponents the base is said to be "**to the ____ power.**"

Avoid the following common error.

Common Error:

Incorrect Solution	Correct Solution
Do NOT multiply the base times the exponent.	**Do** multiply the base times itself.
$10^2 = 10 \cdot 2 = 20$ WRONG	$10^2 = 10 \cdot 10 = 100$ RIGHT
$6^3 = 6 \cdot 3 = 18$ WRONG	$6^3 = 6 \cdot 6 \cdot 6 = 216$ RIGHT

Properties of Exponents

If there is no exponent written with a number, then the exponent is understood to be 1: that is, any number is equal to itself raised to the first power. Thus

$$8 = 8^1, \qquad 6 = 6^1, \qquad \text{and} \qquad 193 = 193^1.$$

Objective 2

Property of Exponents 1

For any whole number a, $a^1 = a$.

When the exponent 0 is used for any base except 0, the value of the power is defined to be 1:

$$2^0 = 1 \qquad 3^0 = 1 \qquad 5^0 = 1 \qquad 46^0 = 1$$

Objective 3

Property of Exponents 2

For any nonzero whole number a, $a^0 = 1$.

To help in understanding 0 as an exponent, we discuss one of the rules for exponents (which will be studied in algebra) that is related to division. To divide $\frac{2^6}{2^2}$ or $\frac{5^4}{5^3}$, we can write

$$\frac{2^6}{2^2} = \frac{2 \cdot 2 \cdot 2 \cdot 2 \cdot 2 \cdot 2}{2 \cdot 2} = 2 \cdot 2 \cdot 2 \cdot 2 = 2^4 \quad \text{or} \quad \frac{2^6}{2^2} = 2^{6-2} = 2^4$$

$$\frac{5^4}{5^3} = \frac{5 \cdot 5 \cdot 5 \cdot 5}{5 \cdot 5 \cdot 5} = 5 \qquad \text{or} \qquad \frac{5^4}{5^3} = 5^{4-3} = 5^1$$

The rule is to **subtract the exponents when dividing numbers with the same base**. So

$$\frac{3^4}{3^4} = 3^{4-4} = 3^0 \quad \text{and} \quad \frac{5^2}{5^2} = 5^{2-2} = 5^0.$$

And,

$$\frac{3^4}{3^4} = \frac{81}{81} = 1 \quad \text{and} \quad \frac{5^2}{5^2} = \frac{25}{25} = 1.$$

For the rules of exponents to make sense, we have

$$3^0 = 1 \qquad \text{and} \qquad 5^0 = 1.$$

Special Note:

The expression 0^0 is not defined.

Example 2

a. $8^1 = 8$

b. $6^0 = 1$

c. $6^2 = 36$ $\quad (6^2 = 6 \cdot 6)$

d. $5^3 = 125$ $\quad (5^3 = 5 \cdot 5 \cdot 5)$

e. $3^3 = 27$ $\quad (3^3 = 3 \cdot 3 \cdot 3)$

f. $2^0 = 1$

Now work exercise 2 in the margin.

2. In each expression, name the exponent and the base. Then find the value of the expression.

a. 8^2

b. 6^3

c. 5^2

d. 2^6

e. 12^0

f. 10^4

Perfect Squares

To improve your speed in factoring and in working with fractions and simplifying expressions, **you should memorize** all the squares of the whole numbers from 1 to 20. These numbers are called **perfect squares**. The following table lists these perfect squares. (**Note**: The square of every whole number is a perfect square. Thus there are an infinite number of perfect squares.)

Objective 4

Number (n)	1	2	3	4	5	6	7	8	9	10
Square (n^2)	1	4	9	16	25	36	49	64	81	100

Number (n)	11	12	13	14	15	16	17	18	19	20
Square (n^2)	121	144	169	196	225	256	289	324	361	400

Practice Problems

Find the value of the expression.

1. 7^2 **2.** 3^3 **3.** 4^0 **4.** 1^5

5. Find a base and exponent form for 1,000,000 without using the exponent 1.

Answers to Practice Problems:

1. 49 **2.** 27 **3.** 1 **4.** 1 **5.** 10^6

Name ______________ Section ______________ Date ______________

Exercises 2.1

*For Exercises 1 – 25, name **a.** the exponent and **b.** the base. Then **c.** find the value of each expression. See Example 2.*

1. 2^3 **2.** 2^5 **3.** 5^2 **4.** 6^2 **5.** 7^0

6. 11^2 **7.** 1^4 **8.** 4^3 **9.** 4^0 **10.** 3^6

11. 3^2 **12.** 2^4 **13.** 5^0 **14.** 1^{50} **15.** 62^1

16. 12^2 **17.** 10^2 **18.** 10^3 **19.** 4^2 **20.** 8^3

21. 10^4 **22.** 5^3 **23.** 6^3 **24.** 10^5 **25.** 19^0

ANSWERS

1. [Respond below exercise.]
2. [Respond below exercise.]
3. [Respond below exercise.]
4. [Respond below exercise.]
5. [Respond below exercise.]
6. [Respond below exercise.]
7. [Respond below exercise.]
8. [Respond below exercise.]
9. [Respond below exercise.]
10. [Respond below exercise.]
11. [Respond below exercise.]
12. [Respond below exercise.]
13. [Respond below exercise.]
14. [Respond below exercise.]
15. [Respond below exercise.]
16. [Respond below exercise.]
17. [Respond below exercise.]
18. [Respond below exercise.]
19. [Respond below exercise.]
20. [Respond below exercise.]
21. [Respond below exercise.]
22. [Respond below exercise.]
23. [Respond below exercise.]
24. [Respond below exercise.]
25. [Respond below exercise.]

26. ______

27. ______

28. ______

29. ______

30. ______

31. ______

32. ______

33. ______

34. ______

35. ______

36. ______

37. ______

38. ______

39. ______

40. ______

41. ______

42. ______

43. ______

44. ______

45. ______

46. ______

47. ______

48. ______

49. ______

50. ______

Find a base and exponent form for each number without using the exponent 1. ***[Hint:*** *the table inside the back cover may be helpful.] See Example 1.*

26. 4 **27.** 25 **28.** 16 **29.** 27

30. 32 **31.** 121 **32.** 49 **33.** 8

34. 9 **35.** 36 **36.** 125 **37.** 81

38. 64 **39.** 144 **40.** 216 **41.** 100

42. 1000 **43.** 10,000 **44.** 100,000 **45.** 243

46. 625 **47.** 225 **48.** 196 **49.** 343

50. 169

Name ______________ Section ______________ Date ______________

Use exponents to rewrite the following products. See Examples 1 and 2.

51. $6 \cdot 6 \cdot 6 \cdot 6 \cdot 6$ **52.** $7 \cdot 7 \cdot 7 \cdot 7$ **53.** $2 \cdot 2 \cdot 7 \cdot 7$

54. $5 \cdot 5 \cdot 9 \cdot 9 \cdot 9$ **55.** $2 \cdot 2 \cdot 3 \cdot 3 \cdot 3$ **56.** $3 \cdot 3 \cdot 5 \cdot 5 \cdot 5$

57. $7 \cdot 7 \cdot 13$ **58.** $11 \cdot 11 \cdot 11$ **59.** $2 \cdot 3 \cdot 3 \cdot 11 \cdot 11$

60. $5 \cdot 5 \cdot 5 \cdot 11 \cdot 11$

Find the value of each of the following squares. Write as many of them as you can from memory. See Examples 1 and 2.

61. 8^2 **62.** 3^2 **63.** 7^2 **64.** 11^2 **65.** 15^2

66. 14^2 **67.** 18^2 **68.** 9^2 **69.** 12^2 **70.** 20^2

71. 10^2 **72.** 16^2 **73.** 30^2 **74.** 40^2 **75.** 50^2

ANSWERS

51. ______________
52. ______________
53. ______________
54. ______________
55. ______________
56. ______________
57. ______________
58. ______________
59. ______________
60. ______________
61. ______________
62. ______________
63. ______________
64. ______________
65. ______________
66. ______________
67. ______________
68. ______________
69. ______________
70. ______________
71. ______________
72. ______________
73. ______________
74. ______________
75. ______________

Writing and Thinking about Mathematics

76. [Respond below exercise.]

77. [Respond below exercise.]

78. [Respond below exercise.]

76. Continue the list of perfect squares by making a list of the squares of the numbers from 21 to 30.

77. Make a list of the numbers that are perfect cubes from 1 to 1000. Memorize as many of these as you can.

78. Use your calculator to find the following values:

a. 86^0 **b.** 623^0 **c.** 9072^0

Discuss, in your own words, what these results indicate.

2.2 Order of Operations

Rules for Order of Operations

To ensure that everyone gets the same correct answer when evaluating an expression with several of the operations (addition, subtraction, multiplication, and division) and possibly grouping symbols such as parentheses and brackets, mathematicians have agreed on a set of rules for order of operations. These rules are used in all branches of mathematics and computer science (including your calculator). For example, evaluate the following expression in whatever way you think is correct:

Objectives

① Know the Rules for Order of Operations.

② Be able to use the Rules for Order of Operations to simplify numerical expressions.

$$6(10-2^3)+12\div 2\cdot 3$$

The correct value of this expression can be found by using the rules for order of operations, as follows:

$$\begin{aligned}6(10-2^3)+12\div 2\cdot 3 &= 6(10-8)+12\div 2\cdot 3\\ &= 6(2)+12\div 2\cdot 3\\ &= 12+12\div 2\cdot 3\\ &= 12+6\cdot 3\\ &= 12+18\\ &= 30\end{aligned}$$

See if you can understand how the following rules for order of operations were applied at each step in the example above.

Objective ①

Rules for Order of Operations

1. First, simplify within grouping symbols, such as parentheses ( ), brackets [ ], or braces { }. Start with the innermost grouping.
2. Second, evaluate any numbers or expressions indicated by exponents.
3. Third, moving from **left to right**, perform any multiplication or division in the order in which it appears.
4. Fourth, moving from **left to right**, perform any addition or subtraction in the order in which it appears.

The rules are very explicit. Read them carefully. Note that in Rule 3, neither multiplication nor division has priority over the other. Whichever of these operations occurs first, moving left to right, is done first. In Rule 4, addition and subtraction are handled in the same way.

A well-known mnemonic device for remembering these rules is the following:

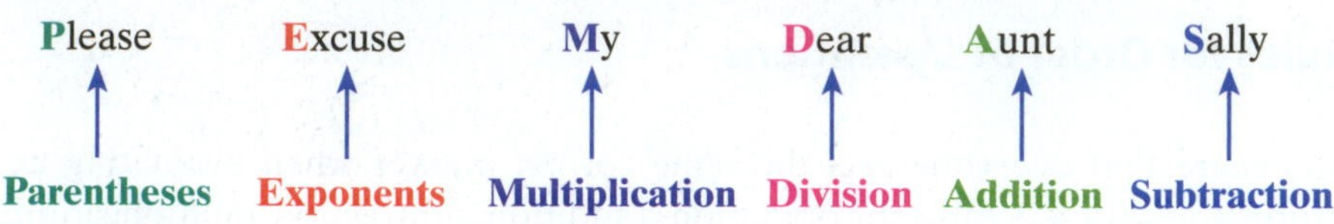

Objective 2

The following examples show how to apply the rules for order of operations. Generally, more than one step can be performed at the same time. You may use the + and – signs to separate the expressions into various parts. The addition and subtraction occur last unless they are within grouping symbols.

1. Evaluate $6 \cdot 2 - 4 \cdot 3 + 8$.

Example 1

Evaluate $14 \div 7 + 3 \cdot 2 - 5$.

Solution

	$14 \div 7 + 3 \cdot 2 - 5$	Divide before multiplying in this case.
$=$	$2 + 3 \cdot 2 - 5$	Multiply before adding or subtracting.
$=$	$2 + 6 - 5$	Add before subtracting in this case.
$=$	$8 - 5$	Subtract.
$=$	3	

Now work exercise 1 in the margin.

2. Evaluate $8 \div 2 \cdot 7 - 5 \cdot 3$.

Example 2

Evaluate $3 \cdot 6 \div 9 - 1 + 4 \cdot 7$.

Solution

	$3 \cdot 6 \div 9 - 1 + 4 \cdot 7$	Multiply.
$=$	$18 \div 9 - 1 + 4 \cdot 7$	Divide.
$=$	$2 - 1 + 4 \cdot 7$	Multiply.
$=$	$2 - 1 + 28$	Subtract and add, left to right.
$=$	$1 + 28$	Add.
$=$	29	

Now work exercise 2 in the margin.

Example 3

Evaluate $(6 + 2) + (8 + 1) \div 9$.

Solution

	$(6 + 2) + (8 + 1) \div 9$	Operate within parentheses.
$=$	$8 + 9 \div 9$	Divide.
$=$	$8 + 1$	Add.
$=$	9	

Now work exercise 3 in the margin.

3. Evaluate $(9 + 3) \div 4 + (9 - 1)$.

Example 4

Evaluate $30 \div 3 \cdot 2 + 3\,(6 - 21 \div 7)$.

Solution

	$30 \div 3 \cdot 2 + 3\,(6 - 21 \div 7)$	Operate within parentheses.
$=$	$30 \div 3 \cdot 2 + 3\,(6 - 3)$	Divide and simplify within parentheses since the two parts are separated by a + sign. Do **not** multiply yet.
$=$	$10 \cdot 2 + 3\,(3)$	Multiply.
$=$	$20 + 9$	Add.
$=$	29	

Now work exercise 4 in the margin.

4. Evaluate $4\,(9 - 32 \div 8) - 5 \cdot 6 \div 2$.

Example 5

Evaluate $2 \cdot 3^2 + 18 \div 3^2$.

Solution

	$2 \cdot 3^2 + 18 \div 3^2$	Find powers.
$=$	$2 \cdot 9 + 18 \div 9$	Multiply and divide since the two parts are separated by a + sign.
$=$	$18 + 2$	Add.
$=$	20	

Now work exercise 5 in the margin.

5. Evaluate $6^2 \div 9 + 2 \cdot 5^2$.

6. Evaluate
$[(7^2 - 9) \div 2^2 - 3](17 - 14)$.

Example 6

Evaluate $[(5 + 2^2) \div 3 + 8]\,(14 - 10)$.

Solution

$$[(5 + 2^2) \div 3 + 8]\,(14 - 10)$$ Begin by simplifying the exponents and then operate within parentheses (twice).

$$= [(5 + 4) \div 3 + 8] \cdot (4)$$ Operate within parentheses.

$$= [9 \div 3 + 8]\,(4)$$ Divide within brackets.

$$= [3 + 8]\,(4)$$ Add within brackets.

$$= [11]\,(4)$$ Multiply.

$$= 44$$

Now work exercise 6 in the margin.

7. Evaluate
$(16 - 12)\,[(7 - 2^2) \cdot 2 + 3]$.

Completion Example 7

Evaluate $(15 - 10)\,[(9 + 3^2) \div 2 + 6]$.

Solution

$$(15 - 10)\,[(9 + 3^2) \div 2 + 6]$$ Operate within parentheses and find the power.

$$= \_\_\_\_\_\_\,[(9 + \_\_\_) \div 2 + 6]$$ Operate within parentheses within the brackets.

$$= \_\_\_\_\_\_\,[(\_\_\_\_\_) \div 2 + 6]$$ Divide within the brackets.

$$= \_\_\_\_\_\_\,[\_\_\_\_\_ + 6]$$ Add within the brackets.

$$= \_\_\_\_\_\_\,[\_\_\_\_\_]$$ Multiply.

$$= \_\_\_\_\_\_$$

Now work exercise 7 in the margin.

Completion Example 8

Evaluate $3(2+2^2)-6-3\cdot 2^2$.

Solution

$3(2+2^2)-6-3\cdot 2^2$

$= 3(2 + \_\_\_\_\_) - 6 - 3 \cdot \_\_\_\_\_$

$= 3(\_\_\_\_\_\_) - 6 - \_\_\_\_\_$

$= \_\_\_\_\_ - 6 - \_\_\_\_\_$

$= \_\_\_\_\_ - \_\_\_\_\_$

$= \_\_\_\_\_$

Now work exercise 8 in the margin.

8. Evaluate
$6^2 - 4(3^2 - 2) - 5 + 4 \cdot 2^2$.

Completion Example Answers

7. $(15-10)[(9+3^2)\div 2+6]$ Operate within parentheses and find the power.
$= \mathbf{(5)}[(9+\mathbf{9})\div 2+6]$ Operate within parentheses within the brackets.
$= \mathbf{(5)}[\mathbf{(18)}\div 2+6]$ Divide within the brackets.
$= \mathbf{(5)}[\mathbf{9}+6]$ Add within the brackets.
$= \mathbf{(5)}[\mathbf{15}]$ Multiply.
$= \mathbf{75}$

8. $3(2+2^2)-6-3\cdot 2^2$
$= 3(2+\mathbf{4})-6-3\cdot\mathbf{4}$
$= 3(\mathbf{6})-6-\mathbf{12}$
$= \mathbf{18}-6-\mathbf{12}$
$= \mathbf{12}-\mathbf{12}$
$= \mathbf{0}$

Practice Problems

Find the value for each of the following expressions by using the Rules for Order of Operations.

1. $15 \div 15 + 10 \cdot 2$

2. $3 \cdot 2^3 - 12 - 3 \cdot 2^2$

3. $4 \div 2^2 + 3 \cdot 2^2$

4. $(5 + 7) \div 3 + 1$

5. $19 - 5(3 - 1)$

Answers to Practice Problems:

1. 21 **2.** 0 **3.** 13 **4.** 5 **5.** 9

Name ______________________ Section ____________ Date __________

ANSWERS

Exercises 2.2

Fill in each blank with the word (or words) that indicates which operation (or operations) is performed at each step in evaluating the expression.

1. [Respond beside exercise.]

2. [Respond beside exercise.]

3. [Respond beside exercise.]

4. [Respond beside exercise.]

1. $5 + \underbrace{6 \cdot 4} \div 2$ ____________________

$= 5 + \underbrace{24 \div 2}$ ____________________

$= \underbrace{5 + 12}$ ____________________

$= 17$

2. $\underbrace{(9-2)} - \underbrace{(3+2)} \div 5$ ____________________

$= 7 - \underbrace{5 \div 5}$ ____________________

$= \underbrace{7 - 1}$ ____________________

$= 6$

3. $(\underbrace{22 \div 2} + 3) \div 7$ ____________________

$= \underbrace{(11 + 3)} \div 7$ ____________________

$= \underbrace{14 \div 7}$ ____________________

$= 2$

4. $3 \cdot \underbrace{2^2} - 6 \div 2 + 5 \cdot 7$ ____________________

$= \underbrace{3 \cdot 4} - \underbrace{6 \div 2} + \underbrace{5 \cdot 7}$ ____________________

$= \underbrace{12 - 3} + 35$ ____________________

$= \underbrace{9 + 35}$ ____________________

$= 44$

5. ________

6. ________

7. ________

8. ________

9. ________

10. ________

11. ________

12. ________

13. ________

14. ________

15. ________

16. ________

17. ________

18. ________

Find the value of each of the following expressions by using the Rules for Order of Operations. See Examples 1 through 8.

5. $4 \div 2 + 7 - 3 \cdot 2$

6. $8 \cdot 3 \div 12 + 13$

7. $6 + 3 \cdot 2 - 10 \div 2$

8. $14 \cdot 3 \div 7 \div 2 + 6$

9. $6 \div 2 \cdot 3 - 1 + 2 \cdot 7$

10. $5 \cdot 1 \cdot 3 - 4 \div 2 + 6 \cdot 3$

11. $72 \div 4 \div 9 - 2 + 3$

12. $14 + 63 \div 3 - 35$

13. $(2 + 3 \cdot 4) \div 7 + 3$

14. $(2 + 3) \cdot 4 \div 5 + 3 \cdot 2$

15. $(7 - 3) + (2 + 5) \div 7$

16. $16(2 + 4) - 90 - 3 \cdot 2$

17. $35 \div (6 - 1) - 5 + 6 \div 2$

18. $22 - 11 \cdot 2 + 15 - 5 \cdot 3$

Name ______________ Section ______________ Date ______________

ANSWERS

19. $(42-2\div 2\cdot 3)\div 13$

20. $18+18\div 2\div 3-3\cdot 1$

21. $4(7-2)\div 10+5$

22. $(33-2\cdot 6)\div 7+3-6$

23. $72\div 8+3\cdot 4-105\div 5$

24. $6(14-6\div 2-11)$

25. $48\div 12\div 4-1+6$

26. $5-1\cdot 2+4(6-18\div 3)$

27. $8-1\cdot 5+6(13-39\div 3)$

28. $(21\div 7-3)42+6$

29. $16-16\div 2-2+7\cdot 3$

30. $(135\div 3+21\div 7)\div 12-4$

31. $(13-5)\div 4+12\cdot 4\div 3-72\div 18\cdot 2+16$

19. ______________
20. ______________
21. ______________
22. ______________
23. ______________
24. ______________
25. ______________
26. ______________
27. ______________
28. ______________
29. ______________
30. ______________
31. ______________

32. ____________

33. ____________

34. ____________

35. ____________

36. ____________

37. ____________

38. ____________

39. ____________

40. ____________

41. ____________

42. ____________

43. ____________

44. ____________

32. $100 \div 10 \div 10 + 1000 \div 10 \div 10 \div 10 - 2$

33. $15 \div 3 + 2 - 6 + (3)(2)(18)(0)(5)$

34. $[(85 + 5) \div 3 \cdot 2 + 15] \div 15$

35. $2 \cdot 5^2 - 4 \div 2 + 3 \cdot 7$

36. $16 \div 2^4 - 9 \div 3^2$

37. $(4^2 - 7) \cdot 2^3 - 8 \cdot 5 \div 10$

38. $4^2 - 2^4 + 5 \cdot 6^2 - 10^2$

39. $(2^5 + 1) \div 11 - 3 + 7(3^3 - 7)$

40. $(6 + 8^2 - 10 \div 2) \div 5 + 5 \cdot 3^2$

41. $(5 + 7) \div 4 + 2$

42. $(2^3 + 2) \div 5 + (7^2 \div 7)$

43. $(5^2 + 7) \div 8 - (14 \div 7 \cdot 2)$

44. $(3 \cdot 2^2 - 5 \cdot 2 + 2) - (2 - 1^2 + 5 \cdot 2 - 10)$

Name ______________________ Section ____________ Date __________

ANSWERS

45. $2^3 \cdot 3^2 \div 24 - 3 + 6^2 \div 4$

46. $2 \cdot 3^2 + 5 \cdot 3^2 + 15^2 - (21 \cdot 3^2 + 6)$

47. $3 \cdot 2^3 - 2^2 + 4 \cdot 2 - 2^4$

48. $2 \cdot 5^2 - 4(21 \div 3 - 7) + 10^3 - 1000$

49. $(4+3)^2 - (2+3)^2$

50. $40 \div 2 \cdot 5 + 1 \cdot 3^2 \cdot 2$

51. $20 - 2(3-1) + 6^2 \div 2 \cdot 3$

52. $8 \div 2 \cdot 4 + 16 \div 4 \cdot 2 + 3 \cdot 2^2$

53. $50 \cdot 10 \div 2 - 2^2 \cdot 5 + 14 - 2 \cdot 7$

54. $(20 \div 2^2 \cdot 5) + (51 \div 17)^2$

55. $[(2+3)(5-1) \div 2](10+1)$

56. $3[4 + (6 \div 3 \cdot 2)]$

57. $5[3^2 + (8 + 2^3)] - 15$

58. $(2^4 - 16)[13 - (5^2 - 20)]$

45. ________
46. ________
47. ________
48. ________
49. ________
50. ________
51. ________
52. ________
53. ________
54. ________
55. ________
56. ________
57. ________
58. ________

59. ____________

60. ____________

61. ____________

62. ____________

63. ____________

64. ____________

65. ____________

59. $100+2[(7^2-9)(5+1)^2]$

60. $(3+5)^2[(2+1)^2-14\div 2]$

61. $2+5\left[10\div 5\cdot 2+3^2-(6+4)\right]$

62. $(2+5)\left[10\div 5\cdot 2+3^2-(6+4)\right]$

63. $(6+2)\left[10^2\cdot 2+(3\cdot 5^2-4^2)\right]$

64. $6+2\left[10^2\cdot 2+(3\cdot 5^2-4^2)\right]$

65. $(10-7)[7^2\div(5+2)^2+13\cdot 5]$

2.3 Tests for Divisibility (2, 3, 5, 6, 9, and 10)

Rules of Tests for Divisibility

In our work with factoring (Section 2.5) and fractions (Chapter 3), we will need to be able to divide quickly by small numbers. We will want to know if a number is **exactly divisible** by some number **before** we divide.

> **Divisibility**
>
> If a number can be divided by another number so that the remainder is 0, then we say:
>
> 1. The first number is **exactly divisible by** (or is **divisible by**) the second.
> 2. The second number **divides** the first.

Objective ①

Objectives

① Understand the terms **exactly divisible**, **divisible**, and **divides**.

② Know the tests for easily checking divisibility by 2, 3, 5, 6, 9, and 10.

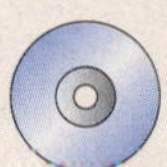

There are simple tests we can use to determine whether a number is divisible by 2, 3, 5, 6, 9, and 10 **without actually dividing**. For example, can you tell (without dividing) if 6495 is divisible by 2? By 3?

The answer is that 6495 is divisible by 3 but not by 2.

```
   2165            3247
3)6495          2)6495
  6               6
  04              04
   3               4
   19              09
   18               8
    15              15
    15              14
     0  remainder    1  remainder (remainder not 0)
```

The following list of rules explains how to test for divisibility by 2, 3, 5, 6, 9, and 10. There are other tests for other numbers such as 4, 7, 8, and 15, but the rules given here are sufficient for our purposes.

> **Tests for Divisibility by 2, 3, 5, 6, 9, and 10**
>
> **For 2:** If the last digit (units digit) of a whole number is 0, 2, 4, 6, or 8 (an even number), then the whole number is divisible by 2.
>
> **For 3:** If the sum of the digits of a whole number is divisible by 3, then the number is divisible by 3.
>
> *Continued on next page...*

Objective ②

1. a. Is 548 divisible by 2? Explain why or why not.

b. Is 7912 divisible by 3? Explain why or why not.

c. Is 1576 divisible by 6? Explain why or why not.

d. Is 6823 divisible by 5? Explain why or why not.

e. Is 4653 divisible by 9? Explain why or why not.

f. Is 8510 divisible by 10? Explain why or why not.

For 5: If the last digit of a whole number is 0 or 5, then the number is divisible by 5.

For 6: If the whole number is divisible by both 2 and 3, then it is divisible by 6.

For 9: If the sum of the digits of a whole number is divisible by 9, then the number is divisible by 9.

For 10: If the last digit of a whole number is 0, then the number is divisible by 10.

Even and Odd Whole Numbers

Even whole numbers are divisible by 2.
(If a whole number is divided by 2 and the remainder is 0, then the whole number is even.)

Odd whole numbers are not divisible by 2.
(If a whole number is divided by 2 and the remainder is 1, then the whole number is odd.)

Example 1

a. 356 is divisible by 2 since the last digit is 6.

b. 6801 is divisible by 3 since $6 + 8 + 0 + 1 = 15$ and 15 is divisible by 3.

c. 1365 is divisible by 5 since 5 is the last digit.

d. 9054 is divisible by both 2 and 3. (It is an even number and $9 + 0 + 4 + 5 = 18$.) Therefore, 9054 is divisible by 6.

e. 9657 is divisible by 9 since $9 + 6 + 5 + 7 = 27$ and 27 is divisible by 9.

f. 3590 is divisible by 10 since 0 is the last digit.

Now work exercise 1 in the margin.

If a number is divisible by 9, then it will be divisible by 3. In Example **1e.**, $9 + 6 + 5 + 7 = 27$ and 27 is divisible by 9 and by 3, so 9657 is divisible by 9 and by 3. We also say that 9 and 3 **divide into** (or **divide exactly into**) 9657 with the implication that the remainder is 0.

However, a number that is divisible by 3 may not be divisible by 9. In Example **1b.**, 6 + 8 + 0 + 1 = 15 and 15 is **not** divisible by 9, so 6801 is divisible by 3 but not by 9.

Similarly, any number divisible by 10 is also divisible by 5, but a number that is divisible by 5 might not be divisible by 10. The number 2580 is divisible by 10 and also by 5, but 4365 is divisible by 5 and not by 10.

In Examples 2, 3, and 4 we use all six tests to determine which of the numbers 2, 3, 5, 6, 9, and 10 (if any) will divide into the given numbers.

Determine which of the numbers 2, 3, 5, 6, 9, and 10 divide into the numbers given in exercises 2 and 3.

2. 6270

3. 1710

Example 2

5712

Solution

a. divisible by 2 (last digit is 2, an even digit)

b. divisible by 3 (5 + 7 + 1 + 2 = 15 and 15 is divisible by 3)

c. not divisible by 5 (last digit is not 0 or 5)

d. divisible by 6 (it is divisible by both 2 and 3).

e. not divisible by 9 (5 + 7 + 1 + 2 = 15 and 15 is not divisible by 9)

f. not divisible by 10 (last digit is not 0)

Now work exercise 2 in the margin.

Example 3

2530

Solution

a. divisible by 2 (last digit is 0, an even digit)

b. not divisible by 3 (2 + 5 + 3 + 0 = 10 and 10 is not divisible by 3)

c. divisible by 5 (last digit is 0)

d. not divisible by 6 (2 + 5 + 3 + 0 = 10 and 10 is not divisible by 3)

e. not divisible by 9 (2 + 5 + 3 + 0 = 10 and 10 is not divisible by 9)

f. divisible by 10 (last digit is 0)

Now work exercise 3 in the margin.

4. Determine which of the numbers 2, 3, 5, 6, 9, and 10 divide into the number 3289.

Example 4

3401

Solution

a. not divisible by 2 (last digit is 1, an odd digit)

b. not divisible by 3 ($3 + 4 + 0 + 1 = 8$ and 8 is not divisible by 3)

c. not divisible by 5 (last digit is not 0 or 5)

d. not divisible by 6 (not divisible by 2 as last digit is 1, an odd digit)

e. not divisible by 9 ($3 + 4 + 0 + 1 = 8$ and 8 is not divisible by 9)

f. not divisible by 10 (last digit is not 0)

Therefore, 3401 is not divisible by any number in this list.

Now work exercise 4 in the margin.

5. Is 255 divisible by

a. 3?

b. 5?

c. 6?

Explain your reasoning.

Completion Example 5

a. 250 is divisible by 10 because

____.

b. 250 is not divisible by 3 because

____.

c. 250 is not divisible by 6 because

____.

Now work exercise 5 in the margin.

6. Is 378 divisible by

a. 2?

b. 5?

c. 9?

Explain your reasoning.

Completion Example 6

a. 512 is divisible by 2 because

____.

b. 512 is not divisible by 9 because

____.

c. 512 is not divisible by 5 because

____.

Now work exercise 6 in the margin.

Checking Divisibility of Products

To emphasize the relationships among the concepts of multiplication, factors, and divisibility, we now discuss how these are related to given products. We will use these concepts as the basis of our discussion of the least common multiple (LCM) in Section 2.6 and with common denominators of fractions in Chapter 3.

Consider the fact that $3 \cdot 4 \cdot 5 \cdot 5 = 300$. Note that if any one factor (or the product of two or more factors) is divided into 300, the quotient will be the product of the remaining factors. For example, if 300 is divided by 3, then the product $4 \cdot 5 \cdot 5 = 100$ will be the quotient. Similarly, various groupings give

$$3 \cdot 4 \cdot 5 \cdot 5 = (3 \cdot 4)(5 \cdot 5) = 12 \cdot 25 = 300$$
$$3 \cdot 4 \cdot 5 \cdot 5 = (4 \cdot 5)(3 \cdot 5) = 20 \cdot 15 = 300$$

and

$$3 \cdot 4 \cdot 5 \cdot 5 = (3 \cdot 4 \cdot 5)(5) = 60 \cdot 5 = 300$$

Thus we can make the following statements:

12 divides 300 25 times and 25 divides 300 12 times
(or, $300 \div 12 = 25$ and $300 \div 25 = 12$)

20 divides 300 15 times and 15 divides 300 20 times
(or, $300 \div 20 = 15$ and $300 \div 15 = 20$)

60 divides 300 5 times and 5 divides 300 60 times
(or, $300 \div 60 = 5$ and $300 \div 5 = 60$)

Example 7

Does 36 divide the product $3 \cdot 4 \cdot 5 \cdot 7 \cdot 9$? If so, how many times?

Solution

Because $36 = 4 \cdot 9$, we have

$$\begin{aligned} 3 \cdot 4 \cdot 5 \cdot 7 \cdot 9 &= (4 \cdot 9)(3 \cdot 5 \cdot 7) \\ &= 36 \cdot 105 \end{aligned}$$

Thus 36 does divide the product, and it divides the product 105 times.

Now work exercise 7 in the margin.

7. Does 24 divide the product $4 \cdot 7 \cdot 3 \cdot 9 \cdot 10$? If so, how many times?

8. Does 21 divide the product $4 \cdot 5 \cdot 6 \cdot 9 \cdot 2$? If so, how many times?

Example 8

Does 15 divide the product $5 \cdot 7 \cdot 2 \cdot 3 \cdot 2$? If so, how many times?

Solution

Because $15 = 3 \cdot 5$, we have

$$\begin{aligned} 5 \cdot 7 \cdot 2 \cdot 3 \cdot 2 &= (3 \cdot 5)(7 \cdot 2 \cdot 2) \\ &= (15)(28) \end{aligned}$$

Thus 15 does divide the product, and it divides the product 28 times.

Now work exercise 8 in the margin.

9. Does 30 divide the product $7 \cdot 5 \cdot 2 \cdot 9$? If so, how many times?

Example 9

Does 35 divide the product $3 \cdot 4 \cdot 5 \cdot 11$?

Solution

We know that $35 = 5 \cdot 7$ and even though 5 is a factor of the product, 7 is not. Therefore, 35 does not divide the product $3 \cdot 4 \cdot 5 \cdot 11$. In other words, $3 \cdot 4 \cdot 5 \cdot 11 = 660$; 660 is not divisible by 35.

Now work exercise 9 in the margin.

10. Does 55 divide the product $3 \cdot 5 \cdot 4 \cdot 11 \cdot 8$? If so, how many times?

Completion Example 10

Does 77 divide the product $3 \cdot 11 \cdot 6 \cdot 7 \cdot 2$? If so, how many times?

Solution

Because 77 = ____ · ____, we have

3·11·6·7·2 = (____ · ____) (____ · ____ · ____)
= (____) (____)

Thus 77 does divide the product, and it divides the product _____ times.

Now work exercise 10 in the margin.

Completion Example Answers

5. **a.** 250 is divisible by 10 because **the units digit is 0**.

b. 250 is not divisible by 3 because **the sum of the digits is 7 and 7 is not divisible by 3**.

c. 250 is not divisible by 6 because **the number 250 is not divisible by 3**.

6. **a.** 512 is divisible by 2 because **the number 2 (the last digit) is divisible by 2**.

b. 512 is not divisible by 9 because **the sum of the digits is 8 and 8 is not divisible by 9**.

c. 512 is not divisible by 5 because **the units digit is not 0 or 5**.

10. Because $77 = 7 \cdot 11$, we have

$$\begin{aligned} 3 \cdot 11 \cdot 6 \cdot 7 \cdot 2 &= (\mathbf{7 \cdot 11})(\mathbf{2 \cdot 3 \cdot 6}) \\ &= (\mathbf{77})(\mathbf{36}) \end{aligned}$$

Thus 77 does divide the product, and it divides the product **36** times.

Practice Problems

Using the techniques of this section, determine which of the numbers 2, 3, 5, 6, 9, and 10 (if any) will divide exactly into each of the following numbers.

1. 842 **2.** 9030 **3.** 4031

4. Does 16 divide the product $3 \cdot 5 \cdot 4 \cdot 7 \cdot 4$? If so, how many times?

Answers to Practice Problems:

1. 2 **2.** 2, 3, 5, 6, 10 **3.** None **4.** Yes, 105 times

Name ______________ Section ______________ Date ______________

Exercises 2.3

Using the tests for divisibility, determine which of the numbers 2, 3, 5, 6, 9, and 10 will divide exactly into each of the following numbers. See Examples 1 through 6.

1. 72 **2.** 81 **3.** 105

4. 333 **5.** 150 **6.** 471

7. 664 **8.** 154 **9.** 372

10. 375 **11.** 443 **12.** 173

13. 567 **14.** 480 **15.** 331

16. 370 **17.** 571 **18.** 466

19. 897 **20.** 695 **21.** 795

22. 777 **23.** 45,000 **24.** 885

ANSWERS

1. ______________

2. ______________

3. ______________

4. ______________

5. ______________

6. ______________

7. ______________

8. ______________

9. ______________

10. ______________

11. ______________

12. ______________

13. ______________

14. ______________

15. ______________

16. ______________

17. ______________

18. ______________

19. ______________

20. ______________

21. ______________

22. ______________

23. ______________

24. ______________

25. ______
26. ______
27. ______
28. ______
29. ______
30. ______
31. ______
32. ______
33. ______
34. ______
35. ______
36. ______
37. ______
38. ______
39. ______
40. ______
41. ______
42. ______
43. ______
44. ______
45. ______
46. ______
47. ______
48. ______
49. ______
50. ______
51. ______

25. 4422 **26.** 1234 **27.** 4321

28. 8765 **29.** 5678 **30.** 402

31. 705 **32.** 732 **33.** 441

34. 555 **35.** 666 **36.** 9000

37. 10,000 **38.** 576 **39.** 549

40. 792 **41.** 5700 **42.** 4391

43. 5476 **44.** 6930 **45.** 4380

46. 510 **47.** 8805 **48.** 1155

49. 8377 **50.** 2222 **51.** 35,622

Name ______________________ Section ______________ Date ___________

ANSWERS

52. 75,495 **53.** 12,324 **54.** 55,555

55. 632,448 **56.** 578,400 **57.** 9,737,001

58. 17,158,514 **59.** 36,762,252 **60.** 20,498,105

Determine whether each of the given numbers divides (or is a factor of) the given product. If it does divide the product, tell how many times. Find each product and make a written statement concerning the divisibility of the product by the given number. See Examples 7 through 10.

61. $6; 2 \cdot 3 \cdot 3 \cdot 5$ **62.** $10; 2 \cdot 3 \cdot 3 \cdot 5$

63. $14; 2 \cdot 3 \cdot 5 \cdot 7$ **64.** $20; 3 \cdot 4 \cdot 5 \cdot 11$

65. $10; 3 \cdot 3 \cdot 5 \cdot 7$ **66.** $25; 2 \cdot 3 \cdot 5 \cdot 7 \cdot 11$

67. $25; 2 \cdot 2 \cdot 3 \cdot 5 \cdot 5$ **68.** $35; 3 \cdot 4 \cdot 5 \cdot 7 \cdot 10$

69. $21; 3 \cdot 3 \cdot 5 \cdot 7 \cdot 11$ **70.** $30; 2 \cdot 3 \cdot 4 \cdot 5 \cdot 13$

52. ______

53. ______

54. ______

55. ______

56. ______

57. ______

58. ______

59. ______

60. ______

61. [Respond below exercise.]

62. [Respond below exercise.]

63. [Respond below exercise.]

64. [Respond below exercise.]

65. [Respond below exercise.]

66. [Respond below exercise.]

67. [Respond below exercise.]

68. [Respond below exercise.]

69. [Respond below exercise.]

70. [Respond below exercise.]

71. [Respond below exercise.] ______

72. [Respond below exercise.] ______

73. [Respond below exercise.] ______

Writing and Thinking about Mathematics

71. If a number is divisible by both 2 and 9, must it be divisible by 18? Explain your reasoning and give several examples to support your answer.

72. If a number is divisible by both 3 and 9, must it be divisible by 27? Explain your reasoning and give several examples to support your answer.

73. With your understanding of factors and divisibility, make up a rule for divisibility by 15 and a rule for divisibility by 21.

REVIEW

1. ______
2. ______
3. ______
4. ______
5. ______
6. ______
7. ______
8. ______
9. ______
10. ______

Review Problems (from Section 1.4)

Round as indicated.

To the nearest ten:	**1.** 847	**2.** 1931
To the nearest hundred:	**3.** 439	**4.** 2563
To the nearest thousand:	**5.** 13,612	**6.** 20,500

First, estimate the answer using rounded numbers; then find the following sums and differences.

7. $\begin{array}{r} 485 \\ 93 \\ +\ 115 \\ \hline \end{array}$

8. $\begin{array}{r} 661 \\ 1730 \\ +\ 2059 \\ \hline \end{array}$

9. $\begin{array}{r} 8752 \\ -\ 3527 \\ \hline \end{array}$

10. $\begin{array}{r} 74,605 \\ -\ 46,083 \\ \hline \end{array}$

2.4 Prime Numbers and Composite Numbers

Prime Numbers and Composite Numbers

Every counting number, except 1, has two or more factors (or divisors). The following list shows examples of the factors of some counting numbers.

Counting Numbers		Factors
3	→	1, 3
14	→	1, 2, 7, 14
17	→	1, 17
18	→	1, 2, 3, 6, 9, 18
21	→	1, 3, 7, 21
23	→	1, 23
36	→	1, 2, 3, 4, 6, 9, 12, 18, 36

Objectives

① Know the definition of a prime number.

② Know the definition of a composite number.

③ Be able to list all the prime numbers less than 50.

④ Be able to determine whether a number is prime or composite.

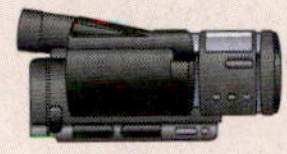

Note that in this list 23, 17, and 3 have **only** two factors. Such numbers are called **prime numbers**.

Prime Number

A **prime number** is a counting number greater than 1 that has only 1 and itself as factors.

OR

A **prime number** is a counting number with exactly two different factors (or divisors).

Composite Number

A **composite number** is a counting number with more than two different factors (or divisors).

Objective ②

Thus in the list discussed, 14, 18, 21, and 36 are **composite numbers**.

Note:

1 is **neither** a prime nor a composite number. $1 = 1 \cdot 1$, and 1 is the only factor of 1. 1 does not have **exactly** two **different** factors, and it does not have more than two different factors.

Determine whether each of the following numbers is prime or composite. Explain your reasoning.

1. a. 13

b. 25

c. 32

2. a. 31

b. 27

c. 19

Example 1

Some prime numbers:

2	2 has exactly two different factors, 1 and 2.
7	7 has exactly two different factors, 1 and 7.
11	11 has exactly two different factors, 1 and 11.
29	29 has exactly two different factors, 1 and 29.

Now work exercise 1 in the margin.

Example 2

Some composite numbers:

12 $1 \cdot 12 = 12$, $2 \cdot 6 = 12$, and $3 \cdot 4 = 12$.
1, 2, 3, 4, 6, and 12 are all factors of 12. Thus 12 has more than two different factors.

33 $1 \cdot 33 = 33$ and $3 \cdot 11 = 33$.
So 1, 3, 11, and 33 are all factors of 33; and 33 has more than two different factors.

Now work exercise 2 in the margin.

The Sieve of Eratosthenes

There is no formula to help us find all the prime numbers. However, there is a technique developed by a Greek mathematician named Eratosthenes. He used the concept of **multiples**. To find the multiples of any counting number, multiply each of the counting numbers by that number.

Counting Numbers:	**1, 2, 3, 4, 5, 6, 7, 8,...**
Multiples of 8:	8, 16, 24, 32, 40, 48, 56, 64,...
Multiples of 2:	2, 4, 6, 8, 10, 12, 14, 16,...
Multiples of 3:	3, 6, 9, 12, 15, 18, 21, 24,...
Multiples of 10:	10, 20, 30, 40, 50, 60, 70, 80,...

None of the multiples of a number, except possibly the number itself, can be prime since they all have that number as a factor. To sift out the prime numbers according to the **Sieve of Eratosthenes**, we proceed by eliminating multiples as the following steps on the next page describe.

Objective 3

1. To find the prime numbers from 1 to 50, list all the counting numbers from 1 to 50 in rows of ten.

1	2	3	4	5	6	7	8	9	10
11	12	13	14	15	16	17	18	19	20
21	22	23	24	25	26	27	28	29	30
31	32	33	34	35	36	37	38	39	40
41	42	43	44	45	46	47	48	49	50

2. Start by crossing out 1 (since 1 is not a prime number). Next, circle 2 and cross out all the other multiples of 2; that is, cross out every second number.

~~1~~	(2)	3	~~4~~	5	~~6~~	7	~~8~~	9	~~10~~
11	~~12~~	13	~~14~~	15	~~16~~	17	~~18~~	19	~~20~~
21	~~22~~	23	~~24~~	25	~~26~~	27	~~28~~	29	~~30~~
31	~~32~~	33	~~34~~	35	~~36~~	37	~~38~~	39	~~40~~
41	~~42~~	43	~~44~~	45	~~46~~	47	~~48~~	49	~~50~~

3. The first number after 2 not crossed out is 3. Circle 3 and cross out all multiples of 3 that are not already crossed out; that is, after 3, every third number should be crossed out. If we proceed this way, we will have all the prime numbers between 1 and 50 circled, and the composite numbers crossed out.

~~1~~	(2)	(3)	~~4~~	5	~~6~~	7	~~8~~	~~9~~	~~10~~
11	~~12~~	13	~~14~~	~~15~~	~~16~~	17	~~18~~	19	~~20~~
~~21~~	~~22~~	23	~~24~~	25	~~26~~	~~27~~	~~28~~	29	~~30~~
31	~~32~~	~~33~~	~~34~~	35	~~36~~	37	~~38~~	~~39~~	~~40~~
41	~~42~~	43	~~44~~	~~45~~	~~46~~	47	~~48~~	49	~~50~~

4. The final table shows that the prime numbers less than 50 are:
2, 3, 5, 7, 11, 13, 17, 19, 23, 29, 31, 37, 41, 43, and 47.

~~1~~	**(2)**	**(3)**	~~4~~	**(5)**	~~6~~	**(7)**	~~8~~	~~9~~	~~10~~
(11)	~~12~~	**(13)**	~~14~~	~~15~~	~~16~~	**(17)**	~~18~~	**(19)**	~~20~~
~~21~~	~~22~~	**(23)**	~~24~~	~~25~~	~~26~~	~~27~~	~~28~~	**(29)**	~~30~~
(31)	~~32~~	~~33~~	~~34~~	~~35~~	~~36~~	**(37)**	~~38~~	~~39~~	~~40~~
(41)	~~42~~	**(43)**	~~44~~	~~45~~	~~46~~	**(47)**	~~48~~	~~49~~	~~50~~

From the table, you should also note that

a. 2 is the only even prime number.

b. All other prime numbers are odd, but not all odd numbers are prime.

Determining Prime Numbers

Computers can be used to determine whether or not very large numbers are prime. The following procedure of dividing by prime numbers can be used to determine whether or not relatively small numbers are prime. If a prime number smaller than the given number is found to be a factor (or divisor), then the given number is composite.

Objective 4

To Determine Whether a Number is Prime:

Divide the number by progressively larger prime numbers (2, 3, 5, 7, 11, and so forth) until:

1. **The remainder is 0.** This means that the prime number is a factor and **the given number is composite;** or
2. **You find a quotient smaller than the prime divisor.** This means that **the given number is prime** because it has no smaller prime factors.

Note: Reasoning that if a composite number were a factor, then one of its prime factors would have been found to be a factor in an earlier division, we divide only by prime numbers (that is, there is no need to divide by a composite number).

3. Is 504 prime? Explain your reasoning.

Example 3

Is 605 prime?

Solution

Since the units digit is 5, the number 605 is divisible by 5 (using the divisibility test in Section 2.3) and is not prime. The number 605 is a composite number. In fact, $605 = 5 \cdot 121 = 5 \cdot 11 \cdot 11$, and 605 has the factors 1, 5, 11, 55, 121, and 605.

Now work exercise 3 in the margin.

Example 4

Is 103 prime?

Solution

Tests for 2, 3, and 5 fail. (The number 103 is not even; $1 + 0 + 3 = 4$ and 4 is not divisible by 3; and the last digit is not 0 or 5.)

Divide by 7:

$$\begin{array}{r} 14 \\ 7\overline{)103} \\ \underline{7} \\ 33 \\ \underline{28} \\ 5 \end{array}$$

Quotient is greater than divisor. Remainder is not 0.

Divide by 11:

$$\begin{array}{r} 9 \\ 11\overline{)103} \\ \underline{99} \\ 4 \end{array}$$

Quotient is less than divisor.

The number 103 is prime.

Now work exercise 4 in the margin.

4. Is 113 prime? Explain your reasoning.

Example 5

Is 221 prime or composite?

Solution

Tests for 2, 3, and 5 fail.

Divide by 7:

$$\begin{array}{r} 31 \\ 7\overline{)221} \\ \underline{21} \\ 11 \\ \underline{7} \\ 4 \end{array}$$

Quotient is greater than divisor. Remainder is not 0.

Divide by 11:

$$\begin{array}{r} 20 \\ 11\overline{)221} \\ \underline{22} \\ 01 \\ \underline{0} \\ 1 \end{array}$$

Quotient is greater than divisor. Remainder is not 0.

Divide by 13:

$$\begin{array}{r} 17 \\ 13\overline{)221} \\ \underline{13} \\ 91 \\ \underline{91} \\ 0 \end{array}$$

Remainder is 0.

The number 221 is composite and not prime.

Note: $221 = 13 \cdot 17$; that is, 13 and 17 are factors of 221.

Now work exercise 5 in the margin.

5. Is 247 prime or composite? Explain your reasoning.

6. Is 299 prime or composite?

Completion Example 6

Is 211 prime or composite?

Solution

Tests for 2, 3, and 5 all fail.

Divide by 7: $7\overline{)211}$ Divide by 11: $11\overline{)211}$

Divide by 13: $13\overline{)211}$ Divide by ___: $\overline{)211}$

211 is __________.

Now work exercise 6 in the margin.

7. Find two factors of 84 such that their product is 84 and their sum is 25.

Example 7

One interesting application of factors of counting numbers (very useful in beginning algebra) involves finding two factors whose sum is some specified number. For example, find two factors of 70 such that their product is 70 and their sum is 19.

Solution

The factors of 70 are 1, 2, 5, 7, 10, 14, 35, and 70, and the pairs whose products are 70 are

$1 \cdot 70 = 70,$ $2 \cdot 35 = 70,$ $5 \cdot 14 = 70,$ $7 \cdot 10 = 70$

Thus the numbers we are looking for are 5 and 14 because

$5 \cdot 14 = 70$ and $5 + 14 = 19.$

Now work exercise 7 in the margin.

Completion Example Answers

6. Tests for 2, 3, and 5 all fail.

Divide by 7:
$$\begin{array}{r} \mathbf{30} \\ 7\overline{)211} \\ \underline{\mathbf{21}} \\ \mathbf{01} \end{array}$$

Divide by 11:
$$\begin{array}{r} \mathbf{19} \\ 11\overline{)211} \\ \underline{\mathbf{11}} \\ \mathbf{101} \\ \underline{\mathbf{99}} \\ \mathbf{2} \end{array}$$

Divide by 13:
$$\begin{array}{r} \mathbf{16} \\ 13\overline{)211} \\ \underline{\mathbf{13}} \\ \mathbf{81} \\ \underline{\mathbf{78}} \\ \mathbf{3} \end{array}$$

Divide by 17:
$$\begin{array}{r} \mathbf{12} \\ 17\overline{)211} \\ \underline{\mathbf{17}} \\ \mathbf{41} \\ \underline{\mathbf{34}} \\ \mathbf{7} \end{array}$$

The number 211 is **prime**.

Practice Problems

Decide whether each of the following numbers is prime or composite.

1. 39 **2.** 79 **3.** 143

4. Find two factors of 36 such that their product is 36 and their sum is 20.

Answers to Practice Problems:

1. Composite **2.** Prime **3.** Composite **4.** 2, 18

Name ______________________ Section ____________ Date __________

ANSWERS

Exercises 2.4

List the first four multiples for each of the following numbers.

1. 5 **2.** 7 **3.** 11 **4.** 13

5. 12 **6.** 9 **7.** 20 **8.** 17

9. 16 **10.** 25

11. Construct a Sieve of Eratosthenes for the numbers from 1 to 100. List the prime numbers from 1 to 100.

Decide whether each of the following numbers is prime or composite. If the number is composite, find at least three factors for the number.

12. 17 **13.** 19 **14.** 28 **15.** 32

16. 47 **17.** 59 **18.** 16 **19.** 63

1. ________

2. ________

3. ________

4. ________

5. ________

6. ________

7. ________

8. ________

9. ________

10. ________

11. [Respond below exercise.]

12. ________

13. ________

14. ________

15. ________

16. ________

17. ________

18. ________

19. ________

20. ______

21. ______

22. ______

23. ______

24. ______

25. ______

26. ______

27. ______

28. ______

29. ______

30. ______

31. ______

32. ______

33. ______

34. ______

35. ______

36. ______

37. ______

38. ______

39. ______

40. ______

41. ______

42. ______

43. ______

20. 14 **21.** 51 **22.** 67 **23.** 89

24. 73 **25.** 61 **26.** 52 **27.** 57

28. 98 **29.** 86 **30.** 53 **31.** 37

Two numbers are given. Find two factors of the first number such that their product is the first number and their sum is the second number. See Example 7.

32. 24, 10 **33.** 12, 7 **34.** 16, 10 **35.** 12, 13

36. 14, 9 **37.** 50, 27 **38.** 20, 9 **39.** 24, 11

40. 48, 19 **41.** 36, 15 **42.** 7, 8 **43.** 63, 24

Name ______________________ Section __________ Date __________

ANSWERS

44. 51, 20 **45.** 25, 10 **46.** 16, 8 **47.** 60, 17

48. 52, 17 **49.** 27, 12 **50.** 72, 22

Writing and Thinking about Mathematics

51. Find the set of all prime numbers less than 1000 that are not odd.

52. Are all odd numbers also prime numbers? Explain your answer.

53. Explain why the number 1 is not prime and not composite.

54. The number 1001 is composite. Find all the factors of 1001.

44. ______

45. ______

46. ______

47. ______

48. ______

49. ______

50. ______

51. [Respond below exercise.]

52. [Respond below exercise.]

53. [Respond below exercise.]

54. [Respond below exercise.]

55. [Respond below exercise.]

56. [Respond below exercise.]

Collaborative Learning

55. In teams of 2 to 4 students, try to find the largest prime number you can within fifteen minutes. The team leader is to explain the team's reasoning to the class.

56. Mathematicians have been interested since ancient times in a search for **perfect numbers**. A **perfect number** is a counting number that is equal to the sum of its proper divisors (divisors not including itself). For example, the first perfect number is 6. The proper divisors of 6 are 1, 2, and 3, and $1 + 2 + 3 = 6$. The teams formed for Exercise 55 are to try to find the second and third perfect numbers. (**Hint**: The second perfect number is between 20 and 30, and the third perfect number is between 450 and 500.)

Name ______________________ Section ____________ Date ____________

REVIEW

Review Problems (from Section 1.5)

Tell what property of multiplication is illustrated.

1. $8 \cdot 7 = 7 \cdot 8$

2. $5(2 \cdot 6) = (5 \cdot 2)6$

Use the technique of multiplying by powers of 10 to find the following products mentally.

3. $40 \cdot 400$

4. $300 \cdot 500$

5. $120 \cdot 7000$

Find the following products.

6. $\begin{array}{r} 314 \\ \times\ 73 \\ \hline \end{array}$

7. $\begin{array}{r} 182 \\ \times\ 466 \\ \hline \end{array}$

8. $\begin{array}{r} 105 \\ \times\ 1700 \\ \hline \end{array}$

1. ____________
2. ____________
3. ____________
4. ____________
5. ____________
6. ____________
7. ____________
8. ____________

2.5 Prime Factorizations

Finding a Prime Factorization

To add and subtract fractions (Chapter 3), we first need to find a common denominator for the fractions. Finding **all** the prime factors of a composite number will help us to accomplish this. For example, factoring 28 gives $28 = 4 \cdot 7$. While 7 is a prime number, 4 is not prime. So, continuing to factor 4 as $2 \cdot 2$, we have all prime factors, and the **prime factorization** of 28: $28 = 2 \cdot 2 \cdot 7$.

We could have started factoring 28 as $28 = 2 \cdot 14$; however, since $14 = 2 \cdot 7$, we still would have had the same end result; $28 = 2 \cdot 2 \cdot 7$. Regardless of the method used or the factors used in the beginning, **there is only one prime factorization for any composite number**. This fact is so important that it is called the **Fundamental Theorem of Arithmetic**.

Objectives

① Know the Fundamental Theorem of Arithmetic.

② Be able to find the prime factorization of a composite number.

③ Know the meaning of the term **prime factorization**.

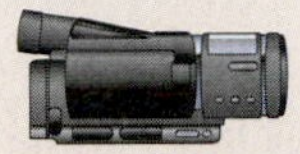

Objective ①

The Fundamental Theorem of Arithmetic

Every composite number has exactly one prime factorization.

Remember that, because multiplication is a commutative operation, the order in which the prime factors are written is not important. That is, a different ordering of the factors is not a different prime factorization. Thus $2 \cdot 2 \cdot 7$ and $2 \cdot 7 \cdot 2$ are the same prime factorizations of 28. What is important is that **all of the factors must be prime numbers**. One procedure for finding the prime factorization of a composite number is outlined in the following box.

Objective ②

To Find the Prime Factorization of a Composite Number:

1. Factor the composite number into any two factors.
2. Factor each factor that is not prime.
3. Continue this process until all factors are prime.

Objective ③

The **prime factorization** of a number is a factorization of that number using only prime factors.

Many times the beginning factors needed to start the process of finding a prime factorization can be found by using the tests for divisibility by 2, 3, 5, 6, 9, and 10 discussed in Section 2.3. This was one reason for developing these tests, and you should review them or write them down for easy reference. They are used in the following examples.

1. Find the prime factorization of 84.

2. Find the prime factorization of 110.

Example 1

Find the prime factorization of 60.

Solution

$60 = 6 \cdot 10$ Since the last digit is 0, we know 10 is a factor.

$= 2 \cdot 3 \quad 2 \cdot 5$ 6 and 10 can both be factored so that each factor is a prime number. This is the prime factorization of 60.

or,

$60 = 3 \cdot 20$ 3 is prime, but 20 is not.

$= 3 \cdot 4 \cdot 5$ 4 is not prime.

$= 3 \cdot 2 \cdot 2 \cdot 5$ All factors are prime.

Since multiplication is commutative, the order of the factors is not important. What is important is that **all of the factors must be prime numbers**.

Writing the factors in order, we see the prime factorization of 60 is $2 \cdot 2 \cdot 3 \cdot 5$ or, using exponents, $2^2 \cdot 3 \cdot 5$.

Now work exercise 1 in the margin.

Example 2

Find the prime factorization of 70.

Solution

$70 = 7 \cdot 10$ 10 is a factor since the last digit is 0.

$= 7 \cdot 2 \cdot 5$

$= 2 \cdot 5 \cdot 7$ Writing the factors in order is not necessary, but it is convenient for comparing answers.

Now work exercise 2 in the margin.

Example 3

Find the prime factorization of each number.

Solution

a. $85 = 5 \cdot 17$ 5 is a factor since the last digit is 5. Since both 5 and 17 are prime, $5 \cdot 17$ is the prime factorization.

b. $72 = 8 \cdot 9$ or $72 = 2 \cdot 36$

$= 2 \cdot 4 \cdot 3 \cdot 3$ $= 2 \cdot 6 \cdot 6$

$= 2 \cdot 2 \cdot 2 \cdot 3 \cdot 3$ $= 2 \cdot 2 \cdot 3 \cdot 2 \cdot 3$

$= 2^3 \cdot 3^2$ $= 2^3 \cdot 3^2$ Using exponents.

c. $245 = 5 \cdot 49$

$= 5 \cdot 7 \cdot 7$

$= 5 \cdot 7^2$

d. $264 = 2 \cdot 132$ or $264 = 4 \cdot 66$

$= 2 \cdot 2 \cdot 66$ $= 2 \cdot 2 \cdot 6 \cdot 11$

$= 2 \cdot 2 \cdot 2 \cdot 33$ $= 2 \cdot 2 \cdot 2 \cdot 3 \cdot 11$

$= 2 \cdot 2 \cdot 2 \cdot 3 \cdot 11$ $= 2^3 \cdot 3 \cdot 11$

$= 2^3 \cdot 3 \cdot 11$

Regardless of your choices for the first two factors, there is only one prime factorization for any composite number.

Now work exercise 3 in the margin.

3. Find the prime factorization of each number.
a. 65
b. 80
c. 192
d. 180

Completion Example 4

Find the prime factorization of 90.

Solution

$90 = 9 \cdot$ ____

$= 3 \cdot 3 \cdot$ ____ $\cdot$ ____

$=$ ____________ Using exponents.

Now work exercise 4 in the margin.

4. Find the prime factorization of 250.

Completion Example 5

Find the prime factorization of 925.

Solution

$925 = 5 \cdot$ ____

$= 5 \cdot$ ____ $\cdot$ ____

$=$ ____________ Using exponents.

Now work exercise 5 in the margin.

5. Find the prime factorization of 775.

6. Find the prime factorization of 165.

Completion Example 6

Find the prime factorization of 196.

Solution

$196 = 2 \cdot 98$

$= 2 \cdot \_\_\_\_ \cdot \_\_\_\_$

$= 2 \cdot \_\_\_\_ \cdot \_\_\_\_ \cdot \_\_\_\_$

$= \_\_\_\_\_\_\_\_$ Using exponents.

Now work exercise 6 in the margin.

Finding Factors of Composite Numbers

Once the prime factorization of a composite number is known, all the factors (or divisors) of that number can be found. For a number to be a factor of a composite number, it must be either 1, the number itself, one of the prime factors, or the product of two or more of the prime factors.

Factors of Composite Numbers

The only factors (or divisors) of a composite number are:

1. 1 and the number itself;

2. Each prime factor; and

3. Products formed by all combinations of the prime factors (including repeated factors).

7. Find all the factors of 42.

Example 7

Find all the factors of 30.

Solution

Since $30 = 2 \cdot 3 \cdot 5$, the factors are

a. 1 and the number itself: 1 and 30.

b. Each prime factor: 2, 3, 5.

c. Products of all combinations of the prime factors:
$2 \cdot 3 = 6,\quad 2 \cdot 5 = 10,\quad 3 \cdot 5 = 15$

The factors are 1, 30, 2, 3, 5, 6, 10, and 15. These are the only factors of 30.

Now work exercise 7 in the margin.

Example 8

Find all factors of 140.

Solution

$$140 = 14 \cdot 10$$
$$= 2 \cdot 7 \cdot 2 \cdot 5$$
$$= 2 \cdot 2 \cdot 5 \cdot 7$$

The factors are

a. 1 and the number itself: 1 and 140.

b. Each prime factor: 2, 5, 7.

c. Products of all combinations of the prime factors:
$2 \cdot 2 = 4$, $2 \cdot 5 = 10$, $2 \cdot 7 = 14$ $5 \cdot 7 = 35$,
$2 \cdot 2 \cdot 5 = 20$, $2 \cdot 2 \cdot 7 = 28$, $2 \cdot 5 \cdot 7 = 70$,

The factors are
1, 140, 2, 5, 7, 4, 10, 14, 35, 20, 28, and 70.

There are no other factors (or divisors) of 140.

Now work exercise 8 in the margin.

8. Find all the factors of 160.

Completion Example Answers

4. $90 = 9 \cdot \mathbf{10}$
$= 3 \cdot 3 \cdot \mathbf{2} \cdot \mathbf{5}$
$= \mathbf{2} \cdot \mathbf{3^2} \cdot \mathbf{5}$ Using exponents.

5. $925 = 5 \cdot \mathbf{185}$
$= 5 \cdot \mathbf{5} \cdot \mathbf{37}$
$= \mathbf{5^2} \cdot \mathbf{37}$ Using exponents.

6. $196 = 2 \cdot 98$
$= 2 \cdot \mathbf{2} \cdot \mathbf{49}$
$= 2 \cdot \mathbf{2} \cdot \mathbf{7} \cdot \mathbf{7}$
$= \mathbf{2^2} \cdot \mathbf{7^2}$ Using exponents.

Practice Problems

Find the prime factorization of each of the following numbers.

1. 42 **2.** 56 **3.** 230

4. Using the prime factorization of 63, find all the factors of 63.

Answers to Practice Problems:

1. $2 \cdot 3 \cdot 7$ **2.** $2^3 \cdot 7$ **3.** $2 \cdot 5 \cdot 23$

4. 1, 63, 3, 7, 9, 21

Name ____________________ Section ____________ Date ____________

ANSWERS

1. ____________
2. ____________
3. ____________
4. ____________
5. ____________
6. ____________
7. ____________
8. ____________
9. ____________
10. ____________
11. ____________
12. ____________
13. ____________
14. ____________
15. ____________
16. ____________
17. ____________
18. ____________
19. ____________
20. ____________

Exercises 2.5

Find the prime factorization for each of the following numbers. Use the tests for divisibility for 2, 3, 5, 6, 9, and 10 whenever they help find beginning factors. See Examples 1 through 6.

1. 24 **2.** 28 **3.** 27 **4.** 16

5. 36 **6.** 60 **7.** 72 **8.** 78

9. 81 **10.** 105 **11.** 125 **12.** 160

13. 75 **14.** 150 **15.** 210 **16.** 40

17. 250 **18.** 93 **19.** 168 **20.** 360

21. ______
22. ______
23. ______
24. ______
25. ______
26. ______
27. ______
28. ______
29. ______
30. ______
31. ______
32. ______
33. ______
34. ______
35. ______
36. ______
37. ______
38. ______
39. ______
40. ______

21. 126 **22.** 48 **23.** 17 **24.** 47

25. 51 **26.** 144 **27.** 121 **28.** 169

29. 225 **30.** 52 **31.** 32 **32.** 98

33. 108 **34.** 103 **35.** 101 **36.** 202

37. 170 **38.** 500 **39.** 10,000 **40.** 100,000

Name ______________________ Section ____________ Date __________

ANSWERS

Using the prime factorization of each number, find all the factors (or divisors) of each number. See Examples 7 and 8.

41. 12 **42.** 18 **43.** 28 **44.** 98 **45.** 121

46. 45 **47.** 105 **48.** 54 **49.** 97 **50.** 144

Collaborative Learning

51. The product of the counting numbers from 1 to a given number is called the **factorial** of that number. For example, the symbol 5! is read "five factorial," and the value is $5! = 1 \cdot 2 \cdot 3 \cdot 4 \cdot 5$.

In teams of 2 to 4 students, analyze the following questions and statements and discuss your conclusions in class.

a. What are the meanings of the symbols 6!, 7!, 8! and 10!?

b. Write the prime factorization of each of these factorials.

c. Determine whether each of the following statements is true or false.

24 divides 6! (T) (F)

28 divides 6! (T) (F)

75 divides 10! (T) (F)

75 divides 8! (T) (F)

41. ________

42. ________

43. ________

44. ________

45. ________

46. ________

47. ________

48. ________

49. ________

50. ________

51. a. [Respond in exercise.]

b. [Respond in exercise.]

c. [Respond in exercise.]

[Respond below
52. exercise.] ______

52. The following expressions are part of formulas used in probability and statistics. For example, the answer to part **a.** indicates the number of committees that can be formed with two people from a group of six people.

a. $\frac{6!}{2! \cdot 4!}$ **b.** $\frac{10!}{2! \cdot 8!}$ **c.** $\frac{20!}{6! \cdot 14!}2$

In teams of 2 to 4 students, evaluate each of these expressions and discuss how you arrived at these results in class. Did you find any shortcuts in the calculations? What are possible "committee" problems related to parts **b.** and **c.**?

REVIEW

Review Problems (from Sections 1.7 and 1.8)

1. ______

1. Find the difference between the product of 42 and 38 and the quotient of 1470 and 70.

2. ______

2. Financing. Chris bought a car for $16,200. The salesperson added $972 for taxes and $520 for license fees. If he made a down payment of $4750 and financed the rest with his bank, how much did he finance?

3. ______

3. Find the average (or mean) of the numbers 103, 114, 98, and 101.

4. a. ______

4. A right triangle (one angle is 90°) has three sides: a base of 12 meters, a height of 5 meters, and a third side of 13 meters.

a. Find the perimeter (distance around) of the triangle.

b. Find the area of the triangle in square meters. (To find the area, multiply the base times height, then divide by 2.)

b. ______

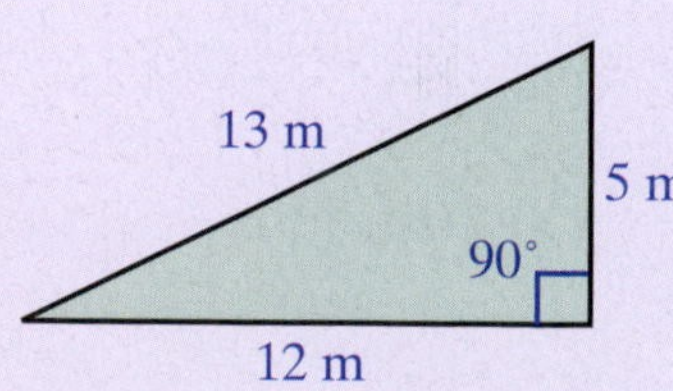

2.6 Least Common Multiple (LCM) with Applications

Finding the LCM of a Set of Counting Numbers

The techniques discussed in this section are used throughout Chapter 3 on fractions. Study these ideas thoroughly because they will make your work with fractions much easier.

Objectives

① Understand the meaning of the term **least common multiple**.

② Use prime factorizations to find the least common multiple of a set of numbers.

③ Recognize the application of the LCM concept in a word problem.

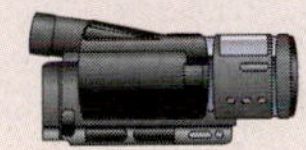

Remember that the multiples of a number are the products of that number with the counting numbers. The first multiple of a number is that number, and all other multiples are larger than that number.

Counting Numbers:	1,	2,	3,	4,	5,	6,	7,	8,	9,	10, . . .
Multiples of 6:	6,	12,	18,	24,	(30),	36,	42,	48,	54,	(60), . . .
Multiples of 10:	10,	20,	(30),	40,	50,	(60),	70,	80,	(90),	100, . . .

The common multiples of 6 and 10 are 30, 60, 90, 120, ...

The **Least Common Multiple (LCM)** is 30.

Least Common Multiple

The **Least Common Multiple (LCM)** of a set of counting numbers is the smallest number common to all the sets of multiples of the given numbers.

Objective ①

Except for the number itself, **factors** (or **divisors**) of the number are smaller than the number, and **multiples** are larger than the number. For example, 6 and 10 are factors of 30, and both are smaller than 30. 30 is a multiple of both 6 and 10 and is larger than either 6 or 10.

Listing all of the multiples as we did for 6 and 10 and then choosing the least common multiple (LCM) is not very efficient. Either of two other techniques, one involving prime factorizations and the other involving division by prime factors, is easier to apply.

Method I for Finding the LCM

To find the LCM of a set of counting numbers:

1. Find the prime factorization of each number.
2. Find the prime factors that appear in **any one** of the prime factorizations.
3. Form the product of these primes using each prime the most number of times it appears in **any one** of the prime factorizations.

Objective ②

1. Find the LCM of 12 and 25.

Example 1

Find the LCM of 20 and 45.

Solution

a. Prime factorizations:

$20 = 4 \cdot 5 = 2 \cdot 2 \cdot 5$ two 2's, one 5

$45 = 9 \cdot 5 = 3 \cdot 3 \cdot 5$ two 3's, one 5

b. 2, 3, and 5 are the only prime factors.

c. Most of each factor in any one factorization:

two 2's (in 20)
two 3's (in 45)
one 5 (one in 30 and one in 45)

So, $\text{LCM} = 2 \cdot 2 \cdot 3 \cdot 3 \cdot 5 = 2^2 \cdot 3^2 \cdot 5 = 180$

180 is the smallest number divisible by both 20 and 45. (Note also that the LCM, 180, contains all the factors of the numbers 20 and 45.)

Now work exercise 1 in the margin.

2. Find the LCM of 18, 21, and 42.

Example 2

Find the LCM of 12, 18, and 48.

Solution

a. Prime factorizations:

$$\left.\begin{aligned} 12 &= 4 \cdot 3 = 2^2 \cdot 3 \\ 18 &= 2 \cdot 9 = 2 \cdot 3^2 \\ 48 &= 16 \cdot 3 = 2^4 \cdot 3 \end{aligned}\right\} \begin{aligned} \text{LCM} &= 2^4 \cdot 3^2 \\ &= 144 \end{aligned}$$

b. The only prime factors that appear are 2 and 3.

c. There are four 2's in 48 and two 3's in 18.

Thus the product $2^4 \cdot 3^2 = 144$ is the least common multiple.

Now work exercise 2 in the margin.

Example 3

Find the LCM of 27, 30, and 42.

Solution

$$\left.\begin{aligned} 27 &= 3\cdot 9 = 3^3 \\ 30 &= 6\cdot 5 = 2\cdot 3\cdot 5 \\ 42 &= 2\cdot 21 = 2\cdot 3\cdot 7 \end{aligned}\right\} \begin{aligned} \text{LCM} &= 2\cdot 3^3\cdot 5\cdot 7 \\ &= 1890 \end{aligned}$$

Now work exercise 3 in the margin.

3. Find the LCM of 36, 45, and 60.

Completion Example 4

Find the LCM of 36, 24, and 48.

Solution

a. Prime factorizations:

36 = __________

24 = __________

48 = __________

b. _____ and _____ are the only prime factors.

c. Most of each factor in any one factorization:

__________ (in 48)

__________ (in 36)

LCM = __________ = _______ = 144

_______ is the smallest number divisible by all the numbers 36, 24, and 48.

Now work exercise 4 in the margin.

4. Find the LCM of 12, 30, and 40.

The second technique for finding the LCM involves division.

> **Method II for Finding the LCM**
>
> To find the LCM of a set of counting numbers:
>
> 1. Write the numbers horizontally and find a prime number that will divide into more than one number, if possible.
> 2. Divide by that prime and write the quotients beneath the dividends. Rewrite any numbers not divided beneath themselves.
> 3. Continue the process until no two numbers have a common prime divisor.
> 4. The LCM is the product of all the prime divisors and the last set of quotients.

5. Find the LCM of 10, 24, 30, and 32.

Example 5

Find the LCM of 20, 25, 18, and 6.

Solution

5)	20	25	18	6
2)	4	5	18	6
3)	2	5	9	3
	2	5	3	1

$$\begin{aligned} LCM &= 5 \cdot 2 \cdot 3 \cdot 2 \cdot 5 \cdot 3 \cdot 1 \\ &= 2^2 \cdot 3^2 \cdot 5^2 \\ &= 900 \end{aligned}$$

Now work exercise 5 in the margin.

6. Find the LCM of 14, 21, and 45.

Example 6

Find the LCM of 15, 49, and 35.

Solution

5)	15	49	35
3)	3	49	7
7)	1	49	7
	1	7	1

$$\begin{aligned} LCM &= 5 \cdot 3 \cdot 7 \cdot 1 \cdot 7 \cdot 1 \\ &= 3 \cdot 5 \cdot 7^2 \\ &= 735 \end{aligned}$$

Now work exercise 6 in the margin.

Finding How Many Times Each Number Divides into the LCM

In Example 3, by using prime factorizations, we found that the LCM for 27, 30, and 42 is 1890. We know that 1890 is a multiple of 27. That is, 27 will divide evenly into 1890. How many times will 27 divide into 1890? You could perform long division:

$$\begin{array}{r} 70 \\ 27\overline{)1890} \\ \underline{189} \\ 00 \\ \underline{0} \\ 0 \end{array}$$

Thus 27 divides 70 times into 1890.

However, we do not need to divide. We can find the same result much more quickly by looking at the prime factorization of 1890 (which we have as a result of finding the LCM). First, find the prime factors for 27. Then the remaining prime factors will have 70 as their product.

$$1890 = 2 \cdot 3^3 \cdot 5 \cdot 7 = \underbrace{(3 \cdot 3 \cdot 3)}_{27} \cdot \underbrace{(2 \cdot 5 \cdot 7)}_{70}$$

$$= 27 \cdot 70$$

The following examples illustrate how to use prime factors to tell how many times each number in a set divides into the LCM for that set of numbers. **This technique is very useful in working with fractions.**

Example 7

Find the LCM for 27, 30, and 42, and tell how many times each number divides into the LCM.

Solution

$$\left.\begin{aligned} 27 &= 3\cdot 9 = 3^3 \\ 30 &= 6\cdot 5 = 2\cdot 3\cdot 5 \\ 42 &= 2\cdot 21 = 2\cdot 3\cdot 7 \end{aligned}\right\} \text{LCM} = 2\cdot 3^3\cdot 5\cdot 7 = 1890$$

$$1890 = 2\cdot 3\cdot 3\cdot 3\cdot 5\cdot 7 = \underbrace{(3\cdot 3\cdot 3)}\cdot\underbrace{(2\cdot 5\cdot 7)} = 27\cdot 70$$

$$1890 = 2\cdot 3\cdot 3\cdot 3\cdot 5\cdot 7 = \underbrace{(2\cdot 3\cdot 5)}\cdot\underbrace{(3\cdot 3\cdot 7)} = 30\cdot 63$$

$$1890 = 2\cdot 3\cdot 3\cdot 3\cdot 5\cdot 7 = \underbrace{(2\cdot 3\cdot 7)}\cdot\underbrace{(3\cdot 3\cdot 5)} = 42\cdot 45$$

So 27 divides into 1890 70 times;
30 divides into 1890 63 times;
42 divides into 1890 45 times.

Now work exercise 7 in the margin.

7. Find the LCM of 15, 35, and 42; then tell how many times each number divides into the LCM.

Example 8

Find the LCM for 12, 18, and 66, and tell how many times each number divides into the LCM.

Solution

$$\left.\begin{aligned} 12 &= 2^2\cdot 3 \\ 18 &= 2\cdot 3^2 \\ 66 &= 2\cdot 3\cdot 11 \end{aligned}\right\} \text{LCM} = 2^2\cdot 3^2\cdot 11 = 396$$

$$396 = 2\cdot 2\cdot 3\cdot 3\cdot 11 = \underbrace{(2\cdot 2\cdot 3)}\cdot\underbrace{(3\cdot 11)} = 12\cdot 33$$

$$396 = 2\cdot 2\cdot 3\cdot 3\cdot 11 = \underbrace{(2\cdot 3\cdot 3)}\cdot\underbrace{(2\cdot 11)} = 18\cdot 22$$

$$396 = 2\cdot 2\cdot 3\cdot 3\cdot 11 = \underbrace{(2\cdot 3\cdot 11)}\cdot\underbrace{(2\cdot 3)} = 66\cdot 6$$

So 12 divides into 396 33 times;
18 divides into 396 22 times;
66 divides into 396 6 times.

Now work exercise 8 in the margin.

8. Find the LCM for 10, 18, and 75, and tell how many times each number divides into the LCM.

9. a. Find the LCM of 20, 25, and 36.

b. State the number of times each number divides into the LCM.

Completion Example 9

a. Find the LCM of 30, 50, and 63.
b. State the number of times each number divides into the LCM.

Solution

a. 30 = ________
50 = ________
63 = ________

LCM = ________
= ________ = 3150

b. 3150 = ________ = (2 · 3 · 5) · (______) = 105 · ______
3150 = ________ = (2 · 5 · 5) · (______) = 63 · ______
3150 = ________ = (3 · 3 · 7) · (______) = 50 · ______

Now work exercise 9 in the margin.

Objective 3

An Application

Example 10

Suppose three weather satellites – A, B, and C – are orbiting the earth. Satellite A takes 18 hours, B takes 14 hours, and C takes 10 hours. If they are directly above each other now, as shown in part (a) of the figure below, in how many hours will they again be directly above each other in the positions shown in part (a) of the figure?

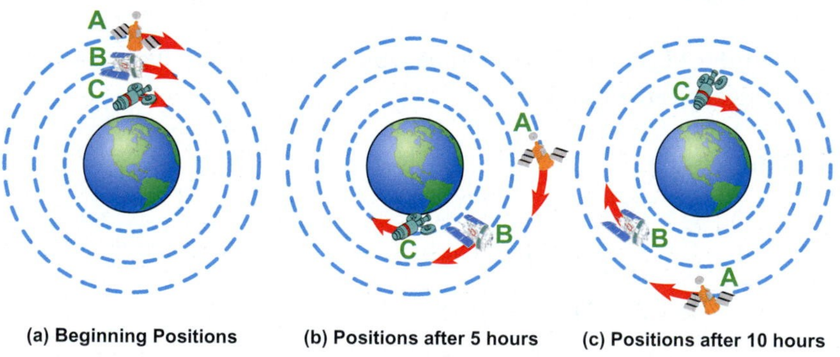

(a) Beginning Positions (b) Positions after 5 hours (c) Positions after 10 hours

Solution

Note that when the three satellites are in this position again, each one will have made some number of complete orbits. Since satellite A takes 18 hours to make one complete orbit of the earth, our solution must be a multiple of 18. Similarly, our solution must also be a multiple of 14 and a multiple of 10 to account for the complete orbits of satellites B and C.

The LCM of 18, 14, and 10 will tell us the next time satellites A, B, and C will all complete an orbit at the same time.

$$18 = 2 \cdot 3^2$$
$$14 = 2 \cdot 7 \quad \text{LCM} = 2 \cdot 3^2 \cdot 5 \cdot 7 = 630$$
$$10 = 2 \cdot 5$$

Thus the satellites will not align again for 630 hours (or 26 days and 6 hours).

Note:

Satellite A will have made 35 orbits, since $630 = 18 \cdot 35$.
Satellite B will have made 45 orbits, since $630 = 14 \cdot 45$.
Satellite C will have made 63 orbits, since $630 = 10 \cdot 63$.

Now work exercise 10 in the margin.

Completion Example 11

The problem stated in **Mathematics at Work!** at the beginning of this chapter is similar to Example 10. Only the orbiting times of the satellites are different. Redraw the figures from page 124 here. In how many hours will they again be directly above each other in the positions shown in the first figure? How many orbits will each satellite have completed at that time?

Solution

They will be in the original position in _____ hours.

Satellite A will have made _____ orbits.
Satellite B will have made _____ orbits.
Satellite C will have made _____ orbits.

Now work exercise 11 in the margin.

10. Three interns spent their day distributing messages in the Statehouse. It took the first intern 30 min to make each round, the second intern 36 min, and the third 18 min. If they began their rounds in the lobby, at 8 a.m., how much time elapsed before all three met again in the lobby? (Assume they made rounds continuously and that each round began and ended in the lobby.)

11. Using the information you calculated in Exercise 10 in the margin, how many rounds will each intern have completed at the time all three interns meet?

Completion Example Answers

4. **a.** Prime factorizations:

$36 = \mathbf{2 \cdot 2 \cdot 3 \cdot 3}$
$24 = \mathbf{2 \cdot 2 \cdot 2 \cdot 3}$
$48 = \mathbf{2 \cdot 2 \cdot 2 \cdot 2 \cdot 3}$

b. **2** and **3** are the only prime factors.

c. Most of each factor in any one factorization:
4 **2's** (in 48)
2 **3's** (in 36)

$\text{LCM} = \mathbf{2 \cdot 2 \cdot 2 \cdot 2 \cdot 3 \cdot 3} = \mathbf{2^4 \cdot 3^2} = 144$

144 is the smallest number divisible by all the numbers 36, 24, and 48.

9. **a.**
$30 = \mathbf{2 \cdot 3 \cdot 5}$
$50 = \mathbf{2 \cdot 5 \cdot 5}$
$63 = \mathbf{3 \cdot 3 \cdot 7}$

$\text{LCM} = \mathbf{2 \cdot 3 \cdot 3 \cdot 5 \cdot 5 \cdot 7}$
$= \mathbf{2 \cdot 3^2 \cdot 5^2 \cdot 7} = 3150$

b. $3150 = \mathbf{2 \cdot 3 \cdot 3 \cdot 5 \cdot 5 \cdot 7} = (2 \cdot 3 \cdot 5) \cdot \mathbf{(3 \cdot 5 \cdot 7)} = 105 \cdot \mathbf{30}$
$3150 = \mathbf{2 \cdot 3 \cdot 3 \cdot 5 \cdot 5 \cdot 7} = (2 \cdot 5 \cdot 5) \cdot \mathbf{(3 \cdot 3 \cdot 7)} = 63 \cdot \mathbf{50}$
$3150 = \mathbf{2 \cdot 3 \cdot 3 \cdot 5 \cdot 5 \cdot 7} = (3 \cdot 3 \cdot 7) \cdot \mathbf{(2 \cdot 5 \cdot 5)} = 50 \cdot \mathbf{63}$

11. They will be in the original position in **72** hours.

Satellite A will have made **3** orbits.
Satellite B will have made **4** orbits.
Satellite C will have made **6** orbits.

Practice Problems

Find the LCM for each of the following sets of numbers.

1. 30, 40, 50 **2.** 28, 70 **3.** 168, 140

Answers to Practice Problems:

1. 600 **2.** 140 **3.** 840

Name ______________ Section ______________ Date ______________

Exercises 2.6

Find the LCM for each of the sets of numbers. See Examples 1 through 6.

1. 8, 12

2. 3, 5, 7

3. 4, 6, 9

4. 3, 5, 9

5. 2, 5, 11

6. 4, 14, 18

7. 6, 10, 12

8. 6, 8, 27

9. 25, 40

10. 40, 75

11. 28, 98

12. 30, 75

13. 30, 80

14. 16, 28

15. 25, 100

16. 20, 50

17. 35, 100

18. 144, 216

19. 36, 42

20. 40, 100

21. 2, 4, 8

22. 10, 15, 35

23. 8, 13, 15

24. 25, 35, 49

ANSWERS

1. ______________
2. ______________
3. ______________
4. ______________
5. ______________
6. ______________
7. ______________
8. ______________
9. ______________
10. ______________
11. ______________
12. ______________
13. ______________
14. ______________
15. ______________
16. ______________
17. ______________
18. ______________
19. ______________
20. ______________
21. ______________
22. ______________
23. ______________
24. ______________

25. ______
26. ______
27. ______
28. ______
29. ______
30. ______
31. ______
32. ______
33. ______
34. ______
35. ______
36. ______
37. ______
38. ______
39. ______
40. ______
41. a. ______
b. ______
42. a. ______
b. ______
43. a. ______
b. ______
44. a. ______
b. ______
45. a. ______
b. ______
46. a. ______
b. ______

25. 6, 12, 15 **26.** 8, 10, 120 **27.** 6, 15, 80

28. 13, 26, 169 **29.** 45, 125, 150 **30.** 34, 51, 54

31. 33, 66, 121 **32.** 36, 54, 72 **33.** 45, 145, 290

34. 54, 81, 108 **35.** 45, 75, 135 **36.** 35, 40, 72

37. 10, 20, 30, 40 **38.** 15, 25, 30, 40 **39.** 24, 40, 48, 56

40. 169, 637, 845

*In Exercises 41 – 50, **a.** find the LCM and **b.** state the number of times each number divides into the LCM. See Examples 7 through 11.*

41. 8, 10, 15 **42.** 6, 15, 30 **43.** 10, 15, 24

44. 8, 10, 120 **45.** 6, 18, 27, 45 **46.** 12, 95, 228

Name ________________ Section ________ Date ________

ANSWERS

47. 45, 63, 98 **48.** 40, 56, 196 **49.** 99, 143, 363

50. 125, 135, 225

51. Running. Two long-distance joggers are running on the same course in the same direction. They meet at the water fountain and say "Hi." One jogger goes around the course in 10 minutes and the other goes around the course in 14 minutes. They continue to jog until they meet again at the water fountain.

a. How many minutes elapsed between meetings at the fountain?

b. How many times will each have jogged around the course?

52. Security. Three night watchmen meet at a doughnut shop before they walk around inspecting buildings at a shopping center. The watchmen take 9, 12, and 14 minutes, respectively, for the inspection trip.

a. If they start at the same time, in how many minutes will they meet again at the doughnut shop?

b. How many inspection trips will each watchman have made?

53. Space Travel. Two astronauts miss connections at their first rendezvous in space.

a. If one astronaut circles the earth every 12 hours and the other every 16 hours, in how many hours will they rendezvous again at the same place?

b. How many orbits will each astronaut have made before the second rendezvous?

54. Trucking. Three truck drivers eat lunch together whenever all three are at the routing station at the same time. The first driver's route takes 5 days, the second driver's takes 15 days, and the third driver's takes 6 days.

a. How often do the three drivers eat lunch together?

b. If the first driver's route is changed to take 6 days, how often would they eat lunch together?

55. Travel. Four book salespersons leave the home office the same day. They take 10 days, 12 days, 15 days, and 18 days, respectively, to travel their own sales regions.

a. In how many days will they all meet again at the home office?

b. How many sales trips will each have made?

47. a. ________

b. ________

48. a. ________

b. ________

49. a. ________

b. ________

50. a. ________

b. ________

51. a. ________

b. ________

52. a. ________

b. ________

53. a. ________

b. ________

54. a. ________

b. ________

55. a. ________

b. ________

56. [Respond below exercise.]

Writing and Thinking about Mathematics

56. From studying Example 10 and working Exercises 51 – 55, write a general statement about the types of word problems (or characteristics of these problems) in which the least common multiple is a related idea.

57. [Respond below exercise.]

Collaborative Learning

57. With the class separated into teams of two to four students, each team is to write one or two paragraphs explaining what topic in this chapter they found to be the most interesting and why. Each team leader is to read the paragraph(s), with classroom discussion to follow.

Chapter 2 Index of Key Terms and Ideas

Section 2.1 Exponents

Exponents pages 124 – 125

An **exponent** is a number that tells how many times its base is to be used as a factor.

Properties of Exponents page 126

1. For any whole number a, $a^1 = a$.
2. For any nonzero whole number a, $a^0 = 1$.

Perfect squares pages 127 – 128

$1^2 = 1, 2^2 = 4, 3^2 = 9, 4^2 = 16, 5^2 = 25, 6^2 = 36, 7^2 = 49, 8^2 = 64, 9^2 = 81,$
$10^2 = 100, 11^2 = 121, 12^2 = 144, 13^2 = 169, 14^2 = 196, 15^2 = 225, 16^2 = 256,$
$17^2 = 289, 18^2 = 324, 19^2 = 361, 20^2 = 400$

Section 2.2 Order of Operations

Rules for Order of Operations page 133

1. First, simplify within grouping symbols, such as parentheses ( ), brackets [ ], or braces { }. Start with the innermost grouping.
2. Second, evaluate any numbers or expressions indicated by exponents.
3. Third, moving from **left to right**, perform any multiplication or division in the order in which it appears.
4. Fourth, moving from **left to right**, perform any addition or subtraction in the order in which it appears.

Section 2.3 Tests for Divisibility (2, 3, 5, 6, 9, and 10)

Divisibility page 145

If a number can be divided by another number so that the remainder is 0, then we say:

1. The first number is **exactly divisible by** (or is **divisible by**) the second.
2. The second number **divides** the first.

Tests for Divisibility by 2, 3, 5, 6, 9, and 10 pages 145 – 146

For 2: If the last digit (units digit) of a whole number is 0, 2, 4, 6, or 8 (an even number), then the whole number is divisible by 2.

For 3: If the sum of the digits of a whole number is divisible by 3, then the number is divisible by 3.

For 5: If the last digit of a whole number is 0 or 5, then the number is divisible by 5.

For 6: If the whole number is divisible by both 2 and 3, then it is divisible by 6.

Continued on next page...

Section 2.3 Tests for Divisibility (2, 3, 5, 6, 9, and 10) (continued)

For 9: If the sum of the digits of a whole number is divisible by 9, then the number is divisible by 9.

For 10: If the last digit of a whole number is 0, then the number is divisible by 10.

Even and Odd Whole Numbers page 146

Even whole numbers are divisible by 2.

(If a whole number is divided by 2 and the remainder is 0, then the whole number is even.)

Odd whole numbers are not divisible by 2.

(If a whole number is divided by 2 and the remainder is 1, then the whole number is odd.)

Section 2.4 Prime Numbers and Composite Numbers

Prime Number page 157

A **prime number** is a counting number greater than 1 that has only 1 and itself as factors.

OR

A **prime number** is a counting number with exactly two different factors (or divisors).

Composite Number page 157

A **composite number** is a counting number with more than two different factors (or divisors).

Prime Numbers less than 50 page 159

2, 3, 5, 7, 11, 13, 17, 19, 23, 29, 31, 37, 41, 43, 47

To Determine Whether a Number is Prime: page 160

Divide the number by progressively larger prime numbers (2, 3, 5, 7, 11, and so forth) until:

1. **The remainder is 0.** This means that the prime number is a factor and **the given number is composite**; or
2. **You find a quotient smaller than the prime divisor.** This means that **the given number is prime** because it has no smaller prime factors.

Section 2.5 Prime Factorizations

The Fundamental Theorem of Arithmetic page 171

Every composite number has exactly one prime factorization.

Continued on next page...

Section 2.5 Prime Factorizations (continued)

To Find the Prime Factorization of a Composite Number: page 171

1. Factor the composite number into any two factors.
2. Factor each factor that is not prime.
3. Continue this process until all factors are prime.

The **prime factorization** of a number is a factorization of that number using only prime factors.

Factors of Composite Numbers page 174

The only factors (or divisors) of a composite number are:

1. 1 and the number itself;
2. Each prime factor; and
3. Products formed by all combinations of the prime factors

Section 2.6 Least Common Multiple (LCM) with Applications

The Least Common Multiple (LCM) page 181

The **Least Common Multiple (LCM)** of a set of counting numbers is the smallest number common to all the sets of multiples of the given numbers.

Method I for finding the LCM page 181

To find the LCM of a set of counting numbers:

1. Find the prime factorization of each number.
2. Find the prime factors that appear in **any one** of the prime factorizations.
3. Form the product of these primes using each prime the most number of times it appears in **any one** of the prime factorizations.

Method II for finding the LCM page 183

To find the LCM of a set of counting numbers:

1. Write the numbers horizontally and find a prime number that will divide into more than one number, if possible.
2. Divide by that prime and write the quotients beneath the dividends. Rewrite any numbers not divided beneath themselves.
3. Continue the process until no two numbers have a common prime divisor.
4. The LCM is the product of all the prime divisors and the last set of quotients.

Name ______________________ Section ____________ Date ____________

ANSWERS

Chapter 2 Review Questions

1. In the expression 3^4, 3 is called the _________, and 4 is called the _________. The value of 3^4 is _______.

2. A prime number is a counting number with exactly two ________.

3. Every composite number has a unique _________ factorization.

4. Since 36 has more than two different factors, 36 is a _________ number.

Find a base and exponent form for each number without using 1 as the exponent.

5. 16
6. 27
7. 169
8. 400

Evaluate each of the following expressions.

9. $7+3\cdot2-1+9\div3$
10. $3\cdot2^5-2\cdot5^2$
11. $14\div2+2\cdot8+30\div5\cdot2$
12. $(16\div2^2+6)\div2+8$
13. $(75-3\cdot5)\div10-4$
14. $(7^2\cdot2+2)\div10\div(2+3)$

In Exercises 15 – 20, etermine which of the numbers 2, 3, 5, 6, 9, and 10 (if any) will divide exactly into each of the follwing numbers.

15. 45
16. 72
17. 479

1. ______________
2. ______________
3. ______________
4. ______________
5. ______________
6. ______________
7. ______________
8. ______________
9. ______________
10. ______________
11. ______________
12. ______________
13. ______________
14. ______________
15. ______________
16. ______________
17. ______________

18. ____________

19. ____________

20. ____________

21. ____________

22. ____________

23. ____________

24. ____________

25. ____________

26. ____________

27. ____________

28. ____________

29. ____________

30. ____________

31. ____________

32. ____________

33. ____________

34. ____________

18. 5040 **19.** 8836 **20.** 575,493

21. List the first 10 multiples of 3. Are any of these multiples prime numbers?

22. Is 223 a prime number? Show your work.

23. List the prime numbers less than 60.

24. Find two factors of 24 whose product is 24 and whose sum is 10.

25. Find two factors of 60 whose product is 60 and whose sum is 17.

Find the prime factorization for each of the following numbers.

26. 150 **27.** 65 **28.** 84 **29.** 92

Find the LCM for each of the following sets of numbers.

30. 8, 14, 24 **31.** 8, 12, 25, 36 **32.** 27, 54, 135

33. Find the LCM for the numbers 18, 39, and 63 and tell how many times each number divides into the LCM.

34. **Racing.** One race car goes around the track every 30 seconds and the other every 35 seconds. If both cars start a race from the same point, in how many seconds will the first car be exactly one lap ahead of the second car? How many laps will each car have made when the first car is two laps ahead of the second?

Name ______________________ Section ______________ Date ____________

ANSWERS

Chapter 2 Test

1. In the expression 7^3, 7 is called the _______ and 3 is called the _______. The value of 7^3 is _______.

2. **a.** List the prime numbers between 6 and 20.
b. List the squares of these prime numbers.

In Exercises 3 – 6, find the value of each expression by using the rules for order of operations.

3. $12+9\div3-2$

4. $60\div4(4-1)+3$

5. $2\cdot3^2-(2^2\cdot3\div2)-3(5-1)$

6. $12\cdot2^3\div4\cdot3$

Using the tests for divisibility, tell which of the numbers 2, 3, 5, 6, 9, and 10 (if any) will divide the numbers in Exercises 7 – 10.

7. 90

8. 221

9. 324

10. 1700

11. List the multiples of 13 that are less than 100.

12. List all of the composite numbers found in the following set: 1, 2, 6, 9, 13, 15, 17, 27, 39, 41, 51, 59.

13. Show, by division, that 81 is a factor of 7452.

1. ____________

2 a. ____________

b. ____________

3. ____________

4. ____________

5. ____________

6. ____________

7. ____________

8. ____________

9. ____________

10. ____________

11. ____________

12. ____________

13. ____________

14. ______

15. ______

16. ______

17. ______

18. ______

19. ______

20. ______

21. ______

22. a. ______

b. ______

14. Does 42 divide the product $2 \cdot 5 \cdot 6 \cdot 7 \cdot 9$? If so, how many times? If not, explain briefly why not.

15. List all the factors of 60.

Find the prime factorization of each number.

16. 124

17. 165

18. Determine whether or not 107 is a prime number. Show each step in your reasoning.

Find the LCM of each set of numbers and tell how many times each number divides into the LCM.

19. 4, 14, 21

20. 6, 15, 60

21. 8, 10, 15, 28

22. Sales Travel. Three salesmen have lunch together each time they are in the home office on the same day.

a. If it takes Salesman A 6 days to cover his territory, Salesman B 9 days, and Salesman C 12 days, how often do they have lunch together?

b. If Salesman A's route is changed so that he takes only 5 days to cover his territory, how often do they have lunch together?

Name ______________________ Section ____________ Date __________

ANSWERS

Cumulative Review: Chapters 1 – 2

1. Write the number 50,732 in **a.** expanded notation and **b.** its English word equivalent.

In Exercises 2 – 5, name the property illustrated.

2. $45 \cdot 2 = 2 \cdot 45$

3. $7 \cdot (3 \cdot 4) = (7 \cdot 3) \cdot 4$

4. $19 + 0 = 19$

5. $25 + 3 = 3 + 25$

6. Round 41,624 to the nearest thousand.

*In Exercises 7 – 10, first **a.** estimate each answer and then **b.** find the answer by performing the indicated operation.*

7.
$$\begin{array}{r} 83 \\ 947 \\ +1035 \\ \hline \end{array}$$

8.
$$\begin{array}{r} 6003 \\ -\ 759 \\ \hline \end{array}$$

9.
$$\begin{array}{r} 74 \\ \times\ 86 \\ \hline \end{array}$$

10. $17\overline{)2210}$

11. Multiply mentally: 90×300.

Use the Rules for Order of Operations to evaluate the expressions in Exercises 12 and 13.

12. $15 + 2\left(8 - 2^3\right) - 3 \cdot 2^2$

13. $75 \div 5 \cdot 3 + 4\left(7 - 2^2\right)$

14. Without actually dividing, determine which of the numbers 2, 3, 5, 6, 9, and 10 will divide into 10,840. Give a brief reason for each decision.

1. a. ______
b. ______
2. ______
3. ______
4. ______
5. ______
6. ______
7. a. ______
b. ______
8. a. ______
b. ______
9. a. ______
b. ______
10. a. ______
b. ______
11. ______
12. ______
13. ______
14. ______

15.

16.

17.

18.

19.

20. a.

b.

21. a.

b.

22.

23.

24.

25. a.

b.

15. Determine whether the number 307 is prime or not. Show all the steps you used.

16. List all the prime numbers less than 20,000 that are even.

17. List all the factors of 65.

18. Find the prime factorization of 475.

19. Does 56 divide the product $2 \cdot 3 \cdot 4 \cdot 6 \cdot 7 \cdot 9$? If so, how many times? If not, explain briefly why not.

*First, find **a.** the LCM of each set of numbers and **b.** tell how many times each number divides into the LCM.*

20. 15, 27, 35

21. 33, 44, 55, 121

22. Geography. The Danube River is 2842 kilometers long and flows into the Black Sea. It is 509 kilometers longer than the Colorado River which flows into the Gulf of California. How long is the Colorado River?

23. A painting is in a rectangular frame that is 36 inches wide and 48 inches high. How many square inches of wall space is covered when the painting is hung?

24. Grading. On a psychology exam, three students scored 75, four students scored 82, two students scored 85, one student scored 87, five students scored 91, and one student scored 95. What was their average score on the exam?

25. Running. Two long-distance runners practice on the same track and start at the same point. One takes 54 seconds and the other takes 63 seconds to go around the track once.

a. If each continues at the same pace, in how many seconds will they meet at the starting point?

b. How many laps will each have run by the time they meet again at the starting point?

3

Fractions

Chapter 3 Fractions

WHAT TO EXPECT IN CHAPTER 3

Chapter 3 deals with understanding fractions (called rational numbers) and operations with fractions. Each section has some word problems to reinforce the concepts as they are developed. The term improper fraction is introduced in Section 3.1, and then used throughout the chapter. Despite the use of the term "improper", you should understand that improper fractions are just as useful as any other type of number.

In Section 3.1 the commutative and associative properties for multiplication (Section 3.2) and addition (Section 3.4) are shown to apply with fractions just as they do with whole numbers. The prime factorization techniques developed in Chapter 2 are used to reduce fractions to lower terms and to raise fractions to higher terms and form the basis for the development of the operations of multiplication (Section 3.2), division (Section 3.3), addition (Section 3.4), and subtraction (Section 3.5). For many students this prime factoring method provides valuable insight into the nature of fractions and helps to build a new confidence in working with fractions.

The rules for order of operations (Section 3.6) are the same for evaluating expressions that contain fractions as they are for evaluating expressions with whole numbers.

Chapter 3 Fractions

Mathematics at Work!

Politics and the political process affect everyone in some way. In local, state or national elections, registered voters make decisions about who will represent them and make choices about various ballot measures.

In major issues at the state and national levels, pollsters use mathematics (in particular, statistics and statistical methods) to indicate attitudes and to predict, within certain percentages, how the voters will vote. When there is an important election in your area, read the papers and magazines and listen to the television reports for mathematically related statements predicting the outcome.

Consider the following situation: There are 8000 registered voters in Brownsville, and $\frac{3}{8}$ of these voters live in neighborhoods on the north-side of town. A survey indicates that $\frac{4}{5}$ of these north-side voters are in favor of a bond measure for constructing a new recreation facility that would largely benefit their neighborhoods. Also, $\frac{7}{10}$ of the registered voters from all other parts of town are in favor of the measure. We might then want to know:

a. How many north-side voters favor the bond measure?
b. How many voters in all favor the bond measure?

(See Example 18, Section 3.1.)

3.1 Basic Multiplication and Changing to Higher Terms

Introduction to Fractions (or Rational Numbers)

Objective 1

In this chapter we will deal with fractions in which the **numerator** (top number) and **denominator** (bottom number) are whole numbers, and the denominator is not 0. Such fractions are called **rational numbers**. (Note that there are other fractions, called **irrational numbers**, that are studied in algebra and other mathematics courses.)

Rational Numbers

A **rational number** is a number that can be written in the fraction form $\frac{a}{b}$ where a is a whole number and b is a nonzero whole number.

$$\frac{a}{b} \begin{array}{l} \leftarrow \text{numerator} \\ \leftarrow \text{denominator} \end{array}$$

Note: The numerator a can be 0, but the denominator b cannot be 0.

Examples of fractions (or rational numbers) are

$\frac{1}{2}$	$\frac{3}{4}$	$\frac{9}{10}$	$\frac{17}{3}$
one-half	*three-fourths*	*nine-tenths*	*seventeen-thirds*

Unless otherwise stated, we will use the terms **fraction** and **rational number** to mean the same thing. Example 1 shows illustrations of several fractions indicating a part of a whole.

Objectives

1. Know the terms **numerator** and **denominator**.
2. Learn two meanings of fractions: to indicate equal parts of a whole and to indicate division.
3. Understand why no denominator can be 0.
4. Learn how to multiply fractions.
5. Determine what to multiply by in order to build a fraction to higher terms.

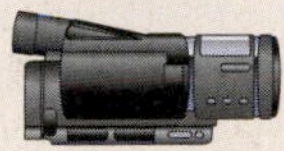
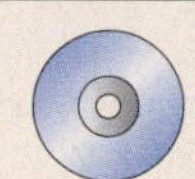

Example 1

a.

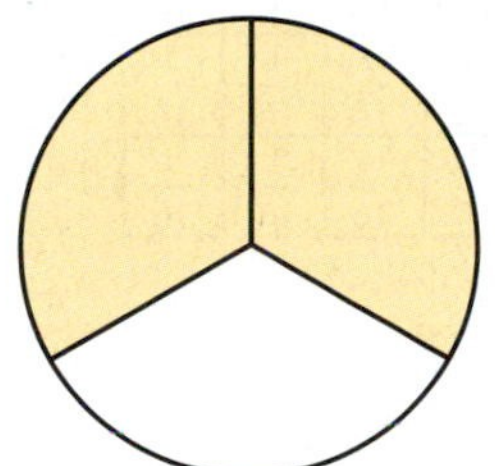

In the circle, 2 of the 3 equal parts are shaded. Thus $\frac{2}{3}$ of the circle is shaded and $\frac{1}{3}$ is not shaded.

b.

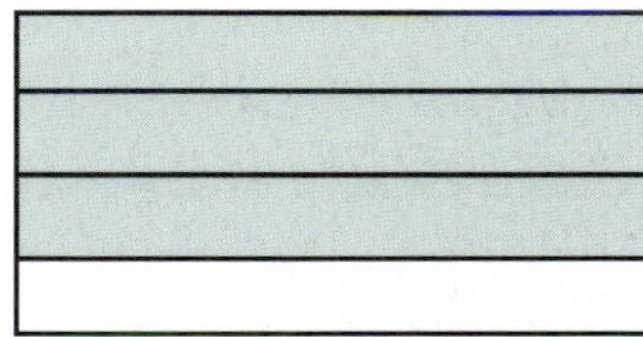

In the rectangle, 3 of the 4 equal parts are shaded. Thus $\frac{3}{4}$ of the rectangle is shaded and $\frac{1}{4}$ is not shaded.

c.

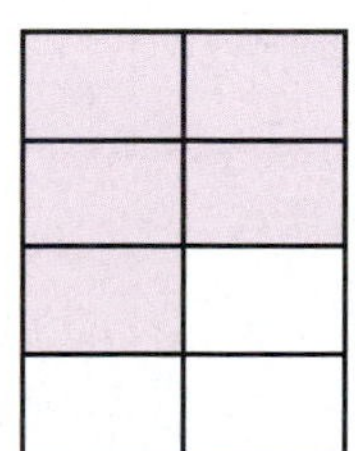

In the rectangle, 5 of the 8 equal parts are shaded. Thus $\frac{5}{8}$ of the rectangle is shaded and $\frac{3}{8}$ is not shaded.

Now work exercise 1 in the margin.

Write a fraction that indicates the shaded parts in each of the following diagrams.

1. a.

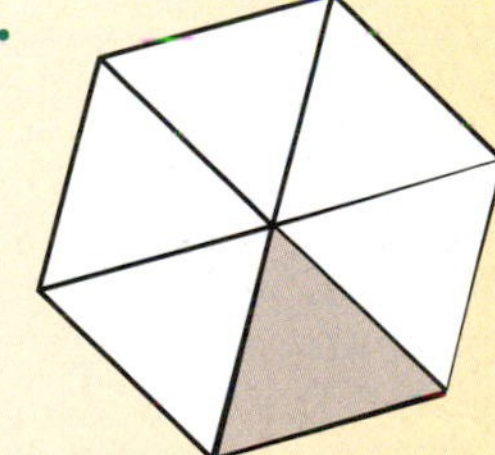

b.

Figure 3.1 shows how a whole may be separated into equal parts in several ways and still represent the same (or equal) part of the whole. In this case each shaded region represents the same part of the whole. **Equal** fractions are said to be **equivalent**.

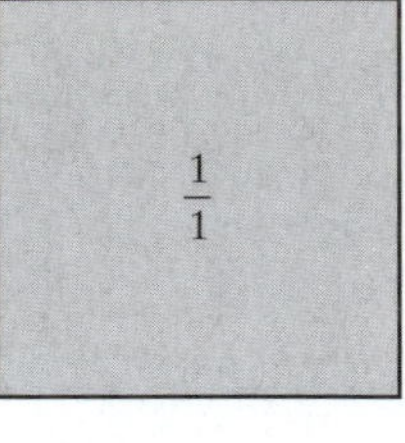

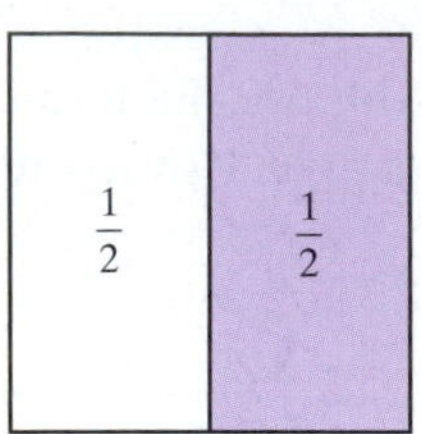

Figure 3.1

From the figure we see that $\frac{1}{2}=\frac{2}{4}=\frac{4}{8}=\frac{8}{16}$.

Fractions can be classified in two categories: proper fractions and improper fractions.

Fractions

A fraction is a **proper fraction** if the numerator is less than the denominator.

A fraction is an **improper fraction** if the numerator is equal to or greater than the denominator.

Examples of proper fractions are: $\frac{1}{2}, \frac{6}{7}, \frac{3}{10}$, and $\frac{5}{13}$.

Examples of improper fractions are: $\frac{11}{4}, \frac{6}{5}, \frac{20}{7}$, and $\frac{8}{1}=8$.

Note about improper fractions: The term improper is somewhat misleading because there is nothing "improper" about such fractions. In fact, in algebra and much of mathematics, the form of improper fractions is preferred over the alternative form of mixed numbers. In Chapter 4, we will discuss how improper fractions can be written in the form of mixed numbers, such as $1\frac{2}{3}$ and $2\frac{1}{6}$.

Fractions can be used in two basic ways:

1. To indicate equal parts of a whole.
2. To indicate division. (The numerator is to be divided by the denominator.)

Write a fraction that indicates the shaded parts in each of the following diagrams.

Example 2

$\frac{3}{4}$ indicates 3 of 4 equal parts.

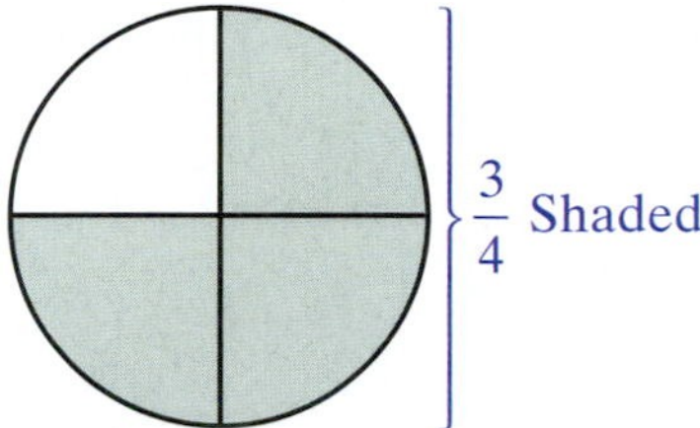

Now work exercise 2 in the margin.

2. 

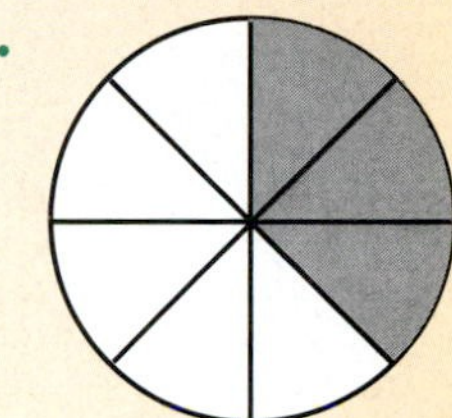

Example 3

$\frac{2}{3}$ indicates 2 of 3 equal parts.

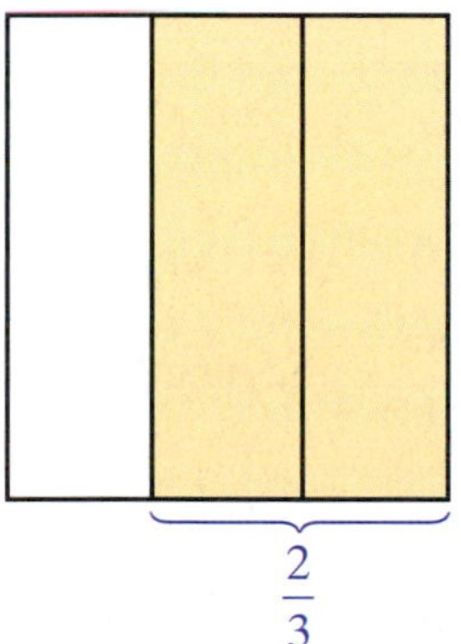

Now work exercise 3 in the margin.

3.

Example 4

$\frac{5}{3}$ indicates 5 of 3 equal parts (more than a whole).

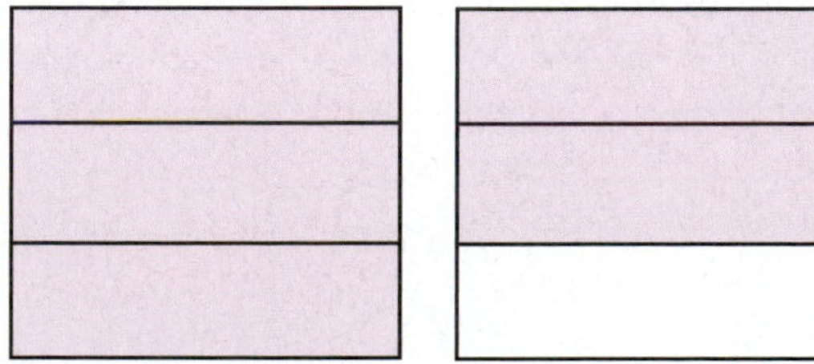

Now work exercise 4 in the margin.

4.

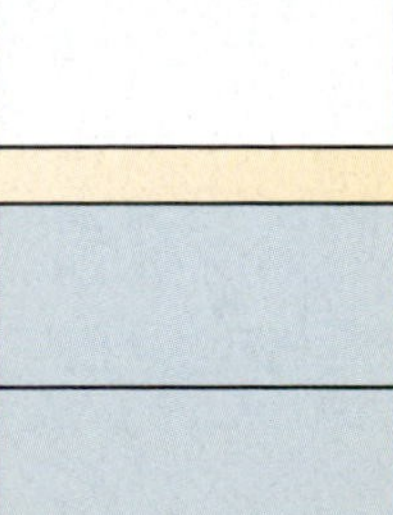

Whole numbers can be thought of as fractions with denominator 1. Thus, in fraction form:

$$0 = \frac{0}{1}, \quad 1 = \frac{1}{1}, \quad 2 = \frac{2}{1}, \quad 3 = \frac{3}{1}, \quad \text{and so on.}$$

Therefore, in working with fractions and whole numbers, whole numbers may be written in fraction form with denominator 1. We will see that this is particularly convenient when multiplying and dividing with fractions.

We stated earlier that fractions can be used to indicate division. For example, the fraction $\frac{24}{8}$ is the same as $24 \div 8$. That is, the numerator is to be divided by the denominator. Because division is related to multiplication, we can make the following statements:

$$\frac{24}{8} = 3 \quad \text{because} \quad 24 = 8 \cdot 3$$

$$\text{and} \quad \frac{0}{5} = 0 \quad \text{because} \quad 0 = 5 \cdot 0$$

This meaning of division for fractions helps to clarify the fact that the denominator can never be 0. Note that the definition of rational numbers states that the denominator b in the form $\frac{a}{b}$ cannot be 0. We say that **division by 0 is undefined**. The following discussion explains why.

Objective 3

Division by 0 is Undefined

Consider $\frac{5}{0} = \boxed{?}$. Then we must have $5 = 0 \cdot \boxed{?}$. But this is impossible because $0 \cdot \boxed{?} = 0$ and $5 \neq 0$. Therefore, $\frac{5}{0}$ is undefined.

$\frac{0}{0}$ is Undefined

Next, consider $\frac{0}{0} = \boxed{?}$. Then we must have $0 = 0 \cdot \boxed{?}$. But $0 = 0 \cdot \boxed{?}$ is true for any value of $\boxed{?}$, and in arithmetic, an operation such as division cannot give more than one answer. So we agree that $\frac{0}{0}$ is undefined.

Thus $\frac{a}{0}$ is undefined for any value of a.

Example 5

a. $\frac{0}{36}=0$ **b.** $\frac{0}{124}=0$

Now work exercise 5 in the margin.

Example 6

a. $\frac{17}{0}$ is undefined. **b.** $\frac{233}{0}$ is undefined.

Now work exercise 6 in the margin.

What is the value, if any, of the following fractions:

5. $\frac{0}{45}$

6. $\frac{10}{0}$

Multiplying Fractions

Objective 4

We want to learn how to perform operations with fractions. That is, we want to be able to add, subtract, multiply, and divide with fractions. We begin with multiplication. Remember that any whole number can be written in fraction form with denominator 1. Also, no denominator can be 0.

To Multiply Fractions:

1. Multiply the numerators.
2. Multiply the denominators.

$$\frac{a}{b}\cdot\frac{c}{d}=\frac{a\cdot c}{b\cdot d}$$

Special Note about Fractions:

We know that any whole number $\boldsymbol{a}$ can be considered as a fraction with denominator 1; that is $a=\frac{a}{1}$. Similarly, any fraction can be considered to be the product of a whole number and a fraction; that is,

$$\frac{a}{b}=\frac{a}{1}\cdot\frac{1}{b}=a\cdot\frac{1}{b}.$$

This idea will be helpful throughout the text in understanding how various forms of whole numbers, fractions, mixed numbers, and decimal numbers are related. For example,

$$\frac{2}{3}=2\cdot\frac{1}{3}\text{ and }\frac{5}{8}=5\cdot\frac{1}{8}.$$

Finding the product of two fractions can be thought of as finding one fractional part **of** another fraction. For example, when we multiply

$$\frac{1}{2}\cdot\frac{3}{10}$$

we are finding

$$\frac{1}{2} \textbf{ of } \frac{3}{10}$$

Thus $\frac{1}{2}$ **of** $\frac{3}{10}$ is $\frac{3}{20}$ because $\frac{1}{2}\cdot\frac{3}{10}=\frac{3}{20}$.

Examples 7 and 8 illustrate how multiplication of two fractions can be interpreted by shading in appropriate parts of a whole (the square being the whole).

7. Find the product of $\frac{1}{4}$ and $\frac{5}{6}$.

Example 7

Find the product of $\frac{2}{3}$ and $\frac{4}{5}$.

Solution

$$\frac{2}{3}\cdot\frac{4}{5}=\frac{2\cdot4}{3\cdot5}=\frac{8}{15}$$

This product can be illustrated graphically. We can think of the multiplication as finding $\frac{2}{3}$ **of** $\frac{4}{5}$.

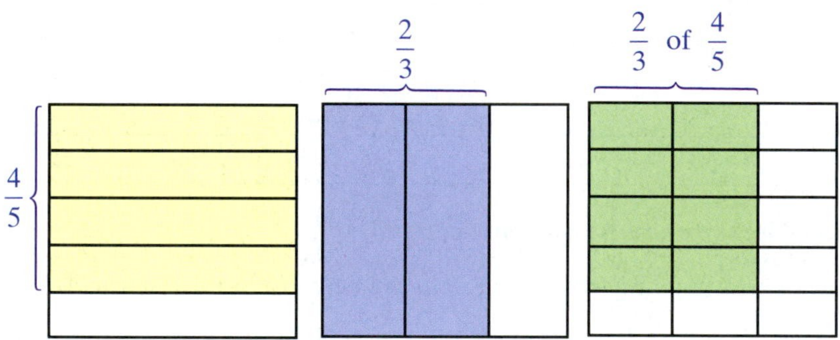

a. Shade $\frac{4}{5}$.

b. Shade $\frac{2}{3}$.

c. Shaded region is $\frac{2}{3}$ of $\frac{4}{5}$ or $\frac{2}{3}\cdot\frac{4}{5}=\frac{8}{15}$.

Now work exercise 7 in the margin.

Example 8

Find the product of $\frac{1}{3}$ and $\frac{2}{5}$ and illustrate the product as in Example 7 by shading parts of a square.

Solution

$$\frac{1}{3}\cdot\frac{2}{5}=\frac{1\cdot2}{3\cdot5}=\frac{2}{15}.$$

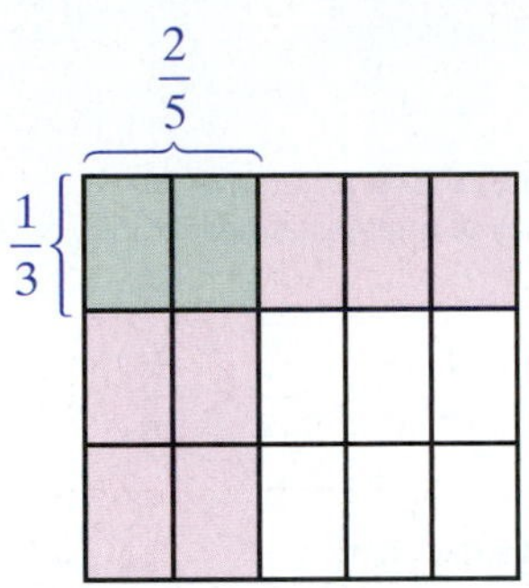

Now work exercise 8 in the margin.

Completion Example 9

Find the following products.

a. $\frac{2}{3}\cdot\frac{5}{7}=\frac{2\cdot 5}{3\cdot 7}=$ ______

b. $\frac{1}{4}\cdot\frac{3}{5}=\frac{1\cdot 3}{4\cdot 5}=$ ______

c. $\frac{7}{5}\cdot 2=\frac{7}{5}\cdot\frac{2}{1}=$ ______

d. $\frac{1}{4}\cdot\frac{3}{5}\cdot\frac{7}{2}=\frac{1\cdot 3\cdot 7}{\_\_\_\_}=$ ______

e. $\frac{2}{3}\cdot\frac{11}{15}\cdot\frac{1}{7}=$ ______

Now work exercise 9 in the margin.

Both the commutative property and the associative property of multiplication apply to fractions.

Commutative Property of Multiplication:

If $\frac{a}{b}$ and $\frac{c}{d}$ are fractions, then

$$\frac{a}{b}\cdot\frac{c}{d}=\frac{c}{d}\cdot\frac{a}{b}.$$

Associative Property of Multiplication:

If $\frac{a}{b}$, $\frac{c}{d}$, and $\frac{e}{f}$ are fractions, then

$$\frac{a}{b}\cdot\frac{c}{d}\cdot\frac{e}{f}=\left(\frac{a}{b}\cdot\frac{c}{d}\right)\cdot\frac{e}{f}=\frac{a}{b}\cdot\left(\frac{c}{d}\cdot\frac{e}{f}\right).$$

8. Find the product of $\frac{1}{2}$ and $\frac{3}{7}$ and illustrate the product as in Example 8 by shading parts of a square.

9. Find the following products.

a. $\frac{1}{7}\cdot\frac{3}{5}$

b. $\frac{0}{4}\cdot\frac{5}{8}$

c. $\frac{3}{14}\cdot 5$

d. $\frac{1}{2}\cdot\frac{3}{4}\cdot\frac{7}{5}$

e. $\frac{7}{8}\cdot\frac{3}{4}\cdot 1$

10. State which property of multiplication is illustrated in each problem.

a. $\frac{1}{2}\cdot\frac{5}{8}=\frac{5}{8}\cdot\frac{1}{2}$

b. $\frac{3}{4}\cdot 6=6\cdot\frac{3}{4}$

11. State which property of multiplication is illustrated:

$$\frac{1}{3}\cdot\left(\frac{1}{2}\cdot\frac{1}{4}\right)=\left(\frac{1}{3}\cdot\frac{1}{2}\right)\cdot\frac{1}{4}$$

Example 10

$\frac{3}{4}\cdot\frac{1}{7}=\frac{1}{7}\cdot\frac{3}{4}$ Illustrates the commutative property of multiplication.

$$\frac{3}{4}\cdot\frac{1}{7}=\frac{3\cdot 1}{4\cdot 7}=\frac{3}{28} \quad\text{and}\quad \frac{1}{7}\cdot\frac{3}{4}=\frac{1\cdot 3}{7\cdot 4}=\frac{3}{28}$$

Now work exercise 10 in the margin

Example 11

$\left(\frac{5}{8}\cdot\frac{3}{2}\right)\cdot\frac{3}{11}=\frac{5}{8}\cdot\left(\frac{3}{2}\cdot\frac{3}{11}\right)$ Illustrates the associative property of multiplication.

$$\left(\frac{5}{8}\cdot\frac{3}{2}\right)\cdot\frac{3}{11}=\left(\frac{5\cdot 3}{8\cdot 2}\right)\cdot\frac{3}{11}=\frac{15}{16}\cdot\frac{3}{11}=\frac{15\cdot 3}{16\cdot 11}=\frac{45}{176} \quad\text{and}$$

$$\frac{5}{8}\cdot\left(\frac{3}{2}\cdot\frac{3}{11}\right)=\frac{5}{8}\cdot\left(\frac{3\cdot 3}{2\cdot 11}\right)=\frac{5}{8}\cdot\frac{9}{22}=\frac{5\cdot 9}{8\cdot 22}=\frac{45}{176}$$

Now work exercise 11 in the margin.

Changing Fractions to Higher Terms

We know that 1 is the multiplicative identity for whole numbers; that is, for any whole number a, $a\cdot 1=a$. The number 1 is also the multiplicative identity for fractions (or rational numbers), since

$$\frac{a}{b}\cdot 1=\frac{a}{b}\cdot\frac{1}{1}=\frac{a\cdot 1}{b\cdot 1}=\frac{a}{b}.$$

Multiplicative Identity:

The number 1 is called the **multiplicative identity**, and for any fraction $\frac{a}{b}$,

$$\frac{a}{b}\cdot 1=\frac{a}{b}.$$

This fact states that a fraction has the same value if it is multiplied by 1. It allows us to find a fraction equal to (or equivalent to) a given fraction with a larger denominator. In this case, the given fraction is said to be **changed to higher terms**. For example, if we multiply $\frac{1}{2}\cdot\frac{3}{3}$, this is the same as $\frac{1}{2}\cdot 1$. So

$$\frac{1}{2}=\frac{1}{2}\cdot 1=\frac{1}{2}\cdot\frac{3}{3}=\frac{3}{6}, \text{ or } \frac{1}{2}=\frac{3}{6}.$$

To Change a Fraction to Higher Terms:

Multiply the numerator and denominator by the same nonzero whole number.

$$\frac{a}{b}=\frac{a}{b}\cdot 1=\frac{a}{b}\cdot\frac{\boldsymbol{k}}{\boldsymbol{k}}=\frac{a\cdot \boldsymbol{k}}{b\cdot \boldsymbol{k}},\text{ where } \boldsymbol{k}\neq 0\ \left(1=\frac{\boldsymbol{k}}{\boldsymbol{k}}\right).$$

Objective 5

The following examples illustrate the use of this technique of changing a fraction to higher terms. Note carefully the importance of the choice of the form of $\frac{k}{k}$.

Example 12

Suppose we want to find a fraction equal to $\frac{3}{4}$ with a denominator of 28.

$\frac{3}{4}=\frac{?}{28}$

We know $4\cdot 7=28$, so use $1=\frac{k}{k}=\frac{7}{7}$.

$\frac{3}{4}=\frac{3}{4}\cdot 1=\frac{3}{4}\cdot\frac{7}{7}=\frac{21}{28}$.

Now work exercise 12 in the margin.

Example 13

Suppose we want to find a fraction equal to $\frac{9}{10}$ with a denominator of 30.

$\frac{9}{10}=\frac{?}{30}$

Since $10\cdot 3=30$, use $1=\frac{k}{k}=\frac{3}{3}$.

$\frac{9}{10}=\frac{9}{10}\cdot 1=\frac{9}{10}\cdot\frac{3}{3}=\frac{27}{30}$.

Now work exercise 13 in the margin.

Example 14

Suppose we want to find a fraction equal to $\frac{7}{8}$ with a denominator of 32.

$\frac{7}{8}=\frac{?}{32}$

Since $8\cdot 4=32$, use $1=\frac{k}{k}=\frac{4}{4}$.

$\frac{7}{8}=\frac{7}{8}\cdot 1=\frac{7}{8}\cdot\frac{4}{4}=\frac{28}{32}$.

Now work exercise 14 in the margin.

Find the missing numerator or denominator that will make the fractions equal.

12. $\frac{2}{3}=\frac{?}{12}$

13. $\frac{5}{8}=\frac{?}{24}$

14. $\frac{1}{6}=\frac{?}{72}$

15. Find the missing numerator that will make the fractions equal.

$\frac{3}{5} = \frac{?}{20}$

16. Find a fraction with denominator 63 equal to $\frac{4}{7}$.

17. Find a fraction with numerator 48 equal to $\frac{4}{5}$.

18. For a certain piece of furniture, $\frac{1}{8}$ of the included hardware is wingnuts. Of these wingnuts, $\frac{1}{20}$ are defective. What fraction of the included hardware is defective?

Example 15

$\frac{11}{8} = \frac{?}{40}$

Since $8 \cdot 5 = 40$, use $1 = \frac{5}{5}$.

$\frac{11}{8} = \frac{11}{8} \cdot \frac{5}{5} = \frac{55}{40}$.

Now work exercise 15 in the margin.

Completion Example 16

Find a fraction with denominator 35 equal to $\frac{3}{5}$.

$\frac{3}{5} = \frac{3}{5} \cdot 1 = \frac{3}{5} \cdot \_\_\_\_\_\_ = \frac{?}{35}$

Now work exercise 16 in the margin.

Completion Example 17

Find a fraction with numerator 28 equal to $\frac{2}{3}$.

$\frac{2}{3} = \frac{2}{3} \cdot 1 = \frac{2}{3} \cdot \_\_\_\_\_\_ = \frac{28}{?}$.

Now work exercise 17 in the margin.

Example 18

In a certain voting district, $\frac{3}{5}$ of the eligible voters are actually registered to vote. Of these registered voters, $\frac{2}{7}$ are independents (have no party affiliation). What fraction of the eligible voters are registered independents?

Solution

Since the independents are a fraction **of** the eligible voters, we multiply:

$$\frac{2}{7} \cdot \frac{3}{5} = \frac{2 \cdot 3}{7 \cdot 5} = \frac{6}{35}$$

Thus $\frac{6}{35}$ of the eligible voters are registered as independents.

Now work exercise 18 in the margin.

Completion Example Answers

9. a. $\frac{2}{3}\cdot\frac{5}{7}=\frac{2\cdot 5}{3\cdot 7}=\mathbf{\frac{10}{21}}$

b. $\frac{1}{4}\cdot\frac{3}{5}=\frac{1\cdot 3}{4\cdot 5}=\mathbf{\frac{3}{20}}$

c. $\frac{7}{5}\cdot 2=\frac{7}{5}\cdot\frac{2}{1}=\mathbf{\frac{7\cdot 2}{5\cdot 1}}=\mathbf{\frac{14}{5}}$

d. $\frac{1}{4}\cdot\frac{3}{5}\cdot\frac{7}{2}=\frac{1\cdot 3\cdot 7}{\mathbf{4\cdot 5\cdot 2}}=\mathbf{\frac{21}{40}}$

e. $\frac{2}{3}\cdot\frac{11}{15}\cdot\frac{1}{7}=\mathbf{\frac{2\cdot 11\cdot 1}{3\cdot 15\cdot 7}}=\mathbf{\frac{22}{315}}$

16. $\frac{3}{5}=\frac{3}{5}\cdot 1=\frac{3}{5}\cdot\frac{\mathbf{7}}{\mathbf{7}}=\frac{\mathbf{21}}{35}$

17. $\frac{2}{3}=\frac{2}{3}\cdot 1=\frac{2}{3}\cdot\frac{\mathbf{14}}{\mathbf{14}}=\frac{28}{\mathbf{42}}$

Practice Problems

Find the products.

1. $\frac{1}{4}\cdot\frac{1}{4}$ **2.** $\frac{3}{5}\cdot\frac{4}{7}$ **3.** $0\cdot\frac{5}{6}\cdot\frac{7}{8}$ **4.** $\frac{1}{2}\cdot\frac{3}{7}\cdot\frac{9}{2}$

5. Find $\frac{1}{2}$ of $\frac{1}{6}$.

Answers to Practice Problems:

1. $\frac{1}{16}$ **2.** $\frac{12}{35}$ **3.** 0 **4.** $\frac{27}{28}$ **5.** $\frac{1}{12}$

Name ______________________ Section ____________ Date __________

Exercises 3.1

What is the value, if any, of each of the following numbers or products? See Examples 5 and 6.

1. $\frac{0}{6}$

2. $\frac{0}{35}$

3. $\frac{0}{6} \cdot \frac{1}{10}$

4. $\frac{0}{5} \cdot \frac{0}{11}$

5. $\frac{13}{0}$

6. $\frac{2}{0}$

7. $\frac{3}{2} \cdot \frac{1}{0}$

8. $\frac{76}{0} \cdot \frac{21}{0}$

9. Explain, in your own words, why no denominator can be 0.

10. What is the meaning of the term **rational number**?

Write a fraction that indicates the shaded parts in each of the following diagrams. See Examples 1 through 4.

11.

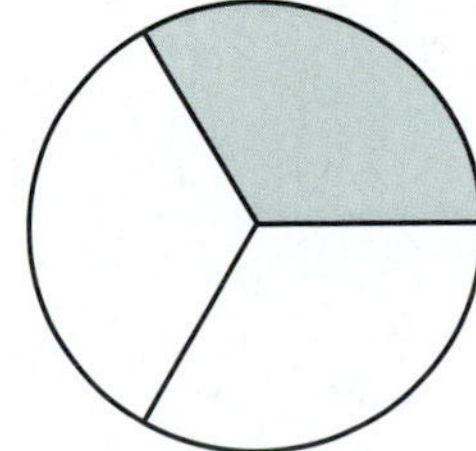

12.

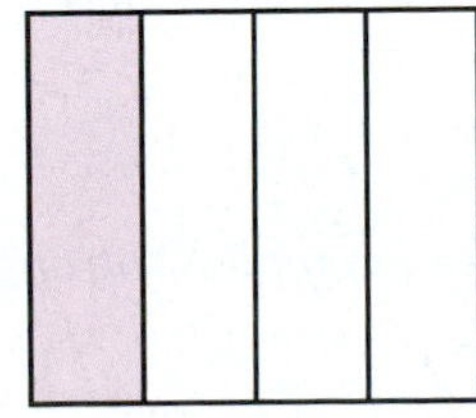

13.

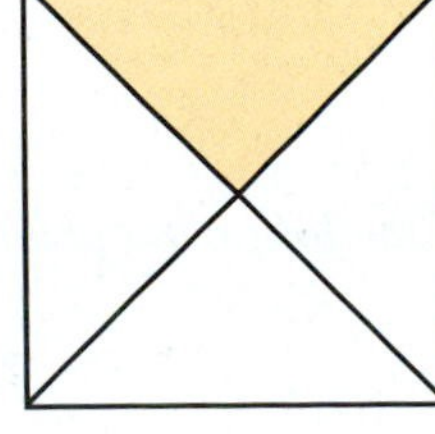

ANSWERS

1. ____________

2. ____________

3. ____________

4. ____________

5. ____________

6. ____________

7. ____________

8. ____________

9. [Respond below exercise.]

10. [Respond below exercise.]

11. ____________

12. ____________

13. ____________

14. ______

15. ______

16. ______

17. [Respond below exercise.]

18. [Respond below exercise.]

19. [Respond below exercise.]

20. [Respond below exercise.]

21. ______

22. ______

23. ______

24. ______

14. 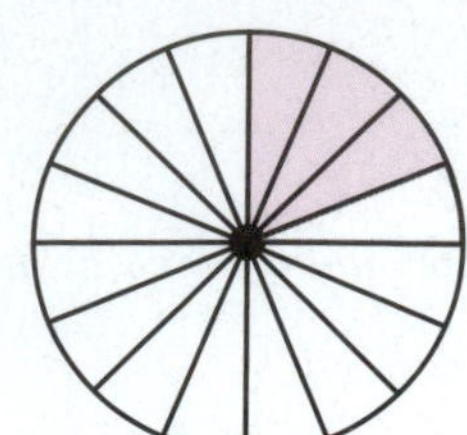

15. 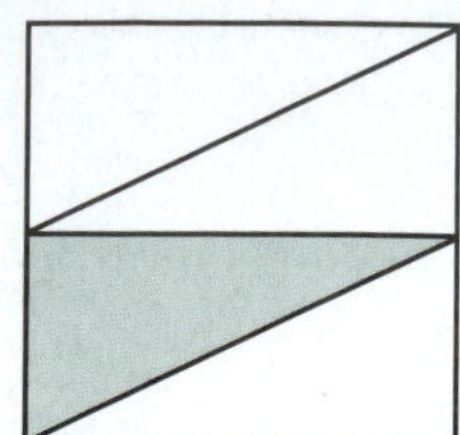

16. 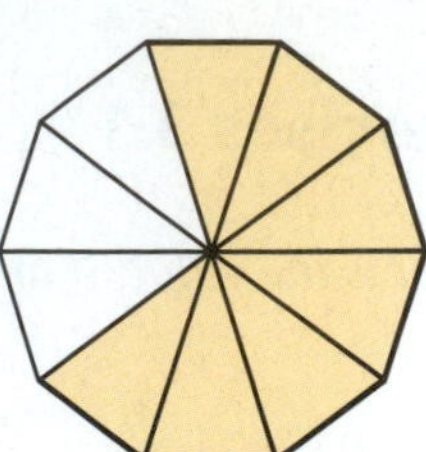

Determine each product and then illustrate each product with appropriate shading in a square. See Examples 7 and 8.

17. $\frac{1}{5}\cdot\frac{3}{4}$

18. $\frac{5}{6}\cdot\frac{5}{6}$

19. $\frac{2}{3}\cdot\frac{4}{7}$

20. $\frac{3}{4}\cdot\frac{3}{4}$

Find the following products. See Examples 7 through 9.

21. $\frac{3}{16}\cdot\frac{1}{2}$

22. $\frac{2}{5}\cdot\frac{2}{5}$

23. $\frac{3}{7}\cdot\frac{3}{7}$

24. $\frac{1}{2}\cdot\frac{3}{4}$

Name ______________________ Section ____________ Date __________

ANSWERS

25. $\frac{5}{8} \cdot \frac{3}{4}$

26. $\frac{1}{9} \cdot \frac{4}{9}$

27. $\frac{0}{3} \cdot \frac{5}{7}$

28. $\frac{0}{4} \cdot \frac{7}{6}$

29. $\frac{7}{6} \cdot \frac{5}{2}$

30. $\frac{4}{1} \cdot \frac{3}{1}$

31. $\frac{2}{1} \cdot \frac{5}{1}$

32. $\frac{14}{1} \cdot \frac{0}{2}$

33. $\frac{15}{1} \cdot \frac{3}{2}$

34. $\frac{6}{5} \cdot \frac{7}{1}$

35. $\frac{8}{5} \cdot \frac{4}{3}$

36. $\frac{5}{6} \cdot \frac{11}{3}$

37. $\frac{9}{4} \cdot \frac{11}{5}$

38. $\frac{1}{5} \cdot \frac{2}{7} \cdot \frac{3}{11}$

39. $\frac{4}{13} \cdot \frac{2}{5} \cdot \frac{6}{7}$

40. $\frac{7}{8} \cdot \frac{7}{9} \cdot \frac{7}{3}$

Find the fractional parts as indicated. See Examples 7 through 9.

41. Find $\frac{1}{5}$ of $\frac{1}{2}$.

42. Find $\frac{1}{3}$ of $\frac{1}{2}$.

43. Find $\frac{1}{4}$ of $\frac{1}{2}$.

44. Find $\frac{2}{3}$ of $\frac{2}{15}$.

45. Find $\frac{4}{7}$ of $\frac{3}{5}$.

46. Find $\frac{1}{3}$ of $\frac{2}{3}$.

25. ____________
26. ____________
27. ____________
28. ____________
29. ____________
30. ____________
31. ____________
32. ____________
33. ____________
34. ____________
35. ____________
36. ____________
37. ____________
38. ____________
39. ____________
40. ____________
41. ____________
42. ____________
43. ____________
44. ____________
45. ____________
46. ____________

47. ____________

Tell which property of multiplication is illustrated in each exercise. See Examples 10 and 11.

47. $\frac{5}{9}\cdot\frac{11}{14}=\frac{11}{14}\cdot\frac{5}{9}$

48. $\frac{3}{5}\cdot\left(\frac{7}{8}\cdot\frac{1}{2}\right)=\left(\frac{3}{5}\cdot\frac{7}{8}\right)\cdot\frac{1}{2}$

48. ____________

49. $\left(\frac{6}{11}\cdot\frac{13}{5}\right)\cdot\frac{7}{5}=\frac{6}{11}\cdot\left(\frac{13}{5}\cdot\frac{7}{5}\right)$

50. $\left(\frac{2}{3}\cdot\frac{4}{5}\right)\cdot\frac{1}{7}=\frac{1}{7}\cdot\left(\frac{2}{3}\cdot\frac{4}{5}\right)$

49. ____________

50. ____________

51. Food Budget. If you had \$20 and you spent \$7 for food and a soft drink, what fraction of your money did you spend? What fraction would you still have?

51. ____________

52. Grade Distribution. In a class of 35 students, 6 students received A's on a mathematics exam. What fraction of the class did not receive an A?

52. ____________

53. Cooking. A recipe calls for $\frac{3}{4}$ cup of flour. How much flour should be used if only half the recipe is to be made?

53. ____________

54. Candy. One of Maria's birthday presents was a box of candy. Half of the candy was chocolate covered and one-fourth of the chocolate covered candy had cherries inside. What fraction of the candy was chocolate covered cherries?

54. ____________

Name ______________________ Section ______________ Date ____________

Find the missing numerators and denominators in changing each fraction to higher terms. See Examples 12 through 17.

55. $\frac{5}{8}=\frac{5}{8}\cdot\frac{?}{?}=\frac{?}{24}$

56. $\frac{1}{16}=\frac{1}{16}\cdot\frac{?}{?}=\frac{?}{64}$

57. $\frac{2}{5}=\frac{2}{5}\cdot\frac{?}{?}=\frac{?}{25}$

58. $\frac{6}{7}=\frac{6}{7}\cdot\frac{?}{?}=\frac{?}{49}$

59. $\frac{1}{9}=\frac{1}{9}\cdot\frac{?}{?}=\frac{5}{?}$

60. $\frac{3}{4}=\frac{3}{4}\cdot\frac{?}{?}=\frac{15}{?}$

61. $\frac{5}{8}=\frac{5}{8}\cdot\frac{?}{?}=\frac{10}{?}$

62. $\frac{6}{5}=\frac{6}{5}\cdot\frac{?}{?}=\frac{?}{45}$

63. $\frac{14}{3}=\frac{14}{3}\cdot\frac{?}{?}=\frac{?}{9}$

64. $\frac{5}{8}=\frac{5}{8}\cdot\frac{?}{?}=\frac{?}{96}$

65. $\frac{9}{16}=\frac{9}{16}\cdot\frac{?}{?}=\frac{?}{96}$

66. $\frac{7}{2}=\frac{7}{2}\cdot\frac{?}{?}=\frac{?}{20}$

67. $\frac{10}{11}=\frac{10}{11}\cdot\frac{?}{?}=\frac{?}{44}$

68. $\frac{3}{16}=\frac{3}{16}\cdot\frac{?}{?}=\frac{?}{80}$

69. $\frac{11}{12}=\frac{11}{12}\cdot\frac{?}{?}=\frac{?}{48}$

70. $\frac{5}{21}=\frac{5}{21}\cdot\frac{?}{?}=\frac{?}{42}$

ANSWERS

55. ____________

56. ____________

57. ____________

58. ____________

59. ____________

60. ____________

61. ____________

62. ____________

63. ____________

64. ____________

65. ____________

66. ____________

67. ____________

68. ____________

69. ____________

70. ____________

71. ______________

72. ______________

73. ______________

74. ______________

75. ______________

76. ______________

77. ______________

78. ______________

79. ______________

80. ______________

71. $\frac{2}{3} = \frac{2}{3} \cdot \frac{?}{?} = \frac{?}{48}$

72. $\frac{1}{13} = \frac{1}{13} \cdot \frac{?}{?} = \frac{?}{39}$

73. $\frac{9}{10} = \frac{9}{10} \cdot \frac{?}{?} = \frac{?}{100}$

74. $\frac{3}{10} = \frac{3}{10} \cdot \frac{?}{?} = \frac{?}{100}$

75. $\frac{7}{10} = \frac{7}{10} \cdot \frac{?}{?} = \frac{?}{70}$

76. $\frac{9}{10} = \frac{9}{10} \cdot \frac{?}{?} = \frac{?}{90}$

77. $\frac{5}{12} = \frac{5}{12} \cdot \frac{?}{?} = \frac{?}{108}$

78. $\frac{8}{5} = \frac{8}{5} \cdot \frac{?}{?} = \frac{?}{30}$

79. $\frac{7}{6} = \frac{7}{6} \cdot \frac{?}{?} = \frac{?}{36}$

80. $\frac{10}{3} = \frac{10}{3} \cdot \frac{?}{?} = \frac{?}{33}$

3.2 Multiplication and Reducing with Fractions

Reducing Fractions

In Section 3.1 we used the multiplicative identity, in the form of $\frac{k}{k}$, to change a fraction to higher terms. We can use this same idea to **change a fraction to lower terms** (to **reduce a fraction**). **A fraction is reduced to lowest terms if the numerator and denominator have no common factors other than 1.**

> **To Reduce a Fraction to Lowest Terms:**
>
> **1.** Factor the numerator and denominator into prime factorizations.
>
> **2.** Use the fact that $\frac{k}{k} = 1$ and **divide out** common factors.

Note that reduced fractions might be improper fractions. Changing a fraction to a mixed number is not the same as reducing it. Mixed numbers will be discussed in Chapter 4. Example 1 illustrates how to reduce fractions to lowest terms.

Objectives

① Learn how to reduce a fraction to lowest terms.

② Recognize when a fraction is in lowest terms.

③ Be able to multiply fractions and reduce at the same time.

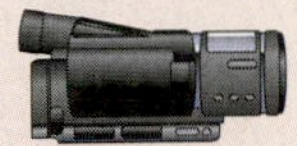

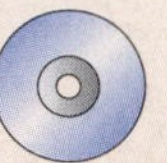

Objective ①

Example 1

a. $\frac{14}{21} = \frac{2 \cdot 7}{3 \cdot 7} = \frac{2}{3} \cdot \frac{7}{7} = \frac{2}{3} \cdot 1 = \frac{2}{3}$

b. $\frac{15}{20} = \frac{3 \cdot 5}{2 \cdot 2 \cdot 5} = \frac{3}{4} \cdot \frac{5}{5} = \frac{3}{4} \cdot 1 = \frac{3}{4}$

Now work exercise 1 in the margin.

1. Reduce the fractions to lowest terms.

a. $\frac{8}{36}$

b. $\frac{18}{33}$

Example 2

$\frac{44}{20}$

We can just divide out common factors (prime or not) with the understanding that $\frac{k}{k} = 1$.

a. $\frac{44}{20} = \frac{\cancel{4} \cdot 11}{\cancel{4} \cdot 5} = \frac{11}{5}$

b. Or, using prime factors,

$$\frac{44}{20} = \frac{\cancel{2} \cdot \cancel{2} \cdot 11}{\cancel{2} \cdot \cancel{2} \cdot 5} = \frac{11}{5}.$$

Now work exercise 2 in the margin.

2. a. Use common factors to reduce $\frac{40}{35}$ to lowest terms.

b. Use prime factors to reduce $\frac{40}{35}$ to lowest terms.

Example 3

$\frac{8}{72}$

Remember that 1 is a factor of any whole number. So, if all of the factors in the numerator or denominator are divided out, 1 must be used as a factor.

Continued on next page...

Reduce each fraction to lowest terms.

3. $\frac{12}{25}$

4. $\frac{66}{44}$

5. $\frac{10}{60}$

6. $\frac{78}{104}$

a. Using prime factors,

$$\frac{8}{72}=\frac{\not{2}\cdot\not{2}\cdot\not{2}\cdot 1}{\not{2}\cdot\not{2}\cdot\not{2}\cdot 3\cdot 3}=\frac{1}{9}$$ 1 is used as a factor in the numerator.

b. Or, if you see that 8 is a common factor, divide it out. But remember that 1 is a factor.

$$\frac{8}{72}=\frac{\not{8}\cdot 1}{\not{8}\cdot 9}=\frac{1}{9}$$

Now work exercise 3 in the margin.

Example 4

$\frac{36}{27}$

a. Using prime factors,

$$\frac{36}{27}=\frac{2\cdot 2\cdot\not{3}\cdot\not{3}}{3\cdot\not{3}\cdot\not{3}}=\frac{4}{3}$$

b. Or, using 3 as a factor, we get an incomplete answer.

$$\frac{36}{27}=\frac{\not{3}\cdot 12}{\not{3}\cdot 9}=\frac{12}{9}$$ not in lowest terms

Objective 2

By using prime factors, you can be certain that the fraction is reduced to lowest terms. You may use larger numbers, but be sure you have the largest common factor.

$$\frac{36}{27}=\frac{4\cdot\not{9}}{\not{9}\cdot 3}=\frac{4}{3}$$

Now work exercise 4 in the margin.

Completion Example 5

Reduce $\frac{12}{18}$ to lowest terms.

a. $\frac{12}{18}=\frac{2\cdot 2\cdot 3}{2\cdot 3\cdot 3}=$ ________

b. Or, $\frac{12}{18}=\frac{2\cdot ?}{3\cdot ?}=$ ________

Now work exercise 5 in the margin.

Completion Example 6

Reduce $\frac{52}{65}$ to lowest terms.

Finding a common factor could be difficult here. Prime factoring helps.

$\frac{52}{65}=\frac{2\cdot 2\cdot ?}{5\cdot ?}=$ ________

Now work exercise 6 in the margin.

Multiplying and Reducing Fractions at the Same Time

Objective 3

Now we can multiply fractions and reduce all in one step by using prime factors (or other common factors). If you have any difficulty understanding how to multiply and reduce, use prime factors. This will help you gain confidence that your answers are correct.

Examples 7 – 10 illustrate how to multiply and reduce at the same time by factoring the numerators and denominators. Note that **if all of the factors in the numerator or denominator divide out, then 1 must be used as a factor**. (See Completion Example 10).

Example 7

$$\frac{15}{28}\cdot\frac{7}{9}$$

Method 1 (Poor Method)

$$\frac{15}{28}\cdot\frac{7}{9}=\frac{15\cdot 7}{28\cdot 9}=\frac{105}{252}$$

Now factor and reduce.

$$\frac{105}{252}=\frac{5\cdot 21}{4\cdot 63}=\frac{5\cdot\cancel{3}\cdot\cancel{7}}{2\cdot 2\cdot 3\cdot\cancel{3}\cdot\cancel{7}}=\frac{5}{12}$$

By multiplying first you have caused extra work for yourself because you have to factor.

Method 2 (Preferred Method)

$$\frac{15}{28}\cdot\frac{7}{9}=\frac{15\cdot 7}{28\cdot 9}=\frac{\cancel{3}\cdot 5\cdot\cancel{7}}{2\cdot 2\cdot\cancel{7}\cdot 3\cdot\cancel{3}}=\frac{5}{2\cdot 2\cdot 3}=\frac{5}{12}$$

Only the preferred method is used to find the products in Examples 8 and 9.

Now work exercise 7 in the margin.

Example 8

$$\frac{9}{10}\cdot\frac{25}{32}\cdot\frac{44}{33}$$

$$\frac{9}{10}\cdot\frac{25}{32}\cdot\frac{44}{33}=\frac{9\cdot 25\cdot 44}{10\cdot 32\cdot 33}$$

$$=\frac{\cancel{3}\cdot 3\cdot\cancel{5}\cdot 5\cdot\cancel{2}\cdot\cancel{2}\cdot\cancel{11}}{2\cdot\cancel{5}\cdot 2\cdot 2\cdot 2\cdot\cancel{2}\cdot\cancel{2}\cdot\cancel{3}\cdot\cancel{11}}$$

$$=\frac{3\cdot 5}{2\cdot 2\cdot 2\cdot 2}$$

$$=\frac{15}{16}$$

Now work exercise 8 in the margin.

Find each product in lowest terms by using prime factors.

7. $\frac{10}{9}\cdot\frac{3}{5}$

8. $\frac{5}{6}\cdot\frac{8}{7}\cdot\frac{14}{10}$

9. Find the product in lowest terms by using prime factors.

$\frac{2}{5} \cdot \frac{8}{5} \cdot \frac{5}{10}$

10. Find $\frac{15}{6} \cdot \frac{4}{19} \cdot \frac{38}{10}$ in lowest terms by using prime factors.

Find the product by dividing the numerators and denominators by common factors.

11. $\frac{10}{9} \cdot \frac{3}{5}$

12. $\frac{14}{15} \cdot \frac{25}{21} \cdot \frac{3}{10}$

Example 9

$\frac{36}{49} \cdot \frac{14}{75} \cdot \frac{15}{18}$

$$\frac{36}{49} \cdot \frac{14}{75} \cdot \frac{15}{18} = \frac{36 \cdot 14 \cdot 15}{49 \cdot 75 \cdot 18}$$

$$= \frac{\cancel{2} \cdot 2 \cdot \cancel{3} \cdot \cancel{3} \cdot 2 \cdot \cancel{7} \cdot \cancel{3} \cdot \cancel{5}}{\cancel{7} \cdot 7 \cdot \cancel{3} \cdot 5 \cdot \cancel{5} \cdot \cancel{2} \cdot \cancel{3} \cdot \cancel{3}}$$

$$= \frac{2 \cdot 2}{7 \cdot 5} = \frac{4}{35}$$

Now work exercise 9 in the margin.

Completion Example 10

$\frac{55}{26} \cdot \frac{8}{44} \cdot \frac{91}{35}$

$$\frac{55}{26} \cdot \frac{8}{44} \cdot \frac{91}{35} = \frac{55 \cdot 8 \cdot 91}{26 \cdot 44 \cdot 35}$$

$= \frac{\_\_\_\_\_\_}{\_\_\_\_\_\_}$

$= \frac{\_\_\_\_}{\_\_\_\_}$

$= \_\_\_\_$

Now work exercise 10 in the margin.

Another method frequently used to multiply and reduce at the same time is to divide numerators and denominators by common factors whether they are prime or not. If these factors are easily determined, then this method is probably faster. But common factors are sometimes missed with this method, whereas they are not missed with the prime factorization method. In either case, be careful and organized.

Examples 7 – 10 are shown again as Examples 11 – 14, this time using the division method.

Example 11

$$\frac{\overset{5}{\cancel{15}}}{\underset{4}{\cancel{28}}} \cdot \frac{\overset{1}{\cancel{7}}}{\underset{3}{\cancel{9}}} = \frac{5}{12}$$

3 divides both 15 and 9.
7 divides both 7 and 28.

Now work exercise 11 in the margin.

Example 12

$$\frac{\overset{3}{\cancel{9}}}{\underset{2}{\cancel{10}}} \cdot \frac{\overset{5}{\cancel{25}}}{\underset{8}{\cancel{32}}} \cdot \frac{\overset{\overset{1}{\cancel{4}}}{\cancel{44}}}{\underset{\underset{1}{\cancel{3}}}{\cancel{33}}} = \frac{15}{16}$$

11 divides both 44 and 33.
5 divides both 25 and 10.
4 divides both 4 and 32.
3 divides both 3 and 9.

Now work exercise 12 in the margin.

Example 13

$$\frac{\overset{2}{\cancel{36}}}{\underset{7}{\cancel{49}}}\cdot\frac{\overset{2}{\cancel{14}}}{\underset{5}{\cancel{75}}}\cdot\frac{\overset{1}{\cancel{15}}}{\underset{1}{\cancel{18}}}=\frac{4}{35}$$

18 divides both 18 and 36.
7 divides both 14 and 49.
15 divides both 15 and 75.

Another approach is to use factors that are not all prime:

$$\frac{36}{49}\cdot\frac{14}{75}\cdot\frac{15}{18}=\frac{4\cdot\cancel{9}\cdot\cancel{2}\cdot\cancel{7}\cdot\cancel{15}}{\cancel{7}\cdot7\cdot5\cdot\cancel{15}\cdot\cancel{2}\cdot\cancel{9}}=\frac{4}{35}$$

Now work exercise 13 in the margin.

Example 14

$$\frac{\overset{1}{\cancel{5}}\;\cancel{55}}{\cancel{26}\;\underset{1}{\cancel{2}}}\cdot\frac{\overset{1}{\cancel{4}}\;\cancel{8}}{\cancel{44}\;\underset{1}{\cancel{4}}}\cdot\frac{\overset{1}{\cancel{7}}\;\cancel{91}}{\cancel{35}\;\underset{1}{\cancel{7}}}=\frac{1}{1}=1$$

11 divides both 55 and 44.
13 divides both 26 and 91.
5 divides both 5 and 35.
7 divides both 7 and 7.
2 divides both 2 and 8.
4 divides both 4 and 4.

Now work exercise 14 in the margin.

Example 15

A study showed that $\frac{5}{8}$ of the members of a public service organization were in favor of a new set of bylaws. If the organization had a membership of 200 people, how many were in favor of the changes in the bylaws?

Solution

We want to find $\frac{5}{8}$ of 200, so we multiply:

$$\frac{5}{8}\cdot200=\frac{5}{8}\cdot\frac{200}{1}=\frac{5\cdot2\cdot10\cdot10}{2\cdot2\cdot2}=\frac{5\cdot\cancel{2}\cdot\cancel{2}\cdot5\cdot\cancel{2}\cdot5}{\cancel{2}\cdot\cancel{2}\cdot\cancel{2}}$$

$$=\frac{5\cdot5\cdot5}{1}=125$$

Thus there are 125 members in favor of the bylaw changes.

Now work exercise 15 in the margin.

13. Find the product by dividing the numerators and denominators by common factors.

$$\frac{16}{27}\cdot\frac{45}{6}\cdot\frac{2}{5}$$

14. Find $\frac{17}{100}\cdot\frac{27}{34}\cdot\frac{25}{9}$ by dividing the numerators and denominators by common factors.

15. A delivery service found that $\frac{3}{4}$ of its drivers wore their seatbelts. If this company employed 500 drivers, how many drivers wore their seatbelts?

Completion Example Answers

5. **a.** $\frac{12}{18}=\frac{\not{2}\cdot 2\cdot \not{3}}{\not{2}\cdot 3\cdot \not{3}}=\mathbf{\frac{2}{3}}$

b. Or, $\frac{12}{18}=\frac{2\cdot \not{6}}{3\cdot \not{6}}=\mathbf{\frac{2}{3}}$

6. $\frac{52}{65}=\frac{2\cdot 2\cdot \not{13}}{5\cdot \not{13}}=\mathbf{\frac{4}{5}}$

10. $\frac{55}{26}\cdot\frac{8}{44}\cdot\frac{91}{35}=\frac{55\cdot 8\cdot 91}{26\cdot 44\cdot 35}=\frac{\not{5}\cdot\not{11}\cdot\not{2}\cdot\not{2}\cdot\not{2}\cdot\not{7}\cdot\not{13}}{\not{2}\cdot\not{13}\cdot\not{2}\cdot\not{2}\cdot\not{11}\cdot\not{7}\cdot\not{5}}=\mathbf{\frac{1}{1}}=\mathbf{1}$

Practice Problems

Reduce to lowest terms.

1. $\frac{25}{55}$

2. $\frac{34}{51}$

Multiply.

3. $\frac{3}{5}\cdot\frac{7}{8}$

4. $\frac{2}{21}\cdot\frac{15}{22}$

5. $\frac{17}{100}\cdot\frac{27}{34}\cdot\frac{25}{9}\cdot 6$ [Hint: Write 6 as $\frac{6}{1}$ and factor before multiplying.]

Answers to Practice Problems:

1. $\frac{5}{11}$ **2.** $\frac{2}{3}$ **3.** $\frac{21}{40}$ **4.** $\frac{5}{77}$ **5.** $\frac{9}{4}$

Name ______________ Section ______________ Date ______________

Exercises 3.2

Reduce each fraction to lowest terms. If it is already in lowest terms, simply rewrite the fraction. See Examples 1 through 6.

1. $\frac{3}{9}$ **2.** $\frac{16}{24}$ **3.** $\frac{9}{12}$ **4.** $\frac{6}{20}$

5. $\frac{16}{40}$ **6.** $\frac{24}{30}$ **7.** $\frac{14}{36}$ **8.** $\frac{5}{11}$

9. $\frac{0}{25}$ **10.** $\frac{75}{100}$ **11.** $\frac{22}{55}$ **12.** $\frac{60}{75}$

13. $\frac{30}{36}$ **14.** $\frac{7}{28}$ **15.** $\frac{26}{39}$ **16.** $\frac{27}{56}$

17. $\frac{34}{51}$ **18.** $\frac{36}{48}$ **19.** $\frac{24}{100}$ **20.** $\frac{16}{32}$

21. $\frac{30}{45}$ **22.** $\frac{28}{42}$ **23.** $\frac{12}{35}$ **24.** $\frac{66}{84}$

ANSWERS

1. ______
2. ______
3. ______
4. ______
5. ______
6. ______
7. ______
8. ______
9. ______
10. ______
11. ______
12. ______
13. ______
14. ______
15. ______
16. ______
17. ______
18. ______
19. ______
20. ______
21. ______
22. ______
23. ______
24. ______

25. ______

26. ______

27. ______

28. ______

29. ______

30. ______

31. ______

32. ______

33. ______

34. ______

35. ______

36. ______

37. ______

38. ______

39. ______

40. ______

41. ______

42. ______

43. ______

44. ______

45. ______

46. ______

47. ______

48. ______

49. ______

25. $\frac{14}{63}$ **26.** $\frac{30}{70}$ **27.** $\frac{25}{76}$ **28.** $\frac{70}{84}$

29. $\frac{50}{100}$ **30.** $\frac{48}{12}$ **31.** $\frac{27}{72}$ **32.** $\frac{18}{40}$

33. $\frac{144}{156}$ **34.** $\frac{150}{135}$ **35.** $\frac{121}{165}$ **36.** $\frac{140}{112}$

37. $\frac{96}{108}$ **38.** $\frac{72}{36}$ **39.** $\frac{84}{42}$ **40.** $\frac{51}{85}$

*Find each product in lowest terms. See Example 7. (**Hint**: Factor before multiplying.)*

41. $\frac{2}{3} \cdot \frac{4}{3}$ **42.** $\frac{1}{5} \cdot \frac{4}{7}$ **43.** $\frac{3}{7} \cdot \frac{5}{3}$

44. $\frac{2}{11} \cdot \frac{3}{2}$ **45.** $\frac{5}{16} \cdot \frac{16}{15}$ **46.** $\frac{7}{8} \cdot \frac{9}{14}$

47. $\frac{10}{18} \cdot \frac{9}{5}$ **48.** $\frac{11}{22} \cdot \frac{6}{8}$ **49.** $\frac{15}{27} \cdot \frac{9}{30}$

Name ______________________ Section ______________ Date ___________

ANSWERS

50. $\frac{35}{20} \cdot \frac{36}{14}$

51. $\frac{25}{9} \cdot \frac{3}{100}$

52. $\frac{30}{42} \cdot \frac{7}{100}$

53. $\frac{18}{42} \cdot \frac{14}{75}$

54. $\frac{42}{70} \cdot \frac{20}{12}$

55. $8 \cdot \frac{5}{12}$

56. $9 \cdot \frac{7}{24}$

57. $\frac{6}{85} \cdot \frac{34}{9}$

58. $\frac{13}{91} \cdot \frac{34}{65}$

*Find each product. See Examples 7 through 10. (**Hint**: Factor before multiplying.)*

59. $\frac{23}{36} \cdot \frac{20}{46}$

60. $\frac{7}{8} \cdot \frac{4}{21}$

61. $\frac{5}{15} \cdot \frac{18}{24}$

62. $\frac{20}{32} \cdot \frac{9}{13} \cdot \frac{26}{7}$

63. $\frac{69}{15} \cdot \frac{30}{8} \cdot \frac{14}{46}$

64. $\frac{42}{52} \cdot \frac{27}{22} \cdot \frac{33}{9}$

65. $\frac{3}{4} \cdot 18 \cdot \frac{7}{2} \cdot \frac{22}{54}$

66. $\frac{9}{10} \cdot \frac{35}{40} \cdot \frac{65}{15}$

67. $\frac{66}{84} \cdot \frac{12}{5} \cdot \frac{28}{33}$

68. $\frac{24}{100} \cdot \frac{36}{48} \cdot \frac{15}{9}$

69. $\frac{17}{10} \cdot \frac{5}{42} \cdot \frac{18}{51} \cdot 4$

70. $\frac{75}{8} \cdot \frac{16}{36} \cdot 9 \cdot \frac{7}{25}$

50. ____
51. ____
52. ____
53. ____
54. ____
55. ____
56. ____
57. ____
58. ____
59. ____
60. ____
61. ____
62. ____
63. ____
64. ____
65. ____
66. ____
67. ____
68. ____
69. ____
70. ____

71. ______________

71. **Falling Object.** Suppose that a ball is dropped from a height of 20 feet. If the ball bounces back to five-eighths the height from which it was dropped, how high will it bounce on its third bounce?

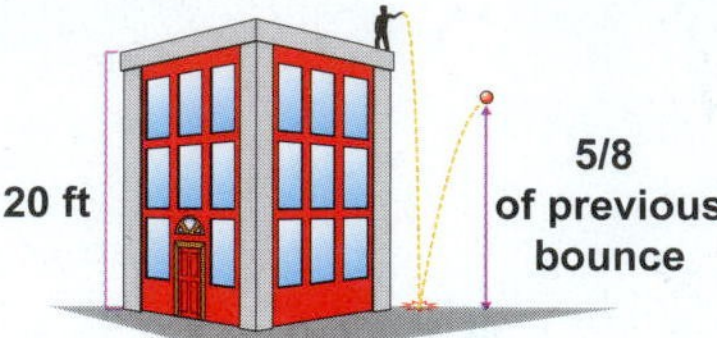

72. ______________

72. **Dominant Hand.** A study showed that one-fifth of the students in an elementary school were left-handed. If the school had an enrollment of 550 students, how many were left-handed?

73. ______________

73. **Construction.** For her sprinkler system, Patricia needs to cut a 20-foot piece of plastic pipe into four pieces. If each new piece of pipe is to be half the length of the previous piece, how long will each of the four pieces be?

74. ______________

74. **Bicycling.** If you go on a 100-mile bicycle trip in the mountains and one-fourth of the trip is downhill, how many miles are uphill?

75. **Division of a Whole.** A pizza pie is to be cut into fourths. Each of these fourths is to be cut into thirds. What fraction of the pie is each of the final pieces?

75. ______________

76. **Baseball.** Major league baseball teams play 162 games each season. If a team has played $\frac{5}{9}$ of its games by the All-Star break (around mid-season), how many games has it played by that time?

76. ______________

77. **Planetary Orbits.** Venus orbits the sun in $\frac{45}{73}$ the time that the earth takes to orbit the sun. Assuming that one "earth-year" is 365 days along, how long is a "Venus-year" in terms of "earth-days"?

77. ______________

Name ______________ Section ______________ Date ______________

ANSWERS

78. [Respond below exercise.]

Check Your Number Sense

78. Consider the following statement:

"If a fraction is less than 1 and greater than 0, the product of this fraction with another number (fraction or whole number) must be less than this other number." Do you agree with this statement? To help you understand this concept, answer the following questions involving the fraction $\frac{3}{4}$.

Is $\frac{3}{4}$ of 12 less than 12?

Is $\frac{3}{4}$ of $\frac{2}{3}$ less than $\frac{2}{3}$?

Is $\frac{3}{4}$ of $\frac{1}{2}$ less than $\frac{1}{2}$?

Is $\frac{3}{4}$ of $\frac{4}{5}$ less than $\frac{4}{5}$?

Answer these same questions using $\frac{2}{3}$ and $\frac{1}{2}$ in place of $\frac{3}{4}$.

Review Problems (from Section 2.5)

Find the prime factorization for each of the following numbers. Use the tests for divisibility by 2, 3, 5, 6, 9, and 10 whenever they help in finding the beginning factors.

1. 45 **2.** 78 **3.** 130

4. 460 **5.** 1,000,000

Find all of the factors (or divisors) of each number.

6. 30 **7.** 48 **8.** 63

REVIEW

1. ______
2. ______
3. ______
4. ______
5. ______
6. ______
7. ______
8. ______

3.3 Division with Fractions

Reciprocals

If the product of two fractions is 1, then the fractions are called reciprocals of each other. (Remember that whole numbers can also be written in fraction form.) For example,

$\frac{5}{8}$ and $\frac{8}{5}$ are reciprocals because $\frac{5}{8}\cdot\frac{8}{5}=\frac{40}{40}=1.$

Objectives

1. Recognize and find the reciprocal of a number.
2. Know that division is accomplished by multiplication by the reciprocal of the divisor.

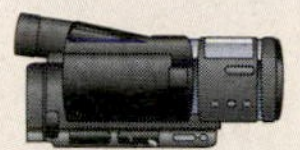
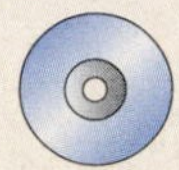

Reciprocals

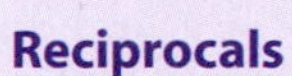

The **reciprocal** of $\frac{a}{b}$ is $\frac{b}{a}$ $(a\neq 0 \text{ and } b\neq 0)$. The product of a nonzero number and its reciprocal is always 1.

$$\frac{a}{b}\cdot\frac{b}{a}=1$$

Note: $0=\frac{0}{1}$, but $\frac{1}{0}$ is undefined. That is, **the number 0 has no reciprocal**.

Find the reciprocal of each number:

1. $\frac{7}{8}$

2. $\frac{1}{10}$

3. 16

Example 1

The reciprocal of $\frac{2}{3}$ is $\frac{3}{2}$.

$$\frac{2}{3}\cdot\frac{3}{2}=\frac{2\cdot 3}{3\cdot 2}=1$$

Now work exercise 1 in the margin.

Example 2

The reciprocal of $\frac{5}{8}$ is $\frac{8}{5}$.

$$\frac{5}{8}\cdot\frac{8}{5}=\frac{5\cdot 8}{8\cdot 5}=1$$

Now work exercise 2 in the margin.

Example 3

The reciprocal of 10 is $\frac{1}{10}$.

$$10\cdot\frac{1}{10}=\frac{10}{1}\cdot\frac{1}{10}=\frac{10\cdot 1}{1\cdot 10}=1$$

Now work exercise 3 in the margin.

Division with Fractions

To develop an understanding of division, we first write a division problem in fraction form with fractions in the numerator and denominator. For example,

$$\frac{2}{3} \div \frac{5}{11} = \frac{\frac{2}{3}}{\frac{5}{11}}$$

Now we multiply the numerator and the denominator (both fractions) by the reciprocal of the denominator. This is the same as multiplying by 1 and does not change the value of the expression. The reciprocal of $\frac{5}{11}$ is $\frac{11}{5}$, so we multiply both the numerator and the denominator by $\frac{11}{5}$.

$$\frac{2}{3} \div \frac{5}{11} = \frac{\frac{2}{3}}{\frac{5}{11}} \cdot \frac{\frac{11}{5}}{\frac{11}{5}} = \frac{\frac{2}{3} \cdot \frac{11}{5}}{\frac{5}{11} \cdot \frac{11}{5}} = \frac{\frac{2}{3} \cdot \frac{11}{5}}{1} = \frac{2}{3} \cdot \frac{11}{5}$$

Now that we have a denominator of 1, a division problem has been changed into a multiplication problem:

To divide, multiply by the reciprocal.

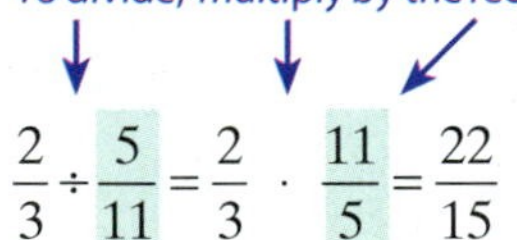

$$\frac{2}{3} \div \frac{5}{11} = \frac{2}{3} \cdot \frac{11}{5} = \frac{22}{15}$$

Objective 2

To Divide Fractions:

To divide by any nonzero number, multiply by its reciprocal. In general,

$$\frac{a}{b} \div \frac{c}{d} = \frac{a}{b} \cdot \frac{d}{c} \quad \text{where } b, c, d \neq 0.$$

As illustrated in the following examples, once a division problem is in the form of a product, we can reduce by factoring.

4. Find the quotient: $\frac{3}{4} \div \frac{1}{4}$.

Example 4

$\frac{5}{6} \div \frac{1}{6}$ How many $\frac{1}{6}$'s are there in $\frac{5}{6}$?

$\frac{5}{6} \div \frac{1}{6} = \frac{5}{\cancel{6}} \cdot \frac{\cancel{6}}{1} = 5$ $\qquad$ $\frac{6}{1}$ is the reciprocal of $\frac{1}{6}$.

Thus there are five $\frac{1}{6}$'s in $\frac{5}{6}$.

Now work exercise 4 in the margin.

Example 5

$\frac{2}{3} \div \frac{3}{4}$ The divisor is $\frac{3}{4}$. Its reciprocal is $\frac{4}{3}$.

$$\frac{2}{3} \div \frac{3}{4} = \frac{2}{3} \cdot \frac{4}{3} = \frac{8}{9}$$

Now work exercise 5 in the margin.

Example 6

$\frac{7}{16} \div 7$ The divisor is 7. Its reciprocal is $\frac{1}{7}$.

$$\frac{7}{16} \div 7 = \frac{7}{16} \cdot \frac{1}{7} = \frac{\cancel{7} \cdot 1}{16 \cdot \cancel{7}} = \frac{1}{16}$$

Now work exercise 6 in the margin.

Example 7

$\frac{16}{27} \div \frac{4}{9}$ The divisor is $\frac{4}{9}$. Its reciprocal is $\frac{9}{4}$.

$$\frac{16}{27} \div \frac{4}{9} = \frac{16}{27} \cdot \frac{9}{4} = \frac{4 \cdot \cancel{4} \cdot \cancel{9}}{\cancel{9} \cdot 3 \cdot \cancel{4}} = \frac{4}{3}$$

Now work exercise 7 in the margin.

Example 8

$\frac{9}{4} \div \frac{9}{2}$ The divisor is $\frac{9}{2}$. Its reciprocal is $\frac{2}{9}$.

$$\frac{9}{4} \div \frac{9}{2} = \frac{9}{4} \cdot \frac{2}{9} = \frac{\cancel{9} \cdot \cancel{2} \cdot 1}{\cancel{2} \cdot 2 \cdot \cancel{9}} = \frac{1}{2}$$

Now work exercise 8 in the margin.

Completion Example 9

$\frac{13}{4} \div \frac{39}{5}$

$$\frac{13}{4} \div \frac{39}{5} = \frac{13}{4} \cdot \underline{\hspace{2cm}} = \frac{13 \cdot \underline{\hspace{1cm}}}{4 \cdot \underline{\hspace{1cm}}} = \underline{\hspace{1.5cm}}$$

Now work exercise 9 in the margin.

Completion Example 10

$\frac{4}{9} \div \frac{4}{9}$

$$\frac{4}{9} \div \frac{4}{9} = \frac{4}{9} \cdot \underline{\hspace{2cm}} = \frac{4 \cdot \underline{\hspace{0.5cm}}}{9 \cdot \underline{\hspace{0.5cm}}} = \underline{\hspace{1.5cm}}$$

Now work exercise 10 in the margin.

Divide as indicated and reduce to lowest terms.

5. $\frac{3}{5} \div \frac{2}{7}$

6. $\frac{1}{2} \div 2$

7. $\frac{25}{48} \div \frac{5}{6}$

8. $\frac{9}{10} \div \frac{3}{4}$

9. $\frac{7}{12} \div \frac{9}{3}$

10. $\frac{23}{33} \div \frac{23}{33}$

11. The result of multiplying two numbers is $\frac{5}{24}$. If one of the numbers is $\frac{7}{8}$, what is the other number?

Example 11

The result of multiplying two numbers is $\frac{7}{16}$. If one of the numbers is $\frac{3}{4}$, what is the other number? (**Hint:** Think in terms of whole numbers to convince yourself that this is a division problem. If the product of two numbers is 24 and one of the numbers is 6, what is the other number? You would divide 24 by 6.)

$$\begin{array}{r} 4 \\ 6\overline{)24} \\ \underline{24} \\ 0 \end{array}$$

The other number is 4.

Solution

$\frac{7}{16}$ is the result of multiplying two numbers. Divide $\frac{7}{16}$ by $\frac{3}{4}$ to find the second number.

$$\frac{7}{16} \div \frac{3}{4} = \frac{7}{16} \cdot \frac{4}{3} = \frac{7 \cdot \cancel{4}}{\cancel{4} \cdot 4 \cdot 3} = \frac{7}{12}$$

The other number is $\frac{7}{12}$.

Check by multiplying: $\frac{3}{4} \cdot \frac{7}{12} = \frac{\cancel{3} \cdot 7}{4 \cdot 4 \cdot \cancel{3}} = \frac{7}{16}$

Now work exercise 11 in the margin.

12. If the product of $\frac{7}{5}$ and another number is $\frac{4}{9}$, what is the other number?

Example 12

If the product of $\frac{3}{2}$ and another number is $\frac{5}{18}$, what is the other number?

Solution

As in Example 11, we know the product of two numbers. So we divide the product by the given number to find the other number.

$$\frac{5}{18} \div \frac{3}{2} = \frac{5}{18} \cdot \frac{2}{3} = \frac{5 \cdot \cancel{2}}{\cancel{2} \cdot 9 \cdot 3} = \frac{5}{27}$$

$\frac{5}{27}$ is the other number.

Check by multiplying: $\frac{3}{2} \cdot \frac{5}{27} = \frac{\cancel{3} \cdot 5}{2 \cdot \cancel{3} \cdot 9} = \frac{5}{18}$

Now work exercise 12 in the margin.

Completion Example Answers

9. $\frac{13}{4} \div \frac{39}{5} = \frac{13}{4} \cdot \frac{\mathbf{5}}{\mathbf{39}} = \frac{\cancel{13} \cdot \mathbf{5}}{4 \cdot \mathbf{3} \cdot \cancel{\mathbf{13}}} = \frac{\mathbf{5}}{\mathbf{12}}$

10. $\frac{4}{9} \div \frac{4}{9} = \frac{4}{9} \cdot \frac{\mathbf{9}}{\mathbf{4}} = \frac{4 \cdot \mathbf{9}}{9 \cdot \mathbf{4}} = \mathbf{1}$

Practice Problems

Find each of the following quotients. Reduce whenever possible.

1. $\frac{5}{8} \div \frac{5}{8}$

2. $\frac{3}{4} \div \frac{1}{4}$

3. $\frac{3}{4} \div 4$

4. $\frac{8}{25} \div \frac{2}{15}$

5. $\frac{6}{7} \div 0$

Answers to Practice Problems:

1. 1

2. 3

3. $\frac{3}{16}$

4. $\frac{12}{5}$

5. Undefined

Name ________________ Section ________ Date ________

ANSWERS

1. ____
2. ____
3. ____
4. ____
5. ____
6. ____
7. ____
8. ____
9. ____
10. ____
11. ____
12. ____
13. ____
14. ____
15. ____
16. ____
17. ____
18. ____
19. ____
20. ____
21. ____
22. ____

Exercises 3.3

1. What is the reciprocal of $\frac{12}{13}$?

2. To divide by any nonzero number, multiply by its__________.

3. Find the quotient of $0 \div \frac{5}{6}$.

4. Find the quotient of $\frac{5}{6} \div 0$.

Find the following quotients. Reduce to lowest terms. See Examples 4 through 10.

5. $\frac{2}{3} \div \frac{3}{4}$

6. $\frac{1}{5} \div \frac{3}{4}$

7. $\frac{3}{7} \div \frac{3}{5}$

8. $\frac{2}{11} \div \frac{2}{3}$

9. $\frac{3}{5} \div \frac{3}{7}$

10. $\frac{2}{3} \div \frac{2}{11}$

11. $\frac{5}{16} \div \frac{15}{16}$

12. $\frac{7}{18} \div \frac{3}{9}$

13. $\frac{3}{14} \div \frac{2}{7}$

14. $\frac{13}{40} \div \frac{26}{35}$

15. $\frac{5}{12} \div \frac{15}{16}$

16. $\frac{12}{27} \div \frac{10}{18}$

17. $\frac{17}{48} \div \frac{51}{90}$

18. $\frac{3}{5} \div \frac{7}{8}$

19. $\frac{13}{16} \div \frac{2}{3}$

20. $\frac{5}{6} \div \frac{3}{4}$

21. $\frac{3}{4} \div \frac{5}{6}$

22. $\frac{14}{15} \div \frac{21}{25}$

23. ______
24. ______
25. ______
26. ______
27. ______
28. ______
29. ______
30. ______
31. ______
32. ______
33. ______
34. ______
35. ______
36. ______
37. ______
38. ______
39. ______
40. ______
41. ______
42. ______
43. ______
44. ______
45. ______
46. ______
47. a. ______
b. ______

23. $\frac{3}{7} \div \frac{3}{7}$

24. $\frac{6}{13} \div \frac{6}{13}$

25. $\frac{16}{27} \div \frac{7}{18}$

26. $\frac{20}{21} \div \frac{15}{42}$

27. $\frac{25}{36} \div \frac{5}{24}$

28. $\frac{17}{20} \div \frac{3}{14}$

29. $\frac{26}{35} \div \frac{39}{40}$

30. $\frac{5}{6} \div \frac{13}{4}$

31. $\frac{7}{8} \div \frac{15}{2}$

32. $\frac{29}{50} \div \frac{31}{10}$

33. $\frac{21}{5} \div \frac{10}{3}$

34. $\frac{35}{17} \div \frac{5}{4}$

35. $\frac{21}{5} \div 3$

36. $\frac{41}{6} \div 2$

37. $3 \div \frac{21}{5}$

38. $2 \div \frac{41}{6}$

39. $5 \div \frac{15}{8}$

40. $14 \div \frac{1}{7}$

41. $\frac{15}{8} \div 5$

42. $\frac{1}{7} \div 14$

43. $56 \div \frac{1}{8}$

44. $24 \div \frac{1}{4}$

45. $\frac{33}{32} \div \frac{11}{4}$

46. $\frac{92}{7} \div \frac{46}{11}$

47. **a.** $\frac{3}{4} \div 2$ **b.** $\frac{3}{4} \div 3$

Name ______________________ Section ____________ Date ___________

ANSWERS

48. **a.** $\frac{4}{7} \div 4$ **b.** $\frac{4}{7} \div \frac{1}{4}$

49. **a.** $\frac{5}{8} \div 2$ **b.** $\frac{5}{8} \div \frac{1}{2}$

50. **a.** $\frac{3}{10} \div 10$ **b.** $\frac{3}{10} \div \frac{1}{10}$

51. **Missing Number.** The product of $\frac{9}{10}$ and another number is $\frac{5}{3}$. What is the other number? [**HINT:** Think $\frac{9}{10} \cdot \square = \frac{5}{3}$.]

52. **Missing Number.** The result of multiplying two numbers is 150. If one of the numbers is $\frac{5}{7}$, what is the other number? [**HINT**: Think $\frac{5}{7} \cdot \square = 150$.]

53. **Enrollment.** A small private college has determined that about $\frac{11}{25}$ of the students that it accepts will actually enroll. If the college wants 550 freshmen to enroll, how many should it accept?

54. **Geology.** The floor of the Atlantic Ocean is spreading apart at an average rate of $\frac{3}{50}$ of a meter per year. About how long will it take for the sea floor to spread 12 meters?

Check Your Number Sense

55. **Vehicle Capacity.** A bus has 45 passengers. This is $\frac{3}{4}$ of its capacity. Is the capacity of the bus more or less than 45? What is the capacity of the bus?

56. **Computers.** A computer printer can print eight pages in one minute. Will the printer print more or fewer than eight pages in 30 seconds? How many pages will the printer print in 15 seconds?

48. a. ____________

b. ____________

49. a. ____________

b. ____________

50. a. ____________

b. ____________

51. ____________

52. ____________

53. ____________

54. ____________

55. ____________

56. ____________

57. [Respond below exercise.]

58.

59.

60.

61.

62.

57. **Salary.** If $\frac{2}{3}$ of your salary is \$1200 per month, is your monthly salary more or less than \$1200? We know that $\frac{1}{10}$ of \$1200 is \$120. If you get a raise of $\frac{1}{10}$ of your monthly salary, will your new monthly salary be more than, less than, or equal to \$1320? Why?

Writing and Thinking about Mathematics

58. Show that the phrases "6 divided by two" and "6 divided by one-half" have different meanings.

59. Show that the phrases "15 divided by three" and "15 divided by one-third" have different meanings.

60. Show that the phrases "12 divided by three" and "12 times one-third" have the same meaning.

61. Show that the phrases "20 divided by four" and "20 times one-fourth" have the same meaning.

62. Is division a commutative operation? Explain briefly and give three examples using fractions to help justify your answer.

Name ______________________ Section ____________ Date __________

ANSWERS

48. **a.** $\frac{4}{7} \div 4$ **b.** $\frac{4}{7} \div \frac{1}{4}$

49. **a.** $\frac{5}{8} \div 2$ **b.** $\frac{5}{8} \div \frac{1}{2}$

50. **a.** $\frac{3}{10} \div 10$ **b.** $\frac{3}{10} \div \frac{1}{10}$

51. **Missing Number.** The product of $\frac{9}{10}$ and another number is $\frac{5}{3}$. What is the other number? [**HINT:** Think $\frac{9}{10} \cdot \square = \frac{5}{3}$.]

52. **Missing Number.** The result of multiplying two numbers is 150. If one of the numbers is $\frac{5}{7}$, what is the other number? [**HINT**: Think $\frac{5}{7} \cdot \square = 150$.]

53. **Enrollment.** A small private college has determined that about $\frac{11}{25}$ of the students that it accepts will actually enroll. If the college wants 550 freshmen to enroll, how many should it accept?

54. **Geology.** The floor of the Atlantic Ocean is spreading apart at an average rate of $\frac{3}{50}$ of a meter per year. About how long will it take for the sea floor to spread 12 meters?

Check Your Number Sense

55. **Vehicle Capacity.** A bus has 45 passengers. This is $\frac{3}{4}$ of its capacity. Is the capacity of the bus more or less than 45? What is the capacity of the bus?

56. **Computers.** A computer printer can print eight pages in one minute. Will the printer print more or fewer than eight pages in 30 seconds? How many pages will the printer print in 15 seconds?

48. a. ____________

b. ____________

49. a. ____________

b. ____________

50. a. ____________

b. ____________

51. ____________

52. ____________

53. ____________

54. ____________

55. ____________

56. ____________

[Respond below
57. exercise.]

58. ______

59. ______

60. ______

61. ______

62. ______

57. **Salary.** If $\frac{2}{3}$ of your salary is \$1200 per month, is your monthly salary more or less than \$1200? We know that $\frac{1}{10}$ of \$1200 is \$120. If you get a raise of $\frac{1}{10}$ of your monthly salary, will your new monthly salary be more than, less than, or equal to \$1320? Why?

Writing and Thinking about Mathematics

58. Show that the phrases "6 divided by two" and "6 divided by one-half" have different meanings.

59. Show that the phrases "15 divided by three" and "15 divided by one-third" have different meanings.

60. Show that the phrases "12 divided by three" and "12 times one-third" have the same meaning.

61. Show that the phrases "20 divided by four" and "20 times one-fourth" have the same meaning.

62. Is division a commutative operation? Explain briefly and give three examples using fractions to help justify your answer.

Name ____________ Section ________ Date ________

Review Problems (from Section 2.6)

Find the least common multiple (LCM) of the following sets of numbers.

1. 30, 65 **2.** 28, 36 **3.** 10, 20, 50

4. 39, 51 **5.** 15, 25, 100 **6.** 44, 88, 121

For each of the following sets of numbers, **a.** *find the LCM, and* **b.** *state the number of times each number divides into the LCM.*

7. 50, 125 **8.** 20, 24, 30

9. 12, 35, 70 **10.** 45, 63, 99

11. Appointments. Two people meet in the optometrist's office and have a pleasant conversation. They agree to have lunch together the next time they are in the optometrist's office on the same day. If their appointments are once every 30 days for one person and once every 45 days for the other person, in how many days will they have lunch together?

REVIEW

1. ______
2. ______
3. ______
4. ______
5. ______
6. ______
7. a. ______
   b. ______
8. a. ______
   b. ______
9. a. ______
   b. ______
10. a. ______
    b. ______
11. ______

Name ____________ Section ________ Date ________

Review Problems (from Section 2.6)

Find the least common multiple (LCM) of the following sets of numbers.

1. 30, 65

2. 28, 36

3. 10, 20, 50

4. 39, 51

5. 15, 25, 100

6. 44, 88, 121

*For each of the following sets of numbers, **a.** find the LCM, and **b.** state the number of times each number divides into the LCM.*

7. 50, 125

8. 20, 24, 30

9. 12, 35, 70

10. 45, 63, 99

11. Appointments. Two people meet in the optometrist's office and have a pleasant conversation. They agree to have lunch together the next time they are in the optometrist's office on the same day. If their appointments are once every 30 days for one person and once every 45 days for the other person, in how many days will they have lunch together?

REVIEW

1. ________

2. ________

3. ________

4. ________

5. ________

6. ________

7. a. ________

b. ________

8. a. ________

b. ________

9. a. ________

b. ________

10. a. ________

b. ________

11. ________

3.4 Addition with Fractions

Addition with Fractions

Figure 3.3 illustrates how the **sum** of the fractions $\frac{3}{7}$ and $\frac{2}{7}$ might be diagrammed.

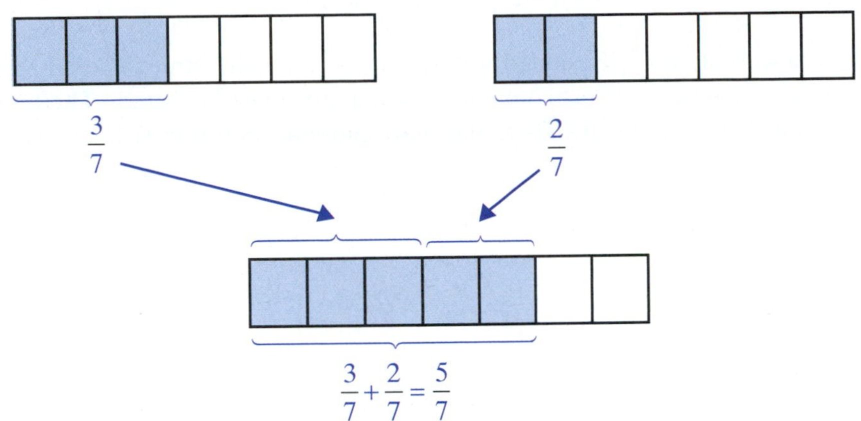

Figure 3.2

Objectives

① Be able to add fractions with the same denominator.

② Recall how to find the LCM.

③ Know how to add fractions with different denominators.

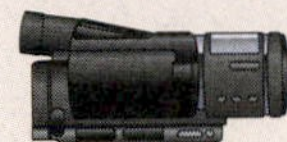

The common denominator "names" each fraction. The sum has this common name. Just as 3 *apples* plus 2 *apples* gives a total of 5 *apples*, we have 3 *sevenths* plus 2 *sevenths* giving a total of 5 *sevenths*.

To Add Two (or More) Fractions with the Same Denominator:

1. Add the numerators.

2. Keep the common denominator.

$$\frac{a}{b}+\frac{c}{b}=\frac{a+c}{b}$$

Objective ①

Example 1

$$\frac{1}{5}+\frac{3}{5}=\frac{1+3}{5}=\frac{4}{5}$$

Now work exercise 1 in the margin.

Example 2

$$\frac{2}{7}+\frac{3}{7}+\frac{1}{7}=\frac{2+3+1}{7}=\frac{6}{7}$$

Now work exercise 2 in the margin.

You may be able to reduce after adding.

Example 3

$$\frac{4}{15}+\frac{6}{15}=\frac{4+6}{15}=\frac{10}{15}=\frac{2\cdot\cancel{5}}{3\cdot\cancel{5}}=\frac{2}{3}$$

Now work exercise 3 in the margin.

Add the fractions.

1. $\frac{4}{7}+\frac{2}{7}$

2. $\frac{5}{13}+\frac{2}{13}+\frac{3}{13}$

3. $\frac{3}{12}+\frac{5}{12}$

4. Add $\frac{9}{10}+\frac{7}{10}+\frac{2}{10}$ and reduce if possible.

5. Find the LCD for $\frac{5}{9}$ and $\frac{7}{18}$.

6. Find the LCD for $\frac{7}{36}$ and $\frac{13}{42}$.

Example 4

$$\frac{1}{8}+\frac{2}{8}+\frac{7}{8}=\frac{1+2+7}{8}=\frac{10}{8}=\frac{\cancel{2}\cdot 5}{\cancel{2}\cdot 4}=\frac{5}{4}$$

Now work exercise 4 in the margin.

Objective 2

Of course, fractions to be added will not always have the same denominator. In these cases, the smallest common denominator must be found. **The least common denominator (LCD) is the least common multiple (LCM) of the denominators**.

Example 5

Find the LCD for $\frac{3}{8}$ and $\frac{11}{12}$.

Solution

Using prime factorization:

$$8 = 2\cdot 2\cdot 2$$
$$12 = 2\cdot 2\cdot 3$$
$$\text{LCD} = 2\cdot 2\cdot 2\cdot 3 = 24$$

Now work exercise 5 in the margin.

Example 6

Find the LCD for $\frac{5}{21}$ and $\frac{5}{28}$.

Solution

Using prime factorization:

$$21 = 3\cdot 7$$
$$28 = 2\cdot 2\cdot 7$$
$$\text{LCD} = 2\cdot 2\cdot 3\cdot 7 = 84$$

Now work exercise 6 in the margin.

Objective 3

To Add Fractions with Different Denominators:

1. Find the least common denominator (LCD).
2. Change each fraction to an equivalent fraction with that denominator.
3. Add the new fractions.
4. Reduce if possible.

3.4 Addition with Fractions

Addition with Fractions

Figure 3.3 illustrates how the **sum** of the fractions $\frac{3}{7}$ and $\frac{2}{7}$ might be diagrammed.

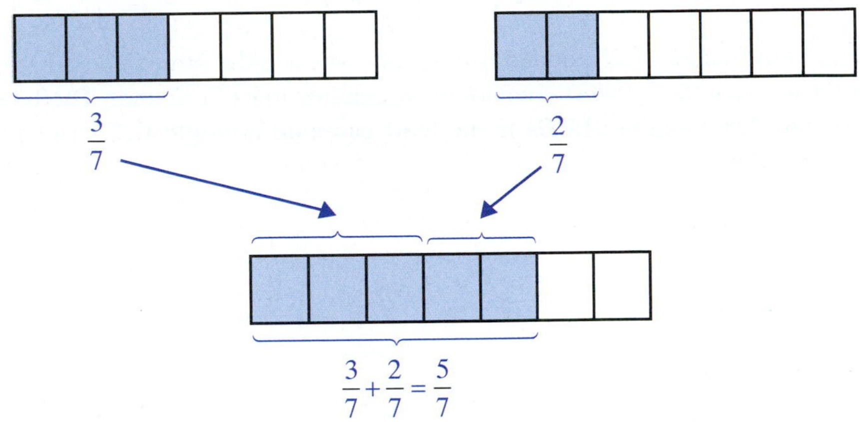

Figure 3.2

Objectives

① Be able to add fractions with the same denominator.

② Recall how to find the LCM.

③ Know how to add fractions with different denominators.

The common denominator "names" each fraction. The sum has this common name. Just as 3 *apples* plus 2 *apples* gives a total of 5 *apples*, we have 3 *sevenths* plus 2 *sevenths* giving a total of 5 *sevenths*.

To Add Two (or More) Fractions with the Same Denominator:

1. Add the numerators.
2. Keep the common denominator.

$$\frac{a}{b}+\frac{c}{b}=\frac{a+c}{b}$$

Objective ①

Add the fractions.

1. $\frac{4}{7}+\frac{2}{7}$

Example 1

$$\frac{1}{5}+\frac{3}{5}=\frac{1+3}{5}=\frac{4}{5}$$

Now work exercise 1 in the margin.

2. $\frac{5}{13}+\frac{2}{13}+\frac{3}{13}$

Example 2

$$\frac{2}{7}+\frac{3}{7}+\frac{1}{7}=\frac{2+3+1}{7}=\frac{6}{7}$$

Now work exercise 2 in the margin.

You may be able to reduce after adding.

3. $\frac{3}{12}+\frac{5}{12}$

Example 3

$$\frac{4}{15}+\frac{6}{15}=\frac{4+6}{15}=\frac{10}{15}=\frac{2\cdot\cancel{5}}{3\cdot\cancel{5}}=\frac{2}{3}$$

Now work exercise 3 in the margin.

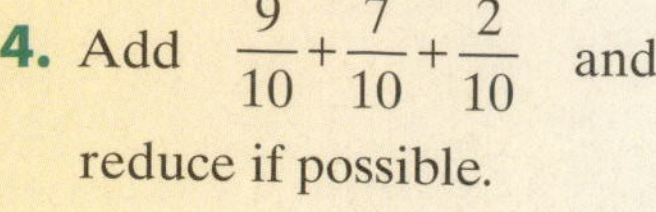

4. Add $\frac{9}{10}+\frac{7}{10}+\frac{2}{10}$ and reduce if possible.

Example 4

$$\frac{1}{8}+\frac{2}{8}+\frac{7}{8}=\frac{1+2+7}{8}=\frac{10}{8}=\frac{\not{2}\cdot 5}{\not{2}\cdot 4}=\frac{5}{4}$$

Now work exercise 4 in the margin.

Objective ②

Of course, fractions to be added will not always have the same denominator. In these cases, the smallest common denominator must be found. **The least common denominator (LCD) is the least common multiple (LCM) of the denominators.**

5. Find the LCD for $\frac{5}{9}$ and $\frac{7}{18}$.

Example 5

Find the LCD for $\frac{3}{8}$ and $\frac{11}{12}$.

Solution

Using prime factorization:

$$\begin{aligned} 8 &= 2\cdot 2\cdot 2 \\ 12 &= 2\cdot 2\cdot 3 \\ \text{LCD} &= 2\cdot 2\cdot 2\cdot 3 = 24 \end{aligned}$$

Now work exercise 5 in the margin.

6. Find the LCD for $\frac{7}{36}$ and $\frac{13}{42}$.

Example 6

Find the LCD for $\frac{5}{21}$ and $\frac{5}{28}$.

Solution

Using prime factorization:

$$\begin{aligned} 21 &= 3\cdot 7 \\ 28 &= 2\cdot 2\cdot 7 \\ \text{LCD} &= 2\cdot 2\cdot 3\cdot 7 = 84 \end{aligned}$$

Now work exercise 6 in the margin.

Objective ③

To Add Fractions with Different Denominators:

1. Find the least common denominator (LCD).
2. Change each fraction to an equivalent fraction with that denominator.
3. Add the new fractions.
4. Reduce if possible.

Example 7

Find the sum $\frac{1}{2}+\frac{3}{5}$.

Solution

a. Find the LCD.

$\text{LCD} = 2 \cdot 5 = 10$

b. Find equivalent fractions with denominator 10.

$$\frac{1}{2}=\frac{1}{2}\cdot\frac{5}{5}=\frac{5}{10}$$

$$\frac{3}{5}=\frac{3}{5}\cdot\frac{2}{2}=\frac{6}{10}$$

c. Add.

$$\frac{1}{2}+\frac{3}{5}=\frac{5}{10}+\frac{6}{10}=\frac{5+6}{10}=\frac{11}{10}.$$

Now work exercise 7 in the margin.

Example 8

Find the sum $\frac{3}{8}+\frac{11}{12}$.

Solution

a. Find the LCD.

$$\left.\begin{aligned} 8 &= 2 \cdot 2 \cdot 2 \\ 12 &= 2 \cdot 2 \cdot 3 \end{aligned}\right\} \text{LCD} = 2 \cdot 2 \cdot 2 \cdot 3 = 24$$

b. Find the equivalent fractions with denominator 24.

$\frac{3}{8}=\frac{3}{8}\cdot\frac{3}{3}=\frac{9}{24}$ Multiply by $\frac{3}{3}$ since $8 \cdot 3 = 24$.

$\frac{11}{12}=\frac{11}{12}\cdot\frac{2}{2}=\frac{22}{24}$ Multiply by $\frac{2}{2}$ since $12 \cdot 2 = 24$.

c. Add.

$$\frac{3}{8}+\frac{11}{12}=\frac{9}{24}+\frac{22}{24}=\frac{9+22}{24}=\frac{31}{24}.$$

Now work exercise 8 in the margin.

Add and reduce if possible.

7. $\frac{1}{7}+\frac{4}{9}$

8. $\frac{3}{8}+\frac{9}{11}$

Add and reduce if possible.

9. $\frac{2}{9}+\frac{1}{6}$

Example 9

Add $\frac{5}{21}+\frac{5}{28}$ and reduce if possible.

Solution

a. Find the LCD.

$$\left.\begin{aligned} 21 &= 3\cdot 7 \\ 28 &= 2\cdot 2\cdot 7 \end{aligned}\right\} \text{LCD} = 2\cdot 2\cdot 3\cdot 7 = 84$$

b. Find the equivalent fractions with denominator 84.

$$\frac{5}{21}=\frac{5}{21}\cdot\frac{4}{4}=\frac{20}{84}$$

$$\frac{5}{28}=\frac{5}{28}\cdot\frac{3}{3}=\frac{15}{84}$$

c. $\frac{5}{21}+\frac{5}{28}=\frac{20}{84}+\frac{15}{84}=\frac{20+15}{84}=\frac{35}{84}.$

d. Now reduce.

$$\frac{35}{84}=\frac{\cancel{7}\cdot 5}{2\cdot 2\cdot 3\cdot \cancel{7}}=\frac{5}{12}$$

Now work exercise 9 in the margin.

10. $\frac{1}{4}+\frac{3}{8}+\frac{7}{10}$

Example 10

Find the sum $\frac{2}{3}+\frac{1}{6}+\frac{5}{12}.$

Solution

a. LCD = 12. You can simply observe this or use prime factorizations.

b. Steps **b.**, **c.** and **d.** can be written together in one process.

$$\begin{aligned} \frac{2}{3}+\frac{1}{6}+\frac{5}{12} &= \frac{2}{3}\cdot\frac{4}{4}+\frac{1}{6}\cdot\frac{2}{2}+\frac{5}{12} \\ &= \frac{8}{12}+\frac{2}{12}+\frac{5}{12} \\ &= \frac{15}{12}=\frac{\cancel{3}\cdot 5}{2\cdot 2\cdot \cancel{3}}=\frac{5}{4} \end{aligned}$$

Or, the numbers can be written vertically. The process is the same.

$$\begin{aligned} \frac{2}{3} &= \frac{2}{3}\cdot\frac{4}{4}=\frac{8}{12} \\ \frac{1}{6} &= \frac{1}{6}\cdot\frac{2}{2}=\frac{2}{12} \\ +\ \frac{5}{12} &= \quad\frac{5}{12}=\frac{5}{12} \\ \hline & \qquad\quad \frac{15}{12}=\frac{\cancel{3}\cdot 5}{2\cdot 2\cdot \cancel{3}}=\frac{5}{4} \end{aligned}$$

Now work exercise 10 in the margin.

Example 11

Add $5+\frac{7}{10}+\frac{3}{1000}$.

Solution

a. LCD = 1000. All the denominators are powers of 10 and 1000 is the largest. We can write 5 as $\frac{5}{1}$.

b. $5+\frac{7}{10}+\frac{3}{1000}=\frac{5}{1}\cdot\frac{1000}{1000}+\frac{7}{10}\cdot\frac{100}{100}+\frac{3}{1000}$

$=\frac{5000}{1000}+\frac{700}{1000}+\frac{3}{1000}$

$=\frac{5703}{1000}$

Now work exercise 11 in the margin.

Common Error

The following common error must be avoided:

Find the sum $\frac{3}{2}+\frac{1}{6}$.

Wrong Solution

$\frac{\overset{1}{\cancel{3}}}{2}+\frac{1}{\underset{2}{\cancel{6}}}=\frac{1}{2}+\frac{1}{2}=1$

You **cannot** cancel across the + sign.

Correct Solution

Use LCD = 6.

$\frac{3}{2}+\frac{1}{6}=\frac{3}{2}\cdot\frac{\mathbf{3}}{\mathbf{3}}+\frac{1}{6}=\frac{9}{6}+\frac{1}{6}=\frac{10}{6}$

NOW reduce.

$\frac{10}{6}=\frac{5\cdot\cancel{2}}{3\cdot\cancel{2}}=\frac{5}{3}$ 2 is a factor in both the numerator and the denominator.

11. Add $3+\frac{3}{10}+\frac{9}{1000}$ and reduce if possible.

12. Find the sum $\frac{3}{4}+\frac{5}{6}+\frac{1}{2}$ and reduce if possible.

Completion Example 12

Find the sum:

$$\frac{2}{3}+\frac{5}{8}+\frac{1}{6}$$

Solution

a. $\left.\begin{array}{l}3=3\\ 8=2\cdot2\cdot2\\ 6=2\cdot3\end{array}\right\}$ LCD $=2\cdot2\cdot2\cdot3=24$

b. $\frac{2}{3}+\frac{5}{8}+\frac{1}{6}=\frac{2}{3}\cdot\underline{\qquad}+\frac{5}{8}\cdot\underline{\qquad}+\frac{1}{6}\cdot\underline{\qquad}$

$=\underline{\qquad}+\underline{\qquad}+\underline{\qquad}$

$=\underline{\qquad}$

Now work exercise 12 in the margin.

Both the commutative and associative properties of addition apply to fractions.

Commutative Property of Addition

If $\frac{a}{b}$ and $\frac{c}{d}$ are fractions, then

$$\frac{a}{b}+\frac{c}{d}=\frac{c}{d}+\frac{a}{b}.$$

Associative Property of Addition

If $\frac{a}{b}$, $\frac{c}{d}$, and $\frac{e}{f}$ are fractions, then

$$\frac{a}{b}+\frac{c}{d}+\frac{e}{f}=\frac{a}{b}+\left(\frac{c}{d}+\frac{e}{f}\right)=\left(\frac{a}{b}+\frac{c}{d}\right)+\frac{e}{f}.$$

Completion Example Answer

12. $\frac{2}{3}+\frac{5}{8}+\frac{1}{6}=\frac{2}{3}\cdot\frac{\mathbf{8}}{\mathbf{8}}+\frac{5}{8}\cdot\frac{\mathbf{3}}{\mathbf{3}}+\frac{1}{6}\cdot\frac{\mathbf{4}}{\mathbf{4}}$

$=\frac{\mathbf{16}}{\mathbf{24}}+\frac{\mathbf{15}}{\mathbf{24}}+\frac{\mathbf{4}}{\mathbf{24}}$

$=\frac{\mathbf{35}}{\mathbf{24}}$

Practice Problems

Find the following sums. Reduce all answers.

1. $\frac{1}{8}+\frac{3}{8}+\frac{2}{8}$

2. $\frac{2}{13}+\frac{5}{2}+\frac{2}{5}$

3. $\frac{7}{10}+\frac{1}{100}+\frac{5}{1000}$

4. $\begin{array}{r} \frac{3}{10} \\ \frac{1}{20} \\ +\frac{7}{30} \\ \hline \end{array}$

5. $\begin{array}{r} \frac{17}{100} \\ +\frac{18}{100} \\ \hline \end{array}$

Answers to Practice Problems:

1. $\frac{3}{4}$ **2.** $\frac{397}{130}$ **3.** $\frac{715}{1000}=\frac{143}{200}$ **4.** $\frac{7}{12}$ **5.** $\frac{7}{20}$

Name ____________ Section ____________ Date ____________

Exercises 3.4

Find the least common denominator (LCD) of the fractions in each exercise. See Examples 5 and 6.

1. $\frac{3}{8}, \frac{5}{16}$

2. $\frac{2}{39}, \frac{1}{3}, \frac{4}{13}$

3. $\frac{2}{27}, \frac{5}{18}, \frac{1}{6}$

4. $\frac{5}{8}, \frac{1}{12}, \frac{5}{9}$

5. $\frac{3}{10}, \frac{1}{100}, \frac{1}{1000}$

Add the following fractions and reduce answers if possible. See Examples 1 through 4.

6. $\frac{6}{10}+\frac{4}{10}$

7. $\frac{3}{14}+\frac{2}{14}$

8. $\frac{1}{20}+\frac{3}{20}$

9. $\frac{3}{4}+\frac{3}{4}$

10. $\frac{5}{6}+\frac{4}{6}$

11. $\frac{7}{5}+\frac{3}{5}$

12. $\frac{11}{15}+\frac{7}{15}$

13. $\frac{7}{9}+\frac{8}{9}$

14. $\frac{3}{25}+\frac{12}{25}$

15. $\frac{7}{90}+\frac{37}{90}+\frac{21}{90}$

16. $\frac{11}{75}+\frac{12}{75}+\frac{62}{75}$

17. $\frac{14}{32}+\frac{7}{32}+\frac{1}{32}$

18. $\frac{4}{100}+\frac{35}{100}+\frac{76}{100}$

19. $\frac{21}{95}+\frac{33}{95}+\frac{3}{95}$

20. $\frac{1}{200}+\frac{17}{200}+\frac{25}{200}$

ANSWERS

1. ____________
2. ____________
3. ____________
4. ____________
5. ____________
6. ____________
7. ____________
8. ____________
9. ____________
10. ____________
11. ____________
12. ____________
13. ____________
14. ____________
15. ____________
16. ____________
17. ____________
18. ____________
19. ____________
20. ____________

21. ______

22. ______

23. ______

24. ______

25. ______

26. ______

27. ______

28. ______

29. ______

30. ______

31. ______

32. ______

33. ______

34. ______

35. ______

36. ______

37. ______

38. ______

39. ______

40. ______

41. ______

42. ______

43. ______

44. ______

45. ______

46. ______

21. $\frac{1}{12}+\frac{2}{3}+\frac{1}{4}$

22. $\frac{3}{8}+\frac{5}{16}$

23. $\frac{2}{5}+\frac{3}{10}+\frac{3}{20}$

24. $\frac{3}{4}+\frac{1}{16}+\frac{6}{32}$

25. $\frac{2}{7}+\frac{4}{21}+\frac{1}{3}$

26. $\frac{1}{6}+\frac{1}{4}+\frac{1}{3}$

27. $\frac{2}{39}+\frac{1}{3}+\frac{4}{13}$

28. $\frac{1}{2}+\frac{3}{10}+\frac{4}{5}$

29. $\frac{1}{27}+\frac{4}{18}+\frac{1}{6}$

30. $\frac{2}{7}+\frac{3}{20}+\frac{9}{14}$

31. $\frac{1}{8}+\frac{1}{12}+\frac{1}{9}$

32. $\frac{2}{5}+\frac{4}{7}+\frac{3}{8}$

33. $\frac{2}{3}+\frac{3}{4}+\frac{9}{14}$

34. $\frac{1}{5}+\frac{7}{30}+\frac{1}{6}$

35. $\frac{1}{5}+\frac{2}{15}+\frac{1}{6}$

36. $\frac{1}{5}+\frac{1}{10}+\frac{1}{4}$

37. $\frac{1}{5}+\frac{7}{40}+\frac{1}{4}$

38. $\frac{1}{3}+\frac{5}{12}+\frac{1}{15}$

39. $\frac{1}{4}+\frac{1}{20}+\frac{8}{15}$

40. $\frac{7}{10}+\frac{3}{25}+\frac{3}{4}$

41. $\frac{5}{8}+\frac{4}{27}+\frac{1}{24}$

42. $\frac{3}{16}+\frac{5}{48}+\frac{1}{32}$

43. $\frac{72}{105}+\frac{2}{45}+\frac{15}{21}$

44. $\frac{2}{15}+\frac{1}{18}+\frac{2}{25}$

45. $\frac{0}{27}+\frac{0}{16}+\frac{1}{5}$

46. $\frac{5}{6}+\frac{0}{100}+\frac{0}{70}+\frac{1}{3}$

Name ____________ Section ____________ Date ____________

ANSWERS

47. $\frac{3}{10}+\frac{1}{100}+\frac{7}{1000}$

48. $\frac{11}{100}+\frac{15}{10}+\frac{1}{10}$

49. $\frac{17}{1000}+\frac{1}{100}+\frac{1}{10,000}$

50. $6+\frac{1}{100}+\frac{3}{10}$

51. $8+\frac{1}{10}+\frac{9}{100}+\frac{1}{1000}$

52. $\frac{1}{10}+\frac{3}{10}+\frac{9}{1000}$

53. $\frac{7}{10}+\frac{5}{100}+\frac{3}{1000}$

54. $\frac{1}{2}+\frac{3}{4}+\frac{1}{100}$

55. $\frac{1}{4}+\frac{1}{8}+\frac{7}{100}$

56. $\frac{9}{1000}+\frac{7}{1000}+\frac{21}{10,000}$

57. $\frac{11}{100}+\frac{1}{2}+\frac{3}{1000}$

58. $\frac{3}{4}+\frac{17}{1000}+\frac{13}{10,000}+2$

59. $5+\frac{1}{10}+\frac{3}{100}+\frac{4}{1000}$

60. $\frac{13}{10,000}+\frac{1}{100,000}+\frac{21}{1,000,000}$

61. $\begin{array}{r}\frac{3}{4}\\ \frac{1}{2}\\ +\frac{5}{12}\\ \hline\end{array}$

62. $\begin{array}{r}\frac{1}{5}\\ \frac{2}{15}\\ +\frac{5}{6}\\ \hline\end{array}$

63. $\begin{array}{r}\frac{7}{8}\\ \frac{2}{3}\\ +\frac{1}{9}\\ \hline\end{array}$

64. $\begin{array}{r}\frac{1}{27}\\ \frac{1}{18}\\ +\frac{4}{9}\\ \hline\end{array}$

65. $\begin{array}{r}\frac{3}{20}\\ \frac{1}{100}\\ +\frac{3}{100}\\ \hline\end{array}$

47. ____________
48. ____________
49. ____________
50. ____________
51. ____________
52. ____________
53. ____________
54. ____________
55. ____________
56. ____________
57. ____________
58. ____________
59. ____________
60. ____________
61. ____________
62. ____________
63. ____________
64. ____________
65. ____________

66. ______

67. ______

68. ______

69. ______

70. ______

71. ______

72. ______

73. ______

74. ______

66. $\begin{array}{r} \dfrac{13}{100} \\ \dfrac{4}{10} \\ +\dfrac{1}{1000} \\ \hline \end{array}$ **67.** $\begin{array}{r} \dfrac{7}{12} \\ \dfrac{1}{9} \\ +\dfrac{2}{3} \\ \hline \end{array}$ **68.** $\begin{array}{r} \dfrac{1}{3} \\ \dfrac{8}{15} \\ +\dfrac{7}{10} \\ \hline \end{array}$ **69.** $\begin{array}{r} \dfrac{9}{16} \\ \dfrac{5}{48} \\ +\dfrac{3}{32} \\ \hline \end{array}$ **70.** $\begin{array}{r} \dfrac{3}{10} \\ \dfrac{1}{20} \\ +\dfrac{1}{25} \\ \hline \end{array}$

71. Postage. Three pieces of mail weigh $\frac{1}{2}$ ounce, $\frac{1}{5}$ ounce, and $\frac{3}{10}$ ounce. What is the total weight of the letters?

72. Measurement. Using a microscope, a scientist measures the diameters of three hairs to be $\frac{1}{1000}$ inch, $\frac{3}{1000}$ inch, and $\frac{1}{100}$ inch. What is the total of these three diameters?

73. Machine Shop. A machinist drills four holes in a straight line. Each hole has a diameter of $\frac{1}{10}$ inch and there is $\frac{1}{4}$ inch between the holes. What is the distance between the outer edges of the first and last holes?

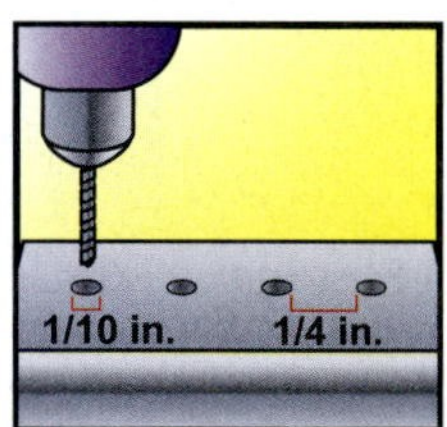

74. Stationery. A notebook contains 30 sheets of paper (each $\frac{1}{100}$ inch thick), 2 pieces of cardboard (each $\frac{1}{16}$ inch thick), and a front and back cover (each $\frac{1}{4}$ inch thick). What is the total thickness of the notebook?

Name ______________________ Section ____________ Date __________

ANSWERS

75. ____________

75. Carpentry. A carpenter is installing baseboard and toe molding. If the baseboard is $\frac{3}{8}$ inch thick and the toe molding (to be put in front of the baseboard) is $\frac{1}{4}$ inch thick, what is the total thickness of the two trim pieces?

1/4 in. 3/8 in.

76. ____________

76. Budgeting. Cynthia budgets $\frac{1}{5}$ of her monthly income for housing, $\frac{1}{4}$ for auto and gas, and $\frac{1}{6}$ for food. What fraction of her income does she budget each month for housing, auto and gas, and food?

77. Investment. Beth's investment strategy is to put $\frac{1}{6}$ of her paycheck into a savings account and another $\frac{1}{9}$ into a retirement account. If she maintains this strategy for 24 paychecks and receives \$900 per paycheck, how much money will she have saved?

77. ____________

78. Ancient number systems. In ancient Egypt, fractions were described as sums with a numerator of 1. What fraction would be described by the sum of $\frac{1}{20}$, $\frac{1}{124}$, and $\frac{1}{155}$?

78. ____________

3.5 Subtraction with Fractions

Subtraction with Fractions

Figure 3.4 shows how the **difference** of the two fractions $\frac{7}{8}$ and $\frac{4}{8}$ might be diagrammed.

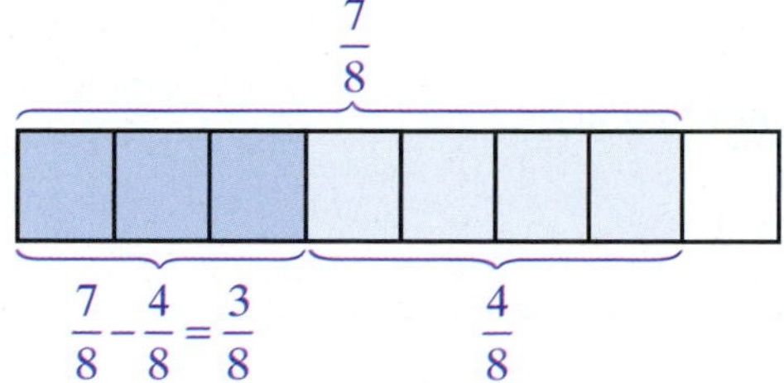

Figure 3.3

From the figure 3.3, we see that $\frac{7}{8}-\frac{4}{8}=\frac{3}{8}$. Just as with addition, the common denominator "names" each fraction. The difference is found by subtracting the numerators and using the common denominator.

Objectives

① Be able to subtract fractions with the same denominator.

② Know how to subtract fractions with different denominators.

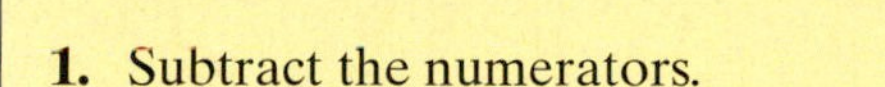

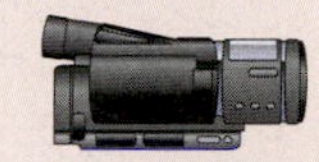

Objective ①

To Subtract Fractions with the Same Denominator:

1. Subtract the numerators.
2. Keep the common denominator.

$$\frac{a}{b}-\frac{c}{b}=\frac{a-c}{b}$$

Example 1

$$\frac{5}{6}-\frac{1}{6}=\frac{5-1}{6}=\frac{4}{6}=\frac{\cancel{2}\cdot 2}{\cancel{2}\cdot 3}=\frac{2}{3}$$

Now work exercise 1 in the margin.

Example 2

$$\frac{9}{10}-\frac{7}{10}=\frac{9-7}{10}=\frac{2}{10}=\frac{\cancel{2}\cdot 1}{\cancel{2}\cdot 5}=\frac{1}{5}$$

Now work exercise 2 in the margin.

Example 3

$$\frac{19}{10}-\frac{11}{10}=\frac{19-11}{10}=\frac{8}{10}=\frac{\cancel{2}\cdot 4}{\cancel{2}\cdot 5}=\frac{4}{5}$$

Now work exercise 3 in the margin.

Example 4

$$\frac{7}{8}-\frac{3}{8}=\frac{7-3}{8}=\frac{4}{8}=\frac{\cancel{4}\cdot 1}{\cancel{4}\cdot 2}=\frac{1}{2}$$

Now work exercise 4 in the margin.

Subtract and reduce if possible.

1. $\frac{6}{10}-\frac{3}{10}$

2. $\frac{27}{30}-\frac{5}{30}$

3. $\frac{11}{14}-\frac{4}{14}$

4. $\frac{8}{9}-\frac{2}{9}$

Objective 2

To Subtract Fractions with Different Denominators:

1. Find the least common denominator (LCD).
2. Change each fraction to an equivalent fraction with that denominator.
3. Subtract the new fractions.
4. Reduce if possible.

Subtract and reduce if possible.

5. $\frac{7}{11}-\frac{3}{5}$

Example 5

$\frac{9}{10}-\frac{2}{15}$

Solution

a. Find the LCD.

$$\left.\begin{array}{l}10=2\cdot 5\\15=3\cdot 5\end{array}\right\}\text{LCD}=2\cdot 3\cdot 5=30$$

b. Find equivalent fractions with denominator 30.

$$\frac{9}{10}=\frac{9}{10}\cdot\frac{3}{3}=\frac{27}{30}$$

$$\frac{2}{15}=\frac{2}{15}\cdot\frac{2}{2}=\frac{4}{30}$$

c. Subtract.

$$\frac{9}{10}-\frac{2}{15}=\frac{27}{30}-\frac{4}{30}=\frac{27-4}{30}=\frac{23}{30}$$

Now work exercise 5 in the margin.

6. $\frac{13}{14}-\frac{3}{35}$

Example 6

$\frac{12}{55}-\frac{2}{33}$

Solution

a. Find the LCD.

$$\left.\begin{array}{l}55=5\cdot 11\\33=3\cdot 11\end{array}\right\}\text{LCD}=3\cdot 5\cdot 11=165$$

b. Find equivalent fractions with denominator 165.

$$\frac{12}{55}=\frac{12}{55}\cdot\frac{3}{3}=\frac{36}{165}$$

$$\frac{2}{33}=\frac{2}{33}\cdot\frac{5}{5}=\frac{10}{165}$$

c. Subtract.

$$\frac{12}{55}-\frac{2}{33}=\frac{36}{165}-\frac{10}{165}=\frac{36-10}{165}=\frac{26}{165}$$

d. Reduce if possible.

$$\frac{26}{165}=\frac{2\cdot 13}{3\cdot 5\cdot 11}=\frac{26}{165}$$ This cannot be reduced because there are no common prime factors in the numerator and denominator.

Now work exercise 6 in the margin.

Example 7

$$\frac{7}{12}-\frac{3}{20}$$

Solution

a. $\left.\begin{matrix}12=2\cdot 2\cdot 3\\ 20=2\cdot 2\cdot 5\end{matrix}\right\}$ LCD $=2\cdot 2\cdot 3\cdot 5=60$

b. Steps **b.**, **c.**, and **d.** of Example 6 can be written as one process.

$$\frac{7}{12}-\frac{3}{20}=\frac{7}{12}\cdot\frac{5}{5}-\frac{3}{20}\cdot\frac{3}{3}=\frac{35}{60}-\frac{9}{60}$$
$$=\frac{35-9}{60}=\frac{26}{60}=\frac{\cancel{2}\cdot 13}{\cancel{2}\cdot 30}=\frac{13}{30}$$

Or, writing the fractions vertically,

$$\frac{7}{12}=\frac{7}{12}\cdot\frac{5}{5}=\frac{35}{60}$$
$$-\frac{3}{20}=\frac{3}{20}\cdot\frac{3}{3}=\frac{9}{60}$$
$$\frac{26}{60}=\frac{\cancel{2}\cdot 13}{\cancel{2}\cdot 30}=\frac{13}{30}$$

Now work exercise 7 in the margin.

Example 8

$$1-\frac{5}{8}$$

Solution

a. LCD = 8. Since 1 can be written as $\frac{1}{1}$, the common denominator is $8\cdot 1=8$.

b. $1-\frac{5}{8}=\frac{1}{1}\cdot\frac{8}{8}-\frac{5}{8}=\frac{8}{8}-\frac{5}{8}=\frac{8-5}{8}=\frac{3}{8}$

Now work exercise 8 in the margin.

Subtract and reduce if possible.

7. $\frac{7}{10}-\frac{2}{15}$

8. $3-\frac{5}{12}$

Subtract and reduce if possible.

9. $\frac{7}{38}-\frac{3}{19}$

10. $\frac{4}{15}-\frac{3}{25}$

11. $5-\frac{3}{8}$

Completion Example 9

$\frac{5}{18}-\frac{1}{6}$

a. LCD = 18.

b. $\frac{5}{18}-\frac{1}{6}=\frac{5}{18}-\frac{1}{6}\cdot\frac{3}{3}$ Use $\frac{3}{3}$ since $6\cdot 3=18$.

$= \_\_\_\_ - \_\_\_\_$

$= \_\_\_\_ = \_\_\_\_$

Now work exercise 9 in the margin.

Completion Example 10

$\frac{11}{12}-\frac{13}{20}$

a. $\left.\begin{array}{l}12=2\cdot 2\cdot 3\\ 20=2\cdot 2\cdot 5\end{array}\right\}$ LCD = ____

b. $\frac{11}{12}-\frac{13}{20}=\frac{11}{12}\cdot\_\_\_\_-\frac{13}{20}\cdot\_\_\_\_$

$= \_\_\_\_ - \_\_\_\_$

$= \_\_\_\_ = \_\_\_\_$

Now work exercise 10 in the margin.

Example 11

$2-\frac{7}{16}$

Solution

a. LCD = 16. Since 2 can be written $\frac{2}{1}$, the common denominator is 16.

b. $2-\frac{7}{16}=\frac{2}{1}\cdot\frac{16}{16}-\frac{7}{16}=\frac{32}{16}-\frac{7}{16}=\frac{32-7}{16}=\frac{25}{16}$

Now work exercise 11 in the margin.

Example 12

The Narragansett Grays baseball team lost 90 games in one season. If $\frac{1}{5}$ of their losses were by 1 or 2 runs and $\frac{4}{9}$ of their losses were by 3 or fewer runs, what fraction of their losses were by exactly 3 runs?

Solution

Before answering the question, we need to analyze the information and decide what arithmetic operation is needed. The losses that were by 3 or fewer runs include those that were by 1 or 2 runs. So, if we subtract the fraction of losses that were by 1 or 2 runs from the fraction of losses that were by 3 or fewer runs, we will be left with the fraction of losses that were by exactly 3 runs. That is, we subtract $\frac{1}{5}$ from $\frac{4}{9}$. (The LCD is 45.)

$$\frac{4}{9}-\frac{1}{5}=\frac{4}{9}\cdot\frac{5}{5}-\frac{1}{5}\cdot\frac{9}{9}=\frac{20}{45}-\frac{9}{45}=\frac{11}{45}$$

Thus the Grays lost $\frac{11}{45}$ of their games by exactly 3 runs.

Now work exercise 12 in the margin.

12. The Cavalier baseball team lost 30 games in one season. If $\frac{1}{3}$ of their losses were by more than 5 runs, and $\frac{2}{5}$ of their losses were by more than 4 runs, what fraction of their losses were by exactly 5 runs?

Completion Example Answers

9. b. $\frac{5}{18}-\frac{1}{6}=\frac{5}{18}-\frac{1}{6}\cdot\frac{3}{3}$

$=\frac{\mathbf{5}}{\mathbf{18}}-\frac{\mathbf{3}}{\mathbf{18}}$

$=\frac{\mathbf{2}}{\mathbf{18}}=\frac{\mathbf{1}}{\mathbf{9}}$

10. a. $\left.\begin{array}{l}12=2\cdot2\cdot3\\20=2\cdot2\cdot5\end{array}\right\}$ LCD $= \mathbf{2\cdot2\cdot3\cdot5=60}$

b. $\frac{11}{12}-\frac{13}{20}=\frac{11}{12}\cdot\frac{\mathbf{5}}{\mathbf{5}}-\frac{13}{20}\cdot\frac{\mathbf{3}}{\mathbf{3}}$

$=\frac{\mathbf{55}}{\mathbf{60}}-\frac{\mathbf{39}}{\mathbf{60}}$

$=\frac{\mathbf{16}}{\mathbf{60}}=\frac{\mathbf{4}}{\mathbf{15}}$

Practice Problems

Find the following differences. Reduce all the answers.

1. $\frac{5}{9}-\frac{1}{9}$ **2.** $\frac{7}{6}-\frac{2}{3}$ **3.** $\frac{7}{10}-\frac{7}{15}$ **4.** $1-\frac{3}{4}$

Answers to Practice Problems:

1. $\frac{4}{9}$ **2.** $\frac{1}{2}$ **3.** $\frac{7}{30}$ **4.** $\frac{1}{4}$

Name ______________________ Section ____________ Date __________

ANSWERS

1. ____________
2. ____________
3. ____________
4. ____________
5. ____________
6. ____________
7. ____________
8. ____________
9. ____________
10. ____________
11. ____________
12. ____________
13. ____________
14. ____________
15. ____________
16. ____________
17. ____________
18. ____________
19. ____________
20. ____________
21. ____________
22. ____________
23. ____________
24. ____________

Exercises 3.5

Subtract and reduce if possible. See Examples 1 through 11.

1. $\frac{4}{7}-\frac{1}{7}$ **2.** $\frac{5}{7}-\frac{3}{7}$ **3.** $\frac{9}{10}-\frac{3}{10}$ **4.** $\frac{11}{10}-\frac{7}{10}$

5. $\frac{5}{8}-\frac{1}{8}$ **6.** $\frac{7}{8}-\frac{5}{8}$ **7.** $\frac{11}{12}-\frac{7}{12}$ **8.** $\frac{7}{12}-\frac{3}{12}$

9. $\frac{13}{15}-\frac{4}{15}$ **10.** $\frac{21}{15}-\frac{11}{15}$ **11.** $\frac{5}{6}-\frac{1}{3}$ **12.** $\frac{5}{6}-\frac{1}{2}$

13. $\frac{11}{15}-\frac{3}{10}$ **14.** $\frac{8}{10}-\frac{3}{15}$ **15.** $\frac{3}{4}-\frac{2}{3}$ **16.** $\frac{2}{3}-\frac{1}{4}$

17. $\frac{15}{16}-\frac{21}{32}$ **18.** $\frac{3}{8}-\frac{1}{16}$ **19.** $\frac{5}{4}-\frac{3}{5}$ **20.** $\frac{5}{12}-\frac{1}{6}$

21. $\frac{14}{27}-\frac{7}{18}$ **22.** $\frac{25}{18}-\frac{21}{27}$ **23.** $\frac{8}{45}-\frac{11}{72}$ **24.** $\frac{46}{55}-\frac{10}{33}$

25. ______

26. ______

27. ______

28. ______

29. ______

30. ______

31. ______

32. ______

33. ______

34. ______

35. ______

36. ______

37. ______

38. ______

39. ______

40. ______

41. ______

42. ______

43. ______

44. ______

45. ______

46. ______

47. ______

48. ______

49. ______

50. ______

25. $\frac{5}{36}-\frac{1}{45}$ **26.** $\frac{5}{1}-\frac{3}{4}$ **27.** $\frac{4}{1}-\frac{5}{8}$ **28.** $2-\frac{9}{16}$

29. $1-\frac{13}{16}$ **30.** $6-\frac{2}{3}$ **31.** $\frac{9}{10}-\frac{3}{100}$ **32.** $\frac{159}{1000}-\frac{1}{10}$

33. $\frac{76}{100}-\frac{7}{10}$ **34.** $\frac{999}{1000}-\frac{99}{100}$ **35.** $\frac{54}{100}-\frac{5}{10}$ **36.** $\frac{7}{24}-\frac{10}{36}$

37. $\frac{31}{40}-\frac{5}{8}$ **38.** $\frac{14}{35}-\frac{12}{30}$ **39.** $\frac{20}{35}-\frac{24}{42}$ **40.** $\frac{3}{10}-\frac{298}{1000}$

41. $1-\frac{9}{10}$ **42.** $1-\frac{7}{8}$ **43.** $1-\frac{2}{3}$ **44.** $1-\frac{1}{16}$

45. $1-\frac{3}{20}$ **46.** $1-\frac{4}{9}$ **47.** $\begin{array}{r}\frac{7}{8}\\ -\frac{2}{3}\\ \hline\end{array}$ **48.** $\begin{array}{r}\frac{14}{15}\\ -\frac{3}{10}\\ \hline\end{array}$

49. $\begin{array}{r}\frac{1}{10}\\ -\frac{8}{100}\\ \hline\end{array}$ **50.** $\begin{array}{r}\frac{3}{100}\\ -\frac{1}{1000}\\ \hline\end{array}$

Name ______________________ Section ______________ Date ____________

ANSWERS

51. Find the sum of $\frac{1}{4}$ and $\frac{3}{16}$. Subtract $\frac{1}{8}$ from the sum. What is the difference?

52. Find the difference between $\frac{2}{3}$ and $\frac{5}{9}$. Add $\frac{1}{12}$ to the difference. What is the sum?

53. Find the sum of $\frac{3}{4}$ and $\frac{5}{8}$. Multiply the sum by $\frac{2}{3}$. What is the product?

54. Find the product of $\frac{9}{10}$ and $\frac{2}{3}$. Divide the product by $\frac{5}{3}$. What is the quotient?

55. Find the quotient of $\frac{3}{4}$ and $\frac{15}{16}$. Add $\frac{3}{10}$ to the quotient. What is the sum?

56. **Astronomy.** About $\frac{1}{2}$ of all incoming solar radiation is absorbed by the earth, $\frac{1}{5}$ is absorbed by the atmosphere, and $\frac{1}{20}$ is scattered by the atmosphere. The rest is reflected by the earth and clouds. What fraction of solar radiation is reflected?

51. ____________

52. ____________

53. ____________

54. ____________

55. ____________

56. ____________

Collaborative Learning

57. With the class separated into teams of two to four students, each team is to write one or two paragraphs explaining what topic in this chapter they found to be most difficult and what techniques they used to learn the related material. Each team leader is to read the paragraph(s) with classroom discussion to follow.

57. [Respond below exercise.]

REVIEW

1. ______
2. ______
3. ______
4. ______
5. ______
6. ______
7. ______
8. ______
9. ______
10. ______

Review Problems (from Section 2.1 and 2.2)

Find the value of each of the following squares. Write as many as you can from memory.

1. 9^2 **2.** 13^2 **3.** 15^2 **4.** 18^2

Find the value of each of the following expressions by using the rules for order of operations.

5. $8 \div 2 + 6 - 5 \cdot 2$

6. $8 \div (2+6) + 14 \cdot 2$

7. $10^2 - 9 \cdot 6 \div 2 + 1^3$

8. $3(4+7) - 4 \cdot 3 - 3 \cdot 7$

9. $2 \cdot 5^2 + 3(16 - 2 \cdot 8) + 5 \cdot 2^2$

10. $(5^2 + 7) \div 4 - (10 \div 5 \cdot 2)$

3.6 Comparisons and Order of Operations with Fractions

Comparing Two or More Fractions

Many times we want to compare two (or more) fractions to see which is smaller or larger. Then we can subtract the smaller from the larger, or possibly make a decision based on the relative sizes of the fractions. Related word problems will be discussed in detail in later chapters.

Objectives

① Compare fractions by finding a common denominator and comparing the numerators.

② Follow the rules for order of operations with fractions.

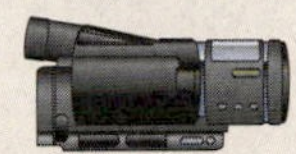

To Compare Two Fractions (to Find Which is Larger or Smaller):

1. Find the least common denominator (LCD).
2. Change each fraction to an equivalent fraction with that denominator.
3. Compare the numerators.

Objective ①

Example 1

Which is larger: $\frac{5}{6}$ or $\frac{7}{8}$? How much larger?

Solution

a. Find the LCD.

$$\left.\begin{array}{l}6 = 2 \cdot 3\\ 8 = 2 \cdot 2 \cdot 2\end{array}\right\} \text{LCD} = 2 \cdot 2 \cdot 2 \cdot 3 = 24$$

b. Find equal fractions with denominator 24.

$$\frac{5}{6} = \frac{5}{6} \cdot \frac{4}{4} = \frac{20}{24} \text{ and } \frac{7}{8} = \frac{7}{8} \cdot \frac{3}{3} = \frac{21}{24}$$

c. $\frac{7}{8}$ is larger than $\frac{5}{6}$, since 21 is larger than 20.

d. $\frac{7}{8} - \frac{5}{6} = \frac{21}{24} - \frac{20}{24} = \frac{1}{24}$

$\frac{7}{8}$ is larger by $\frac{1}{24}$.

Now work exercise 1 in the margin.

1. Which is larger: $\frac{9}{22}$ or $\frac{10}{33}$? How much larger?

Example 2

Which is larger: $\frac{8}{9}$ or $\frac{11}{12}$? How much larger?

Solution

a. LCD $= 2 \cdot 2 \cdot 3 \cdot 3 = 36$

2. Which is larger: $\frac{7}{9}$ or $\frac{19}{24}$? How much larger?

b. $\frac{8}{9} = \frac{8}{9} \cdot \frac{4}{4} = \frac{32}{36}$ and $\frac{11}{12} = \frac{11}{12} \cdot \frac{3}{3} = \frac{33}{36}$

c. $\frac{11}{12}$ is larger than $\frac{8}{9}$, since 33 is larger than 32.

d. $\frac{11}{12} - \frac{8}{9} = \frac{33}{36} - \frac{32}{36} = \frac{1}{36}$

$\frac{11}{12}$ is larger by $\frac{1}{36}$.

Now work exercise 2 in the margin.

3. Arrange $\frac{7}{12}$, $\frac{5}{9}$, and $\frac{2}{3}$ in order, from smallest to largest.

Example 3

Arrange $\frac{2}{3}$, $\frac{7}{10}$, and $\frac{9}{15}$ in order, from smallest to largest. Then find the difference between the smallest and the largest.

Solution

a. LCD = 30

b. $\frac{2}{3} = \frac{2}{3} \cdot \frac{10}{10} = \frac{20}{30}$; $\frac{7}{10} = \frac{7}{10} \cdot \frac{3}{3} = \frac{21}{30}$; $\frac{9}{15} = \frac{9}{15} \cdot \frac{2}{2} = \frac{18}{30}$

c. Smallest to largest: $\frac{9}{15}, \frac{2}{3}, \frac{7}{10}$.

d. $\frac{7}{10} - \frac{9}{15} = \frac{21}{30} - \frac{18}{30} = \frac{3}{30} = \frac{1}{10}$

Now work exercise 3 in the margin.

Objective 2

Using the Rules for Order of Operations with Fractions

An expression with fractions may involve more than one arithmetic operation. To simplify such expressions, use the rules for order of operations just as they were discussed in Chapter 2 for whole numbers. Of course, all the rules for fractions must be followed, too. That is, to add or subtract, you need a common denominator; to divide, you multiply by the reciprocal of the divisor.

Rules for Order of Operations

1. First, simplify within grouping symbols, such as parentheses ( ), brackets [ ], and braces { }. Start with the innermost grouping.
2. Second, evaluate any numbers or expressions indicated by exponents.
3. Third, moving from **left to right**, perform any multiplications or divisions in the order in which they appear.
4. Fourth, moving from **left to right**, perform any additions or subtractions in the order in which they appear.

Evaluate each the following expressions:

4. $\frac{2}{5}\cdot\frac{1}{4}+\frac{1}{9}\div\frac{2}{5}$

Example 4

Evaluate the expression $\frac{1}{2}\div\frac{3}{4}+\frac{5}{6}\cdot\frac{1}{5}$

Solution

$$\frac{1}{2}\div\frac{3}{4}+\frac{5}{6}\cdot\frac{1}{5}$$

$$=\frac{1}{\cancel{2}}\cdot\frac{\overset{2}{\cancel{4}}}{3}+\frac{5}{6}\cdot\frac{1}{5}$$ Divide first.

$$=\frac{2}{3}+\frac{\cancel{5}}{6}\cdot\frac{1}{\cancel{5}}$$ Now multiply.

$$=\frac{2}{3}+\frac{1}{6}$$ Now add (LCD = 6).

$$=\frac{2}{3}\cdot\frac{2}{2}+\frac{1}{6}$$

$$=\frac{4}{6}+\frac{1}{6}$$

$$=\frac{5}{6}$$

Now work exercise 4 in the margin.

Example 5

Evaluate the expression $\left(\frac{3}{4}-\frac{5}{8}\right)\div\left(\frac{15}{16}-\frac{1}{2}\right)$.

Solution

$$\left(\frac{3}{4}-\frac{5}{8}\right)\div\left(\frac{15}{16}-\frac{1}{2}\right)$$ Work inside the parentheses.

$$=\left(\frac{6}{8}-\frac{5}{8}\right)\div\left(\frac{15}{16}-\frac{8}{16}\right)$$

$$=\left(\frac{1}{8}\right)\div\left(\frac{7}{16}\right)$$ Now divide.

Continued on next page...

Evaluate each of the following expressions.

5. $\frac{5}{4}\div\left(1-\frac{1}{3}\right)$

6. $\frac{1}{4}+\left(\frac{1}{3}\right)^2\div\frac{7}{3}$

7. $\left(\frac{1}{2}-\frac{1}{6}\right)\div\left(4+\frac{2}{9}\right)$

$=\frac{1}{\cancel{8}}\cdot\frac{\overset{2}{\cancel{16}}}{7}$

$=\frac{2}{7}$

Now work exercise 5 in the margin.

Example 6

Evaluate the expression $\frac{9}{10}-\left(\frac{1}{4}\right)^2+\frac{1}{2}$.

Solution

$\frac{9}{10}-\left(\frac{1}{4}\right)^2+\frac{1}{2}=\frac{9}{10}-\frac{1}{16}+\frac{1}{2}$ Simplify the exponent first. Remember, $\left(\frac{1}{4}\right)\cdot\left(\frac{1}{4}\right)=\left(\frac{1}{4}\right)^2$

$=\frac{9}{10}\cdot\frac{8}{8}-\frac{1}{16}\cdot\frac{5}{5}+\frac{1}{2}\cdot\frac{40}{40}$ Now add and subtract. (LCD = 80)

$=\frac{72}{80}-\frac{5}{80}+\frac{40}{80}$

$=\frac{107}{80}$

Now work exercise 6 in the margin.

Example 7

Evaluate the expression $\left(\frac{3}{4}+\frac{1}{2}\right)\div\left(1-\frac{1}{3}\right)$.

Solution

Simplifying inside the first set of parentheses gives:

$$\frac{3}{4}+\frac{1}{2}=\frac{3}{4}+\frac{1}{2}\cdot\frac{2}{2}=\frac{3}{4}+\frac{2}{4}=\frac{5}{4}.$$

Simplifying inside the second set of parentheses gives:

$$1-\frac{1}{3}=\frac{3}{3}-\frac{1}{3}=\frac{2}{3}.$$

Therefore, $\left(\frac{3}{4}+\frac{1}{2}\right)\div\left(1-\frac{1}{3}\right)=\left(\frac{5}{4}\right)\div\left(\frac{2}{3}\right)=\frac{5}{4}\cdot\frac{3}{2}=\frac{15}{8}$.

Now work exercise 7 in the margin.

Completion Example 8

Evaluate the expression $\frac{1}{2}\cdot\frac{5}{6}+\frac{7}{15}\div 2$

Solution

$$\frac{1}{2}\cdot\frac{5}{6}+\frac{7}{15}\div 2=\frac{5}{12}+\frac{7}{15}\div 2$$
$$=\frac{5}{12}+\frac{7}{15}\cdot\frac{\quad}{\underline{\quad\quad}}$$
$$=\frac{5}{12}+\frac{\quad}{\underline{\quad\quad}}$$
$$=\frac{5}{12}\cdot\frac{\quad}{\underline{\quad\quad}}+\frac{\quad}{\underline{\quad\quad}}\cdot\frac{\quad}{\underline{\quad\quad}} \qquad \text{(LCD = \_\_\_)}$$
$$=\frac{\quad}{\underline{\quad\quad}}+\frac{\quad}{\underline{\quad\quad}}$$
$$=\frac{\quad}{\underline{\quad\quad}}=\frac{\quad}{\underline{\quad\quad}}=\frac{\quad}{\underline{\quad\quad}}$$

Now work exercise 8 in the margin.

8. Evaluate the expression $\frac{2}{3}\div\frac{7}{9}+2\cdot\frac{4}{21}$.

Completion Example 9

Evaluate the expression $\left(\frac{7}{8}-\frac{7}{10}\right)\div\frac{7}{2}$.

Solution

$$\left(\frac{7}{8}-\frac{7}{10}\right)\div\frac{7}{2}=\left(\frac{35}{40}-\frac{\quad}{\underline{\quad\quad}}\right)\div\frac{7}{2}$$
$$=\left(\frac{\quad}{\underline{\quad\quad}}\right)\div\frac{7}{2}$$
$$=\left(\frac{\quad}{\underline{\quad\quad}}\right)\cdot\frac{\quad}{\underline{\quad\quad}}$$
$$=\frac{\quad}{\underline{\quad\quad}}=\frac{\quad}{\underline{\quad\quad}}$$

Now work exercise 9 in the margin.

9. Evaluate the expression $\frac{2}{3}\div\left(\frac{4}{5}-\frac{4}{7}\right)$.

Completion Example Answers

8. $\frac{1}{2}\cdot\frac{5}{6}+\frac{7}{15}\div 2=\frac{5}{12}+\frac{7}{15}\div 2$

$=\frac{5}{12}+\frac{7}{15}\cdot\frac{\mathbf{1}}{\mathbf{2}}$

$=\frac{5}{12}+\frac{\mathbf{7}}{\mathbf{30}}$

$=\frac{5}{12}\cdot\frac{\mathbf{5}}{\mathbf{5}}+\frac{7}{\mathbf{30}}\cdot\frac{\mathbf{2}}{\mathbf{2}}$ (LCD = **60**)

$=\frac{\mathbf{25}}{\mathbf{60}}+\frac{\mathbf{14}}{\mathbf{60}}$

$=\frac{\mathbf{39}}{\mathbf{60}}=\frac{\cancel{3}\cdot\mathbf{13}}{\cancel{3}\cdot\mathbf{20}}=\frac{\mathbf{13}}{\mathbf{20}}$

9. $\left(\frac{7}{8}-\frac{7}{10}\right)\div\frac{7}{2}=\left(\frac{35}{40}-\frac{\mathbf{28}}{\mathbf{40}}\right)\div\frac{7}{2}$

$=\left(\frac{\mathbf{7}}{\mathbf{40}}\right)\div\frac{7}{2}$

$=\left(\frac{\mathbf{7}}{\mathbf{40}}\right)\cdot\frac{\mathbf{2}}{\mathbf{7}}$

$=\frac{\cancel{7}\cdot\cancel{2}}{\cancel{2}\cdot\mathbf{20}\cdot\cancel{7}}=\frac{\mathbf{1}}{\mathbf{20}}$

Practice Problems

1. Which is larger: $\frac{2}{3}$ or $\frac{8}{9}$? How much larger?

2. Arrange $\frac{3}{5}$, $\frac{5}{7}$, and $\frac{7}{10}$ in order, from smallest to largest.

Evaluate the following expressions.

3. $\frac{1}{6}\div\frac{3}{5}+\frac{3}{4}\cdot\frac{2}{7}$ **4.** $\frac{3}{4}\div\frac{2}{3}-\frac{1}{2}\cdot\frac{5}{6}$ **5.** $\left(\frac{2}{3}+\frac{7}{8}\right)\div\left(2-\frac{1}{5}\right)$

Answers to Practice Problems:

1. $\frac{8}{9}$ is $\frac{2}{9}$ larger. **2.** Smallest to largest: $\frac{3}{5}$, $\frac{7}{10}$, $\frac{5}{7}$

3. $\frac{31}{63}$ **4.** $\frac{17}{24}$ **5.** $\frac{185}{216}$

Name ____________________ Section ____________ Date ____________

ANSWERS

Exercises 3.6

Find the larger number of each pair and state how much larger it is. See Examples 1 and 2.

1. $\frac{2}{3}, \frac{3}{4}$

2. $\frac{7}{10}, \frac{8}{15}$

3. $\frac{4}{5}, \frac{17}{20}$

4. $\frac{4}{10}, \frac{3}{8}$

5. $\frac{13}{20}, \frac{5}{8}$

6. $\frac{13}{16}, \frac{21}{25}$

7. $\frac{14}{35}, \frac{12}{30}$

8. $\frac{10}{36}, \frac{7}{24}$

9. $\frac{17}{80}, \frac{11}{48}$

10. $\frac{37}{100}, \frac{24}{75}$

Arrange the numbers in order, from smallest to largest. Then find the difference between the largest and smallest numbers. See Example 3.

11. $\frac{2}{3}, \frac{3}{5}, \frac{7}{10}$

12. $\frac{8}{9}, \frac{9}{10}, \frac{11}{12}$

13. $\frac{7}{6}, \frac{11}{12}, \frac{19}{20}$

14. $\frac{1}{3}, \frac{5}{42}, \frac{3}{7}$

15. $\frac{1}{2}, \frac{1}{3}, \frac{1}{4}$

16. $\frac{2}{3}, \frac{3}{4}, \frac{5}{8}$

17. $\frac{7}{9}, \frac{31}{36}, \frac{13}{18}$

18. $\frac{17}{12}, \frac{40}{36}, \frac{31}{24}$

19. $\frac{1}{100}, \frac{3}{1000}, \frac{20}{10,000}$

20. $\frac{32}{100}, \frac{298}{1000}, \frac{3}{10}$

1. ____________
2. ____________
3. ____________
4. ____________
5. ____________
6. ____________
7. ____________
8. ____________
9. ____________
10. ____________
11. ____________
12. ____________
13. ____________
14. ____________
15. ____________
16. ____________
17. ____________
18. ____________
19. ____________
20. ____________

21. ______

22. ______

23. ______

24. ______

25. ______

26. ______

27. ______

28. ______

29. ______

30. ______

31. ______

32. ______

33. ______

34. ______

35. ______

36. ______

37. ______

38. ______

39. ______

40. ______

41. ______

42. ______

43. ______

44. ______

45. ______

46. ______

47. ______

Evaluate each expression using the rules for order of operations. See Examples 4 through 9.

21. $\frac{1}{2} \div \frac{7}{8} + \frac{1}{7} \cdot \frac{2}{3}$

22. $\frac{3}{5} \cdot \frac{1}{6} + \frac{1}{5} \div 2$

23. $\frac{1}{2} \div \frac{1}{2} + \frac{2}{3} \cdot \frac{2}{3}$

24. $5 - \frac{3}{4} \div 3$

25. $6 - \frac{5}{8} \div 4$

26. $\frac{2}{15} \cdot \frac{1}{4} \div \frac{3}{5} + \frac{1}{27}$

27. $\frac{5}{8} \cdot \frac{1}{10} \div \frac{3}{4} + \frac{1}{6}$

28. $\left(\frac{7}{15} + \frac{8}{21}\right) \div \frac{3}{35}$

29. $\left(\frac{1}{2} - \frac{1}{3}\right) \div \left(\frac{5}{8} + \frac{3}{16}\right)$

30. $\left(\frac{1}{3} + \frac{1}{5}\right) \cdot \left(\frac{3}{4} - \frac{1}{6}\right)$

31. $\left(\frac{1}{2}\right)^2 - \left(\frac{1}{4}\right)^3$

32. $\frac{2}{3} + \frac{3}{4} + \left(\frac{1}{2}\right)^2$

33. $\left(\frac{1}{3}\right)^2 + \left(\frac{1}{6}\right)^2 + \frac{2}{3}$

34. $\frac{1}{2} \div \frac{2}{3} + \left(\frac{1}{3}\right)^2$

35. $\left(\frac{3}{4} - \frac{1}{2}\right) \div \left(1 + \frac{1}{3}\right)$

36. $\left(\frac{1}{8} + \frac{1}{2}\right) \div \left(1 - \frac{2}{5}\right)$

37. $\left(\frac{1}{5} + \frac{1}{6}\right) \div \left(2 + \frac{1}{3}\right)$

38. $\left(\frac{5}{6} - \frac{1}{3}\right) \div \left(\frac{1}{2} + \frac{1}{5}\right)$

39. $\left(\frac{2}{3} - \frac{1}{4}\right) \div \left(\frac{3}{5} - \frac{1}{4}\right)$

40. $\left(\frac{5}{6} - \frac{2}{3}\right) \div \left(\frac{5}{8} - \frac{1}{16}\right)$

41. $\left(\frac{7}{8} - \frac{3}{16}\right) \div \left(\frac{1}{3} - \frac{1}{4}\right)$

42. $\left(\frac{2}{5} + \frac{5}{6}\right) \div \left(\frac{1}{4} + \frac{1}{10}\right)$

43. $\left(\frac{5}{8} - \frac{1}{8}\right) \div \left(\frac{1}{2} - \frac{1}{4}\right)$

44. $16 + \left(\frac{1}{3} \div \frac{2}{3}\right)$

45. $\left(\frac{2}{3}\right)^2 + \left(\frac{1}{3}\right)^2 - \frac{5}{9}$

46. $\left(\frac{1}{2} - \frac{1}{3}\right)^2 + \left(\frac{1}{4} - \frac{1}{6}\right)^2$

47. $\left(\frac{2}{5} + \frac{1}{4}\right) \cdot \left(\frac{3}{4} - \frac{1}{2}\right)$

Name ______________________ Section ____________ Date __________

ANSWERS

48. ______________

49. a. ______________

b. ______________

50. a. ______________

b. ______________

51. ______________

52. ______________

48. At Cerritos College, $\frac{3}{5}$ of the students take a math course. Of the students taking math, $\frac{1}{4}$ take statistics. What fraction of the Cerritos College students take statistics?

Check Your Number Sense

49. **a.** If two fractions are between 0 and 1, can their sum be more than 1? Explain.

b. If two fractions are between 0 and 1, can their product be more than 1? Explain.

50. Consider the fraction $\frac{1}{2}$.

a. If this fraction is divided by 2, will the quotient be more or less than $\frac{1}{2}$?

b. If this fraction is divided by 3, will the quotient be more or less than the quotient in part **a.**?

51. Will the quotient always get smaller and smaller when a nonzero number is divided by larger and larger numbers? Can you think of a case in which this is not true? What happens when 0 is divided by larger and larger numbers?

52. Consider any fraction between 0 and 1, not including 0 or 1. If you square this number, will the result be larger or smaller than the original number? Is this always the case? Explain your answer.

Chapter 3 Index of Key Terms and Ideas

Section 3.1 Basic Multiplication and Changing to Higher Terms

Rational Numbers page 204

A **rational number** is a number that can be written in the fraction form $\frac{a}{b}$ where a is a whole number and b is a nonzero whole number.

$$\frac{a \leftarrow \text{numerator}}{b \leftarrow \text{denominator}}$$

Note: The numerator a can be 0, but the denominator b cannot be 0.

Fractions page 206

A fraction is a **proper fraction** if the numerator is less than the denominator.

A fraction is an **improper fraction** if the numerator is equal to or greater than the denominator.

Fractions can be Used in Two Basic Ways page 207

1. To indicate equal parts of a whole.
2. To indicate division. (The numerator is to be divided by the denominator.)

Division by 0 is Undefined. $\frac{0}{0}$ is undefined. page 208

To Multiply Fractions: pages 209 – 210

1. Multiply the numerators.
2. Multiply the denominators.

$$\frac{a}{b}\cdot\frac{c}{d}=\frac{a\cdot c}{b\cdot d}$$

Properties of Multiplication pages 211– 212

Commutative property: $\frac{a}{b}\cdot\frac{c}{d}=\frac{c}{d}\cdot\frac{a}{b}$.

Associative property: $\frac{a}{b}\cdot\frac{c}{d}\cdot\frac{e}{f}=\left(\frac{a}{b}\cdot\frac{c}{d}\right)\cdot\frac{e}{f}=\frac{a}{b}\cdot\left(\frac{c}{d}\cdot\frac{e}{f}\right)$.

Identity: $\frac{a}{b}\cdot 1=\frac{a}{b}$.

To Change a Fraction to Higher Terms: pages 212 – 213

Multiply the numerator and denominator by the same nonzero whole number.

$$\frac{a}{b}=\frac{a}{b}\cdot 1=\frac{a}{b}\cdot\frac{\boldsymbol{k}}{\boldsymbol{k}}=\frac{a\cdot \boldsymbol{k}}{b\cdot \boldsymbol{k}}, \text{ where } \boldsymbol{k}\neq 0 \left(1=\frac{\boldsymbol{k}}{\boldsymbol{k}}\right).$$

Section 3.2 Multiplication and Reducing with Fractions

Lowest Terms page 223

A fraction is reduced to **lowest terms** if the numerator and denominator have no common factors other than 1. You can check for lowest terms by using prime factors.

Continued on next page...

Section 3.2 Multiplication and Reducing with Fractions (continued)

To Reduce a Fraction to Lowest Terms: page 223

1. Factor the numerator and denominator into prime factorizations.
2. Use the fact that $\frac{k}{k} = 1$ and **divide out** common factors.

Multiplying and Reducing Fractions at the Same Time pages 225–226

Section 3.3 Division with Fractions

Reciprocals page 235

The **reciprocal** of $\frac{a}{b}$ is $\frac{b}{a}$ $(a \neq 0 \text{ and } b \neq 0)$. The product of a nonzero number and its reciprocal is always 1.

$$\frac{a}{b} \cdot \frac{b}{a} = 1$$

To Divide Fractions: page 236

To divide by any nonzero number, multiply by its reciprocal. In general,

$$\frac{a}{b} \div \frac{c}{d} = \frac{a}{b} \cdot \frac{d}{c} \text{ where } b, c, d \neq 0.$$

Section 3.4 Addition with Fractions

To Add Two (or More) Fractions with the Same Denominator: page 247

1. Add the numerators.
2. Keep the common denominator.

$$\frac{a}{b} + \frac{c}{b} = \frac{a+c}{b}$$

To Add Fractions with Different Denominators: page 248

1. Find the least common denominator (LCD).
2. Change each fraction to an equivalent fraction with that denominator.
3. Add the new fractions.
4. Reduce if possible.

Properties of Addition page 252

Commutative property: $\frac{a}{b} + \frac{c}{d} = \frac{c}{d} + \frac{a}{b}$.

Associative property: $\frac{a}{b} + \frac{c}{d} + \frac{e}{f} = \frac{a}{b} + \left(\frac{c}{d} + \frac{e}{f}\right) = \left(\frac{a}{b} + \frac{c}{d}\right) + \frac{e}{f}$.

Section 3.5 Subtraction with Fractions

To Subtract Fractions with the Same Denominator: page 261

1. Subtract the numerators.
2. Keep the common denominator.

$$\frac{a}{b} - \frac{c}{b} = \frac{a-c}{b}$$

Continued on next page...

Section 3.5 Subtraction with Fractions (continued)

To Subtract Fractions with Different Denominators: page 262

1. Find the least common denominator (LCD).
2. Change each fraction to an equivalent fraction with that denominator.
3. Subtract the new fractions.
4. Reduce if possible.

Section 3.6 Comparisons and Order of Operations with Fractions

To Compare Two Fractions (to Find Which is Larger or Smaller): page 271

1. Find the least common denominator (LCD).
2. Change each fraction to an equivalent fraction with that denominator.
3. Compare the numerators.

Rules for Order of Operations page 272 – 273

1. First, simplify within grouping symbols, such as parentheses ( ), brackets [ ], and braces { }. Start with the innermost grouping.
2. Second, evaluate any numbers or expressions indicated by exponents.
3. Third, moving from **left to right**, perform any multiplications or divisions in the order in which they appear.
4. Fourth, moving from **left to right**, perform any additions or subtractions in the order in which they appear.

Name ____________ Section ____________ Date ____________

ANSWERS

Chapter 3 Review Questions

1. The denominator of a rational number cannot be ____________.

2. $\frac{0}{7} = 0$, but $\frac{7}{0}$ is ____________.

3. The reciprocal of $\frac{2}{3}$ is _____, and the reciprocal of $\frac{3}{2}$ is _______.

4. Which property of addition is illustrated by the following statement?

$$\frac{1}{3} + \left(\frac{5}{6} + \frac{1}{2}\right) = \left(\frac{1}{3} + \frac{5}{6}\right) + \frac{1}{2}$$

5. Find $\frac{2}{3}$ of $\frac{2}{5}$.

Multiply and reduce all answers.

6. $\frac{1}{3} \cdot \frac{1}{2} \cdot \frac{1}{5}$

7. $\frac{1}{7} \cdot \frac{3}{7}$

8. $\frac{35}{56} \cdot \frac{4}{15} \cdot \frac{5}{10}$

Fill in the missing terms so that each equation is true.

9. $\frac{1}{6} = \frac{?}{12}$

10. $\frac{9}{10} = \frac{?}{60}$

11. $\frac{15}{13} = \frac{?}{65}$

1. ____________
2. ____________
3. ____________
4. ____________
5. ____________
6. ____________
7. ____________
8. ____________
9. ____________
10. ____________
11. ____________

12. ____________

13. ____________

14. ____________

15. ____________

16. ____________

17. ____________

18. ____________

19. ____________

20. ____________

21. ____________

22. ____________

23. ____________

24. ____________

25. ____________

26. ____________

27. ____________

28. ____________

29. ____________

30. ____________

Reduce each fraction to its lowest terms.

12. $\frac{15}{30}$ **13.** $\frac{99}{88}$ **14.** $\frac{0}{4}$ **15.** $\frac{150}{120}$

Add or subtract as indicated and reduce all answers.

16. $\frac{3}{7}+\frac{2}{7}$ **17.** $\frac{5}{6}-\frac{1}{6}$ **18.** $\frac{5}{8}-\frac{3}{8}$

19. $\frac{1}{12}+\frac{5}{36}+\frac{11}{24}$ **20.** $\frac{13}{22}-\frac{9}{33}$ **21.** $\frac{5}{27}+\frac{5}{18}$

22. $1-\frac{13}{20}$ **23.** $\frac{3}{4}-\frac{5}{12}$ **24.**

$$\begin{array}{r} \frac{2}{3} \\ \frac{1}{8} \\ +\frac{1}{12} \\ \hline \end{array}$$

Divide and reduce all answers.

25. $\frac{2}{3}\div 6$ **26.** $1\div\frac{3}{5}$ **27.** $\frac{7}{12}\div\frac{7}{12}$

28. $\frac{15}{16}\div\frac{3}{4}$ **29.** $\frac{3}{4}\div\frac{15}{16}$

30. Which is larger, $\frac{2}{3}$ or $\frac{4}{5}$? How much larger?

Name ________________ Section ____________ Date __________ **ANSWERS**

31. Arrange the following fractions in order, from smallest to largest.

$\frac{7}{12}, \frac{5}{9}, \frac{11}{20}$

Evaluate each expression using the rules for order of operations.

32. $\frac{5}{8}\cdot\frac{3}{10}+\frac{1}{14}\div 2$

33. $\left(\frac{3}{5}-\frac{1}{3}\right)\div\left(\frac{1}{6}\div\frac{7}{8}\right)$

34. $\frac{7}{15}+\frac{5}{9}\div\frac{2}{3}-\frac{2}{3}$

35. $\left(\frac{2}{3}\right)^2-\left(\frac{1}{3}\right)^2+\frac{1}{18}$

36. $\left(\frac{3}{8}+\frac{1}{2}\right)\div\left(\frac{1}{2}-\frac{1}{10}\right)$

37. Sue bought 3 bags of candy weighing $\frac{1}{4}$ pound each. She gave one bag to Tom. How many pounds of candy did she have left?

38. Ms. Clarke teaches statistics. This semester, $\frac{2}{3}$ of her statistics class of 42 students are planning to be elementary school teachers. How many of her students do not plan to be elementary school teachers?

31. ____________

32. ____________

33. ____________

34. ____________

35. ____________

36. ____________

37. ____________

38. ____________

Name ______________________ Section ____________ Date __________

ANSWERS

Chapter 3 Test

1. The reciprocal of $\frac{5}{8}$ is _______.

2. The fraction equivalent to $\frac{7}{16}$ with denominator 80 is _______.

3. The equation $\frac{3}{4}\cdot\frac{7}{2}=\frac{7}{2}\cdot\frac{3}{4}$ illustrates the _________ property of _________.

Reduce to lowest terms.

4. $\frac{90}{108}$ **5.** $\frac{77}{55}$ **6.** $\frac{117}{156}$

7. Find the product of $\frac{2}{7}$ and $\frac{14}{15}$.

8. From the sum of $\frac{3}{8}$ and $\frac{5}{12}$, subtract the sum of $\frac{1}{6}$ and $\frac{5}{9}$.

9. Find the quotient when 4 is divided by $\frac{4}{9}$.

Perform the indicated operations and reduce all answers to lowest terms. Follow the rules for order of operations.

10. $\frac{3}{7}\cdot\frac{14}{27}$ **11.** $10\div\frac{2}{5}$ **12.** $\frac{3}{5}\cdot\frac{1}{2}\cdot\frac{3}{8}$ **13.** $\frac{5}{11}\div\frac{4}{5}$

1. ____________
2. ____________
3. ____________
4. ____________
5. ____________
6. ____________
7. ____________
8. ____________
9. ____________
10. ____________
11. ____________
12. ____________
13. ____________

14. ______

15. ______

16. ______

17. ______

18. ______

19. ______

20. ______

21. ______

22. ______

23. ______

24. ______

25. ______

26. ______

27. ______

14. $2-\frac{14}{11}$

15. $\frac{4}{35}+\frac{2}{7}+\frac{1}{10}$

16. $\frac{5}{16}+\left(\frac{1}{4}\right)^2$

17. $\left(\frac{51}{16}-3\right)\div\frac{3}{8}$

18. $\left(\frac{54}{17}-\frac{3}{17}\right)\div\left(\frac{1}{2}+\frac{3}{4}\right)$

19. $\frac{17}{19}+\frac{6}{19}+\frac{15}{19}$

20. $\frac{2}{7}\cdot\frac{3}{4}\cdot\frac{7}{9}$

21. $\left(\frac{1}{2}\right)^3-\left(\frac{1}{4}\right)^2$

22. $\frac{4}{5}-\left(\frac{1}{2}\right)^2+\frac{1}{10}$

23. $\left(\frac{3}{4}-\frac{1}{5}\right)\div\frac{3}{2}$

24. $\frac{1}{3}\cdot\frac{5}{6}+\left(\frac{1}{3}\right)^2\div 2$

25. Arrange the numbers $\frac{7}{8}$, $\frac{3}{4}$, and $\frac{2}{3}$ in order, from smallest to largest. Show how you arrived at this ordering.

26. The result of multiplying two numbers is $\frac{3}{5}$. If one of the numbers is $\frac{9}{10}$, what is the other number?

27. At Bio Hondo College, $\frac{3}{4}$ of the students have jobs while going to school. Of those who have jobs, $\frac{3}{5}$ work more than 20 hours per week. What fraction of the students work more than 20 hours per week?

Name ______________________ Section ____________ Date __________

ANSWERS

Cumulative Review: Chapters 1 – 3

1. Round 2,549,700 to the nearest hundred thousand.

2. Give two illustrations of each of the named properties.
a. Associative property of addition.
b. Identity property for multiplication

*For each of the following, **a.** estimate each answer; then **b.** find the answer by performing the indicated operation.*

3. $\begin{array}{r} 961 \\ 1745 \\ 87 \\ +\ 620 \\ \hline \end{array}$

4. $\begin{array}{r} 224 \\ \times\ 108 \\ \hline \end{array}$

5. $\begin{array}{r} 9048 \\ -8052 \\ \hline \end{array}$

6. $15\overline{)4750}$

7. Multiply mentally: 50×7000.

8. Without actually dividing, determine which of the numbers 2, 3, 5, 6, 9, and 10 will divide into 8190. Give a brief reason for each decision.

9. Determine whether or not the number 431 is prime. Show all the steps you use.

10. Find the prime factorization of 780.

1. ____________

2. a. ____________

b. ____________

3. a. ____________

b. ____________

4. a. ____________

b. ____________

5. a. ____________

b. ____________

6. a. ____________

b. ____________

7. ____________

8. [Respond below exercise.]

9. ____________

10. ____________

11. [Respond below exercise.] ________

12. [Respond below exercise.] ________

13. ________

14. ________

15. ________

16. ________

17. ________

18. ________

19. ________

20. ________

21. ________

In Exercises 11 and 12, find the LCM of each set of numbers and tell how many times each number divides into the LCM.

11. 18, 42, 90

12. 36, 60, 84, 96

13. Use the rules for order of operations to evaluate the expression $12\cdot 9\div 3^2+2(5^2-3\cdot 5)$.

Perform the indicated operations and reduce if possible.

14. $\frac{5}{9}\left(\frac{3}{28}\right)\left(\frac{14}{20}\right)$

15. $\frac{3}{20}+\frac{7}{8}+\frac{2}{15}$

16. $\frac{27}{34}-\frac{1}{51}$

17. $\left(\frac{3}{4}\right)^2+\frac{1}{2}\div\frac{8}{3}\cdot 3$

18. Find $\frac{3}{4}$ of $\frac{12}{7}$.

19. Find the average of the numbers 97, 85, 76, and 122.

20. Banking. Samantha opened her checking account with a deposit of \$5380. She wrote checks for \$95, \$265, \$107, and \$1573 and made another deposit of \$340. What was the new balance in her account?

21. If the quotient of 119 and 17 is subtracted from the product of 23 and 34, what is the difference?

Name ______________________ Section ____________ Date __________

ANSWERS

22. **Computers.** The base of a laser jet printer is in the shape of a rectangle 25 centimeters wide and 32 centimeters long. What area of a desk top does the printer cover?

23. **Travel.** The distance between New York and Berlin is approximately $\frac{2}{3}$ the distance between Mexico City and Berlin. If the distance between Mexico City and Berlin is approximately 6000 miles, about how far apart are New York and Berlin?

24. **Geography.** The area of Switzerland is approximately 16,000 square miles; the area of France is approximately 210,000 square miles; and the area of Germany is approximately 140,000 square miles. **a.** About what fraction (reduced) of the area of France is the area of Switzerland? **b.** About what fraction (reduced) of the area of Germany is the area of Switzerland? **c.** About what fraction (reduced) of the area of France is the area of Germany?

25. **Soccer.** A soccer field is in the shape of a rectangle and has dimensions 55 yards by 100 yards. Find **a.** the perimeter and **b.** the area of the field.

26. **Geometry.** A triangle has dimensions as shown in the figure. Find **a.** the perimeter and **b.** the area of the triangle.

5 in. 5 in. 4 in. 6 in.

27. **Geometry.** A right triangle has sides of length 10 feet, 24 feet, and 26 feet. **a.** Draw a sketch and label the sides of the triangle. Find **b.** the perimeter and **c.** the area of the triangle.

22. ____________

23. ____________

24. a. ____________

b. ____________

c. ____________

25. a. ____________

b. ____________

26. a. ____________

b. ____________

27. a. ____________

b. ____________

c. ____________

4

Mixed Numbers

Chapter 4 Mixed Numbers

WHAT TO EXPECT IN CHAPTER 4

In Chapter 4 we will discuss the meaning of mixed numbers and develop techniques for operating with them. All of the concepts and skills developed in Chapter 1 (Whole Numbers), Chapter 2 (Exponents, Prime Numbers, & LCM), and Chapter 3 (Fractions) are an integral part of the topics in Chapter 4, and they should be reviewed often, perhaps even on a daily basis.

One interesting fact that will be made clear in Section 4.1 is that a mixed number indicates the sum of a whole number and a fraction. This idea leads to an understanding of how to change mixed numbers to improper fractions and how to change improper fractions to mixed numbers. Section 4.1 also makes the distinction between the two concepts of reducing an improper fraction and changing an improper fraction to a mixed number.

The importance of the relationship between improper fractions and mixed numbers is emphasized in Section 4.2, where multiplication and division with mixed numbers are accomplished by using the numbers in improper fraction form. Sections 4.3 and 4.4 deal with addition and subtraction with mixed numbers. In subtraction, we can sometimes allow the fraction part of a mixed number to be larger than 1. Section 4.5 explains how the rules for order of operations can be applied with mixed numbers. The extension of mixed numbers into a real world setting, denominate numbers, is discussed in Section 4.6.

Chapter 4 Mixed Numbers

Mathematics at Work!

Carpenters construct all kinds of things we use every day: desks, bookshelves, cabinets, end tables, tool storage sheds, and a host of other useful items. For example, to build bookshelves in a corner of a room, a carpenter will need to take measurements and buy lumber of the type and quality needed. Many times, the measurements of the pieces he needs to cut are not whole numbers, but mixed numbers (whole numbers with fraction parts). For example, pieces might need to be $2\frac{1}{2}$ feet long or $56\frac{1}{2}$ inches long. Consider the following problem:

A carpenter wants to build a single stand-alone bookshelf that will involve three shelves – a top, a bottom, and two sides. For each bookshelf, he will need three pieces of lumber each $3\frac{1}{4}$ feet long (including two shelves and a bottom), one piece $3\frac{3}{4}$ feet long for the top and two pieces each $3\frac{1}{2}$ feet long for the sides. At the local home supply store the type of lumber needed comes in lengths of 8 feet and 12 feet. How many pieces of length 12 feet will he need to buy to build the bookshelf and how much lumber will be left over? Would he have been better off to buy lengths of 8 feet?

4.1 Introduction to Mixed Numbers

Changing Mixed Numbers to Fraction Form

Objective ①

A **mixed number** is the sum of a whole number and a fraction with the fraction part less than 1. By convention, we usually write the whole number and the fraction side by side without the plus sign. For example,

$$7+\frac{3}{4}=7\frac{3}{4}$$ Read "seven and three-fourths."

$$10+\frac{1}{2}=10\frac{1}{2}$$ Read "ten and one-half."

Most people are familiar with mixed numbers and use them daily, as in: "I rode my bicycle $10\frac{1}{2}$ miles today" or, "This recipe calls for $2\frac{1}{4}$ cups of flour." However, as we will see in Section 4.2, this form of mixed numbers is not convenient for multiplication and division with mixed numbers. These operations are more easily accomplished by changing the mixed numbers to improper fractions first.

Objective ②

Note: To change a mixed number to an improper fraction, add the whole number and the fraction. Remember that a whole number can be written in fraction form with denominator 1.

Example 1

Change each mixed number to an improper fraction.

a. $3\frac{4}{5}=3+\frac{4}{5}=\frac{3}{1}\cdot\frac{5}{5}+\frac{4}{5}=\frac{15}{5}+\frac{4}{5}=\frac{19}{5}$

b. $2\frac{7}{8}=2+\frac{7}{8}=\frac{2}{1}\cdot\frac{8}{8}+\frac{7}{8}=\frac{16}{8}+\frac{7}{8}=\frac{23}{8}$

Now work exercise 1 in the margin.

1. Change each mixed number to an improper fraction.

a. $7\frac{1}{8}$

b. $11\frac{3}{4}$

There is a pattern to changing mixed numbers to improper fractions that leads to a familiar shortcut. Since the denominator of the whole number is always 1, the LCD is always the denominator of the fraction part. Therefore, in Example **1a.**, the LCD was 5 and we multiplied the whole number 3 by the common denominator 5. Similarly, in Example **1b.**, we multiplied the whole number 2 by the common denominator 8. After each multiplication, we added that product to the numerator of the fraction part and used the common denominator. This process is summarized as follows.

Objectives

① Understand that a mixed number is the sum of a whole number and a fraction.

② Be able to change a mixed number to the form of an improper fraction.

③ Be able to change an improper fraction to the form of a mixed number.

④ Know the difference between reducing an improper fraction and changing it to a mixed number.

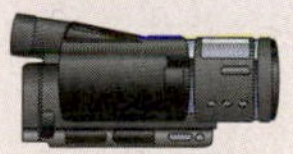

Shortcut for Changing Mixed Numbers to Fraction Form:

1. Multiply the whole number by the denominator of the fraction part.
2. Add the numerator of the fraction part to this product.
3. Write this sum over the denominator of the fraction.

With Example **1b.** as a guide, this shortcut can be diagrammed as follows:

16 + 7

$$2\frac{7}{8}=\frac{2\cdot 8+7}{8}=\frac{16+7}{8}=\frac{23}{8}$$

2 · 8

Example 2

Change $5\frac{2}{3}$ to an improper fraction.

Solution

Multiply $5\cdot 3=15$ and add 2: $15+2=17$

Write 17 over the denominator 3: $5\frac{2}{3}=\frac{17}{3}$

Now work exercise 2 in the margin.

2. Change $7\frac{9}{11}$ to an improper fraction.

Change each mixed number to improper fraction form.

3. $8\frac{1}{9}$

4. $5\frac{1}{2}$

Example 3

Change $6\frac{9}{10}$ to an improper fraction.

Solution

Multiply $10 \cdot 6 = 60$ and add 9: $60 + 9 = 69$

Write 69 over the denominator 10: $6\frac{9}{10} = \frac{69}{10}$

Now work exercise 3 in the margin.

Completion Example 4

Change $11\frac{5}{6}$ to an improper fraction.

Solution

Multiply $11 \cdot 6 =$ _____ and add 5: _____ + _____ = _____.

Write this sum, _____, over the denominator, _____.

Therefore,

$11\frac{5}{6} =$ _____.

Now work exercise 4 in the margin.

Changing Improper Fractions to Mixed Numbers

To reverse the process (that is, to change an improper fraction to a mixed number), we use the fact that division is one of the meanings of a fraction.

Objective 3

To Change an Improper Fraction to a Mixed Number:

1. Divide the numerator by the denominator to find the whole number part of the mixed number.
2. Write the remainder over the denominator as the fraction part of the mixed number. (Note that this fraction part will always be less than 1.)

Example 5

Change $\frac{29}{4}$ to a mixed number.

Solution

Divide 29 by 4:

$$\begin{array}{r} 7 \\ 4\overline{)29} \\ \underline{28} \\ 1 \end{array} \rightarrow \frac{29}{4} = 7 + \frac{1}{4} = 7\frac{1}{4}$$

Now work exercise 5 in the margin.

Example 6

Change $\frac{59}{3}$ to a mixed number.

Solution

Divide 59 by 3:

$$\begin{array}{r} 19 \\ 3\overline{)59} \\ \underline{3} \\ 29 \\ \underline{27} \\ 2 \end{array} \rightarrow \frac{59}{3} = 19 + \frac{2}{3} = 19\frac{2}{3}$$

Now work exercise 6 in the margin.

Note: Changing an improper fraction to a mixed number is not the same as reducing an improper fraction. **Reducing** involves finding common factors in the numerator and denominator. **Changing to a mixed number** involves division of the numerator by the denominator. Common factors are not involved. In any case, the fraction part of a mixed number should be in reduced form. To ensure this, we can follow either of the following two procedures:

1. Reduce the improper fraction first, and then change this fraction to a mixed number.
2. Change the improper fraction to a mixed number first, and then reduce the fraction part.

Change each improper fraction to mixed number form with the fraction part reduced.

5. a. $\frac{18}{7}$

b. $\frac{20}{16}$

6. a. $\frac{25}{2}$

b. $\frac{35}{15}$

Objective 4

7. A customer went to a fish market and bought the following amounts of seafood: $\frac{3}{4}$ pound of shrimp, $\frac{2}{3}$ pound of blue crabs, $\frac{5}{8}$ pound of squid, and $\frac{1}{2}$ pound of red snapper. How many pounds of seafood were purchased? (Express the answer as a mixed number.)

Example 7

A customer at a supermarket deli ordered the following amounts of sliced meats: $\frac{1}{3}$ pound of roast beef, $\frac{3}{4}$ pound of turkey, $\frac{3}{8}$ pound of salami, and $\frac{1}{2}$ pound of boiled ham. What was the total amount of meat purchased? (Express the answer as a mixed number.)

Solution

To find the total amount of meat purchased, we **add** the individual amounts. From Section 3.4, we know that to add fractions, we need a common denominator. The LCD is the least common multiple of the numbers 3, 4, 8, and 2: $\text{LCD} = 2 \cdot 2 \cdot 2 \cdot 3 = 24$.

$$\frac{1}{3}+\frac{3}{4}+\frac{3}{8}+\frac{1}{2}=\frac{1}{3}\cdot\frac{8}{8}+\frac{3}{4}\cdot\frac{6}{6}+\frac{3}{8}\cdot\frac{3}{3}+\frac{1}{2}\cdot\frac{12}{12}$$

$$=\frac{8}{24}+\frac{18}{24}+\frac{9}{24}+\frac{12}{24}$$

$$=\frac{47}{24}=1\frac{23}{24}$$

The total purchase was $1\frac{23}{24}$ pounds of meat.

Now work exercise 7 in the margin.

Completion Example Answers

4. Multiply $11 \cdot 6 = \mathbf{66}$ and add 5: $\mathbf{66} + \mathbf{5} = \mathbf{71}$.
Write this sum, **71**, over the denominator, **6**.

Therefore, $11\frac{5}{6} = \frac{\mathbf{71}}{\mathbf{6}}$.

Practice Problems

Reduce to lowest terms.

1. $\frac{18}{16}$ **2.** $\frac{35}{15}$

Change to a mixed number.

3. $\frac{51}{34}$ **4.** $\frac{35}{15}$

Change to an improper fraction.

5. $6\frac{2}{3}$ **6.** $7\frac{1}{100}$

Answers to Practice Problems:

1. $\frac{9}{8}$ **2.** $\frac{7}{3}$ **3.** $1\frac{1}{2}$ **4.** $2\frac{1}{3}$ **5.** $\frac{20}{3}$ **6.** $\frac{701}{100}$

Name ______________________ Section ____________ Date __________

Exercises 4.1

Reduce each of the following improper fractions to lowest terms.

1. $\frac{24}{18}$ **2.** $\frac{25}{10}$ **3.** $\frac{16}{12}$ **4.** $\frac{10}{8}$

5. $\frac{39}{26}$ **6.** $\frac{48}{32}$ **7.** $\frac{35}{25}$ **8.** $\frac{18}{16}$

9. $\frac{80}{64}$ **10.** $\frac{75}{60}$

Change each of the following to a mixed number with the fraction part reduced. See Examples 5 and 6.

11. $\frac{100}{24}$ **12.** $\frac{25}{10}$ **13.** $\frac{16}{12}$ **14.** $\frac{10}{8}$

15. $\frac{39}{26}$ **16.** $\frac{42}{8}$ **17.** $\frac{43}{7}$ **18.** $\frac{34}{16}$

19. $\frac{45}{6}$ **20.** $\frac{75}{12}$ **21.** $\frac{56}{18}$ **22.** $\frac{31}{15}$

23. $\frac{36}{12}$ **24.** $\frac{48}{16}$ **25.** $\frac{72}{16}$ **26.** $\frac{70}{34}$

ANSWERS

1. ________
2. ________
3. ________
4. ________
5. ________
6. ________
7. ________
8. ________
9. ________
10. ________
11. ________
12. ________
13. ________
14. ________
15. ________
16. ________
17. ________
18. ________
19. ________
20. ________
21. ________
22. ________
23. ________
24. ________
25. ________
26. ________

27. ______

28. ______

29. ______

30. ______

31. ______

32. ______

33. ______

34. ______

35. ______

36. ______

37. ______

38. ______

39. ______

40. ______

41. ______

42. ______

43. ______

44. ______

45. ______

46. ______

47. ______

48. ______

49. ______

50. ______

51. ______

27. $\frac{45}{15}$ **28.** $\frac{60}{36}$ **29.** $\frac{35}{20}$ **30.** $\frac{185}{100}$

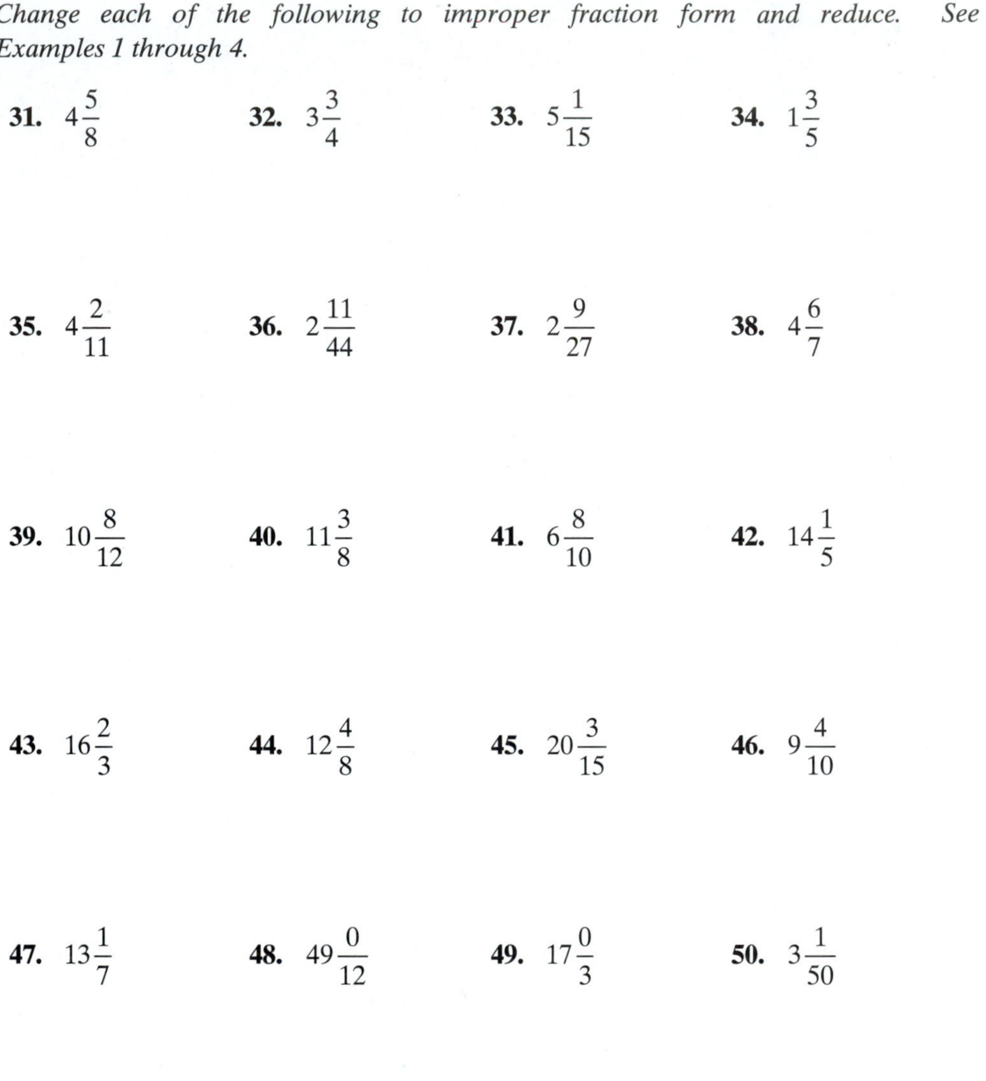

Change each of the following to improper fraction form and reduce. See Examples 1 through 4.

31. $4\frac{5}{8}$ **32.** $3\frac{3}{4}$ **33.** $5\frac{1}{15}$ **34.** $1\frac{3}{5}$

35. $4\frac{2}{11}$ **36.** $2\frac{11}{44}$ **37.** $2\frac{9}{27}$ **38.** $4\frac{6}{7}$

39. $10\frac{8}{12}$ **40.** $11\frac{3}{8}$ **41.** $6\frac{8}{10}$ **42.** $14\frac{1}{5}$

43. $16\frac{2}{3}$ **44.** $12\frac{4}{8}$ **45.** $20\frac{3}{15}$ **46.** $9\frac{4}{10}$

47. $13\frac{1}{7}$ **48.** $49\frac{0}{12}$ **49.** $17\frac{0}{3}$ **50.** $3\frac{1}{50}$

In each of the following exercises, write the answer in mixed number form. See Example 7.

51. Forestry. A tree in Yosemite National Forest grew $\frac{2}{3}$ foot, $\frac{3}{4}$ foot, $\frac{7}{8}$ foot, and $\frac{1}{2}$ foot in four consecutive years. How many feet did the tree grow during these four years?

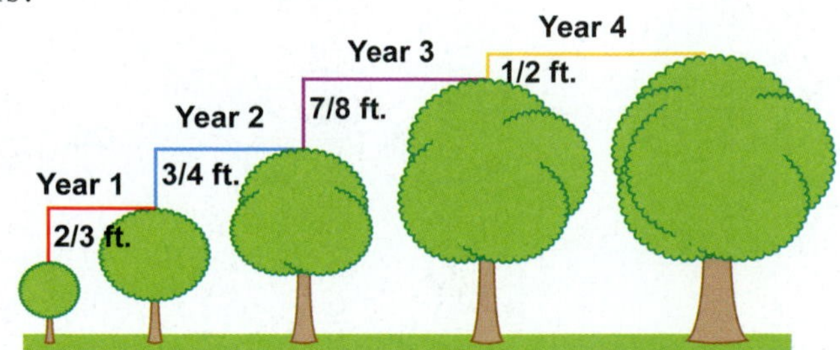

Name ______________________ Section ____________ Date __________

ANSWERS

52. Human Growth. A baby grew $\frac{1}{4}$ inch, $\frac{3}{8}$ inch, $\frac{3}{4}$ inch, and $\frac{9}{16}$ inch over a four-month period. How many inches did the baby grow over this time period?

52. ____________

53. Stock Market. During five days in one week the price of Xerox stock rose $\frac{1}{4}$, rose $\frac{7}{8}$, rose $\frac{3}{4}$, fell $\frac{1}{2}$, and rose $\frac{3}{8}$ of a dollar. What was the net gain (or loss) in the price of Xerox stock over this five-day period?

53. ____________

54. Measurement. After typing the manuscript for this textbook, Dr. Wright measured the height (or thickness) of the stack of pages for each of the first four chapters as $\frac{7}{8}$ inch, $\frac{1}{2}$ inch, $\frac{3}{4}$ inch, and $\frac{5}{8}$ inch. What was the total height of the first four chapters in typed form?

54. ____________

Writing and Thinking about Mathematics

55. You were probably familiar with the shortcut for changing a mixed number to an improper fraction (page 297) before you read this section. In your own words, explain how this method works and why it gives the correct improper fraction every time.

55. [Respond below exercise.]

56. Explain, in your own words, why the use of the word **improper** is somewhat misleading when referring to improper fractions. Can you think of another word used in a special sense (other than its normal English usage) that could be misleading to a person not familiar with the new meaning? (**Note:** This could lead to a lively class discussion about how and why some words are given special meanings, other than their normal meanings, in technical fields.)

56. [Respond below exercise.]

REVIEW

1. ________

2. ________

3. ________

4. ________

5. ________

6. ________

7. ________

8. ________

Review Problems (from Section 3.2 and 3.3)

Reduce each fraction to lowest terms.

1. $\frac{35}{45}$ **2.** $\frac{128}{320}$ **3.** $\frac{300}{100}$ **4.** $\frac{102}{221}$

Find each of the following products reduced to lowest terms.

5. $\frac{2}{3} \cdot \frac{15}{16} \cdot \frac{8}{25}$ **6.** $\frac{13}{24} \cdot \frac{15}{39} \cdot \frac{18}{20}$

Find each of the following quotients reduced to lowest terms.

7. $\frac{25}{36} \div \frac{35}{28}$ **8.** $\frac{65}{22} \div \frac{91}{33}$

4.2 Multiplication and Division with Mixed Numbers

Multiplication with Mixed Numbers

In Sections 4.3 and 4.4 we will see that addition and subtraction with mixed numbers both rely on the fact that (as we discussed in Section 4.1) a mixed number is the sum of a whole number and a fraction. However, in this section, we will see that multiplication and division are easier to perform if we change the mixed numbers to improper fraction form rather than treat them as sums. Thus multiplication and division with mixed numbers are the same as multiplication and division with fractions. Just as with fractions, we can use prime factorizations (or find other common factors) in both the numerator and the denominator and then reduce.

Objectives

① Learn how to multiply with mixed numbers.

② Learn how to divide with mixed numbers.

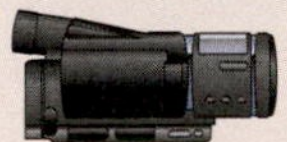

To Multiply Mixed Numbers:

1. Change each number to fraction form.
2. Multiply by factoring numerators and denominators; then reduce.
3. Change the answer to a mixed number or leave it in fraction form. (The choice sometimes depends on what use is to be made of the answer.)

Objective ①

Example 1

Find the product: $\left(\frac{5}{6}\right)\left(3\frac{3}{10}\right)$.

Solution

$$\frac{5}{6}\cdot 3\frac{3}{10} = \frac{5}{6}\cdot\frac{33}{10} = \frac{\cancel{5}\cdot\cancel{3}\cdot 11}{2\cdot\cancel{3}\cdot 2\cdot\cancel{5}} = \frac{11}{4} \quad \text{or} \quad 2\frac{3}{4}$$

Now work exercise 1 in the margin.

1. Find the product:

$\frac{1}{12}\cdot 18\frac{3}{4}$

Example 2

Multiply and reduce to lowest terms: $4\frac{1}{2}\cdot 1\frac{1}{6}\cdot 3\frac{1}{3}$.

Solution

$$4\frac{1}{2}\cdot 1\frac{1}{6}\cdot 3\frac{1}{3} = \frac{9}{2}\cdot\frac{7}{6}\cdot\frac{10}{3} = \frac{\cancel{3}\cdot\cancel{3}\cdot 7\cdot\cancel{2}\cdot 5}{\cancel{2}\cdot 2\cdot\cancel{3}\cdot\cancel{3}} = \frac{35}{2} \quad \text{or} \quad 17\frac{1}{2}$$

Now work exercise 2 in the margin.

2. Multiply and reduce to lowest terms:

$2\frac{5}{7}\cdot 5\frac{2}{5}\cdot 1\frac{1}{9}$

Large mixed numbers can be multiplied in the same way that we multiplied in Examples 1 and 2. The numerators and their products will be relatively large, but the technique of changing to improper fractions and then multiplying and reducing is the same.

3. The free-throw lane on a basketball court is $18\frac{5}{6}$ feet long and 12 feet wide. What is the area inside the lane?

Example 3

Linda is framing a rectangular antique circus poster that measures $24\frac{3}{8}$ inches wide by $45\frac{1}{4}$ inches long. What is the area of the glass needed to cover the poster?

Solution

Find the area (measured in square inches) of the rectangle by multiplying the width by the length.

a. To multiply $24\frac{3}{8} \cdot 45\frac{1}{4}$, first change both numbers to improper fractions.

$$\begin{array}{r} 24 \\ \times\ 8 \\ \hline 192 \end{array} \quad \left.\begin{array}{r} 192 \\ +\ 3 \\ \hline 195 \end{array}\right\} \quad 24\frac{3}{8} = \frac{195}{8}$$

$$\begin{array}{r} 45 \\ \times\ 4 \\ \hline 180 \end{array} \quad \left.\begin{array}{r} 180 \\ +\ 1 \\ \hline 181 \end{array}\right\} \quad 45\frac{1}{4} = \frac{181}{4}$$

b. Now multiply the improper fractions; then change the product back to a mixed number.

$$24\frac{3}{8} \cdot 45\frac{1}{4} = \frac{195}{8} \cdot \frac{181}{4} = \frac{35,295}{32} = 1102\frac{31}{32}$$

$$\begin{array}{r} 195 \\ \times\ 181 \\ \hline 195 \\ 15\ 60 \\ 19\ 5 \\ \hline 35,295 \end{array} \qquad \begin{array}{r} 1102\frac{31}{32} \\ 32\overline{)35,295} \\ \underline{32} \\ 3\ 2 \\ \underline{3\ 2} \\ 09 \\ \underline{0} \\ 95 \\ \underline{64} \\ 31 \end{array}$$

Thus the area of the glass is $1102\frac{31}{32}$ square inches.

Now work exercise 3 in the margin.

Completion Example 4

Find the product and write it as a mixed number with the fraction part in lowest terms:

$$\left(4\frac{3}{8}\right)\left(3\frac{1}{5}\right)\left(\frac{1}{34}\right).$$

Solution

$$\left(4\frac{3}{8}\right)\left(3\frac{1}{5}\right)\left(\frac{1}{34}\right)=\frac{35}{8}\cdot\_\_\_\_\_\_\_\_\_\cdot\_\_\_\_\_\_\_\_\_$$

$$=\frac{5\cdot 7\cdot\_\_\_\_\_\_\_\_\_}{2\cdot 2\cdot 2\cdot\_\_\_\_\_\_\_\_\_}=\_\_\_\_\_\_\_\_\_$$

Now work exercise 4 in the margin.

4. Find the product $4\frac{1}{5}\cdot 2\frac{3}{4}\cdot\frac{10}{33}$ and write it as a mixed number with the fraction part in lowest terms.

Completion Example 5

Multiply: $2\frac{3}{5}\cdot 8\frac{1}{4}$.

Solution

$$2\frac{3}{5}\cdot 8\frac{1}{4}=\frac{13}{5}\cdot\_\_\_\_\_\_\_$$

$$=\_\_\_\_\_\_\_=\_\_\_\_\_\_\_$$

Now work exercise 5 in the margin.

5. Find the product $\left(2\frac{1}{2}\right)\left(5\frac{3}{5}\right)$ and write the product in mixed number form.

In Section 3.1 we discussed the fact that finding a fractional part of another number requires multiplication. The key word is **of**, and we will find in later chapters that this is also true for decimals and percents. We emphasize this concept again here with mixed numbers.

> **Note:** To find a fraction **of** a number means to **multiply** the number by the fraction.

Example 6

Find $\frac{3}{4}$ of 40.

Solution

$$\frac{3}{4}\cdot 40=\frac{3}{\cancel{4}_{1}}\cdot\frac{\cancel{40}^{10}}{1}=30$$

Now work exercise 6 in the margin.

6. Find $\frac{3}{10}$ of 60.

Example 7

Find $\frac{2}{3}$ of $5\frac{1}{4}$.

Solution

$$\frac{2}{3}\cdot 5\frac{1}{4}=\frac{2}{3}\cdot\frac{21}{4}=\frac{\cancel{2}\cdot\cancel{3}\cdot 7}{\cancel{3}\cdot\cancel{2}\cdot 2}=\frac{7}{2}\text{ or }3\frac{1}{2}$$

Now work exercise 7 in the margin.

7. Find $\frac{1}{2}$ of $3\frac{1}{5}$.

Division with Mixed Numbers

Division with mixed numbers is the same as division with fractions, as discussed in Section 3.3. Simply change each mixed number to an improper fraction before dividing. Recall that, **to divide by any nonzero number, we multiply by its reciprocal**. That is, for $c \neq 0$ and $d \neq 0$, the **reciprocal** of $\frac{c}{d}$ is $\frac{d}{c}$ and

$$\frac{a}{b} \div \frac{c}{d} = \frac{a}{b} \cdot \frac{d}{c}$$

Objective 2

To Divide Mixed Numbers:

1. Change each number to fraction form.
2. Write the reciprocal of the divisor.
3. Multiply by factoring numerators and denominators; then reduce.

8. Divide: $3\frac{3}{4} \div 2\frac{2}{5}$

Example 8

Divide: $3\frac{1}{4} \div 7\frac{4}{5}$.

Solution

$$3\frac{1}{4} \div 7\frac{4}{5} = \frac{13}{4} \div \frac{39}{5} = \frac{13}{4} \cdot \frac{5}{39}$$

$$= \frac{\cancel{13} \cdot 5}{4 \cdot \cancel{13} \cdot 3} = \frac{5}{12}$$

Note that the divisor is $\frac{39}{5}$, and we multiply by its reciprocal, $\frac{5}{39}$.

Now work exercise 8 in the margin.

9. Find the quotient: $4\frac{2}{3} \div 4\frac{2}{3}$.

Example 9

Find the quotient: $7\frac{7}{8} \div 6$.

Solution

$$7\frac{7}{8} \div 6 = \frac{63}{8} \div \frac{6}{1} = \frac{63}{8} \cdot \frac{1}{6}$$

$$= \frac{\cancel{3} \cdot 3 \cdot 7}{8 \cdot 2 \cdot \cancel{3}} = \frac{21}{16} \text{ or } 1\frac{5}{16}$$

Now work exercise 9 in the margin.

Completion Example 10

Divide: $10\frac{1}{2} \div 3\frac{3}{4}$.

Solution

$$10\frac{1}{2} \div 3\frac{3}{4} = \frac{21}{2} \div \underline{\qquad} = \frac{21}{2} \cdot \underline{\qquad}$$

$$= \frac{3 \cdot 7 \cdot \underline{\qquad}}{2 \cdot \underline{\qquad}} = \underline{\qquad} \quad \text{or} \quad \underline{\qquad}$$

Now work exercise 10 in the margin.

10. Divide: $16\frac{1}{3} \div 2\frac{1}{8}$

If we know two numbers, say 2 and 7, we can find their product, 14, by multiplying the two numbers. But, if we know that the product of two numbers is 24 and one of the numbers is 3, how do we find the other number? Recall from our work with whole numbers in Chapter 1 that the answer is found by dividing the product by the known number. Thus the missing number is $24 \div 3 = 8$. Example 11 illustrates this same idea (finding a number by dividing the product of two numbers by the known number) with mixed numbers.

Example 11

The product of $2\frac{1}{3}$ with another number is $5\frac{1}{6}$. What is the other number?

Solution

$$2\frac{1}{3} \cdot ? = 5\frac{1}{6}$$

To find the missing number, *divide* $5\frac{1}{6}$ by $2\frac{1}{3}$.

$$5\frac{1}{6} \div 2\frac{1}{3} = \frac{31}{6} \div \frac{7}{3} = \frac{31}{\cancel{6}_2} \cdot \frac{\cancel{3}^1}{7} = \frac{31}{14} \text{ or } 2\frac{3}{14}$$

The other number is $2\frac{3}{14}$.

Check: $2\frac{1}{3} \cdot 2\frac{3}{14} = \frac{\cancel{7}^1}{3} \cdot \frac{31}{\cancel{14}_2} = \frac{31}{6} = 5\frac{1}{6}$

Now work exercise 11 in the margin.

11. The product of $3\frac{2}{7}$ with another number is $9\frac{1}{2}$. What is the other number?

Completion Example Answers

4. $\left(4\frac{3}{8}\right)\left(3\frac{1}{5}\right)\left(\frac{1}{34}\right) = \frac{35}{8}\cdot\frac{\mathbf{16}}{\mathbf{5}}\cdot\frac{\mathbf{1}}{\mathbf{34}}$

$= \frac{5\cdot 7\cdot \mathbf{2\cdot 2\cdot 2\cdot 2\cdot 1}}{2\cdot 2\cdot 2\cdot \mathbf{5\cdot 2\cdot 17}} = \frac{\mathbf{7}}{\mathbf{17}}$

5. $2\frac{3}{5}\cdot 8\frac{1}{4} = \frac{13}{5}\cdot\frac{\mathbf{33}}{\mathbf{4}}$

$= \frac{\mathbf{429}}{\mathbf{20}} = \mathbf{21}\frac{\mathbf{9}}{\mathbf{20}}$

10. $10\frac{1}{2} \div 3\frac{3}{4} = \frac{21}{2} \div \frac{\mathbf{15}}{\mathbf{4}} = \frac{21}{2}\cdot\frac{\mathbf{4}}{\mathbf{15}}$

$= \frac{3\cdot 7\cdot \mathbf{2\cdot 2}}{2\cdot \mathbf{3\cdot 5}} = \frac{\mathbf{14}}{\mathbf{5}}$ or $\mathbf{2\frac{4}{5}}$

Practice Problems

Find the indicated products.

1. $4\frac{1}{3}\cdot\frac{2}{13}$

2. $5\frac{1}{2}\cdot 3\frac{1}{3}$

3. $\frac{5}{8}\cdot 2\frac{3}{5}\cdot 5\frac{1}{3}$

4. $16\frac{1}{2}\cdot 20\frac{2}{3}$

5. Find $\frac{9}{10}$ of 70.

Find the indicated quotients.

6. $2\frac{1}{3} \div \frac{7}{3}$

7. $5\frac{1}{4} \div 3\frac{1}{2}$

8. $12 \div \frac{3}{4}$

9. $6\frac{3}{5} \div 11$

Answers to Practice Problems:

1. $\frac{2}{3}$ **2.** $\frac{55}{3}$ or $18\frac{1}{3}$ **3.** $\frac{26}{3}$ or $8\frac{2}{3}$ **4.** 341 **5.** 63

6. 1 **7.** $\frac{3}{2}$ or $1\frac{1}{2}$ **8.** 16 **9.** $\frac{3}{5}$

Name ______________________ Section ____________ Date __________

Exercises 4.2

Find the indicated products. See Examples 1, 2, 4, and 5.

1. $\left(2\frac{1}{3}\right)\left(3\frac{1}{4}\right)$

2. $\left(1\frac{1}{5}\right)\left(1\frac{1}{7}\right)$

3. $4\frac{1}{2}\left(2\frac{1}{3}\right)$

4. $3\frac{1}{3}\left(2\frac{1}{5}\right)$

5. $6\frac{1}{4}\left(3\frac{3}{5}\right)$

6. $5\frac{1}{3}\left(2\frac{1}{4}\right)$

7. $\left(8\frac{1}{2}\right)\left(3\frac{2}{3}\right)$

8. $\left(9\frac{1}{3}\right)2\frac{1}{7}$

9. $\left(6\frac{2}{7}\right)1\frac{3}{11}$

10. $\left(11\frac{1}{4}\right)1\frac{1}{15}$

11. $6\frac{2}{3}\cdot 4\frac{1}{2}$

12. $4\frac{3}{8}\cdot 2\frac{2}{7}$

13. $9\frac{3}{4}\cdot 2\frac{6}{26}$

14. $7\frac{1}{2}\cdot\frac{2}{15}$

15. $\frac{3}{4}\cdot 1\frac{1}{3}$

16. $3\frac{4}{5}\cdot 2\frac{1}{7}$

17. $12\frac{1}{2}\cdot 2\frac{1}{5}$

18. $9\frac{3}{5}\cdot 1\frac{1}{16}$

19. $6\frac{1}{8}\cdot 3\frac{1}{7}$

20. $5\frac{1}{4}\cdot 11\frac{1}{3}$

21. $\frac{1}{4}\cdot\frac{2}{3}\cdot\frac{6}{7}$

22. $\frac{7}{8}\cdot\frac{24}{25}\cdot\frac{5}{21}$

23. $\frac{3}{16}\cdot\frac{8}{9}\cdot\frac{3}{5}$

24. $\frac{2}{5}\cdot\frac{1}{5}\cdot\frac{4}{7}$

25. $2\frac{1}{4}\cdot 6\frac{3}{8}\cdot 1\frac{5}{27}$

26. $1\frac{3}{32}\cdot 1\frac{1}{7}\cdot 1\frac{1}{25}$

27. $1\frac{5}{16}\cdot 1\frac{1}{3}\cdot 1\frac{1}{5}$

28. $4\frac{1}{15}\cdot 5\frac{1}{6}$

ANSWERS

1. ____
2. ____
3. ____
4. ____
5. ____
6. ____
7. ____
8. ____
9. ____
10. ____
11. ____
12. ____
13. ____
14. ____
15. ____
16. ____
17. ____
18. ____
19. ____
20. ____
21. ____
22. ____
23. ____
24. ____
25. ____
26. ____
27. ____
28. ____

29. ______

30. ______

31. ______

32. ______

33. ______

34. ______

35. ______

36. ______

37. ______

38. ______

39. ______

40. ______

41. ______

42. ______

43. ______

44. ______

45. ______

46. ______

47. ______

48. ______

49. ______

50. ______

51. ______

52. ______

53. ______

54. ______

55. ______

56. ______

29. $7\frac{3}{5}\cdot 5\frac{1}{16}$ **30.** $4\frac{5}{6}\cdot 3\frac{1}{7}$ **31.** $5\frac{1}{13}\cdot 4\frac{1}{26}$ **32.** $6\frac{3}{4}\cdot 7\frac{5}{12}$

33. Find $\frac{2}{3}$ of 60. **34.** Find $\frac{1}{4}$ of 80. **35.** Find $\frac{1}{5}$ of 100. **36.** Find $\frac{3}{5}$ of 100.

37. Find $\frac{1}{2}$ of $2\frac{5}{8}$. **38.** Find $\frac{1}{6}$ of $1\frac{3}{4}$. **39.** Find $\frac{9}{10}$ of $3\frac{5}{7}$. **40.** Find $\frac{7}{8}$ of $6\frac{4}{5}$.

Find the indicated quotients. See Examples 8 through 10.

41. $\frac{2}{21}\div\frac{2}{7}$ **42.** $\frac{9}{32}\div\frac{5}{8}$ **43.** $\frac{5}{12}\div\frac{3}{4}$ **44.** $\frac{6}{17}\div\frac{6}{17}$

45. $\frac{5}{6}\div 3\frac{1}{4}$ **46.** $\frac{7}{8}\div 7\frac{1}{2}$ **47.** $\frac{29}{50}\div 3\frac{1}{10}$ **48.** $4\frac{1}{5}\div 3\frac{1}{3}$

49. $2\frac{1}{17}\div 1\frac{1}{4}$ **50.** $5\frac{1}{6}\div 3\frac{1}{4}$ **51.** $2\frac{2}{49}\div 3\frac{1}{14}$ **52.** $6\frac{5}{6}\div 2$

53. $4\frac{1}{5}\div 3$ **54.** $4\frac{5}{8}\div 4$ **55.** $6\frac{5}{6}\div\frac{1}{2}$ **56.** $4\frac{5}{8}\div\frac{1}{4}$

Name ____________ Section ________ Date ________

ANSWERS

57. $4\frac{1}{5} \div \frac{1}{3}$ **58.** $1\frac{1}{32} \div 3\frac{2}{3}$ **59.** $7\frac{5}{11} \div 4\frac{1}{10}$ **60.** $13\frac{1}{7} \div 4\frac{2}{11}$

61. Commuting. A man drives $17\frac{7}{10}$ miles one way to work, five days a week. How many miles does he drive each week going to and from work?

62. Plumbing. A length of pipe is $27\frac{3}{4}$ feet. What would be the total length if $36\frac{1}{2}$ of these pipe sections were laid end-to-end?

63. Reading. JoAnn reads $\frac{1}{6}$ of a book in 3 hours. If the book contains 540 pages, how many pages does she read in 3 hours? How long will she take to read the entire book?

64. Geography. Three towns (A, B, and C) are located on the same highway (assume a straight section of highway). Towns A and B are 53 kilometers apart. Town C is $45\frac{9}{10}$ kilometers from town B. How far apart are towns A and C? There is more than one correct answer. (The sketch shows that there are two possible situations to consider.)

53 km

A C B

A B C

53 km

57. ____________

58. ____________

59. ____________

60. ____________

61. ____________

62. ____________

63. ____________

64. ____________

65. ____________

65. **Engineering.** A telephone pole is 32 feet long. If $\frac{5}{16}$ of the pole must be underground and $\frac{11}{16}$ of the pole must be above ground, how much of the pole is underground? How much is above ground?

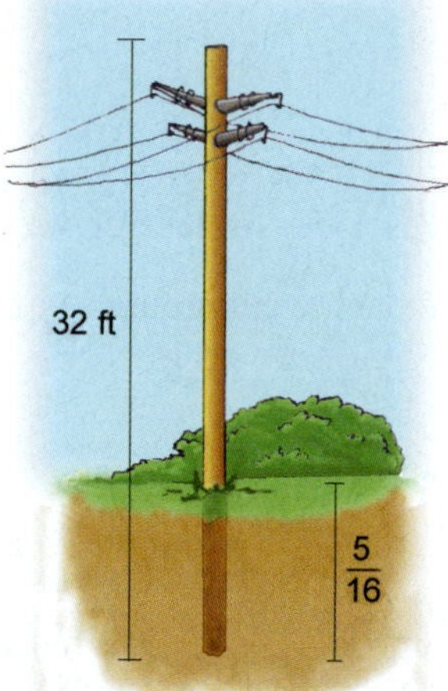

66. ____________

66. **Geometry.** The total distance around a square (its perimeter) is found by multiplying the length of one side by 4. Find the perimeter of a square if the length of one side is $5\frac{1}{16}$ inches.

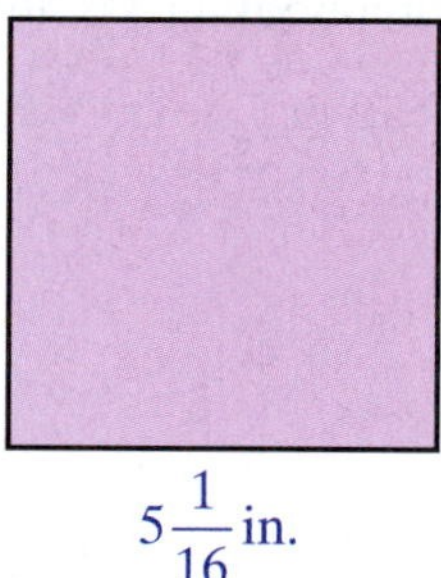

67. ____________

67. The product of $\frac{9}{10}$ with another number is $1\frac{2}{3}$. What is the other number?

68. ____________

68. The result of multiplying two numbers is $10\frac{1}{3}$. If one of the numbers is $7\frac{1}{6}$, what is the other number?

Name ______________________ Section ____________ Date ___________

ANSWERS

69. Airline Operations. An airplane is carrying 150 passengers. This is $\frac{6}{7}$ of its capacity. What is the capacity of the airplane?

69. ____________

70. Shopping. The sale price of a coat is \$135. This is $\frac{3}{4}$ of the original price. What was the original price (price before the sale)?

70. ____________

71. Shopping. The sale price of a new computer system (including a printer) is \$2400. This is $\frac{3}{4}$ of the original price. What was the original price (price before the sale)?

71. ____________

72. ____________

72. Shopping. A used car is advertised at a special price of \$3500. If this sale price is $\frac{4}{5}$ of the original price, what was the original price (price before the sale)?

73. a. ____________

73. Auto Mileage. Your car averages 26 miles per gallon of gas.

a. How many miles can your car travel on $17\frac{1}{2}$ gallons of gas?

b. If each gallon costs $202\frac{9}{10}$ cents, what would you pay (in cents) for $17\frac{1}{2}$ gallons (round to the nearest cent)?

b. ____________

74. Geometry. An equilateral triangle is one in which all three sides are the same length. What is the perimeter of an equilateral triangle with sides of length $15\frac{7}{10}$ centimeters?

74. ____________

75. ______

75. Geography. On a road map, 1 inch represents 50 miles. How many miles are represented by $3\frac{1}{4}$ inches?

76. ______

76. Diet. One tablespoon of butter contains 36 milligrams of cholesterol. How many milligrams of cholesterol are there in $2\frac{3}{4}$ tablespoons of butter?

Check Your Number Sense

77. a. ______

77. Suppose the product of $5\frac{7}{10}$ and some other number is $10\frac{1}{2}$. Answer the following questions without doing any calculations.

a. Do you think that this other number is more than 1 or less than 1? Why?

b. Do you think that this other number is more than 2 or less than 2? Why? Find the other number.

b. ______

REVIEW

1. ______

2. ______

3. ______

4. ______

5. ______

6. ______

7. ______

8. ______

Review Problems (from Section 3.4 and 3.5)

Find each of the following sums and reduce to lowest terms.

1. $\frac{2}{3}+\frac{1}{7}$ **2.** $\frac{5}{8}+\frac{9}{10}$ **3.** $\frac{3}{5}+\frac{7}{20}+\frac{1}{12}$ **4.** $\frac{11}{18}+\frac{2}{5}+\frac{3}{10}$

Find each of the following differences and reduce to lowest terms.

5. $\frac{7}{8}-\frac{9}{20}$ **6.** $\frac{19}{21}-\frac{3}{28}$ **7.** $\frac{17}{32}-\frac{17}{64}$ **8.** $\frac{20}{39}-\frac{9}{26}$

4.3 Addition with Mixed Numbers

Addition with Mixed Numbers

From Section 4.1, we know that a mixed number is the sum of a whole number and a fraction. In other words, we learned that the sum of mixed numbers is the sum of whole numbers and fractions. Keeping this relationship and the commutative and associative properties of addition in mind, we add mixed numbers by treating the whole numbers and the fraction parts separately.

Objectives

① Know how to add mixed numbers when the fraction parts have the same denominator.

② Know how to add mixed numbers when the fraction parts have different denominators.

③ Be able to write the sum of mixed numbers in the form of a mixed number with the fraction part less than 1.

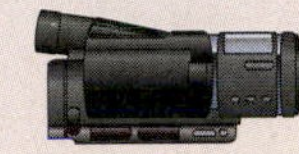

To Add Mixed Numbers:

1. Add the fraction parts.
2. Add the whole numbers.
3. Write the sum as a mixed number so that the fraction part is less than 1. (If the sum of the fraction parts is more than 1, rewrite it as a mixed number and add it to the sum of the whole numbers.)

Objective ①

Example 1

Find the sum: $4\frac{2}{7}+6\frac{3}{7}$.

Solution

We can write each number as a sum and then use the commutative and associative properties of addition to treat the whole numbers and fraction parts separately.

$$\begin{aligned} 4\frac{2}{7}+6\frac{3}{7} &= 4+\frac{2}{7}+6+\frac{3}{7} \\ &= (4+6)+\left(\frac{2}{7}+\frac{3}{7}\right) \\ &= 10+\frac{5}{7} \\ &= 10\frac{5}{7} \end{aligned}$$

1. Find the sum: $2\frac{1}{9}+3\frac{7}{9}$.

Now work exercise 1 in the margin.

Objective ②

Example 2

Add: $25\frac{1}{6}+3\frac{7}{18}$.

Solution

$$\begin{aligned} 25\frac{1}{6}+3\frac{7}{18} &= 25+\frac{1}{6}+3+\frac{7}{18} \\ &= (25+3)+\left(\frac{1}{6}+\frac{7}{18}\right) \end{aligned}$$

Continued on next page...

2. Add: $14\frac{7}{10}+22\frac{1}{5}$.

$$=28+\left(\frac{1}{6}\cdot\frac{3}{3}+\frac{7}{18}\right)$$
$$=28+\left(\frac{3}{18}+\frac{7}{18}\right)$$
$$=28\frac{10}{18}=28\frac{5}{9}$$

Or, vertically,

$$\begin{array}{rcccl} 25\frac{1}{6} = & 25\frac{1}{6}\cdot\frac{3}{3} = & 25\frac{3}{18} & & \text{LCD} = 18 \\ +\,3\frac{7}{18} = & 3\frac{7}{18} = & 3\frac{7}{18} & & \\ \hline & & 28\frac{10}{18} = 28\frac{5}{9} & & \end{array}$$

Now work exercise 2 in the margin.

Objective 3

3. Add the mixed numbers: $4\frac{5}{8}+10\frac{7}{10}$

Example 3

Add the mixed numbers: $7\frac{2}{3}+9\frac{4}{5}$.

Solution

$$\begin{array}{rcl} 7\frac{2}{3} = 7\frac{2}{3}\cdot\frac{5}{5} = & 7\frac{10}{15} & \quad \text{LCD} = 15 \\ +\,9\frac{4}{5} = 9\frac{4}{5}\cdot\frac{3}{3} = & 9\frac{12}{15} & \\ \hline & 16\frac{22}{15} = 16+1\frac{7}{15} = 17\frac{7}{15} & \end{array}$$

Fraction part is greater than 1. (points to $16\frac{22}{15}$)

Change it to a mixed number. (points to $1\frac{7}{15}$)

Now work exercise 3 in the margin.

Example 4

A triangle has sides measuring $3\frac{1}{3}$ meters, $6\frac{4}{5}$ meters, and $6\frac{14}{15}$ meters.

a. Find the perimeter of (total distance around) the triangle.

b. Find the area of the triangle if the height of the triangle is $3\frac{1}{5}$ meters and the base is $6\frac{14}{15}$ meters.

Solution

a. We find the perimeter by adding the lengths of the three sides.

$$3\frac{1}{3} = 3\frac{5}{15}$$
$$6\frac{4}{5} = 6\frac{12}{15}$$
$$+\,6\frac{14}{15} = 6\frac{14}{15}$$
$$15\frac{31}{15} = 15 + 2\frac{1}{15} = 17\frac{1}{15}\text{ meters}$$

The perimeter is $17\frac{1}{15}$ meters.

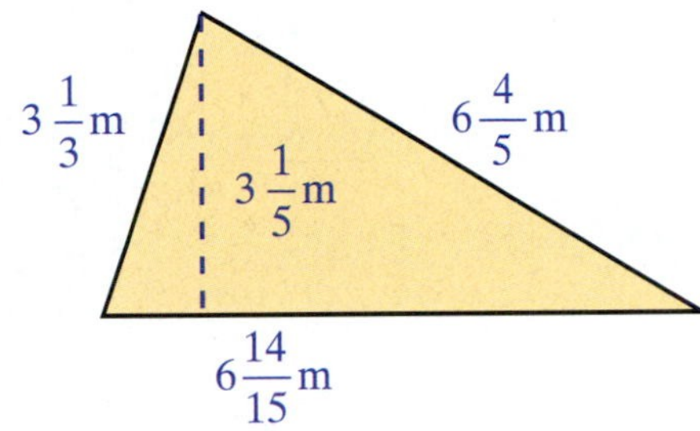

b. Let the base $= 6\frac{14}{15}$ meters and the height $= 3\frac{1}{5}$ meters.

To find the area of the triangle, multiply the base times the height and divide by 2.

$$A = \frac{1}{2}\cdot 6\frac{14}{15}\cdot 3\frac{1}{5} = \frac{1}{2}\cdot\frac{104}{15}\cdot\frac{16}{5} = \frac{832}{75} = 11\frac{7}{75}$$

$$\begin{array}{r} 11\frac{7}{75} \\ 75\overline{)832} \\ \underline{750} \\ 82 \\ \underline{75} \\ 7 \end{array}$$

Thus the area of the triangle is $11\frac{7}{75}$ square meters.

Now work exercise 4 in the margin.

4. A triangle has sides measuring $4\frac{1}{4}$ meters, $6\frac{1}{4}$ meters, and $6\frac{1}{2}$ meters.

a. Find the perimeter of the triangle.

b. Find the area of the triangle if the height of the triangle is 6 m. (**Hint:** the base is $4\frac{1}{4}$ meters.)

Practice Problems

Find the indicated sums. Write the answer as a mixed number.

1. $\begin{array}{r} 11\frac{1}{7} \\ +\ 9\frac{3}{4} \\ \hline \end{array}$

2. $5\frac{1}{3}+\frac{2}{13}$

3. $7\frac{3}{4}+\frac{1}{5}$

4. $\frac{7}{8}+1\frac{3}{5}+7\frac{1}{3}$

5. $\begin{array}{r} 1\frac{2}{7} \\ 7\frac{1}{2} \\ +\ 2\frac{3}{5} \\ \hline \end{array}$

Answers to Practice Problems:

1. $20\frac{25}{28}$ **2.** $5\frac{19}{39}$ **3.** $7\frac{19}{20}$ **4.** $9\frac{97}{120}$ **5.** $11\frac{27}{70}$

Name ______________________ Section ______________ Date ____________

Exercises 4.3

Find the indicated sums. See Examples 1 through 3. Write the answer as a mixed number.

1. $10 + 4\frac{5}{7}$

2. $6\frac{1}{2} + 3\frac{1}{2}$

3. $5\frac{1}{3} + 2\frac{2}{3}$

4. $8 + 7\frac{3}{11}$

5. $12 + \frac{3}{4}$

6. $15 + \frac{9}{10}$

7. $3\frac{1}{4} + 5\frac{5}{12}$

8. $5\frac{11}{12} + 3\frac{5}{6}$

9. $21\frac{1}{3} + 13\frac{1}{18}$

10. $9\frac{2}{3} + \frac{5}{6}$

11. $4\frac{1}{2}+3\frac{1}{6}$

12. $3\frac{1}{4}+7\frac{1}{8}$

13. $25\frac{1}{10}+17\frac{1}{4}$

14. $5\frac{1}{7}+3\frac{1}{3}$

15. $6\frac{5}{12}+4\frac{1}{3}$

16. $5\frac{3}{10}+2\frac{1}{14}$

17. $8\frac{2}{9}+4\frac{1}{27}$

18. $11\frac{3}{4}+2\frac{5}{16}$

19. $6\frac{4}{9}+12\frac{1}{15}$

20. $4\frac{1}{6}+13\frac{9}{10}$

21. $21\frac{3}{4}+6\frac{3}{4}$

22. $3\frac{5}{8}+3\frac{5}{8}$

23. $7\frac{3}{5}+2\frac{1}{8}$

24. $9\frac{1}{8}+3\frac{7}{12}$

ANSWERS

1. ____________
2. ____________
3. ____________
4. ____________
5. ____________
6. ____________
7. ____________
8. ____________
9. ____________
10. ____________
11. ____________
12. ____________
13. ____________
14. ____________
15. ____________
16. ____________
17. ____________
18. ____________
19. ____________
20. ____________
21. ____________
22. ____________
23. ____________
24. ____________

25. ______

26. ______

27. ______

28. ______

29. ______

30. ______

31. ______

32. ______

33. ______

34. ______

35. ______

36. ______

37. ______

25. $3\frac{1}{3}+4\frac{1}{4}+5\frac{1}{5}$

26. $\frac{3}{7}+2\frac{1}{14}+2\frac{1}{6}$

27. $20\frac{5}{8}+42\frac{5}{6}$

28. $25\frac{2}{3}+1\frac{1}{16}$

29. $32\frac{1}{64}+4\frac{1}{24}+17\frac{3}{8}$

30. $3\frac{1}{20}+7\frac{1}{15}+2\frac{3}{10}$

31. $\begin{array}{r} 4\frac{1}{3} \\ 8\frac{3}{8} \\ +\ 6\frac{1}{6} \\ \hline \end{array}$

32. $\begin{array}{r} 3\frac{2}{3} \\ 14\frac{1}{10} \\ +\ 5\frac{1}{5} \\ \hline \end{array}$

33. $\begin{array}{r} 13\frac{5}{8} \\ 13\frac{1}{12} \\ +10\frac{1}{4} \\ \hline \end{array}$

34. $\begin{array}{r} 5\frac{1}{8} \\ 1\frac{1}{5} \\ +3\frac{1}{40} \\ \hline \end{array}$

35. $\begin{array}{r} 27\frac{2}{3} \\ 30\frac{5}{8} \\ +31\frac{5}{6} \\ \hline \end{array}$

36. Travel. A bus trip is made in three parts. The first takes part $2\frac{1}{3}$ hours, the second part takes $2\frac{1}{2}$ hours, and the third part takes $3\frac{3}{4}$ hours. How long does the entire trip take?

37. Construction. A construction company was contracted to build three sections of highway. One section was $20\frac{7}{10}$ kilometers, the second section was $3\frac{4}{10}$ kilometers, and the third section was $11\frac{6}{10}$ kilometers. What was the total length of highway built?

Name ______________________ Section ____________ Date __________

ANSWERS

38. Geometry. A triangle has sides that measure $42\frac{3}{4}$ feet, $23\frac{1}{2}$ feet, and $22\frac{7}{8}$ feet. Find the perimeter of the triangle.

38. ____________

39. Geometry. A quadrilateral (four-sided figure) has sides that measure $3\frac{1}{2}$ inches, $2\frac{1}{4}$ inches, $3\frac{5}{8}$ inches, and $2\frac{3}{4}$ inches. What is the total distance around the quadrilateral?

39. ____________

40. Construction. A carpenter buys boards of five widths at the lumber yard $3\frac{1}{2}$, $9\frac{1}{4}$, $11\frac{1}{4}$, $7\frac{1}{4}$, and $5\frac{1}{2}$ inches. If he bought one of each of these boards, what would be the total width of these boards?

40. ____________

41. Movie Rentals. Among the top 50 movie rentals in 2004, "The Day After Tomorrow" earned $\$52\frac{7}{25}$ million, "The Last Samurai" earned $\$44\frac{3}{20}$ million, "Starsky and Hutch" earned $\$38\frac{2}{25}$ million, and "Gothika" earned $\$40\frac{7}{20}$ million. What was the total of the rental earnings of these four films in 2004?

41. ____________

42. Construction. A carpenter wants to build a single stand-alone bookshelf that will involve three shelves – a top, a bottom, and two sides. He will need three pieces of lumber each $3\frac{1}{4}$ feet long (including two shelves and a bottom), one piece $3\frac{3}{4}$ feet long for the top and two pieces each $3\frac{1}{2}$ feet long for the sides. At the local home supply store, the type of lumber needed comes in lengths of 8 feet and 12 feet.

a. How many pieces of length 12 feet will he need to buy to build the bookshelf and how much lumber will be left over?

b. Would he have been better off to buy lengths of 8 feet?

42. a. ____________

b. ____________

43. a. ______

b. ______

c. ______

44. ______

Check Your Number Sense

43. Truncated mixed numbers are the whole number part of a mixed number. For example, 4 is the truncated mixed number of $4\frac{3}{5}$. Using the idea of truncated mixed numbers, estimate each of the following products mentally.

a. $2\frac{1}{7} \cdot 3\frac{5}{8}$ **b.** $20\frac{1}{7} \cdot 30\frac{5}{8}$ **c.** $200\frac{1}{7} \cdot 300\frac{5}{8}$

44. Of your three estimated answers in Exercise 43, which one do you think is closest to the actual product? Why?

4.4 Subtraction with Mixed Numbers

Subtraction with Mixed Numbers

Subtraction with mixed numbers involves working with the whole numbers and the fraction parts separately in a manner similar to addition. With subtraction, though, we have another concern related to the sizes of the fraction parts. We will see that when the fraction part being subtracted is larger than the other fraction part, the smaller fraction part must be changed to an improper fraction by **borrowing** 1 from its whole number.

Objectives

① Know how to subtract mixed numbers when the fraction part of the number being subtracted is smaller than or equal to the fraction part of the other number.

② Know how to subtract mixed numbers when the fraction part of the number being subtracted is larger than the fraction part of the other number.

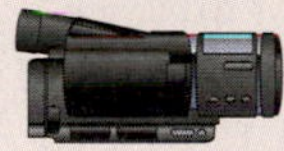 

To Subtract Mixed Numbers:

1. Subtract the fraction parts.
2. Subtract the whole numbers.

Objective ①

Example 1

Find the difference: $4\frac{3}{5}-1\frac{2}{5}$.

Solution

$$4\frac{3}{5}-1\frac{2}{5}=(4-1)+\left(\frac{3}{5}-\frac{2}{5}\right)$$
$$=3+\frac{1}{5}$$
$$=3\frac{1}{5}$$

Or, vertically,

$$\begin{array}{rcrr} 4\frac{3}{5} & & 4 & \frac{3}{5} \\ -1\frac{2}{5} & = & -1 & -\frac{2}{5} \\ \hline 3\frac{1}{5} & = & 3 & +\ \frac{1}{5} = 3\frac{1}{5} \end{array}$$

Now work exercise 1 in the margin.

1. Find the difference: $8\frac{6}{7}-2\frac{4}{7}$.

Example 2

Subtract: $10\frac{3}{5}-6\frac{3}{20}$.

Solution

$$\begin{array}{rcrcr} 10\frac{3}{5} & = & 10\frac{3}{5}\cdot\frac{4}{4} & = & 10\frac{12}{20} \\ -6\frac{3}{20} & = & -6\frac{3}{20} & = & -6\frac{3}{20} \\ \hline & & & & 4\frac{9}{20} \end{array}$$

LCD = 20

Now work exercise 2 in the margin.

2. Subtract: $9\frac{7}{10}-3\frac{1}{5}$.

When the fraction part of the number being subtracted is larger than the fraction part of the first number, we **borrow** 1 from the whole-number part and rewrite the first number as a whole number plus an improper fraction. Then the subtraction of the fraction parts can proceed as before.

Objective 2

If the Fraction Part Being Subtracted Is Larger than the First Fraction:

1. **Borrow** the whole number 1 from the first whole number.
2. Add this 1 to the first fraction. (This will always result in an improper fraction that is larger than the fraction being subtracted.)
3. Then subtract.

Find the difference for each expression.

3. $\begin{array}{r} 15\frac{1}{10} \\ -\ 7\frac{2}{10} \\ \hline \end{array}$

Example 3

Find the difference: $7\frac{2}{5} - 4\frac{3}{5}$.

Solution

Since $\frac{3}{5}$ is larger than $\frac{2}{5}$, **borrow** the whole number 1 from 7. We write 7 as 6 + 1; then we write $1+\frac{2}{5}$ as $\frac{7}{5}$.

$$\begin{array}{rcrcrcr} 7\frac{2}{5} & = & 6+1+\frac{2}{5} & = & 6+1\frac{2}{5} & = & 6\frac{7}{5} \\ -4\frac{3}{5} & = & 4+\quad\frac{3}{5} & = & 4+\quad\frac{3}{5} & = & 4\frac{3}{5} \\ \hline & & & & & & 2\frac{4}{5} \end{array}$$

Now work exercise 3 in the margin.

4. $6\frac{3}{14} - 2\frac{3}{4}$

Example 4

Subtract: $19\frac{2}{3} - 5\frac{3}{4}$.

Solution

First, change the fraction parts so that they have the same denominator, 12. Then **borrow** the whole number 1 from 19.

$$\begin{array}{rcrcrcr} 19\frac{2}{3} & = & 19\frac{2}{3}\cdot\frac{4}{4} & = & 19\frac{8}{12} & = & 18\frac{20}{12} \\ -5\frac{3}{4} & = & 5\frac{3}{4}\cdot\frac{3}{3} & = & 5\frac{9}{12} & = & 5\frac{9}{12} \\ \hline & & & & & & 13\frac{11}{12} \end{array}$$

$\leftarrow 1+\frac{8}{12}=\frac{12}{12}+\frac{8}{12}=\frac{20}{12}$

Now work exercise 4 in the margin.

Example 5

Find the difference: $7-4\frac{2}{3}$.

Solution

In this case the whole number 7 has no fraction part. We still **borrow** the whole number 1. We write 1 as $\frac{3}{3}$ so that its denominator will be the same as that of the other fraction part, $\frac{2}{3}$.

$$\begin{array}{rr} 7 & =6\frac{3}{3} \\ -4\frac{2}{3} & =4\frac{2}{3} \\ \hline & 2\frac{1}{3} \end{array}$$

Here $7=6+1=6+\frac{3}{3}=6\frac{3}{3}$.

Now work exercise 5 in the margin.

5. Find the difference:

$17-9\frac{6}{11}$.

Completion Example 6

Subtract: $10-6\frac{5}{8}$.

Solution

The whole number 10 has no fraction part. We still need to **borrow** the whole number 1 from 10.

$$\begin{array}{rr} 10 & =9\underline{\quad} \\ -6\frac{5}{8} & =6\frac{5}{8} \\ \hline & \underline{\quad} \end{array}$$

Now work exercise 6 in the margin.

6. Subtract: $\begin{array}{r} 13 \\ -\ 8\frac{5}{6} \\ \hline \end{array}$

Completion Example 7

Find the difference as indicated.

$$\begin{array}{rll} 14\frac{3}{4} & =14\underline{\quad} & =13\underline{\quad} \\ -5\frac{9}{10} & =\ 5\underline{\quad} & =\ 5\underline{\quad} \\ \hline & & \underline{\quad} \end{array}$$

Now work exercise 7 in the margin.

7. Find the difference as indicated.

$$\begin{array}{r} 16\frac{3}{8} \\ -10\frac{4}{5} \\ \hline \end{array}$$

8. When he first began a jogging program, Ted could run a mile in $8\frac{1}{4}$ minutes. Now he can run a mile in $6\frac{3}{5}$ minutes. How much has he cut from his timed mile?

Example 8

When he was first riding a trail with his new bicycle, Karl found that he was taking about $25\frac{1}{2}$ minutes to ride the 4 mile loop. Now he takes about $18\frac{3}{5}$ minutes to ride the same 4 miles. By how many minutes has he improved in riding 4 miles? In riding 1 mile?

Solution

a. To find by how many minutes Karl has improved in riding 4 miles, we simply find the difference between the two times.

$$\begin{aligned} 25\frac{1}{2} &= 25\frac{1}{2}\cdot\frac{5}{5} = 25\frac{5}{10} = 24\frac{15}{10} \\ -\ 18\frac{3}{5} &= 18\frac{3}{5}\cdot\frac{2}{2} = 18\frac{6}{10} = 18\frac{6}{10} \\ & \qquad\qquad\qquad\qquad\quad 6\frac{9}{10} \end{aligned}$$

So, in riding 4 miles, Karl has improved by about $6\frac{9}{10}$ minutes.

(**Note**: With truncated numbers, we could have estimated his improvement time as $25 - 18 = 7$ minutes.)

b. To find his improvement time in riding one mile, we divide the answer in part **a.** by 4.

$$6\frac{9}{10} \div 4 = \frac{69}{10} \div \frac{4}{1} = \frac{69}{10}\cdot\frac{1}{4} = \frac{69}{40} = 1\frac{29}{40}$$

He has improved by about $1\frac{29}{40}$ minutes per mile.

Now work exercise 8 in the margin.

Completion Example Answers

6.
$$\begin{aligned} 10 &= 9\mathbf{\frac{8}{8}} \\ -\ 6\frac{5}{8} &= 6\frac{5}{8} \\ & \quad \mathbf{3\frac{3}{8}} \end{aligned}$$

7.
$$\begin{aligned} 14\frac{3}{4} &= 14\mathbf{\frac{15}{20}} = 13\mathbf{\frac{35}{20}} \\ -\ 5\frac{9}{10} &= 5\mathbf{\frac{18}{20}} = 5\mathbf{\frac{18}{20}} \\ & \qquad\qquad\quad \mathbf{8\frac{17}{20}} \end{aligned}$$

Practice Problems

Find the following differences.

1. $\begin{array}{r} 3\frac{3}{4} \\ \underline{-2} \end{array}$

2. $\begin{array}{r} 7\frac{9}{10} \\ \underline{-4\frac{1}{7}} \end{array}$

3. $5\frac{2}{5}-3\frac{7}{10}$

4. $5\frac{7}{12}-2\frac{1}{6}$

5. $17\frac{3}{7}-5\frac{3}{10}$

Answers to Practice Problems:

1. $1\frac{3}{4}$ **2.** $3\frac{53}{70}$ **3.** $1\frac{7}{10}$ **4.** $3\frac{5}{12}$ **5.** $12\frac{9}{70}$

Name ____________ Section ________ Date ________

Exercises 4.4

Find each of the following differences. See Examples 1 through 7.

1. $5\frac{1}{2} - 1$

2. $7\frac{3}{4} - 2$

3. $4\frac{5}{12} - 3$

4. $3\frac{5}{8} - 2$

5. $6\frac{1}{2} - 2\frac{1}{2}$

6. $9\frac{1}{4} - 5\frac{1}{4}$

7. $5\frac{3}{4} - 2\frac{1}{4}$

8. $7\frac{9}{10} - 3\frac{3}{10}$

9. $14\frac{5}{8} - 11\frac{3}{8}$

10. $20\frac{7}{16} - 15\frac{5}{16}$

11. $4\frac{7}{8} - 1\frac{1}{4}$

12. $9\frac{5}{16} - 2\frac{1}{4}$

13. $5\frac{11}{12} - 1\frac{1}{4}$

14. $10\frac{5}{6} - 4\frac{2}{3}$

15. $8\frac{5}{6} - 2\frac{1}{4}$

16. $15\frac{5}{8} - 11\frac{3}{4}$

17. $14\frac{6}{10} - 3\frac{4}{5}$

18. $8\frac{3}{32} - 4\frac{3}{16}$

19. $5\frac{9}{10} - 2$

20. $7 - 6\frac{2}{3}$

ANSWERS

1. ____________
2. ____________
3. ____________
4. ____________
5. ____________
6. ____________
7. ____________
8. ____________
9. ____________
10. ____________
11. ____________
12. ____________
13. ____________
14. ____________
15. ____________
16. ____________
17. ____________
18. ____________
19. ____________
20. ____________

21. ______

22. ______

23. ______

24. ______

25. ______

26. ______

27. ______

28. ______

29. ______

30. ______

31. a. ______

b. ______

32. ______

33. ______

34. ______

21. $12-4\frac{1}{5}$ **22.** $75-17\frac{5}{6}$ **23.** $4\frac{9}{16}-2\frac{7}{8}$ **24.** $3\frac{7}{10}-2\frac{5}{6}$

25. $20\frac{3}{6}-3\frac{4}{8}$ **26.** $17\frac{3}{12}-12\frac{2}{8}$ **27.** $18\frac{2}{7}-4\frac{1}{3}$ **28.** $13\frac{5}{8}-6\frac{11}{20}$

29. $18\frac{7}{8}-2\frac{2}{3}$ **30.** $10\frac{3}{10}-2\frac{1}{2}$

31. Painting. Sara can paint a room in $3\frac{3}{5}$ hours, and Emily can paint a room of the same size in $4\frac{1}{5}$ hours.

a. How many hours are saved by having Sara paint a room of this size?
b. How many minutes are saved?

32. Grading. A teacher graded two sets of test papers. The first set took $3\frac{3}{4}$ hours to grade, and the second set took $2\frac{3}{5}$ hours. How much faster did she grade the second set?

33. Cleaning. Mike takes $1\frac{1}{2}$ hours to clean a pool, and Tom takes $2\frac{1}{3}$ hours to clean the same pool. How much longer does Tom take?

34. Running. When she first started training, a long-distance runner could run 10 miles in $70\frac{3}{10}$ minutes. Three months later she ran the same 10 miles in $63\frac{7}{10}$ minutes. By how much did her time improve?

Name ______________________________ Section ______________ Date ___________

ANSWERS

35. **Stock Market.** A certain stock was selling for $43\frac{7}{8}$ dollars per share. One month later it was selling for $48\frac{1}{2}$ dollars per share. By how much did the stock increase in price?

35. ____________

36. **Weight Loss.** You need to lose 10 pounds. If you weigh 180 pounds now and you lose $3\frac{1}{4}$ pounds during the first week and $3\frac{1}{2}$ pounds during the second week, how much more weight do you need to lose?

36. ____________

37. **Weight Loss.** Mr. Johnson originally weighed 240 pounds. During each week of six weeks of dieting, he lost $5\frac{1}{2}$ pounds, $2\frac{3}{4}$ pounds, $4\frac{5}{16}$ pounds, $1\frac{3}{4}$ pounds, $2\frac{5}{8}$ pounds, and $3\frac{1}{4}$ pounds. What did he weigh at the end of the six weeks?

37. ____________

38. **Travel.** A salesman drove $5\frac{3}{4}$ hours one day and $6\frac{1}{2}$ hours the next day. How much more time did he spend driving on the second day?

38. ____________

39. **Physiology.** On average, the air that we inhale includes $1\frac{1}{4}$ parts water, and the air we exhale includes $5\frac{9}{10}$ parts water. How many more parts water are in exhaled air?

39. ____________

40. **Running.** A person who is running will burn about $14\frac{7}{10}$ calories each minute, and a person who is walking will burn about $5\frac{1}{2}$ calories each minute. How many more calories does a runner burn in a minute than a walker?

40. ____________

41. [Respond below exercise.] ______

42. a. ______

b. ______

c. ______

43. a. ______

b. ______

c. ______

Writing and Thinking about Mathematics

41. In your own words, describe the technique for estimating answers when mixed numbers are involved in calculations.

42. Using the rule you just described in Exercise 41, mentally estimate each of the following differences.

a. $15\frac{6}{7}-11\frac{3}{4}$ **b.** $136\frac{17}{40}-125\frac{23}{30}$ **c.** $945\frac{1}{10}-845\frac{3}{100}$

43. Consider the following problem:

The product of two numbers is $16\frac{1}{2}$. If one of the numbers is $2\frac{3}{4}$, what is the other number?

a. Rewrite the problem using truncated mixed numbers and solve this new problem.

b. Does this technique make it easier for you to understand the original problem? Why or why not?

c. What is the solution to the original problem?

REVIEW

1. ______

2. ______

3. ______

4. ______

Review Problems (from Section 2.2 and 3.6)

Use the rules for order of operations to simplify each of the following expressions.

1. $6+2(13-5\cdot 2)+12\div 2$ **2.** $5^3+6^2-3\cdot 5\cdot 2^2$

3. $\left(\frac{1}{3}\right)^2+\frac{5}{8}\div\frac{1}{4}$ **4.** $\frac{3}{4}\div\frac{1}{2}\cdot 6-\frac{1}{5}(45)$

4.5 Order of Operations with Mixed Numbers

Objective

① Be able to follow the rules for order of operations with mixed numbers.

Order of Operations

We have discussed the rules for order of operations with whole numbers (see Section 2.2) and with fractions (see Section 3.6). These same rules are valid when using mixed numbers. In fact, some calculators are programmed to use these rules when operating with any type of number, including decimal numbers. The rules are restated here for easy reference.

Rules for Order of Operations

1. First, simplify within grouping symbols, such as parentheses ( ), brackets [ ], and braces { }. Start with the innermost grouping.
2. Second, evaluate any numbers or expressions indicated by exponents.
3. Third, moving from **left to right**, perform any multiplications or divisions in the order in which they appear.
4. Fourth, moving from **left to right**, perform any additions or subtractions in the order in which they appear.

Objective ①

The Rules for Order of Operations are used in each of the following examples. Carefully study the steps in each example. You should note that, in general, when mixed numbers are changed to improper fractions, working with mixed numbers is the same as working with fractions.

Example 1

Evaluate the expression $3\frac{2}{5} \div \left(\frac{1}{4} + \frac{3}{5}\right)$.

Solution

$3\frac{2}{5} = \frac{17}{5}$ Change the mixed number to an improper fraction.

$\frac{1}{4} + \frac{3}{5} = \frac{1}{4} \cdot \frac{5}{5} + \frac{3}{5} \cdot \frac{4}{4} = \frac{5}{20} + \frac{12}{20} = \frac{17}{20}$ Add the numbers inside parentheses.

Now divide:

$$3\frac{2}{5} \div \left(\frac{1}{4} + \frac{3}{5}\right) = \frac{17}{5} \div \frac{17}{20} = \frac{17}{5} \cdot \frac{20}{17} = \frac{\cancel{17} \cdot 4 \cdot \cancel{5}}{\cancel{5} \cdot \cancel{17}} = \frac{4}{1} = 4$$

Now work exercise 1 in the margin.

1. Evaluate the expression $3\frac{1}{4} \div \left(\frac{3}{4} + \frac{7}{8}\right)$.

Evaluate each of the following expressions.

2. $\left(4\frac{1}{3}+2\frac{2}{3}\right)\div\left(3\frac{2}{5}-\frac{2}{5}\right)$

3. $3\frac{1}{5}\div\frac{2}{3}+1\frac{1}{2}\cdot2\frac{1}{3}$

Example 2

Evaluate the expression $\left(5\frac{2}{3}-2\frac{1}{3}\right)\div\left(1\frac{1}{2}+\frac{1}{2}\right)$.

Solution

Work inside each set of parentheses:

$$5\frac{2}{3}-2\frac{1}{3}=3\frac{1}{3}=\frac{10}{3}\text{ and }1\frac{1}{2}+\frac{1}{2}=2$$

Now divide:

$$\left(5\frac{2}{3}-2\frac{1}{3}\right)\div\left(1\frac{1}{2}+\frac{1}{2}\right)=\frac{10}{3}\div2=\frac{10}{3}\cdot\frac{1}{2}=\frac{5\cdot\cancel{2}\cdot1}{3\cdot\cancel{2}}=\frac{5}{3}=1\frac{2}{3}$$

Now work exercise 2 in the margin.

Example 3

Evaluate the expression $2\frac{1}{2}\cdot1\frac{1}{6}+7\div\frac{3}{4}$.

Solution

$$\begin{aligned}2\frac{1}{2}\cdot1\frac{1}{6}+7\div\frac{3}{4}&=\frac{5}{2}\cdot\frac{7}{6}+\frac{7}{1}\cdot\frac{4}{3}\\&=\frac{35}{12}+\frac{28}{3}&&\text{Multiply and divide from left to right.}\\&=\frac{35}{12}+\frac{28}{3}\cdot\frac{4}{4}&&\text{Now add.}\\&=\frac{35}{12}+\frac{112}{12}\\&=\frac{147}{12}=12\frac{3}{12}=12\frac{1}{4}\end{aligned}$$

Or, working with separate parts, we can write

$$2\frac{1}{2}\cdot1\frac{1}{6}=\frac{5}{2}\cdot\frac{7}{6}=\frac{35}{12}=2\frac{11}{12}\qquad\text{multiplying}$$

$$7\div\frac{3}{4}=\frac{7}{1}\cdot\frac{4}{3}=\frac{28}{3}=9\frac{1}{3}\qquad\text{dividing}$$

$$\begin{aligned}2\frac{11}{12}&=2\frac{11}{12}\\+\ 9\frac{1}{3}&=9\frac{4}{12}&&\text{adding the results}\\&\ 11\frac{15}{12}=12\frac{3}{12}=12\frac{1}{4}\end{aligned}$$

Now work exercise 3 in the margin.

Example 4

Evaluate the expression $\left(2\frac{1}{2}\right)^2 + \frac{1}{5}\cdot\frac{1}{6} \div \frac{1}{15}$.

Solution

$$\begin{aligned}\left(2\frac{1}{2}\right)^2 + \frac{1}{5}\cdot\frac{1}{6} \div \frac{1}{15} &= \left(\frac{5}{2}\right)^2 + \frac{1}{5}\cdot\frac{1}{6} \div \frac{1}{15}\\ &= \frac{25}{4} + \frac{1}{30}\cdot\frac{15}{1}\\ &= \frac{25}{4} + \frac{1}{2}\\ &= \frac{25}{4} + \frac{2}{4}\\ &= \frac{27}{4}\\ &= 6\frac{3}{4}\end{aligned}$$

Now work exercise 4 in the margin.

4. Evaluate the expression: $3\frac{5}{7} \div \left(\left(\frac{1}{2}\right)^2 + 1\frac{3}{4}\right)$.

Example 5

Find the average of the mixed numbers $1\frac{1}{2}$, $2\frac{3}{4}$, and $3\frac{5}{8}$. You may want to refer back to Section 1.7.

Solution

We find the sum first and then divide by 3.

$$\begin{aligned}1\frac{1}{2} &= 1\frac{4}{8}\\ 2\frac{3}{4} &= 2\frac{6}{8}\\ +3\frac{5}{8} &= 3\frac{5}{8}\\ \hline 6\frac{15}{8} &= 7\frac{7}{8}\end{aligned}$$

$$7\frac{7}{8} \div 3 = \frac{63}{8}\cdot\frac{1}{3} = \frac{\cancel{3}\cdot 3\cdot 7}{8\cdot\cancel{3}} = \frac{21}{8} = 2\frac{5}{8}$$

Therefore, the average is $2\frac{5}{8}$.

Now work exercise 5 in the margin.

5. Find the average of the mixed numbers $3\frac{1}{4}$, $4\frac{1}{12}$, and $2\frac{5}{6}$.

Practice Problems

Use the rules for order of operations to evaluate the expressions.

1. $2 \div \left(\frac{1}{2} + \frac{3}{7}\right)$

2. $\left(\frac{4}{5} - \frac{2}{3}\right) \div 3\frac{1}{2}$

3. $\frac{2}{7} \cdot \frac{9}{10} - \frac{1}{8} \div 2$

4. $5\frac{1}{2} \div \left(2 - \frac{1}{3}\right)^2$

5. $\left(4\frac{1}{3} + 2\frac{5}{6}\right) \div \left(3\frac{1}{7} - 2\frac{1}{4}\right)$

6. Find the average of the numbers $3\frac{1}{2}$, $5\frac{1}{6}$, and $7\frac{2}{7}$.

Answers to Practice Problems:

1. $\frac{28}{13}$ or $2\frac{2}{13}$ **2.** $\frac{4}{105}$ **3.** $\frac{109}{560}$

4. $\frac{99}{50}$ or $1\frac{49}{50}$ **5.** $\frac{602}{75}$ or $8\frac{2}{75}$ **6.** $\frac{335}{63}$ or $5\frac{20}{63}$

Name ____________ Section ____________ Date ____________

Exercises 4.5

Use the rules for order of operations to evaluate the expressions. See Examples 1 through 4.

1. $1 \div \left(\frac{1}{12}+\frac{1}{6}\right)$

2. $1 \div \left(\frac{1}{15}+\frac{2}{15}\right)$

3. $\left(\frac{3}{10}+\frac{1}{6}\right) \div 3$

4. $\left(\frac{7}{12}+\frac{1}{15}\right) \div 5$

5. $\left(2+\frac{1}{5}\right) \div 2\frac{1}{5}$

6. $\left(4+\frac{1}{2}\right) \div 4\frac{1}{2}$

7. $\left(4+\frac{1}{3}\right) \div \left(6+\frac{1}{4}\right)$

8. $\left(2-\frac{1}{3}\right) \div \left(1-\frac{1}{3}\right)$

9. $\left(\frac{5}{8}+\frac{5}{8}\right) \div \left(2\frac{1}{2}\right)^2$

10. $\left(\frac{3}{10}\right)^2 \div \left(1+\frac{2}{3}\right)^2$

11. $\left(6\frac{1}{100}+5\frac{3}{100}\right) \div \left(2\frac{1}{10}+3\frac{1}{10}\right)$

12. $5\frac{1}{100} \div \left(4\frac{7}{10}-2\frac{9}{10}\right)$

13. $\left(7\frac{1}{5}-2\frac{1}{3}\right)-\left(6\frac{1}{10}-2\right)$

14. $\left(5\frac{2}{3}-1\frac{1}{6}\right)-\left(3\frac{1}{2}+\frac{1}{6}\right)$

15. $\frac{3}{5}\cdot\frac{1}{6}+\frac{1}{5}\div 2$

16. $\frac{1}{2}\div\frac{1}{2}+1-\frac{2}{3}\cdot 3$

17. $3\frac{1}{2}\cdot 5\frac{1}{3}+\frac{5}{12}\div\frac{15}{16}$

ANSWERS

1. ____________
2. ____________
3. ____________
4. ____________
5. ____________
6. ____________
7. ____________
8. ____________
9. ____________
10. ____________
11. ____________
12. ____________
13. ____________
14. ____________
15. ____________
16. ____________
17. ____________

18. ________

19. ________

20. ________

21. ________

22. ________

23. ________

24. ________

25. ________

26. ________

27. ________

28. ________

29. ________

18. $2\frac{1}{4}+1\frac{1}{5}+2\div\frac{20}{21}$ **19.** $\frac{5}{8}-\frac{1}{3}\cdot\frac{2}{5}+6\frac{1}{10}$ **20.** $1\frac{1}{6}\cdot1\frac{2}{19}\div\frac{7}{8}+\frac{1}{38}$

21. $\frac{3}{10}+\frac{5}{6}\div\frac{1}{4}\cdot\frac{1}{8}-\frac{7}{60}$ **22.** $5\frac{1}{7}\div\left(2+1\frac{1}{3}\right)^2$ **23.** $\left(2-\frac{1}{3}\right)\div\left(1-\frac{1}{3}\right)^2$

24. $\left(2\frac{4}{9}+1\frac{1}{18}\right)\div\left(1\frac{2}{9}-\frac{1}{6}\right)$

25. If the product of $5\frac{1}{2}$ and $2\frac{1}{4}$ is added to the quotient of $\frac{9}{10}$ and $\frac{3}{4}$, what is the sum?

26. If the quotient of $\frac{5}{8}$ and $\frac{1}{2}$ is subtracted from the product of $2\frac{1}{4}$ and $3\frac{1}{5}$, what is the difference?

27. If $\frac{9}{10}$ of 70 is divided by $\frac{3}{4}$ of 10, what is the quotient?

28. If $\frac{2}{3}$ of $4\frac{1}{4}$ is added to $\frac{5}{8}$ of $6\frac{1}{3}$, what is the sum?

29. Find the average of the numbers $\frac{7}{8}$, $\frac{9}{10}$, and $1\frac{3}{4}$.

Name ____________________ Section ____________ Date __________

ANSWERS

30. Find the average of the numbers $\frac{5}{6}$, $\frac{1}{15}$, and $\frac{17}{30}$.

30. __________

31. Find the average of the numbers $5\frac{1}{8}$, $7\frac{1}{2}$, $4\frac{3}{4}$, and $10\frac{1}{2}$.

31. __________

32. Find the average of the numbers $4\frac{7}{10}$, $3\frac{9}{10}$, $5\frac{1}{100}$, and $11\frac{3}{20}$.

32. __________

33. A video cassette tape has a running time of 2 hours at standard speed, 4 hours at long-play speed, and 6 hours at super long-play speed. If $\frac{1}{2}$ of the tape is recorded at standard speed, $\frac{1}{4}$ at long-play speed, and $\frac{1}{4}$ at super long-play speed, what is the total playing time of the entire recording?

33. __________

Collaborative Learning

34. With the class separated into teams of two to four students, each team is to write one or two paragraphs on the following two topics. Then the team leader is to read the paragraphs with classroom discussion to follow.

a. What topic have you found to be the most interesting in the text so far? Why?

b. What topic have you found to be the most useful in the text so far? Why?

34. a. __________

b. __________

4.6 Denominate Numbers

Simplifying Mixed Denominate Numbers

Abstract numbers are numbers with no units of measure attached. **Denominate numbers** are numbers with units of measure attached.

Objective ①

Abstract Numbers	Denominate Numbers
25	25 centimeters
1	1 ounce
$\frac{1}{2}$	$\frac{1}{2}$ inch
$5\frac{3}{4}$	$5\frac{3}{4}$ feet

Denominate numbers with two or more related units are called **mixed denominate numbers**. Examples of commonly used mixed denominate numbers are

5 ft 8 in., 3 lb 4 oz, and 1 hr 45 min.

With mixed numbers we make sure that the fraction part is less than 1. Thus we write $6\frac{1}{2}$, not $5\frac{3}{2}$. Similarly, in **simplified mixed denominate numbers**, the number of smaller units is less than 1 of the larger unit. Table 4.1 shows equivalent measures of length, weight, liquid volume, and time used in the U.S. customary system of measure. (Metric tables are included in the Appendix for interested students.)

Objectives

① Know the difference between abstract numbers and denominate numbers.

② Be able to simplify mixed denominate numbers.

③ Know how to add like denominate numbers.

④ Know how to subtract like denominate numbers.

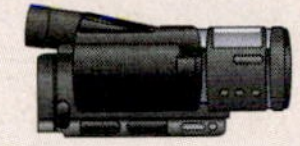

Table 4.1 U.S. Customary Units of Measure

Length		***Liquid Volume***	
1 foot (ft)	= 12 inches (in.)	1 pint (pt)	= 16 fluid ounces (fl oz)
1 yard (yd)	= 3 ft	1 quart (qt)	= 2 pt = 32 fl oz
1 mile (mi)	= 5280 ft	1 gallon (gal)	= 4 qt = 128 fl oz
Weight		***Time***	
1 pound (lb)	= 16 ounces (oz)	1 minute (min)	= 60 seconds (sec)
1 ton (t)	= 2000 lb	1 hour (hr)	= 60 min
		1 day	= 24 hr

To simplify mixed denominate numbers, either you must have a table of equivalent values (such as Table 4.1) with you or you must memorize the basic equivalent values. Most people know at least some of these values from experience. The examples that follow illustrate the technique for simplifying mixed denominate numbers. Remember that the number of smaller units cannot be greater than 1 whole unit of the next larger unit. See Examples 1, 2, and 3.

Objective 2

Simplify each of the following mixed denominate numbers.

1. 10 ft 13 in.

2. 2 lb 22 oz

3. 3 hr 75 min

Example 1

Simplify the mixed denominate number 3 ft 14 in.

Solution

Since 14 in. is more than 1 ft (there are 12 in. in 1 ft), we write

$$\begin{aligned} 3\text{ ft }14\text{ in.} &= 3\text{ ft} + 12\text{ in.} + 2\text{ in.} \\ &= \underbrace{3\text{ ft} + 1\text{ ft}} + 2\text{ in.} \\ &= 4\text{ ft} + 2\text{ in.} \\ &= 4\text{ ft }2\text{ in.} \end{aligned}$$

Now work exercise 1 in the margin.

Example 2

Simplify the mixed denominate number 5 lb 30 oz.

Solution

Since 30 oz is more than 1 lb, we write

$$\begin{aligned} 5\text{ lb }30\text{ oz} &= 5\text{ lb} + 16\text{ oz} + 14\text{ oz} \\ &= \underbrace{5\text{ lb} + 1\text{ lb}} + 14\text{ oz} \\ &= 6\text{ lb} + 14\text{ oz} \\ &= 6\text{ lb }14\text{ oz} \end{aligned}$$

Now work exercise 2 in the margin.

Example 3

Simplify the mixed denominate number 2 hr 70 min.

Solution

Since 70 min is more than 1 hr, we write

$$\begin{aligned} 2\text{ hr }70\text{ min} &= 2\text{ hr} + 60\text{ min} + 10\text{ min} \\ &= \underbrace{2\text{ hr} + 1\text{ hr}} + 10\text{ min} \\ &= 3\text{ hr} + 10\text{ min} \\ &= 3\text{ hr }10\text{ min} \end{aligned}$$

Now work exercise 3 in the margin.

Adding and Subtracting Like Denominate Numbers

Understanding how to simplify mixed denominate numbers helps in both adding and subtracting such numbers. **Like denominate numbers** are denominate numbers that have the same units or that can be written with the same units. For example,

5 ft 10 in.	and	2 ft 3 in.	are like denominate numbers, and
3 hr 5 min	and	4 hr 15 min	are like denominate numbers.

To Add Like Denominate Numbers:

1. Write the numbers in column form so that like units are aligned.
2. Add the numbers in each column.
3. Simplify the resulting mixed denominate number, if necessary.

Objective 3

Examples 4 – 6 illustrate addition with like denominate numbers and simplifying the sum.

Example 4

$$\begin{array}{r} 3\text{ ft}\;\;2\text{ in.} \\ 2\text{ ft}\;\;8\text{ in.} \\ +\;\;5\text{ ft}\;\;5\text{ in.} \\ \hline 10\text{ ft}\;\;15\text{ in.} \end{array}$$

$$10\text{ ft }15\text{ in.} = 10\text{ ft} + 12\text{ in.} + 3\text{ in.} = 11\text{ ft }3\text{ in.}$$

Now work exercise 4 in the margin.

4. Add and simplify the following sum:

$$\begin{array}{r} 5\text{ ft}\;\;6\text{ in.} \\ 3\text{ ft}\;\;4\text{ in.} \\ +\;\;9\text{ ft}\;\;3\text{ in.} \\ \hline \end{array}$$

Example 5

$$\begin{array}{r} 2\text{ hr }15\text{ min} \\ +\;\;4\text{ hr }50\text{ min} \\ \hline 6\text{ hr }65\text{ min} \end{array}$$

$$6\text{ hr }65\text{ min} = 6\text{ hr} + 60\text{ min} + 5\text{ min} = 7\text{ hr }5\text{ min}$$

Now work exercise 5 in the margin.

5. Add and simplify the following sum:

$$\begin{array}{r} 20\text{ min }45\text{ sec} \\ +\;\;3\text{ min }25\text{ sec} \\ \hline \end{array}$$

Example 6

In three different trips to the supermarket, Dan bought the following amounts of milk: 2 gal 2 qt, 1 gal 3 qt, and 1 gal 2 qt. What was the total amount of milk he bought on these three trips?

Solution

To find the total amount, we add the denominate numbers and simplify.

$$\begin{array}{r} 2\text{ gal }2\text{ qt} \\ 1\text{ gal }3\text{ qt} \\ +\;1\text{ gal }2\text{ qt} \\ \hline 4\text{ gal }7\text{ qt} \end{array}$$

$$4\text{ gal }7\text{ qt} = 4\text{ gal} + 4\text{ qt} + 3\text{ qt} = 5\text{ gal }3\text{ qt}$$

The total amount of milk that Dan bought was 5 gal 3 qt.

Now work exercise 6 in the margin.

6. Dennis filled up three buckets – 1 gal 1 qt, 2 gal 2 qt, and 1 gal 3 qt. What was the total amount of water in the buckets?

Objective 4

To Subtract Like Denominate Numbers:

1. Write the numbers in column form so that like units are aligned.
2. If necessary for subtraction, borrow 1 of the larger units and rewrite the top numbers.
3. Subtract the like units.

Examples 7 – 9 illustrate subtraction with like denominate numbers.

7. Subtract:

$$\begin{array}{r} 9 \text{ lb } 12 \text{ oz} \\ -7 \text{ lb } \ \ 6 \text{ oz} \\ \hline \end{array}$$

Example 7

$$\begin{array}{r} 8 \text{ lb } 14 \text{ oz} \\ -\ 3 \text{ lb } 10 \text{ oz} \\ \hline 5 \text{ lb } \ \ 4 \text{ oz} \end{array}$$

Now work exercise 7 in the margin.

8. Subtract:

$$\begin{array}{r} 13 \text{ ft } 2 \text{ in.} \\ -\ \ 4 \text{ ft } 9 \text{ in.} \\ \hline \end{array}$$

Example 8

$$\begin{array}{r} 6 \text{ ft } 5 \text{ in.} \\ -2 \text{ ft } 8 \text{ in.} \\ \hline \end{array}$$

Here 5 in. is smaller than 8 in., and we cannot subtract directly. So we **borrow** 1 ft = 12 in. from 6 ft.

$$\begin{array}{rcr} 6 \text{ ft } 5 \text{ in.} & = & 5 \text{ ft } \ 17 \text{ in.} \\ -2 \text{ ft } 8 \text{ in.} & = & 2 \text{ ft } \ \ 8 \text{ in.} \\ \hline & & 3 \text{ ft } \ \ 9 \text{ in.} \end{array}$$

(12 in. + 5 in. = 17 in.)

Now work exercise 8 in the margin.

9. Two baseball games on consecutive days lasted 3 hr 25 min and 2 hr 55 min. What was the difference in duration between the two games?

Example 9

On the first day of a long business trip (including both airplane and car travel) a salesman traveled for 13 hr 20 min. On the second day he traveled for 10 hr 50 min. What was the difference in his travel time for the two days?

Solution

To find the difference, we subtract the two times. In this case, we need to know that there are 60 minutes in 1 hour.

$$\begin{array}{r} 13 \text{ hr } 20 \text{ min} \\ -\ 10 \text{ hr } 50 \text{ min} \\ \hline \end{array}$$

Since 20 min is smaller than 50 min, we cannot subtract directly. So we **borrow** 1 hr = 60 min from 13 hr.

$$\begin{array}{rcr} 13 \text{ hr } 20 \text{ min} & = & 12 \text{ hr } 80 \text{ min} \\ -10 \text{ hr } 50 \text{ min} & = & 10 \text{ hr } 50 \text{ min} \\ \hline & & 2 \text{ hr } 30 \text{ min} \end{array}$$

(60 min + 20 min = 80 min)

So the difference in travel time for the two days was 2 hr 30 min.

Now work exercise 9 in the margin.

Practice Problems

Simplify the following mixed denominate numbers.

1. 10 min 130 sec **2.** 6 ft 13 in. **3.** 1 lb 34 oz

Add and simplify if necessary.

4.
3 ft 4 in.
2 ft 9 in.
+ 5 ft 6 in.

5.
3 hr 45 min 10 sec
+ 2 hr 25 min 55 sec

Subtract.

6.
16 lb 8 oz
− 12 lb 14 oz

7.
10 gal 2 qt 9 fl oz
− 8 gal 1 qt 11 fl oz

Answers to Practice Problems:

1. 12 min 10 sec **2.** 7 ft 1 in. **3.** 3 lb 2 oz **4.** 11 ft 7 in.
5. 6 hr 11 min 5 sec **6.** 3 lb 10 oz **7.** 2 gal 30 fl oz

Name ______________________ Section ____________ Date __________

Exercises 4.6

Simplify the following mixed denominate numbers. See Examples 1 through 3.

1. 3 ft 20 in. **2.** 4 ft 18 in. **3.** 6 lb 20 oz **4.** 3 lb 24 oz

5. 5 min 80 sec **6.** 14 min 90 sec **7.** 2 days 30 hr **8.** 5 days 36 hr

9. 8 gal 5 qt **10.** 2 gal 6 qt **11.** 4 pt 20 fl oz **12.** 3 pt 24 fl oz

Add and simplify if necessary. See Examples 4 and 5.

13.
2 ft 8 in.
5 ft 4 in.
+1 ft 7 in.

14.
3 ft 5 in.
6 ft 5 in.
+2 ft 3 in.

15.
10 lb 10 oz
+ 7 lb 8 oz

16.
4 lb 5 oz
+ 4 lb 7 oz

17.
8 min 35 sec
+ 9 min 35 sec

18.
5 min 10 sec
+ 14 min 35 sec

19.
2 hr 15 min 45 sec
+ 1 hr 55 min 30 sec

20.
5 hr 20 min 30 sec
+ 2 hr 35 min 40 sec

21.
2 days 20 hr 50 min
+ 3 days 5 hr 45 min

22.
1 day 15 hr 40 min
+ 2 days 10 hr 25 min

ANSWERS

1. ______
2. ______
3. ______
4. ______
5. ______
6. ______
7. ______
8. ______
9. ______
10. ______
11. ______
12. ______
13. ______
14. ______
15. ______
16. ______
17. ______
18. ______
19. ______
20. ______
21. ______
22. ______

23. ______

24. ______

25. ______

26. ______

27. ______

28. ______

29. ______

30. ______

31. ______

32. ______

33. ______

34. ______

35. ______

36. ______

37. ______

38. ______

39. ______

40. ______

41. ______

42. ______

23. 5 gal 2 qt
3 gal 2 qt
+ 4 gal 3 qt

24. 4 gal 3 qt
1 gal 2 qt
+ 3 gal 1 qt

25. 4 gal 3 qt 10 fl oz
+ 2 gal 3 qt 10 fl oz

26. 5 gal 3 qt 15 fl oz
+1 gal 2 qt 8 fl oz

27. 5 yd 2 ft 8 in.
+ 6 yd 2 ft 10 in.

28. 3 yd 1 ft 7 in.
+2 yd 2 ft 3 in.

Subtract. See Examples 7 and 8.

29. 5 yd 1 ft 7 in.
−2 yd 2 ft 5 in.

30. 8 yd 2 ft 3 in.
−7 yd 1 ft 8 in.

31. 9 gal 2 qt 4 fl oz
−5 gal 3 qt 6 fl oz

32. 20 gal 1 qt 13 fl oz
− 14 gal 2 qt 10 fl oz

33. 15 hr 30 min
− 12 hr 45 min

34. 6 hr 20 min
−4 hr 40 min

35. 15 min 20 sec
− 10 min 30 sec

36. 30 min 15 sec
−20 min 25 sec

37. 8 lb 4 oz
− 3 lb 12 oz

38. 20 lb 10 oz
− 10 lb 14 oz

39. 6 ft 5 in.
−2 ft 9 in.

40. 3 ft 8 in.
− 1 ft 10 in.

41. What is 2 ft 5 in. more than 4 ft 8 in.?

42. Find the sum of 6 hr 30 min, 9 hr 15 min, and 4 hr 45 min.

Name ______________________ Section ____________ Date __________

ANSWERS

43. What is the difference between 14 hr and 10 hr 20 min?

43. ____________

44. What is 3 lb 11 oz less than 6 lb 8 oz?

44. ____________

45. What is 7 qt more than 5 gal 2 qt?

45. ____________

46. What is the sum of 4 gal 3 qt, 2 gal 3 qt, and 1 gal 2 qt?

46. ____________

47. **Construction.** The plans for a new bookcase call for 3 shelves that are each 3 ft 8 in. long and 4 shelves that are each 1 ft long. What total length of shelf board is required for these shelves?

47. ____________

48. **Race-Walking.** Mr. and Mrs. Gonzalez spent the following amounts of time race-walking for their health during one week: 1 hr 10 min, 1 hr 30 min, 45 min, 1 hr 5 min, 50 min, 1 hr 40 min, and 2 hr. How much time did they spend race-walking that week?

48. ____________

Writing and Thinking about Mathematics

49. In your own words, explain the difference between abstract numbers and denominate numbers. Write a sentence in which each type of number is used at least once.

49. [Respond below exercise.]

50. Explain how to simplify the denominate number 23 hr 59 min 60 sec.

50. [Respond below exercise.]

Chapter 4 Index of Key Terms and Ideas

Section 4.1 Introduction to Mixed Numbers

Mixed Numbers page 296

A **mixed number** is the sum of a whole number and a fraction with the fraction part less than 1.

Shortcut for Changing Mixed Numbers to Fraction Form: page 297

1. Multiply the whole number by the denominator of the fraction part.
2. Add the numerator of the fraction part to this product.
3. Write this sum over the denominator of the fraction.

To Change an Improper Fraction to a Mixed Number: page 298

1. Divide the numerator by the denominator to find the whole number part of the mixed number.
2. Write the remainder over the denominator as the fraction part of the mixed number. (Note that this fraction part will always be less than 1.)

Reducing Fractions Versus Changing Fractions to Mixed Numbers page 299

Reducing involves finding common factors in the numerator and denominator.

Changing to a mixed number involves division of the numerator by the denominator. Common factors are not involved.

Section 4.2 Multiplication and Division with Mixed Numbers

To Multiply Mixed Numbers: page 305

1. Change each number to fraction form.
2. Multiply by factoring numerators and denominators; then reduce.
3. Change the answer to a mixed number or leave it in fraction form. (The choice sometimes depends on what use is to be made of the answer.)

To Divide Mixed Numbers: page 308

1. Change each number to fraction form.
2. Write the reciprocal of the divisor.
3. Multiply by factoring numerators and denominators; then reduce.

Section 4.3 Addition with Mixed Numbers

To Add Mixed Numbers: page 317

1. Add the fraction parts.
2. Add the whole numbers.
3. Write the sum as a mixed number so that the fraction part is less than 1. (If the sum of the fraction parts is more than 1, rewrite it as a mixed number and add it to the sum of the whole numbers.)

Section 4.4 Subtraction with Mixed Numbers

To Subtract Mixed Numbers: page 325

1. Subtract the fraction parts.
2. Subtract the whole numbers.

If the Fraction Part Being Subtracted is Larger than the First Fraction page 326

1. **Borrow** the whole number 1 from the first whole number.
2. Add this 1 to the first fraction. (This will always result in an improper fraction that is larger than the fraction being subtracted.)
3. Subtract.

Section 4.5 Order of Operations with Mixed Numbers

Rules for Order of Operations page 335

1. First, simplify within grouping symbols, such as parentheses ( ), brackets [ ], or braces { }. Start with the innermost grouping.
2. Second, evaluate any numbers or expressions indicated by exponents.
3. Third, moving from **left to right**, perform any multiplications or divisions in the order in which they appear.
4. Fourth, moving from **left to right**, perform any additions or subtractions in the order in which they appear.

Section 4.6 Denominate Numbers

Abstract and Denominate Numbers page 343

Abstract numbers are numbers with no units of measure attached. **Denominate numbers** are numbers with units of measure attached.

To Add Like Denominate Numbers: page 345

1. Write the numbers in column form so that like units are aligned.
2. Add the numbers in each column.
3. Simplify the resulting mixed denominate number, if necessary.

To Subtract Like Denominate Numbers: page 346

1. Write the numbers in column form so that like units are aligned.
2. If necessary for subtraction, borrow 1 of the larger units and rewrite the top numbers.
3. Subtract the like units.

Name ______________________ Section ______________ Date ____________

ANSWERS

Chapter 4 Review Questions

Change each of the following to mixed numbers with the fraction reduced to lowest terms.

1. $\frac{53}{8}$ **2.** $\frac{91}{13}$ **3.** $\frac{342}{100}$

Change each of the following to an improper fraction.

4. $5\frac{1}{10}$ **5.** $2\frac{11}{12}$ **6.** $13\frac{2}{5}$

Perform the indicated operations. Reduce all fractions to lowest terms.

7. $27\frac{1}{4}+3\frac{1}{2}$ **8.** $5\frac{2}{5}-4\frac{2}{3}$ **9.** $4\frac{5}{7}\cdot 2\frac{6}{11}$

10. $\frac{5}{6}\div 3\frac{3}{4}$ **11.** $4\frac{5}{8}+2\frac{3}{4}$ **12.** $\frac{5}{12}\left(6\frac{3}{10}\right)\left(7\frac{1}{9}\right)$

13. $7\frac{1}{11}\left(2\frac{3}{4}\right)\left(5\frac{1}{3}\right)$ **14.** $2\frac{4}{5}\div 4$ **15.** $12\frac{5}{6}-6\frac{1}{4}$

16. $4\frac{7}{8}-3\frac{1}{4}$

17.
$$\begin{array}{r} 14\frac{5}{12} \\ -\ 3\frac{3}{4} \\ \hline \end{array}$$

18.
$$\begin{array}{r} 42\frac{7}{9} \\ 53\frac{4}{15} \\ +\ 24\frac{9}{10} \\ \hline \end{array}$$

1. ____________
2. ____________
3. ____________
4. ____________
5. ____________
6. ____________
7. ____________
8. ____________
9. ____________
10. ____________
11. ____________
12. ____________
13. ____________
14. ____________
15. ____________
16. ____________
17. ____________
18. ____________

19. ______________

20. ______________

21. ______________

22. ______________

23. ______________

24. ______________

25. ______________

26. ______________

27. ______________

19. $4\frac{2}{7} \div 3\frac{3}{5} + 4\frac{1}{6} \cdot 2\frac{4}{5}$

20. $2\frac{3}{10} - 5\frac{3}{5} \div 4\frac{2}{3} + 2\frac{1}{6}$

21. $\left(\frac{1}{5} + \frac{1}{6}\right) \div 2\frac{1}{3}$

22. $\left(\frac{7}{8} - \frac{3}{16}\right) \div \left(\frac{1}{3} - \frac{1}{4}\right)$

23. Find the average of $\frac{3}{4}, \frac{5}{8}$, and $\frac{9}{10}$.

24. Find the average of $1\frac{3}{10}, 2\frac{1}{5}$, and $1\frac{3}{4}$.

25. Find $\frac{2}{3}$ of 96.

26. Find $\frac{2}{5}$ of $6\frac{1}{4}$.

27. The product of $20\frac{2}{3}$ with some other number is $6\frac{1}{2}$. What is the other number?

Name ______________________ Section ____________ Date __________ **ANSWERS**

28. Diet. While dieting, Ken lost $4\frac{1}{3}$ pounds the first week, $1\frac{3}{4}$ pounds the second week, and $2\frac{1}{6}$ pounds the third week. What was his average weekly weight loss?

28. ____________

29. Sale price. The sale price of a television is \$375. This is $\frac{3}{5}$ of the original price. What was the original price?

29. ____________

30. Unknown number. The sum of $12\frac{7}{8}$ and some other number is $20\frac{1}{4}$. What is the other number?

30. ____________

31. Dressmaking. Rachel wants to make matching dresses for her three daughters. The pattern requires $1\frac{7}{8}$ yards of material for a size 2 dress, $2\frac{1}{4}$ yards for a size 5, and $3\frac{1}{2}$ yards for a size 8. What is the least amount of material Rachel should buy if she makes one dress of each size?

31. ____________

32. Lottery. Jason won \$7500 in the state lottery. If $\frac{1}{5}$ of his prize money was withheld for taxes, how much money did he actually receive?

32. ____________

33. Construction. Michael is building a fence $4\frac{1}{2}$ feet high. It is recommended that $\frac{1}{4}$ of the length of each fence post should be underground. How long must each fence post be?

33. ____________

34. ______

35. ______

36. ______

37. ______

38. ______

34. Attendance. An English class has 42 students enrolled. If $\frac{6}{7}$ of the students enrolled are present in class, how many are absent?

35. Bicycling. A bicyclist rode his bicycle $37\frac{9}{10}$ miles each day for five days. How far did he ride his bicycle during those five days?

36.
```
  35 min 25 sec
+ 10 min 35 sec
```

37.
```
  8 ft 9 in.
  3 ft 4 in.
+ 1 ft 3 in.
```

38.
```
  2 days 2 hr 30 min
+ 3 days 8 hr 50 min
```

Name ______________________ Section ____________ Date __________

ANSWERS

1. ____________

2. a. ____________

b. ____________

3. a. ____________

b. ____________

4. ____________

5. ____________

6. ____________

7. ____________

8. ____________

9. ____________

Chapter 4 Test

1. The expression $\frac{13}{0}$ is undefined, but $\frac{0}{13}$ = _____.

2. Change each improper fraction to a mixed number.

a. $\frac{17}{5}$ **b.** $\frac{100}{33}$

3. Change each mixed number to an improper fraction.

a. $6\frac{5}{8}$ **b.** $4\frac{3}{10}$

In each of the following exercises, perform the operations indicated. Express your answers as either whole numbers or mixed numbers with the fraction part reduced.

4. Find the sum of $2\frac{3}{4}$ and $3\frac{5}{6}$.

5. Find the difference between $7\frac{1}{10}$ and $4\frac{4}{15}$.

6. Find the sum of the product of $3\frac{3}{5}$ and $6\frac{2}{3}$ with the quotient of $2\frac{1}{3}$ and 14.

In each of the following exercises, perform the indicated operations.

7. $\begin{array}{r} 9\frac{3}{8} \\ +\ \frac{5}{6} \\ \hline \end{array}$ **8.** $\begin{array}{r} 7 \\ -5\frac{3}{5} \\ \hline \end{array}$ **9.** $4\frac{3}{10}+2\frac{3}{4}$

10. ______

11. ______

12. ______

13. ______

14. ______

15. ______

16. ______

17. ______

18. ______

19. ______

20. ______

21. ______

22. ______

10. $6\frac{1}{6}-4\frac{2}{7}$

11. $\frac{3}{5}+1\frac{7}{10}+2\frac{1}{8}$

12. $6\frac{2}{5}\cdot 3\frac{1}{8}$

13. $2\frac{2}{3}\cdot\frac{5}{8}\cdot 2\frac{1}{4}$

14. $\frac{5}{6}\div 2\frac{1}{2}$

15. $4\frac{7}{8}\div 3\frac{1}{4}$

16. $\left(3\frac{1}{4}\right)\left(1\frac{2}{13}\right)\left(3\frac{1}{9}\right)$

17. $\frac{4}{5}\div 1\frac{1}{3}+\frac{2}{15}\cdot\frac{10}{7}$

18. $\left(6\frac{1}{5}+3\frac{7}{10}\right)\div\left(5\frac{3}{10}-4\frac{1}{2}\right)^2$

19.
  22 min 15 sec
+ 15 min 55 sec

20.
  32 gal 2 qt 8 fl oz
− 12 gal 3 qt 12 fl oz

21. Travel. A businesswoman flew to Japan in 14 hours and 10 minutes. Her flight to Europe lasted 8 hours and 30 minutes. What was the difference in flight times?

22. Carpentry. A carpenter measured the lengths of three boards. The lengths were 2 ft 3 in., 6 ft 8 in., and 10 ft 4 in. What was the total length of the three boards?

Name ______________________ Section ______________ Date ____________

ANSWERS

23. Geometry. The lengths of the three sides of a triangle are $5\frac{4}{5}$ centimeters, $3\frac{3}{5}$ centimeters, and $7\frac{1}{5}$ centimeters. The height of the triangle is $2\frac{3}{5}$ cm.

a. Find the perimeter of the triangle.

b. Find the area of the triangle.

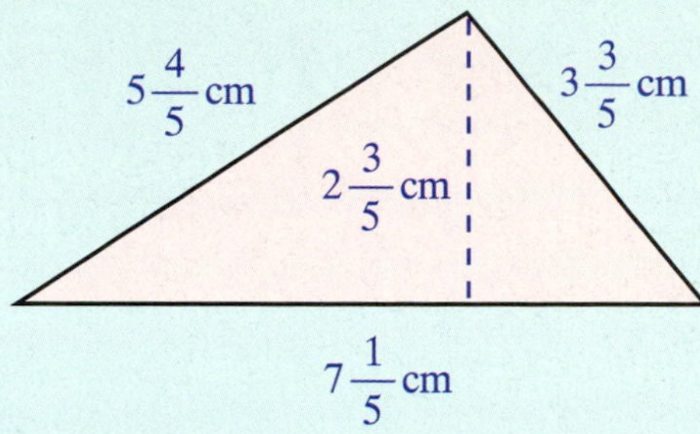

23. a. ____________

b. ____________

24. Running. During three days of practice, a high school cross country team member ran $8\frac{3}{10}$ miles, $14\frac{1}{5}$ miles, and $6\frac{1}{4}$ miles.

a. How far did she run during those three days?

b. What was her average distance per day?

24. a. ____________

b. ____________

25. a. Geometry. Find the perimeter of a rectangle that is 8 inches wide and $11\frac{1}{2}$ inches long.

b. Find the area of the rectangle. (The area of a rectangle is found by multiplying length times width.)

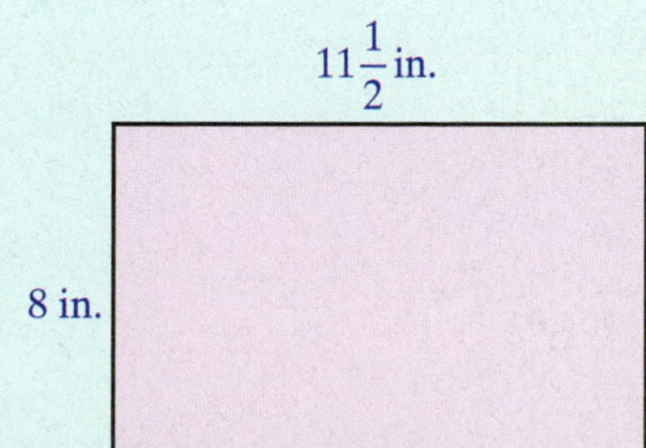

25. a. ____________

b. ____________

Name ______________________ Section ____________ Date __________

ANSWERS

Cumulative Review: Chapters 1 – 4

1. [Respond below exercise.]
2. ______________
3. ______________
4. ______________
5. ______________
6. ______________
7. ______________
8. a. ______________
   b. ______________
9. ______________
10. ______________

1. Write 53,460 in expanded notation and in its English word equivalent.

2. Round 265,400 to the nearest ten thousand.

3. Name the property of addition illustrated: $16 + 35 = 35 + 16$.

4. Match each expression with its best estimate.

_______	(a) $165 + 92 + 86 + 131$	A.	200
_______	(b) $8467 \div 43$	B.	500
_______	(c) $33 \cdot 46$	C.	1000
_______	(d) $6476 - 5392$	D.	1500

5. Multiply mentally: 800(7000).

6. Use the rules for order of operations to evaluate the following expression: $7^2 + 2(12 \cdot 4 \div 2 \cdot 3) - 7 \cdot 5$

7. List all the prime numbers less than 30.

8. Find the prime factorization of each number:
a. 170 **b.** 305

9. The value of $0 \div 36$ is _____, while $36 \div 0$ is _____.

10. Find the average of 44, 35, 53, and 40.

11. ______

12. ______

13. ______

14. ______

15. ______

16. ______

17. ______

18. ______

19. ______

20. ______

21. ______

22. ______

23. ______

24. ______

25. a. ______

b. ______

Perform the indicated operations. Reduce all fractions to lowest terms.

11. $\begin{array}{r} 8597 \\ +\ 4653 \\ \hline \end{array}$

12. $\begin{array}{r} 3782 \\ -\ 1255 \\ \hline \end{array}$

13. $\begin{array}{r} 732 \\ \times\ 36 \\ \hline \end{array}$

14. $13\overline{)2639}$

15. $\frac{7}{8} \div 6$

16. $\frac{7}{18}\left(\frac{3}{10}\right)\left(\frac{12}{28}\right)$

17. $\frac{7}{8} - \frac{7}{12}$

18. $\frac{9}{10} + \frac{4}{5} + \frac{1}{6}$

19. $\begin{array}{r} 13\frac{4}{5} \\ -\ 8\frac{4}{7} \\ \hline \end{array}$

20. $4\frac{3}{8} \div 2\frac{1}{2}$

21. $5\left(5\frac{2}{5}\right)\left(5\frac{5}{6}\right)$

22. $\left(1\frac{3}{10} + 2\frac{3}{5}\right) \div \left(2\frac{4}{5} - 1\frac{1}{2}\right)$

23. $\left(1\frac{1}{3}\right)^2 + 6\frac{4}{5} \div 2\frac{1}{8} - 2\frac{3}{10}$

24. Arrange the following fractions in order from smallest to largest.

$\frac{5}{6}, \frac{7}{10}, \frac{8}{9}, \frac{3}{4}$

25. Find the following sums and simplify if possible.

a. $\begin{array}{r} \text{3 hr 15 min 35 sec} \\ +\ \text{1 hr 55 min 40 sec} \\ \hline \end{array}$

b. $\begin{array}{r} \text{20 lb 10 oz} \\ +\ \text{7 lb 12 oz} \\ \hline \end{array}$

Name ______________________ Section ____________ Date __________

ANSWERS

26. Banking. On Tuesday morning Kevin had $950 in his checking account. That day he deposited $1750 in his account and wrote checks for the following amounts: $92, $345, and $750. What was his new balance on Wednesday morning?

26. ____________

27. Auto maintenance. If a car has a 22-gallon gas tank, how many gallons of gas will it take to fill the tank when it is $\frac{1}{4}$ full?

27. ____________

28. Shopping. Paula bought two pairs of shoes for $75 per pair, three blouses for $40 each, a pair of slacks for $65, and a sweater for $70. What was the average price per item?

28. ____________

29. Shopping. The discount price of a new DVD player is $180. If this price is $\frac{4}{5}$ of the original price, what was the original price? Round your answer to the nearest cent.

29. ____________

30. Sociology. In September 2002, boys watched TV for an average of 2 hours and 56 minutes per day. During the same month, girls watched TV for an average of 2 hours and 36 minutes per day. What was the difference between the boys' and the girls' average TV viewing times for one day in September 2002? What was the difference for the entire month?

30. ____________

5

Decimal Numbers

Chapter 5 Decimal Numbers

WHAT TO EXPECT IN CHAPTER 5

In Chapter 5, the concept of decimal numbers is developed and students will learn to add, subtract, multiply, and divide with decimal numbers. Also, because of the current emphasis on working with calculators, the topic of square roots is included.

Section 5.1 covers reading, writing, and rounding decimal numbers. The student should note that the word *and* is used to indicate the placement of the decimal point. Operations with decimal numbers are covered in Sections 5.2 through 5.4, as well as the use of rounded decimal numbers to estimate answers.

Section 5.5 discusses the relationships between fractions and decimal numbers. Sections 5.6 and 5.7 involve square roots and geometric concepts (the Pythagorean Theorem, circles, and volume). A calculator will be very useful for these sections.

Chapter 5 Decimal Numbers

Mathematics at Work!

As illustrated in the figure below, the shape of a baseball infield is a square 90 feet on each side. (A square is a four-sided plane figure in which all four sides are the same length and adjoining sides meet at 90° angles.) Do you think that the distance from home plate to second base is more than 180 feet or less than 180 feet? The distance from the pitcher's mound to home plate is 60½ feet. Is the pitcher's mound exactly halfway between home plate and second base? Do the two diagonals of the square (shown as dashed lines in the figure) intersect at the pitcher's mound? What is the distance from home plate to second base (to the nearest tenth of a foot)? (See Exercise 65 in Section 5.6.)

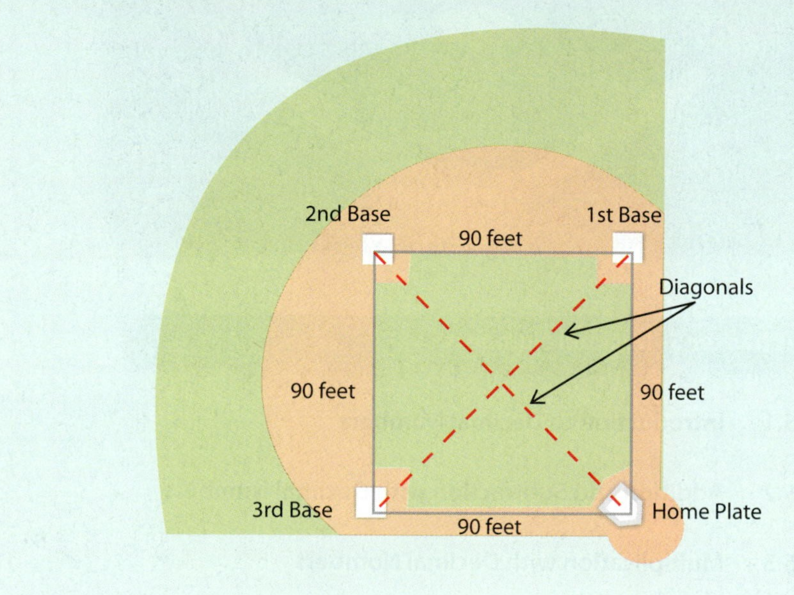

5.1 Introduction to Decimal Numbers

Introduction to Decimal Numbers

In Section 1.1, we discussed whole numbers and their representation using a **place value system** and the ten digits 0, 1, 2, 3, 4, 5, 6, 7, 8, and 9. The value of each place (for whole numbers) in this system (called the **decimal system**) is a power of ten:

1, 10, 100, 1000, 10,000, 100,000, 1,000,000 and so on.

In this system, the decimal point is thought of as the beginning point and the value of each place (moving to the left from the decimal point) is ten times the value of the place to its right, as shown in Figure 5.1.

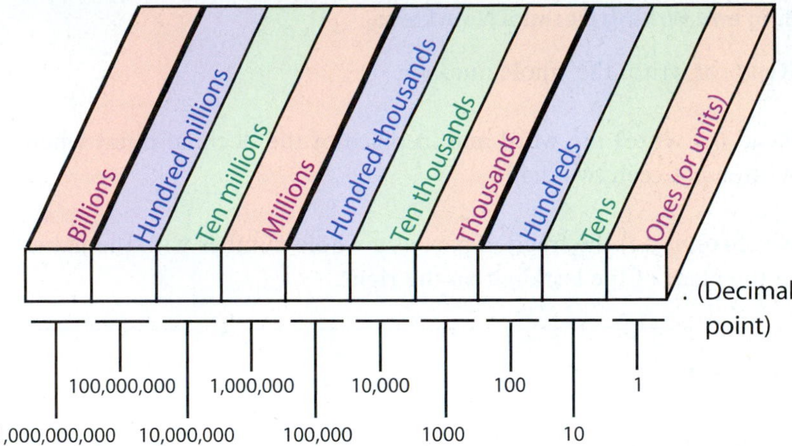

Figure 5.1

Objectives

1. Learn to read and write decimal numbers.
2. Know the importance of **and** in reading and writing decimal numbers.
3. Understand that **th** at the end of a word indicates a fraction.
4. Know how to round a decimal number to an indicated place of accuracy.

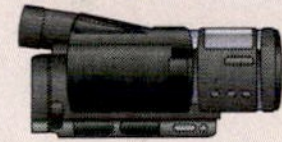
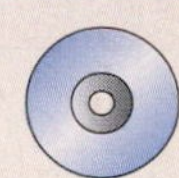

However, the decimal system is used to represent fractions as well as whole numbers. You are certainly familiar with dollars and cents as you pay for food, gasoline, and clothes. For example, the price of a pound of meat might be \$3.60 or you might pay \$45.25 for a new shirt. The 60 cents and 25 cents are fractions of a dollar. Thus, fractions can be indicated by writing digits to the right of the decimal point. Thus, in **decimal notation**:

$$\frac{60}{100} = 0.60 \quad \text{and} \quad \frac{25}{100} = 0.25$$

The common **decimal notation** uses a decimal point, with whole numbers written to the left of the decimal point and fractions written to the right of the decimal point. Numbers written in decimal notation are called **decimal numbers**, or simply **decimals**. The values of several places in this decimal system are shown in Figure 5.2. Note that fractions less than 1 are indicated to the right of the decimal point and each place has a value $\frac{1}{10}$ of the value of the place to its left.

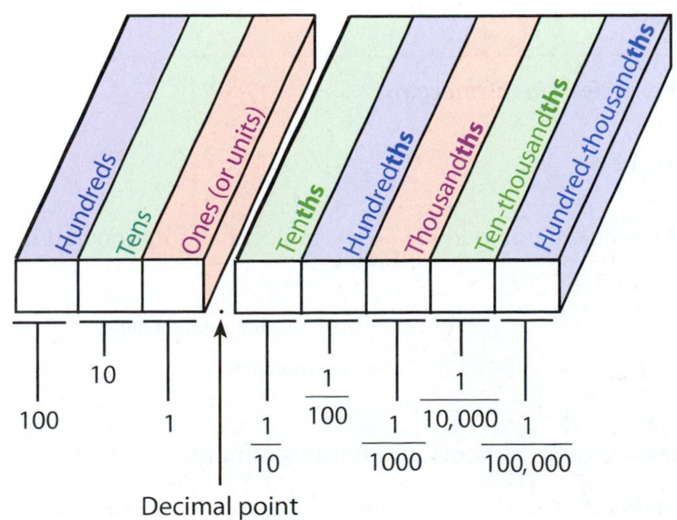

Figure 5.2

Objective 1

Reading and Writing Decimal Numbers:

1. Read (or write) the whole number.
2. Read (or write) the word **and** in place of the decimal point when written between two digits.
3. Read (or write) the fraction part as a whole number with the name of the place of the last digit on the right.

Write each mixed number in decimal notation and in words.

1. $19\frac{3}{10}$

Example 1

Write the mixed number $48\frac{6}{10}$ in decimal notation and in words.

Solution

4 8 . 6 — in decimal notation

forty-eight and six tenths — in words

And indicates the decimal point; the digit 6 is in the tenths position.

Now work exercise 1 in the margin.

Objective 2

2. $8\frac{743}{1000}$

Example 2

Write the mixed number $5\frac{398}{1000}$ in decimal notation and in words.

Solution

5 . 3 9 8 — in decimal notation

five and three hundred ninety-eight thousandths — in words

And indicates the decimal point; the digit 8 is in the thousandths position.

Now work exercise 2 in the margin.

3. $39\frac{184}{10,000}$

Example 3

Write the mixed number $12\frac{75}{10,000}$ in decimal notation and in words.

Solution

Two 0's must be inserted as placeholders.

1 2 . 0 0 7 5 — in decimal notation

twelve and seventy-five ten-thousandths — in words

And indicates the decimal point; the digit 5 is in the ten-thousandths position.

Now work exercise 3 in the margin.

Special Notes:

1. The letters **th** at the end of a word indicate a fraction part (a part to the right of the decimal point).
   six hundred = 600
   six hundred**ths** = 0.06

2. The hyphen (-) indicates one word.
   four hundred thousand = 400,000
   four hundred-thousand**ths** = 0.00004

Objective 3

Write the following in decimal notation:

Example 4

Write **seventeen thousandths** in decimal notation.

Solution

0 is inserted as placeholder.

0. 0 1 7

The digit 7 is in the thousandths position.

Now work exercise 4 in the margin.

4. ninety-one hundredths

Compare Examples 5 and 6 very carefully. Note how deletion of the word **and** in Example 5 completely changes the number from that in Example 6.

Example 5

Write **six hundred and five thousandths** in decimal notation.

Solution

Two 0's are inserted as placeholders.

6 0 0 . 0 0 5

The digit 5 is in the thousandths position.

Now work exercise 5 in the margin.

5. two hundred and twenty-one thousandths

Example 6

Write **six hundred five thousandths** in decimal notation.

Solution

0.605

Now work exercise 6 in the margin.

6. two hundred twenty-one thousandths

Note:

As a safety measure when writing a check, the amount is written in number form and in word form to avoid problems of spelling and poor penmanship. If there is a discrepancy between the number and the written form, the bank will honor the written form. (See Example 7.)

7. If you fill out a check in the amount of $100 but mistakenly write "Two hundred" in word form, which amount will be honored?

Example 7

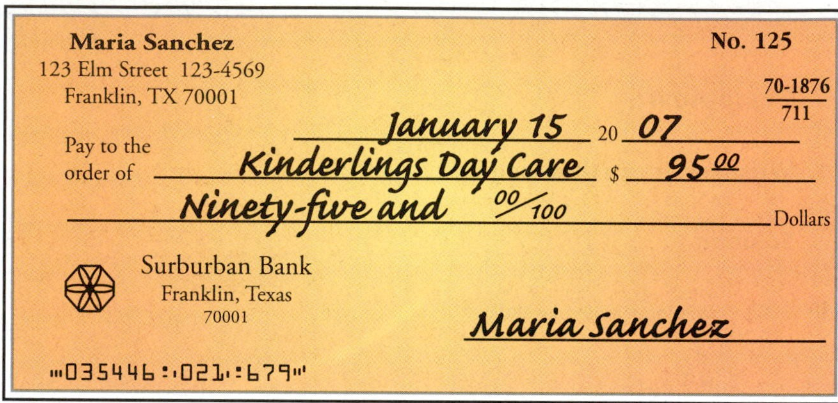

When writing a check, the written form of the amount is written on the line next to "Dollars" in decimal notation. This check is written for $95.

Now work exercise 7 in the margin.

Rounding Decimal Numbers

Measuring devices such as rulers, meter sticks, speedometers, micrometers, and surveying transits give only approximate measurements (See Figure 5.3). This is because, whether the units are large (such as miles and kilometers) or small (such as inches and centimeters), there are always smaller, more accurate units (such as eighths of an inch and millimeters) that could be used to indicate a particular measurement. We are constantly dealing with approximate (or rounded) numbers in our daily lives. If a recipe calls for 1.5 cups of flour and the cook puts in 1.47 cups (or 1.53 cups), the result will be about the same. In fact, the cook will never put exactly the same amount of flour in the mixture, no matter how many times the recipe is used.

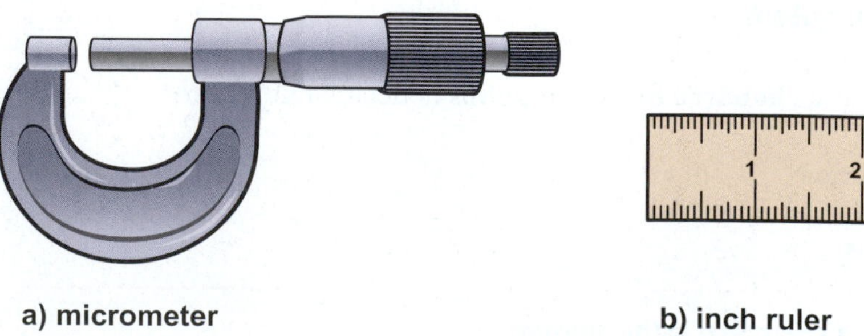

a) micrometer   b) inch ruler

Figure 5.3

Rounding a given number means finding another number close to the given number. The desired place of accuracy must be stated. For example, looking at the number line below, we can see that 5.77 is closer to 5.80 than it is to 5.70. So (to the nearest tenth), 5.77 rounds to 5.80, or 5.8.

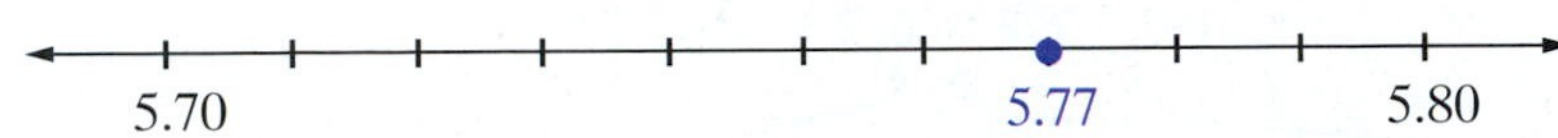

There are several generally accepted rules for rounding decimal numbers. For example, the IRS allows rounding to the nearest dollar on income tax forms. Sometimes, in cases where many numbers are involved, a decimal number is rounded to the nearest even digit at some particular place value. The rule chosen depends on the use of the numbers and whether or not there might be a penalty for an error.

In this text, we will use the following rules for rounding:

Rules for Rounding Decimal Numbers (to the right of the decimal point):

1. If the digit to the right of the given place value is less than 5: leave the digit in the given place as is and drop the remaining digits.

2. If the digit to the right of the given place value is 5 or more: increase the digit in the given place value by 1 and drop the remaining digits.

(Important Reminder: In rounding whole numbers, digits to the left of the decimal point that are dropped must be replaced by 0's.)

Objective 4

Example 8

Round 6.749 to the nearest tenth.

Solution

a. 6.749

7 is in the tenths position. The next digit is 4.

b. Because 4 is less than 5, leave 7 as it is and drop 4 and 9.

c. Thus, 6.749 rounds to **6.7** (to the nearest tenth).

Now work exercise 8 in the margin.

8. Round 13.2541 to the nearest hundredth.

9. Round 58.94479 to the nearest thousandth.

Example 9

Round 13.73962 to the nearest thousandth.

Solution

a. 1 3 . 7 3 9 6 2

9 is in the thousandths position. The next digit is 6.

b. Because 6 is greater than 5, increase 9 by 1 and drop 6 and 2.

c. Thus, 13.73962 rounds to **13.740** (to the nearest thousandth).

(Note carefully that increasing 9 by 1 gives 10 and this affects the digit 3 as well. Think of 39 becoming 40.)

Now work exercise 9 in the margin.

10. Round 21.92355 to the nearest ten-thousandth.

Completion Example 10

Round 8.00241 to the nearest ten-thousandth.

Solution

a. The digit in the ten-thousandths position is _____.

b. The next digit to the right is _____.

c. Because _____ is less than 5, leave _____ as it is and drop the _____.

d. Thus, 8.00241 rounds to _________ (to the nearest ___________).

Now work exercise 10 in the margin.

11. Round 40,387 to the nearest thousand.

Completion Example 11

Round 7361 to the nearest hundred.

Solution

a. The decimal point is understood to be to the right of _____.

b. The digit in the hundreds position is _____.

c. The next digit to the right is _____.

d. Because _____ is more than 5, increase _____ by 1 and replace _____ and ____ with 0's.

e. Thus, 7361 rounds to ________ (to the nearest hundred).

Now work exercise 11 in the margin.

Completion Example Answers

10. **a.** The digit in the ten-thousandths position is **4**.

b. The next digit to the right is **1**.

c. Because **1** is less than 5, leave **4** as it is and drop the **1**.

d. Thus, 8.00241 rounds to **8.0024** (to the nearest **ten-thousandth**).

11. **a.** The decimal point is understood to be to the right of **1**.

b. The digit in the hundreds position is **3**.

c. The next digit to the right is **6**.

d. Because **6** is more than 5, increase **3** by 1 and replace **6** and **1** with 0's.

e. Thus, 7361 rounds to **7400** (to the nearest hundred).

Practice Problems

1. Write 20.7 in words.

2. Write 18.051 in words.

3. Write $4\frac{6}{100}$ in decimal notation.

4. Write eight hundred and three tenths in decimal notation.

Round as indicated.

5. 572.3 (to the nearest ten)

6. 6.749 (to the nearest tenth)

7. 7558 (to the nearest thousand)

8. 0.07961 (to the nearest thousandth)

9. A penny dated anytime from 1959 through 1982 had an original weight of 3.11 grams. A penny dated 1983 or later had an original weight of 2.5 grams. Write the numbers representing weights in words.

Answers to Practice Problems:

1. twenty and seven tenths **2.** eighteen and fifty-one thousandths
3. 4.06 **4.** 800.3 **5.** 570
6. 6.7 **7.** 8000 **8.** 0.080
9. three and eleven hundredths; two and five tenths

Name ______________________ Section ____________ Date __________

ANSWERS

1. ______
2. ______
3. ______
4. ______
5. ______
6. ______
7. ______
8. ______
9. ______
10. ______
11. ______
12. ______
13. ______
14. ______
15. ______
16. ______
17. ______
18. ______
19. ______

Exercises 5.1

Write the following mixed numbers in decimal notation. See Examples 1 through 3.

1. $37\frac{498}{1000}$ **2.** $18\frac{76}{100}$ **3.** $4\frac{11}{100}$

4. $87\frac{3}{1000}$ **5.** $95\frac{2}{10}$ **6.** $56\frac{3}{100}$

7. $62\frac{7}{10}$ **8.** $100\frac{25}{100}$

Write the following decimal numbers in mixed number form.

9. 82.56 **10.** 93.07 **11.** 10.576

12. 100.6 **13.** 65.003 **14.** 172.35

Write the following numbers in decimal notation. See Examples 4 through 6.

15. three tenths **16.** fourteen thousandths

17. seventeen hundredths **18.** six and twenty-eight hundredths

19. sixty and twenty-eight thousandths

20. ____________

21. ____________

22. ____________

23. ____________

24. ____________

25. ____________

26. ____________

27. ____________

28. ____________

29. ____________

30. ____________

31. a. ____________

b. ____________

c. ____________

d. ____________

20. seventy-two and three hundred ninety-two thousandths

21. eight hundred fifty and thirty-six ten-thousandths

22. seven hundred and seventy-seven thousandths

Write the following decimal numbers in words.

23. 0.5

24. 0.93

25. 5.06

26. 35.078

27. 7.003

28. 607.607

29. 10.4638

30. 600.615

Fill in the blanks to correctly complete each statement. See Examples 8 through 11.

31. Round 8.4732 to the nearest hundredth.

a. The digit in the hundredths position is _____.

b. The next digit to the right is _____.

c. Because _____ is less than 5, leave _____ as it is and drop ____ and ____.

d. So, 8.4732 rounds to ________ (to the nearest hundredth).

Name ____________________ Section __________ Date __________

ANSWERS

32. Round 1.00659 to the nearest thousandth.

a. The digit in the thousandths position is _____.

b. The next digit to the right is _____.

c. Because _____ is 5 or more, increase _____ by 1 and drop _____ and _____.

d. So, 1.00659 rounds to _______ (to the nearest thousandth).

Round each of the numbers as indicated in Exercises 33 – 90. See Examples 8 through 11.

To the nearest tenth:

33. 4.763 **34.** 5.031 **35.** 76.349

36. 76.352 **37.** 89.015 **38.** 7.555

39. 18.009 **40.** 14.33382

To the nearest hundredth:

41. 0.385 **42.** 0.296 **43.** 5.7226

44. 8.9874 **45.** 6.99613 **46.** 13.13465

47. 0.0782 **48.** 6.0035

32. a. ______

b. ______

c. ______

d. ______

33. ______

34. ______

35. ______

36. ______

37. ______

38. ______

39. ______

40. ______

41. ______

42. ______

43. ______

44. ______

45. ______

46. ______

47. ______

48. ______

49. ______
50. ______
51. ______
52. ______
53. ______
54. ______
55. ______
56. ______
57. ______
58. ______
59. ______
60. ______
61. ______
62. ______
63. ______
64. ______
65. ______
66. ______
67. ______
68. ______
69. ______
70. ______
71. ______
72. ______

To the nearest thousandth:

49. 0.0672 **50.** 0.05550 **51.** 0.6338

52. 7.6666 **53.** 32.4785 **54.** 9.4302

55. 0.00191 **56.** 20.76962

To the nearest whole number (or nearest unit, or nearest one):

57. 479.23 **58.** 6.8 **59.** 19.999

60. 382.48 **61.** 649.66 **62.** 439.78

63. 6333.11 **64.** 8122.825

To the nearest ten:

65. 5163 **66.** 6475 **67.** 495

68. 572.5 **69.** 998.5 **70.** 378.92

71. 92,540.9 **72.** 7007.7

Name ______________ Section ______________ Date ______________

To the nearest thousand:

73. 7398

74. 62,275

75. 47,823.4

76. 103,499

77. 217,480.2

78. 9872.5

79. 4,500,762

80. 573,333.3

81. 0.0005783 (nearest hundred-thousandth)

82. 0.5449 (nearest hundredth)

83. 94.6 (nearest ten)

84. 5.00632 (nearest thousandth)

85. 473.8 (nearest hundred)

86. 5750 (nearest thousand)

87. 3.2296 (nearest thousandth)

88. 15.548 (nearest tenth)

89. 78,419 (nearest ten thousand)

90. 78,419 (nearest ten)

ANSWERS

73. ______________

74. ______________

75. ______________

76. ______________

77. ______________

78. ______________

79. ______________

80. ______________

81. ______________

82. ______________

83. ______________

84. ______________

85. ______________

86. ______________

87. ______________

88. ______________

89. ______________

90. ______________

91. ____________

92. ____________

93. ____________

94. ____________

95. ____________

96. ____________

97. ____________

98. ____________

99. ____________

In each of the following exercises, write the decimal numbers that are not whole numbers in words. See Examples 1 through 3.

91. One inch is equal to 2.54 centimeters.

92. One gallon of water weighs 8.33 pounds.

93. The winning car in the 2005 Indianapolis 500 race averaged 157.603 mph.

94. In 2005, the U.S. dollar was worth 1.3144 Swiss francs.

95. The number π is approximately equal to 3.14159.

96. The number **e** (used in higher-level mathematics) is approximately equal to 2.71828.

97. One acre is approximately equal to 0.405 hectares.

98. One liter is approximately equal to 0.264 gallons.

Write sample checks for the amounts indicated. See Example 7.

99. $356.45

Maria Sanchez No. 126
123 Elm Street 123-4569
Franklin, TX 70001
70-1876/711

February 1 20 07

Pay to the order of Management Systems $ ________

________ Dollars

Surburban Bank
Franklin, Texas
70001

Maria Sanchez

⑆035446⑆021⑈679⑈

Name ______________ Section ______ Date ______

ANSWERS

100. $651.50

Maria Sanchez 123 Elm Street 123-4569 Franklin, TX 70001	**No. 127** **70-1876** **711**
	February 9 20 *07*
Pay to the order of *McCann's Furniture*	$ ______
______ Dollars	
Surburban Bank Franklin, Texas 70001	*Maria Sanchez*
⑈035446⑆021⑆679⑈	

100. ______

101. ______

101. $2506.64

Maria Sanchez 123 Elm Street 123-4569 Franklin, TX 70001	**No. 128** **70-1876** **711**
	February 15 20 *07*
Pay to the order of *Eastern Securities, Inc.*	$ ______
______ Dollars	
Surburban Bank Franklin, Texas 70001	*Maria Sanchez*
⑈035446⑆021⑆679⑈	

102. [Respond below exercise.]

Writing and Thinking about Mathematics

102. In your own words, state why the word **and** is so commonly misused when numbers are spoken and/or written. Bring an example of this from a newspaper, magazine, or television show to class for discussion.

103. [Respond below exercise.]

103. Why are hyphens used when some words are written? Give examples of the use of hyphens when writing decimal numbers in word form.

REVIEW

1. a. ____________

b. ____________

2. a. ____________

b. ____________

3. a. ____________

b. ____________

4. a. ____________

b. ____________

5. a. ____________

b. ____________

6. a. ____________

b. ____________

7. ____________

Review Problems (from 1.2, 1.3, and 1.4)

*First **a.** estimate each sum; then **b.** find each sum.*

1. $\begin{array}{r} 2093 \\ +\ 7014 \\ \hline \end{array}$

2. $\begin{array}{r} 875 \\ 573 \\ +\ 107 \\ \hline \end{array}$

3. $\begin{array}{r} 127,484 \\ 382,567 \\ +\ \ 52,661 \\ \hline \end{array}$

*First **a.** estimate each difference; then **b.** find each difference.*

4. $\begin{array}{r} 357 \\ -\ 169 \\ \hline \end{array}$

5. $\begin{array}{r} 7844 \\ -\ 3843 \\ \hline \end{array}$

6. $\begin{array}{r} 90,006 \\ -\ \ 5\ 115 \\ \hline \end{array}$

7. What number should be added to 850 in order to get a sum of 1200?

5.2 Addition and Subtraction with Decimal Numbers

Addition with Decimal Numbers

Objectives

① Be able to add with decimal numbers.

② Be able to subtract with decimal numbers.

③ Know how to estimate sums and differences with rounded decimal numbers.

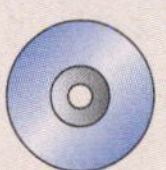

Addition with decimals can be accomplished by writing the decimal numbers in a vertical format with the decimal points aligned. In this way, whole numbers are added to whole numbers, tenths to tenths, hundredths to hundredths, and so on. The decimal point in the sum is in line with the decimal points in the addends. For example,

Decimal points are aligned vertically.

$$\begin{array}{r} 6.15 \\ +3.42 \\ \hline 9.57 \end{array}$$

Any number of 0's may be written to the right of the last digit in the fraction part of a number to help keep the digits in the correct alignment. This will not change the value of any number or the value of the sum.

To Add Decimal Numbers:

1. Write the addends in a vertical column.
2. Keep the decimal points aligned vertically.
3. Keep digits with the same position value aligned. (Zeros may be written in to help keep the digits aligned properly.)
4. Add the numbers, keeping the decimal point in the sum aligned with the other decimal points.

Objective ①

Example 1

Find the sum 6.3 + 5.42 + 14.07.

Solution

$$\begin{array}{r} 6.30 \\ 5.42 \\ +\ 14.07 \\ \hline 25.79 \end{array}$$

0 may be written in to help keep the digits aligned.

Now work exercise 1 in the margin.

1. Find the sum: 14.86 + 5 + 9.1.

2. Find the sum:
23.8 + 4.2567 + 11 + 3.01

Example 2

Find the sum: $9 + 4.86 + 37.479 + 0.6$

Solution

$$\begin{array}{r} 9.000 \\ 4.860 \\ 37.479 \\ +\ 0.600 \\ \hline 51.939 \end{array}$$

The decimal point is understood to be to the right of 9, as in 9.0, 9.00, or 9.000.

0's are written in to help keep the digits aligned properly.

Now work exercise 2 in the margin.

3. Add: 1.093 + 37 + 24.58

Example 3

Add:

$$\begin{array}{r} 56.2 \\ 85.75 \\ +29.001 \\ \hline \end{array}$$

Solution

You can write

$$\begin{array}{r} 56.200 \\ 85.750 \\ +\ 29.001 \\ \hline 170.951 \end{array}$$

0's written in to keep the digits in line.

Now work exercise 3 in the margin.

4. Mr. Riley went to the furniture store and bought a rug for \$50.79, a lamp for \$33.68 and a nightstand for \$72.11. How much did he spend? (Tax was included in the prices.) Estimate his expenses mentally before calculating his actual expenses.

Example 4

Mrs. Finn went to the local store and bought a pair of shoes for \$42.50, a blouse for \$25.60, and a skirt for \$37.55. How much did she spend? (Tax was included in the prices.) Estimate her expenses mentally before calculating her actual expenses.

Solution

$$\begin{array}{r} \$\ 42.50 \\ 25.60 \\ +\ \ 37.55 \\ \hline \$105.65 \end{array}$$

She spent \$105.65. (Did you estimate \$110?)

Now work exercise 4 in the margin.

Subtraction with Decimal Numbers

Objective 2

To Subtract Decimal Numbers:

1. Write the addends in a vertical column.
2. Keep the decimal points aligned vertically.
3. Keep digits with the same position value aligned. (Zeros may be written in as aids.)
4. Subtract, keeping the decimal point in the difference aligned with the other decimal points.

Find each difference.

Example 5

Find the difference: 16.715 – 4.823

Solution

$$\begin{array}{r} {\scriptstyle 5\ 16\ 1} \\ 1\not{6}.\not{7}15 \\ -4.823 \\ \hline 11.892 \end{array}$$

5. 9.262 – 3.07

Now work exercise 5 in the margin.

Example 6

Find the difference: 21.2 – 13.716

Solution

$$\begin{array}{r} {\scriptstyle 1\ 10\ 11\ 9\ 1} \\ \not{2}\not{1}.\not{2}\not{0}0 \\ -13.716 \\ \hline 7.484 \end{array}$$ ← Write in 0's.

6. 170.44 – 28.973

Now work exercise 6 in the margin.

Estimating Sums and Differences

We can estimate a sum (or difference) by rounding each number to the place of the **leftmost nonzero digit** and then adding (or subtracting) these rounded numbers. This technique of estimating answers is especially helpful when working with decimal numbers, where the placement of the decimal point is so important.

Note that, depending on the position of the leftmost nonzero digit, the rounded numbers may be whole numbers or decimal fractions. Also, different numbers within one problem might be rounded to different places of accuracy.

7. Estimate and then find the sum:
6.68 + 103.5 + 21.94

Example 7

First
a. Estimate the sum 74 + 3.529 + 52.61; then
b. Find the actual sum.

Solution

a. Estimate by adding rounded numbers.

74	rounds to →	70
3.529	rounds to →	4
52.61	rounds to →	+ 50
		124 ← estimate

b. Find the actual sum.

$$\begin{array}{r} 74.000 \\ 3.529 \\ +\ 52.610 \\ \hline 130.139 \end{array}$$ ← actual sum

Now work exercise 7 in the margin.

Practice Problems

Find each indicated sum or difference.

1. 46.2 + 3.07 + 2.6

2. 9 + 5.6 + 0.58

3. 6.4 – 3.7

4. 18 – 0.4384

Answers to Practice Problems:

1. 51.87 **2.** 15.18 **3.** 2.7 **4.** 17.5616

Name ______________________ Section ____________ Date __________

Exercises 5.2

Find each of the indicated sums. Estimate your answers, either mentally or on paper, before doing the actual calculations. Check to see that your sums are close to the estimated values. See Examples 1 through 4 and 7.

1. $0.6 + 0.4 + 1.3$

2. $5 + 6.1 + 0.4$

3. $0.59 + 6.91 + 0.05$

4. $3.488 + 16.593 + 25.002$

5. $37.02 + 25 + 6.4 + 3.89$

6. $4.0086 + 0.034 + 0.6 + 0.05$

7. $43.766 + 9.33 + 17 + 206$

8. $52.3 + 6 + 21.01 + 4.005$

9. $2.051 + 0.2006 + 5.4 + 37$

10. $5 + 2.37 + 463 + 10.88$

ANSWERS

1. ____________

2. ____________

3. ____________

4. ____________

5. ____________

6. ____________

7. ____________

8. ____________

9. ____________

10. ____________

11. ____________

12. ____________

13. ____________

14. ____________

15. ____________

16. ____________

17. ____________

18. ____________

19. ____________

20. ____________

21. ____________

22. ____________

23. ____________

24. ____________

25. ____________

11.
```
   47.3
   42.03
+ 29.003
```

12.
```
   1.007
  20.063
+  0.49
```

13.
```
  4.128
  0.02
+ 3.0
```

14.
```
  5.0015
  2.443
+ 0.0469
```

15.
```
  75.2
   3.682
+ 14.995
```

16.
```
  107.39
    5.061
   23.54
+  64.9801
```

17.
```
  34.967
  50.6
   8.562
+  9.3
```

18.
```
   4.156
   3.7
  25.682
+ 13.405
```

19.
```
  74.0
   3.529
  52.62
+  7.001
```

20.
```
  983.4
   47.518
  805.411
+ 300.766
```

Find each of the indicated differences. First estimate the differences mentally. See Examples 5 and 6.

21. 5.2 – 3.76 **22.** 17.83 – 8.9 **23.** 29.5 – 13.61

24. 1.0057 – 0.03 **25.** 78.015 – 13.068

Name ______________ Section ______________ Date ______________

26. $\begin{array}{r} 22.418 \\ -\ 17.523 \\ \hline \end{array}$

27. $\begin{array}{r} 4.8 \\ -\ 0.0026 \\ \hline \end{array}$

28. $\begin{array}{r} 31.009 \\ -\ 0.534 \\ \hline \end{array}$

29. $\begin{array}{r} 4.0 \\ -\ 1.0566 \\ \hline \end{array}$

30. $\begin{array}{r} 40.718 \\ -\ 6.532 \\ \hline \end{array}$

31. Goods and services. Theresa got a haircut for \$30.00 and a manicure for \$10.50. If she tipped the stylist \$5, how much change did she receive from a \$50 bill?

32. Goods and services. Mr. Johnson bought the following items at a department store: slacks, \$32.50; shoes, \$43.75; shirt, \$18.60.

a. How much did he spend?

b. What was his change if he gave the clerk a \$100 bill? (Tax was included in the prices.)

33. Construction. An architect's scale drawing shows a rectangular lot 2.38 inches on one side and 3.76 inches on the other side. What is the perimeter of (distance around) the rectangle in the drawing?

ANSWERS

26. ______________

27. ______________

28. ______________

29. ______________

30. ______________

31. ______________

32. a. ______________

b. ______________

33. ______________
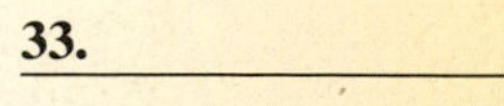

34. a. ______

b. ______

35. a. ______

b. ______

36. a. ______

b. ______

c. ______

37. a. ______

b. ______

c. ______

34. **Goods and services.** Mrs. Jones bought the following items at a department store: dress, $47.25; shoes, $35.75; purse, $12.50.
   a. How much did she spend?
   b. What was her change if she gave the clerk a $100 bill? (Tax was included in the prices.)

35. **Banking.** Suppose your checking account shows a balance of $280.35 at the beginning of the month. During the month, you make deposits of $310.50, $200, and $185.50, and you write checks for $85.50, $210.20, and $600.
   a. **Estimate** the balance at the end of the month.
   b. Find the balance at the end of the month.

36. **Income and expenses.** Your income for one month was $2800. You paid $670 for the rent, $50 for electricity, $20 for water, and $35 for gas.
   a. **Estimate** what remained of your month's income after you paid those bills.
   b. Calculate the actual amount that remained.
   c. How good was your estimate?

37. **Automobile purchase.** Martin wants to buy a car for $24,500. His credit union will loan him $21,750, but he must pay $280 for a license fee and $900 for taxes.
   a. **Estimate** the amount of cash he needs in order to buy the car.
   b. Calculate the actual amount of cash he needs.
   c. How good was your estimate?

Name ______________________ Section ____________ Date __________

ANSWERS

38. ______________

38. Agricultural production. In 2004 U.S. farmers produced 115.935 million bushels of oats, 279.253 million bushels of barley, and 2158.245 million bushels of wheat. What was the total combined amount of these grains produced in 2004?

39. ______________

39. Advertising. In 2004 Procter & Gamble spent $5.504 billion on advertising. In the same year, Johnson & Johnson spent $1.9 billion and Bristol-Myers spent $1.411 billion on advertising. How much more did Procter & Gamble spend on advertising than the other two companies combined?

40. ______________

40. Astronomy. The eccentricity of a planet's orbit is the measure of how much the orbit varies from a perfectly circular pattern. Mercury's orbit has an eccentricity of 0.205630, while Earth's orbit has an eccentricity of 0.016711. How much greater is Mercury's eccentricity of orbit than Earth's?

41. ______________

41. Weather. In one year, Boston, MA receives an average of 41.63 inches of rain and 40.9 inches of snow, while Burlington, VT receives an average of 39.37 inches of rain and 78.8 inches of snow. How much more rain is there in Boston than in Burlington? How much more snow is there in Burlington than in Boston?

[Respond below
42. a. exercise.] ____

[Respond below
b. exercise.] ____

[Respond below
c. exercise.] ____

[Respond below
43. a. exercise.] ____

[Respond below
b. exercise.] ____

[Respond below
c. exercise.] ____

Check Your Number Sense

42. Suppose that you were to add two decimal numbers with 0 as their whole number parts. Can their sum possibly be

a. more than 2? Explain briefly.
b. more than 1? Explain briefly.
c. less than 1? Explain briefly.

43. Suppose that you were to subtract two decimal numbers with 0 as their whole number parts. Can their difference possibly be

a. more than 1? Explain briefly.
b. less than 1? Explain briefly.
c. equal to 1? Explain briefly.

REVIEW

1. ____

2. ____

3. ____

4. ____

5. ____

Review Problems (from Section 1.5)

Find each of the following products.

1. 50×7000

2. $\begin{array}{r} 35 \\ \times\ 15 \\ \hline \end{array}$

3. $\begin{array}{r} 195 \\ \times\ 36 \\ \hline \end{array}$

4. First estimate the following product, and then find the product.

$$\begin{array}{r} 4571 \\ \times\ 375 \\ \hline \end{array}$$

5. Find the sum of seventy-five and ninety-three and the difference between one hundred sixty-four and eighty-five. Then, find the product of the sum and the difference.

5.3 Multiplication with Decimal Numbers

Multiplying Decimal Numbers

The method for multiplying decimal numbers is similar to that for multiplying whole numbers. The difference is in determining where to place the decimal point in the product. To illustrate how to place the decimal point, several products are shown in both fraction form and decimal form. (Note that we will generally indicate multiplication with the × sign or parentheses rather than the raised dot · to avoid confusion with the dot and the decimal point.)

Objectives

① Be able to multiply decimal numbers and place the decimal point correctly in the product.

② Be able to multiply decimal numbers mentally by powers of 10.

③ Know how to estimate products with rounded decimal numbers.

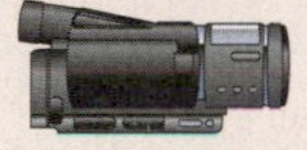 

Products in Fraction Form	Products in Decimal Form
$\frac{1}{10} \times \frac{1}{100} = \frac{1}{1000}$	$\begin{array}{r} 0.1 \\ \times\ 0.01 \\ \hline 0.001 \end{array}$ ← 1 place, ← 2 places; total of 3 places (thousandths)
$\frac{3}{10} \times \frac{5}{100} = \frac{15}{1000}$	$\begin{array}{r} 0.3 \\ \times\ 0.05 \\ \hline 0.015 \end{array}$ ← 1 place, ← 2 places; total of 3 places (thousandths)
$\frac{6}{100} \times \frac{4}{1000} = \frac{24}{100,000}$	$\begin{array}{r} 0.004 \\ \times\ 0.06 \\ \hline 0.00024 \end{array}$ ← 3 places, ← 2 places; total of 5 places (hundred thousandths)

As these illustrations indicate, there is **no need to keep the decimal points lined up for multiplication**. The following rule states how to multiply two decimal numbers and place the decimal point in the product.

Objective ①

To Multiply Decimal Numbers:

1. Multiply the two numbers as if they were whole numbers.
2. Count the total number of places to the right of the decimal points in both numbers being multiplied.
3. Place the decimal point in the product so that the number of places to the right of it is the same as that found in Step 2.

Find each product:

1. (0.8)(0.2)

2. (5.6)(0.04)

3. $\begin{array}{r} 3.781 \\ \times \quad 3.01 \\ \hline \end{array}$

Example 1

Multiply: 2.432×5.1.

Solution

$$\begin{array}{r} 2.432 \\ \times \quad 5.1 \\ \hline 2432 \\ 12\,160 \\ \hline 12.4032 \end{array}$$

2.432 ← 3 places, 5.1 ← 1 place: total of 4 places

12.4032 ← 4 places in the product

Now work exercise 1 in the margin.

Example 2

Multiply: 4.35×12.6.

Solution

$$\begin{array}{r} 4.35 \\ \times \quad 12.6 \\ \hline 2\,610 \\ 8\,70 \\ 43\,5 \\ \hline 54.810 \end{array}$$

4.35 ← 2 places, 12.6 ← 1 place: total of 3 places

54.810 ← 3 places in the product

Now work exercise 2 in the margin.

Example 3

Multiply: (0.046)(0.007).

Solution

$$\begin{array}{r} 0.046 \\ \times \quad 0.007 \\ \hline 0.000322 \end{array}$$

0.046 ← 3 places, 0.007 ← 3 places: total of 6 places

0.000322 ← 6 places in the product

This means that three 0's had to be inserted between the 3 and the decimal point.

Now work exercise 3 in the margin.

Completion Example 4

Multiply: 3.4×5.8.

Solution

```
    3.4  ← ___ place(s) ┐
 ×  5.8  ← ___ place(s) ┘ total of ___ place(s)
   2 72
  17 0
  ______    ___ place(s) in the product.
```

Now work exercise 4 in the margin.

4. (0.29)(11.4)

Completion Example 5

Multiply 0.003×0.03.

Solution

```
   0.003  ← ___ places ┐
 ×  0.03  ← ___ places ┘ total of ___ places

  _______      ___ places in the product
                  ___ 0's had to be inserted to place the decimal point.
```

Now work exercise 5 in the margin.

5. 3.06×0.019

Multiplying Decimal Numbers by Powers of 10

In Section 1.5, we discussed multiplication of whole numbers by powers of ten by placing 0's to the right of the number. Now that decimal numbers have been introduced, we can see that inserting 0's in a whole number has the effect of moving the decimal point to the right. (**Note:** Even though we do not always write the decimal point in a whole number, it is understood to be just to the right of the rightmost digit.) The following more general guidelines can be used to multiply all decimal numbers by powers of 10.

Objective 2

To Multiply a Decimal Number by a Power of 10:

1. Move the decimal point to the right.
2. Move it the same number of places as the number of 0's in the power of 10.

Multiplication by **10** moves the decimal point **one** place **to the right**.

Multiplication by **100** moves the decimal point **two** places **to the right**.

Multiplication by **1000** moves the decimal point **three** places **to the right**.

And so on.

6. Multiply the following:

a. 10(0.13)

b. 1000(8.78)

Example 6

The following products illustrate multiplication by powers of 10.

a. $10(1.59) = 15.9$ Move the decimal point 1 place to the right.

b. $100(2.68) = 268. = 268$ Move the decimal point 2 places to the right.

c. $1000(0.9653) = 965.3$ Move the decimal point 3 places to the right.

d. $1000(7.2) = 7200. = 7200$ Move the decimal point 3 places to the right.

e. $10^2(3.5149) = 351.49$ Move the decimal point 2 places to the right. The exponent tells how many places to move the decimal point.

Now work exercise 6 in the margin.

Metric Units of Length

The **meter** is the basic unit of length in the metric system. Smaller and larger units are named by putting a prefix in front of the basic unit:

for example, **centi**meter (cm) and **kilo**meter (km).

Units of length are set up so that there are 10 of one unit in the next larger unit. Thus, there are:

10 millimeters (mm) in 1 centimeter (cm),
10 centimeters (cm) in 1 decimeter (dm),
10 decimeters (dm) in 1 meter (m),
10 meters (m) in 1 dekameter (dam),
10 dekameters in 1 hectometer (hm),
10 hectometers in 1 kilometer (km).

Therefore, **to change from any number of a particular unit of length to a smaller unit of length**, we multiply by some power of 10 (or simply move the decimal point to the right). A visual representation of this is shown in Figure 5.4. This makes sense because there will be more of the smaller unit. Table 5.1 illustrates some relationships between the meter and other units of length in the metric system. (See Appendix I for a more detailed discussion of the metric system and other units of measure for weight and volume.)

1 millimeter (mm): about the width of the wire in a paper clip

1 centimeter (cm): about the width of a paper clip

Figure 5.4

Table 5.1 Basic Units of Measure of Length in the Metric System		
1 meter (m)	=	1000 **milli**meters (mm)
1 meter (m)	=	100 **centi**meters (cm)
1 meter (m)	=	10 **deci**meters (dm)
1 **kilo**meter (km)	=	1000 meters (m)

Changing Metric Measures of Length:

To change to a measure that is:

	Example
One unit smaller, multiply by **10**.	3 cm = 30 mm
Two units smaller, multiply by **100**.	5 m = 500 cm
Three units smaller, multiply by **1000**.	14 m = 14,000 mm

And so on. The chart in the margin shows a visual representation of this technique using columns to show placement of the decimals.

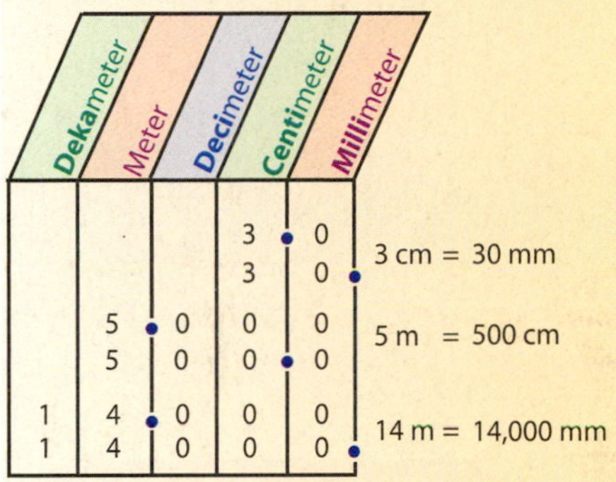

Example 7

The following example illustrates how to change from larger to smaller units of length in the metric system. (Refer to Table 5.1 if you need help.)

a. 4.32 m = 100 (4.32) cm = 432 cm

b. 4.32 m = 1000 (4.32) mm = 4320 mm

c. 14.6 cm = 10 (14.6) mm = 146 mm

d. 3.51 km = 1000 (3.51) m = 3510 m

Now work exercise 7 in the margin.

7. Determine the equivalent measures in the metric system.

a. 5.6 m = _______cm

b. 35.25 cm = _______mm

c. 16.43 km = _______m

As an aid in understanding relative lengths in the metric system, we show some comparisons to the U.S. customary unit system here.

1 meter is about 39.36 inches.

1 meter is about 3.28 feet.

1 centimeter is about 0.394 inches.

1 kilometer is about 0.62 miles.

meterstick

1 meter = 10 decimeters = 100 centimeters

yardstick

1 yard = 3 feet = 36 inches

ruler

1 foot = 12 inches

Objective 3

Estimating Products of Decimal Numbers

Estimating products can be done by rounding each number to the place of the last nonzero digit on the left and multiplying these rounded numbers. This technique is particularly helpful in correctly placing the decimal point in the actual product.

8. Estimate and then find the product: 11.287×3.4.

Example 8

First **a.** estimate the product (0.356)(6.1); then **b.** find the product and use the estimation to help place the decimal point.

Solution

a. Estimate by multiplying rounded numbers.

$$\begin{array}{rl} 0.4 & \text{(0.356 rounded)} \\ \times\ 6 & \text{(6.1 rounded)} \\ \hline 2.4 & \leftarrow \text{estimate} \end{array}$$

b. Find the actual product.

$$\begin{array}{rl} 0.356 & \\ \times\ 6.1 & \\ \hline 356 & \\ 2136 & \\ \hline 2.1716 & \leftarrow \text{actual product} \end{array}$$

The estimated product 2.4 helps place the decimal point correctly in the product 2.1716.

Now work exercise 8 in the margin.

Example 9

First **a.** find the product (19.9)(2.3); then **b.** estimate the product and use this estimation as a check.

Solution

a. Find the actual product.

$$\begin{array}{r} 19.9 \\ \times\ \ 2.3 \\ \hline 5.97 \\ 39.8\ \ \\ \hline 45.77 \end{array} \leftarrow \text{actual product}$$

b. Estimate the product and use this estimation to check that the actual product is reasonable.

$$\begin{array}{rl} 20.0 & \text{(19.9 rounded)} \\ \times\ \ \ 2 & \text{(2.3 rounded)} \\ \hline 40.0 & \leftarrow \text{estimated product} \end{array}$$

The estimated product of 40.0 indicates that the actual product is reasonable and the placement of the decimal point is correct.

Now work exercise 9 in the margin.

9. Find the product (31.5)(0.64) and then estimate to check your answer.

Note that although estimated products (or estimated sums or estimated differences) are close to the actual value, they do not guarantee the absolute accuracy of this actual value. The estimates help only in placing decimal points and checking the reasonableness of answers. Experience and understanding are needed to judge whether or not a particular answer is reasonably close to an estimate or vice versa.

Word Problems

Some word problems may involve several operations with decimal numbers. The problems do not usually say directly to add, subtract, or multiply. Experience and reasoning abilities are needed to decide which operation (if any) to perform with the given numbers. In Example 10, we illustrate a problem that involves several steps, and we show how estimating can provide a check for a reasonable answer.

Example 10

You can buy a car for $15,000 cash or you can make a down payment of $3750 and then pay $1093.33 each month for 12 months. How much can you save by paying cash?

Solution

a. Find the amount paid in monthly payments by multiplying the amount of each payment by 12. In this case, judgment dictates that we use 12 and do not round to 10, since we do not want to lose two full monthly payments in our estimate. (See Exercise 81 to understand what happens if 10 is used in the estimated calculations.)

Continued on next page...

10. How much will you save on a new car by paying \$500 a month for 24 months as opposed to \$1050 a month for 12 months? Assume the down payment is the same.

Estimate		Actual Amount	
\$1000		\$1093.33	
× 12		× 12	
2000		2186 66	
10,000		10,933 30	
\$12,000	estimated monthly payments	\$13,119.96	paid in monthly payments

b. Find the total amount paid by adding the down payment to the answer in part **a.**

Estimate		Actual Amount	
\$ 4 000	down payments	\$3 750.00	down payments
+12, 000	monthly payments	+13,119.96	monthly payments
\$16,000	estimated total	\$16,869.96	total paid

c. Find the savings by subtracting \$15,000 (the cash price) from the answer to part **b.**

Estimate		Actual Amount	
\$16,000	estimated total	\$16,869.96	total paid
−15,000	cash price	−15,000.00	cash price
\$ 1 000	estimated savings	\$ 1 869.96	savings by paying cash

The estimated \$1000 saved by paying cash is reasonably close to the actual savings of \$1869.96.

Now work exercise 10 in the margin.

Completion Example Answers

4.

```
   3.4  ← 1 place  }
 × 5.8  ← 1 place  } total of 2 places
   272
  170
 19.72   2 places in the product
```

5.

```
   0.003  ← 3 places }
 ×  0.03  ← 2 places } total of 5 places
 0.00009   5 places in the product
           4 0's had to be inserted to place the decimal point
```

Practice Problems

Multiply the following.

1. 1.23×0.7

2. 9.5×6.884

3. $(0.08)(0.542)$

4. 1000×0.0079

Perform the following conversions.

5. 8930 m = ______ mm

6. 7.002 dm ______ mm

Answers to Practice Problems:

1. 0.861 **2.** 65.398 **3.** 0.04336 **4.** 7.9

5. 8,930,000 mm **6.** 700.2 mm

Name ________________ Section ____________ Date ___________

ANSWERS

Exercises 5.3

1. Match each indicated product with the best estimate of that product.

________(a)	0.7×0.8	A. 0.06
________(b)	34.5×0.11	B. 1.0
________(c)	0.63×9.81	C. 3.0
________(d)	0.34×0.18	D. 5.0
________(e)	4.6×1.2	E. 6.0

2. Match each indicated product with the best estimate of that product.

________(a)	1.75×0.04	A. 0.008
________(b)	1.75×0.004	B. 0.08
________(c)	17.5×0.04	C. 0.8
________(d)	1.75×4	D. 8.0

Find each of the indicated products. See Examples 1 through 5.

3. 0.6×0.7
4. 0.3×0.8
5. 0.2×0.2
6. 0.3×0.3
7. 8×2.7
8. 4×9.6
9. 1.4×0.3
10. 1.5×0.6
11. 0.2×0.02
12. 0.3×0.03
13. 5.4×0.02
14. 7.3×0.01
15. 0.23×0.12
16. 0.15×0.15
17. 8.1×0.006
18. 7.1×0.008
19. 0.06×0.01
20. 0.25×0.01

1. [Respond below exercise.] ________
2. [Respond below exercise.] ________
3. ________
4. ________
5. ________
6. ________
7. ________
8. ________
9. ________
10. ________
11. ________
12. ________
13. ________
14. ________
15. ________
16. ________
17. ________
18. ________
19. ________
20. ________

21. ______
22. ______
23. ______
24. ______
25. ______
26. ______
27. ______
28. ______
29. ______
30. ______
31. ______
32. ______
33. ______
34. ______
35. ______
36. ______
37. ______
38. ______
39. ______
40. ______
41. ______
42. ______
43. ______
44. ______
45. ______
46. ______
47. ______
48. ______
49. ______
50. ______
51. ______
52. ______
53. ______
54. ______

21. 3×0.125 **22.** 4×0.375 **23.** 1.6×0.875

24. 5.3×0.75 **25.** 6.9×0.25 **26.** 4.8×0.25

27. 0.83×6.1 **28.** 0.27×0.24 **29.** 0.16×0.5

30. 0.28×0.5

Find each of the following products mentally by using your knowledge of multiplication by powers of ten. See Example 6.

31. 100×3.46 **32.** 100×20.57 **33.** 100×7.82

34. 100×6.93 **35.** 100×16.1 **36.** 100×38.2

37. 10×0.435 **38.** 10×0.719 **39.** 10×1.86

40. 1000×4.1782 **41.** 1000×0.38 **42.** 1000×0.47

43. $10{,}000 \times 0.005$ **44.** $10{,}000 \times 0.00615$ **45.** $10{,}000 \times 7.4$

Find the equivalent measures in the metric system. See Example 7. (Refer to Table 5.1 for help.)

46. 5 cm = _____ mm **47.** 13 cm = _____ mm **48.** 3 m = _____ cm

49. 15 m = _____ cm **50.** 3.2 m = _____ mm **51.** 6.17 m= _____ mm

52. 6.5 km = _____ m **53.** 16 km = _____ m **54.** 0.5 km = _____ m

Name ______________ Section ______________ Date ______________

ANSWERS

55. 0.6 km = ________ m = ________ cm = ________ mm

56. 2.53 km = ________ m = ________ cm = ________ mm

57. 0.02 km = ________ m = ________ cm = ________ mm

58. 10.7 km = ________ m = ________ cm = ________ mm

First, estimate the product, and then find the actual product. See Examples 8 and 9.

59. 0.106×0.09

60. 1.07×0.5

61. 5.08×0.4

62. 0.0106×0.087

63. 0.0213×0.065

64. 83.105×0.111

65. 17.002×0.101

66. 86.1×0.057

67. 7.83×0.18

68. 95.62×0.57

69. 6.02×0.57

70. 8.034×0.29

71. Car payments. To buy a car, you can pay \$2036.50 in cash, or you can put down \$400 and make 18 monthly payments of \$104.30. How much would you save by paying cash?

72. Property tax. Suppose a tax assessor figures tax at 0.06 of the assessed value of a home and the assessed value is determined at a rate of 0.42 times the market value.

a. Estimate the taxes to be paid on a home worth a market value of \$236,500.

b. What are the exact taxes paid?

55. ________
56. ________
57. ________
58. ________
59. ________
60. ________
61. ________
62. ________
63. ________
64. ________
65. ________
66. ________
67. ________
68. ________
69. ________
70. ________
71. ________
72. a. ________
b. ________

73. a. ______________

b. ______________

74. ______________

75. ______________

76. ______________

77. ______________

78. ______________

79. ______________

80. ______________

73. **Sales tax.**
   a. If the sale price of a new refrigerator is \$583 and sales tax is figured at 0.06 times the price, approximately what amount is paid for the refrigerator?
   b. What is the exact amount paid for the refrigerator?

74. **Overtime pay.** If you were paid a salary of \$350 per week and \$13.75 for each hour you worked over 40 hours in a week, how much would you make if you worked 45 hours in one week?

75. Find the perimeter and area of a square with sides 3.2 ft long.

76. Find the area of a triangle with base 10.5 cm and height 2.8 cm.

77. Find the perimeter and area of a rectangle with length 18.04 m and width 6.25 m.

78. Find the area of a square with sides 4.7 mm long.

79. **Library funding.** In 2003 Ohio led the nation in per capita funding for its public libraries with \$39.87 spent for each person. How much funding would have been received that year by a library that served a town of 23,500 people?

80. **Agriculture.** In July of 2004 the average price per pound received by U.S. farmers for cattle was \$0.92. At this price, what would be the value of 42,500 pounds of cattle?

Name ______________________ Section ____________ Date ___________

ANSWERS

81. [Respond below exercise.]

82. [Respond below exercise.]

83. [Respond below exercise.]

Writing and Thinking about Mathematics

81. Automobile purchase. In Example 10, we stated the following problem:

> You can buy a car for $15,000 cash or you can make a down payment of $3750 and pay $1093.33 each month for 12 months. How much can you save by paying cash?

Estimate the savings by using rounded numbers. (This includes rounding 12 months to 10 months.) Explain why this estimated savings does not seem reasonable, and explain why we must be careful about using rounded numbers in practical applications.

82. Estimation. Suppose you are interested only in a rounded answer for a product. Would there be any difference in the products produced by the following two procedures?

(a) First multiply the two numbers as they are and then round the product to the desired place of accuracy.

(b) First round each number to the desired place of accuracy and then multiply the rounded numbers.

Explain why you think these two procedures would produce the same result or different results.

83. Measurement. We stated in the text that 1 meter is about 39.36 inches. Use each of the techniques discussed in Exercise 82 to find (to the nearest tenth of an inch) how many inches are in 17.523 meters. Discuss why the results are different (or the same).

REVIEW

1. ______

2. ______

3. ______

4. ______

5. a. ______

b. ______

Review Problems (from Section 1.6)

Find the whole number quotients and remainders for each of the following.

1. $15\overline{)120}$

2. $20\overline{)315}$

3. $50\overline{)4057}$

4. $230\overline{)46,790}$

5. **Geography.** The state of Alaska has a population of about 640,000 people and a land area of about 570,000 square miles (or about [illegible],480,000 square kilometers).

a. Approximately how many square miles are there in Alaska for each person?

b. About how many square kilometers are there for each person?

5.4 Division with Decimal Numbers

Dividing Decimal Numbers

The process of division (called the **division algorithm**) with decimal numbers is, in effect, the same as that for division with whole numbers. This is reasonable because whole numbers are decimal numbers; that is, we can always write the decimal point to the right of a whole number if we so choose. As the following example illustrates, division with whole numbers gives a quotient and possibly a remainder.

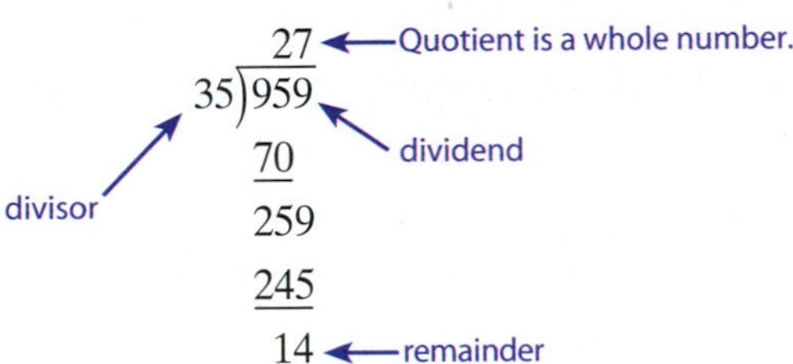

Objectives

1. Be able to divide decimal numbers and place the decimal point correctly in the quotient.
2. Be able to divide decimal numbers mentally by powers of 10.
3. Know how to estimate quotients with rounded decimal numbers.

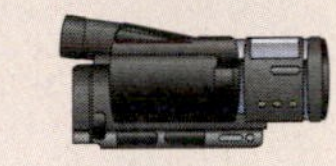
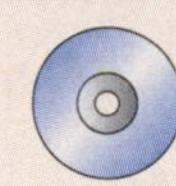

Now that we have decimal numbers, we can continue to divide and get a quotient that is a decimal number other than a whole number. Zeros are added onto the dividend if they are needed.

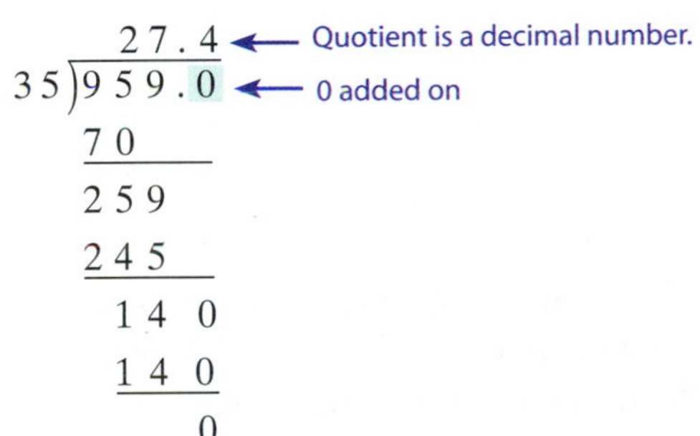

If the divisor is a decimal number other than a whole number, multiply both the divisor and the dividend by a power of 10 so that the divisor is a whole number. For example, we can write

$$4.9\overline{)51.45} \quad \text{as} \quad \frac{51.45}{4.9} \cdot \frac{10}{10} = \frac{514.5}{49} \quad \text{or} \quad 49\overline{)514.5}$$

By multiplying both the divisor (4.9) and the dividend (51.45) by 10, we have a new divisor (49) that is a whole number. In a similar manner, we can write

$$1.36\overline{)5.1} \quad \text{as} \quad \frac{5.1}{1.36} \cdot \frac{100}{100} = \frac{510}{136} \quad \text{or} \quad 136\overline{)510.}$$

In this case, both the divisor (1.36) and the dividend (5.1) are multiplied by 100 so that we have a new whole number divisor (136).

This discussion leads to the following procedure for dividing decimal numbers.

Objective 1

To Divide Decimal Numbers:

1. Move the decimal point in the divisor to the right so that the divisor is a whole number.
2. Move the decimal point in the dividend the same number of places to the right.
3. Place the decimal point in the quotient directly above the new decimal point in the dividend. (**Note:** Be sure to do this before dividing.)
4. Divide just as you would with whole numbers. (**Note:** 0's may be added in the dividend as needed to be able to continue the division process.)

1. Divide: $14.03 \div 2.3$

Example 1

Divide: $51.45 \div 4.9$

Solution

a. Write the problem as follows:

$4.9\overline{)51.45}$

b. To move the decimal point so that the divisor is a whole number, move each decimal point one place. This makes the whole number 49 the divisor. Place the decimal point in the quotient before dividing.

. ← decimal point in quotient

$4.9.\overline{)51.4.5}$

c. Divide as with whole numbers.

$$\begin{array}{r} 10.5 \\ 49.\overline{)514.5} \\ \underline{49} \\ 24 \\ \underline{0} \\ 24\,5 \\ \underline{24\,5} \\ 0 \end{array}$$

Now work exercise 1 in the margin.

Example 2

Divide: $5.1 \div 1.36$

Solution

a. Write the problem as follows: $1.36\overline{)5.1}$

b. Move the decimal points so the divisor is a whole number. Add 0's in the dividend if needed. Place the decimal point in the quotient before dividing.

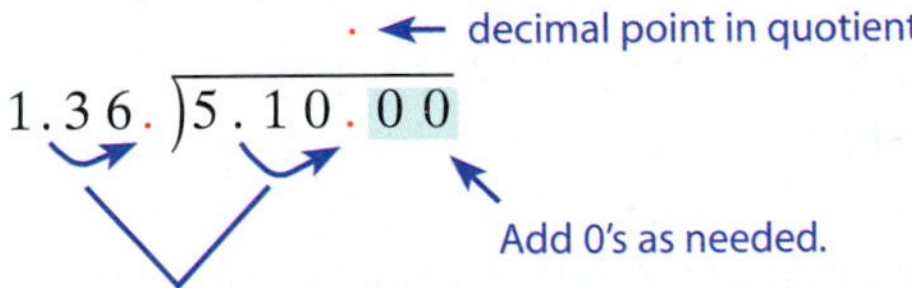

c. Divide.

```
          3.75
 136.)510.00
      408
      1020
       952
        680
        680
          0
```

Now work exercise 2 in the margin.

2. Divide: $85.12 \div 6.4$

Example 3

Divide: $6.3252 \div 6.3$

Solution

```
           1.0 0 4
 6.3.)6.3.2 5 2
      6 3
        0 2
          0
          2 5
            0
          2 5 2
          2 5 2
              0
```

Note: There must be a digit to the right of the decimal point in the quotient above every digit to the right of the decimal point in the dividend.

Now work exercise 3 in the margin.

3. Divide: $33.524 \div 2.9$

4. Divide: $14.872 \div 2.86$

5. Find $9.86 \div 2.1$ (to the nearest tenth).

Completion Example 4

Divide: $24.225 \div 4.25$

Solution

$$4.25.\overline{)24.22.5}$$ quotient: 5.

Now work exercise 4 in the margin.

In each of Examples 1 – 4, the remainder was 0. Certainly, this will not always be the case. In general, some place of accuracy for the quotient is agreed upon before the division is performed. If the remainder is not 0 by the time this place of accuracy is reached, then we divide one more place and round the quotient.

When the Remainder is not 0:

1. Decide first how many decimal places are to be in the quotient.
2. Divide until the quotient is **one digit past the place of desired accuracy**.
3. Using this last digit, round the quotient to the desired place of accuracy.

Example 5

Find $8.24 \div 2.9$ to the nearest tenth.

Solution

Divide until the quotient is in hundredths (one more place than tenths); then round to tenths.

hundredths
read **approximately**
rounded to tenths

$2.84 \approx 2.8$

```
      2.84
2.9.)8.2.40
     5 8
     2 4 4
     2 3 2
       1 2 0
       1 1 6
           4
```

$8.24 \div 2.9 \approx 2.8$ accurate to the nearest tenth

Now work exercise 5 in the margin.

Example 6

Find 1.83 ÷ 4.1 to the nearest hundredth.

Solution

Divide until the quotient is in thousandths (one more place than hundredths); then round to hundredths.

read **approximately**

$$
\begin{array}{r}
0.446 \approx 0.45 \\
4.1\overline{)1.8300} \\
\underline{164} \\
190 \\
\underline{164} \\
260 \\
\underline{246} \\
14
\end{array}
$$

Add as many 0's as needed.

$1.83 \div 4.1 \approx 0.45$ accurate to the nearest hundredth

Now work exercise 6 in the margin.

6. Find $14.393 \div 9.3$ (to the nearest hundredth).

Example 7

Find $17 \div 3.3$ to the nearest thousandth.

Solution

Divide until the quotient is in ten-thousandths; then round to thousandths.

$$
\begin{array}{r}
5.1515 \approx 5.152 \\
3.3\overline{)17.00000} \\
\underline{165} \\
50 \\
\underline{33} \\
170 \\
\underline{165} \\
50 \\
\underline{33} \\
170 \\
\underline{165} \\
5
\end{array}
$$

← Add as many 0's as needed.

$17 \div 3.3 \approx 5.152$ accurate to thousandths

Now work exercise 7 in the margin.

7. Find $43.721 \div 0.06$ (to the nearest thousandth).

Dividing Decimal Numbers by Powers of 10

In Section 5.3, we found that multiplication by powers of 10 can be accomplished by moving the decimal point to the right (making a number larger than the original number). Division by powers of 10 can be accomplished by moving the decimal point to the left (making a number smaller than the original number).

Objective 2

To Divide a Decimal Number by a Power of 10:

1. Move the decimal point to the left.
2. Move it the same number of places as the number of 0's in the power of 10.

Division by **10** moves the decimal point **one** place **to the left.**

Division by **100** moves the decimal point **two** places **to the left.**

Division by **1000** moves the decimal point **three** places **to the left.**

And so on.

Two general guidelines will help you understand working with powers of 10.

1. Multiplication by a power of 10 will make a number larger, so move the decimal point to the right.
2. Division by a power of 10 will make a number smaller, so move the decimal point to the left.

8. Find the quotient by using your knowledge of division by powers of 10.
$\frac{57.6}{10,000}$

Example 8

The following quotients illustrate division by powers of 10.

a. $4.16 \div 100 = \frac{4.16}{100} = 0.0416$ Move the decimal point two places to the left.

b. $782 \div 10 = \frac{782}{10} = 78.2$ Move the decimal point one place to the left.

c. $5.933 \div 1000 = \frac{5.933}{1000} = 0.005933$ Move the decimal point three places to the left.

Now work exercise 8 in the margin.

Metric Units of Length

Now, in the metric system, **to change from any number of a particular unit of length to a larger unit of length**, we divide by some power of 10 (or simply move the decimal point to the left). This makes sense because there will be fewer of the larger unit. Refer to Table 5.1 to review the basic relationships between units of length in the metric system.

Changing Metric Measures of Length

To change to a measure that is:

	Example
One unit larger, divide by **10**.	4 mm = 0.4 cm
Two units larger, divide by **100**.	6 mm = 0.06 dm
Three units larger, divide by **1000**.	8 mm = 0.008 m

And so on. The chart in the margin shows a visual representation of this technique using columns to show placement of the decimals.

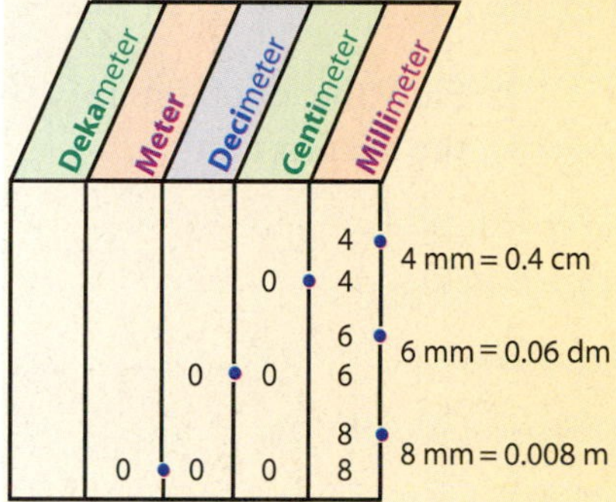

Example 9

The following examples illustrate changing from smaller to larger units of length in the metric system.
(Refer to Table 5.1 if you need help.)

a. $564 \text{ cm} = \frac{564}{100} \text{ m} = 5.64 \text{ m}$

b. $564 \text{ mm} = \frac{564}{1000} \text{ m} = 0.564 \text{ m}$

c. $1030 \text{ m} = \frac{1030}{1000} \text{ km} = 1.03 \text{ km}$

Now work exercise 9 in the margin.

9. Find the equivalent measures in the metric system.

a. 350 cm = _____ m

b. 68 mm = _____ cm

c. 3952 mm = _____ m

Objective 3

Estimating Quotients of Decimal Numbers

As with addition, subtraction, and multiplication, we can use estimating with division to help in placing the decimal point in the quotient and to verify the reasonableness of the quotient. In order to estimate with division, round both the divisor and the dividend to the place of the last nonzero digit on the left and then divide with these rounded values.

10. a. Estimate the quotient $13.18 \div 2.41$.

b. Find the quotient to the nearest tenth.

Example 10

First

a. Estimate the quotient $6.2 \div 0.302$, and then

b. Find the quotient to the nearest tenth.

Solution

a. Estimate the quotient using $6.2 \approx 6$ and $0.302 \approx 0.3$.

```
       20.
0.3.)6.0.
     6
     0 0
       0
       0
```

b. Find the quotient to the nearest tenth.

```
          20.52 ≈ 20.5
0.302.)6.200.00
       6 04
         160
           0
         1600
         1510
           900
           604
```

The estimated value 20 is very close to the rounded quotient 20.5.

Now work exercise 10 in the margin.

We know from Section 1.7 that the **average** (or **arithmetic average**) of a set of numbers can be found by adding the numbers, then dividing the sum by the number of addends. The term **average** is used in phrases such as "average speed of 43 miles per hour" or "the average price of a pair of shoes." These averages can also be found by division. If the total amount of a quantity (distance, dollars, gallons of gas, etc.) and a number of units (time, items bought, miles, etc.) are known, then we can find the **average amount per unit** by dividing the amount by the number of units.

Example 11

The gas tank of a car holds 17 gallons of gasoline. Approximately how many miles per gallon does the car average if it will go 470 miles on one tank of gas?

Solution

The question calls only for an approximate answer. Thus, we can use rounded values:

$17 \approx 20$ gal and $470 \approx 500$ miles

Now divide to approximate the **average** number of miles per gallon:

$$\begin{array}{r} 25 \\ 20\overline{)500} \\ \underline{40} \\ 100 \\ \underline{100} \\ 0 \end{array} \quad \text{miles per gallon}$$

The car averages about 25 miles per gallon.

Now work exercise 11 in the margin.

11. If a sprinter runs 100 meters in 10.13 seconds, what is his approximate average speed in meters per second?

If an average amount per unit is known, then a corresponding total amount can be found by multiplying. For example, if you ride your bicycle at an average speed of 15.2 miles per hour, then the distance you travel can be found by multiplying your average speed by the time you spend riding.

Example 12

If you ride your bicycle at an average speed of 15.2 miles per hour, how far will you ride in 3.5 hours?

Solution

Multiply the average speed by the number of hours.

$$\begin{array}{rl} 15.2 & \text{miles per hour} \\ \times\quad 3.5 & \text{hours} \\ \hline 7\,60 & \\ 45\,6 & \\ \hline 53.20 & \text{miles} \end{array}$$

You will ride 53.2 miles in 3.5 hours.

Now work exercise 12 in the margin.

12. A secretary can type 72 words per minute on average. How many words will she be able to type in 15 minutes?

Completion Example Answer

4.
$$\begin{array}{r} 5.7 \\ 4.25.\overline{)24.22.5} \\ \underline{21\,25} \\ 2\,975 \\ \underline{2\,975} \\ 0 \end{array}$$

Practice Problems

Find each indicated quotient or product.

1. $4\overline{)1.83}$ (to the nearest hundredth)
2. $0.06\overline{)43.721}$ (to the nearest thousandth)
3. $\dfrac{42.31}{1000}$
4. $87.96 \div 1000$

Find the equivalent measures in the metric system.

5. 98 mm = _____ cm
6. 3.4 cm = ______ m

Answers to Practice Problems:

1. 0.46 **2.** 728.683 **3.** 0.04231
4. 0.08796 **5.** 9.8 cm **6.** 0.034 m

Name ______________________ Section __________ Date __________

ANSWERS

Exercises 5.4

1. *Match the indicated quotient with the best estimate of that quotient. See Example 10.*

_______ (a) $3.1\overline{)6.386}$ A. 0.02

_______ (b) $0.1\overline{)216.5}$ B. 2

_______ (c) $3.7\overline{)281.6}$ C. 5

_______ (d) $18.5\overline{)127.9}$ D. 75

_______ (e) $4.1\overline{)0.0884}$ E. 2000

2. *Match the indicated quotient with the best estimate of that quotient. See Example 10.*

_______ (a) $27.58 \div 0.003$ A. 100

_______ (b) $27.58 \div 0.03$ B. 10

_______ (c) $27.58 \div 0.3$ C. 10,000

_______ (d) $27.58 \div 3$ D. 1000

Divide. See Examples 1 through 3.

3. $4.68 \div 2$

4. $1.71 \div 3$

5. $4.95 \div 5$

6. $1.62 \div 9$

7. $0.064 \div 0.8$

8. $0.63 \div 0.7$

9. $82.24 \div 0.04$

10. $16.02 \div 0.03$

11. $48 \div 2.4$

12. $28 \div 5.6$

1. [Respond in exercise.] ________

2. [Respond in exercise.] ________

3. ________

4. ________

5. ________

6. ________

7. ________

8. ________

9. ________

10. ________

11. ________

12. ________

13. ______

14. ______

15. ______

16. ______

17. ______

18. ______

19. ______

20. ______

21. ______

22. ______

23. ______

24. ______

25. ______

26. ______

27. ______

28. ______

29. ______

30. ______

31. ______

32. ______

33. ______

34. ______

Find each quotient to the nearest tenth. See Example 5.

13. $8\overline{)455}$ **14.** $4\overline{)263}$ **15.** $9.4\overline{)6.538}$

16. $4.6\overline{)5}$ **17.** $7.05\overline{)0.4977}$ **18.** $0.37\overline{)4.683}$

19. $1.62\overline{)34}$ **20.** $1.33\overline{)75}$

Find each quotient to the nearest hundredth. See Example 6.

21. $24\overline{)0.1463}$ **22.** $1.23\overline{)14.91129}$ **23.** $0.075\overline{)0.42753}$

24. $2.7\overline{)2.583}$ **25.** $23\overline{)62.949}$ **26.** $9\overline{)2}$

27. $13\overline{)65.476}$ **28.** $3.181\overline{)6}$

Find each quotient mentally by using your knowledge of division by powers of 10. See Example 8.

29. $78.4 \div 100$ **30.** $16.4963 \div 100$ **31.** $50.36 \div 100$

32. $45.621 \div 1000$ **33.** $73.85 \div 1000$ **34.** $18.6 \div 1000$

Name ____________________ Section ____________ Date __________

ANSWERS

35. $\frac{167}{10}$ **36.** $\frac{138.1}{10}$ **37.** $\frac{7.85}{10}$

38. $\frac{1.54}{10,000}$ **39.** $\frac{169.9}{10,000}$ **40.** $\frac{10.413}{10,000}$

Find the equivalent measures in the metric system. See Example 9.

41. 5 mm = _____ cm **42.** 11 mm = _____ cm

43. 83 cm = _____ m **44.** 95 cm = _____ m

45. 344 mm = _____ m **46.** 255 mm = _____ m

47. 1500 m = _____ km **48.** 2400 m = _____ km

49. 97.2 mm = _____ cm **50.** 18.5 mm = _____ cm

51. 32 mm = _____ cm = _____ dm = _____ m

52. 560 mm = _____ cm = _____ dm = _____ m

35. ____________
36. ____________
37. ____________
38. ____________
39. ____________
40. ____________
41. ____________
42. ____________
43. ____________
44. ____________
45. ____________
46. ____________
47. ____________
48. ____________
49. ____________
50. ____________
51. ____________
52. ____________

53. ________

54. ________

55. ________

56. ________

57. ________

58. ________

59. a. ________

b. ________

60. a. ________

b. ________

61. a. ________

b. ________

62. a. ________

b. ________

In Exercises 53 – 58, first estimate each quotient; then find the quotient to the nearest thousandth. See Example 10.

53. $23\overline{)71}$ **54.** $69\overline{)293}$ **55.** $85.3\overline{)24.31}$

56. $2.57\overline{)0.4961}$ **57.** $16.2\overline{)0.11623}$ **58.** $25.7\overline{)6.27}$

Read each problem carefully before you decide which operation(s) are required. See Examples 11 and 12.

59. Fuel economy.
a. If a car averages 24.6 miles per gallon, estimate how far it will go on 18 gallons of gas?
b. Exactly how many miles will it go on 18 gallons of gas?

60. Bicycling.
a. If Lance Armstrong rode 250.6 miles in 13.2 hours, estimate how fast he was riding per hour.
b. What was his average speed in miles per hour (to the nearest tenth)?

61. Wholesale purchasing. A quarter section of beef can be bought cheaper than the same amount of meat purchased a few pounds at a time.
a. Estimate the cost per pound if 150 pounds costs $187.50.
b. What is the cost per pound?

62. Automobile travel.
a. If you drive 9.5 hours at an average speed of 52.2 miles per hour, about how far will you drive?
b. Exactly how far will you drive?

Name ______________________ Section ____________ Date ____________

ANSWERS

63. Sales tax. If new tires cost $56.50 per tire and tax is figured at 0.06 times the cost of each tire, what will you pay for four new tires?

63. ____________

64. Sales tax. If you bought 10 books for a total price of $225 plus tax, and tax is figured at 0.06 times the price, what average amount did you pay per book, including tax?

64. ____________

65. Sales tax. If the total price of a stereo was $312.70 including a tax of 0.06 times the list price, you can find the list price by dividing the total price by 1.06. What was the list price? (**Note:** 1.06 represents the list price plus 0.06 times the list price.)

65. ____________

66. Mortgages. Suppose that the total interest paid on a 30-year mortgage for a home loan of $60,000 will be $189,570. What will be the payment each month if the payments are to pay off both the loan and the interest?

66. ____________

67. Baseball. In 2004 the Cleveland Indians had a team batting average of 0.276 and had 1565 base hits. Find the number of team at bats, to the nearest whole number, by dividing base hits by batting average.

67. ____________

68. Baseball. In a recent year, Chris Carpenter of the St. Louis Cardinals led his team with 21 wins and an 0.808 winning percentage. Find Carpenter's total number of pitching decisions (games won or lost), to the nearest whole number, by dividing wins by winning percentage.

68. ____________

69.

69. Basketball. In a recent year, Reggie Miller of the Indiana Pacers basketball team led the NBA with a 0.933 free-throw percentage. He successfully made 250 free throws. Find his number of attempted free throws, to the nearest whole number, by dividing free throws made by free-throw percentage.

70.

70. Baseball. At Turner Field in Atlanta, home of the Braves baseball team, the distance down the right-field foul line is 330 feet. Convert this distance to meters, to the nearest tenth of a meter. (Use 1 m = 3.28 ft.)

71.

71. Running. A marathon footrace is 26.219 miles in length. Convert this distance to kilometers, to the nearest tenth of a kilometer. (Use 1 km = 0.621 mi.)

72.

Check Your Number Sense

72. Publishing. When a textbook is made, it is often printed and bound in sets of 16 pages, called signatures. (Each signature is actually one large sheet of paper with the individual pages laid out in a rectangular arrangement. It is then folded several times to create the "booklet," or signature, of 16 pages.) Which of the following are possible textbook lengths if only whole 16-page signatures are used? (a) 256 pages (b) 500 pages (c) 368 pages (d) 1264 pages (e) 648 pages. Explain, briefly, why you chose the values you did.

73. a.

b.

73. Baseball. The public address announcer at SBC Park in San Francisco earns $75 for announcing each Giants baseball game.

a. If she announced 72 games in one particular season, about how much did she earn: $3500, $4500, $5500, or $6500?

b. If there are 81 home games in one year, approximately what is the maximum amount that she can earn announcing games: $3500, $4500, $5500, or $6500?

Name ______________________ Section ____________ Date __________

Writing and Thinking about Mathematics

74. In your own words, explain why, when converting units in the metric system, we **multiply** by a power of 10 to change to a **smaller** unit of measure, and we **divide** by a power of 10 to change to a **larger** unit of measure.

ANSWERS

74. [Respond below exercise.]

5.5 Decimals and Fractions

Changing from Decimals to Fractions

The decimals and fractions we have discussed are simply different forms of the same type of number—namely, **rational numbers**. To understand how these numbers are related, we show how to change from one form to the other. From Section 5.1, we know that decimal numbers can be written in fraction form with denominators that are powers of 10. For example,

$$0.35 = \frac{35}{100} \quad \text{and} \quad 0.017 = \frac{17}{1000}$$

The rightmost digit, 5, is in the hundredths position, so the fraction has 100 in the denominator.

The rightmost digit, 7, is in the thousandths position, so the fraction has 1000 in the denominator.

In each case, the denominator is the value of the position of the rightmost digit. Thus we proceed as follows to change a decimal number to fraction form.

Changing from Decimals to Fractions:

A finite (or terminating) decimal number can be written in fraction form by writing a fraction with the following:

1. A **numerator** that consists of the whole number formed by all the digits of the decimal number, and
2. A **denominator** that is the power of ten that names the position of the rightmost digit.

In Examples 1 – 6, each decimal number is changed to fraction form and then reduced, if possible, by using the factoring techniques discussed in Chapter 3 for reducing fractions.

Example 1

$$0.25 = \frac{25}{100} = \frac{\cancel{5} \cdot \cancel{5} \cdot 1}{2 \cdot \cancel{5} \cdot 2 \cdot \cancel{5}} = \frac{1}{4}$$

hundredths

Now work exercise 1 in the margin.

Example 2

$$0.32 = \frac{32}{100} = \frac{\cancel{4} \cdot 8}{\cancel{4} \cdot 25} = \frac{8}{25}$$

hundredths

Now work exercise 2 in the margin.

Objectives

① Know how to change decimal numbers to fraction form and/or mixed number form.

② Know how to change fractions and mixed numbers to decimal form.

③ Recognize both terminating and non-terminating decimals.

④ Understand that working with decimals and fractions in the same problem may involve rounded numbers and approximate answers.

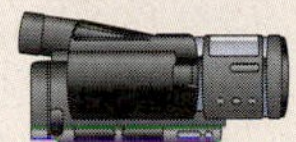

Objective ①

Convert to fraction form and then reduce:

1. 0.48

2. 0.95

Convert to fraction form and then reduce:

3. 0.762

4. 0.03

5. 15.8

6. 7.75

Example 3

$$0.131 = \frac{131}{1000}$$

↑ thousandths

Now work exercise 3 in the margin.

Example 4

$$0.075 = \frac{75}{1000} = \frac{\cancel{25} \cdot 3}{\cancel{25} \cdot 40} = \frac{3}{40}$$

↑ thousandths

Now work exercise 4 in the margin.

Example 5

$$2.6 = \frac{26}{10} = \frac{\cancel{2} \cdot 13}{\cancel{2} \cdot 5} = \frac{13}{5}$$

↑ tenths

or, as a mixed number,

$$2.6 = 2\frac{6}{10} = 2\frac{3}{5}$$

Now work exercise 5 in the margin.

Example 6

$$1.42 = \frac{142}{100} = \frac{\cancel{2} \cdot 71}{\cancel{2} \cdot 50} = \frac{71}{50}$$

↑ hundredths

or, as a mixed number,

$$1.42 = 1\frac{42}{100} = 1\frac{21}{50}$$

To use the TI-84 plus calculator to change a decimal to a fraction, follow the steps given below.

Step 1: Enter 1.42.

Step 2: Press 

.

Step 3: Press 

.

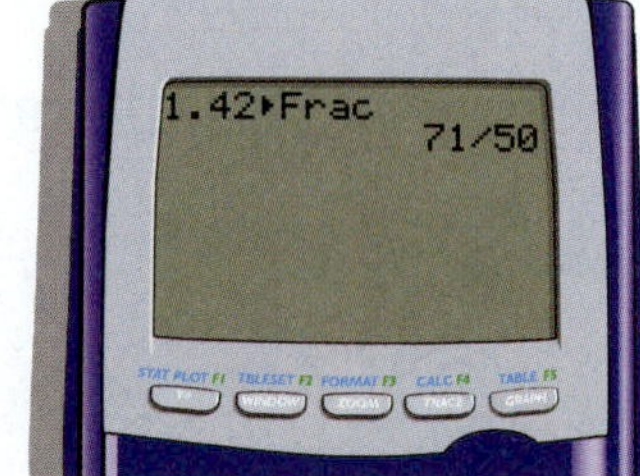

To the right is the display after pressing enter.

Now work exercise 6 in the margin.

Changing from Fractions to Decimals

Changing from Fractions to Decimals:

We have seen that fractions indicate parts of a whole and that fractions indicate division. To change a fraction to decimal form, we use the meaning that the numerator is to be divided by the denominator.

Objective 2

1. If the remainder is 0, the decimal is said to be **terminating**.

Objective 3

2. If the remainder is not 0, the decimal is said to be **non-terminating**.

The following examples illustrate fractions that convert to terminating decimals.

Example 7

$$\frac{3}{8} \rightarrow 8\overline{)3.000} \rightarrow \frac{3}{8} = 0.375$$

$$\begin{array}{r} 0.375 \\ 8\overline{)3.000} \\ \underline{2\,4} \\ 60 \\ \underline{56} \\ 40 \\ \underline{40} \\ 0 \end{array}$$

Now work exercise 7 in the margin.

Example 8

$$\frac{3}{4} \rightarrow 4\overline{)3.00} \rightarrow \frac{3}{4} = 0.75$$

$$\begin{array}{r} 0.75 \\ 4\overline{)3.00} \\ \underline{2\,8} \\ 20 \\ \underline{20} \\ 0 \end{array}$$

Now work exercise 8 in the margin.

Example 9

$$\frac{4}{5} \rightarrow 5\overline{)4.0} \rightarrow \frac{4}{5} = 0.8$$

$$\begin{array}{r} 0.8 \\ 5\overline{)4.0} \\ \underline{4\,0} \\ 0 \end{array}$$

Now work exercise 9 in the margin.

Convert the following to terminating decimals:

7. $\frac{2}{5}$

8. $\frac{13}{20}$

9. $\frac{6}{12}$

Non-terminating decimals can be **repeating** or **non-repeating**. A non-terminating repeating decimal has a repeating pattern to its digits. Every fraction with a whole number numerator and nonzero denominator is either a terminating decimal or a repeating decimal. (Such numbers are called **rational numbers**.) Non-terminating, non-repeating decimals are called **irrational numbers** and are discussed in Section 5.6 with square roots.

Convert the following to repeating decimals:

10. $\frac{3}{11}$

11. $\frac{2}{15}$

The following examples illustrate fractions that convert to non-terminating repeating decimals.

Example 10

$$\frac{1}{3} \rightarrow 3\overline{)1.000} = 0.333$$

← The 3 will repeat without end.

$$\begin{array}{r} 0.333 \\ 3\overline{)1.000} \\ \underline{9} \\ 10 \\ \underline{9} \\ 10 \\ \underline{9} \\ 1 \end{array}$$

← Continuing to divide will give a remainder of 1 each time.

We write $\frac{1}{3} = 0.333...$ The three dots mean "and so on" or to continue without stopping.

Now work exercise 10 in the margin.

Example 11

$$\frac{7}{12} \rightarrow \begin{array}{r} 0.5833 \\ 12\overline{)7.0000} \\ \underline{6\,0} \\ 1\,00 \\ \underline{96} \\ 40 \\ \underline{36} \\ 40 \\ \underline{36} \\ 4 \end{array}$$

← The 3 will repeat without end.

← Continuing to divide will give a remainder of 4 each time

We write $\frac{7}{12} = 0.58333...$

Now work exercise 11 in the margin.

Example 12

$$\frac{1}{7} \rightarrow 7\overline{)1.000000}$$

Quotient: 0.142857 ← The six digits will repeat in the same pattern without end.

```
  0.142857
7)1.000000
   7
   30
   28
    20
    14
     60
     56
      40
      35
       50
       49
        1
```

← The remainder will repeat in sequence 1, 3, 2, 6, 4, 5, 1, and so on. Therefore, the digits in the quotient will also repeat in sequence.

We write $\frac{1}{7} = 0.142857142857142857...$

Now work exercise 12 in the margin.

12. $\frac{4}{9}$

Another way of writing repeating decimals is to write a **bar** over the repeating digits. Thus, in Examples 10, 11, and 12, we can write

$$\frac{1}{3} = 0.\overline{3} \quad \text{and} \quad \frac{7}{12} = 0.58\overline{3} \quad \text{and} \quad \frac{1}{7} = 0.\overline{142857}$$

We may choose to round the quotient to some decimal place just as we did with division in Section 5.4. Perform the division one place past the desired rounding position.

Example 13

$$\frac{5}{11} \rightarrow 11\overline{)5.000}$$

0.454 ≈ 0.45 (nearest hundredth)

```
   0.454
11)5.000
   4 4
     60
     55
      50
      44
```

Now work exercise 13 in the margin.

13. $\frac{1}{27}$

Convert the following to repeating decimals:

14. $\frac{1}{12}$

Example 14

$$\frac{5}{6} \rightarrow 6\overline{)5.000} = 0.833 \approx 0.83 \text{ (nearest hundredth)}$$

$$\begin{array}{r} 0.833 \\ 6\overline{)5.000} \\ \underline{4\,8} \\ 20 \\ \underline{18} \\ 20 \\ \underline{18} \\ \vdots \end{array}$$

Now work exercise 14 in the margin.

Objective 4

Operating with Both Fractions and Decimals

As the following examples illustrate, we can perform operations and comparisons with both fractions and decimal numbers by changing the fractions to decimal form.

> **Note:** In some cases this may involve rounding the decimal form of a number and settling for an approximate answer. To have a more accurate answer, we may need to change the decimals to fraction form and then perform the operations.

15. Find the sum $2.88 + \frac{1}{4} + 13.9$ in decimal form.

Example 15

Find the sum $10\frac{1}{2} + 7.32 + 5\frac{3}{5}$ in decimal form.

Solution

$$\begin{array}{rcrl} 10\frac{1}{2} & = & 10.50 & \left(\frac{1}{2} = 0.50\right) \\ 7.32 & = & 7.32 & \\ +\,5\frac{3}{5} & = & 5.60 & \left(\frac{3}{5} = 0.60\right) \\ \hline & & 23.42 & \end{array}$$

Now work exercise 15 in the margin.

Example 16

Determine whether $\frac{3}{16}$ is larger than 0.18 by changing $\frac{3}{16}$ to decimal form and then comparing the two numbers. Find the difference.

Solution

Divide first.

$$\begin{array}{r} 0.1875 \\ 16\overline{)3.0000} \\ \underline{1\ 6} \\ 1\ 40 \\ \underline{1\ 28} \\ 120 \\ \underline{112} \\ 80 \\ \underline{80} \\ 0 \end{array}$$

So $\frac{3}{16} = 0.1875.$

Now subtract.

$$\begin{array}{r} 0.1875 \\ \underline{-0.1800} \\ 0.0075 \end{array} \quad \text{difference}$$

Thus $\frac{3}{16}$ is larger than 0.18 and their difference is 0.0075.

To use the TI-84 Plus calculator to change $\frac{3}{16}$ to a decimal, enter the fraction in the calculator and press ENTER. The display will be as follows:

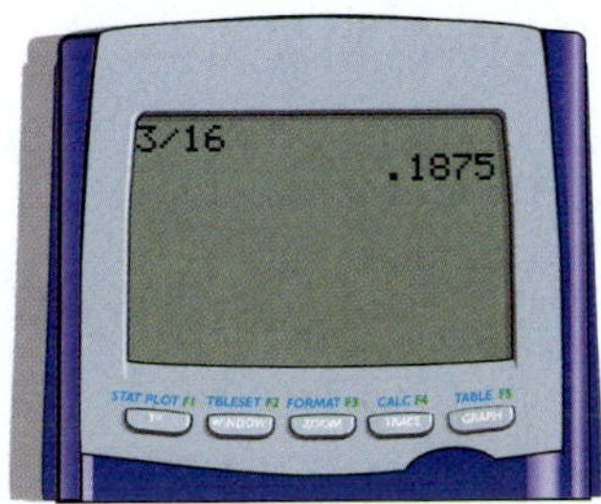

Now work exercise 16 in the margin.

16. Determine whether $\frac{9}{16}$ is larger than 0.52 by changing $\frac{9}{16}$ to decimal form and then comparing the two numbers. Find the difference.

Practice Problems

Convert the following to fraction form.

1. 0.475

2. 2.9

3. 0.003

4. 1.17

Convert the following to decimal form. Write repeating decimals using bar notation.

5. $\frac{13}{16}$

6. $\frac{1}{20}$

7. $\frac{2}{7}$

8. $\frac{18}{37}$

Anwers to Practice Problems:

1. $\frac{19}{40}$ **2.** $\frac{29}{10}$ **3.** $\frac{3}{1000}$ **4.** $\frac{117}{100}$

5. 0.8125 **6.** 0.05 **7.** $0.\overline{285714}$ **8.** $0.\overline{486}$

Name ______________________ Section ____________ Date __________

Exercises 5.5

Change each decimal to fraction form. Do not reduce.

1. 0.9 **2.** 0.3 **3.** 0.5 **4.** 0.8

5. 0.62 **6.** 0.38 **7.** 0.57 **8.** 0.41

9. 0.526 **10.** 0.625 **11.** 0.016 **12.** 0.012

13. 5.1 **14.** 7.2 **15.** 8.15 **16.** 6.35

Change each decimal to fraction form (or mixed number form) and reduce if possible. See Examples 1 through 6.

17. 0.125 **18.** 0.36 **19.** 0.18 **20.** 0.375

21. 0.225 **22.** 0.455 **23.** 0.17 **24.** 0.029

ANSWERS

1. __________
2. __________
3. __________
4. __________
5. __________
6. __________
7. __________
8. __________
9. __________
10. __________
11. __________
12. __________
13. __________
14. __________
15. __________
16. __________
17. __________
18. __________
19. __________
20. __________
21. __________
22. __________
23. __________
24. __________

25. ______
26. ______
27. ______
28. ______
29. ______
30. ______
31. ______
32. ______
33. ______
34. ______
35. ______
36. ______
37. ______
38. ______
39. ______
40. ______
41. ______
42. ______
43. ______
44. ______
45. ______
46. ______
47. ______
48. ______
49. ______
50. ______

25. 3.2 **26.** 1.25 **27.** 6.25 **28.** 2.75

Change each fraction to decimal form. If the decimal is non-terminating, write it using the bar notation over the repeating pattern of digits. See Examples 7 through 12.

29. $\frac{2}{3}$ **30.** $\frac{5}{16}$ **31.** $\frac{7}{11}$ **32.** $\frac{3}{11}$

33. $\frac{11}{16}$ **34.** $\frac{9}{16}$ **35.** $\frac{3}{7}$ **36.** $\frac{5}{7}$

37. $\frac{1}{6}$ **38.** $\frac{5}{18}$ **39.** $\frac{5}{9}$ **40.** $\frac{2}{9}$

Change each fraction to decimal form rounded to the nearest thousandth. See Examples 13 and 14.

41. $\frac{7}{24}$ **42.** $\frac{16}{33}$ **43.** $\frac{5}{12}$ **44.** $\frac{13}{16}$

45. $\frac{1}{32}$ **46.** $\frac{1}{14}$ **47.** $\frac{16}{13}$ **48.** $\frac{20}{9}$

49. $\frac{30}{21}$ **50.** $\frac{40}{3}$

Name ______________________ Section ____________ Date __________

ANSWERS

Perform the indicated operations by writing all the numbers in decimal form. Round to the nearest thousandth, if necessary. See Example 15.

51. $\frac{1}{4}+0.25+\frac{1}{5}$

52. $\frac{3}{4}+\frac{1}{10}+3.55$

53. $\frac{5}{8}+\frac{3}{5}+0.41$

54. $6+2\frac{37}{100}+3\frac{11}{50}$

55. $2\frac{53}{100}+5\frac{1}{10}+7.35$

56. $37.02+25+6\frac{2}{5}+3\frac{89}{100}$

57. $1\frac{1}{4}-0.125$

58. $2\frac{1}{2}-1.75$

59. $36.71-23\frac{1}{5}$

60. $3.1-2\frac{1}{100}$

61. $\left(\frac{35}{100}\right)^2(0.73)$

62. $\left(5\frac{1}{10}\right)^2(2.25)$

63. $\left(1\frac{3}{8}\right)(3.1)(2.6)$

64. $\left(1\frac{3}{4}\right)\left(2\frac{1}{2}\right)(5.35)$

65. $5\frac{54}{100}\div 2.1$

66. $72.16\div\frac{2}{5}$

67. $13.65\div\frac{1}{2}$

68. $91.7\div\frac{1}{4}$

51. ______________

52. ______________

53. ______________

54. ______________

55. ______________

56. ______________

57. ______________

58. ______________

59. ______________

60. ______________

61. ______________

62. ______________

63. ______________

64. ______________

65. ______________

66. ______________

67. ______________

68. ______________

69.

70.

71.

72.

73.

74.

75.

76.

77.

78.

79.

80.

81.

82.

In each of the following exercises, change any fraction to decimal form; then determine which number is larger and find the difference between the two numbers. Use a calculator. See Example 16.

69. $2\frac{1}{4}$, 2.3 **70.** $\frac{7}{8}$, 0.878 **71.** 0.28, $\frac{3}{11}$ **72.** $\frac{1}{3}$, 0.3

73. $\frac{22}{7}$, 3.3 **74.** $\frac{4}{9}$, 0.5 **75.** 3.5, $3\frac{2}{3}$ **76.** $5\frac{3}{4}$, 5.5

Write the three numbers in order, smallest to largest. (Hint: Use a calculator to change each fraction to decimal form.) See Example 16.

77. 0.76, $\frac{3}{4}$, $\frac{7}{10}$ **78.** 0.63, $\frac{5}{8}$, 0.64

79. $\frac{5}{16}$, 0.3126, 0.314 **80.** 0.083, $\frac{41}{500}$, $\frac{2}{25}$

Change any decimal number (not a whole number) into fraction or mixed number form. See Examples 1 through 6.

81. By 2005 census estimates, there were 83.8 people per square mile in the United States.

82. The average weight for a one-year-old girl is 9.1 kg.

Name ______________________ Section __________ Date __________

ANSWERS

83. The median age for men at the beginning of their first marriage is 26.3 years. The median age for women at the beginning of their first marriage is 24.1 years.

83. ____________

84. The maximum speed of a giant tortoise on land is about 0.17 mph.

84. ____________

85. There are about 21.5 students per teacher in California public schools.

85. ____________

86. The surface gravity on Mars is about 0.38 times the gravity on Earth. The atmospheric pressure on Mars is about 0.01 times the atmospheric pressure on Earth.

86. ____________

In Exercises 87 – 90, change each fraction or mixed number (not a whole number) into decimal form. Round each number to the nearest hundredth. See Examples 7 through 14.

87. In 2000 it was estimated that $\frac{7}{50}$ of U.S. households viewed 26 or more TV stations regularly.

87. ____________

88. In a recent year about $\frac{8}{57}$ of the advertising budgets in the automotive industry was spent on newspaper ads.

88. ____________

89. In 2005 the average price of unleaded gasoline in California was $6\frac{6}{7}$ times the price it was in 1970.

89. ____________

90. In the 2004 presidential election, George W. Bush received $\frac{143}{269}$ of the popular vote.

90. ____________

5.6 Square Roots and the Pythagorean Theorem

Square Roots and Irrational Numbers

A number is **squared** when it is multiplied by itself. If a whole number is squared, the result is called a **perfect square**. For example, squaring 7 gives $7^2 = 49$, and 49 is a perfect square. Table 5.2 shows the perfect square numbers found by squaring the whole numbers from 1 to 20. A more complete table (of powers, roots, and prime factorizations) is located in the back of the book.

Objectives

1. Memorize the squares of the whole numbers from 1 to 20.
2. Understand the terms **square root**, **radical sign**, **radicand**, and **radical**.
3. Memorize the perfect square whole numbers from 1 to 400.
4. Know how to use a calculator to find square roots.
5. Understand the terms **right triangle**, **hypotenuse**, and **leg**.
6. Know and be able to use the Pythagorean Theorem.

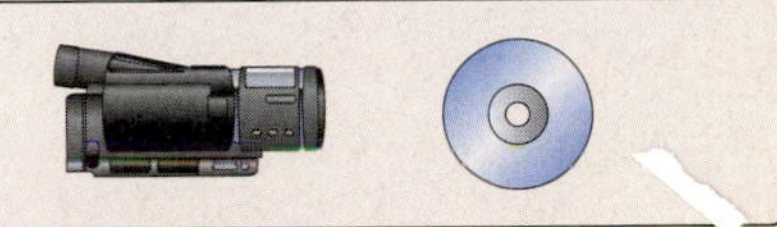

Table 5.2 Squares of Whole Numbers from 1 to 20

$1^2 = 1$	$5^2 = 25$	$9^2 = 81$	$13^2 = 169$	$17^2 = 289$
$2^2 = 4$	$6^2 = 36$	$10^2 = 100$	$14^2 = 196$	$18^2 = 324$
$3^2 = 9$	$7^2 = 49$	$11^2 = 121$	$15^2 = 225$	$19^2 = 361$
$4^2 = 16$	$8^2 = 64$	$12^2 = 144$	$16^2 = 256$	$20^2 = 400$

Since $5^2 = 25$, the number 5 is called the **square root** of 25. We write $\sqrt{25} = 5$. (Read "the square root of 25 is 5.") Similarly,

Objective 1
(See Table 5.2 above.)

$$\sqrt{49} = 7 \quad \text{since} \quad 7^2 = 49$$
$$\sqrt{1} = 1 \quad \text{since} \quad 1^2 = 1$$
$$\sqrt{0} = 0 \quad \text{since} \quad 0^2 = 0$$

Objective 2

The symbol $\sqrt{}$ is called a **radical sign**. The number under the radical sign is called the **radicand**. The complete expression, such as $\sqrt{25}$, is called a **radical**.

Table 5.3 contains the square roots of the perfect square numbers from 1 to 400. (Notice that Table 5.3 is just another way of looking at Table 5.2.) Both tables should be memorized.

Objective 3

Table 5.3 Square Roots of Perfect Squares from 1 to 400

$\sqrt{1} = 1$	$\sqrt{25} = 5$	$\sqrt{81} = 9$	$\sqrt{169} = 13$	$\sqrt{289} = 17$
$\sqrt{4} = 2$	$\sqrt{36} = 6$	$\sqrt{100} = 10$	$\sqrt{196} = 14$	$\sqrt{324} = 18$
$\sqrt{9} = 3$	$\sqrt{49} = 7$	$\sqrt{121} = 11$	$\sqrt{225} = 15$	$\sqrt{361} = 19$
$\sqrt{16} = 4$	$\sqrt{64} = 8$	$\sqrt{144} = 12$	$\sqrt{256} = 16$	$\sqrt{400} = 20$

1. Find the following:

a. 18^2

b. $\sqrt{169}$

Example 1

Use your memory of the results in Tables 5.2 and 5.3 to answer the following

a. 15^2 **b.** 11^2 **c.** $\sqrt{256}$ **d.** $\sqrt{81}$

Solution

a. $15^2 = 225$ **b.** $11^2 = 121$ **c.** $\sqrt{256} = 16$ **d.** $\sqrt{81} = 9$

Now work exercise 1 in the margin.

Most numbers are not perfect squares and the square roots of these numbers are not found as easily as those in Example 1 and the preceding tables. In fact, most square roots are **irrational numbers** (**non-repeating infinite decimals**); that is, most square roots can be only approximated with decimals.

Decimal approximations of $\sqrt{2}$ are given here to emphasize this idea and to help you understand that there is no finite decimal number whose square is 2.

1.4	1.414	1.41421	1.41422
× 1.4	× 1.414	× 1.41421	× 1.41422
56	5656	141421	282844
14	1414	282842	282844
1.96	5656	565684	565688
	1 414	141421	141222
	1.999396	565684	565688
		1 41421	1 41422
		1.9999899241	2.0000182084

So, $\sqrt{2}$ is between 1.41421 and 1.41422.

To find (or approximate) a square root with a calculator, use the key labeled $\boxed{\sqrt{x}}$ or $\boxed{\sqrt{\ }}$.

Objective 4

Examples 2 – 4 show how finding square roots (or approximating square roots) can be accomplished by using a TI-84 Plus calculator. (**Note**: You may have a different calculator and will need to read the manual for that calculator for directions on how to find square roots.)

Example 2

Use a calculator to find $\sqrt{2}$ accurate to four decimal places.

Solution

Step 1: Press the 2ND key.

Step 2: Press the x^2 key (the radical symbol $\sqrt{\ }$ is above the x^2 key).

Step 3: Enter the number 2.

Step 4: Press the right parenthesis key).

Step 5: Press ENTER.

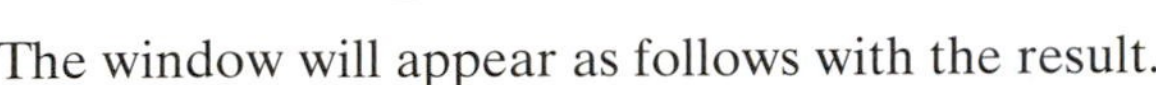

The window will appear as follows with the result.

(**Note**: The right parenthesis is optional. The calculator will give the result with or without the right parenthesis.)

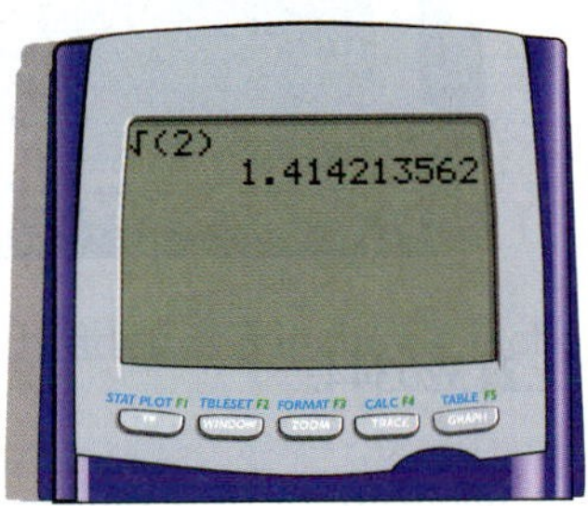

The display shows 1.414213562.

Thus $\sqrt{2} \approx 1.4142$ accurate to four places.

Now work exercise 2 in the margin.

2. Find $\sqrt{5}$ accurate to four decimal places.

Example 3

Find $\sqrt{18}$ accurate to four decimal places.

Solution

Following the steps as in Example 2 gives:

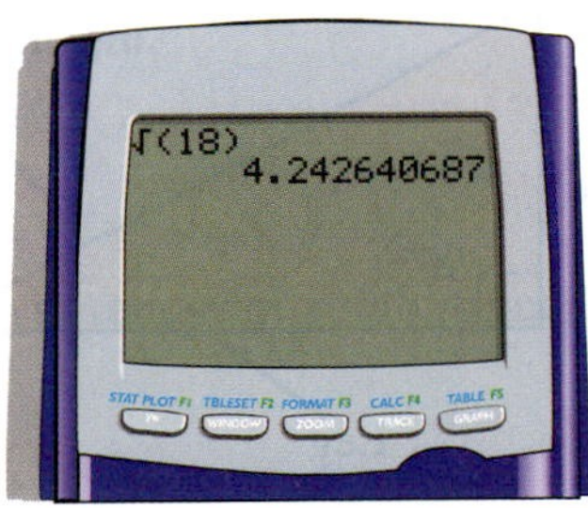

The display shows 4.242640687.

Thus, $\sqrt{18} \approx 4.2426$ accurate to four places.

Now work exercise 3 in the margin.

3. Find $\sqrt{7.5}$ accurate to four decimal places.

38. ______

39. ______

40. ______

41. ______

42. ______

43. ______

44. ______

45. ______

46. ______

47. ______

48. ______

49. ______

50. ______

51. ______

52. ______

53. ______

54. ______

55. ______

56. ______

38. $\sqrt{17}$ **39.** $\sqrt{45}$ **40.** $\sqrt{5}$ **41.** $\sqrt{1.5129}$ **42.** $\sqrt{4.6225}$

43. $\sqrt{9.0601}$ **44.** $\sqrt{1030.41}$ **45.** $\sqrt{800}$ **46.** $\sqrt{500}$ **47.** $\sqrt{0.003}$

48. $\sqrt{0.004}$ **49.** $\sqrt{0.0009}$ **50.** $\sqrt{0.000025}$

Use the Pythagorean Theorem to determine whether or not each of the triangles is a right triangle. See Example 5.

51.

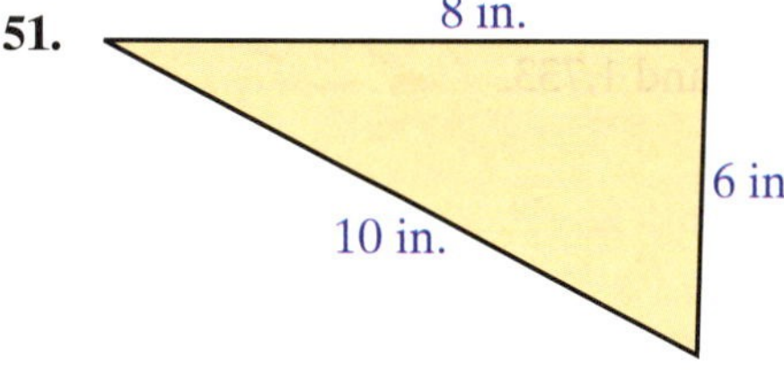

52.

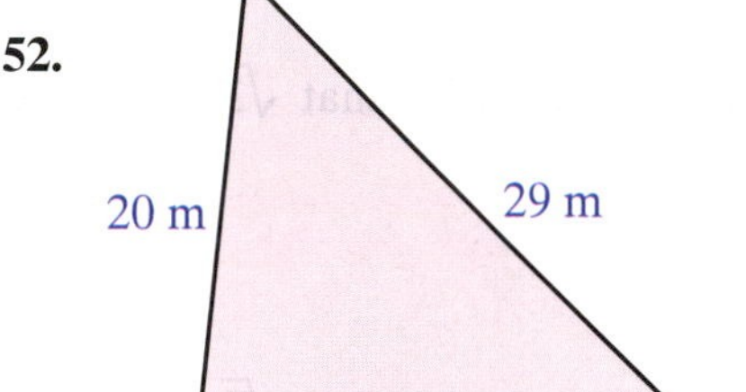

53.

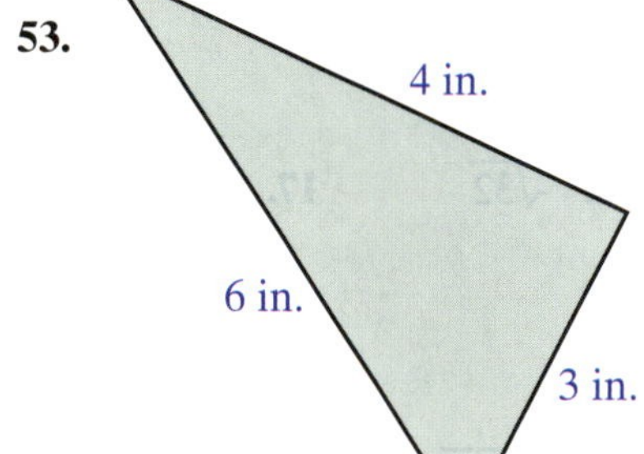

54.

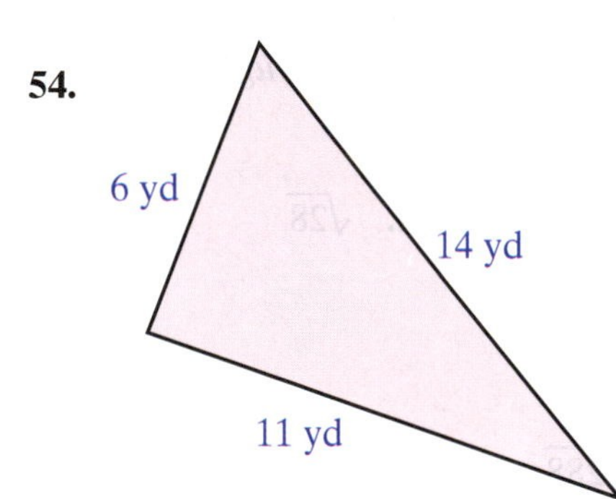

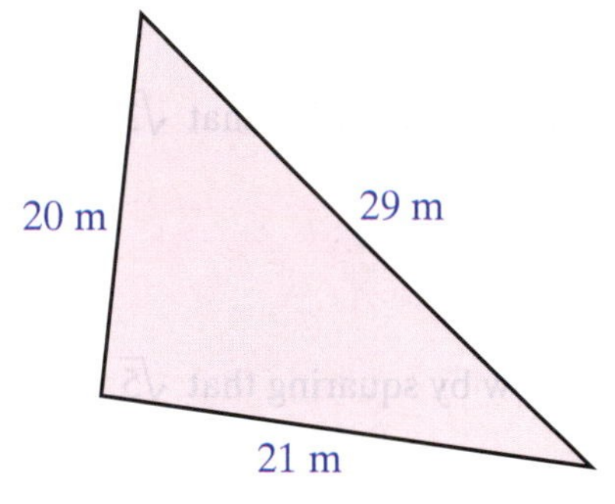

Find the length of the hypotenuse (accurate to four decimal places) in the right triangles. See Examples 6 through 8.

55.

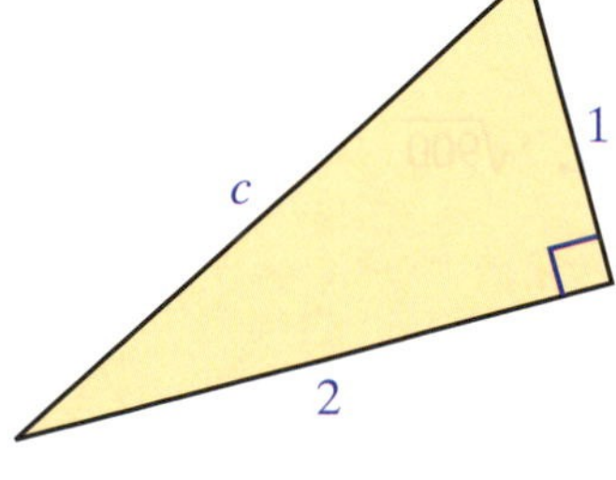

56.

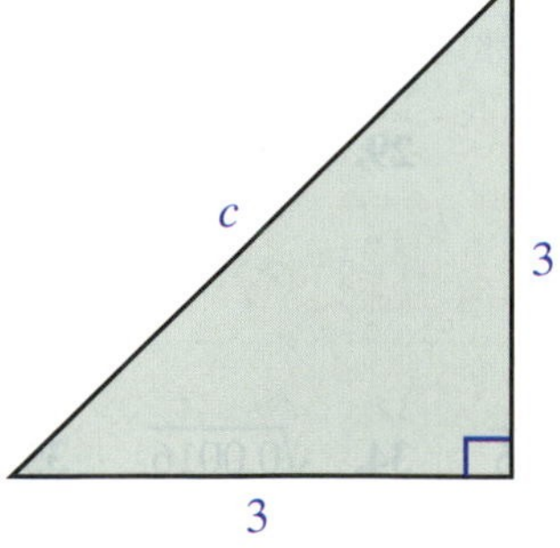

Name ______________________ Section ____________ Date __________

ANSWERS

57.

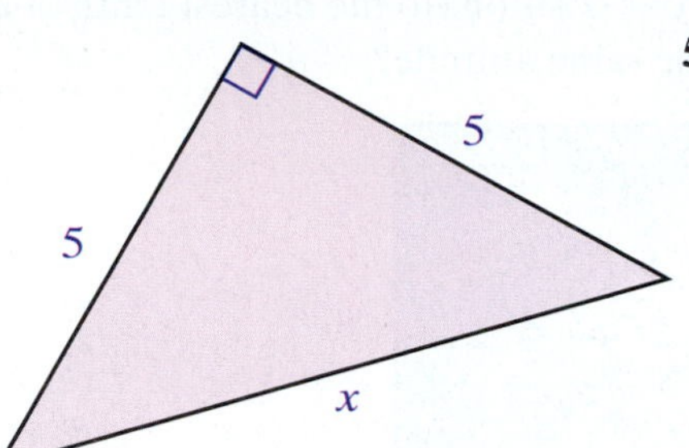

58.

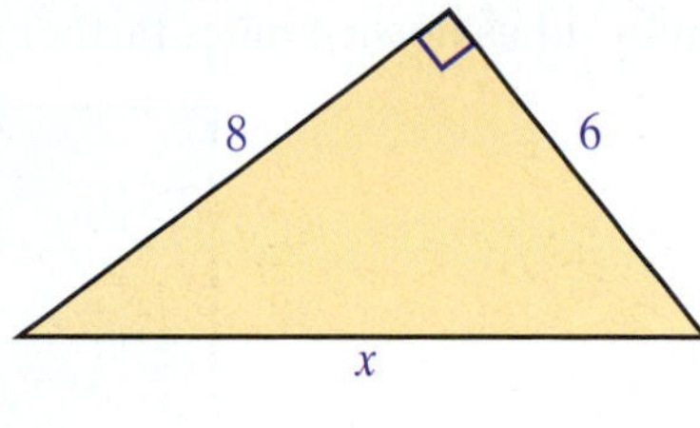

59.

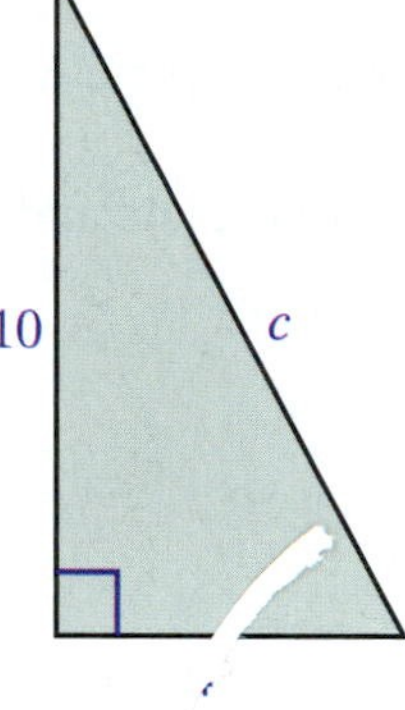

60.

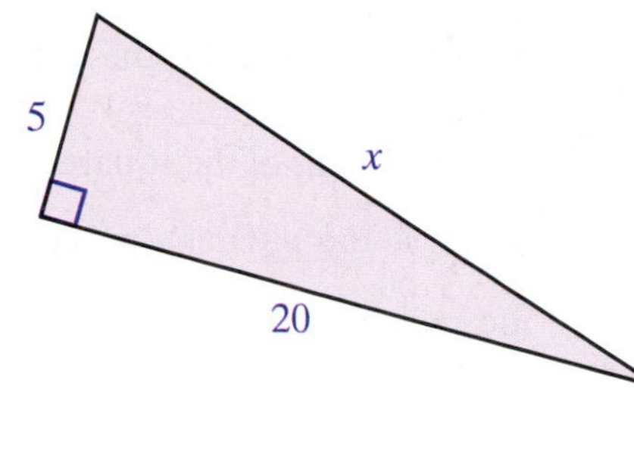

57. ______
58. ______
59. ______
60. ______

61. Ladder and building. The base of a fire-engine ladder is 20 feet from a building and reaches to a fourth floor window 60 feet above the ground level. How far is the ladder extended (to the nearest tenth of a foot)?

61. ______

62. Sailing. A forestay that helps support a ship's mast reaches from the top of the mast, which is 20 meters high, to a point on the deck 10 meters from the base of the mast. What is the length of the stay (to the nearest tenth of a meter)?

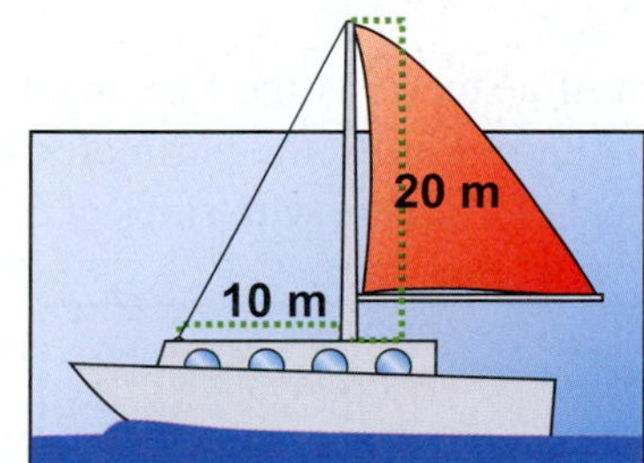

62. ______

63. Architecture. The Xerox Center building in Chicago is 500 feet tall. At a certain time of day, it casts a shadow that is 150 feet long. At that time of day, what is the distance (to the nearest tenth of a foot) from the tip of the shadow to the top of the Xerox building?

63. ______

64.

65. a.

b.

66. a.

b.

c.

64. Airplane distance. If an airplane passes directly over your head at an altitude of 1 mile, how far is the airplane from your position (to the nearest tenth of a mile) after it has flown 4 miles farther at the same altitude?

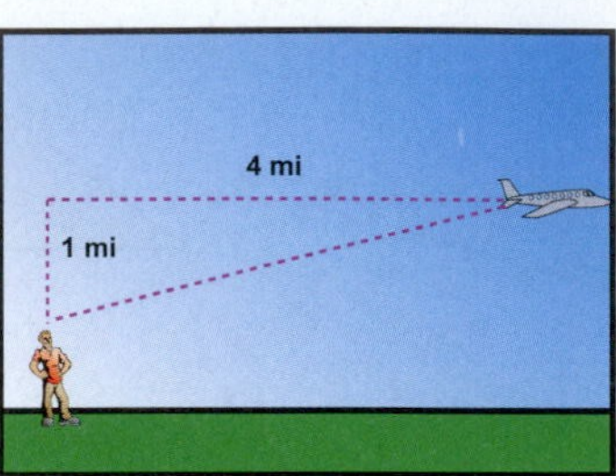

65. Baseball diamond. The shape of a baseball infield is a square with sides 90 feet long.

a. Find the distance (to the nearest tenth of a foot) from home plate to second base.

b. The diagonals of the square intersect halfway between home plate and second base. If the pitcher's mound is $60\frac{1}{2}$ feet from home plate, is the pitcher's mound closer to home plate or to second base?

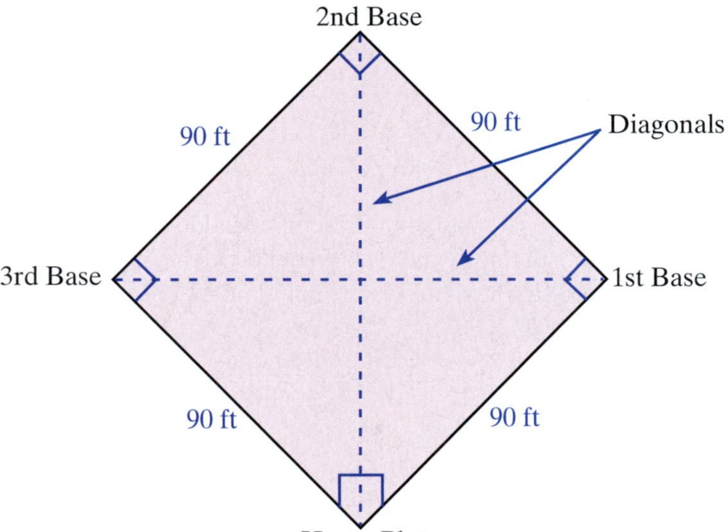

66. Geometry. A diagonal of a square is the line segment from one corner to an opposite corner of the square. Find **a.** the perimeter, **b.** the area of a square, and **c.** the length of a diagonal of a square with sides of 5 inches.

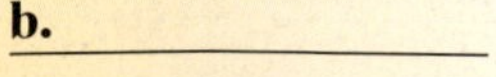

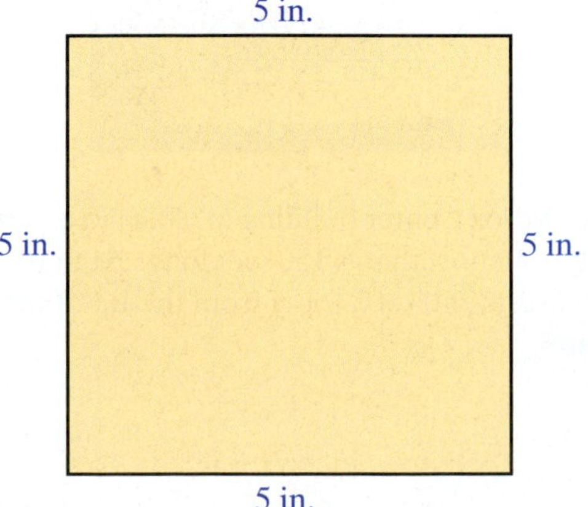

Name ______________________ Section ____________ Date __________

ANSWERS

67. ______________

68. ______________

69. ______________

70. ______________

67. Painting. Before painting a picture on canvas, an artist must stretch the canvas on a rectangular wooden frame. To be sure that the corners of the canvas are true right angles, the artist can measure the diagonals of the stretched canvas. What should be the diagonal measure, to the nearest tenth of an inch, of a canvas whose sides are 24 inches and 38 inches in length?

68. Construction. While installing windows in a new home, a builder measures the diagonals of rectangular window casements to verify that their corners are true right angles. What should be the diagonal measure, to the nearest tenth of an inch, of a window casement with dimensions 36 inches by 54 inches?

69. Quilting. To create a square inside a square, a quilting pattern requires four triangular pieces like the one shaded in the figure shown here. If the square in the center measures 10 cm on a side, and the two legs of each triangle are of equal length, how long are the legs of each triangle, to the nearest tenth of a centimeter?

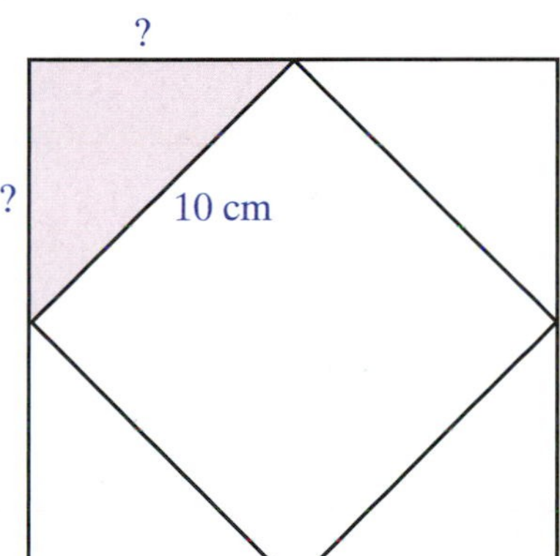

70. Sports. The shape of home plate in the game of baseball can be created by cutting off two triangular pieces at the corners of a square, as shown in the figure. If each of the triangular pieces has a hypotenuse of 12 inches and legs of equal length, what is the length of one side of the original square, to the nearest tenth of an inch?

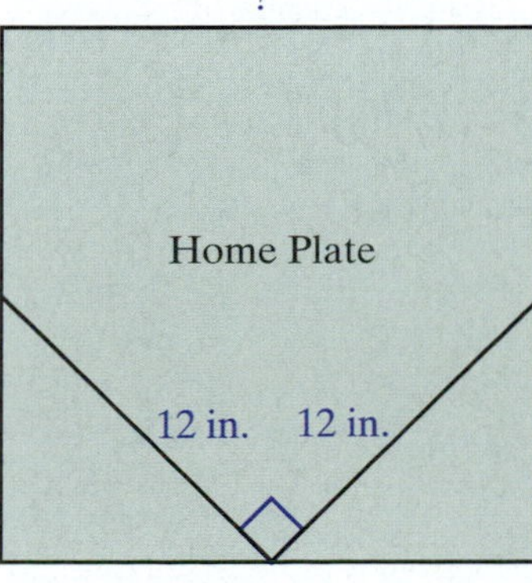

[Respond below
71. exercise.]

Writing and Thinking about Mathematics

71. If three whole numbers satisfy the Pythagorean Theorem, these three numbers are called a Pythagorean triple. For example, 3, 4, and 5 are a Pythagorean triple because

$$3^2 + 4^2 = 5^2 \text{ (or } 9 + 16 = 25).$$

Another Pythagorean triple is 5, 12, and 13 because

$$5^2 + 12^2 = 13^2 \text{ (or } 25 + 144 = 169).$$

Complete the following table by finding ***a***, ***b***, and ***c***, and tell which sets of these three numbers (if any) constitute a Pythagorean triple. The first one is done for you.

m	n	$a = 2mn$	$b = m^2 - n^2$	$c = m^2 + n^2$	Pythagorean triple?
5	1	**10**	**24**	**26**	*yes:* $\mathbf{10^2 + 24^2 = 26^2}$
7	1				
3	2				
7	2				
5	3				
11	3				
13	7				

Extension: Choose some of your own numbers for m and n. Are your results Pythagorean triples? (**Note**: m must be larger than n so that $m^2 - n^2$ will be positive.)

5.7 Geometry: Circles and Volume

Finding the Circumference and Area of a Circle

Recall that the concepts of perimeter and area were discussed in Section 1.8 and formulas related to squares, triangles, and rectangles were given and used in the exercises. Also, recall that a **formula** is a general statement (usually an equation) that relates two or more variables. In this section, we will learn the following definitions and terms related to circles.

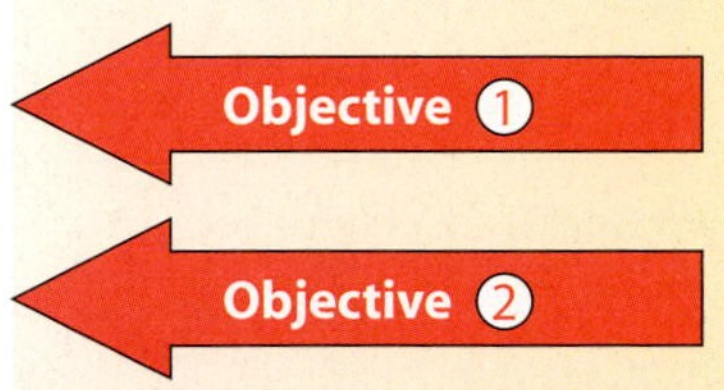

Objectives

① Know the definition of a circle.

② Know the meanings of the terms **center**, **radius**, and **diameter**.

③ Be able to find the circumference and area of a circle.

④ Understand the concept of volume.

⑤ Be able to find the volumes of various geometric figures.

Important Terms and Definitions for Circles:

Circle: The set of all points in a plane that are some fixed distance from a fixed point called the **center** of the circle.

Radius: The fixed distance from the center of a circle to any point on the circle. (The letter r is used to represent the radius of a circle.)

Diameter: The distance from one point on a circle through the center to the point directly opposite it. (The letter d is used to represent the diameter of a circle.)

Circumference: The perimeter of a circle.

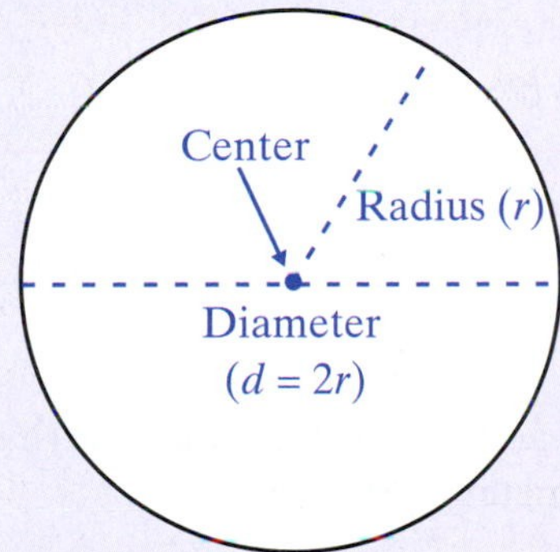

For the formulas for circumference (perimeter) and area related to circles we need to note that a diameter is twice as long as a radius. That is, $\boldsymbol{d = 2r}$.

Objective ③

Formulas for Circles:

For circumference: $C = 2\pi r$ and $C = \pi d$

For area: $A = \pi r^2$

Note: π is the symbol used for the constant 3.1415926535.... This number is an infinite non-repeating decimal (an **irrational number**). For our purposes, we will use $\pi = 3.14$ (accurate to hundredths). However, you should always be aware that 3.14 is only an approximation for π and that the related answers are only approximations.

1. Find the circumference and area of a circle with a radius of 11 meters.

Example 1

Find **a.** the circumference and **b.** the area of a circle with a radius of 6 ft.

Solution

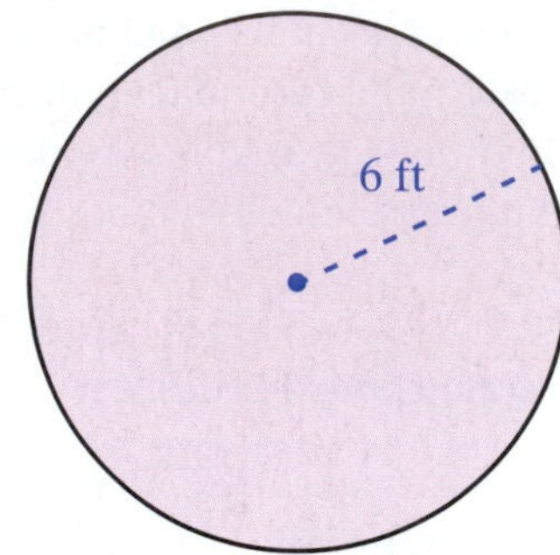

a. Using the formula for circumference:

$C = 2\pi r$

$C = 2 \cdot 3.14 \cdot 6 = 37.68$

The circumference is 37.68 ft.

b. Using the formula for area:

$A = \pi r^2$

$A = 3.14 \cdot 6^2 = 3.14 \cdot 36 = 113.04$

The area is 113.04 ft^2.

Now work exercise 1 in the margin.

2. Find the perimeter of a semicircle which has a diameter of 26 inches.

Example 2

Find the perimeter of a figure that is a semicircle (half of a circle) including its diameter. The diameter is 20 cm long.

Solution

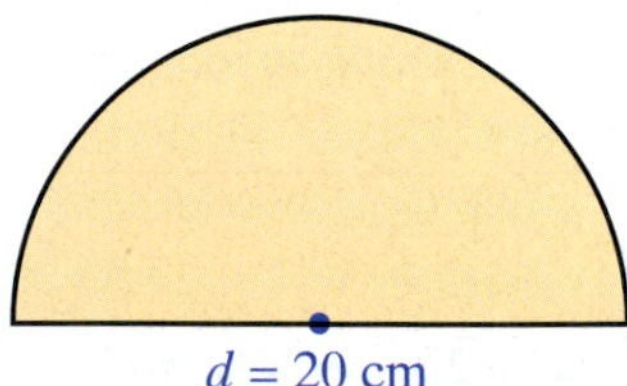

Now we find the perimeter of the figure by adding the length of the semicircle (which is half of the circumference of a circle) to the length of the diameter.

Length of semicircle: $\frac{1}{2}C = \frac{1}{2}\pi d = \frac{1}{2} \cdot 3.14 \cdot 20 = 3.14 \cdot 10 = 31.4$

Perimeter of figure: $P = 31.4 + 20 = 51.4 \text{ cm}$

Now work exercise 2 in the margin.

Example 3

Find **a.** the circumference and **b.** the area of a circle with a diameter of 5.2 in.

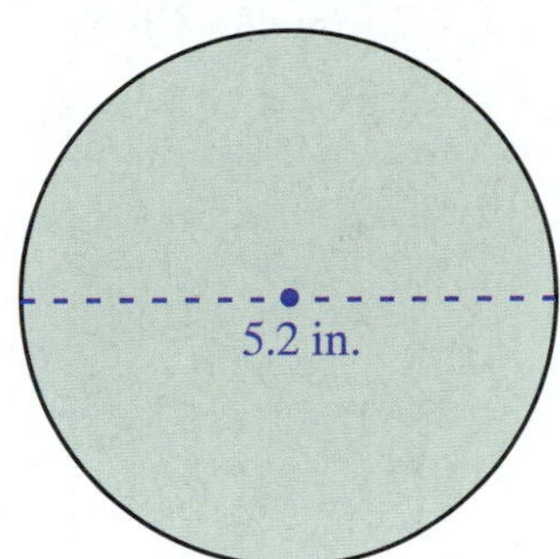

Solution

a. Using the formula for circumference:

$C = \pi d$

$C = \pi d = 3.14 \cdot 5.2 = 16.328$

The circumference is 16.328 in.

b. Using the formula for area:

$A = \pi r^2$ (In this case $r = \frac{1}{2} \cdot 5.2 = 2.6$)

$A = 3.14 \cdot (2.6)^2 = 3.14 \cdot 6.76 = 21.2264$

The area is 21.2264 in.^{2}

Now work exercise 3 in the margin.

3. Find the circumference and area of a circle with a diameter of 15 feet.

Example 4

Find the area of the washer (shaded portion) with dimensions as shown in the figure.

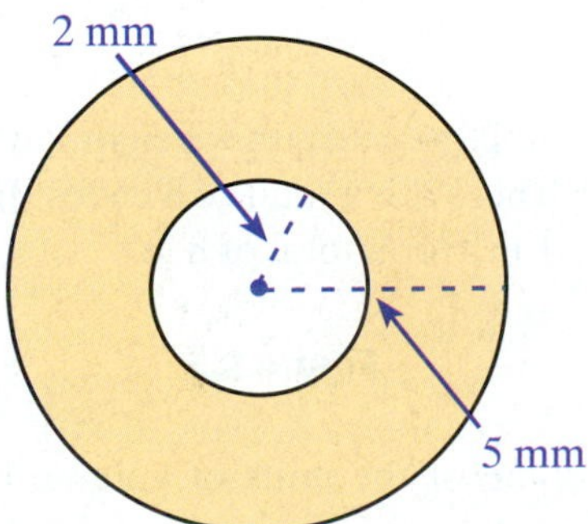

Continued on next page...

4. Find the area of the shaded portion of the figure below. Round to two decimal places.

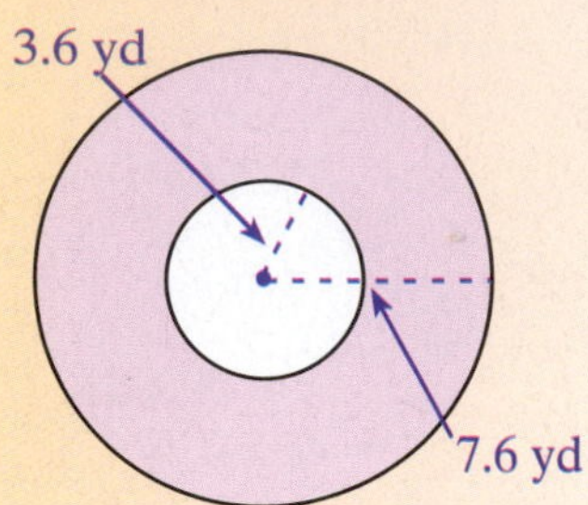

Solution

Subtract the area of the inside (smaller) circle from the area of the outside (larger) circle.

Larger Circle

$A = \pi r^2$

$A = 3.14(5^2)$

$= 3.14(25)$

$= 78.50\,\text{mm}^2$

Smaller Circle

$A = \pi r^2$

$A = 3.14(2^2)$

$= 3.14(4)$

$= 12.56\,\text{mm}^2$

Washer

$$\begin{array}{r} 78.50\,\text{mm}^2 \\ -12.56\,\text{mm}^2 \\ \hline 65.94\,\text{mm}^2 \end{array}$$ area of washer

Now work exercise 4 in the margin.

Objective ④

The Concept of Volume

Volume is the measure of the space enclosed by a three-dimensional figure and is measured in **cubic units**. The concept of volume is illustrated in Figure 5.5 in terms of cubic inches.

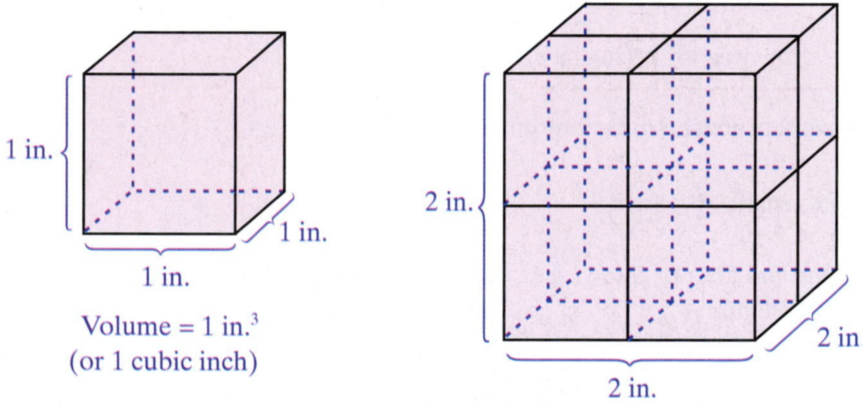

Figure 5.5

In the metric system, some of the units of volume are cubic meters (m^3), cubic decimeters (dm^3), cubic centimeters (cm^3), and cubic millimeters (mm^3). In the U.S. customary system, some of the units of volume are cubic feet (ft^3), cubic inches (in.^3), and cubic yards (yd^3).

Formulas for Volume

The following formulas for the volumes of the common geometric solids shown are valid regardless of the measurement system used. Always be sure to label your answers with the correct units.

Objective 5

Five Geometric Solids and the Formulas for Their Volumes:

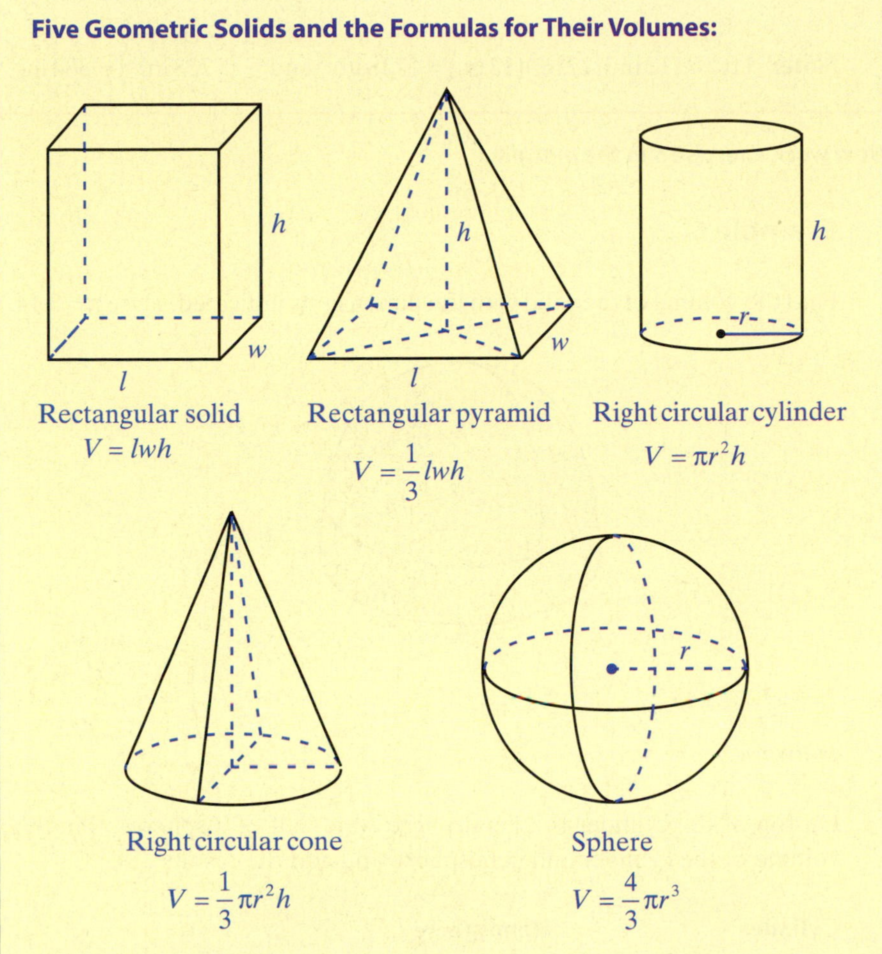

Rectangular solid
$V = lwh$

Rectangular pyramid
$V = \frac{1}{3}lwh$

Right circular cylinder
$V = \pi r^2 h$

Right circular cone
$V = \frac{1}{3}\pi r^2 h$

Sphere
$V = \frac{4}{3}\pi r^3$

Example 5

Find the volume of the rectangular solid with length 8 in., width 4 in., and height 1 ft. Write your answer in cubic inches and in cubic feet.

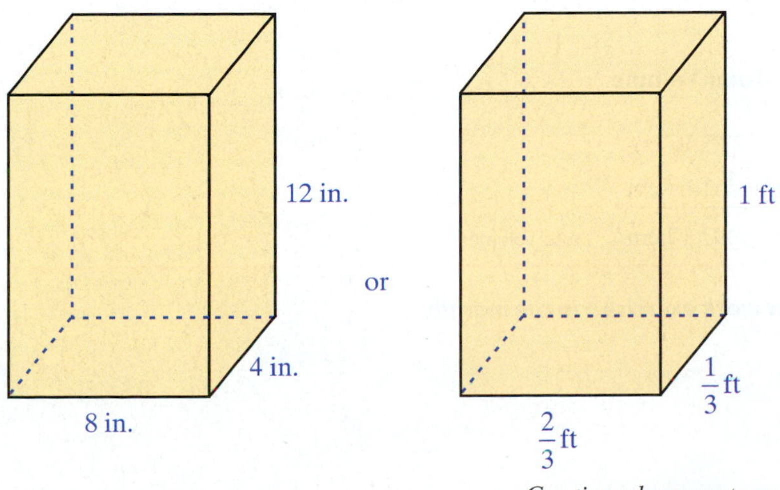

Continued on next page...

5. Find the volume of the solid below.

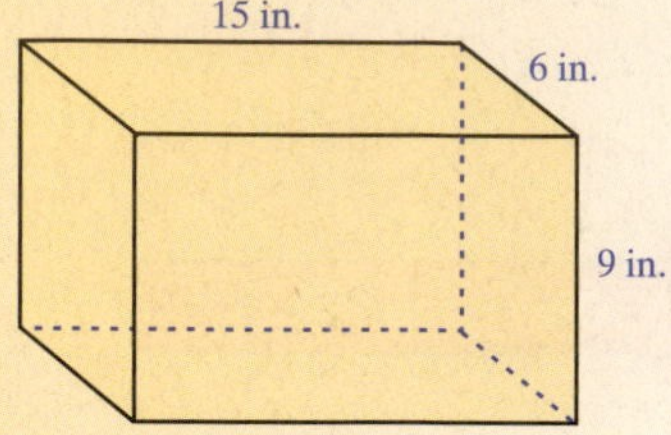

6. Find the volume of the solid shown below.

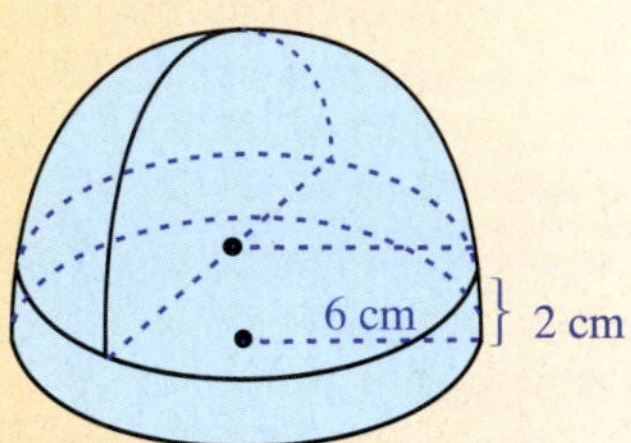

Solution

$$V = lwh \qquad V = lwh$$

$$V = 8 \cdot 4 \cdot 12 \qquad V = \frac{2}{3} \cdot \frac{1}{3} \cdot 1$$

$$= 384 \text{ in.}^3 \qquad = \frac{2}{9} \text{ ft}^3$$

Note: $1 \text{ ft}^3 = (12 \text{ in.})(12 \text{ in.})(12 \text{ in.}) = 1728 \text{ in.}^3$ and $\frac{2}{9}(1728 \text{ in.}^3) = 384 \text{ in.}^3$

Now work exercise 5 in the margin.

Example 6

Find the volume of the solid with the dimensions indicated. (Use $\pi = 3.14$.)

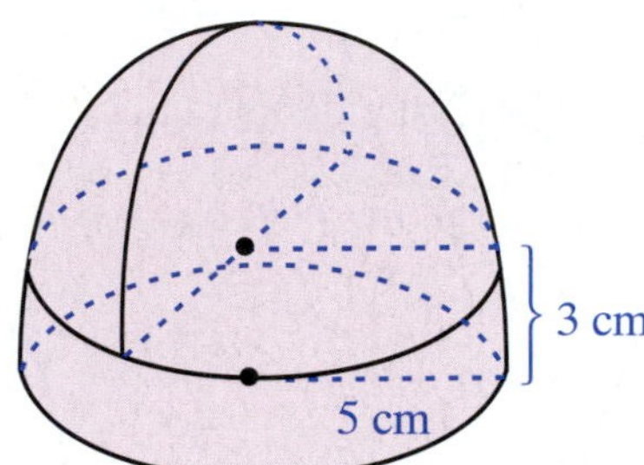

Solution

On top of the cylinder is a hemisphere (one-half of a sphere). Find the volume of the cylinder and hemisphere and add the results.

Cylinder

$$V = \pi r^2 h$$

$$V = 3.14(5^2)(3)$$

$$= 235.5 \text{ cm}^3$$

Hemisphere

$$V = \frac{1}{2} \cdot \frac{4}{3} \pi r^3 \quad \text{(one-half volume of a sphere)}$$

$$V = \frac{2}{3}(3.14)(5^3)$$

$$= 261.67 \text{ cm}^3$$

Total Volume

$$\begin{array}{r} 235.50 \text{ cm}^3 \\ + \ 261.67 \text{ cm}^3 \\ \hline 497.17 \text{ cm}^3 \end{array} \quad \text{total volume}$$

Now work exercise 6 in the margin.

Practice Problems

Find the perimeter and area of the circle with the given radius length.

1. 11 mm

2. 3.8 yd

Find the volume of the following figures.

3. A cube with sides measuring 5 cm.

4. A rectangular solid with length 1.5 in., width 4 in., and height 10 in.

5. A sphere with radius 3 cm.

6. A right circular cone with radius 20 mm and height 9 mm.

Answers to Practice Problems:

1. 69.08 mm; 379.94 mm^2

2. 23.864 yd; 45.3416 yd^2

3. 125 cm^3

4. 60 $in.^3$

5. 113.04 cm^3

6. 3768 mm^3

Name ______________________ Section ____________ Date __________

Exercises 5.7

1. Find **a.** the circumference and **b.** the area of a circle with radius 5 ft.

2. Find **a.** the circumference and **b.** the area of a circle with radius 70 cm.

3. Find **a.** the circumference and **b.** the area of a circle with diameter 6.2 yd.

4. Find **a.** the circumference and **b.** the area of a circle with diameter 14 m.

5. Find **a.** the circumference and **b.** the area of a circle of radius 1 ft.

6. Find **a.** the circumference and **b.** the area of a circle of diameter 1 ft.

*Find **a.** the perimeter and **b.** the area of each figure. See Examples 2 and 3.*

7.

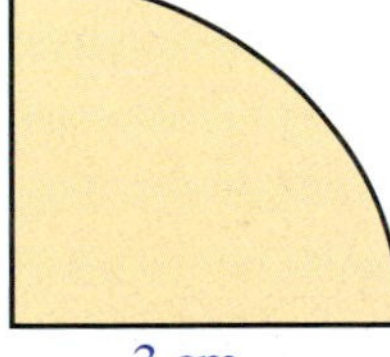

8.

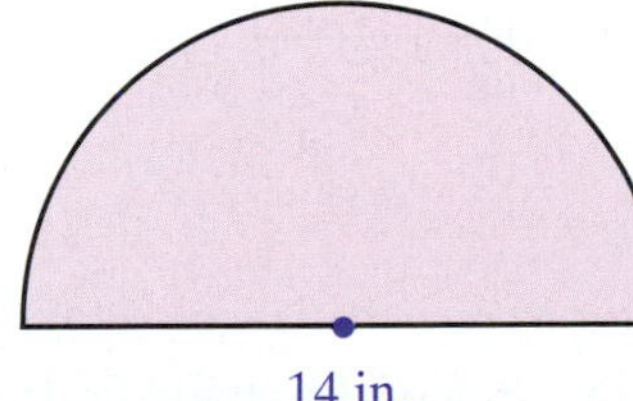

ANSWERS

1. a. ____________
   b. ____________
2. a. ____________
   b. ____________
3. a. ____________
   b. ____________
4. a. ____________
   b. ____________
5. a. ____________
   b. ____________
6. a. ____________
   b. ____________
7. a. ____________
   b. ____________
8. a. ____________
   b. ____________

9. a. ______

b. ______

10. a. ______

b. ______

11. a. ______

b. ______

12. a. ______

b. ______

13. a. ______

b. ______

14. a. ______

b. ______

15. ______

9.

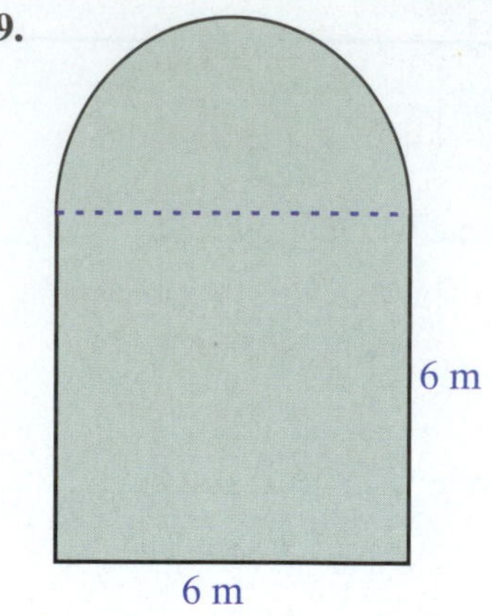

10.

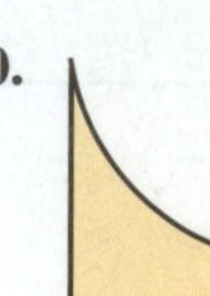

11.

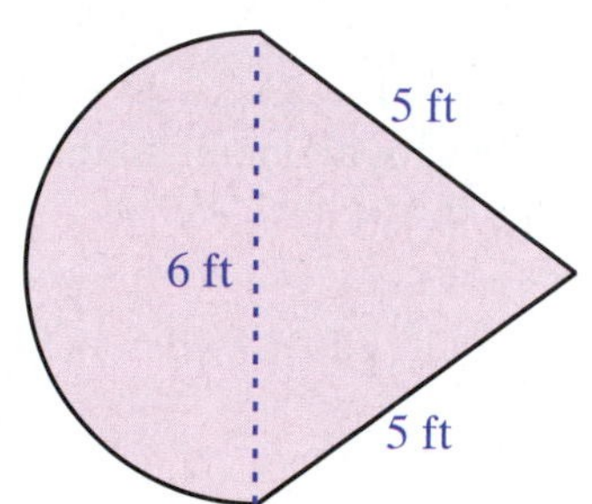

12.

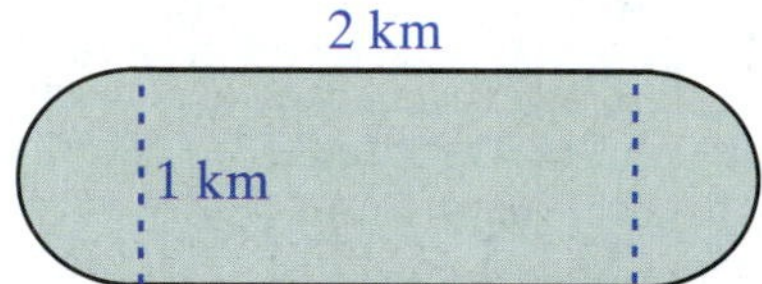

13.

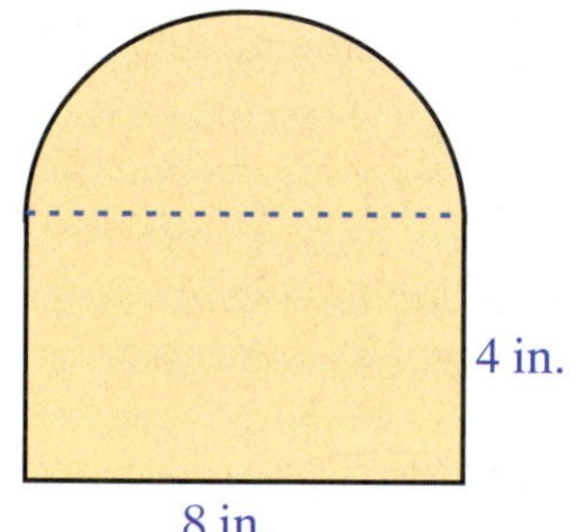

14.

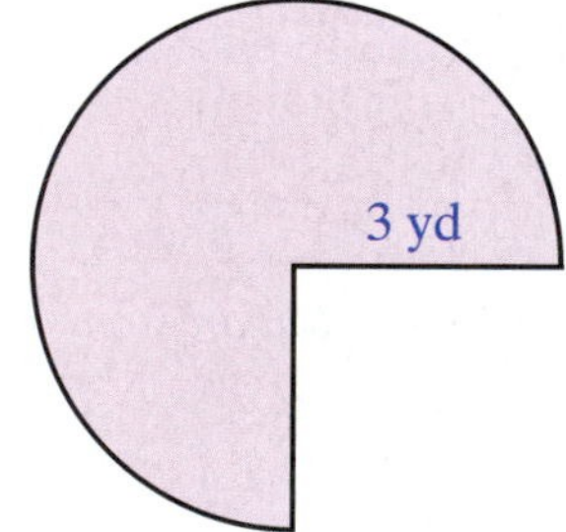

15. Find the area of the shaded portion in the figure.

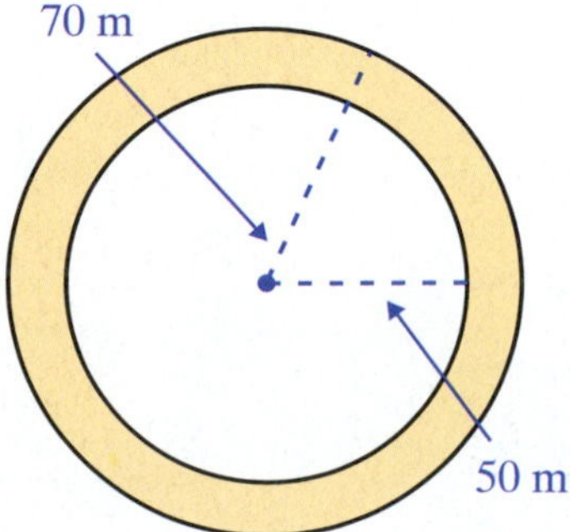

Name ______________________ Section ____________ Date __________

ANSWERS

16. [Respond below exercise]

17. ______________

18. ______________

19. ______________

20. ______________

21. ______________

22. ______________

16. Match each formula for the volume to its corresponding geometric figure.

_______ (a) rectangular solid — A. $V = \frac{4}{3}\pi r^3$

_______ (b) rectangular pyramid — B. $V = \frac{1}{3}\pi r^2 h$

_______ (c) right circular cylinder — C. $V = lwh$

_______ (d) right circular cone — D. $V = \pi r^2 h$

_______ (e) sphere — E. $V = \frac{1}{3}lwh$

Find the volume of each of the following solids in a convenient unit. See Examples 5 and 6. (Use $\pi = 3.14$.)

17. A rectangular solid with length 5 in., width 2 in., and height 7 in.

18. A right circular cylinder 15 in. high and 1 ft in diameter.

19. A sphere with radius 4.5 cm.

20. A sphere with diameter 12 ft.

21. A right circular cone 3 dm high with a 2 dm radius.

22. A rectangular pyramid with length 8 cm, width 10 mm, and height 3 dm.

Find the volume of each of the following solids with the dimensions indicated. See Examples 5 and 6.

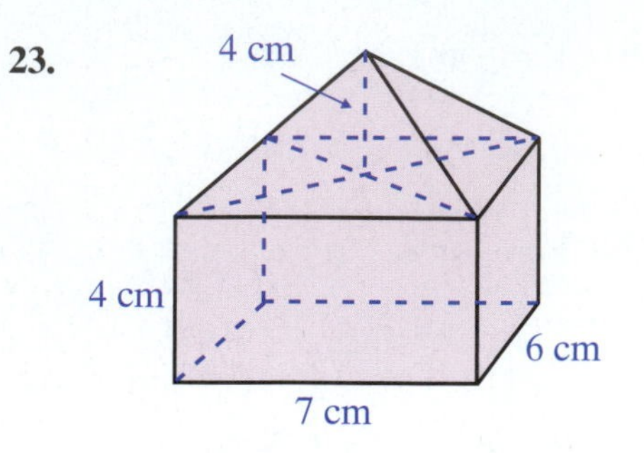

23.

4 cm

4 cm

6 cm

7 cm

24. 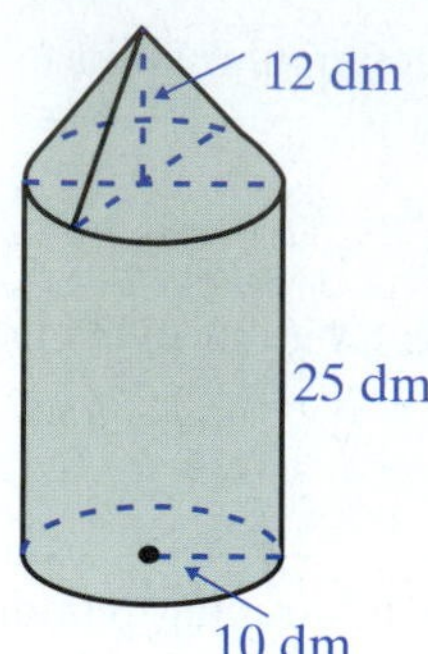

25. 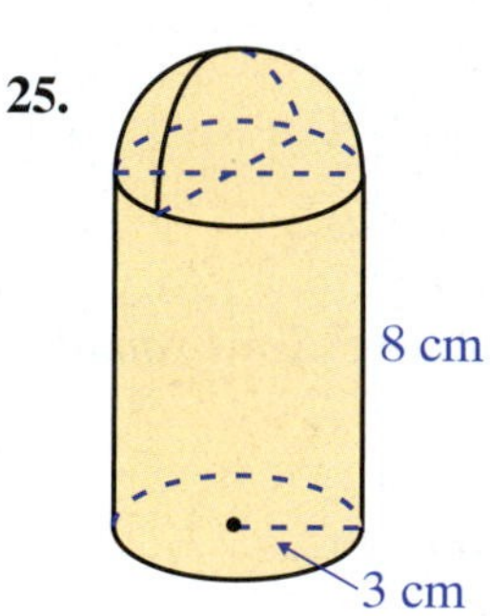

26. 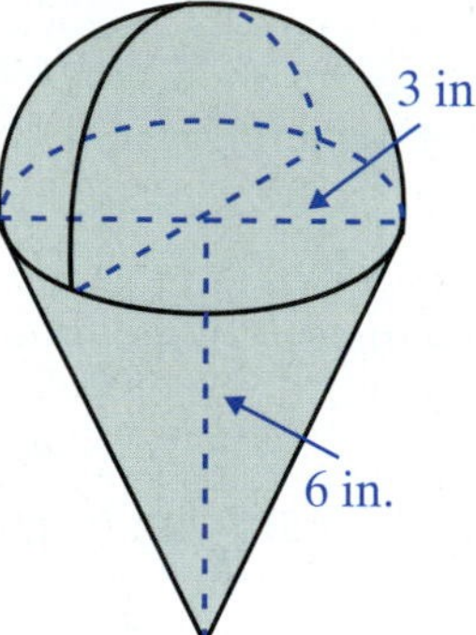

27. 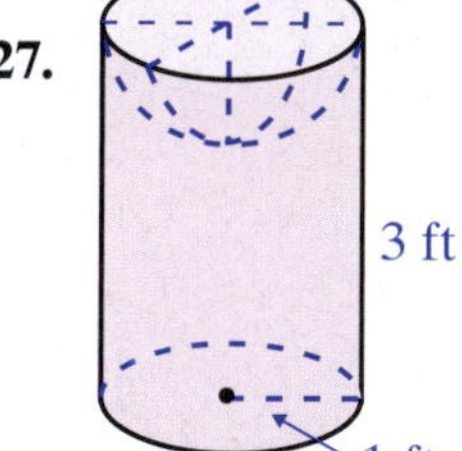

28. 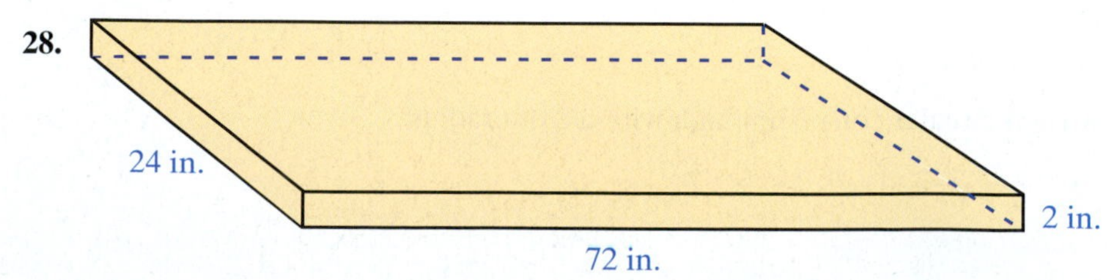

23. __________

24. __________

25. __________

26. __________

27. __________

28. __________

Chapter 5 Index of Key Terms and Ideas

Section 5.1 Introduction to Decimal Numbers

Decimal Notation and Decimal Numbers pages 368 – 369

Decimal notation uses a decimal point, with whole numbers written to the left of the decimal point and fractions written to the right of the decimal point. Numbers written in decimal notation are called **decimal numbers**.

Reading and Writing Decimal Numbers page 370

1. Read (or write) the whole number.
2. Read (or write) the word **and** in place of the decimal point.
3. Read (or write) the fraction part as a whole number with the name of the place of the last digit on the right.

Comments on Notation page 371

1. The letters **th** at the end of a word indicate a fraction part (a part to the right of the decimal point).
2. The hyphen (-) indicates one word.

Rules for Rounding Decimal Numbers (to the right of the decimal point): page 373

1. If the digit to the right of the given place value is less than 5: leave the digit in the given place as is and drop the remaining digits.
2. If the digit to the right of the given place value is 5 or more: increase the digit in the given place value by 1 and drop the remaining digits.

(Important Reminder: In rounding whole numbers, digits to the left of the decimal point that are dropped must be replaced by 0's.)

Section 5.2 Addition and Subtraction with Decimal Numbers

To Add Decimal Numbers page 385

1. Write the addends in a vertical column.
2. Keep the decimal points aligned vertically.
3. Keep digits with the same position value aligned. (Zeros may be written in to help keep the digits aligned properly.)
4. Add the numbers, keeping the decimal point in the sum aligned with the other decimal points.

To Subtract Decimal Numbers page 387

1. Write the addends in a vertical column.
2. Keep the decimal points aligned vertically.
3. Keep digits with the same position value aligned. (Zeros may be written in as aids.)
4. Subtract, keeping the decimal point in the difference aligned with the other decimal points.

Estimating Sums and Differences page 387

To estimate a sum (or difference), round each number to the place of the leftmost nonzero digit and then add (or subtract) these rounded numbers.

Section 5.3 Multiplication with Decimal Numbers

To Multiply Decimal Numbers page 395

1. Multiply the two numbers as if they were whole numbers.
2. Count the total number of places to the right of the decimal points in both numbers being multiplied.
3. Place the decimal point in the product so that the number of places to the right of it is the same as that found in Step 2.

To Multiply a Decimal Number by a Power of 10 page 397

1. Move the decimal point to the right.
2. Move it the same number of places as the number of 0's in the power of 10.

Metric Units of Length pages 398 – 399

In the metric system, to change from any number of a particular unit of length to a smaller unit of length, we multiply by some power of 10 (or simply move the decimal point to the right).

Estimating Products page 400 – 401

Estimating products can be done by rounding each number to the place of the last nonzero digit on the left and multiplying these rounded numbers.

Section 5.4 Division with Decimal Numbers

To Divide Decimal Numbers page 412

1. Move the decimal point in the divisor to the right so that the divisor is a whole number.
2. Move the decimal point in the dividend the same number of places to the right.
3. Place the decimal point in the quotient directly above the new decimal point in the dividend. (**Note:** Be sure to do this before dividing.)
4. Divide just as you would with whole numbers. (**Note:** 0's may be added in the dividend as needed to be able to continue the division process.)

When the Remainder is not 0 page 414

1. Decide first how many decimal places are to be in the quotient.
2. Divide until the quotient is **one digit past the place of desired accuracy**.
3. Using this last digit, round the quotient to the desired place of accuracy.

To Divide a Decimal Number by a Power of 10 page 416

1. Move the decimal point to the left.
2. Move it the same number of places as the number of 0's in the power of 10.

Metric Units of Length page 417

In the metric system, to change from any number of a particular unit of length to a larger unit of length, we divide by some power of 10 (or simply move the decimal point to the left).

Continued on next page...

Section 5.4 Division with Decimal Numbers (continued)

Estimating Quotients page 418

Estimating quotients can be done by rounding both the divisor and the dividend to the place of the last nonzero digit on the left and then dividing with these rounded values.

Section 5.5 Decimals and Fractions

Changing from Decimals to Fractions page 429

A finite (or terminating) decimal number can be written in fraction form by writing a fraction with the following:

1. A **numerator** that consists of the whole number formed by all the digits of the decimal number, and
2. A **denominator** that is the power of ten that names the position of the rightmost digit.

Changing from Fractions to Decimals page 431

We have seen that fractions indicate parts of a whole and that fractions indicate division. To change a fraction to decimal form, we use the meaning that the numerator is to be divided by the denominator.

1. If the remainder is 0, the decimal is said to be **terminating**.
2. If the remainder is not 0, the decimal is said to be **non-terminating**.

Repeating Decimals pages 432–433

One way to write repeating decimals is to write a bar over the repeating digits.

Section 5.6 Square Roots and the Pythagorean Theorem

Squares of Whole Numbers from 1 to 20 page 443

(See Table 5.2)

Terminology of Square Roots page 443

The **square root** of a number equals another number that when squared results in the original number. The symbol $\sqrt{}$ is called a **radical sign**. The number under the radical sign is called the **radicand**. The complete expression is called a **radical**.

Square Roots of Perfect Squares from 1 to 400 page 443

(See Table 5.3)

Right Triangle page 446

A right triangle is a triangle containing a right angle (90°).

Hypotenuse page 446

The hypotenuse is the longest side of a right triangle; the side opposite the right angle.

Continued on next page...

Section 5.6 Square Roots and the Pythagorean Theorem (continued)

Leg page 446

A leg of a triangle is either of the other two sides of a right triangle (the sides that are not the hypotenuse).

The Pythagorean Theorem page 447

In a right triangle, the square of the hypotenuse is equal to the sum of the squares of the two legs: $c^2 = a^2 + b^2$.

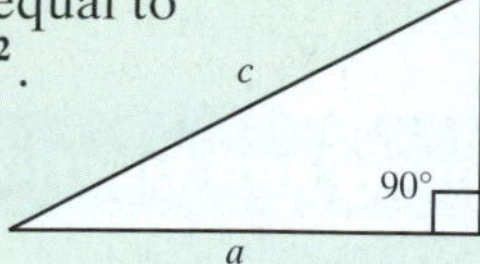

Section 5.7 Geometry: Circles and Volume

Circle page 457

A circle is the set of all points in a plane that are some fixed distance from a fixed point called the **center** of the circle.

Radius page 457

The radius is the fixed distance from the center of a circle to any point on the circle. (The letter r is used to represent the radius of a circle.)

Diameter page 457

The diameter distance from one point on a circle through the center to the point directly opposite it. (The letter d is used to represent the diameter of a circle.)

Circumference page 457

The circumference is the perimeter of a circle.

Formulas for Circles page 457

Circumference: $C = 2\pi r$ and $C = \pi d$ Area: $A = \pi r^2$

Volume page 460

Volume is the measure of the space enclosed by a three-dimensional figure and is measured in cubic units.

Formulas for Volume page 461

Rectangular solid: $V = lwh$ Rectangular pyramid: $V = \frac{1}{3}lwh$

Right circular cylinder: $V = \pi r^2 h$ Right circular cone: $V = \frac{1}{3}\pi r^2 h$

Sphere: $V = \frac{4}{3}\pi r^3$

Name ______________________ Section ____________ Date __________

ANSWERS

Chapter 5 Review Questions

1. A __________ number is a fraction, or a mixed number, with a power of 10 in the denominator.

Write the following decimal numbers in words.

2. 0.4 3. 7.08 4. 92.137 5. 18.5526

Write the following decimal numbers in mixed number form.

6. 81.47 7. 100.03 8. 9.592 9. 200.5

Write the following numbers in decimal notation.

10. two and seventeen hundredths

11. eighty-four and seventy-five thousandths

12. three thousand three and three thousandths

Round as indicated.

13. 5863 (to the nearest hundred)

14. 7.649 (to the nearest tenth)

1. __________
2. __________
3. __________
4. __________
5. __________
6. __________
7. __________
8. __________
9. __________
10. __________
11. __________
12. __________
13. __________
14. __________

15. ____________

16. ____________

17. ____________

18. ____________

19. ____________

20. ____________

21. ____________

22. ____________

23. ____________

24. ____________

25. ____________

26. ____________

27. ____________

28. ____________

29. ____________

30. ____________

31. ____________

15. 0.0385 (to the nearest thousandth)

16. 2.069876 (to the nearest hundred-thousandth)

Add or subtract as indicated.

17. $5.4 + 7.34 + 14.08$

18. $34.967 + 40.8 + 9.451 + 8.2$

19. $16.92 - 7.9$

20. $5 - 1.0377$

21. $\begin{array}{r} 78.6 \\ 9.683 \\ +15.989 \\ \hline \end{array}$

22. $\begin{array}{r} 42.008 \\ -19.3 \\ \hline \end{array}$

Multiply as indicated.

23. $(0.8)(0.9)$

24. $(0.02)(0.32)$

25. $100(2.35)$

26. $10(0.17632)$

27. $10^3(5.9641)$

28. $\begin{array}{r} 1.08 \\ \times\ 1.6 \\ \hline \end{array}$

29. $\begin{array}{r} 36.5 \\ \times\ 4.7 \\ \hline \end{array}$

Divide as indicated. (Round to the nearest hundredth.)

30. $4\overline{)2.83}$

31. $0.06\overline{)52.832}$

Name ______________________ Section ____________ Date ____________

ANSWERS

Divide by using your knowledge of division by powers of 10.

32. $\frac{296.1}{100}$

33. $\frac{5.67}{10^3}$

34. Find the average (to the nearest tenth) of 16.5, 23.4, and 30.7.

Find equivalent measures in the metric system.

35. 182 cm = __________ mm

36. 1.35 m = __________ cm

37. 350 mm = __________ cm

38. 1800 m = __________ km

Change each decimal number to fraction (or mixed number) form. Reduce if possible.

39. 0.07

40. 2.025

41. 0.015

Change each fraction to decimal form. If the decimal is non-terminating, write it using a bar over the repeating digits.

42. $\frac{1}{3}$

43. $\frac{5}{8}$

44. $2\frac{4}{9}$

Change each fraction to decimal form rounded to the nearest thousandth.

45. $\frac{15}{17}$

46. $\frac{99}{101}$

32. __________

33. __________

34. __________

35. __________

36. __________

37. __________

38. __________

39. __________

40. __________

41. __________

42. __________

43. __________

44. __________

45. __________

46. __________

47. ______

48. ______

49. ______

50. ______

51. ______

52. ______

53. ______

54. ______

55. a. ______

b. ______

47. The buyer for a company purchased 17 cars at a price of $33,450 each. How much did he pay for the cars?

48. You are going to make a down payment of $275 on a new computer and ten equal monthly payments of $62.50. How much will you pay for the computer?

In Exercises 49 – 52, use your calculator to find the square roots accurate to four decimal places.

49. $\sqrt{60}$ **50.** $\sqrt{10.24}$ **51.** $\sqrt{92}$ **52.** $\sqrt{0.3969}$

53. Determine whether or not a triangle with sides of length 10 in., 24 in., and 26 in. is a right triangle.

54. Find the length (accurate to four decimal places) of the hypotenuse of a right triangle with two equal legs of length 4 m.

55. Find **a.** the circumference and **b.** the area of a circle with diameter 12 cm.

Name ______________________ Section ____________ Date __________

ANSWERS

56. Find the area of the shaded region between the two circles shown.

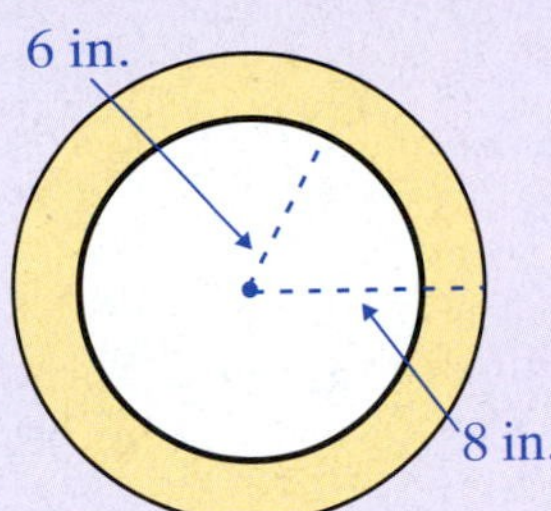

56. ______________

Find the perimeter of each figure in Exercises 57 – 58. Round your answer to the nearest hundredth.

57.

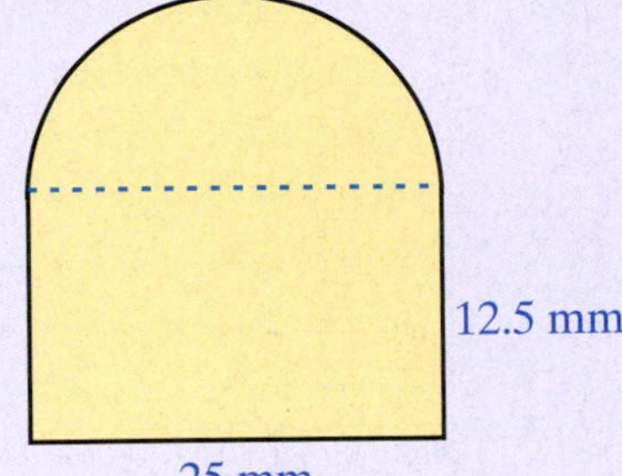

58.

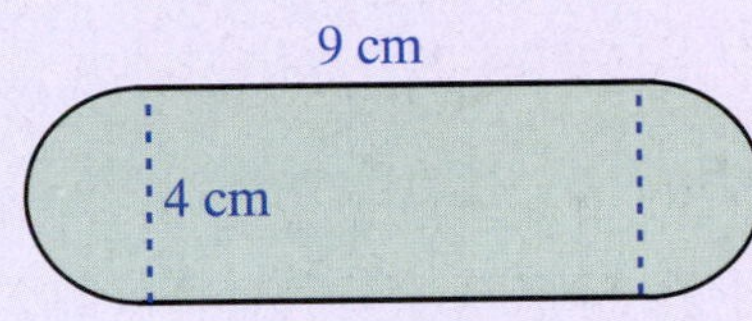

57. ______________

58. ______________

Find the volume of each of the figures shown in Exercises 59 – 62. Round your answer to the nearest hundredth.

59.

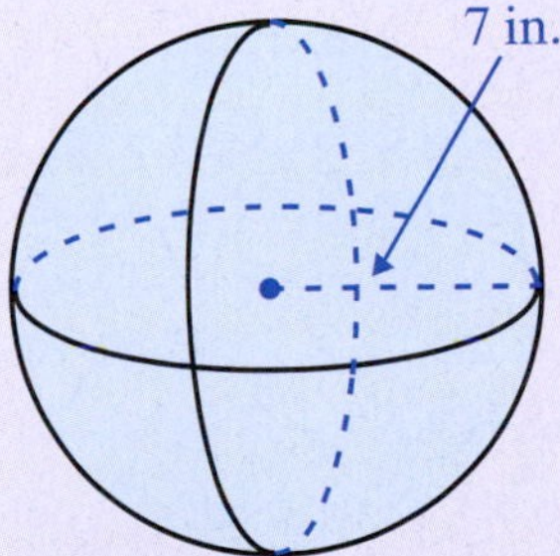

60.

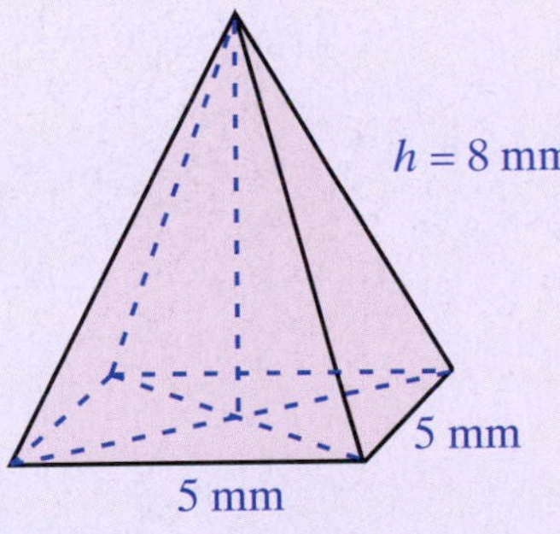

61.

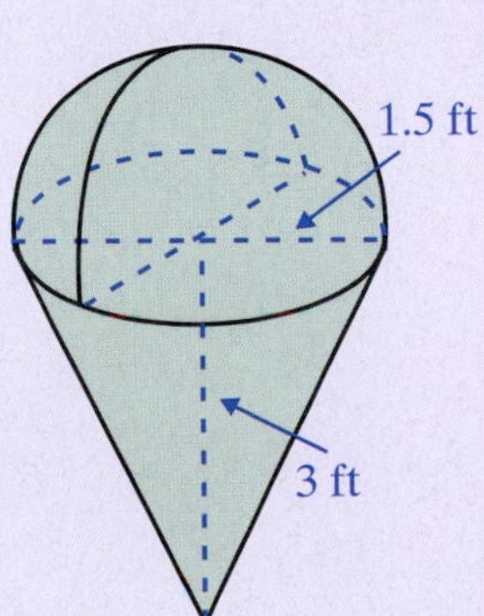

62.

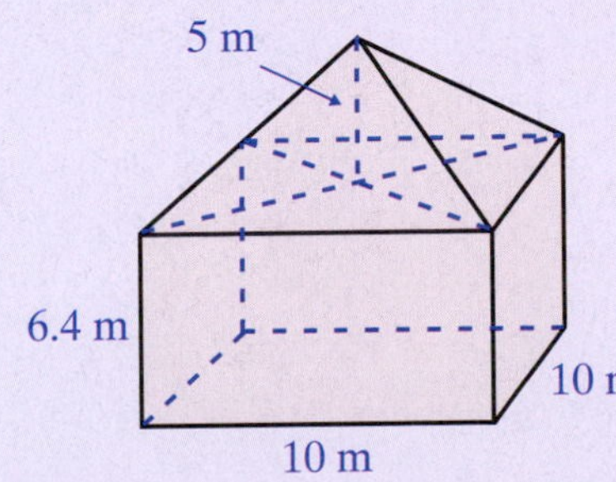

59. ______________

60. ______________

61. ______________

62. ______________

Name ______________________ Section ____________ Date __________

ANSWERS

Chapter 5 Test

1. Write the decimal number 30.657 in words.

2. Write the decimal number 0.375 in fraction form reduced to lowest terms.

3. Change $\frac{5}{16}$ to decimal form.

Round as indicated.

4. 203.018 (to the nearest hundredth)

5. 1972.8046 (to the nearest thousandth)

6. 0.0693 (to the nearest tenth)

Perform the indicated operations. Write your answer with decimal notation.

7. $85.815 + 17.943$

8. $95.6 - 93.712$

9. $82 + 14.91 + 25.2$

10. $100.64 - 82.495$

11. $13\frac{2}{5} + 6 + 17.913$

12. $1\frac{1}{1000} - 0.09705$

1. ____________

2. ____________

3. ____________

4. ____________

5. ____________

6. ____________

7. ____________

8. ____________

9. ____________

10. ____________

11. ____________

12. ____________

13. ____________

14. ____________

15. ____________

16. ____________

17. ____________

18. ____________

19. ____________

20. ____________

21. ____________

22. a. ____________

b. ____________

c. ____________

23. ____________

13. (0.35)(0.84)

14. $\begin{array}{r} 16.31 \\ \times\, 0.785 \\ \hline \end{array}$

15. (1.92)(1000)

16. $3.614 \div 100$

Find each quotient to the nearest hundredth.

17. $82 \div 4.6$

18. $0.13\overline{)8.617}$

Find the equivalent measures in the metric system.

19. 0.681 km = __________ m = __________ cm = __________ mm

20. 355 mm = __________ cm = __________ m

21. An automobile gets 18.3 miles per gallon and the tank holds 21.4 gallons. How far (to the nearest mile) can the car travel on a full tank?

22. Use your calculator to find the following square roots accurate to four decimal places.

a. $\sqrt{6}$ **b.** $\sqrt{800}$ **c.** $\sqrt{1.755625}$

23. A football field is rectangular in shape—100 yards long and 40 yards wide. What is the distance (to the nearest tenth of a yard) from one corner of the field to another corner? (That is, what is the length of a diagonal of the rectangle?)

Name ______________________ Section ____________ Date __________

ANSWERS

24. Find **a.** the circumference and **b.** the area of a circle with diameter 6 inches.

24. a. ______________

b. ______________

25. If a soccer ball (full of air) measures 22 cm in diameter, what is the volume of air (to the nearest cubic centimeter) that the ball contains?

25. ______________

22 cm

Name ______________ Section ______________ Date ______________

ANSWERS

Cumulative Review: Chapters 1 – 5

1. Name the property illustrated.

a. $15 + 6 = 6 + 15$ **b.** $2(6\cdot7) = (2\cdot6)7$ **c.** $51 + 0 = 51$

Evaluate the expressions.

2. $2^2 + 3^2$

3. $18 + 5(2 + 3^2) \div 11 \cdot 2$

4. $24 + [6 + (5^2 \cdot 1 - 3)]$

5. $36 \div 4 + 9 \cdot 2^2$

6. Find the prime factorization of 420.

7. Find the LCM for each set of numbers.

a. 28, 70 **b.** 8, 16, 32, 64

Perform the indicated operations.

8. $\frac{14}{15} - \frac{9}{10}$

9. $4\frac{5}{8} + 3\frac{9}{10}$

10.
$$\begin{array}{r} 306\frac{3}{8} \\ -250\frac{13}{20} \\ \hline \end{array}$$

11.
$$\begin{array}{r} 10\frac{1}{2} \\ 5\frac{3}{4} \\ +16\frac{1}{10} \\ \hline \end{array}$$

12. $\left(\frac{3}{4} - \frac{1}{8}\right) \div \left(\frac{13}{16} - \frac{1}{2}\right)$

13. $\left(\frac{1}{2}\right)^2\left(\frac{14}{15}\right) + \frac{4}{15} \div 2$

14. $22\frac{1}{2} \div 3\frac{3}{4}$

1. a. ______________

b. ______________

c. ______________

2. ______________

3. ______________

4. ______________

5. ______________

6. ______________

7. a. ______________

b. ______________

8. ______________

9. ______________

10. ______________

11. ______________

12. ______________

13. ______________

14. ______________

15. ____________

16. ____________

17. ____________

18. ____________

19. a. ____________

b. ____________

20. ____________

21. a. ____________

b. ____________

c. ____________

d. ____________

22. ____________

15. $\left(\frac{1}{2}+\frac{5}{6}\right)\div\left(1-\frac{5}{8}\right)$

16. Find $\frac{3}{4}$ of 96.

17. Find the quotient $32.1 \div 1.7$ (to the nearest tenth).

18. Find the product of 21.6 and 0.35 and subtract the sum of 1.375 and 4.

19. The business office at a university purchased five vans at a price of $41,540 each, including tax.
a. Approximately how much was paid for the vans?
b. Exactly how much was paid for the vans?

20. The sale price of a new sofa was $752. This was $\frac{4}{5}$ of the original price. What was the original price of the sofa?

21. Use your calculator to find the following square roots accurate to four decimal places.

a. $\sqrt{11}$ **b.** $\sqrt{441}$ **c.** $\sqrt{2000}$ **d.** $\sqrt{3.56}$

22. Use the Pythagorean Theorem to show that a triangle with sides of 12 cm, 16 cm, and 20 cm is a right triangle.

Name ________________________ Section ____________ Date __________

ANSWERS

23. Find the length of the hypotenuse (to the nearest tenth of an inch) of a right triangle with legs of length 36 in. and 50 in.

23. ____________

24. Find **a.** the perimeter and **b.** the area of the figure shown:

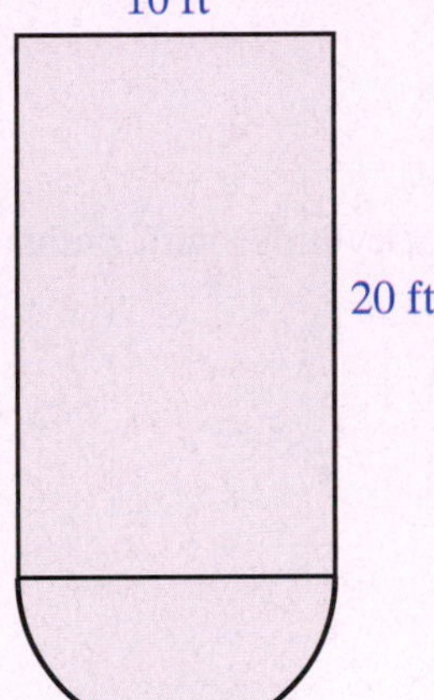

24. a. ____________

b. ____________

25. a. ____________

25. Find **a.** the perimeter and **b.** the area of the figure shown.

b. ____________

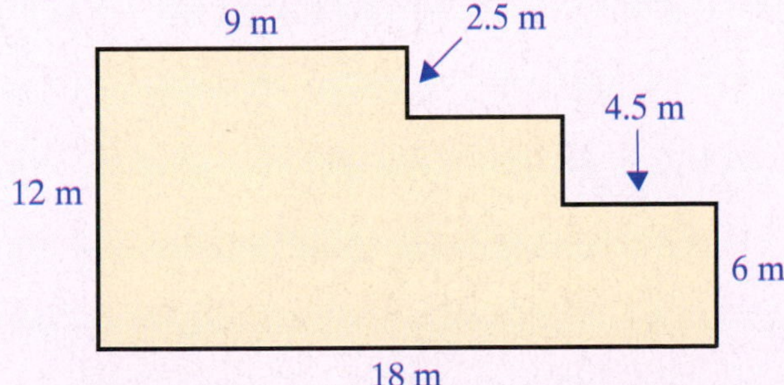

26. a. ____________

b. ____________

26. Is the figure with dimensions as shown a right triangle? Why or why not? If the figure is a right triangle, find **a.** the perimeter and **b.** the area of the figure.

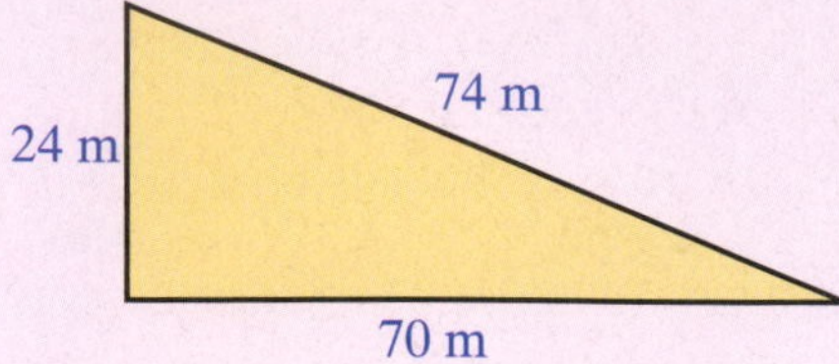

27. ____________

27. Find the volume of a circular cylinder with radius 10 ft and height 4 ft.

28. ____________

28. Find the volume of a circular cone with diameter 8 cm and height 14 cm.

6

Ratios and Proportions

Chapter 6 Ratios and Proportions

WHAT TO EXPECT IN CHAPTER 6

Chapter 6 provides another meaning for fractions (ratios) and introduces the very important topic of solving equations (in the form of proportions). Section 6.1 gives the definition of a ratio and shows three different ways to write a ratio. Also, in Section 6.1, ratios are used in a way familiar to shoppers: price per unit. Proportions (equations stating that two ratios are equal) are introduced in Section 6.2; and then, in Section 6.3, we show how to find the value of one unknown term in a proportion. The unknown term is called a variable, and the methods for finding the value of this term provide an introduction to equation solving. (Note: Solving different forms of equations will be discussed in Chapter 9 and in other courses in mathematics.)

In Section 6.4, we illustrate how proportions can be used in applications. We conclude the chapter with topics from geometry, namely the measurement of angles and the concept of similar triangles.

Chapter 6 Ratios and Proportions

Mathematics at Work!

Homeowners are familiar with trips to the store to buy such things as paint, fertilizer, and grass seed for repairs and maintenance. In many cases a homeowner must purchase goods in quantities greater than what is needed at one time for one particular job. The manufacturer's or distributor's packaging units will likely vary from their exact needs, and the store will not sell part of a can of paint or part of a bag of fertilizer. A little application of mathematics can help stretch the homeowner's dollar by minimizing any excess amounts that must be bought. Consider the following situation.

You want to treat your lawn with fungicide, insecticide, and fertilizer mix. The local nursery sells bags of this mix for $18 per bag, each bag contains 16 pounds of mix, and the recommended coverage for one bag is 2000 square feet. If your lawn consists of two rectangular shapes, one 30 feet by 100 feet and the other 50 feet by 80 feet, how many pounds of fertilizer mix would cover your lawn? How many bags do you need to buy? How much would you pay (not including tax)? (See Exercises 36 – 39, Section 6.4.)

6.1 Ratios and Price per Unit

Objective 1

Understanding Ratios

We know two meanings for fractions.

1. To indicate a part of a whole:

$$\frac{7}{8} \quad \text{means} \quad \frac{\text{7 pieces of cherry pie}}{\text{8 pieces in the whole pie}}$$

2. To indicate division:

$$\frac{3}{8} \quad \text{means} \quad 3 \div 8 \quad \text{or} \quad \begin{array}{r} 0.375 \\ 8\overline{)3.000} \\ \underline{2\,4} \\ 60 \\ \underline{56} \\ 40 \\ \underline{40} \\ 0 \end{array}$$

A third use of fractions is to compare two quantities. Such a comparison is called a **ratio**. For example,

$$\frac{3}{4} \quad \text{might mean} \quad \frac{\text{3 dollars}}{\text{4 dollars}} \quad \text{or} \quad \frac{\text{3 hours}}{\text{4 hours}}$$

Ratio

A **ratio** is a comparison of two quantities by division.
The ratio of a to b can be written as

$$\frac{a}{b} \quad \text{or} \quad a:b \quad \text{or} \quad a \text{ to } b.$$

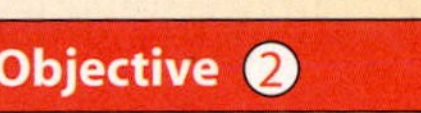

Objectives

1. Understand the meaning of a ratio.
2. Know several notations for ratios.
3. Understand the concept of price per unit.
4. Know how to set up a price per unit as a ratio.
5. Be able to calculate a price per unit by using division.

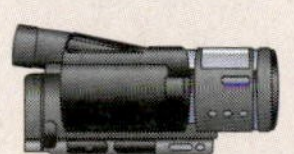
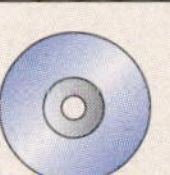

Ratios have the following characteristics:

1. Ratios can be reduced, just as fractions can be reduced.
2. Whenever the units of the numbers in a ratio are the same, then the ratio has no units. We say the ratio is an **abstract number**.
3. When the numbers in a ratio have different units, then the numbers must be labeled to clarify what is being compared. Such a ratio is called a **rate**.

For example, the ratio of 55 miles: 1 hour $\left(\text{or } \frac{55 \text{ miles}}{1 \text{ hour}}\right)$ is a rate of 55 miles per hour (or 55 mph).

Example 1

Compare the quantities 30 students and 40 chairs as a ratio.

Solution

Since the units (students and chairs) are not the same, the units must be written in the ratio. We can write

a. $\frac{30 \text{ students}}{40 \text{ chairs}}$

b. 30 students : 40 chairs

c. 30 students **to** 40 chairs

Furthermore, the same ratio can be simplified by reducing. Since $\frac{30}{40} = \frac{3}{4}$, we can write the ratio as $\frac{3 \text{ students}}{4 \text{ chairs}}$.

The reduced ratio can also be written as

3 students : 4 chairs or 3 students **to** 4 chairs.

Now work exercise 1 in the margin.

1. Compare 12 apples and 15 oranges as a ratio.

Write as a ratio reduced to lowest terms:

2. 3 quarters to 1 dollar

3. Inventory shows 500 washers and 4000 bolts. What is the ratio of washers to bolts?

4. 36 inches to 5 feet.

Example 2

Write the comparison of 2 feet to 3 yards as a ratio.

Solution

a. We can write the ratio as $\frac{2 \text{ feet}}{3 \text{ yards}}$.

b. Another procedure is to change to common units. Because 1 yd = 3 ft, we have 3 yd = 9 ft and we can write the ratio as an abstract number:

$$\frac{2 \text{ feet}}{3 \text{ yards}} = \frac{2 \text{ feet}}{9 \text{ feet}} = \frac{2}{9} \quad \text{or} \quad 2:9 \quad \text{or} \quad 2 \text{ to } 9.$$

Now work exercise 2 in the margin.

Example 3

During baseball season, major league players' batting averages are published in the newspapers. Suppose a player has a batting average of 0.250. What does this indicate?

Solution

A batting average is a ratio (or rate) of hits to times at bat. Thus, a batting average of 0.250 means

$$0.250 = \frac{250 \text{ hits}}{1000 \text{ times at bat}}.$$

Reducing gives

$$0.250 = \frac{250}{1000} = \frac{250 \cdot 1}{250 \cdot 4} = \frac{1}{4} = \frac{1 \text{ hit}}{4 \text{ times at bat}}.$$

This means that we can expect this player to hit successfully at a rate of 1 hit for every 4 times he comes to bat.

Now work exercise 3 in the margin.

Example 4

What is the reduced ratio of 300 centimeters (cm) to 2 meters (m)? (In the metric system, there are 100 centimeters in 1 meter.)

Solution

Since 1 m = 100 cm, we have 2 m = 200 cm. Thus the ratio is

$$\frac{300 \text{ cm}}{2 \text{ m}} = \frac{300 \text{ cm}}{200 \text{ cm}} = \frac{3}{2}.$$

We can also write the ratio as 3 : 2 or 3 to 2.

Now work exercise 4 in the margin.

Price per Unit

Objective 3

When you buy a new battery for your car, you usually buy just one battery; or, if you buy flashlight batteries, you probably buy two or four. So, in cases such as these, the price for one unit (one battery) is clearly marked or understood. However, when you buy groceries, the same item may be packaged in two (or more) different sizes. Since you want to get the most for your money, you want the better (or best) buy. The box below explains how to calculate the **price per unit** (or **unit price**).

Objective 4

To Find the Price per Unit:

1. Set up a ratio (usually in fraction form) of price to units.
2. Divide the price by the number of units.

Note: In the past, many consumers did not understand the concept of price per unit or know how to determine such a number. Now most states have a law that grocery stores must display the price per unit for certain goods they sell so that consumers can be fully informed.

In the following examples, we show the division process as you would write it on paper. However, you may choose to use a calculator to perform these operations. In either case, your first step should be to write each ratio so you can clearly see what you are dividing and that all ratios are comparing the same types of units.

Notice that the comparisons being made in the examples and exercises are with different amounts of the same brand and quality of goods. Any comparison of a relatively expensive brand of high quality with a cheaper brand of lower quality would be meaningless.

Example 5

Objective 5

A 16-ounce jar of grape jam is priced at \$3.99 and a 9.5-ounce jar of the same jam is \$2.69. Which is the better buy?

Solution

We write two ratios of price to units, divide, and compare the results to decide which buy is better. In this problem, we convert dollars to cents (\$3.99 = 399¢ and \$2.69 = 269¢) so that the results will be ratios of cents per ounce.

5. Which is a better buy: a 2-liter bottle of soda for $1.09 or a 3-liter bottle for $1.49?

a. $\frac{399¢}{16 \text{ oz}} \rightarrow$ $16\overline{)399.00}$ = 24.93 or 24.9¢ per ounce

$$\begin{array}{r} 24.93 \\ 16\overline{)399.00} \\ \underline{32} \\ 79 \\ \underline{64} \\ 150 \\ \underline{144} \\ 60 \\ \underline{48} \end{array}$$

b. $\frac{269¢}{9.5 \text{ oz}} \rightarrow$ $9.5\overline{)269.0.00}$ = 28.31 or 28.3¢ per ounce

$$\begin{array}{r} 28.31 \\ 9.5\overline{)269.0.00} \\ \underline{190} \\ 790 \\ \underline{760} \\ 300 \\ \underline{285} \\ 150 \\ \underline{95} \end{array}$$

Thus, the larger jar (16 ounces for $3.99) is the better buy because the price per ounce is less.

Now work exercise 5 in the margin.

Example 6

Pancake syrup comes in three different sized bottles:

36 fluid ounces for $5.29

24 fluid ounces for $4.69

12 fluid ounces for $3.99

Find the price per fluid ounce for each size of bottle and tell which is the best buy.

Solution

After each ratio is set up, the numerator is changed from dollars and cents to just cents so that the division will yield cents per ounce.

a. $\dfrac{\$5.29}{36\text{ oz}} = \dfrac{529¢}{36\text{ oz}} \longrightarrow$

$$\begin{array}{r} 14.69 \\ 36\overline{)529.00} \\ \underline{36} \\ 169 \\ \underline{144} \\ 250 \\ \underline{216} \\ 340 \\ \underline{324} \\ 16 \end{array}$$

or 14.7¢ per ounce or 14.7¢/oz

(**Note :** "14.7¢ /oz" is read "14.7¢ per ounce".)

b. $\dfrac{\$4.69}{24\text{ oz}} = \dfrac{469¢}{24\text{ oz}} \longrightarrow$

$$\begin{array}{r} 19.54 \\ 24\overline{)469.00} \\ \underline{24} \\ 229 \\ \underline{216} \\ 130 \\ \underline{120} \\ 100 \\ \underline{96} \end{array}$$

or 19.5¢/oz

c. $\dfrac{\$3.99}{12\text{ oz}} = \dfrac{399¢}{12\text{ oz}} \longrightarrow$

$$\begin{array}{r} 33.25 \\ 12\overline{)399.00} \\ \underline{36} \\ 39 \\ \underline{36} \\ 30 \\ \underline{24} \\ 60 \\ \underline{60} \end{array}$$

or 33.3¢/oz

The largest container (36 fluid ounces) is the best buy.

Now work exercise 6 in the margin.

In each of the examples, the larger (or largest) amount of units was the better (or best) buy. In general, larger amounts are less expensive because the manufacturer wants you to buy more of the product. However, the larger amount is not always the better buy because of other considerations such as packaging or the consumer's individual needs. For example, people who do not use much pancake syrup may want to buy a smaller bottle. Even though they pay more per unit, it's more economical in the long run not to have to throw any away.

6. Which is a better buy: a 12.5-ounce box of cereal for \$2.50 or a 16-ounce box for \$3.00? What is the price per ounce (to the nearest tenth of a cent) for each?

Special comment on the term *per*: The student should be aware that the term **per** can be interpreted to mean **divided by**. For example,

cents **per** ounce	means	cents **divided by** ounces
dollars **per** pound	means	dollars **divided by** pounds
miles **per** hour	means	miles **divided by** hours
miles **per** gallon	means	miles **divided by** gallons

Practice Problems

Write the following comparisons as ratios reduced to lowest terms. ***Note:*** *there is more than one form of the correct answer.*

1. 2 teachers to 24 students

2. 90 wins to 72 losses

3. 147 miles to 3 hours

4. 5 minutes to 200 seconds

*Find the price per unit for both **a.** and **b.** and determine which is the better deal.*

5. **a.** A 6-pack of cola for $2.99
b. A 12-pack of cola for $5.49

6. **a.** 5 pairs of socks for $12.50
b. 3 pairs of socks for $8.25.

Answers to Practice Problems:

1. 1 teacher : 12 students

2. 5 wins : 4 losses

3. 49 miles : 1 hour, or 49 mph

4. 1 minute : 40 seconds, or 3 : 2

5. **a.** 49.8¢/can; **b.** 45.8¢/can. **b.** is the better deal.

6. **a.** $2.50/pair; **b.** $2.75/pair. **a.** is the better deal.

Name ______________________ Section ____________ Date __________

Exercises 6.1

Write the following comparisons as ratios reduced to lowest terms. Use common units in the numerator and denominator whenever possible. See Examples 1 through 4.

1. 1 dime to 4 nickels

2. 5 nickels to 3 quarters

3. 5 dollars to 5 quarters

4. 6 dollars to 50 dimes

5. 250 miles to 5 hours

6. 270 miles to 4.5 hours

7. 50 miles to 2 gallons of gas

8. 60 miles to 5 gallons of gas

9. 30 chairs to 25 people

10. 25 people to 30 chairs

11. 18 inches to 2 feet

12. 36 inches to 2 feet

13. 8 days to 1 week

14. 21 days to 4 weeks

15. \$200 in profit to \$500 invested

16. \$200 in profit to \$1000 invested

17. 100 centimeters to 1 meter

18. 10 centimeters to 1 millimeter

19. 125 hits to 500 times at bat

20. 100 hits to 500 times at bat

ANSWERS

1. ____________
2. ____________
3. ____________
4. ____________
5. ____________
6. ____________
7. ____________
8. ____________
9. ____________
10. ____________
11. ____________
12. ____________
13. ____________
14. ____________
15. ____________
16. ____________
17. ____________
18. ____________
19. ____________
20. ____________

21. ____________

22. ____________

23. ____________

24. ____________

25. ____________

26. ____________

27. ____________

28. ____________

29. ____________

30. ____________

31. ____________

32. ____________

33. ____________

21. Blood Types. About 28 out of every 100 African-Americans have type-A blood. Express this fact as a ratio in lowest terms.

22. Nutrition. A serving of four home-baked chocolate chip cookies weighs 40 grams and contains 12 grams of fat. What is the ratio, in lowest terms, of fat grams to total grams?

23. Standardized Testing. In recent years, 18 out of every 100 students taking the SAT (Scholastic Aptitude Test) have scored 600 or above on the mathematics portion of the test. Write the ratio, in lowest terms, of the number of scores 600 or above to the number of scores below 600.

24. Weather. In a recent year, Albany, NY, reported a total of 60 clear days, the rest being cloudy or partly cloudy. For a 365-day year, write the ratio, in lowest terms, of clear days to cloudy or partly cloudy days.

Find the unit price (to the nearest tenth of a cent) of each of the following items and tell which is the better (or best) buy. See Examples 5 and 6.

25. sugar
4 lb at \$2.79
10 lb at \$5.89

26. sliced bologna
8 oz at \$1.79
12 oz at \$1.99

27. sour cream
48 oz at \$5.99
24 oz at \$3.59

28. coffee beans
1.75 oz at \$1.99
12 oz at \$7.99

29. coffee
11.5 oz at \$3.99
39 oz at \$8.99

30. coffee
13 oz at \$3.69
39 oz at \$9.39

31. frozen orange juice
16 fl oz at \$2.99
12 fl oz at \$2.19

32. liquid dish soap
75 oz at \$4.99
45 oz at \$3.39

33. powdered dish soap
155 oz at \$6.59
75 oz at \$4.99

Name ________________ Section ____________ Date __________

34. sliced ham
12 oz at $3.99
8 oz at $2.89

35. coffee
13 oz at $4.19
39 oz at $9.99

36. boxed dry milk
9 oz at $3.19
16 oz at $5.69

37. aluminum foil
200 sq ft at $7.39
75 sq ft at $3.69
50 sq ft at $3.19
25 sq ft at $1.49

38. mayonnaise
8 oz at $1.59
16 oz at $2.69
32 oz at $3.89
64 oz at $6.79

39. liquid bleach
48 fl oz at $1.99
96 fl oz at $2.35
174 fl oz at $3.99

40. sliced cheese
8 oz at $3.19
12 oz at $4.59

41. cottage cheese
16 oz at $2.69
32 oz at $4.89

42. large trash bags
18 at $4.99
28 at $6.59

43. kitchen trash bags
20 at $4.99
40 at $6.59

44. baked beans
31 oz at $1.69
53 oz at $2.79

45. pinto beans
29 oz at $1.09
40 oz at $1.59

46. frozen waffles
16 at $3.99
10 at $2.59

47. honey
12 oz at $3.49
24 oz at $7.99
40 oz at $10.99

48. apple sauce
16 oz at $1.69
23 oz at $2.09
48 oz at $3.19

49. mandarin oranges
11 oz at $1.29
15 oz at $1.39

50. spaghetti
32 oz at $2.79
16 oz at $1.50

51. spaghetti sauce
14 oz at $1.79
26 oz at $2.79

ANSWERS

34. ____________

35. ____________

36. ____________

37. ____________

38. ____________

39. ____________

40. ____________

41. ____________

42. ____________

43. ____________

44. ____________

45. ____________

46. ____________

47. ____________

48. ____________

49. ____________

50. ____________

51. ____________

52. ____

53. ____

54. ____

55. ____

56. ____

57. ____

58. ____

59. ____

60. ____

52. pasta sauce
11 oz at $1.29
15 oz at $1.39

53. laundry soap
87 oz at $8.99
175 oz at $16.19

54. laundry soap
73 oz at $6.89
157 oz at $14.89

55. pork and beans
15 oz at $0.89
31 oz at $1.69

56. parmesan cheese
3 oz at $2.49
8 oz at $4.59
16 oz at $7.79

57. diced tomatoes
14.5 oz at $1.29
28 oz at $1.89

58. peanut butter
18 oz at $2.99
28 oz at $4.39
40 oz at $5.89

59. mustard
9 oz at $1.19
14 oz at $1.99

60. catsup
14 oz at $1.99
32 oz at $2.99
64 oz at $4.89

REVIEW

1. ____
2. ____
3. ____
4. ____
5. ____
6. ____
7. ____
8. ____

Review Problems (from Sections 3.2, 4.2, 5.3, and 5.4)

Perform the indicated operations and simplify if possible.

1. $3\frac{1}{2} \cdot 3\frac{1}{7}$

2. $4\frac{5}{8} \div 7\frac{3}{5}$

3. $\frac{14}{32} \cdot \frac{12}{28} \cdot \frac{4}{15}$

4. $20 \div \frac{1}{4}$

5. $(19.2)(7.8)$

6. $14.12(0.0015)$

7. $13.4\overline{)73.7}$

8. $0.25\overline{)100}$

6.2 Proportions

Understanding Proportions

Objective ①

Consider the equation $\frac{3}{6} = \frac{4}{8}$. This statement (or equation) says that two ratios are equal. Such an equation is called a **proportion**. As we will see, proportions may be true or false.

Objectives

① Understand that a proportion is an equation.

② Be familiar with the terms **means** and **extremes**.

③ Know that in a true proportion, the product of the means is equal to the product of the extremes.

Proportion

A **proportion** is a statement that two ratios are equal. In symbols,

$$\frac{a}{b} = \frac{c}{d} \text{ is a proportion.}$$

A proportion has four **terms:**

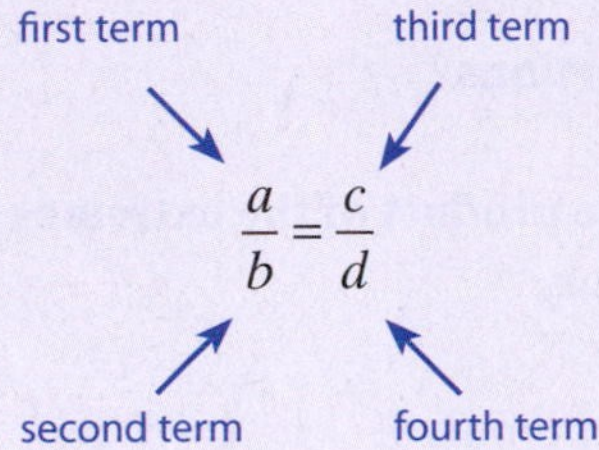

The first and fourth terms (a and d) are called the **extremes**.
The second and third terms (b and c) are called the **means**.

Objective ②

To help in remembering which terms are the extremes and which terms are the means, think of a general proportion written with colons, as shown here.

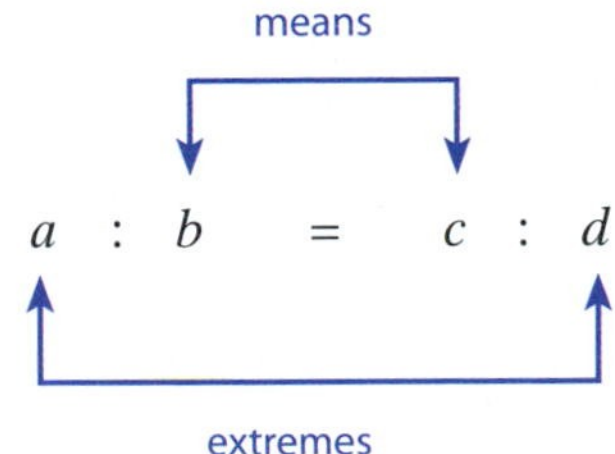

With this form, you can see that a and d are the two end terms; thus, the name **extremes** might seem more reasonable.

Example 1

In the proportion $\frac{8.4}{4.2} = \frac{10.2}{5.1}$, tell which numbers are the extremes and which are the means.

Solution

8.4 and 5.1 are the extremes.

4.2 and 10.2 are the means.

Now work exercise 1 in the margin.

Identify the means and extremes in the following proportion:

1. $\frac{1.3}{1.5} = \frac{1.82}{2.1}$

Identify the means and extremes in the following proportion:

2. $\dfrac{5\frac{1}{2}}{2} = \dfrac{8\frac{1}{4}}{3}$

Completion Example 2

In the proportion $\dfrac{2\frac{1}{2}}{10} = \dfrac{3\frac{1}{4}}{13}$, tell which numbers are the extremes and which are the means.

Solution

________ and ________ are the extremes.

________ and ________ are the means.

Now work exercise 2 in the margin.

Identifying True Proportions

Objective 3

Identifying True Proportions

In a true proportion, the product of the extremes is equal to the product of the means. In symbols,

$$\frac{a}{b} = \frac{c}{d} \quad \text{if and only if} \quad a \cdot d = b \cdot c$$

where $b \neq 0$ and $d \neq 0$.

Determine whether the following proportion is true or false.

3. $\dfrac{3.2}{5} = \dfrac{4}{6.25}$

Note that in a proportion the terms can be any of the types of numbers that we have studied: whole numbers, fractions, mixed numbers, or decimals. Thus, all the related techniques for multiplying these types of numbers should be reviewed at this time.

Example 3

Determine whether the proportion $\dfrac{9}{13} = \dfrac{4.5}{6.5}$ is true or false.

Solution

$$\begin{array}{r} 6.5 \\ \times\ 9 \\ \hline 58.5 \end{array} \leftarrow \text{product of extremes} \qquad \begin{array}{r} 4.5 \\ \times\ 13 \\ \hline 13.5 \\ 45 \\ \hline 58.5 \end{array} \leftarrow \text{product of means}$$

Since 9(6.5) = 13(4.5), the proportion is true.

Now work exercise 3 in the margin.

Example 4

Is the proportion $\frac{5}{8}=\frac{7}{10}$ true or false?

Solution

$5\cdot 10=50$ and $8\cdot 7=56$

Since $50\neq 56$, the proportion is false.

Now work exercise 4 in the margin.

Example 5

Is the proportion $\frac{\frac{3}{5}}{\frac{3}{4}}=\frac{12}{15}$ true or false?

Solution

The extremes are $\frac{3}{5}$ and 15, and their product is

$$\frac{3}{5}\cdot 15=\frac{3}{5}\cdot\frac{15}{1}=9.$$

The means are $\frac{3}{4}$ and 12, and their product is

$$\frac{3}{4}\cdot 12=\frac{3}{4}\cdot\frac{12}{1}=9.$$

Since the product of the extremes is equal to the product of the means, the proportion is true.

Now work exercise 5 in the margin.

Completion Example 6

Is the proportion $\frac{1}{4}:\frac{2}{3}=9:24$ true or false?

Solution

a. The extremes are __________ and __________.
The means are __________ and __________.

b. The product of the extremes is __________.
The product of the means is __________.

c. The proportion is __________ because the products are __________.

Now work exercise 6 in the margin.

Determine whether each proportion is true or false.

4. $\frac{2}{6}=\frac{3}{9}$

5. $\frac{\frac{4}{5}}{\frac{1}{3}}=\frac{4}{3}$

6. $4:18=10:45$

Completion Example Answers

2. $2\frac{1}{2}$ and **13** are the extremes.

$\mathbf{10}$ and $3\frac{1}{4}$ are the means.

6. a. The extremes are $\frac{1}{4}$ and **24**.

The means are $\frac{2}{3}$ and **9**.

b. The product of the extremes is

$\frac{1}{4} \cdot 24 = 6$

The product of the means is

$\frac{2}{3} \cdot 9 = 6$

c. The proportion is **true** because the products are **equal**.

Practice Problems

In the following proportions, identify the extremes and the means.

1. $\frac{2}{3} = \frac{6}{9}$

2. $\frac{80}{100} = \frac{4}{5}$

Determine whether the following proportions are true or false.

3. $\frac{3.7}{11.1} = \frac{5.6}{16.8}$

4. $\frac{9}{8} = \frac{70}{64}$

5. $\frac{12}{19} = \frac{3}{4.75}$

Answers to Practice Problems:

1. The extremes are 2 and 9; the means are 3 and 6.

2. The extremes are 80 and 5; the means are 100 and 4.

3. True **4.** False **5.** True

Name ______________________ Section __________ Date __________

Exercises 6.2

1. In the proportion $\frac{7}{8} = \frac{476}{544}$,

   a. the extremes are ________ and ________ .
   b. the means are ________ and ________ .

2. In the proportion $\frac{x}{y} = \frac{w}{z}$,

   a. the extremes are ________ and ________ .
   b. the means are ________ and ________ .
   c. the two terms ________ and ________ cannot be 0.

Determine whether each proportion is true or false by comparing the product of the means with the product of the extremes. See Examples 3 through 6.

3. $\frac{5}{6} = \frac{10}{12}$

4. $\frac{2}{7} = \frac{5}{17}$

5. $\frac{7}{21} = \frac{4}{12}$

6. $\frac{6}{15} = \frac{2}{5}$

7. $\frac{5}{8} = \frac{12}{17}$

8. $\frac{12}{15} = \frac{20}{25}$

9. $\frac{5}{3} = \frac{15}{9}$

10. $\frac{6}{8} = \frac{15}{20}$

11. $\frac{2}{5} = \frac{4}{10}$

12. $\frac{3}{5} = \frac{6}{10}$

13. $\frac{125}{1000} = \frac{1}{8}$

14. $\frac{3}{8} = \frac{375}{1000}$

15. $\frac{1}{4} = \frac{25}{100}$

16. $\frac{7}{8} = \frac{875}{1000}$

17. $\frac{3}{16} = \frac{9}{48}$

ANSWERS

1. a. ________
b. ________
2. a. ________
b. ________
c. ________
3. ________
4. ________
5. ________
6. ________
7. ________
8. ________
9. ________
10. ________
11. ________
12. ________
13. ________
14. ________
15. ________
16. ________
17. ________

18. ______

19. ______

20. ______

21. ______

22. ______

23. ______

24. ______

25. ______

26. ______

27. ______

28. ______

29. ______

30. ______

31. ______

32. ______

33. ______

34. ______

35. ______

36. ______

37. ______

38. ______

39. ______

40. ______

41. ______

42. ______

43. ______

18. $\frac{2}{3}=\frac{66}{100}$

19. $\frac{1}{3}=\frac{33}{100}$

20. $\frac{14}{6}=\frac{21}{8}$

21. $\frac{4}{9}=\frac{7}{12}$

22. $\frac{19}{16}=\frac{20}{17}$

23. $\frac{3}{6}=\frac{4}{8}$

24. $\frac{12}{18}=\frac{14}{21}$

25. $\frac{5}{6}=\frac{7}{8}$

26. $\frac{7.5}{10}=\frac{3}{4}$

27. $\frac{6.2}{3.1}=\frac{10.2}{5.1}$

28. $\frac{8\frac{1}{2}}{2\frac{1}{3}}=\frac{4\frac{1}{4}}{1\frac{1}{6}}$

29. $\frac{6\frac{1}{5}}{1\frac{1}{7}}=\frac{3\frac{1}{10}}{\frac{8}{14}}$

30. $\frac{6}{24}=\frac{10}{48}$

31. $\frac{7}{16}=\frac{3\frac{1}{2}}{8}$

32. $\frac{10}{17}=\frac{5}{8\frac{1}{2}}$

33. $3:6=5:10$

34. $1\frac{1}{4}:1\frac{1}{2}=\frac{1}{4}:\frac{1}{2}$

35. $210:7=20:\frac{2}{3}$

36. $8.5:6.5=4.5:3.5$

37. $3:5=60:100$

38. $12:1.09=36:3.27$

39. $2\frac{1}{2}:\frac{3}{4}=1\frac{1}{2}:\frac{2}{3}$

40. $6:16=9:24$

41. $6:1.56=2:0.52$

42. $3\frac{1}{5}:1=3\frac{3}{5}:1\frac{2}{5}$

43. $3.75:3=7.5:6$

Name ______________________ Section ____________ Date __________

ANSWERS

44. [Respond below exercise.]

Writing and Thinking about Mathematics

44. Consider the proportion $\frac{16}{5} = \frac{22.4}{7}$. Show why multiplying both sides of the equation by 35 gives the same effect as setting the product of the extremes equal to the product of the means. Do you think that this same technique (of multiplying both sides of the equation by the least common multiple of the denominators) will work with all true proportions? Why or why not?

6.3 Finding the Unknown Term in a Proportion

Understanding the Meaning of an Unknown Term

Proportions can be used to solve certain types of word problems. In these problems, a proportion is set up in which one of the terms in the proportion is not known and the solution to the problem is the value of this unknown term. In this section, we will use the fact that in a true proportion, the product of the extremes is equal to the product of the means as a method for finding the unknown term in a proportion.

Objectives

① Learn how to find the unknown term in a proportion.

② Recall the skills used in multiplying and dividing with whole numbers, fractions, mixed numbers, and decimals.

Variable

A **variable** is a symbol (generally a letter of the alphabet) that is used to represent an unknown number or any one of several numbers. The set of possible values for a variable is called its **replacement set**.

Finding the Unknown Term in a Proportion

Objective ①

To Find the Unknown Term in a Proportion:

1. In the proportion, the unknown term is represented with a variable (some letter such as x, y, w, A, B, etc.).
2. Write an equation that sets the product of the extremes equal to the product of the means.
3. Divide both sides of the equation by the number multiplying the variable. (This number is called the **coefficient** of the variable.)

The resulting equation will have a coefficient of 1 for the variable and will give the missing value for the unknown term in the proportion.

Important Note: As you work through each problem, be sure to write each new equation below the previous equation in the same format shown in the examples. (Arithmetic that cannot be done mentally should be performed to one side, and the results written in the next equation.) This format carries over into solving all types of equations at all levels of mathematics. Also, when a number is written next to a variable, such as in $3x$ or $4y$, the meaning is to multiply the number by the value of the variable. That is, $3x = 3 \cdot x$ and $4y = 4 \cdot y$. The number 3 is the coefficient of x and 4 is the coefficient of y.

Note About the Form of Answers: In general, if the numbers in the problem are in fraction form, then the answer will be in fraction form. If the numbers are in decimal form, then the answer will be in decimal form.

1. Find x if $\frac{12}{x} = \frac{9}{15}$.

Example 1

Find the value of x if $\frac{3}{6} = \frac{5}{x}$.

Solution

In this case, you may be able to see that the correct value of x is 10 since $\frac{3}{6}$ reduces to $\frac{1}{2}$ and $\frac{1}{2} = \frac{5}{10}$.

However, not all proportions involve such simple ratios, and the following general method of solving for the unknown is important.

a. $\frac{3}{6} = \frac{5}{x}$ Write the proportion.

b. $3 \cdot x = 6 \cdot 5$ Write the product of the extremes equal to the product of the means.

c. $\frac{3 \cdot x}{3} = \frac{30}{3}$ Divide both sides by 3, the coefficient of the variable.

d. $\frac{\cancel{3} \cdot x}{\cancel{3}} = \frac{30}{3}$ Reduce both sides to find the solution.

e. $x = 10$

Now work exercise 1 in the margin.

Example 2

Find the value of y if $\frac{6}{16} = \frac{y}{24}$.

Solution

Note that the variable may appear on the right side of the equation as well as on the left side of the equation (as shown in Example 1). In either case, we **divide both sides of the equation by the coefficient of the variable**.

a. $\frac{6}{16} = \frac{y}{24}$ Write the proportion.

b. $6 \cdot 24 = 16 \cdot y$ Write the product of the extremes equal to the product of the means.

c. $\frac{6 \cdot 24}{16} = \frac{16 \cdot y}{16}$ Divide both sides by 16, the coefficient of y.

d. $\frac{6 \cdot 24}{16} = \frac{\cancel{16} \cdot y}{\cancel{16}}$ Reduce both sides to find the value of y.

e. $9 = y$

Second Solution

Reduce the fraction $\frac{6}{16}$ before solving the proportion.

a. $\frac{6}{16}=\frac{y}{24}$ Write the proportion.

b. $\frac{3}{8}=\frac{y}{24}$ Reduce the fraction: $\frac{6}{16}=\frac{3}{8}$.

c. $3\cdot 24=8\cdot y$ Proceed to solve as before.

d. $\frac{3\cdot 24}{8}=\frac{\cancel{8}\cdot y}{\cancel{8}}$

e. $9=y$

Now work exercise 2 in the margin.

Solve each proportion for the unknown term:

2. $\frac{3}{10}=\frac{R}{100}$

Example 3

Find w if $\frac{w}{7}=\frac{20}{\frac{2}{3}}$.

Solution

a. $\frac{w}{7}=\frac{20}{\frac{2}{3}}$ Write the proportion.

b. $\frac{2}{3}\cdot w=7\cdot 20$ Set the product of the extremes equal to the product of the means.

c. $\frac{\frac{\cancel{2}}{\cancel{3}}\cdot w}{\frac{\cancel{2}}{\cancel{3}}}=\frac{7\cdot 20}{\frac{2}{3}}$ Divide each side by the coefficient $\frac{2}{3}$.

d. $w=\frac{7}{1}\cdot\frac{20}{1}\cdot\frac{3}{2}$ Simplify. Remember: to divide by a fraction, multiply by its reciprocal.

e. $w=210$

Now work exercise 3 in the margin.

3. $\frac{\frac{1}{2}}{6}=\frac{5}{x}$

Completion Example 4

Find A if $\frac{A}{3}=\frac{7.5}{6}$.

Solution

a. $\frac{A}{3}=\frac{7.5}{6}$

b. $6\cdot A=$ ________

c. $\frac{\cancel{6}\cdot A}{\cancel{6}}=$ ________

d. $A=$ ________

Now work exercise 4 in the margin.

4. $\frac{x}{2.3}=\frac{3.2}{9.2}$

Solve each proportion for the unknown term:

5. $\frac{z}{6} = \frac{4\frac{1}{2}}{7\frac{1}{2}}$.

6. $\frac{\frac{8}{3}}{5} = \frac{\frac{1}{5}}{z}$

As illustrated in Examples 5 and 6, when the coefficient of the variable is a fraction, we can multiply both sides of the equation by the reciprocal of this coefficient. This is the same as dividing by the coefficient, but fewer steps are involved.

Example 5

Find x if $\frac{x}{1\frac{1}{2}} = \frac{1\frac{2}{3}}{3\frac{1}{3}}$.

Solution

a. $\frac{x}{1\frac{1}{2}} = \frac{1\frac{2}{3}}{3\frac{1}{3}}$ Write the proportion.

b. $\frac{10}{3} \cdot x = \frac{\cancel{3}}{2} \cdot \frac{5}{\cancel{3}}$ Write each mixed number as an improper fraction.

c. $\frac{10}{3} \cdot x = \frac{5}{2}$

d. $\frac{\cancel{3}}{\cancel{10}} \cdot \frac{\cancel{10}}{\cancel{3}} \cdot x = \frac{3}{10} \cdot \frac{5}{2}$ Multiply each side by $\frac{3}{10}$, the reciprocal of $\frac{10}{3}$.

e. $x = \frac{3}{\underset{2}{\cancel{10}}} \cdot \frac{\cancel{5}}{2} = \frac{3}{4}$

Now work exercise 5 in the margin.

Completion Example 6

Find y if $\frac{2\frac{1}{2}}{6} = \frac{3}{y}$.

Solution

a. $\frac{2\frac{1}{2}}{6} = \frac{3}{y}$

b. $\frac{5}{2} \cdot y =$ ______________ $\left(2\frac{1}{2} = \frac{5}{2}\right)$

c. $\frac{\cancel{2}}{\cancel{5}} \cdot \frac{\cancel{5}}{\cancel{2}} \cdot y = \frac{2}{5} \cdot$ ______________

d. $y =$ ______________

Now work exercise 6 in the margin.

Completion Example Answers

4. a. $\dfrac{A}{3} = \dfrac{7.5}{6}$

b. $6 \cdot A = \mathbf{3 \cdot 7.5}$

c. $\dfrac{\cancel{6} \cdot A}{\cancel{6}} = \dfrac{\mathbf{22.5}}{\mathbf{6}}$

d. $A = \mathbf{3.75}$

6. a. $\dfrac{2\frac{1}{2}}{6} = \dfrac{3}{y}$

b. $\dfrac{5}{2} \cdot y = \mathbf{3 \cdot 6}$

c. $\dfrac{\cancel{2}}{\cancel{5}} \cdot \dfrac{\cancel{5}}{\cancel{2}} \cdot y = \dfrac{2}{5} \cdot \mathbf{18}$

d. $y = \dfrac{\mathbf{36}}{\mathbf{5}} \left(\textbf{or } \mathbf{7\frac{1}{5}}\right)$

Practice Problems

Solve the following proportions.

1. $\dfrac{3}{5} = \dfrac{R}{100}$

2. $\dfrac{2\frac{1}{2}}{8} = \dfrac{25}{x}$

3. $\dfrac{3}{2} = \dfrac{B}{4.5}$

4. $\dfrac{x}{1.50} = \dfrac{11}{2.75}$

Answers to Practice Problems:

1. $R = 60$ **2.** $x = 80$ **3.** $B = 6.75$ **4.** $x = 6$

Name ______________________ Section ____________ Date __________

Exercises 6.3

Solve for the variable in each of the following proportions. See Examples 1 through 6.

1. $\frac{3}{6}=\frac{6}{x}$

2. $\frac{7}{21}=\frac{y}{6}$

3. $\frac{5}{7}=\frac{x}{28}$

4. $\frac{4}{10}=\frac{5}{x}$

5. $\frac{8}{B}=\frac{6}{30}$

6. $\frac{7}{B}=\frac{5}{15}$

7. $\frac{1}{2}=\frac{x}{100}$

8. $\frac{3}{4}=\frac{x}{100}$

9. $\frac{A}{3}=\frac{7}{2}$

10. $\frac{x}{100}=\frac{1}{20}$

11. $\frac{3}{5}=\frac{60}{D}$

12. $\frac{3}{16}=\frac{9}{x}$

13. $\frac{\frac{1}{2}}{x}=\frac{5}{10}$

14. $\frac{\frac{2}{3}}{3}=\frac{y}{127}$

15. $\frac{\frac{1}{3}}{x}=\frac{5}{9}$

16. $\frac{\frac{3}{4}}{7}=\frac{3}{z}$

17. $\frac{\frac{1}{8}}{6}=\frac{\frac{1}{2}}{w}$

18. $\frac{\frac{1}{6}}{5}=\frac{5}{w}$

19. $\frac{1}{4}=\frac{1\frac{1}{2}}{y}$

20. $\frac{1}{5}=\frac{x}{2\frac{1}{2}}$

21. $\frac{1}{5}=\frac{x}{7\frac{1}{2}}$

22. $\frac{2}{5}=\frac{R}{100}$

23. $\frac{3}{5}=\frac{R}{100}$

24. $\frac{A}{4}=\frac{75}{100}$

ANSWERS

1. ________
2. ________
3. ________
4. ________
5. ________
6. ________
7. ________
8. ________
9. ________
10. ________
11. ________
12. ________
13. ________
14. ________
15. ________
16. ________
17. ________
18. ________
19. ________
20. ________
21. ________
22. ________
23. ________
24. ________

25. ______

26. ______

27. ______

28. ______

29. ______

30. ______

31. ______

32. ______

33. ______

34. ______

35. ______

36. ______

37. ______

38. ______

39. ______

40. ______

41. ______

42. ______

43. ______

44. ______

45. ______

46. ______

47. ______

48. ______

49. ______

50. ______

25. $\frac{A}{4} = \frac{50}{100}$ **26.** $\frac{20}{B} = \frac{1}{4}$ **27.** $\frac{30}{B} = \frac{25}{100}$ **28.** $\frac{A}{20} = \frac{15}{100}$

29. $\frac{1}{3} = \frac{R}{100}$ **30.** $\frac{2}{3} = \frac{R}{100}$ **31.** $\frac{9}{x} = \frac{4\frac{1}{2}}{11}$ **32.** $\frac{y}{6} = \frac{2\frac{1}{2}}{12}$

33. $\frac{x}{4} = \frac{1\frac{1}{4}}{5}$ **34.** $\frac{5}{x} = \frac{2\frac{1}{4}}{27}$ **35.** $\frac{x}{3} = \frac{16}{3\frac{1}{5}}$ **36.** $\frac{6.2}{5} = \frac{x}{15}$

37. $\frac{3.5}{2.6} = \frac{10.5}{B}$ **38.** $\frac{4.1}{3.2} = \frac{x}{6.4}$ **39.** $\frac{7.8}{1.3} = \frac{x}{0.26}$ **40.** $\frac{7.2}{y} = \frac{4.8}{14.4}$

41. $\frac{150}{300} = \frac{R}{100}$ **42.** $\frac{19.2}{96} = \frac{R}{100}$ **43.** $\frac{12}{B} = \frac{25}{100}$ **44.** $\frac{13.5}{B} = \frac{15}{100}$

45. $\frac{A}{42} = \frac{65}{100}$ **46.** $\frac{A}{244} = \frac{18}{100}$ **47.** $\frac{A}{850} = \frac{30}{100}$ **48.** $\frac{A}{595} = \frac{6}{100}$

49. $\frac{5684}{B} = \frac{98}{100}$ **50.** $\frac{24}{27} = \frac{R}{100}$

Name ____________ Section ________ Date ________

ANSWERS

51. ________

52. ________

53. ________

54. ________

55. ________

56. ________

Check Your Number Sense

Now that you are familiar with proportions and the techniques for finding the unknown term in a proportion, check your general understanding by choosing the answer (using mental calculations only) that seems the most reasonable to you in each of the following exercises. After you have checked the answers in the back of the text, work out each problem that you missed to help you develop a better understanding of proportions.

51. Given the proportion $\frac{x}{100}=\frac{1}{4}$, which of the following values seems the most reasonable for x?

(a) 10 (b) 25 (c) 50 (d) 75

52. Given the proportion $\frac{x}{200}=\frac{1}{10}$, which of the following values seems the most reasonable for x?

(a) 10 (b) 20 (c) 30 (d) 40

53. Given the proportion $\frac{3}{5}=\frac{60}{D}$, which of the following values seems the most reasonable for D?

(a) 50 (b) 80 (c) 100 (d) 150

54. Given the proportion $\frac{4}{10}=\frac{20}{x}$, which of the following values seems the most reasonable for x?

(a) 10 (b) 30 (c) 40 (d) 50

55. Given the proportion $\frac{1.5}{3}=\frac{x}{6}$, which of the following values seems the most reasonable for x?

(a) 1.5 (b) 2.5 (c) 3.0 (d) 4.5

56. Given the proportion $\frac{2\frac{1}{3}}{x}=\frac{4\frac{2}{3}}{10}$, which of the following values seems the most reasonable for x?

(a) $4\frac{2}{3}$ (b) 5 (c) 20 (d) 40

REVIEW

1. ______

2. ______

3. ______

4. ______

5. ______

6. ______

Review Problems (from Sections 5.2 and 5.5)

Perform the indicated operations by using decimal forms of the fractions.

1. $\frac{3}{4}+\frac{3}{5}+7.16$

2. $5\frac{1}{2}+20.3+16.8$

3. $27\frac{5}{8}-13.925$

4. $8.13\left(2\frac{1}{4}\right)$

5. $6\frac{7}{10}+2\frac{1}{2}(6.1)$

6. $5.8\left(3\frac{1}{10}\right)-3\div\frac{1}{5}$

6.4 Problem Solving with Proportions

When to Use Proportions

A proportion is a statement that two ratios are equal. Thus, problems that involve two ratios are the type that can be solved by using proportions.

Objective 1

To Solve a Word Problem Using Proportions:

1. Identify the unknown quantity and use a variable to represent this quantity.

Objective 2

2. Set up a proportion in which the units are compared as in Pattern A or Pattern B shown here.

   Suppose that a motorcycle will travel 352 miles on 11 gallons of gas. How far would you expect it to travel on 15 gallons of gas? (Let x = the unknown number of miles.)

   Pattern A Each ratio has different units, but they are in the same order. For example,

   $$\frac{352\ \textbf{miles}}{11\ \text{gallons}} = \frac{x\ \textbf{miles}}{15\ \text{gallons}}.$$

   Pattern B Each ratio has the same units, the numerators correspond, and the denominators correspond. For example,

   $$\frac{352\ \textbf{miles}}{x\ \textbf{miles}} = \frac{11\ \text{gallons}}{15\ \text{gallons}}.$$

   (352 miles corresponds to 11 gallons, and x miles corresponds to 15 gallons.)

3. Solve the proportion.

Objective 3

Objectives

1. Recognize the type of problem that can be solved by using a proportion.
2. Learn how to identify the unknown quantity in a problem.
3. Be able to set up a proportion with an unknown term in one of the appropriate patterns that will lead to a solution of the problem.

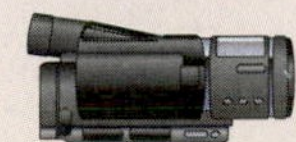

Problem Solving with Proportions

Example 1

You drove your car 500 miles and used 20 gallons of gasoline. How many miles would you expect to drive on 30 gallons of gasoline?

Solution

a. Let x represent the unknown number of miles.

b. Set up a proportion using either Pattern A or Pattern B. **Label the numerators and denominators to be sure the units are in the same order.**

$$\frac{500\text{ miles}}{20\text{ gallons}} = \frac{x\text{ miles}}{30\text{ gallons}}$$

Pattern A Each ratio has different units but the numerators are the same units and the denominators are the same units.

Continued on the next page...

1. One pound of candy costs \$4.75. How many pounds of candy can be bought for \$28.50?

c. Solve the proportion.

$$\frac{500}{20} = \frac{x}{30}$$
$$500 \cdot 30 = 20 \cdot x$$
$$\frac{15{,}000}{20} = \frac{\cancel{20} \cdot x}{\cancel{20}}$$
$$750 = x$$

You would expect to drive 750 miles on 30 gallons of gas.

Note: *Any* of the following proportions yield the same answer.

$$\frac{20 \text{ gallons}}{500 \text{ miles}} = \frac{30 \text{ gallons}}{x \text{ miles}}$$

Pattern A Each ratio has different units but the numerators are the same units and the denominators are the same units.

$$\frac{500 \text{ miles}}{x \text{ miles}} = \frac{20 \text{ gallons}}{30 \text{ gallons}}$$

Pattern B Each ratio has the same units, numerators correspond, and denominators correspond.

$$\frac{x \text{ miles}}{500 \text{ miles}} = \frac{30 \text{ gallons}}{20 \text{ gallons}}$$

Pattern B Each ratio has the same units, numerators correspond, and denominators correspond.

Now work exercise 1 in the margin.

2. Precinct 1 has 520 registered voters and Precinct 2 has 630 registered voters. If the ratio of actual voters to registered voters was the same in both precincts, how many voted in Precinct 2 in the last election if 104 voted in Precinct 1?

Example 2

An architect draws the plans for a building using a scale of $\frac{1}{2}$ inch to represent 10 feet. How many feet would 6 inches represent?

Solution

a. Let y represent the unknown number of feet.

b. Set up a proportion labeling the numerators and denominators so that $\frac{1}{2}$ inch corresponds to 10 feet.

$$\frac{\frac{1}{2} \text{ inch}}{6 \text{ inches}} = \frac{10 \text{ feet}}{y \text{ feet}}$$

c. Solve the proportion $\frac{\frac{1}{2}}{6} = \frac{10}{y}$.

$$\frac{1}{2} \cdot y = 6 \cdot 10$$
$$\frac{\cancel{\frac{1}{2}} \cdot y}{\cancel{\frac{1}{2}}} = \frac{60}{\frac{1}{2}}$$
$$y = \frac{60}{1} \cdot \frac{2}{1}$$
$$y = 120$$

6 inches would represent 120 feet.

Now work exercise 2 in the margin.

Completion Example 3

A recommended mixture of weed killer is 3 capfuls for 2 gallons of water. How many capfuls should be mixed with 5 gallons of water?

Solution

a. Let x = unknown number of capfuls of weed killer.

b. $\frac{x \text{ capfuls}}{5 \text{ gallons}} = \frac{3 \text{ capfuls}}{2 \text{ gallons}}$

$\_\_\_\_\_ \cdot x = 5 \cdot \_\_\_\_\_\_\_\_\_$

$\frac{\_\_\_\_\_ \cdot x}{\_\_\_\_\_\_\_\_\_} = \frac{15}{\_\_\_\_\_\_\_\_\_}$

$x = \_\_\_\_\_\_\_\_\_$

__________ capfuls of weed killer should be mixed with 5 gallons of water.

Now work exercise 3 in the margin.

Completion Example 4

A jelly manufacturer puts 2.5 ounces of sugar into every 6-ounce jar of jelly. How many ounces of jelly can be made with 300 ounces of sugar?

Solution

a. Let A = unknown amount of jelly.

b. $\frac{2.5 \text{ oz sugar}}{6 \text{ oz jelly}} = \frac{300 \text{ oz sugar}}{A \text{ oz jelly}}$

$\_\_\_\_\_ \cdot \_\_\_\_\_\_ = 1800$

$\frac{\_\_\_\_\_ \cdot \_\_\_\_\_}{\_\_\_\_\_} = \frac{1800}{\_\_\_\_\_}$

$A = \_\_\_\_\_\_\_\_\_$

300 ounces of sugar will make ____________ ounces of jelly.

Now work exercise 4 in the margin.

Nurses and doctors work with proportions when prescribing medicine and giving injections. Medical texts write proportions in the form

$2:40::x:100$ instead of $\frac{2}{40} = \frac{x}{100}$.

Using either notation, the solution is found by setting the product of the extremes equal to the product of the means and solving the equation for the unknown quantity.

3. A painter can paint one room of a house in 2 hours and 20 minutes. How long would it take him to paint three more rooms, assuming they are the same size as the first?

4. 4 out of 10 students in a certain college class are female. If there are a total of 35 students in the class, how many are female?

5. Solve the proportion:
81 sq ft : 9 sq yd ::

15 sq ft : x sq yd

Example 5

Solve the proportion.

2 ounces : 40 grams :: x ounces : 100 grams

Solution

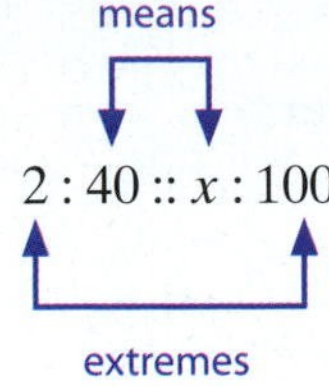

$$2 \cdot 100 = 40 \cdot x$$

$$\frac{200}{40} = \frac{\cancel{40} \cdot x}{\cancel{40}}$$

$$5 = x$$

The solution is $x = 5$ ounces.

Now work exercise 5 in the margin.

Completion Example Answers

3. a. Let x = unknown number of capfuls of weed killer.

b. $\frac{x \text{ capfuls}}{5 \text{ gallons}} = \frac{3 \text{ capfuls}}{2 \text{ gallons}}$

$$\mathbf{2} \cdot x = 5 \cdot \mathbf{3}$$

$$\frac{\cancel{\mathbf{2}} \cdot x}{\cancel{\mathbf{2}}} = \frac{15}{\mathbf{2}}$$

$$x = \mathbf{7.5}$$

7.5 capfuls of weed killer should be mixed with 5 gallons of water.

4. a. Let A = unknown amount of jelly.

b. $\frac{2.5 \text{ oz sugar}}{6 \text{ oz jelly}} = \frac{300 \text{ oz sugar}}{A \text{ oz jelly}}$

$$\mathbf{2.5} \cdot \mathbf{A} = 1800$$

$$\frac{\cancel{\mathbf{2.5}} \cdot \mathbf{A}}{\cancel{\mathbf{2.5}}} = \frac{1800}{\mathbf{2.5}}$$

$$A = \mathbf{720}$$

300 ounces of sugar will make **720** ounces of jelly.

Practice Problems

Solve the following problems using proportions.

1. If a cat normally catches 3 mice in a day, how many would you expect it to catch in a week?

2. If 1 inch represents 35 miles on a map, how many miles do 4.4 inches represent?

3. A baseball player gets 22 hits in 4 weeks. How many hits would you expect him to get in 6 weeks?

4. You can buy 3 onions for $2.89 at the grocery store. What would be the cost of 9 onions?

5. If an airplane flies 2120 miles in 4 hours, how far will it fly in five hours?

Answers to Practice Problems:

1. 21 mice **2.** 154 miles **3.** 33 hits
4. $8.67 **5.** 2650 miles

Name ______________________ Section ____________ Date __________

ANSWERS

Exercises 6.4

Solve the following word problems using proportions. See Examples 1 through 4.

1. **Cartography.** A mapmaker uses a scale of 2 inches to represent 30 miles. How many miles are represented by 3 inches?

2. **Gas prices.** If gasoline sells for $2.05 per gallon, what will 10 gallons cost?

3. **Gas prices.** If gasoline sells for $2.49 per gallon, how many gallons can be bought with $17.43?

4. **Odds.** If the odds on a horse are $5 to win on a $2 bet, how much can be won with a $5 bet?

5. **Investing.** An investor thinks she should make $12 for every $100 she invests. How much would she expect to make on a $1500 investment?

6. **Eggs.** If one dozen (12) eggs cost $1.09, what would three dozen eggs cost?

7. **Fabric.** The price of a certain fabric is $1.75 per yard. How many yards can be bought with $35 (not including tax)?

8. **Painting.** An artist figures she can paint 3 portraits every two weeks. At this rate, how long will it take her to paint 18 portraits?

9. **Chemistry.** Two units of a certain gas weigh 175 grams. What is the weight of 5 units of this gas?

10. **Sports equipment.** A baseball team bought 8 bats for $96. What would they pay for 10 bats?

1. ____________
2. ____________
3. ____________
4. ____________
5. ____________
6. ____________
7. ____________
8. ____________
9. ____________
10. ____________

11.

11. Retail profit. A store owner expects to make a profit of $2 on an item that sells for $10. How much profit will he expect to make on a larger but similar item that sells for $60?

12.

12. Commissions. A saleswoman makes $8 for every $100 worth of the product that she sells. What will she make if she sells $5000 worth of the product?

13.

13. Taxes. If property taxes are figured at $1.50 for every $100 in evaluation, what taxes will be paid on a home valued at $85,000?

14.

14. Taxes. A condominium owner pays property taxes of $2000 per year. If taxes are figured at a rate of $1.25 for every $100 in value, what is the value of his condominium?

15.

15. Taxes. Sales tax is figured at 6¢ for every $1.00 of merchandise purchased. What was the purchase price on an item that had a sales tax of $2.04?

16.

16. Building plans. An architect drew plans for a city park using a scale of $\frac{1}{4}$ inch to represent 25 feet. How many feet would 2 inches represent?

17.

17. Shadows. A building 14 stories high casts a shadow 30 feet long at a certain time of day. What is the length of the shadow of a 20-story building at the same time of day in the same city?

18.

18. Numerical comparisons. Two numbers are in the ratio of 4 to 3. The number 10 is in that same ratio to a fourth number. What is the fourth number?

19.

19. Acceleration. A car is traveling at 45 miles per hour. Its speed is increased by 3 miles per hour every 2 seconds. By how much will its speed increase in 5 seconds? How fast will the car be traveling?

20.

20. Deceleration. A truck is traveling at 55 miles per hour and the driver brakes to slow down 2 miles per hour every 3 seconds. How long will it take the truck to slow to a speed of 45 miles per hour?

Name ______________________ Section ____________ Date ___________

ANSWERS

21. Travel. A salesman figured he drove 560 miles every two weeks. How far would he drive in three months (12 weeks)?

22. Travel. Driving steadily, a woman made a trip of 200 miles in $4\frac{1}{2}$ hours. How long would she take to drive 500 miles at the same rate of speed?

23. Travel. If you can drive 286 miles in $5\frac{1}{2}$ hours, how long will it take you to drive 468 miles at the same rate of speed?

24. Gas prices. How much will 21 gallons of gasoline cost if gasoline costs $2.05 per gallon?

25. Gas prices. If diesel fuel costs $2.27 per gallon, how much diesel fuel will $24.53 buy?

26. Fans. An electric fan makes 180 revolutions per minute. How many revolutions will the fan make if it runs for 24 hours?

27. Investing. An investor made $144 in one year on a $1000 investment. What would she have earned if her investment had been $4500?

28. Typing. A typist can type 8 pages of manuscript in 56 minutes. How long will this typist take to type 300 pages?

29. Maps. On a map, $1\frac{1}{2}$ inches represent 40 miles. How many inches represent 50 miles?

30. Measurement. In the metric system, there are 2.54 centimeters in 1 inch. How many centimeters are there in 1 foot?

31. Lawn care. If 40 pounds of fertilizer are used on 2400 square feet of lawn, how many pounds of fertilizer are needed for a lawn of 5400 square feet?

21. ____________

22. ____________

23. ____________

24. ____________

25. ____________

26. ____________

27. ____________

28. ____________

29. ____________

30. ____________

31. ____________

32. ____________

33. ____________

34. ____________

35. ____________

36. a. ____________

b. ____________

c. ____________

37. a. ____________

b. ____________

38. a. ____________

b. ____________

c. ____________

32. Grading. An English teacher must read and grade 27 essays. If the teacher takes 20 minutes to read and grade 3 essays, how much time will he need to grade all 27 essays?

33. Cooking. If 2 cups of flour are needed make 12 biscuits, how much flour will be needed to make 9 of the same kind of biscuits?

34. Cooking. If 2 cups of flour are needed to make 12 biscuits, how many of the same kind of biscuits can be made with 3 cups of flour?

35. Measurement. There are one thousand grams in one kilogram. How many grams are there in four and seven tenths kilograms?

36. Lawn care. You want to treat your lawn with a mix of fungicide, insecticide, and fertilizer. The local nursery sells bags of this mix for $14 per bag. Each bag contains 16 pounds of mix, and the recommended coverage for one bag is 2000 square feet. **a.** If your lawn consists of two rectangular shapes, one 30 feet by 100 feet and the other 50 feet by 80 feet, how many pounds of fertilizer mix would cover your lawn? **b.** How many bags do you need to buy? **c.** How much would you pay (not including tax)?

37. Lawn care. One bag of Weed Killer & Fertilizer contains 18 pounds of fertilizer and weed treatment with a recommended coverage of 5000 square feet. **a.** If your lawn is in the shape of a rectangle 150 feet by 220 feet, how many pounds of Weed Killer & Fertilizer do you need to cover the lawn? **b.** If the cost of one bag is $12, what will you pay for this fertilizer (not including tax)?

38. House painting. The local paint store recommends that you use one gallon of paint to cover 400 square feet of properly prepared wall space in your home. You want to paint the walls of three rooms, one room measuring 10 feet by 12 feet, another 12 feet by 13 feet, and the third 16 feet by 15 feet. **a.** If each room has 8-foot ceilings, how many gallons of paint do you need to buy? **b.** Will you have any paint left over? **c.** How much? (**Note**: For this exercise, ignore window space and door area.)

Name ______________________ Section ____________ Date __________

ANSWERS

39. Lawn care. One bag of dichondra lawn food contains 20 pounds of fertilizer and its recommended coverage is 4000 square feet. **a.** If you want to cover a lawn that is in the shape of a rectangle 120 feet by 160 feet, how many pounds of lawn food do you need? **b.** How many bags of lawn food would you need to buy?

39. a. ____________

b. ____________

Check Your Number Sense

In each of the following exercises, use mental calculations and your judgment to choose the best answer. After you have checked your answers in the back of the text, work out any problems that you missed to help reinforce your understanding.

40. Computer parts. A computer microchip manufacturer expects that 3 out of every 100 microchips it produces will be defective. If 5000 microchips are produced in one production run, about how many would be expected to be defective in that run?

(a) 15 (b) 50 (c) 150 (d) 300

40. ____________

41. Baseball. A professional baseball player has hit 12 home runs in the first 55 games of the season. At this rate, about how many home runs would you expect him to hit over a complete season of 162 games?

(a) 15 (b) 25 (c) 35 (d) 45

41. ____________

42. Surveys. A food distributor is mailing a survey to shoppers to study their grocery purchasing habits. From experience, the company expects about 6 out of every 100 addresses to return the survey. If the company would like at least 1000 responses, what is the minimum number of surveys that they should mail?

(a) 6000 (b) 10,000 (c) 20,000 (d) 60,000

42. ____________

43. Rainforest destruction. The world's tropical rainforests are being destroyed at a rate of 96,000 acres per day. About how many acres of rainforest are being destroyed every hour?

(a) 4000 (b) 8000 (c) 500,000 (d) 1,000,000

43. ____________

Collaborative Learning

44. ______________

With the class separated into teams of two to four, each team is to analyze the following two exercises and decide on the best of the given choices for the answer. The team leader is to discuss the team's choice and the reason for this choice with the class.

44. Killer Whales. Marine biologists at a killer whale feeding ground photograph and identify 35 whales during a research trip. The next year they return to the same location and identify 40 whales, 8 of which were ones they had identified the year before. About how many whales would you estimate are in the group that these biologists are studying?

(a) 40 (b) 75 (c) 100 (d) 175

45. ______________

45. Wildlife management. Forest rangers are concerned about the population of deer in a certain region. To get a good estimate of the deer population, they find, tranquilize, and tag 50 deer. One month later, they again locate 50 deer and of these 50, 5 are deer that were previously tagged. On the basis of these procedures and results, what do they estimate to be the deer population of that region?

(a) 100 (b) 150 (c) 500 (d) 1000

REVIEW

Review Problems (from Sections 4.2 to 4.5)

1. ______________

1. Consider the product $4\frac{1}{2}\cdot 1\frac{1}{6}\cdot 1\frac{1}{5}$. Do you think that this product is (a) close to 6, (b) close to 20, or (c) close to 100? Find the product.

2. ______________

2. Consider the sum $6\frac{1}{3}+5\frac{3}{10}+8\frac{1}{15}$. Do you think that this sum is (a) close to 10, (b) close to 20, or (c) close to 100? Find the sum.

3. ______________

3. Consider the quotient $6\frac{2}{3}\div 1\frac{1}{3}$. Do you think that this quotient is (a) more than 6 or (b) less than 6? Find the quotient.

4. ______________

4. Use the rules for order of operations to find the value of the following expression: $\frac{1}{2}\cdot 3\frac{1}{2}+5\frac{1}{10}\cdot\frac{10}{17}-3\frac{5}{12}$.

6.5 Geometry: Angles and Triangles

Measuring Angles

We begin the discussion of angles with two definitions using the undefined terms **point** and **line**.

Objective ①

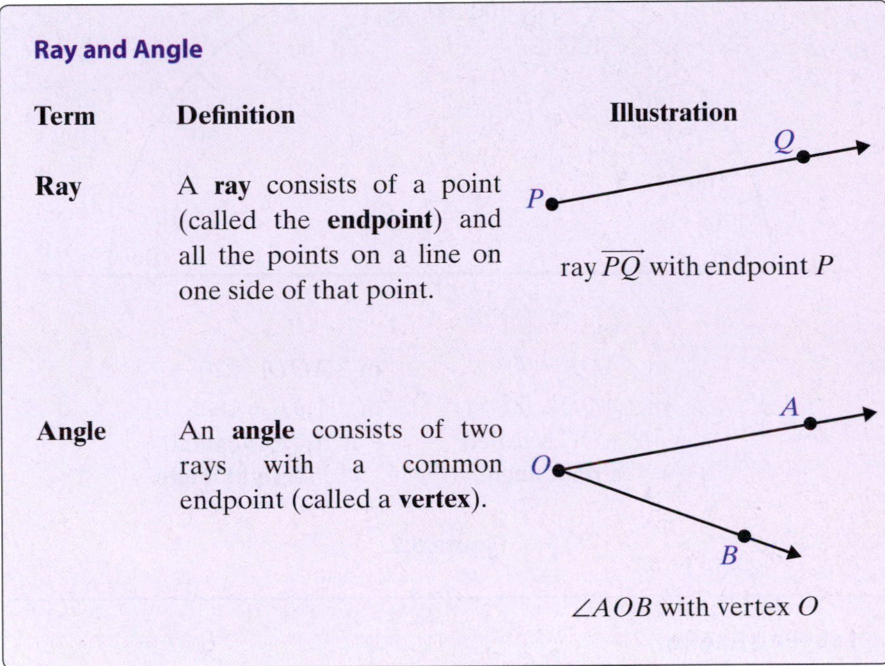

Ray and Angle

Term	Definition	Illustration
Ray	A **ray** consists of a point (called the **endpoint**) and all the points on a line on one side of that point.	ray $\overrightarrow{PQ}$ with endpoint P
Angle	An **angle** consists of two rays with a common endpoint (called a **vertex**).	$\angle AOB$ with vertex O

In an angle, the two rays are called the **sides** of the angle.

Every angle has a **measurement** or **measure** associated with it. Suppose that a circle is divided into 360 equal arcs. If two rays are drawn from the center of the circle through two consecutive points of division on the circle, then that angle is said to **measure one degree** (symbolized 1°). For example, in Figure 6.1, a device called a protractor shows that the measure of $\angle AOB$ is 60 degrees. (We write $m\angle AOB = 60°$.)

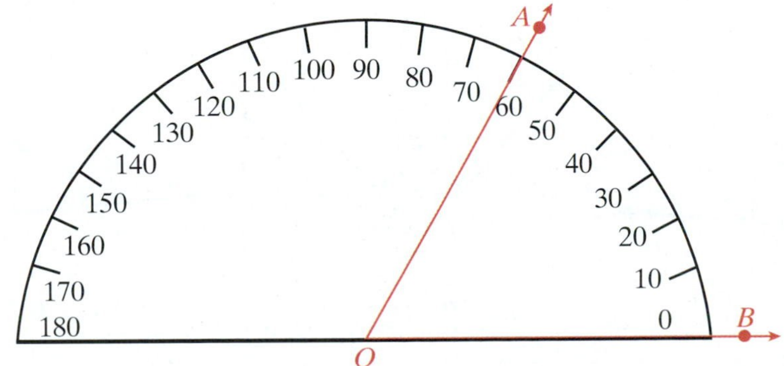

The protractor shows $m\angle AOB = 60°$

Figure 6.1

Objectives

① Know the definition of a **ray** and an **angle**.

② Learn how to classify an angle by its measure as **acute**, **right**, **obtuse**, or **straight**.

③ Know the meanings of the terms **complementary angles** and **supplementary angles**.

④ Learn how to classify a triangle by its sides: **scalene**, **isosceles**, or **equilateral**.

⑤ Learn how to classify a triangle by its angles: **acute**, **right**, or **obtuse**.

⑥ Know the following two important statements about any triangle:
- **a.** The sum of the measures of the angles is 180°.
- **b.** The sum of the lengths of any two sides must be greater than the length of the third side.

⑦ Know that in similar triangles:
- **a.** Corresponding angles have the same measure.
- **b.** Corresponding sides are proportional.

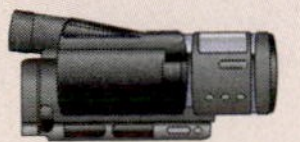

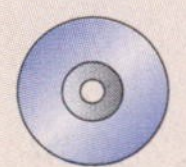

To measure an angle with a protractor, lay the bottom edge of the protractor along one side of the angle with the vertex at the marked center point. Then read the measure from the protractor where the other side of the angle crosses it. (See Figure 6.2.)

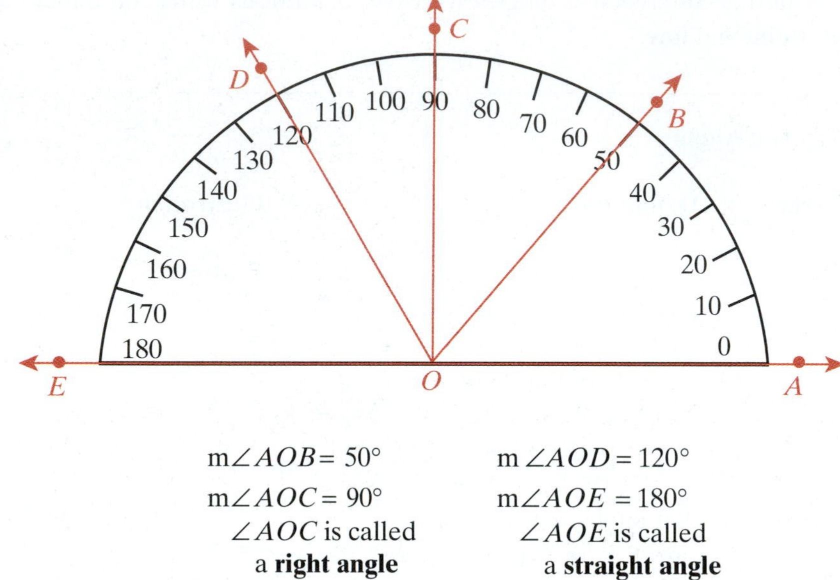

$m\angle AOB = 50°$ | $m\angle AOD = 120°$
$m\angle AOC = 90°$ | $m\angle AOE = 180°$
$\angle AOC$ is called a **right angle** | $\angle AOE$ is called a **straight angle**

Figure 6.2

Labeling Angles

Three common ways of labeling angles (Figure 6.3) are:

1. Using three capital letters with the vertex as the middle letter.
2. Using single numbers such as 1, 2, 3.
3. Using the single capital letter at the vertex when the meaning is clear.

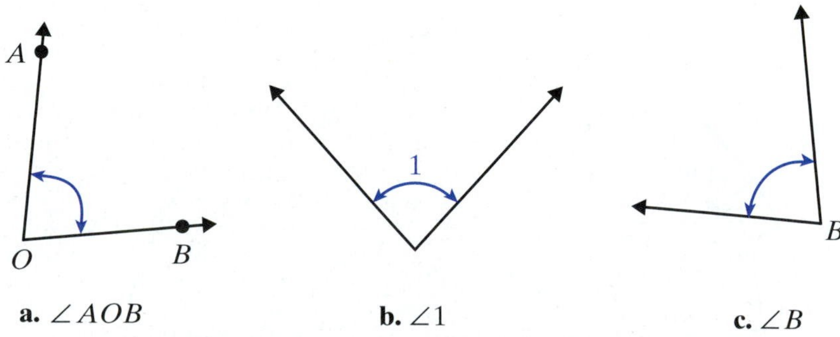

a. $\angle AOB$ **b.** $\angle 1$ **c.** $\angle B$

Figure 6.3: Three ways of labeling angles

Classifying Angles

Angles can be classified (or named) according to their measures.

> **Note:** We will use the two inequality symbols
>
> $<$ (read "is less than")
>
> $>$ (read "is greater than")
>
> to indicate the relative sizes of the measures of angles.

Objective 2

Angles Classified by Measure:

Name	**Measure**	**Illustration**
1. Acute	$0° < m\angle A < 90°$	$\angle A$ is an acute angle. 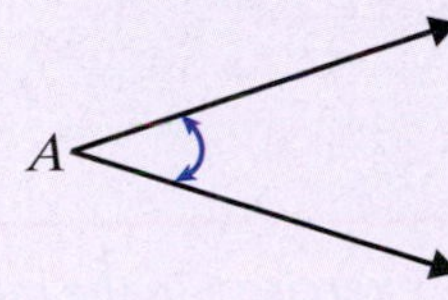
2. Right	$m\angle A = 90°$	$\angle A$ is a right angle. 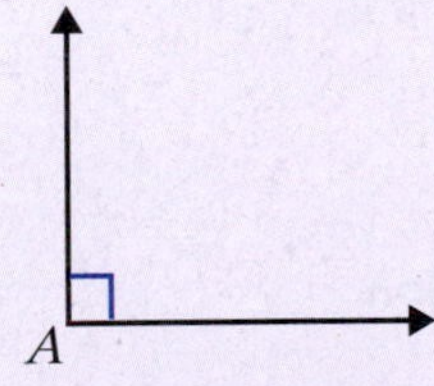
3. Obtuse	$90° < m\angle A < 180°$	$\angle A$ is an obtuse angle. 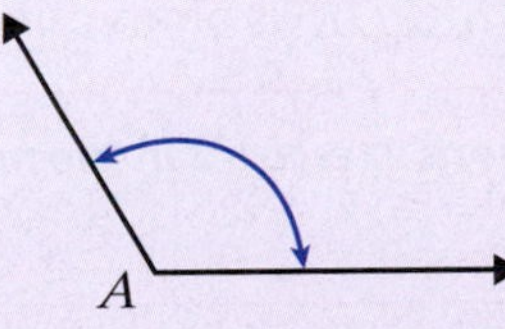
4. Straight	$m\angle A = 180°$	The rays are in opposite directions. $\angle A$ is a straight angle. 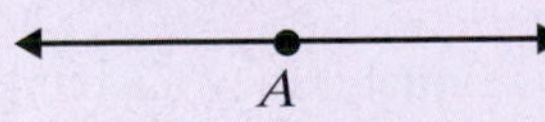

1. 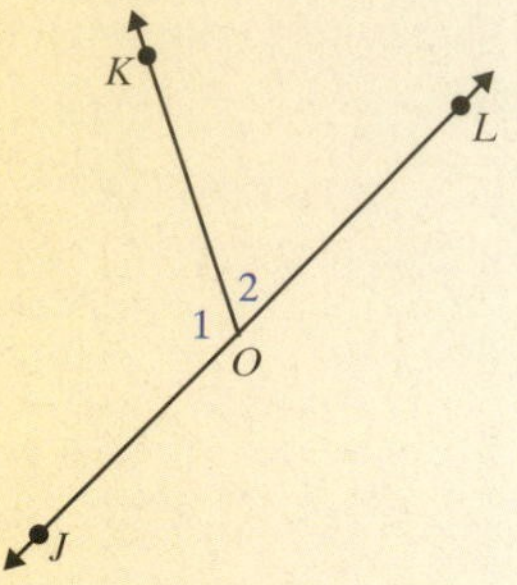

Check the measures of $\angle 1$ and $\angle 2$ above using a protractor.

2. Identify the following (from Examples 1 and 2) as acute, right, obtuse, or straight:
a. $\angle 3$
b. $\angle 4$
c. $\angle SOQ$

The following figure is used for Examples 1 and 2.

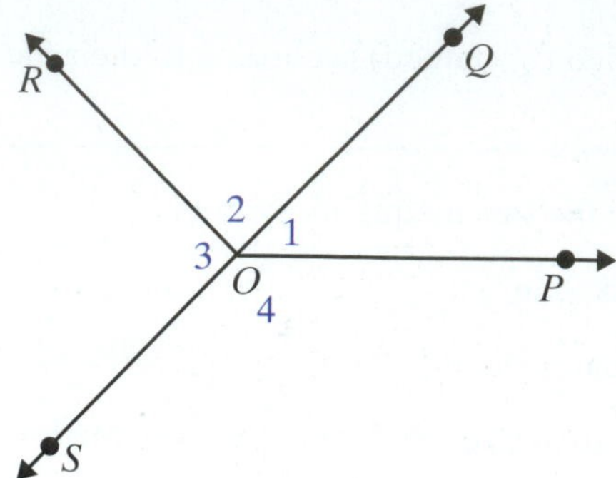

Example 1

Use a protractor to check the measures of the above angles.

Solution

a. $m\angle 1 = 45°$ **b.** $m\angle 2 = 90°$

c. $m\angle 3 = 90°$ **d.** $m\angle 4 = 135°$

Now work exercise 1 in the margin.

Example 2

Tell whether each of the following angles is acute, right, obtuse, or straight.

a. $\angle 1$ **b.** $\angle 2$ **c.** $\angle POR$

Solution

a. $\angle 1$ is acute since $0° < m\angle 1 < 90°$.

b. $\angle 2$ is a right angle since $m\angle 2 = 90°$.

c. $\angle POR$ is obtuse since $m\angle POR = 45° + 90° = 135° > 90°$.

Now work exercise 2 in the margin.

Objective 3

Two Angles are:

1. **Complementary** if the sum of their measures is 90°.
2. **Supplementary** if the sum of their measures is 180°.
3. **Equal** if they have the same measure.

Example 3

In the figure shown,

a. $\angle 1$ and $\angle 2$ are complementary since $m\angle 1 + m\angle 2 = 90°$;

b. $\angle COD$ and $\angle COA$ are supplementary since $m\angle COD + m\angle COA = 70° + 110° = 180°$;

c. $\angle AOD$ is a straight angle since $m\angle AOD = 180°$;

d. $\angle BOA$ and $\angle BOD$ are supplementary; and in this case $m\angle BOD = 90°$.

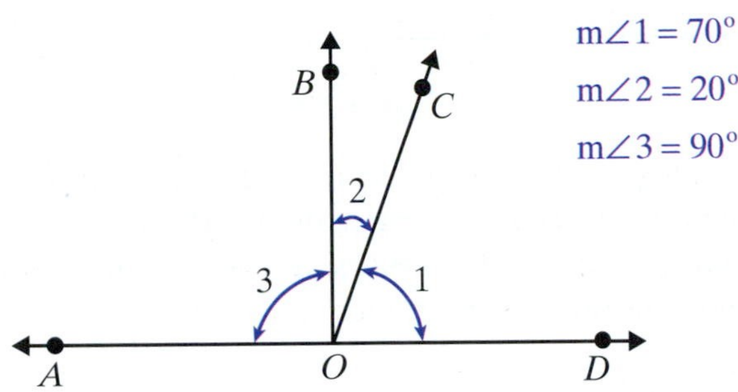

Now work exercise 3 in the margin.

3. In the figure shown in Example 3, are $\angle 2$ and $\angle 3$ supplementary, complementary, or neither? Explain your answer.

Example 4

In the figure below, $\overleftrightarrow{PS}$ is a straight line and $m\angle QOP = 30°$. Find the measures of

a. $\angle QOS$ and **b.** $\angle SOP$.

c. Are any pairs supplementary?

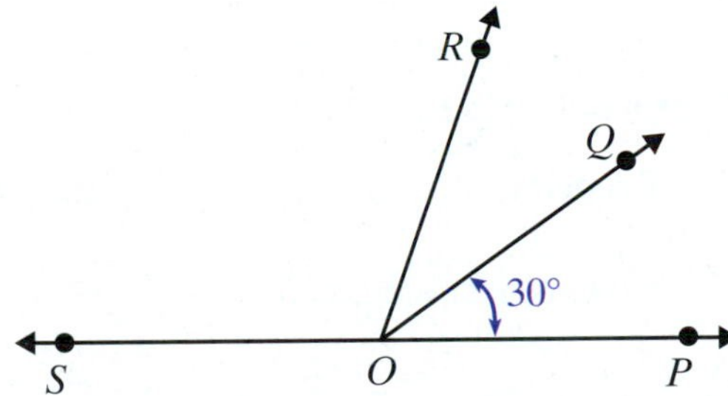

Solution

a. $m\angle QOS = 150°$.

b. $m\angle SOP = 180°$.

c. Yes, $\angle QOP$ and $\angle QOS$ are supplementary and $\angle ROP$ and $\angle ROS$ are supplementary.

Now work exercise 4 in the margin.

4. In the figure shown in Example 4, assume $m\angle ROQ = 40°$ and answer the following questions:

a. What is the measure of $\angle SOR$?

b. Are $\angle SOR$ and $\angle ROP$ equal?

Classifying Triangles

A **triangle** consists of the three line segments that join three points that do not lie on a straight line. The line segments are called the **sides** of the triangle, and the points are called the **vertices** of the triangle. If the points are labeled $A, B,$ and C, the triangle is symbolized $\triangle ABC$. (See Figure 6.4.)

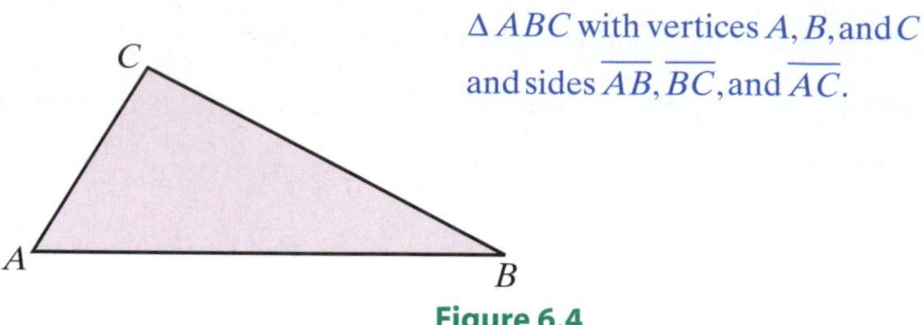

Figure 6.4

The sides of a triangle are said to determine three angles, and these angles are labeled by the vertices. Thus the angles of $\triangle ABC$ are $\angle A$, $\angle B$, and $\angle C$. (Since the definition of an angle involves rays, we can think of the sides of the triangles extended as rays that form these angles.)

Triangles are classified in two ways:

1. according to the lengths of their sides, and
2. according to the measures of their angles.

The corresponding names and properties are listed in the following tables.

> **Special Note:** The line segment with endpoints A and B is indicated by placing a bar over the letters, as in $\overline{AB}$. The length of the segment is indicated by writing only the letters, as in AB.

Objective 4

Triangles Classified by Sides:

Name	**Property**	**Example**
1. Scalene	No two sides are equal.	$\triangle ABC$ is scalene since no two sides are equal.
2. Isosceles	At least two sides are equal.	$\triangle PQR$ is isosceles since $PR = QR$.

3. Equilateral All three sides are equal.

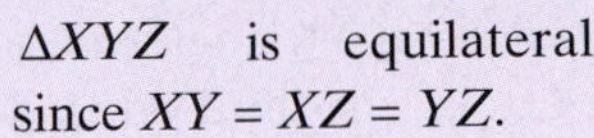

ΔXYZ is equilateral since $XY = XZ = YZ$.

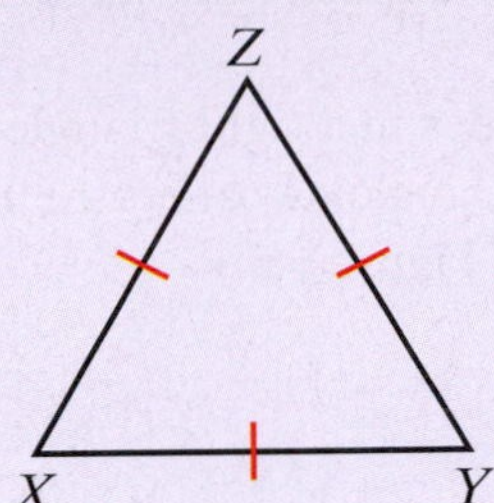

Triangles Classified by Angles:

Objective 5

Name	Property	Example
1. Acute	All three angles are acute.	$\angle A$, $\angle B$, and $\angle C$ are all acute so ΔABC is acute.

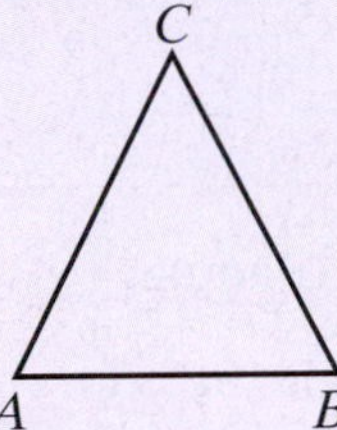

Name	Property	Example
2. Right	One angle is a right angle.	$m\angle P = 90°$ so ΔPQR is a right triangle.

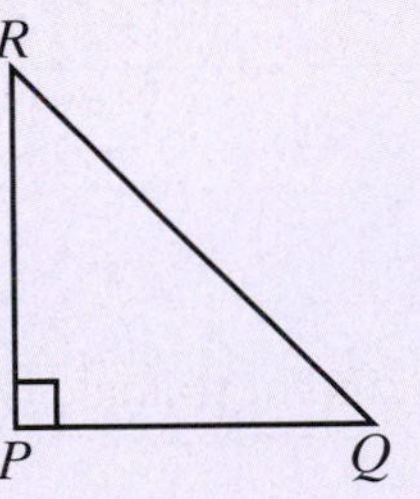

Name	Property	Example
3. Obtuse	One angle is obtuse.	$\angle X$ is obtuse so ΔXYZ is an obtuse triangle.

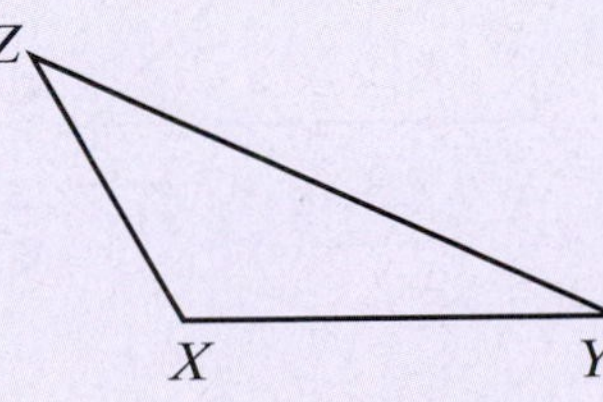

Every triangle is said to have six parts–namely, three angles and three sides. Two sides of a triangle are said to **include** the angle at their common endpoint or vertex. The third side is said to be **opposite** this angle.

As we know from Section 5.6, the sides in a right triangle have special names. The longest side, opposite the right angle, is called the **hypotenuse**, and the other two sides are called **legs**. (See Figure 6.5.)

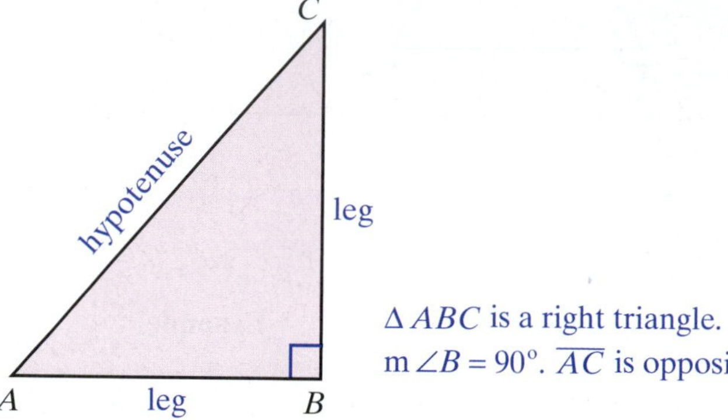

Figure 6.5

Objective 6

Two Important Statements about any Triangle:

1. The sum of the measures of the angles is 180°.
2. The sum of the lengths of any two sides must be greater than the length of the third side.

5. In ΔDEF, $DE = 5$, $EF = 7$, and $DF = 8$. What kind of triangle is ΔDEF?

Example 5

In ΔABC below, $AB = AC$. What kind of triangle is ΔABC?

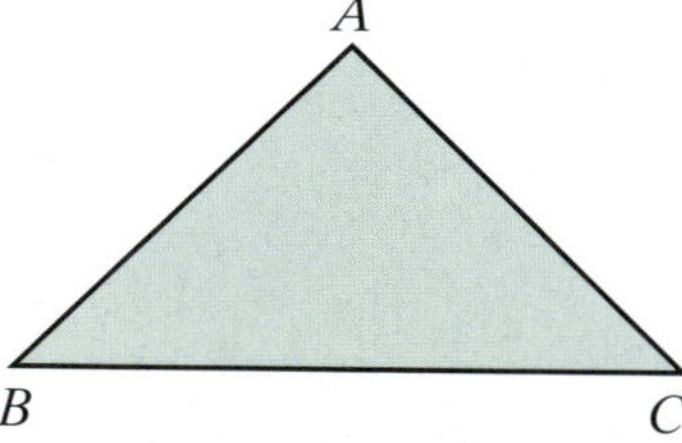

Solution

ΔABC is isosceles because two sides are equal.

Now work exercise 5 in the margin.

Example 6

Suppose the lengths of the sides of ΔPQR are as shown in the figure below. Is this possible?

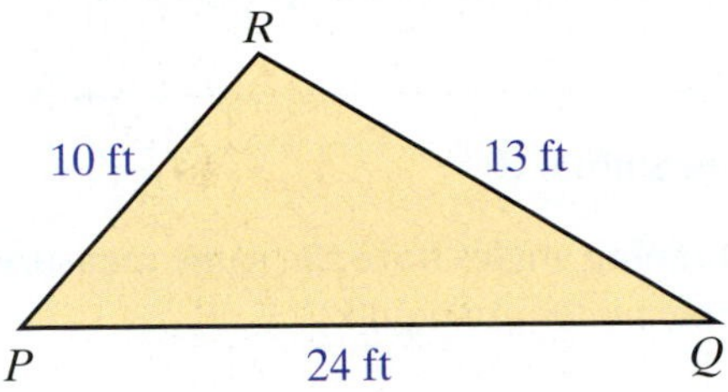

Solution

This is not possible because $PR + QR = 10 \text{ ft} + 13 \text{ ft} = 23 \text{ ft}$ and $PQ = 24 \text{ ft}$, which is longer than the sum of the other two sides. In a triangle, the sum of the lengths of any two sides must be greater than the length of the third side.

Now work exercise 6 in the margin.

6. Is it possible to have a triangle with sides of length 15 in., 37 in. and 51 in.?

Example 7

In ΔBOR below, $m\angle B = 50°$ and $m\angle O = 70°$.

a. What is $m\angle R$?

b. What kind of triangle is ΔBOR?

c. Which side is opposite $\angle R$?

d. Which sides include $\angle R$?

e. Is ΔBOR a right triangle? Why or why not?

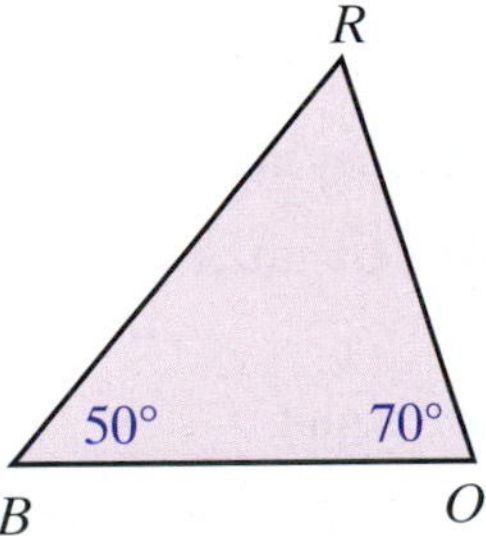

Solution

a. The sum of the measures of the angles must be 180°.
Since $50° + 70° = 120°$, $m\angle R = 180° - 120° = 60°$

b. ΔBOR is an acute triangle since all the angles are acute. Also, ΔBOR is scalene because no two sides are equal.

c. $\overline{BO}$ is opposite $\angle R$.

d. $\overline{RB}$ and $\overline{RO}$ include $\angle R$.

e. ΔBOR is not a right triangle because none of the angles is a right angle.

Now work exercise 7 in the margin.

7. Now assume that in ΔBOR, $m\angle B = 35°$ and $m\angle O = 55°$. Answer the questions **a.** - **e.** as given in Example 7.

Similar Triangles

Two triangles are said to be **similar triangles** if they have the same "shape." They may or may not have the same "size." More formally, two triangles are **similar** if they have the following two properties.

Objective 7

Two Triangles are Similar If:

1. The **corresponding angles have the same measure**. (We say that the corresponding angles are equal.)
2. The **corresponding sides are proportional**. (See Figure 6.6.)

In similar triangles, **corresponding sides** are those sides opposite the equal angles (angles with the same measure) in the respective triangles. (See Figure 6.6 and the following discussion.)

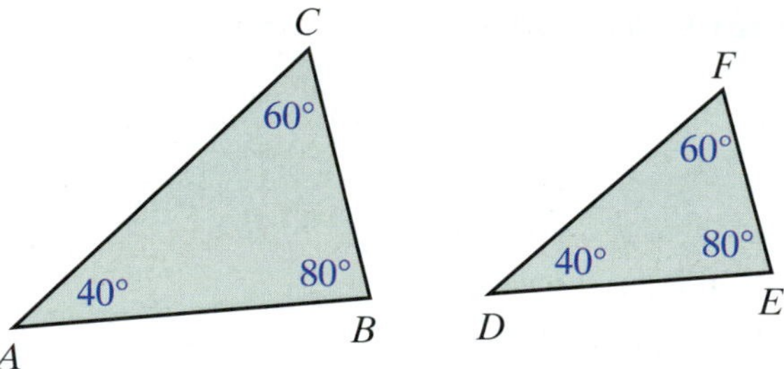

Figure 6.6

For the similar triangles in Figure 6.6, the following relationships are true.

$\angle A$ corresponds to $\angle D$; $m\angle A = m\angle D$.

$\angle B$ corresponds to $\angle E$; $m\angle B = m\angle E$.

$\angle C$ corresponds to $\angle F$; $m\angle C = m\angle F$.

The corresponding angles are equal.

$\overline{AB}$ corresponds to $\overline{DE}$.

$\overline{BC}$ corresponds to $\overline{EF}$.

$\overline{AC}$ corresponds to $\overline{DF}$.

$$\frac{AB}{DE} = \frac{BC}{EF} = \frac{AC}{DF}$$

The corresponding sides are proportional.

We write $\Delta ABC \sim \Delta DEF$. (~ is read "is similar to")

Note: The notation for similar triangles indicates the respective correspondences of angles and sides by the order in which the vertices of each triangle are identified. For example, we could have written $\Delta BCA \sim \Delta EFD$ or $\Delta CAB \sim \Delta FDE$. Be sure to follow this pattern when indicating similar triangles.

Example 8

Given the two triangles ΔABC and ΔAXY with $m\angle ABC = m\angle AXY = 90°$, as shown in the figure, determine whether or not $\Delta ABC \sim \Delta AXY$.

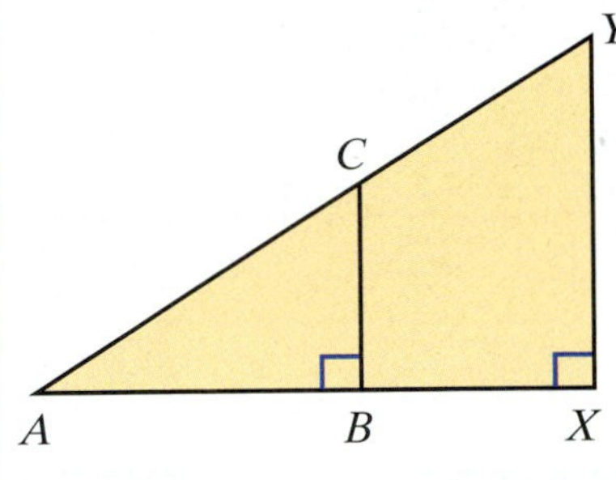

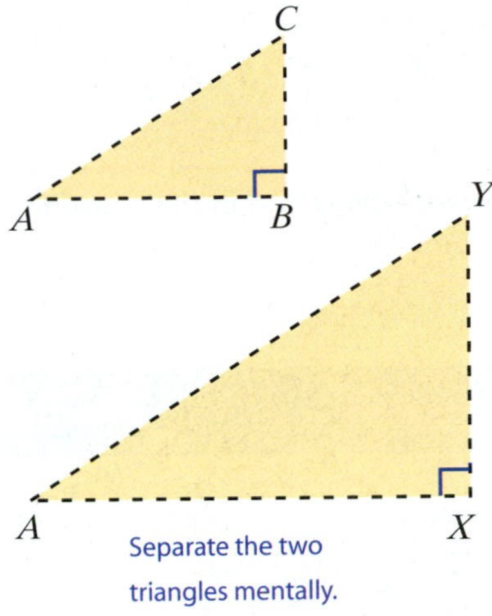

Separate the two triangles mentally.

Solution

We can show that the corresponding angles are equal as follows:

$m\angle CAB = m\angle YAX$ because they are the same angle.

$m\angle CBA = m\angle YXA$ because both are right angles (90°).

$m\angle BCA = m\angle XYA$ because the sum of the measures of the angles in each triangle must be 180°.

Therefore, the corresponding angles are equal, and the triangles are similar.

Now work exercise 8 in the margin.

Example 9

Refer to the figure used in Example 8. If $AB = 4$ centimeters, $BX = 2$ centimeters, and $BC = 3$ centimeters, find XY.

Solution

From Example 8 we know that the two triangles are similar: therefore, their corresponding sides are proportional. Since $\overline{AB}$ and $\overline{AX}$ are corresponding sides (they are opposite equal angles) and $\overline{BC}$ and $\overline{XY}$ are corresponding sides (they are opposite equal angles), the following proportion is true:

$$\frac{AB}{AX} = \frac{BC}{XY}$$

But, $AX = AB + BX = 4 + 2 = 6$ cm.

Continued on the next page...

8. Which of the following statements is correct in reference to the figure given in Example 8?

(a) $\Delta BAC \sim \Delta XYA$

(b) $\Delta CBA \sim \Delta YXA$

9. Using the information given in Examples 8 and 9, and given that $AC = 5$ cm, find CY.

Thus,

$$\frac{4\text{ cm}}{6\text{ cm}} = \frac{3\text{ cm}}{XY}$$

$$4 \cdot XY = 3 \cdot 6$$

$$\frac{\cancel{4} \cdot XY}{\cancel{4}} = \frac{18}{4}$$

$$XY = 4.5\text{ cm}$$

Now work exercise 9 in the margin.

Practice Problems

Identify each given angle as acute, right, obtuse, or straight.

1. $m\angle A = 83°$

2. $m\angle B = 149°$

3. $m\angle C = 180°$

4. $m\angle D = 7°$

Determine whether it is possible for a triangle to have the following side lengths. If such a triangle exists, classify it as scalene, isosceles, or equilateral.

5. 8 in., 13 in., 21 in.

6. 32 mm, 5 mm, 32 mm

Answers to Practice Problems:

1. acute

2. obtuse

3. straight

4. acute

5. does not exist

6. isosceles

Name ______________ Section __________ Date __________

Exercises 6.5

1. Assume that $\angle 1$ and $\angle 2$ are complementary.

   a. If $m\angle 1 = 15°$, what is $m\angle 2$?

   b. If $m\angle 1 = 3°$, what is $m\angle 2$?

   c. If $m\angle 1 = 45°$, what is $m\angle 2$?

   d. If $m\angle 1 = 75°$, what is $m\angle 2$?

2. Assume that $\angle 3$ and $\angle 4$ are supplementary.

   a. If $m\angle 3 = 45°$, what is $m\angle 4$?

   b. If $m\angle 3 = 90°$, what is $m\angle 4$?

   c. If $m\angle 3 = 110°$, what is $m\angle 4$?

   d. If $m\angle 3 = 135°$, what is $m\angle 4$?

3. **a.** What is the supplement of a right angle?

   b. What is the supplement of an obtuse angle?

   c. What is the supplement of an acute angle?

4. In the figure shown below, $\overleftrightarrow{DC}$ is a straight line and $m\angle BOA = 90°$.

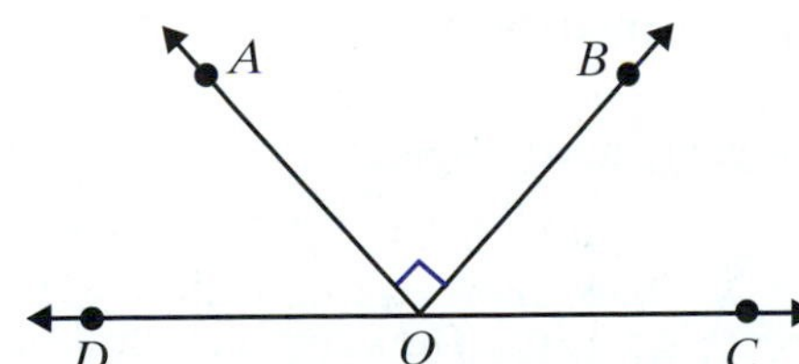

   a. What type of angle is $\angle AOC$?

   b. What type of angle is $\angle BOC$?

   c. What type of angle is $\angle BOA$?

   d. Name a pair of complementary angles.

   e. Name two pairs of supplementary angles.

ANSWERS

1. a. ______
b. ______
c. ______
d. ______

2. a. ______
b. ______
c. ______
d. ______

3. a. ______
b. ______
c. ______

4. a. ______
b. ______
c. ______
d. ______
e. ______

5. ____________

6. ____________

7. ____________

8. ____________

9. ____________

10. ____________

11. ____________

12. a. [Respond below exercise.] ____________

b. [Respond below exercise.] ____________

5. An **angle bisector** is a ray that divides an angle into two angles with equal measures. If $\overrightarrow{OX}$ bisects $\angle COD$ and $m\angle COD = 50°$, what is the measure of each of the equal angles formed?

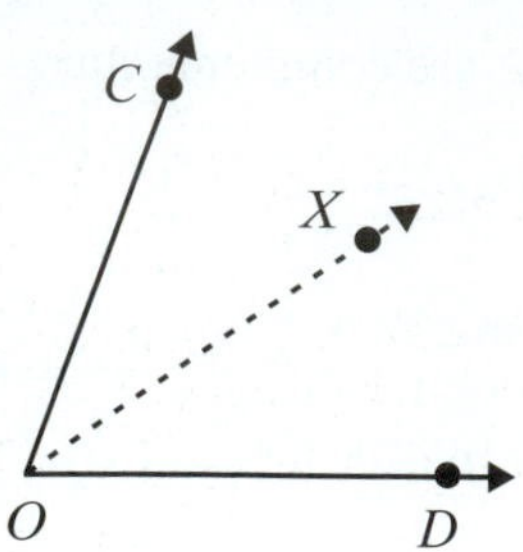

In the figure shown below, $m\angle AOB = 30°$, *and* $m\angle BOC = 80°$ *and* $\overrightarrow{OX}$ *and* $\overrightarrow{OY}$ *are angle bisectors. Find the measures of the following angles. See Example 4.*

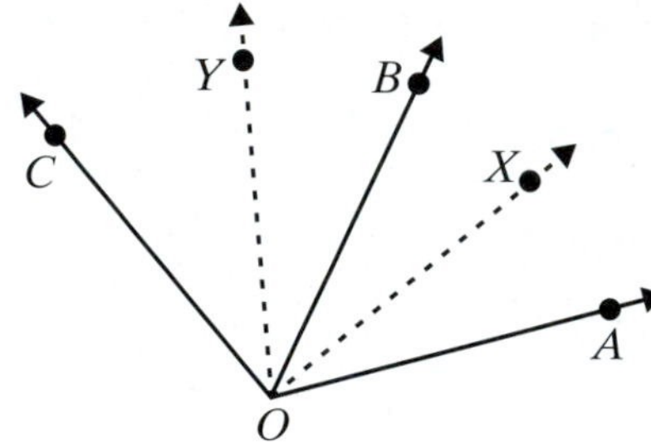

6. $\angle AOX$

7. $\angle BOX$

8. $\angle COX$

9. $\angle BOY$

10. $\angle AOY$

11. $\angle AOC$

12. In the figure shown below:

   a. Name all the pairs of supplementary angles.

   b. Name all the pairs of complementary angles.

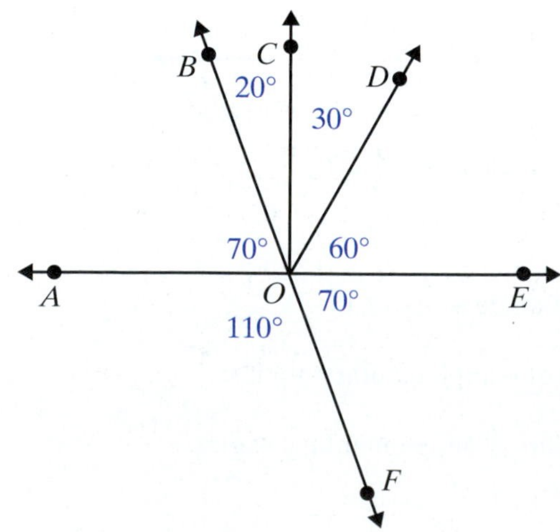

Name ______________ Section ______________ Date ______________

Use a protractor to measure all the angles in each figure. Each line segment may be extended as a ray to form the side of an angle. See Example 1.

13.

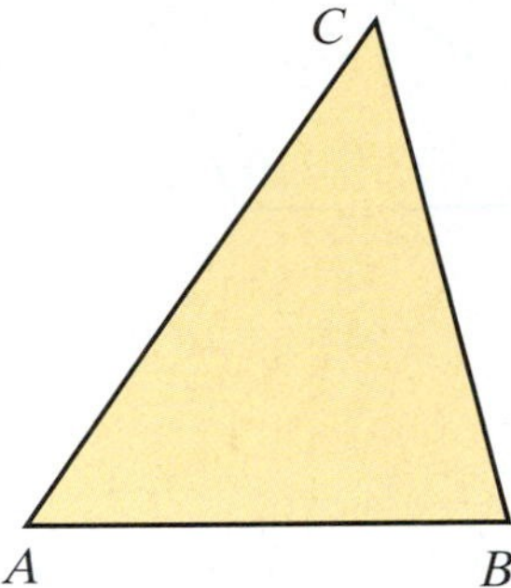

14.

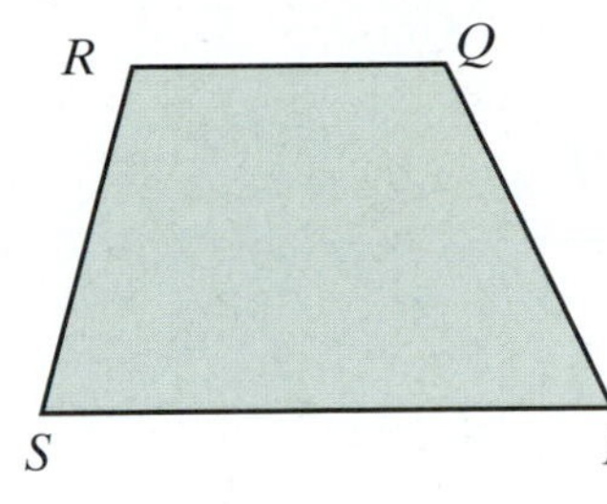

15.

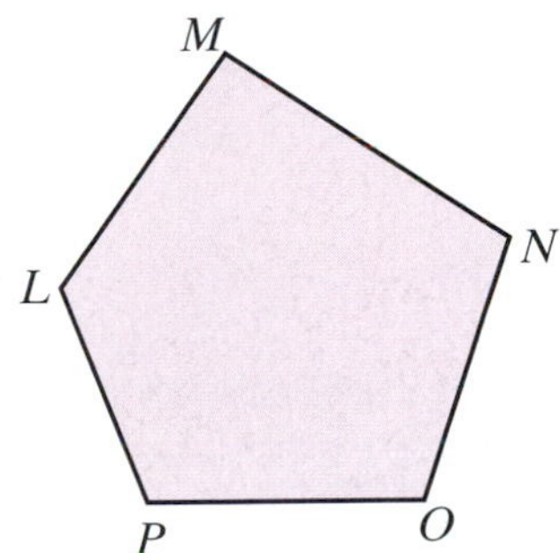

16.

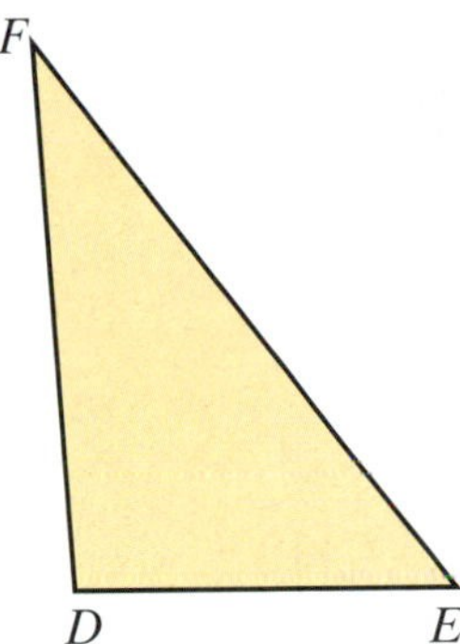

17. Name the type of angle formed by the hands on a clock
a. at six o'clock; **b.** at three o'clock; **c.** at five o'clock

18. What is the measure of each angle formed by the hands of the clock in Exercise 17?

Name each of the following triangles in the most precise way possible, given the indicated measures of angles and lengths of sides. See Examples 5 through 7.

19.

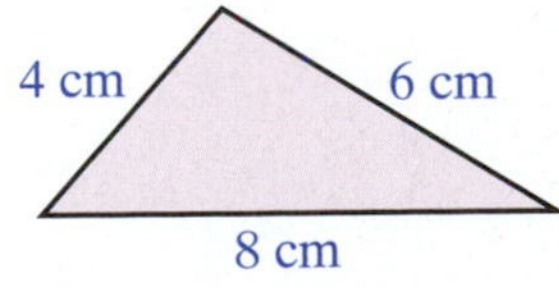

20.

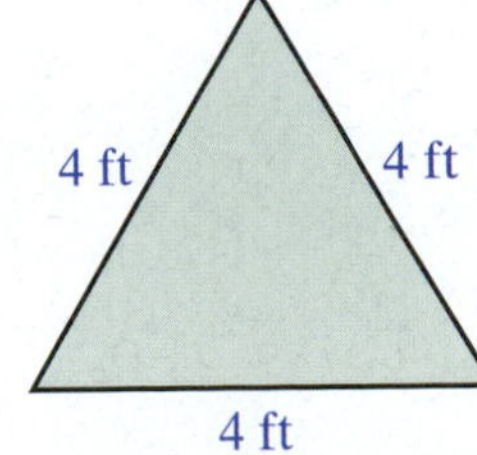

ANSWERS

13. [Respond below exercise.] ______

14. [Respond below exercise.] ______

15. [Respond below exercise.] ______

16. [Respond below exercise.] ______

17. a. ______

b. ______

c. ______

18. a. ______

b. ______

c. ______

19. ______

20. ______

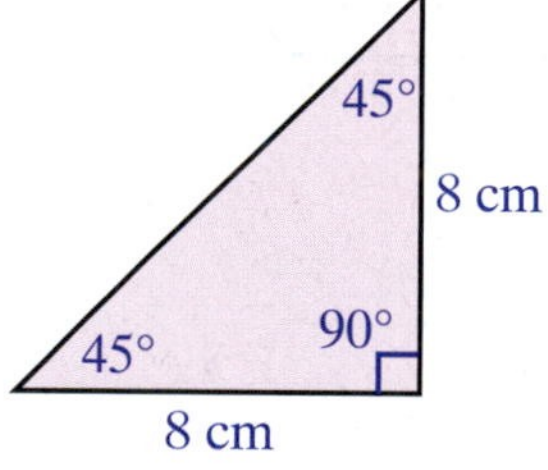

21. ____________

22. ____________

23. ____________

24. ____________

25. ____________

26. ____________

27. ____________

28. ____________

21.

90°

22.

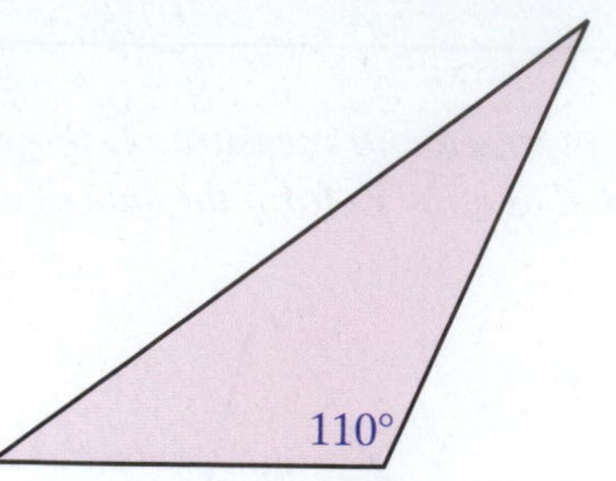

23.

3 in.

6 in.

6 in.

24.

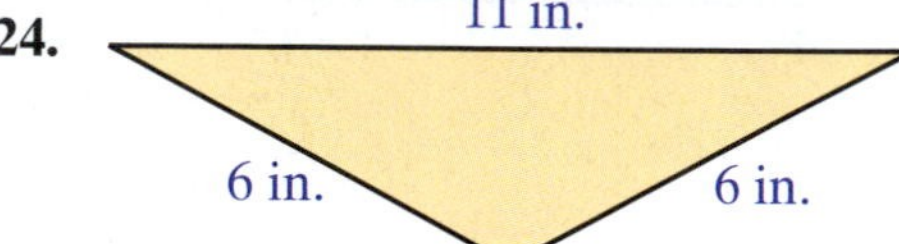

25.

45°

8 cm

45°

90°

8 cm

26.

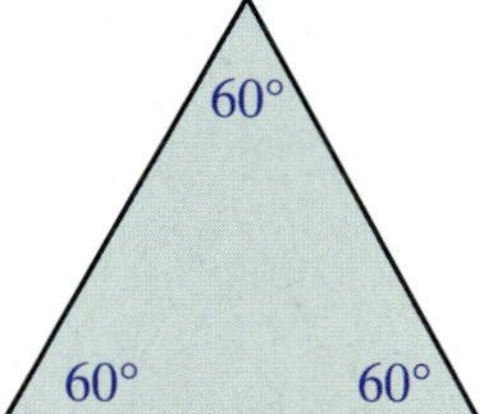

27.

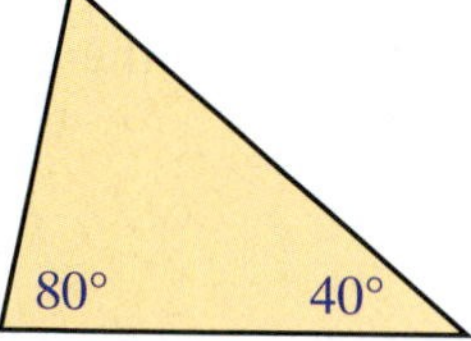

28.

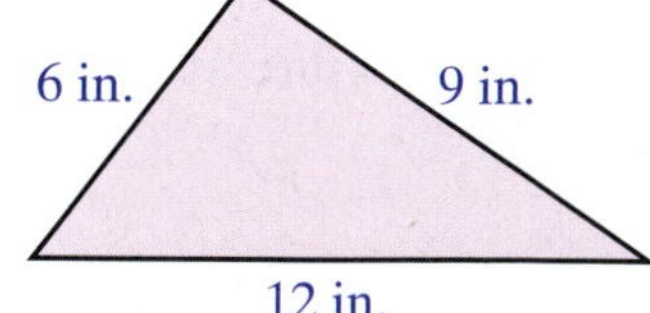

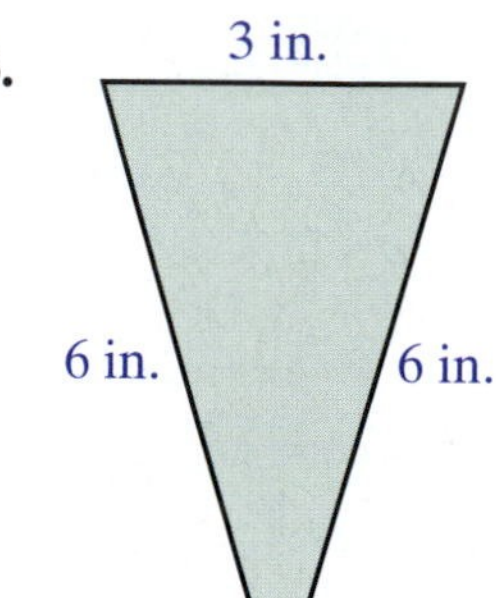

Name ______________________ Section ____________ Date __________

ANSWERS

29.

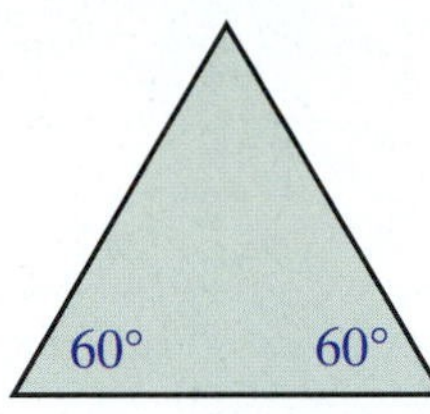

30.

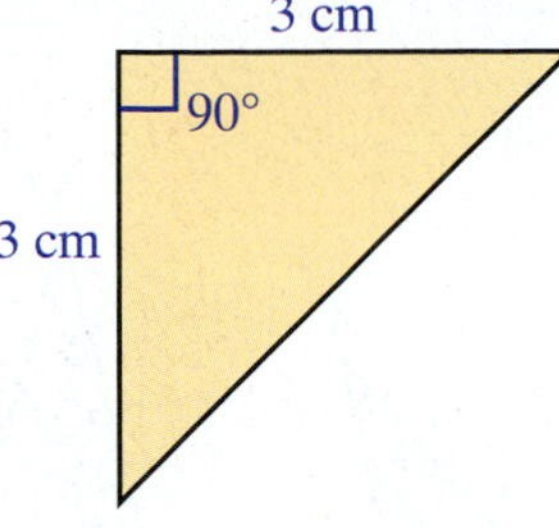

29. ____________

30. ____________

In Exercises 31 – 34, assume that the given triangles are similar and find the values for x and y. See Examples 8 and 9.

31.

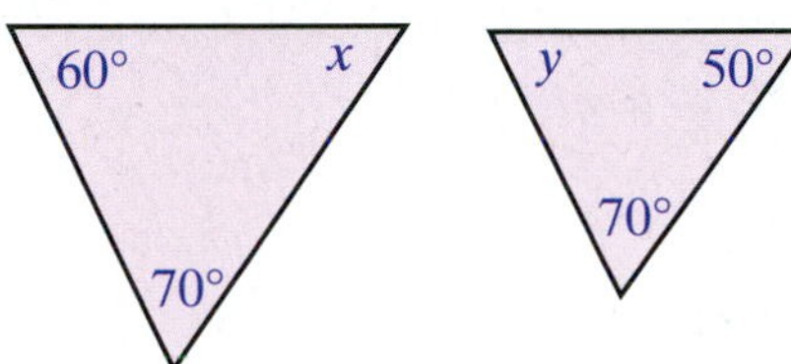

31. ____________

32.

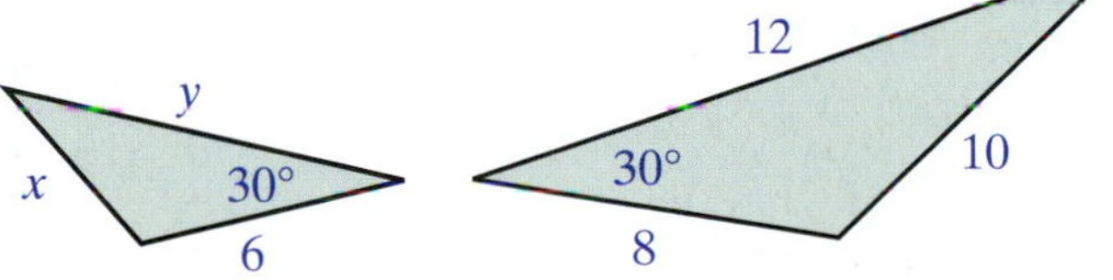

32. ____________

33.

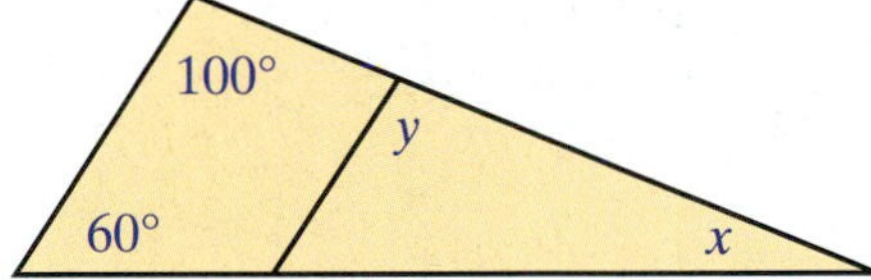

33. ____________

34.

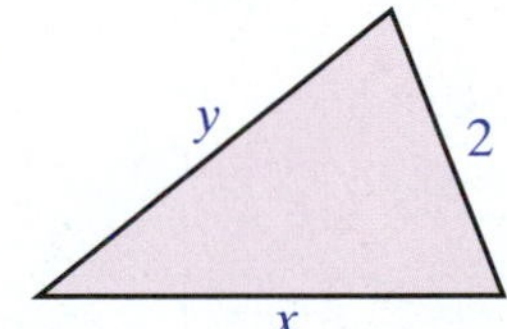

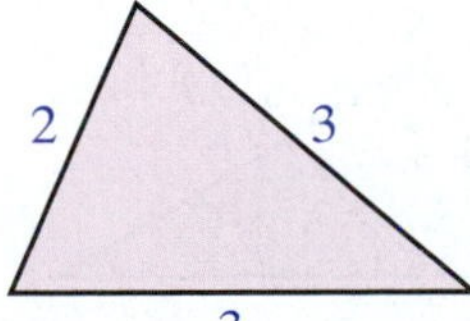

34. ____________

In Exercises 35 – 38, determine whether each pair of triangles is similar. If the triangles are similar, explain why and indicate the similarity by using the ~ symbol. See Example 8.

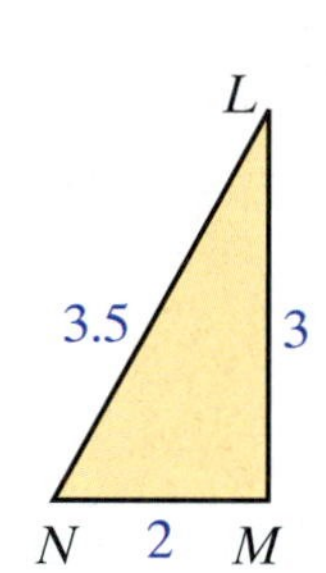

35. ______________

35.

A
4 3
B 2 C

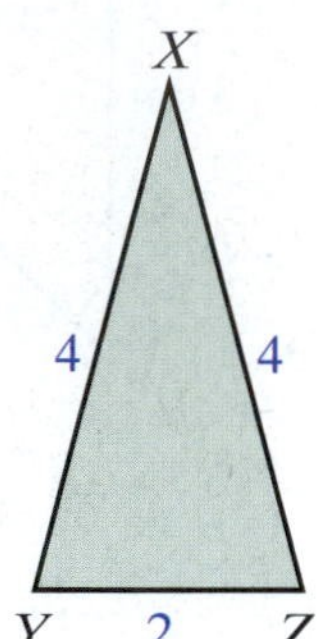

36. ______________

36.

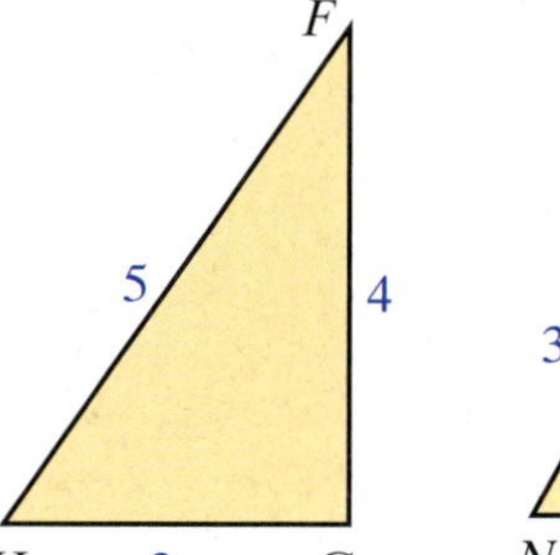

37. ______________

37.

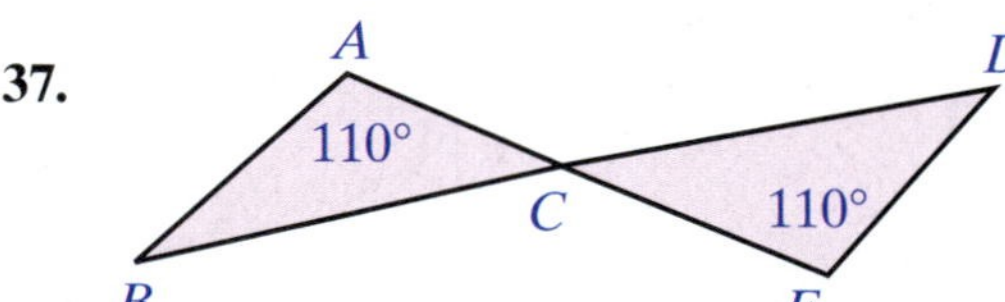

38. ______________

38.

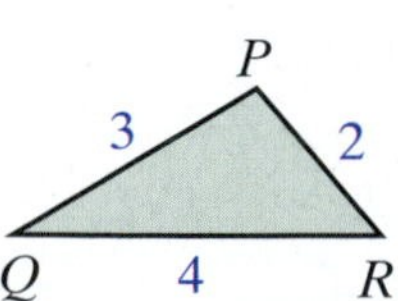

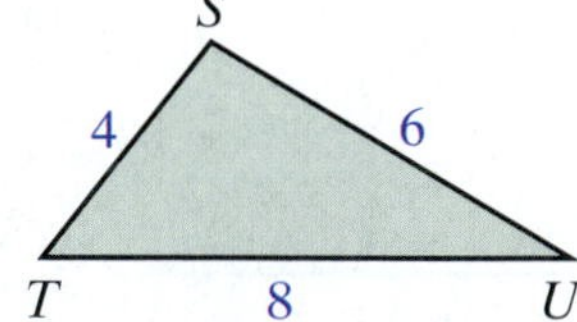

Name ______________________ Section ____________ Date __________

ANSWERS

39. In ΔXYZ and ΔUVW below, $m\angle Z = 30°$ and $m\angle W = 30°$.

a. If both triangles are isosceles, what are the measures of the other four angles? (In an isosceles triangle, the angles opposite the equal sides must be equal.)

b. Are the triangles similar? Explain.

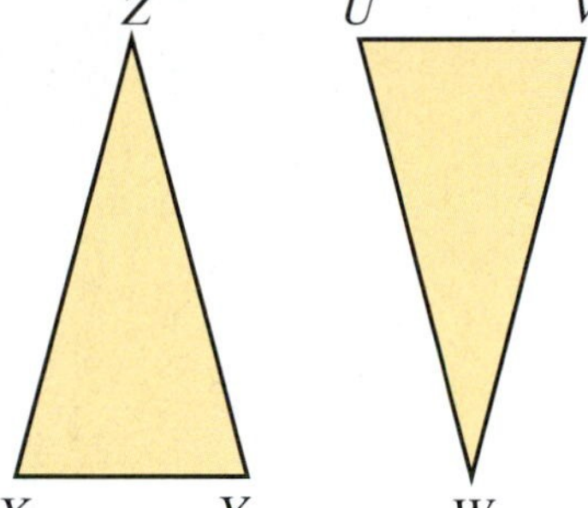

39. a. ______________

b. ______________

40. a. ______________

b. ______________

c. ______________

41. a. ______________

b. ______________

Writing and Thinking about Mathematics

40. a. Is it possible to form a triangle, ΔSTV, if $ST = 12$ centimeters, $TV = 9$ centimeters, and $SV = 15$ centimeters? Explain your reasoning.

b. If so, what kind of triangle is ΔSTV?

c. Use a ruler to draw this triangle, if possible.

41. a. Is it possible to form a triangle, ΔABC if $AB = 6$ inches, $BC = 8$ inches, and $CA = 14$ inches? Explain your reasoning.

b. Use a ruler to draw this triangle, if possible.

Chapter 6 Index of Key Terms and Ideas

Section 6.1 Ratios and Price per Unit

A **ratio** is a comparison of two quantities by division. The ratio of a to b can be written as

$$\frac{a}{b} \quad \text{or} \quad a:b \quad \text{or} \quad a \text{ to } b.$$

To calculate the **price per unit** (or unit price):

1. Set up a ratio (usually in fraction form) of price to units.

2. Divide the price by the number of units.

Section 6.2 Proportions

A **proportion** is a statement that two ratios are equal. In symbols,

$\frac{a}{b} = \frac{c}{d}$ is a proportion.

A proportion has four **terms:**

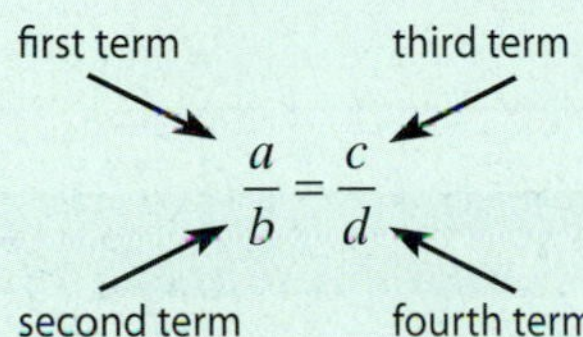

The first and fourth terms (a and d) are called the **extremes**.

The second and third terms (b and c) are called the **means**.

In a true proportion, the product of the extremes is equal to the product of the means.

Section 6.3 Finding the Unknown Term in a Proportion

A **variable** is a symbol (generally a letter of the alphabet) that is used to represent an unknown number or any one of several numbers. The set of possible values for a variable is called its **replacement set**.

Continued on next page...

Section 6.3 Finding the Unknown Term in a Proportion (continued)

To Find the Unknown Term in a Proportion page 507

1. In the proportion, the unknown term is represented with a variable (some letter such as x, y, w, A, B, etc.).
2. Write an equation that sets the product of the extremes equal to the product of the means.
3. Divide both sides of the equation by the number multiplying the variable. (This number is called the **coefficient** of the variable.)

The resulting equation will have a coefficient of 1 for the variable and will give the missing value for the unknown term in the proportion.

Section 6.4 Problem Solving with Proportions

To Solve a Word Problem Using Proportions page 517

1. Identify the unknown quantity and use a variable to represent this quantity.
2. Set up a proportion in which the units are compared as in Pattern A or Pattern B.

   Pattern A Each ratio has different units, but they are in the same order.

   Pattern B Each ratio has the same units, the numerators correspond, and the denominators correspond.
3. Solve the proportion.

Section 6.5 Geometry: Angles and Triangles

Rays page 529

A **ray** consists of a point (called the **endpoint**) and all the points on a line on one side of that point.

Angles page 529

An **angle** consists of two rays with a common endpoint (called a **vertex**).

Angles Classified by Measure page 531

1. If $\angle A$ is an **acute** angle, then $0° < m\angle A < 90°$.
2. If $\angle A$ is a **right** angle, then $m\angle A = 90°$.
3. If $\angle A$ is an **obtuse** angle, then $90° < m\angle A < 180°$.
4. If $\angle A$ is a **straight** angle, then $m\angle A = 180°$.

More Classification of Angles page 532

Two angles are:

1. **Complementary** if the sum of their measures is 90°.
2. **Supplementary** if the sum of their measures is 180°.
3. **Equal** if they have the same measure.

Triangles page 534

A **triangle** consists of the three line segments that join three points that do not lie on a straight line. The line segments are called the **sides** of the triangle, and the points are called the **vertices** of the triangle.

Continued on next page...

Section 6.5 Geometry: Angles and Triangles (continued)

Triangles Classified by Sides page 534–535

1. **Scalene** - No two sides are equal.
2. **Isosceles** - At least two sides are equal.
3. **Equilateral** - All three sides are equal.

Triangles Classified by Angles page 535

1. **Acute** - All three angles are acute.
2. **Right** - One angle is a right angle.
3. **Obtuse** - One angle is obtuse.

Important Characteristics of a Triangle page 536

1. The sum of the measures of the angles is 180°.
2. The sum of the lengths of any two sides must be greater than the length of the third side.

Two Triangles are Similar if: page 538

1. The **corresponding angles have the same measure**. (We say that the corresponding angles are equal.)
2. Their **corresponding sides are proportional**.

Name ____________ Section ____________ Date ____________

ANSWERS

Chapter 6 Review Questions

Write the following comparisons as ratios. Reduce to lowest terms and use common units whenever possible.

1. 2 dimes to 5 nickels

2. 20 inches to 1 yard

3. 10 centimeters to 2 meters

4. 51 miles to 3 gallons of gas

5. 18 hours to 2 days

6. 26 girls to 39 boys

Find the unit prices and tell which is the better buy.

7. Packaged cheese: 6 oz at $1.29
"Deli" cheese: 1 lb (16 oz) at $3.69

8. 1 qt (32 fl oz) milk at $1.99
1 gal (128 fl oz) milk at $3.29

9. 1 lb loaf of bread at $1.20
$1\frac{1}{2}$ lb loaf of bread at $1.65

10. 6 oz bologna at $1.99
8 oz bologna at $2.49

1. ____________
2. ____________
3. ____________
4. ____________
5. ____________
6. ____________
7. ____________
8. ____________
9. ____________
10. ____________

11. ______

12. ______

13. ______

14. ______

15. ______

16. ______

17. ______

18. ______

19. ______

20. ______

21. ______

Name the means and the extremes for each proportion.

11. $\frac{4}{5}=\frac{16}{20}$ **12.** $\frac{\frac{1}{6}}{3}=\frac{\frac{1}{9}}{2}$ **13.** $3:7=0.75:1.75$

Determine whether the following proportions are true or false.

14. $\frac{3}{5}=\frac{9}{15}$ **15.** $\frac{15}{20}=\frac{18}{24}$ **16.** $\frac{6.5}{14}=\frac{8}{16}$

Solve the following proportions. Reduce all fractions.

17. $\frac{10}{12}=\frac{x}{6}$ **18.** $\frac{1.7}{5.1}=\frac{100}{y}$

19. $\frac{7\frac{1}{2}}{3\frac{1}{3}}=\frac{w}{2\frac{1}{4}}$ **20.** $a:7=\frac{1}{3}:5$

21. Fuel economy. A motorcycle averages 42.8 miles per gallon of gas. How many miles can the motorcycle travel on 3.5 gallons of gas?

Name ______________________ Section ____________ Date __________

ANSWERS

22. Cooking. If 2 cups of sugar are needed to make 24 cookies, how many cups are needed to make 32 cookies?

22. ____________

23. Maps. On a certain map, 1 inch represents 35.5 miles. What distance is represented by 4.7 inches?

23. ____________

24. Salary. A part-time clerk earned \$420 the first month on a new job. This was 0.6 times what he had anticipated. How much had he anticipated making?

24. ____________

25. Manufacturing. If a machine produces 5000 safety pins in 2 hours, how many will it produce in two 8-hour days?

25. ____________

26. Acceleration/Deceleration. An automobile was slowing down at the rate of 5 miles per hour (mph) for every 3 seconds. If the automobile was going 65 mph when it began to slow down, how fast was it going at the end of 12 seconds?

26. ____________

27. Homebuilding. An architect draws house plans using a scale of $\frac{3}{4}$ inch to represent 10 feet. How many feet are represented by 2 inches?

27. ____________

28. ______

29. ______

30. ______

31. a. ______

b. ______

c. ______

d. ______

32. ______

28. Travel. If you can drive 200 miles in $4\frac{1}{2}$ hours, how far could you drive (at the same rate) in 6 hours?

29. Travel. The ratio of kilometers to miles is about 8 to 5. Find the rate in kilometers per hour that is equivalent to 40 miles per hour.

30. Demographics. In a certain hospital, 55 out of every 100 children born are boys. In a year when 1040 children are born in the hospital, how many are girls?

31. Name the type of each of the following triangles based on the measures and lengths shown.

a.

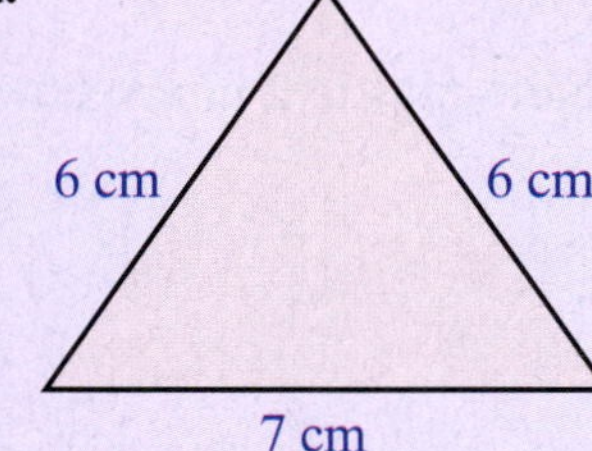

b.

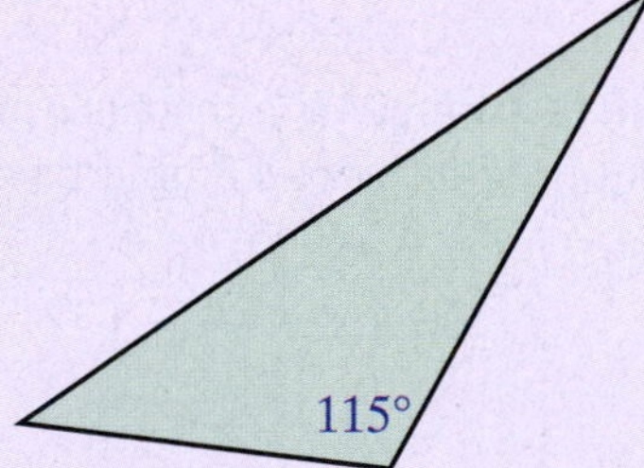

c.

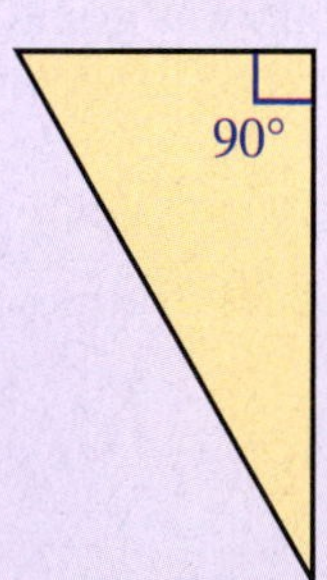

d.

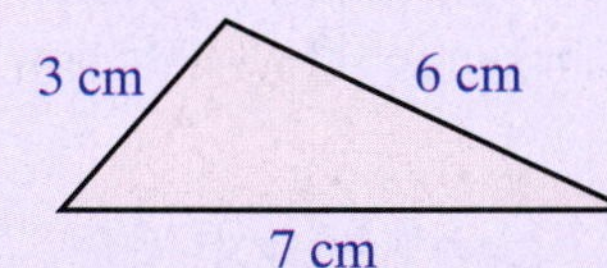

32. Given that $\triangle ABC \sim \triangle ADE$, find x and y.

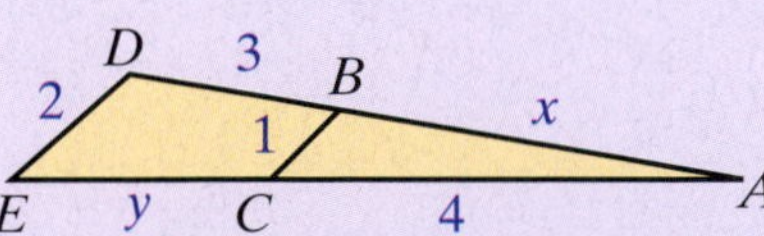

Name ______________________________ Section ______________ Date ___________

ANSWERS

1. ______________

2. ______________

3. ______________

4. ______________

5. ______________

6. ______________

7. ______________

8. ______________

9. ______________

10. ______________

Chapter 6 Test

1. The ratio 7 to 6 can be expressed as the fraction __________.

2. In the proportion $\frac{3}{4} = \frac{75}{100}$, 3 and 100 are called the ____________.

3. Is the proportion $\frac{4}{6} = \frac{9}{14}$ true or false? Give a reason for your answer.

Write the following comparisons as ratios reduced to lowest terms. Use common units whenever possible.

4. 3 weeks to 35 days

5. 6 nickels to 3 quarters

6. 220 miles to 4 hours

7. Find the unit prices (to the nearest tenth of a cent) and tell which is the better buy for a jar of jelly:
9.5 oz at \$2.69 or 16 oz at \$3.99.

Solve the following proportions.

8. $\frac{9}{17} = \frac{x}{51}$

9. $\frac{3}{5} = \frac{10}{y}$

10. $\frac{50}{x} = \frac{0.5}{0.75}$

11. ____________

12. ____________

13. ____________

14. ____________

15. ____________

16. ____________

17. ____________

18. ____________

19. ____________

11. $\frac{2.25}{y} = \frac{1.5}{13}$

12. $\frac{\frac{3}{8}}{x} = \frac{9}{20}$

13. $\frac{\frac{1}{3}}{x} = \frac{5}{\frac{1}{6}}$

14. $9:4::y:2$

15. 9 is to 4 as x is to 16

16. Fuel economy. How far (to the nearest tenth of a mile) can you drive on 5 gallons of gas if you can drive 120 miles on 3.5 gallons of gas?

17. Maps. On a certain map, 2 inches represents 15 miles. How far apart are two towns that are $3\frac{1}{5}$ inches apart on the map?

18. Tires. If you can buy 4 tires for \$236, what will be the cost of 5 of the same type of tires?

19. Travel. If you drive 165 miles in $3\frac{2}{3}$ hours, what is your average speed in miles per hour?

Name ______________________ Section ____________ Date __________

ANSWERS

20. Painting. A house painter knows that he can paint 3 houses every five weeks. At this rate, how long will it take him to paint 15 houses?

20. ______________

21. Weight. If 2 units of water weigh 124.8 pounds, what is the weight of 5 of these units of water?

21. ______________

22. Wholesale profit. A manufacturing company expects to make a profit of $3 on a product that it sells for $8. How much profit does the company expect to make on a similar product that it sells for $20?

22. ______________

23. In the figure shown below, $\overleftrightarrow{AD}$ and $\overleftrightarrow{BE}$ are straight lines.

a. What type of angle is $\angle BOC$?

b. What type of angle is $\angle AOE$?

c. Name two pairs of supplementary angles.

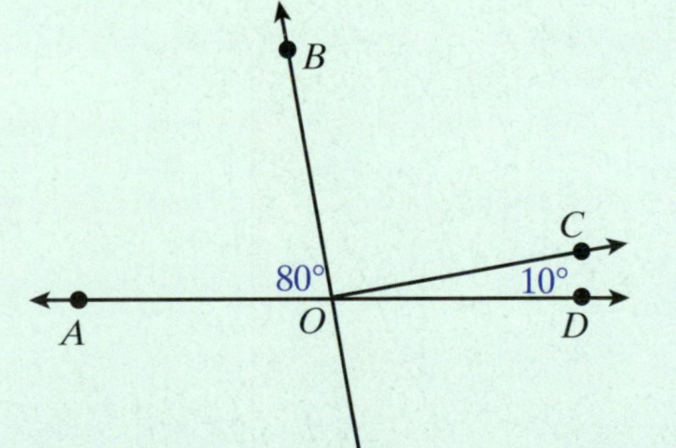

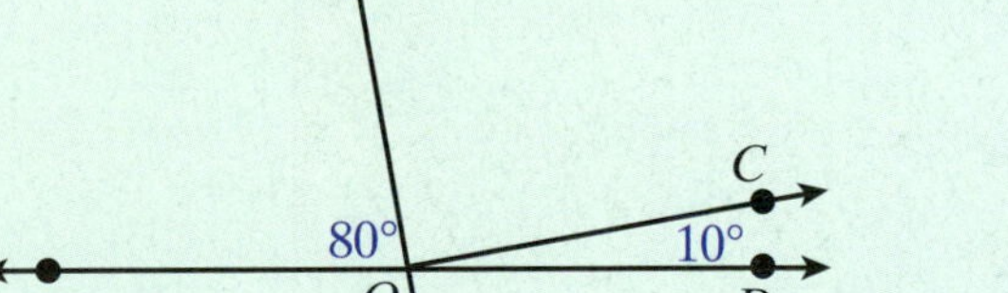

23. a. ______________

b. ______________

c. ______________

24. a. ______________

b. ______________

25. a. ______________

b. ______________

c. ______________

26. ______________

24. **a.** If $\angle 1$ and $\angle 2$ are complementary and $m\angle 1 = 35°$, what is $m\angle 2$?
b. If $\angle 3$ and $\angle 4$ are supplementary and $m\angle 3 = 15°$, what is $m\angle 4$?

25. **a.** Find the measure of $\angle R$.
b. What kind of triangle is ΔRST?
c. Which side is opposite $\angle S$?

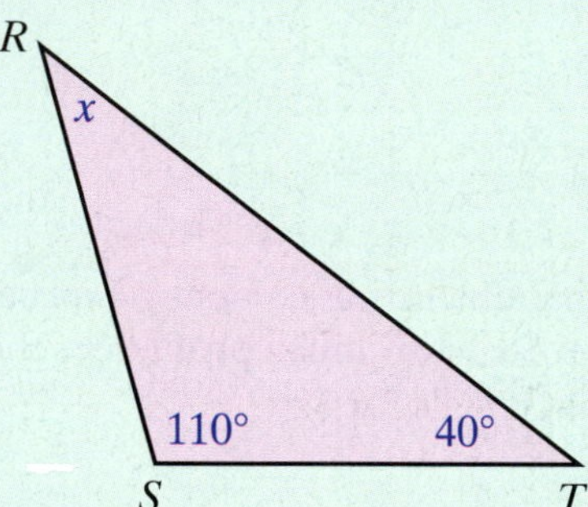

26. Find the value of x given that $\Delta ABC \sim \Delta ADE$.

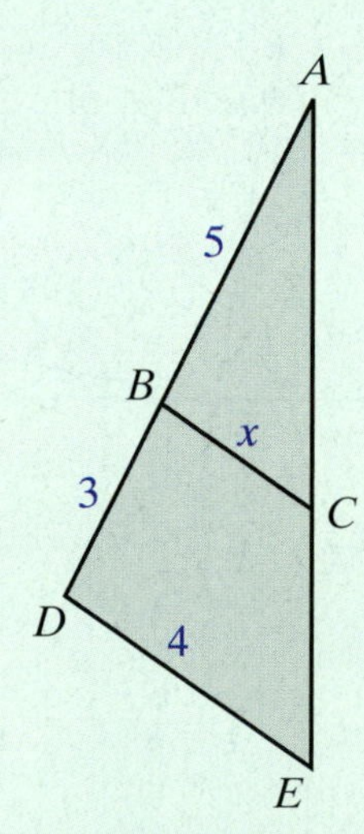

Name ______________ Section ______________ Date ______________

ANSWERS

Cumulative Review: Chapters 1 – 6

1. Write five thousand seven hundred forty in the form of a decimal number.

2. Write the decimal number 3.075 in words.

3. Find the prime factorization of 4510.

Evaluate each of the following expressions.

4. $3 \cdot 5^2 + 20 \div 5$

5. $(80+10) \div 5 \cdot 2 - 14 + 2^2$

6. Find the LCM for 30, 35, and 42.

7. Find the average of 50, 54, 60, and 76.

Perform the indicated operations and reduce all answers.

8. $1 - \frac{15}{16}$

9. $4\frac{5}{8} \div 2$

10. $\left(1\frac{2}{3}\right)\left(2\frac{4}{5}\right)\left(3\frac{1}{7}\right)$

1. ______________
2. ______________
3. ______________
4. ______________
5. ______________
6. ______________
7. ______________
8. ______________
9. ______________
10. ______________

11. ______

12. ______

13. ______

14. ______

15. ______

16. a. ______

b. ______

17. ______

18. ______

19. ______

20. ______

11. $\left(\frac{1}{5}+\frac{1}{10}\right)\div\left(\frac{1}{3}-\frac{1}{4}\right)$

12. $\begin{array}{r} 6 \\ -2\frac{3}{4} \\ \hline \end{array}$

13. $\begin{array}{r} 15\frac{1}{2} \\ +22\frac{3}{4} \\ \hline \end{array}$

14. Find $\frac{3}{4}$ of 76.

15. Round 0.03572 to the nearest thousandth.

16. Write each of the numbers in words.

a. 600.006

b. 0.606

Perform the indicated operations.

17. $\begin{array}{r} 847.8 \\ 436.92 \\ +354.718 \\ \hline \end{array}$

18. $\begin{array}{r} 32.007 \\ -15.835 \\ \hline \end{array}$

19. $\begin{array}{r} 566.3 \\ \times\ 7.2 \\ \hline \end{array}$

20. Divide $1.028 \div 1.3$ and find the quotient to the nearest hundredth.

Name ______________________ Section ____________ Date __________

ANSWERS

21. Use a calculator to find the square roots accurate to four decimal places.

a. $\sqrt{0.72}$ **b.** $\sqrt{95}$

21. a. ______________

b. ______________

Solve each of the following proportions.

22. $\frac{5}{7} = \frac{x}{3\frac{1}{2}}$ **23.** $\frac{0.5}{1.6} = \frac{0.1}{A}$

22. ______________

23. ______________

24. **a. Speed.** If you drove your car 250 miles in 5.5 hours, what was your average speed per hour (to the nearest tenth of a mile)?

b. How far could you drive at this average speed in 7 hours?

24 a. ______________

b. ______________

25. **Taxes.** In December, Mario was told he would receive a bonus of $4000. However, $\frac{1}{10}$ of the bonus was withheld for state taxes, $\frac{1}{5}$ was withheld for federal taxes, and $200 was withheld for Social Security. How much cash did Mario receive after these deductions?

25. ______________

26. **Maps.** The scale on a map indicates that $1\frac{3}{4}$ inches represents 50 miles. How many miles apart are two cities marked 2 inches on the map?

26. ______________

27. Find the length of the hypotenuse (to four decimal places) of a right triangle with both legs measuring 15 feet.

27. ______________

28. a. ______

b. ______

29. ______

30. ______

31. ______

32. ______

28. **a.** Find the perimeter of a rectangle that is $8\frac{1}{2}$ inches wide and $11\frac{2}{3}$ inches long.

b. Find the area of the rectangle.

29. Find the area of a circle that has a diameter of 14 inches.

30. Find the area of a right triangle with height 4 centimeters and base 5 centimeters.

31. Find the volume of the cylinder shown here with the given dimensions. (Use $\pi = 3.14$.)

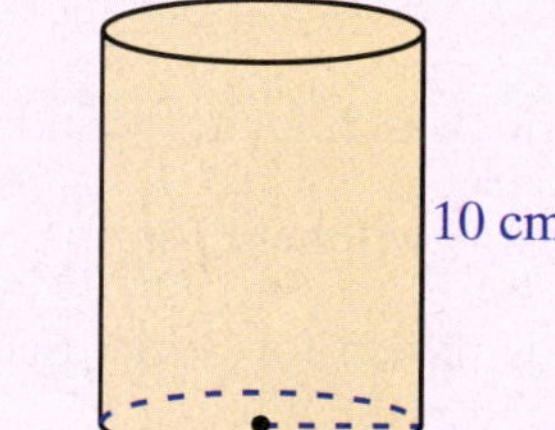

32. Find the volume of the rectangular solid shown here with the given dimensions.

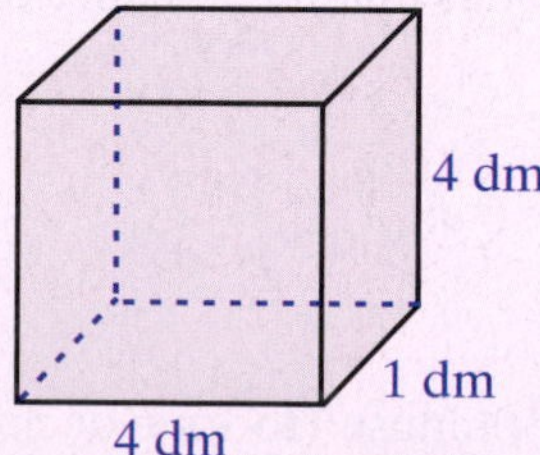

7

PERCENT

Chapter 7 Percent

WHAT TO EXPECT IN CHAPTER 7

Chapter 7 deals with ideas and applications related to percent, one of the most useful and important mathematical concepts in our daily lives. Section 7.1 introduces percent by explaining that percent means hundredths (or per hundred) and that, therefore, percents can be represented in fraction form with denominator 100. Percent of profit and the relationship between percents and decimals are also included in Section 7.1. Section 7.2 discusses the relationships among percents, fractions, and decimals. Sections 7.3 and 7.4 lay the groundwork for applications with percents in Sections 7.5 and 7.6 by helping students realize that all percent problems are basically one of three types of problems. Both the method using the proportion $\frac{P}{100}=\frac{A}{B}$ (Section 7.3) and the method using the formula $R \times B = A$ (Section 7.4) are used in the development.

In Section 7.5 and 7.6 we emphasize Pólya's four-step process for solving problems:

1. Understand the problem.
2. Devise a plan.
3. Carry out the plan.
4. Look back over the results.

Chapter 7 Percent

Mathematics at Work!

The concept of percent is an important part of our daily lives. Income tax is based on a percent of income; property tax is based on a percent of the value of a home; a professional athlete's skills are measured as percents (batting average, free-throw percentage, pass completion percentage); profit is calculated as a percent of investment; commissions are calculated as percents of sales; and tips for service are figured as a percent of the bill.

Even understanding a newspaper or magazine requires some knowledge of percents. Below are some recent news headlines from around the world:

About current U.S. unemployment:
Unemployment Rate Falls to 4.9%

Google to Buy a 5% Stake in AOL

L.A. Major Crime Down 14% in 2005

Best Buy Says December Sales Rose 12% Overall

Greek Budget Deficit Down 26.1%

About National Hockey League player salaries:
NHL Escrow Reduced to 4%

7.1 Decimals and Percents

Objective 1

Understanding Percent

The word **percent** comes from the Latin *per centum*, meaning **per hundred**. So **percent means hundredths**, or the **ratio of a number to 100**. The symbol % is called the **percent sign**. As we shall see, this sign has the same meaning as the fraction $\frac{1}{100}$. For example,

$$\frac{35}{100} = 35\left(\frac{1}{100}\right) = 35\% \quad \text{and} \quad \frac{60}{100} = 60\left(\frac{1}{100}\right) = 60\%$$

In Figure 7.1, the large square is partitioned into 100 small squares, and each small square represents $\frac{1}{100}$, or 1%, of the large square. Therefore, in Figure 7.1, the portion of the large square that is shaded is $\frac{27}{100} = 27\left(\frac{1}{100}\right) = 27\%$.

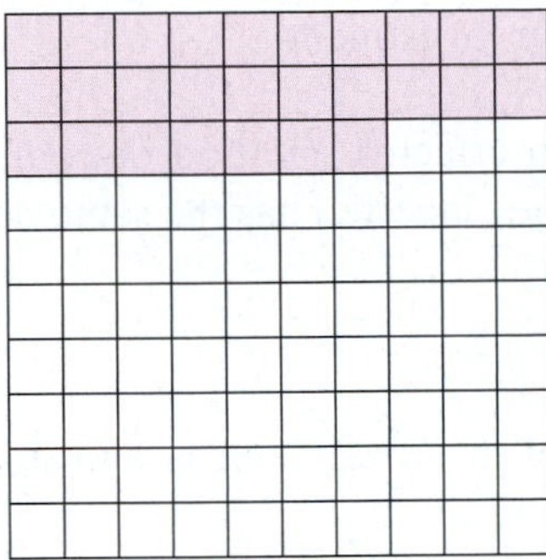

$\frac{27}{100} = 27\%$

Figure 7.1

If a fraction has a denominator of 100, then (with no change in the numerator) the numerator can be read as a percent by dropping the denominator and adding the % sign.

Objectives

1. Understand that percent means hundredths.
2. Relate percent to fractions with denominator 100.
3. Compare profit to investment as a percent.
4. Be able to change percents to decimals and decimals to percents.

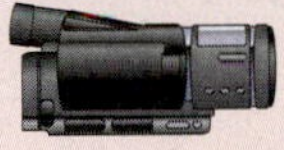

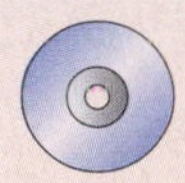

Example 1

Each fraction is changed to a percent.

a. $\frac{20}{100} = 20 \cdot \frac{1}{100} = 20\%$

Remember that **percent** means **hundredths**, and the % sign indicates hundredths, or $\frac{1}{100}$.

b. $\frac{85}{100} = 85\%$

c. $\frac{6.4}{100} = 6.4\%$

Note that the decimal point is not moved; that is, the numerator is unchanged.

d. $\frac{3\frac{1}{2}}{100} = 3\frac{1}{2}\%$

Note that since the denominator is 100, the numerator is unchanged even though it contains a fraction.

e. $\frac{240}{100} = 240\%$

If the numerator is larger than 100, then the number is larger than 1 and the percent is more than 100%.

Now work exercise 1 in the margin.

Objective 2

1. Change each fraction to a percent.

a. $\frac{9}{100}$

b. $\frac{1.25}{100}$

c. $\frac{125}{100}$

d. $\frac{6\frac{1}{4}}{100}$

Percent of Profit

Objective 3

Percent of profit is the ratio of money made to money invested. Generally, two investments do not involve the same amount of money and, therefore, the comparative success of each investment cannot be based on the **amount** of profit. In comparing investments, the investment with the greater percent of profit is considered the better investment.

The use of percent gives an effective method of comparison because each ratio of money made to money invested has the same denominator (100).

2. Which is the better investment: (a) $1000 investment which yields a $120 profit, or (b) $1500 investment with a $200 profit?

Example 2

Calculate the percent of profit for both **a.** and **b.** and tell which is the better investment.

a. $150 profit made by investing $300.

b. $200 profit made by investing $500.

Solution

In each case, find the ratio of dollars profit to dollars invested and reduce the ratio so that it has a denominator of 100. Do not reduce to lowest terms.

a. $\frac{\$150 \text{ profit}}{\$300 \text{ invested}} = \frac{3 \cdot 50}{3 \cdot 100} = \frac{50}{100} = 50\%$

b. $\frac{\$200 \text{ profit}}{\$500 \text{ invested}} = \frac{5 \cdot 40}{5 \cdot 100} = \frac{40}{100} = 40\%$

Investment **a.** is better investment than **b.** because 50% is larger than 40%. Obviously, $200 profit is more than $150 profit, but the $500 risked as an investment in **b.** is considerably more than the $300 risked in **a.**

Now work exercise 2 in the margin.

3. Which is the better investment: (a) Making $250 from an $800 investment, or (b) making $300 from a $1000 investment?

Completion Example 3

Which is the better investment?

a. Making $40 profit by investing $200.

b. Making $75 profit by investing $300.

Solution

Write each ratio as hundredths and compare the percents.

a. $\frac{\$40 \text{ profit}}{\$200 \text{ invested}} = \frac{2 \cdot \ }{2 \cdot 100} = \frac{\ }{100} = \_\_\_\_\%$

b. $\frac{\$75 \text{ profit}}{\$300 \text{ invested}} = \frac{3 \cdot \ }{3 \cdot 100} = \frac{\ }{100} = \_\_\_\_\%$

Investment ________ is better since ______ % is more than ______ %.

Now work exercise 3 in the margin.

Decimals and Percents

Objective 4

One way to change a decimal to a percent is to change the decimal to fraction form with denominator 100 and then change the fraction to percent form. For example,

Decimal Form		Fraction Form		Percent Form
0.47	=	$\frac{47}{100}$	=	47%
0.93	=	$\frac{93}{100}$	=	93%

In both of these illustrations the decimals are in hundredths and changing to fraction form is relatively easy. If the decimal is not in hundredths, we can multiply by 1 in the form of $\frac{100}{100}$, as follows:

$$0.325 = 0.325 \cdot \frac{100}{100} = 0.325(100) \cdot \frac{1}{100} = \frac{32.5}{100} = 32.5\%$$

Now, looking at the fact that $0.325 = 32.5\%$, we can see that the decimal point was moved two places to the right and the % sign was written. This leads to the following general rule.

To Change a Decimal to a Percent:

Step 1: Move the decimal point two places to the right.

Step 2: Write the % sign.

(These two steps have the effect of multiplying by 100 and then dividing by 100.)

The following relationships between decimals and percents will serve as helpful guidelines.

Relationships Between Decimals and Percents

A decimal number that is

1. less than 0.01 is less than 1%.
2. between 0.01 and 0.10 is between 1% and 10%.
3. between 0.10 and 1.00 is between 10% and 100%.
4. more than 1 is more than 100%.

4. Change each decimal to a percent.

a. 0.34

b. 0.0082

c. 1

d. 5.799

Example 4

Convert each decimal to an equivalent percent.

a. 0.253 **b.** 0.905 **c.** 2.65 **d.** 0.7 **e.** 0.002

Solution

a. 0.253 = 25.3%

decimal point moved two places to the right; % sign added

b. 0.905 = 90.5%

c. 2.65 = 265% Note that this is more than 100%.

d. 0.7 = 70% The decimal point is not written here. We could write 70% or 70.0%.

e. 0.002 = 0.2% Note that this is less than 1%.

decimal point moved two places to the right; % sign added

Now work exercise 4 in the margin.

To change percents to decimals, we reverse the procedure for changing decimals to percents. For example,

$$39\% = 39\left(\frac{1}{100}\right) = \frac{39}{100} = 0.39$$

Remember, the decimal point is moved two places to the left when dividing by 100.

To Change a Percent to a Decimal:

Step 1: Move the decimal point two places to the left.

Step 2: Delete the % sign.

Example 5

Convert each percent to an equivalent decimal.

a. 76% **b.** 18.5% **c.** 100% **d.** 1.3% **e.** 0.25%

Solution

a. 76% = 0.76 ← % sign deleted

(76%: understood decimal point; 0.76: decimal point moved two places left)

b. 18.5% = 0.185 **c.** 100% = 1.00 = 1

d. 1.3% = 0.013 **e.** 0.25% = 0.0025

Note that 0.25% is less than 1% and its decimal equivalent is less than 0.01.

Now work exercise 5 in the margin.

5. Change each percent to a decimal.

a. 40%

b. 211%

c. 0.6%

d. 29.37%

Completion Example Answers

3. a. $\frac{\$40 \text{ profit}}{\$200 \text{ invested}} = \frac{2 \cdot \mathbf{20}}{2 \cdot 100} = \frac{\mathbf{20}}{100} = \mathbf{20}\%$

b. $\frac{\$75 \text{ profit}}{\$300 \text{ invested}} = \frac{3 \cdot \mathbf{25}}{3 \cdot 100} = \frac{\mathbf{25}}{100} = \mathbf{25}\%$

Investment **b.** is better since **25%** is more than **20%**.

Practice Problems

Convert each fraction to a percent.

1. $\frac{73}{100}$ **2.** $\frac{16}{100}$

Determine the percent of profit for the following.

3. A \$700 investment yielding a \$147 profit.

4. A \$120 profit from a \$400 investment.

Convert each decimal to a percent.

5. 0.368 **6.** 4.15

Convert each percent to a decimal.

7. 29% **8.** 5.3%

Answers to Practice Problems:

1. 73% **2.** 16%
3. 21% **4.** 30%
5. 36.8% **6.** 415%
7. 0.29 **8.** 0.053

Name ______________________ Section ____________ Date __________

Exercises 7.1

Find the percent of each square that is shaded.

1. 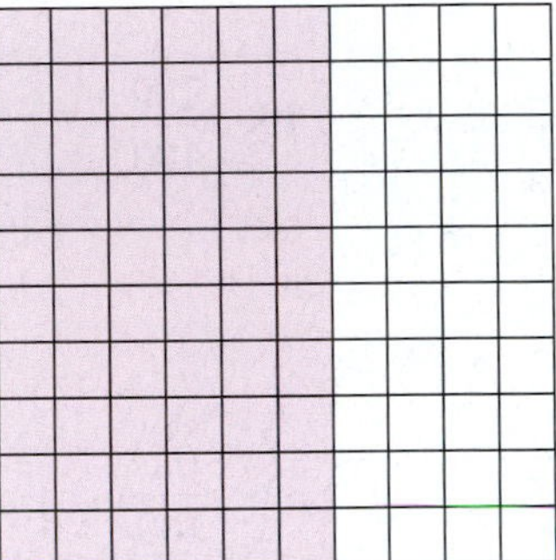

2. 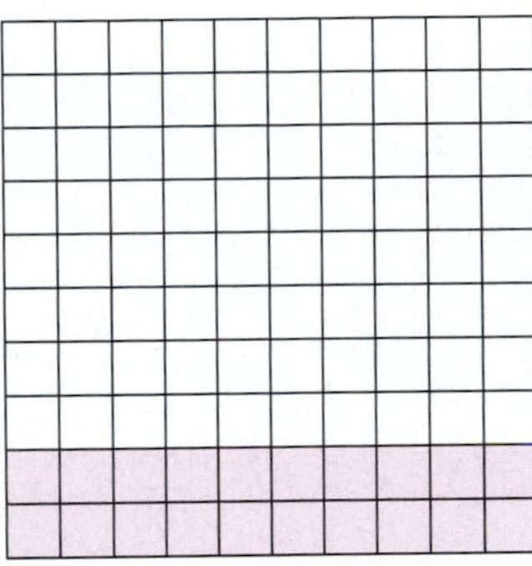

3. 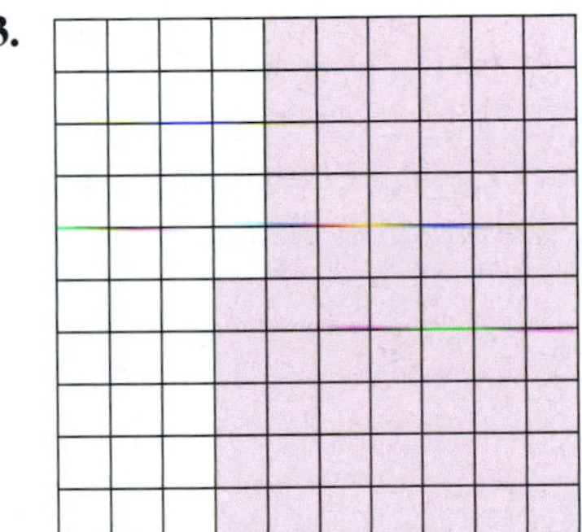

4. 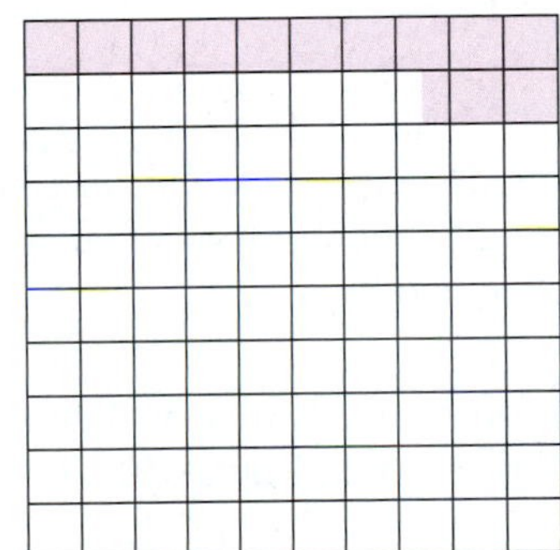

5. 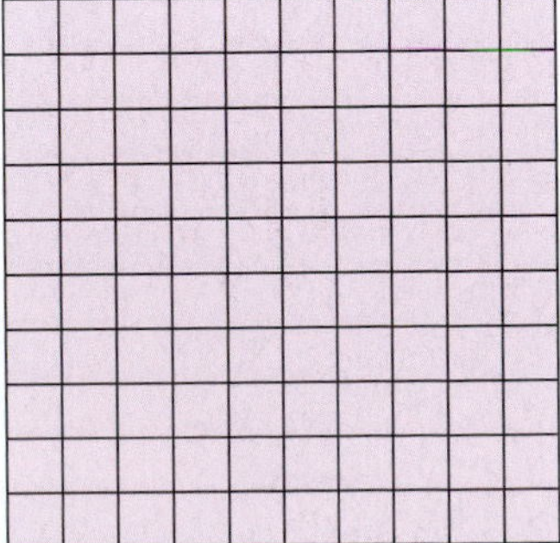

6. 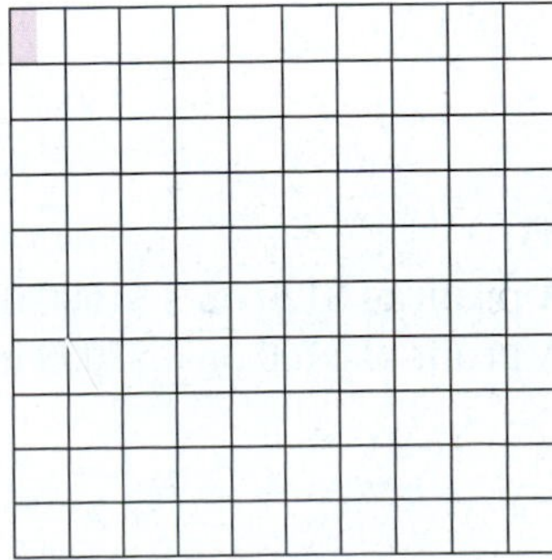

ANSWERS

1. ______________

2. ______________

3. ______________

4. ______________

5. ______________

6. ______________

7. ______

8. ______

9. ______

10. ______

11. ______

12. ______

13. ______

14. ______

15. ______

16. ______

17. ______

18. ______

19. ______

20. ______

21 a. ______

b. ______

22 a. ______

b. ______

23 a. ______

b. ______

24 a. ______

b. ______

Change the following fractions to percents. See Example 1.

7. $\frac{20}{100}$ **8.** $\frac{9}{100}$ **9.** $\frac{15}{100}$ **10.** $\frac{62}{100}$

11. $\frac{53}{100}$ **12.** $\frac{68}{100}$ **13.** $\frac{125}{100}$ **14.** $\frac{200}{100}$

15. $\frac{336}{100}$ **16.** $\frac{13.4}{100}$ **17.** $\frac{0.48}{100}$ **18.** $\frac{0.5}{100}$

19. $\frac{2.14}{100}$ **20.** $\frac{1.62}{100}$

In Exercises 21 – 24, write the ratio of profit to investment as hundredths and then as percents. Compare the percents and tell which investment is better, ***a.*** *or* ***b.*** *See Examples 2 and 3.*

21. a. A profit of $36 on a $200 investment
b. A profit of $51 on a $300 investment

22. a. A profit of $40 on a $400 investment
b. A profit of $60 on a $600 investment

23. a. A profit of $120 on a $1500 investment
b. A profit of $160 on a $2000 investment

24. a. A profit of $300 on a $2000 investment
b. A profit of $360 on a $3000 investment

Name ____________ Section ____________ Date ____________

Change the following decimals to percents. See Example 4.

25. 0.02 **26.** 0.09 **27.** 0.1 **28.** 0.7

29. 0.36 **30.** 0.52 **31.** 0.40 **32.** 0.65

33. 0.025 **34.** 0.035 **35.** 0.055 **36.** 0.004

37. 1.10 **38.** 1.75 **39.** 2 **40.** 2.3

Change the following percents to decimals. See Example 5.

41. 2% **42.** 7% **43.** 18% **44.** 20%

45. 30% **46.** 80% **47.** 0.26% **48.** 0.52%

49. 125% **50.** 120% **51.** 232% **52.** 215%

53. 17.3% **54.** 10.1% **55.** 13.2% **56.** 6.5%

ANSWERS

25. ____________
26. ____________
27. ____________
28. ____________
29. ____________
30. ____________
31. ____________
32. ____________
33. ____________
34. ____________
35. ____________
36. ____________
37. ____________
38. ____________
39. ____________
40. ____________
41. ____________
42. ____________
43. ____________
44. ____________
45. ____________
46. ____________
47. ____________
48. ____________
49. ____________
50. ____________
51. ____________
52. ____________
53. ____________
54. ____________
55. ____________
56. ____________

57. ______

57. Sales tax. Suppose that sales tax is figured at 7.25%. Change 7.25% to a decimal.

58. ______

58. Interest. The interest rate on a loan is 6.4%. Change 6.4% to a decimal.

59. Commission. The sales commission for the clerk in a retail store is figured at 8.5%. Change 8.5% to a decimal.

59. ______

60. Commission. In calculating his sales commission, Mr. Howard multiplies by the decimal 0.12. Change 0.12 to a percent.

60. ______

61. Mortgages. To calculate what your maximum house payment should be, a banker multiplied your income by 0.28. Change 0.28 to a percent.

61. ______

62. Discount. The discount you earn by paying cash is found by multiplying the amount of your purchase by 0.02. Change 0.02 to a percent.

62. ______

63. ______

63. Profit. At Mr. Jones' sporting goods store, the rate of profit based on sales price is 45%. Change 45% to a decimal.

64. ______

64. Discount. The discount during a special sale on dresses is 30%. Change 30% to a decimal.

65. ______

65. License fee. Suppose the state license fee is figured by multiplying the cost of your car by 0.065. Change 0.065 to a percent.

Name ______________________________ Section ____________ Date ___________

ANSWERS

66. Health insurance. With 27.8% of its residents under 65 years of age having no health coverage, Texas was recently ranked as the state with the highest percentage of health care uninsured. Write 27.8% as a decimal.

66. ____________

67. Property taxes. As of 2005, per capita property taxes in New Jersey were 476% higher than those in Alabama. Write 476% as a decimal.

67. ____________

68. Recycling. According to the Aluminum Association, the percentage of aluminum cans recycled in 2003 was 50%, and this percentage rose to 51.2% in 2004. Write 50% and 51.2% as decimals.

68. ____________

Review Problems (from Section 6.3)

Solve for the variable in each of the following proportions.

1. $\frac{R}{100} = \frac{4}{5}$ **2.** $\frac{R}{100} = \frac{2}{3}$ **3.** $\frac{23}{100} = \frac{A}{500}$

4. $\frac{75}{100} = \frac{A}{60}$ **5.** $\frac{90}{100} = \frac{36}{B}$ **6.** $\frac{10}{100} = \frac{7.4}{B}$

REVIEW

1. ____________

2. ____________

3. ____________

4. ____________

5. ____________

6. ____________

7.2 Fractions and Percents

Changing Fractions and Mixed Numbers to Percents

As we discussed in Section 7.1, if a fraction has denominator 100, we can change it to a percent by writing the numerator and adding the % sign. If the denominator is a factor of 100 (2, 4, 5, 10, 20, 25, or 50), we can write the fraction in an equivalent form with denominator 100 and then change it to a percent. For example,

$$\frac{1}{4} = \frac{1}{4} \cdot \frac{25}{25} = \frac{25}{100} = 25\%$$

$$\frac{3}{5} = \frac{3}{5} \cdot \frac{20}{20} = \frac{60}{100} = 60\%$$

$$\frac{17}{50} = \frac{17}{50} \cdot \frac{2}{2} = \frac{34}{100} = 34\%$$

Objectives

① Be able to change fractions and mixed numbers to percents.

② Be able to change percents to mixed numbers and fractions.

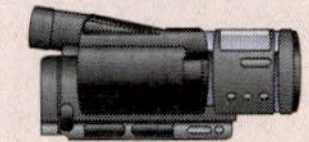

However, most fractions do not have denominators that are factors of 100. A more general approach (easily applied with calculators) is to change the fraction to decimal form by dividing the numerator by the denominator, and then change the decimal to a percent.

Objective ①

To Change a Fraction to a Percent:

Step 1: Change the fraction to a decimal. (Divide the numerator by the denominator.)

Step 2: Change the decimal to a percent.

Example 1

Change $\frac{5}{8}$ to a percent.

Solution

a. Divide:

$$\begin{array}{r} 0.625 \\ 8\overline{)5.000} \\ \underline{4\,8} \\ 20 \\ \underline{16} \\ 40 \\ \underline{40} \\ 0 \end{array}$$

b. Change 0.625 to a percent: $\frac{5}{8} = 0.625 = 62.5\%$

Now work exercise 1 in the margin.

1. Change $\frac{13}{20}$ to a percent.

2. Change $\frac{1}{4}$ to a percent.

Example 2

Change $\frac{18}{20}$ to a percent.

Solution

a. Divide:

$$\begin{array}{r} 0.9 \\ 20\overline{)18.0} \\ \underline{18\,0} \\ 0 \end{array}$$

b. Change 0.9 to a percent: $\frac{18}{20} = 0.9 = 90\%$

Now work exercise 2 in the margin.

3. Change $1\frac{1}{2}$ to a percent.

Example 3

Change $2\frac{1}{4}$ to a percent.

Solution

a. $2\frac{1}{4} = \frac{9}{4}$

$$\begin{array}{r} 2.25 \\ 4\overline{)9.00} \\ \underline{8} \\ 1\,0 \\ \underline{8} \\ 20 \\ \underline{20} \\ 0 \end{array}$$

b. $2\frac{1}{4} = \frac{9}{4} = 2.25 = 225\%$

Now work exercise 3 in the margin.

Agreement for Rounding Decimal Quotients:

In this text,

1. Decimal quotients that are exact with four decimal places (or less) will be written with four decimal places (or less).
2. Decimal quotients that are not exact will be divided to the fourth place, and the quotient will be rounded to the third place (thousandths).

Using a Calculator:

Most calculators will give answers accurate to 8 or 9 decimal places. So, if you use a calculator to perform the long division when changing a fraction to a decimal, be sure to follow the agreement in statement two in the preceding box.

Example 4

Change $\frac{1}{3}$ to a percent.

a. Using a calculator, $\frac{1}{3} = 0.3333333.$

Rounding the decimal quotient to the third decimal place:

$\frac{1}{3} = 0.333 = 33.3\%$ The answer is rounded and not exact.

b. Without a calculator, we can divide and use fractions:

$$\begin{array}{r} 0.33\frac{1}{3} \\ 3\overline{)1.00} \\ \underline{9} \\ 10 \\ \underline{9} \\ 1 \end{array}$$

$$\frac{1}{3} = 0.33\frac{1}{3} = 33\frac{1}{3}\% \text{ or } 33.\bar{3}\%$$

$33\frac{1}{3}\%$ and $33.\bar{3}\%$ are exact, and 33.3% is rounded.

Now work exercise 4 in the margin.

Note: Any one of these answers is acceptable, but be aware that 33.3% is a rounded answer.

Example 5

During the years 1921 to 2005, the New York Yankees baseball team played in 39 World Series Championships and won 26 of them. What percent of these championships did the Yankees win?

Solution

a. The percent won can be found by using a calculator and changing the fraction $\frac{26}{39}$ to decimal form and then changing the decimal to a percent. Using a calculator,

$\frac{26}{39} = 0.6666666 \approx 0.667 = 66.7\%$

b. Without a calculator, we can divide and use fractions:

$$\begin{array}{r} 0.66\frac{2}{3} \\ 3\overline{)2.00} \\ \underline{18} \\ 20 \\ \underline{18} \\ 2 \end{array}$$

$$\frac{26}{39} = \frac{\cancel{13}\cdot 2}{\cancel{13}\cdot 3} = \frac{2}{3} = 0.66\frac{2}{3} = 66\frac{2}{3}\% \text{ or } 66.\bar{6}\%$$

The Yankees won $66\frac{2}{3}\%$ of these championships.

Now work exercise 5 in the margin.

4. Use a calculator to change $\frac{3}{11}$ to a percent. Round this answer to the third decimal place. Then, without a calculator, divide and use fractions to give the exact answer.

5. The Braves little league baseball team won 14 of their 15 games this season. Find the percentage of games won this season.

Note: Any one of these answers– $66\frac{2}{3}\%$, $66.\bar{6}\%$, or 66.7% –is acceptable.

6. Use a calculator to change $\frac{7}{9}$ to a percent. Round this answer to the third decimal place. Then, without a calculator, divide and use fractions to give the exact answer.

Example 6

Using a calculator, $\frac{1}{7} = 0.1428571.$

a. Rounding the decimal quotient to the third decimal place:

$\frac{1}{7} = 0.143 = 14.3\%$

b. Using long division, you can write the answer with a fraction or continue to divide and round the decimal answer. The choice is yours.

$$\begin{array}{r} 0.14\frac{2}{7} = 14\frac{2}{7}\% \\ 7\overline{)1.00} \\ \underline{7} \\ 30 \\ \underline{28} \\ 2 \end{array} \quad \text{or} \quad \begin{array}{r} 0.1428 = 14.3\% \\ 7\overline{)1.0000} \\ \underline{7} \\ 30 \\ \underline{28} \\ 20 \\ \underline{14} \\ 60 \\ \underline{56} \\ 4 \end{array}$$

Now work exercise 6 in the margin.

Changing Percents to Fractions and Mixed Numbers

Objective 2

To Change a Percent to a Fraction or a Mixed Number:

Step 1: Write the percent as a fraction with denominator 100 and drop the % sign.

Step 2: Reduce the fraction.

Example 7

Change each percent to an equivalent fraction or mixed number in reduced form.

a. 60% **b.** $7\frac{1}{4}\%$ **c.** 130%

Solution

a. $60\% = \frac{60}{100} = \frac{3 \cdot \cancel{20}}{5 \cdot \cancel{20}} = \frac{3}{5}$

b. $7\frac{1}{4}\% = \dfrac{7\frac{1}{4}}{100} = \dfrac{\frac{29}{4}}{100} = \dfrac{29}{4} \cdot \dfrac{1}{100} = \dfrac{29}{400}$

c. $130\% = \dfrac{130}{100} = \dfrac{13 \cdot \cancel{10}}{10 \cdot \cancel{10}} = \dfrac{13}{10} = 1\dfrac{3}{10}$

Now work exercise 7 in the margin.

A Common Misunderstanding:

The fractions $\frac{1}{4}$ and $\frac{1}{2}$ are often confused with the percents $\frac{1}{4}\%$ and $\frac{1}{2}\%$. The differences can be clarified by using decimals.

PERCENT	DECIMAL	FRACTION
$\frac{1}{4}\%$ (or 0.25%)	0.0025	$\frac{1}{400}$
$\frac{1}{2}\%$ (or 0.5%)	0.005	$\frac{1}{200}$
25%	0.25	$\frac{1}{4}$
50%	0.50	$\frac{1}{2}$

Thus

$\frac{1}{4} = 0.25$ and $\frac{1}{4}\% = 0.0025$

$\mathbf{0.25 \neq 0.0025}$

Similarly,

$\frac{1}{2} = 0.50$ and $\frac{1}{2}\% = 0.005$

$\mathbf{0.50 \neq 0.005}$

You can think of $\frac{1}{4}$ as being one-fourth of a dollar (a quarter) and $\frac{1}{4}\%$ as being one-fourth of a penny. $\frac{1}{2}$ can be thought of as one-half of a dollar and $\frac{1}{2}\%$ as one-half of a penny.

7. Change each percent to a fraction or mixed number.

a. 80%

b. 16%

c. $5\frac{1}{2}\%$

d. 235%

Some percents are so common that their decimal and fraction equivalents should be memorized. Their fractional values are particularly easy to work with, and many times calculations involving these fractions can be done mentally.

Common Percent - Decimal - Fraction Equivalents

$1\% = 0.01\% = \frac{1}{100}$ $\quad 33\frac{1}{3}\% = 0.33\frac{1}{3} = \frac{1}{3}$ $\quad 12\frac{1}{2}\% = 0.125 = \frac{1}{8}$

$25\% = 0.25 = \frac{1}{4}$ $\quad 66\frac{2}{3}\% = 0.66\frac{2}{3} = \frac{2}{3}$ $\quad 37\frac{1}{2}\% = 0.375 = \frac{3}{8}$

$50\% = 0.50 = \frac{1}{2}$ $\quad 62\frac{1}{2}\% = 0.625 = \frac{5}{8}$

$75\% = 0.75 = \frac{3}{4}$ $\quad 87\frac{1}{2}\% = 0.875 = \frac{7}{8}$

$100\% = 1.00 = 1$

Practice Problems

Convert each fraction to a percent. Round to the third place if necessary.

1. $\frac{6}{25}$ **2.** $\frac{56}{64}$

3. $\frac{1}{18}$ **4.** $\frac{5}{7}$

Convert each percent to a fraction or mixed number reduced to lowest terms.

5. 43% **6.** 292%

Answers to Practice Problems:

1. 24% **2.** 87.5%

3. 5.6% **4.** 71.4%

5. $\frac{43}{100}$ **6.** $\frac{73}{25}$ or $2\frac{23}{25}$

Name ____________ Section ____________ Date ____________

Exercises 7.2

Change the following fractions and mixed numbers to percents. See Examples 1 through 6.

1. $\frac{3}{100}$ **2.** $\frac{16}{100}$ **3.** $\frac{7}{100}$ **4.** $\frac{29}{100}$

5. $\frac{1}{2}$ **6.** $\frac{3}{4}$ **7.** $\frac{1}{4}$ **8.** $\frac{1}{20}$

9. $\frac{11}{20}$ **10.** $\frac{7}{10}$ **11.** $\frac{3}{10}$ **12.** $\frac{3}{5}$

13. $\frac{1}{5}$ **14.** $\frac{2}{5}$ **15.** $\frac{4}{5}$ **16.** $\frac{1}{50}$

17. $\frac{13}{50}$ **18.** $\frac{1}{25}$ **19.** $\frac{12}{25}$ **20.** $\frac{24}{25}$

21. $\frac{1}{8}$ **22.** $\frac{5}{8}$ **23.** $\frac{7}{8}$ **24.** $\frac{1}{9}$

25. $\frac{5}{9}$ **26.** $\frac{39}{50}$ **27.** $\frac{17}{20}$ **28.** $\frac{5}{6}$

ANSWERS

1. ____________
2. ____________
3. ____________
4. ____________
5. ____________
6. ____________
7. ____________
8. ____________
9. ____________
10. ____________
11. ____________
12. ____________
13. ____________
14. ____________
15. ____________
16. ____________
17. ____________
18. ____________
19. ____________
20. ____________
21. ____________
22. ____________
23. ____________
24. ____________
25. ____________
26. ____________
27. ____________
28. ____________

29. ______
30. ______
31. ______
32. ______
33. ______
34. ______
35. ______
36. ______
37. ______
38. ______
39. ______
40. ______
41. ______
42. ______
43. ______
44. ______
45. ______
46. ______
47. ______
48. ______
49. ______
50. ______
51. ______
52. ______

29. $\frac{7}{11}$ **30.** $\frac{5}{11}$ **31.** $1\frac{5}{16}$ **32.** $1\frac{1}{6}$

33. $1\frac{1}{20}$ **34.** $1\frac{1}{4}$ **35.** $1\frac{3}{4}$ **36.** $1\frac{1}{5}$

37. $1\frac{3}{8}$ **38.** $2\frac{1}{2}$ **39.** $2\frac{1}{10}$ **40.** $2\frac{1}{15}$

Change the following percents to fractions or mixed numbers. See Example 7.

41. 10% **42.** 5% **43.** 15% **44.** 17%

45. 25% **46.** 30% **47.** 50% **48.** $12\frac{1}{2}\%$

49. $37\frac{1}{2}\%$ **50.** $16\frac{2}{3}\%$ **51.** $33\frac{1}{3}\%$ **52.** $66\frac{2}{3}\%$

Name ______________ Section ______________ Date ______________

53. 33% **54.** $\frac{1}{2}\%$ **55.** $\frac{1}{4}\%$ **56.** 1%

57. 100% **58.** 125% **59.** 120% **60.** 150%

61. 0.3% **62.** 2.5% **63.** 62.5% **64.** 0.2%

65. 0.75%

Find the missing forms of each number. See Examples 1 through 7.

	Fraction form	Decimal form	Percent form
66.	$\frac{5}{8}$	**a.** ______	**b.** ______
67.	$\frac{11}{20}$	**a.** ______	**b.** ______
68.	**a.** ______	0.09	**b.** ______
69.	**a.** ______	1.75	**b.** ______
70.	**a.** ______	**b.** ______	36%
71.	**a.** ______	**b.** ______	10.5%

ANSWERS

53. ______
54. ______
55. ______
56. ______
57. ______
58. ______
59. ______
60. ______
61. ______
62. ______
63. ______
64. ______
65. ______
66. a. ______
b. ______
67. a. ______
b. ______
68. a. ______
b. ______
69. a. ______
b. ______
70. a. ______
b. ______
71. a. ______
b. ______

72. ____________

73. ____________

74. ____________

75. a. ____________

b. ____________

c. ____________

76. ____________

77. ____________

72. Discounts. A department store offers a 30% discount during a special sale on men's suits. Change 30% to a fraction reduced to lowest terms.

73. Student government. In a sophomore class of 250 students, 10 represent the sophomore class on the student council. What percent of the class is on the student council?

74. Travel. Malcolm planned to drive 360 miles on a trip. After 24 miles, what percent of the trip had he driven?

75. **a.** There are 12 inches in a foot. What percent of a foot is an inch?
b. There are 4 quarts in a gallon. What percent of a gallon is a quart?
c. There are 16 ounces in a pound. What percent of a pound is an ounce?

76. Historic items. A desk once belonging to George Washington was recently sold at auction. It was expected to bring $80,000 but actually sold for $165,000. Write the actual selling price as a percent of the expected price.

77. Population. According to the 2000 census, about 59,000,000 Americans are between the ages of 20 and 34. If the total population is 300,000,000 people, what percentage (to the nearest percent) of the population is between the ages of 20 and 34?

Name ______________ Section ______________ Date ______________

Check Your Number Sense

78. Three pairs of fractions are given. In each case, tell which fraction would be easier to change to a percent mentally and explain your reasoning.

a. $\frac{7}{20}$ or $\frac{2}{9}$ **b.** $\frac{3}{10}$ or $\frac{5}{16}$ **c.** $\frac{1}{12}$ or $\frac{1}{25}$

Review Problems (from Section 6.4)

Solve the following problems.

1. **Merchandise.** A salesman makes $9 for every $100 worth of merchandise he sells. What will he make if he sells $80,000 worth of merchandise?

2. **Architecture.** An architect drew up plans for a new building using a scale of $\frac{1}{2}$ inch to represent 36 feet. How many feet would 3 inches represent?

3. **Gas costs.** If gasoline costs $2.08 per gallon, how many gallons can be bought for $16.80? Round your answer to the nearest hundredth.

4. **Lawn care.** A certain type of fertilizer is sold in bags containing 10 pounds. The recommended coverage is 10 pounds per 3000 square feet.
   a. If your lawn is in the shape of a rectangle 90 feet by 80 feet, how many pounds of this fertilizer do you need to cover your lawn?
   b. How many bags do you need to buy?

ANSWERS

78. a. ______________

b. ______________

c. ______________

REVIEW

1. ______________

2. ______________

3. ______________

4. a. ______________

b. ______________

7.3 Solving Percent Problems Using the Proportion $\frac{P}{100}=\frac{A}{B}$

The Basic Proportion $\frac{P}{100}=\frac{A}{B}$

Many people have difficulty working with percent simply because they do not know whether to add, subtract, multiply, or divide. **There are only three basic types of problems using percent**. In this section, we will develop a technique for solving all three types of problems by using just one basic proportion of the form $\frac{P}{100}=\frac{A}{B}$. The terms represented in this basic proportion are explained in the following box.

Objectives

① Recognize the three types of percent problems.

② Solve percent problems using the proportion $\frac{P}{100}=\frac{A}{B}$.

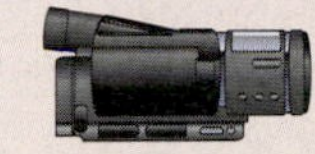 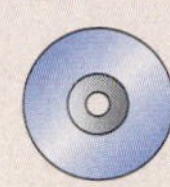

Percent Terms

$P\%$ = Percent

B = Base (number for which we are finding the percent)

A = Amount or percentage (a part of the base)

The relationship among P, B, and A is given in the basic proportion

$$\frac{P}{100}=\frac{A}{B}.$$

As an example of how to set up a basic proportion, consider the statement "25% of 60 is 15."

We can set up this statement in the form of the basic proportion as follows:

25% of 60 is 15.
↑ percent ↑ base ↑ amount

The basic proportion

$$\frac{P}{100}=\frac{A}{B} \quad \text{becomes} \quad \frac{25}{100}=\frac{15}{60}.$$

Using the Basic Proportion

If any two of the values P, A, and B are known, then the third can be found by substituting the known values into the basic proportion and solving for the one unknown term.

Objective 1

The Three Basic Types of Percent Problems and the Proportion $\frac{P}{100} = \frac{A}{B}$

Type 1: Find the amount, given the base and the percent.

What is 65% of 500? $\frac{65}{100} = \frac{A}{500}$

P and B are known. The object is to find A.

Type 2: Find the base, given the percent and the amount.

57% of what number is 51.3? $\frac{57}{100} = \frac{51.3}{B}$

P and A are known. The object is to find B.

Type 3: Find the percent, given the base and the amount.

What percent of 170 is 204? $\frac{P}{100} = \frac{204}{170}$

A and B are known. The object is to find P.

The following examples illustrate how to substitute into the basic proportion and how to solve the resulting proportion for the unknown term by using the methods discussed in Chapter 6. **Remember that *B* is the number of which you are taking the percent**. B is the number that follows the word **of**.

Objective 2

Find the unknown quantity.

1. 15% of 80 is _____ .

Example 1

What is 65% of 500?

Solution

In this problem, $P\% = 65\%$ and $B = 500$. We want to find the value of A. Substitution in the basic proportion gives

$$\frac{65}{100} = \frac{A}{500}$$

$$65 \cdot 500 = 100 \cdot A$$ The product of the extremes must equal the product of the means.

$$\frac{32{,}500}{100} = \frac{\cancel{100} \cdot A}{\cancel{100}}$$

$$325 = A$$

So 65% of 500 is <u>325</u>.

Now work exercise 1 in the margin.

Example 2

57% of what number is 51.3?

Solution

In this problem, $P\% = 57\%$ and $A = 51.3$. We want to find the value of B. Substitution in the basic proportion gives

$$\frac{57}{100} = \frac{51.3}{B}$$

$$57 \cdot B = 100 \cdot 51.3$$ The product of the extremes must equal the product of the means.

$$\frac{\cancel{57} \cdot B}{\cancel{57}} = \frac{5130}{57}$$

$$B = 90$$

So 57% of <u>90</u> is 51.3.

Now work exercise 2 in the margin.

2. 86% of _____ is 430.

Example 3

What percent of 170 is 204?

Solution

In this problem, $B = 170$ and $A = 204$. We want to find the value of P. (Do you expect $P\%$ to be more than 100% or less than 100%?) Substitution in the basic proportion gives

$$\frac{P}{100} = \frac{204}{170}$$

$$170 \cdot P = 100 \cdot 204$$

$$\frac{\cancel{170} \cdot P}{\cancel{170}} = \frac{20,400}{170}$$

$$P = 120 \quad (\text{or } P\% = 120\%)$$

So <u>120%</u> of 170 is 204.

Now work exercise 3 in the margin.

3. _____% of 90 is 19.8.

4. If you eat a meal which contains 300 calories from fat, what percentage of the recommended maximum of 750 is that amount?

Example 4

Many food product labels now list total calories and number of calories from fat. Dietary experts believe that a healthy diet has at most 30% of its calories derived from fat. Following this guideline, if an adult consumes 2500 calories per day, how many calories can be from fat?

Solution

In this problem, $P\% = 30\%$ and $B = 2500$. We want to find the value of A. Substitution in the basic proportion gives

$$\frac{30}{100} = \frac{A}{2500}$$
$$30 \cdot 2500 = 100 \cdot A$$
$$\frac{75{,}000}{100} = \frac{100 \cdot A}{100}$$
$$750 = A$$

So in a healthy diet of 2500 calories, no more than 750 calories should be derived from fat.

Now work exercise 4 in the margin.

5. In November, a hockey player scored on 12 of his shots on goal. The following month, he scored on 15 of his shots. What was the percent increase from November to December?

Example 5

Ben's last two exam scores in algebra were 80 and 88. What was the percent increase from the first exam to the second?

Solution

In this problem, P is unknown. The increase in score from the first exam to the second is $88 - 80 = 8 = A$. $B = 80$ since it is the number that was increased. Substitution in the basic proportion gives

$$\frac{P}{100} = \frac{8}{80}$$
$$80 \cdot P = 8 \cdot 100$$
$$\frac{\cancel{80} \cdot P}{\cancel{80}} = \frac{800}{80}$$
$$P = 10$$

So Ben's score increased by 10%.

Now work exercise 5 in the margin.

Practice Problems

1. What is 85% of 60?

2. What percent of 14 is 42?

3. 62.5% of what number is 27?

4. What is 52% of 225?

Answers to Practice Problems:

1. 51 **2.** 300% **3.** 43.2 **4.** 117

Name ______________ Section ______ Date ______

Exercises 7.3

Use the basic proportion $\frac{P}{100} = \frac{A}{B}$ *to solve each of the following problems for the unknown quantity. See Examples 1 through 4.*

1. 10% of 90 is ______ .

2. 5% of 72 is ______ .

3. 15% of 50 is ______ .

4. 25% of 60 is ______ .

5. 75% of 32 is ______ .

6. 60% of 40 is ______ .

7. 100% of 47 is ______ .

8. 80% of 80 is ______ .

9. 2% of ______ is 5.

10. 20% of ______ is 14.

11. 3% of ______ is 27.

12. 30% of ______ is 45.

13. 100% of ______ is 62.

14. 50% of ______ is 35.

15. 150% of ______ is 69.

16. 110% of ______ is 440.

17. ______% of 50 is 75.

18. ______% of 120 is 48.

19. ______% of 70 is 21.

20. ______% of 15 is 5.

21. ______% of 44 is 66.

22. ______% of 50 is 10.

23. ______% of 54 is 18.

24. ______% of 100 is 38.

ANSWERS

1. ______
2. ______
3. ______
4. ______
5. ______
6. ______
7. ______
8. ______
9. ______
10. ______
11. ______
12. ______
13. ______
14. ______
15. ______
16. ______
17. ______
18. ______
19. ______
20. ______
21. ______
22. ______
23. ______
24. ______

25. ______

26. ______

27. ______

28. ______

29. ______

30. ______

31. ______

32. ______

33. ______

34. ______

35. ______

36. ______

37. ______

38. ______

39. ______

40. ______

41. ______

42. ______

43. ______

44. ______

45. ______

46. ______

47. ______

48. ______

49. ______

50. ______

Each of the following problems is one of the three types discussed, with slightly changed wording. Remember that B (the base) is the number ***of which*** *you are finding the percent.*

25. ______ is 50% of 35.

26. ______ is 31% of 85.

27. 32 is 20% of ______ .

28. 79 is 100% of ______ .

29. 23 is ______% of 10.

30. 18 is ______% of 10.

31. 40 is $33\frac{1}{3}$% of ______ .

32. 76.8 is 15% of ______ .

33. 142.6 is 23% of ______ .

34. 20.25 is 25% of ______ .

35. 48 is ______% of 24.

36. 8 is ______% of 12.

37. ______ is 96% of 35.

38. ______ is 84% of 52.

39. ______ is 18% of 425.

40. ______ is 28% of 640.

41. Find 18% of 345.

42. Find 13.5% of 95.

43. Find 150% of 70.

44. What percent of 32 is 24?

45. 200 is 125% of what number?

46. 450 is 150% of what number?

47. What percent of 100 is 66.5?

48. Find 86.5% of 100.

49. What percent of 96 is 16?

50. 160 is 80% of what number?

Name ______________________ Section ____________ Date __________

ANSWERS

51. ______________

52. ______________

53. ______________

54. ______________

51. Population. According to U.S. Census information from 1990 and 2000, Las Vegas, Nevada had the largest population growth among the nation's 100 most populated cities. Between 1990 and 2000, the population of Las Vegas increased by 83.3%, to 1,563,282. What was Las Vegas' population in 1990 (to the nearest whole number)?

52. Cereal sales. Due to the increasing cost of breakfast cereals, more and more people are buying private-label brands rather than national brands. In a recent year, the sale of private-label cereals rose from 170 million boxes to 180 million boxes. What was the percent increase in sales (to the nearest tenth of a percent)?

53. Baseball attendance. In 2004 the New York Yankees led the major leagues in home attendance, drawing an average of 47,788 fans to their home games. This figure represented 83.3% of the capacity of Yankee Stadium. Estimate how many fans the stadium can hold (to the nearest ten) when it is filled to capacity.

54. Baseball attendance. The same study noted in Exercise 53 found that Yankees attendance increased by an average of 2,711 fans per game from 2004 to 2005. What was the percentage increase (as compared to the 2004 total)? Round your answer to the nearest thousandth.

7.4 Solving Percent Problems Using the Equation $R \times B = A$

The Basic Equation $R \times B = A$

Objectives

① Recognize the three types of percent problems.

② Solve percent problems using the equation $R \times B = A$.

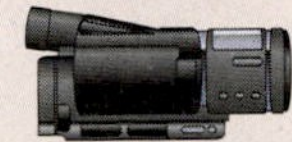

In Section 7.3 we introduced the three types of percent problems and the use of the basic proportion $\frac{P}{100} = \frac{A}{B}$ to solve these problems. If both sides of this proportion are multiplied by B, and P is changed to decimal form R (or fraction form), we have the following result.

$$\frac{P}{100} = \frac{A}{B}$$

$$R = \frac{A}{B}$$

$$R \times B = \frac{A}{B} \times B$$

or

$$R \times B = A,$$

with R equal to the value of $P\%$ **in decimal form**.

We will call this last equation, $\boldsymbol{R \times B = A}$, the basic equation for solving percent problems.

Now consider the statement

25% of 60 is 15.

We can set this statement in the form of the basic equation as follows:

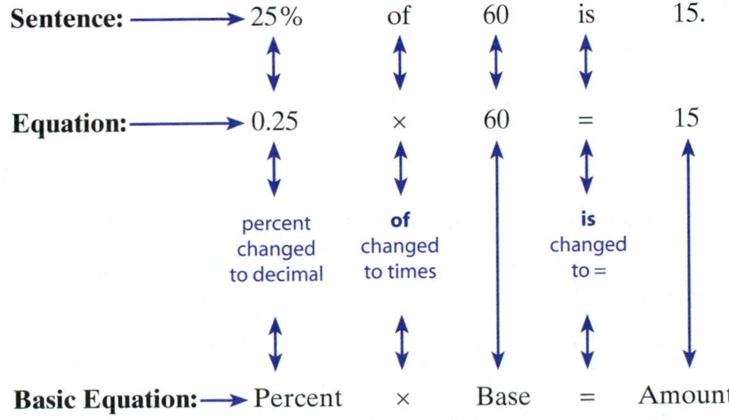

The terms represented in the basic equation are explained in the box on the next page.

Terms

R = Rate or percent (as a decimal or fraction)

B = Base (number for which we are finding the percent)

A = Amount or percentage (a part of the base)

The relationship among R, B, and A is given in the basic equation

$$\boldsymbol{R \times B = A}.$$

Using the Basic Equation

Even though there are just **three basic types of problems that involve percent**, many people have a difficult time differentiating among them and determining whether to multiply or divide in a particular problem. Most of these difficulties can be avoided by using the basic equation $R \times B = A$ and the equation-solving skills learned in Chapter 6. If the values of two of the quantities in the equation $R \times B = A$ are known, these values can be substituted into the equation, and the third value can be determined by solving the resulting equation.

Objective 1

The Three Basic Types of Percent Problems and the Equation $R \times B = A$

Type 1: Find the amount, given the base and the percent.

What is 45% of 70?

$$R \times B = A$$
$$\swarrow \quad \downarrow \quad \downarrow$$
$$0.45 \times 70 = A$$

R and B are known. The object is to find A.

Type 2: Find the base, given the percent and the amount.

30% of what number is 18.36?

$$R \times B = A$$
$$\swarrow \quad \downarrow \quad \searrow$$
$$0.30 \times B = 18.36$$

R and A are known. The object is to find B.

Type 3: Find the percent, given the base and the amount.

What percent of 84 is 16.8?

$$R \times B = A$$
$$\downarrow \quad \downarrow \quad \searrow$$
$$R \times 84 = 16.8$$

A and B are known. The object is to find R.

Remember:

(a) **Of** means to multiply when used with decimal or fractions.

(b) **Is** means =.

(c) The percent is changed to decimal or fraction form.

Problem Type 1: Finding the Value of *A*

Objective 2

Example 1

What is 45% of 70?

Solution

Here $R = 45\% = 0.45$ and $B = 70$.

Substituting into the equation $R \times B = A$ gives

$$0.45 \times 70 = A$$
$$31.5 = A$$

Multiplying 70 by 0.45 gives us 31.5.

$$\begin{array}{r} 70 \\ \times\ 0.45 \\ \hline 3\ 50 \\ 28\ 0\ \ \\ \hline 31.\ 50 \end{array}$$

So 45% of 70 is __31.5__.

Now work exercise 1 in the margin.

Find the unknown quantity.

1. 10% of 137 is ______.

Note: The multiplication can be done by hand, as shown here, or with a calculator. In either case, the equations should be written so the = signs are aligned one above the other, as shown in the solution.

Example 2

What is 18% of 200?

Solution

In this problem $R = 18\% = 0.18$ and $B = 200$.

Substituting into the equation $R \times B = A$ gives

$$0.18 \times 200 = A$$
$$36 = A$$

So 18% of 200 is __36__.

Now work exercise 2 in the margin.

2. 60% of 111 is ______.

3. 8% of _______ is 2.

4. 50% of _______ is 73.

Problem Type 2: Finding the Value of *B*

Example 3

30% of what number is 18.36?

Solution

In this case $R = 30\% = 0.30$ and $A = 18.36$.

Substituting into the equation $R \times B = A$ gives

$$0.30 \times B = 18.36$$

$$\frac{\cancel{0.30} \times B}{\cancel{0.30}} = \frac{18.36}{0.30} \quad \text{Divide both sides by 0.30, the coefficient of } B.$$

$$B = 61.2$$

$$\begin{array}{r} 61.2 \\ .30.\overline{)18.36.0} \\ \underline{18\,0} \\ 36 \\ \underline{30} \\ 60 \\ \underline{60} \\ 0 \end{array}$$

So 30% of <u>61.2</u> is 18.36.

Now work exercise 3 in the margin.

Example 4

82% of ________ is 246.

Solution

Here $R = 82\% = 0.82$ and $A = 246$.

Substituting into the equation $R \times B = A$ gives

$$0.82 \times B = 246$$

$$\frac{\cancel{0.82} \times B}{\cancel{0.82}} = \frac{246}{0.82}$$

Divide both sides by 0.82, the coefficient of *B*. The division can be performed with a calculator.

$$B = 300$$

So 82% of <u>300</u> is 246.

Now work exercise 4 in the margin.

Problem Type 3: Finding the Value of *R*

Example 5

What percent of 84 is 16.8?

Solution

In this problem $B = 84$ and $A = 16.8$.

Substituting into the equation $R \times B = A$ gives

$$R \times 84 = 16.8$$
$$\frac{R \times \cancel{84}}{\cancel{84}} = \frac{16.8}{84}$$
$$R = 0.2 = 20\%$$

Divide both sides by 84, the coefficient of *R*.

$$\begin{array}{r} 0.2 \\ 84\overline{)16.8} \\ \underline{16.8} \\ 0 \end{array}$$

So $\underline{\ 20\%\ }$ of 84 is 16.8.

5. ______% of 56 is 67.2.

Now work exercise 5 in the margin.

Example 6

______% of 195 is 234.

Solution

In this case $B = 195$ and $A = 234$.

Substituting into the equation $R \times B = A$ gives

$$R \times 195 = 234$$
$$\frac{R \times \cancel{195}}{\cancel{195}} = \frac{234}{195}$$
$$R = 1.2 = 120\%$$

Divide both sides by 195, the coefficient of *R*.

R is more than 100% because *A* is larger than *B*.

So $\underline{\ 120\%\ }$ of 195 is 234.

6. ______% of 1000 is 761.

Now work exercise 6 in the margin.

When working with applications, the wording of a problem is unlikely to be exactly as shown in Examples 1 – 6. The following examples show some alternative wordings, but each problem is still one of the three basic types.

Example 7

Find 65% of 42.

Solution

Here the percent = 65% and the base = 42. This is a Type 1 problem:

65% of 42 is _____.

$$0.65 \times 42 = A$$
$$27.3 = A$$

Multiplying 42 by 0.65 gives us 27.3.

$$\begin{array}{r} 0.65 \\ \underline{\times \quad 42} \\ 1.30 \\ \underline{26.0\ } \\ 27.30 \end{array}$$

So 65% of 42 is $\underline{\ 27.3\ }$.

7. Find 33% of 180.

Now work exercise 7 in the margin.

8. ______% of 400 is 624.

Example 8

What percent of 125 gives a result of 31.25?

Solution

Here the percent is unknown. The base = 125 and the amount = 31.25. This is a Type 3 problem:

_____% of 125 is 31.25.

$$R \times 125 = 31.25$$

$$\frac{R \times \cancel{125}}{\cancel{125}} = \frac{31.25}{125}$$

$$R = 0.25 = 25\%$$

Divide both sides by 125, the coefficient of R.

$$\begin{array}{r} 0.25 \\ 125\overline{)31.25} \\ \underline{25\,0} \\ 6\,25 \\ \underline{6\,25} \\ 0 \end{array}$$

Now work exercise 8 in the margin.

The following examples show how to use fraction equivalents for percents to simplify your work.

9. Find 50% of 420.

Example 9

What is 25% of 24?

Solution

25% of 24 is ___. **Type 1 Problem**

$$\frac{1}{\cancel{4}} \times \overset{6}{\cancel{24}} = 6 = A$$

$25\% = \frac{1}{4}$

So 25% of 24 is <u>6</u>.

Now work exercise 9 in the margin.

10. Find 150% of 60.

Example 10

Find 75% of 36.

Solution

75% of 36 is _____. **Type 1 Problem**

$$\frac{3}{\cancel{4}} \times \overset{9}{\cancel{36}} = 27 = A$$

$75\% = \frac{3}{4}$

So 75% of 36 is <u>27</u>.

Now work exercise 10 in the margin.

Example 11

What is the value of $33\frac{1}{3}\%$ of 72?

Solution

$\frac{1}{3} \times 72 = A$

Type 1 Problem

$33\frac{1}{3}\% = \frac{1}{3}$

$\frac{1}{\cancel{3}} \times \overset{24}{\cancel{72}} = 24 = A$

So $33\frac{1}{3}\%$ of 72 is __24__.

Now work exercise 11 in the margin.

11. Find the value of $66\frac{2}{3}\%$ of 189.

Example 12

$37\frac{1}{2}\%$ of what number is 300?

Solution

Type 2 Problem

$\frac{3}{8} \times B = 300$

$37\frac{1}{2}\% = \frac{3}{8}$

$\frac{\cancel{8}}{\cancel{3}} \times \frac{\cancel{3}}{\cancel{8}} \times B = \frac{8}{\cancel{3}} \times \overset{100}{\cancel{300}}$

$\frac{8}{3}$ is the reciprocal of $\frac{3}{8}$.

$B = 800$

So $37\frac{1}{2}\%$ of __800__ is 300.

Now work exercise 12 in the margin.

12. $87\frac{1}{2}\%$ of a number is 14. Determine the number.

Remember that a percent is just another form of a fraction and that, when you find a percent of a given number, the amount will be:

a. smaller than the given number if the percent is less than 100%;

b. greater than the given number if the percent is more than 100%.

Practice Problems

Answer the following using the equation $R \times B = A$.

1. What is 40% of 73?

2. What is 150% of 12.6?

3. 87.5% of what number is 21?

4. 12% of what number is 15?

5. What percent of 88 is 55?

6. What percent of 4 is 7?

Answers to Practice Problems:

1. 29.2
2. 18.9
3. 24
4. 125
5. 62.5%
6. 175%

Name ______________ Section __________ Date __________

Exercises 7.4

Solve each problem for the unknown quantity. You may use a calculator to perform calculations; however, you should first write the basic equation and other resulting equations, one below the other, with = signs aligned. See Examples 1 through 6.

1. 10% of 70 is ______.

2. 5% of 62 is ______.

3. 15% of 60 is ______.

4. 25% of 72 is ______.

5. 75% of 12 is ______.

6. 60% of 30 is ______.

7. 100% of 36 is ______.

8. 80% of 50 is ______.

9. 2% of ______ is 3.

10. 20% of ______ is 17.

11. 3% of ______ is 21.

12. 30% of ______ is 21.

13. 100% of ______ is 75.

14. 50% of ______ is 42.

15. 150% of ______ is 63.

16. 110% of ______ is 330.

17. ______% of 60 is 90.

18. ______% of 150 is 60.

19. ______% of 75 is 15.

20. ______% of 12 is 4.

21. ______% of 34 is 17.

22. ______% of 30 is 6.

23. ______% of 48 is 16.

24. ______% of 100 is 35.

ANSWERS

1. ______

2. ______

3. ______

4. ______

5. ______

6. ______

7. ______

8. ______

9. ______

10. ______

11. ______

12. ______

13. ______

14. ______

15. ______

16. ______

17. ______

18. ______

19. ______

20. ______

21. ______

22. ______

23. ______

24. ______

25. ______

26. ______

27. ______

28. ______

29. ______

30. ______

31. ______

32. ______

33. ______

34. ______

35. ______

36. ______

37. ______

38. ______

39. ______

40. ______

41. ______

42. ______

43. ______

44. ______

45. ______

46. ______

47. ______

Each of the following problems is one of the three types discussed written in a different way. Remember that B is the number for which you are finding the percent. See Examples 7 through 12.

25. _____ is 50% of 25.

26. _____ is 31% of 76.

27. 22 is 20% of _____.

28. 86 is 100% of _____.

29. 18 is _____% of 10.

30. 15 is _____ % of 10.

31. 24 is $33\frac{1}{3}\%$ of _____ .

32. 92.1 is 15% of _____.

33. 119.6 is 23% of _____ .

34. 9.5 is 25% of _____.

35. 36 is _____% of 18.

36. 60 is _____% of 40.

37. _____ is 96% of 17.

38. _____ is 84% of 32.

39. _____ is 18% of 325.

40. _____ is 28% of 460.

41. Find 18% of 244.

42. Find 15.2% of 75.

43. Find 120% of 60.

44. What percent of 32 is 8?

45. 100 is 125% of what number?

Use fractions to solve the following problems. Do the work mentally, if you can, after you have set up the related equation. See Examples 9 through 12.

46. Find 50% of 32.

47. Find $66\frac{2}{3}\%$ of 60.

Name ______________________ Section ____________ Date __________

ANSWERS

48. What is $12\frac{1}{2}\%$ of 80?

49. What is $62\frac{1}{2}\%$ of 16?

50. $33\frac{1}{3}\%$ of 75 is what number?

51. 25% of 150 is what number?

52. 75% of what number is 21?

53. 50% of what number is 35?

54. $37\frac{1}{2}\%$ of what number is 61.2?

55. 100% of what number is 76.3?

56. Magazine subscriptions. Magazine publishers will often give significant discounts to subscribers. The cover price of *Newsweek* is $79.65 for 6 months, but a 6-month subscription costs only $18.57! What is the percent savings (to the nearest tenth) for someone who subscribes to *Newsweek*?

57. Casualties of War. Only 27% of American deaths in the Revolutionary War occurred in battle. All other mortalities were the result of things such as exposure, disease, and starvation. Of the estimated 25,300 deaths, how many men died in battle (to the nearest hundred)?

58. Presidential vetoes. During his presidency from 1945 to 1953, Harry Truman vetoed 250 congressional bills, and 12 of those vetoes were overridden. What percent of Truman's vetoes were overridden?

59. Supreme Court justices. William O. Douglas served as an associate justice on the Supreme Court for 36 years, from 1939 to 1975, the longest tenure of any Supreme Court justice in history. He lived to be 82. What percent of his life (to the nearest one percent) did he spend on the Supreme Court?

60. School teachers. The decade from 1960 to 1970 saw the largest 10-year increase in the number of male elementary and high school teachers in our nation's history-from 402,000 to 691,000. What was the percent increase from 1960 to 1970 (to the nearest one percent)?

48. __________

49. __________

50. __________

51. __________

52. __________

53. __________

54. __________

55. __________

56. __________

57. __________

58. __________

59. __________

60. __________

Check Your Number Sense

61. ______

62. ______

63. ______

64. ______

65. ______

66. ______

67. ______

68. ______

In each of the following exercises, one of the choices is the correct answer to the problem. Work through all the exercises by reasoning and calculating mentally. After you have checked your answers, you should work out the answers to any problems you missed and try to understand why your reasoning was incorrect.

61. Write 7% as a fraction.

(a) $\frac{7}{1}$ (b) $\frac{7}{10}$ (c) $\frac{7}{100}$ (d) $\frac{7}{1000}$

62. Find 10% of 730.

(a) 7.3 (b) 73 (c) 14.6 (d) 146

63. Find 20% of 730.

(a) 7.3 (b) 73 (c) 14.6 (d) 146

64. What is 100% of 9.03?

(a) 0.0903 (b) 0.903 (c) 9.03 (d) 90.3

65. 10% of what number is 29?

(a) 2.9 (b) 29 (c) 290 (d) 2900

66. 1% of what number is 29?

(a) 0.29 (b) 2.9 (c) 290 (d) 2900

67. What percent of 100 is 25?

(a) 0.25% (b) 2.5% (c) 25% (d) 250%

68. What percent of 1000 is 25?

(a) 0.25% (b) 2.5% (c) 25% (d) 250%

Name ______________________ Section ____________ Date __________

ANSWERS

69. 245 is approximately 50% of what number?

(a) 24 (b) 120 (c) 500 (d) 1200

69. ______________

70. 72 is approximately 30% of what number?

(a) 25 (b) 75 (c) 150 (d) 250

70. ______________

7.5 Applications with Percent: Discount, Sales Tax, and Tipping

Pólya's Four-Step Process for Solving Problems

George Pólya (1887–1985), a famous professor at Stanford University, studied the process of discovery learning. Among his many accomplishments, he developed a four-step process as an approach to problem solving:

1. Understand the problem.
2. Devise a plan.
3. Carry out the plan.
4. Look back over the results.

There are a variety of types of applications discussed throughout this text and in subsequent courses in mathematics, and you will find these four steps helpful as guidelines for understanding and solving all of them. *Applying the necessary skills to solve exercises, such as adding fractions or solving equations, is not the same as accumulating the knowledge with which to solve problems.* Problem solving can involve careful reading, reflection, and some original or independent thought.

Objectives

① Learn Pólya's Four-Step Process for Solving Problems.

② Apply skills related to solving percent problems to solving discount problems.

③ Apply skills related to solving percent problems to solving sales tax problems.

④ Apply skills related to solving percent problems to solving problems involving leaving a tip.

Objective ①

Basic Steps for Problem Solving:

1. Understand the problem. For example,
   a. Read the problem carefully and identify the key words.
   b. Understand what information is given and what is to be found.

2. Devise a plan using, for example, one or all of the following:
   a. Guess, estimate, or make a list of possibilities.
   b. Draw a picture or diagram.
   c. Use a variable and form an equation.

3. Carry out the plan. For example,
   a. Try all the possibilities you have listed.
   b. Solve any equation that you may have set up.

4. Look back over the results. For example,
   a. Can you see an easier way to solve the problem?
   b. Does your solution actually work? Does it seem reasonable?
   c. If there is an equation, check your answer in the equation.

In this section we will be concerned with applications that involve percent and the types of percent problems that have been discussed in the first four sections of Chapter 7. **The use of a calculator is recommended, but keep in mind that it is a tool to enhance, not replace, the necessary skills and abilities related to problem solving.** Study the following examples carefully, as they illustrate some basic plans for problem solving with percent.

Objective 2

Discount

A **discount** is a reduction in the **original price** (or **marked price**) of an item. The new reduced price is called the **sale price** and the discount is the difference between the original price and the sale price. The **rate of discount** (or **percent of discount**) is a percent that represents what part the discount is of the original price.

1. A pair of shoes originally priced at \$52 is now discounted 25%.

a. What is the amount of the discount?

b. What is the sale price?

Example 1

A refrigerator that regularly sells for \$975 is on sale at a 30% discount.

a. What is the amount of the discount?

b. What is the sale price?

Solution

First, find the amount of the discount, and then find the sale price by subtracting the discount from the original price.

a. Since the rate of discount is 30%, we find 30% of \$975.

30% of \$975 is__________ .

$$0.30 \times 975 = A$$
$$292.50 = A$$

\$975	original price
× .30	rate of discount
\$292.50	discount

The discount is \$292.50.

b. Find the sale price by subtracting the discount from the original price.

\$975.00	original price
− 292.50	discount
\$682.50	sales price

The sale price is \$682.50.

Now work exercise 1 in the margin.

Objective 3

Sales Tax

Sales tax is a tax charged on goods sold by retailers, and it is assessed by states for income to operate state services. The rate of the sales tax varies from state to state (or even city to city in some cases). In fact, some states do not have a sales tax.

Example 2

If the sales tax rate is 6%, what would be the final cost of the refrigerator in Example 1?

Solution

First find the amount of the sales tax, and then add this tax to the sale price found in Example 1 to find the final cost.

a. Find the amount of sales tax.
6% of $682.50 is ________ .

$0.06 \times 682.50 = A$

$40.95 = A$

$682.50	sale price
× 0.06	tax rate
$40.9500	sales tax

The sales tax is $40.95.

b. Add the sales tax to the sale price to find the final cost.

$682.50	sale price
+ 40.95	sales tax
$723.45	final cost

The final cost of the refrigerator is $723.45.

Now work exercise 2 in the margin.

2. Assuming a 7% sales tax rate, what would be the final cost of the discounted shoes in exercise 1?

Example 3

An auto dealer paid $7566 for a large order of a special part. This was **not** the original price. He received a 3% discount off the original price because he paid cash. What was the original price?

Solution

Since $7566 is not the original price, do **not** take 3% of $7566. In fact, we are to find the original price. Therefore, we must reason that if a price is discounted by 3%, then 97% of the original price is what remains (100% – 3% = 97%). Thus $7566 represents 97% of the original price.

97% of ________ is $7566.

$0.97 \times B = 7566$

$\frac{\cancel{0.97} \times B}{\cancel{0.97}} = \frac{7566}{0.97}$

$B = 7800$

$$\begin{array}{r} 7800. \\ .97.\overline{)7566.00.} \\ \underline{679} \\ 776 \\ \underline{776} \\ 00 \\ \underline{0} \\ 00 \\ \underline{0} \\ 0 \end{array}$$

The original price was $7800.

Now work exercise 3 in the margin.

3. You buy a watch for $90, which is a 40% discount off the original price. What did the watch cost originally?

> **Note:** To check this result, take 3% of $7800 to find the discount. Subtract this discount from $7800 to get the discounted price of $7566.

Objective 4

Tipping

Tipping (or **leaving a tip**) is the custom of leaving a percent of a bill (usually at a restaurant) as a payment to the waiter or waitress for providing good service. The amount of the tip is usually 10 – 20% of the amount charged (some include tax in this calculation, others do not), depending on the quality of the service. The "usual" tip is 15%, but in the case of particularly bad service, no tip is given.

Since most people do not carry a calculator with them when dining out, the calculation of a 15% tip can be an interesting exercise in percents. The following rule of thumb uses the basic facts that 5% is half of 10% and that 10% of a decimal number can be found by moving the decimal point one place to the left.

Rule of Thumb for Calculating a 15% Tip:

1. For ease of calculation, round the amount of the bill to the nearest whole dollar.
2. Find 10% of the rounded amount by moving the decimal point one place to the left.
3. Divide the answer in step 2 by 2. (This represents 5% of the rounded amount, or one-half of 10%.)
4. Add the two amounts found in steps 2 and 3. This sum is the amount of the tip.

> **Note:** There are other rules of thumb that people use to calculate the amount of a tip. For example, you might always round up instead of rounding to the nearest whole dollar, or you might round to the nearest ten dollars, or you might leave a tip that makes the total of your expenses a whole-dollar amount. Consider two bills of $7.95 and $7.20. Some might round each bill to $10.00 and leave a tip of 15% as $1.00 + $0.50 = $1.50 in each case. However, according to the rule in the text, we will calculate the tips as follows:
>
> Round $7.95 to $8.00 and calculate 15% as $0.80 + $0.40 = $1.20.
>
> Round $7.20 to $7.00 and calculate 15% as $0.70 + $0.35 = $1.05.

Example 4

You and a friend dine out and the bill comes to \$28.40, including tax. If you plan to leave a 15% tip, what should be the amount of the tip?

Solution

For ease of computation, round \$28.40 to \$28.00.
Find 10% of \$28.00 by moving the decimal point one place to the left:

$$10\% \text{ of } 28.00 = 2.80$$

Now find 5% of \$28.00 by dividing \$2.80 by 2:

$$2.80 \div 2 = 1.40$$

Adding gives:

$$\begin{array}{r} \$2.80 \\ +\ 1.40 \\ \hline \$4.20 \end{array} \quad \text{amount of tip}$$

We will say that the tip should be \$4.20 as a textbook answer. However, practically speaking (depending on things such as how much change you have in your pocket and how good the service was), you might leave either \$4.25 or \$4.50.

Now work exercise 4 in the margin.

4. Your total bill is \$37.81 and you want to leave a 15% tip. How much tip should you leave?

Practice Problems

1. A television which normally sells for \$300 is priced at a 10% discount. Find **a.** the amount of the discount and **b.** the sale price.

2. If the sales tax rate is 6.5%, what is the tax on an \$800 purchase?

3. Your dinner bill comes to \$18.84. If you plan to tip 15%, how much tip should you leave?

Answers to Practice Problems:

1.a. \$30 **b.** \$270
2. \$52 **3.** \$2.85

Name ______________ Section ____________ Date __________

ANSWERS

1. a. ______

b. ______

2. a. ______

b. ______

3. a. ______

b. ______

4. a. ______

b. ______

5. a. ______

b. ______

c. ______

6. a. ______

b. ______

Exercises 7.5

The following problems may involve several calculations, and calculators are recommended. Follow Pólya's four-step process for problem solving. After you have read the problem so that you understand it, write down the known information and your plan for solving the problem. Then follow through with your plan. See Examples 1 through 3.

1. **Discounts.** A store owner received a 3% discount from the manufacturer when she bought $15,500 worth of dresses.
   a. What was the amount of the discount?
   b. What did she pay for the dresses?

2. **Sales tax.** If the sales tax in a certain state is figured at 6%:
   a. How much tax is there on a purchase of $30.20?
   b. What is the total amount paid for the purchase?

3. **Sales tax.** If sales tax was figured at 6%:
   a. How much tax was paid on the purchase of three textbooks priced at $55.00, $25.50, and $34.95?
   b. What would be the total cost of all three books?

4. **Discounts.** A new briefcase was priced at $275. If it was to be marked down 30%:
   a. What would be the discount?
   b. What would be the new price?

5. **Discounts.** The discount on a fur coat was $150. This was a 20% discount.
   a. What was the original selling price of the coat?
   b. What was the sale price?
   c. What was the total amount paid for the coat if a 6% sales tax was added to the sale price?

6. **Discounts.** The discount on men's suits was $50, and they were on sale for $200.
   a. What was the original selling price?
   b. What was the rate of discount?

7. a. ____________

b. ____________

8. ____________

9. ____________

10. ____________

11. ____________

12. ____________

13. ____________

7. Discounts. In order to get more subscribers, a book club offered three books for a total price of $7.02. The total selling price was originally $17.55 for all three books.

a. What was the amount of the discount?

b. Based on the original selling price, what was the rate of the discount on these three books?

8. Discounts. An auto supply store received a shipment of auto parts and a bill for $845.30. Some of the parts were not as ordered, and they were returned immediately. The value of the parts returned was $175.50. The terms of the billing provided the store with a 2% discount if it paid cash (for the parts it kept) within two weeks. What did the store pay for the parts it kept if it paid cash within two weeks?

9. Discounts. Towels were on sale at a discount of 30%. If the sale price was $3.01, what was the original price?

10. Discounts. Computer disks were on sale for $5.24 per box. What was the original price per box if the sale price represents a discount of 20%?

In Exercises 11 – 16, use the rule of thumb stated in the text to calculate a 15% tip. See Example 4.

11. Tipping. You invited your friend to lunch at the local coffeehouse, and the bill totaled $12.60. Your friend offered to leave the tip. What amount did he leave as a 15% tip?

12. Tipping. A lawyer took four clients to dinner. The bill for the meals and drinks was $150.00 plus a 6% sales tax. What amount did she leave as a tip (rounding up to the nearest dollar) if she left 15%?

13. Tipping. A bill for a family dining at a restaurant was $38.40. What would a 15% tip have been? What total amount should they have left on the table if they wanted to go before the waiter came to pick up the money?

Name ______________ Section ______________ Date ______________

ANSWERS

14. Tipping. On a recent business trip, Ken and Joe had breakfast at the restaurant next to the motel where they were staying. Ken's breakfast bill was \$6.95 and Joe's was \$8.75. How much should each man have left as a tip if each tipped 15%?

14. ______________

15. ______________

15. Tipping. Mrs. Chung had two large pizzas delivered to her home for \$26.00. If she tipped the delivery person 15%, what total amount did she pay the driver?

16. ______________

16. Tipping. Juan took his date out for dinner before the senior prom. The total bill, including tax, was \$42.00. How much did he tip the waitress if he left 15%? What were his total expenses for the meal?

17. a. ______________

17. Discounts and tax. Linda is enrolled in a calculus course. She has the choice of buying the text in hardback form for \$60.00 or in paperback form for \$46.50. Tax is figured at 5% of the selling price. The bookstore buys back hardback books for 40% of the selling price and paperback books for 30% of the selling price.

a. Which book is the more economical buy for Linda if she sells her book back to the bookstore at the end of the semester?

b. How much does she save?

b. ______________

18. Test grades. A student missed 3 problems on a mathematics test and received a grade of 85%. If all the problems were of equal value, how many problems were on the test?

18. ______________

19. Real estate. Suppose you sell your home for \$100,000 and you owe the Savings and Loan \$60,000 on the first trust deed. You pay a real estate agent a commission of 6% of the selling price and other fees and taxes totaling \$1200. How much cash do you have from the sale?

19. ______________

20. ________

21. a. ________

b. ________

22. ________

23. a. ________

b. ________

24. ________

20. Sheets are marked \$22.50 and pillowcases \$7.50. What is the sale price of each item if each item is discounted 25% off the marked price?

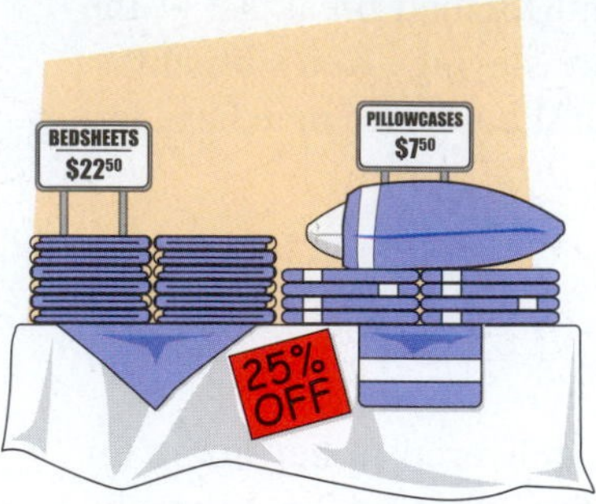

21. In the roofing business, shingles are sold by the "square," which is enough material to cover a 10 ft by 10 ft square (or 100 square feet). A roofing supplier has a closeout on shingles at a 30% discount.

a. If the original price was \$230 per square, what is the sale price per square?

b. How much would a roofer pay for 34 squares?

Writing and Thinking about Mathematics

22. Make up your own rule of thumb for leaving a 20% tip.

23. Weight loss and gain. A man weighed 220 pounds. He lost 22 pounds in three months. Then he gained back 22 pounds over the next four months.

a. What percent of his weight did he lose in the first three months?

b. What percent of his weight did he gain back? The loss and gain are the same amount, but the two percents are different. Explain why.

Collaborative Learning

24. With the class separated into teams of two to four students, each team is to analyze the following discussion on discount and decide how to answer the related questions. Then each team leader is to present the team's answers and related ideas to the class for general discussion.

In Example 1 on page 616, we calculated a 30% discount to find the sale price of a refrigerator. Then, in Example 2, we calculated the sales tax at 6% of the sale price. The final cost of the refrigerator was then found by adding the sales tax to the sale price. Could most of these calculations have been avoided by simply discounting the original price by 24% to determine the final cost (30% − 6% = 24%)? Explain your reasoning.

7.6 Applications with Percent: Commission, Profit, and Others

Commission

Objective 1

A **commission** is a fee paid to an agent or salesperson for a service. Commissions are usually a percent of a negotiated contract or a percent of sales.

Objectives

1. Be familiar with and understand the term **commission**.
2. Understand that percent of profit is a ratio and can be based on cost or on selling price.
3. Know how to calculate percent of profit.

Example 1

A saleswoman earns a fixed salary of \$900 a month plus a commission of 8% on whatever amount she sells over \$6500 in merchandise. What did she earn the month she sold \$11,800 in merchandise?

Solution

First find the base of her commission by subtracting \$6500 from the total amount she sold. Then find 8% of this base and add her monthly salary.

a. Subtract \$6500 from \$11,800.

\$11,800	total amount she sold
− 6500	amount on which she does not earn a commission
\$ 5300	base of commission

b. Find 8% of this base.

8% of \$5300 is ______

$0.08 \times 5300 = A$

$424 = A$

Multiplying the base of commission by the rate gives us total commission.

\$5300	base of commission
× 0.08	rate of commission
\$ 424	commission

c. Her monthly pay is the sum of her salary and her commission.

\$900.00	fixed salary
+ 424.00	commission
\$1324.00	total pay for the month

She earned \$1324 for the month.

Now work exercise 1 in the margin.

1. What is the value of a 6% commission on a \$3200 sale?

Percent of Profit

In business, a company's **profit** is the difference between income (or revenue) and costs, such as wages, materials, and rent. However, manufacturers and retailers are also concerned with the profit on each item produced or sold, which is simply the difference between its selling price to the customer and the cost to the company. This is the type of profit that will be discussed in this section.

Terms Related to Profit

Profit: The difference between selling price and cost (profit = selling price – cost)

Objective 2

Percent of profit: There are two types; both are ratios with profit in the numerator.

1. Percent of profit **based on cost** is the ratio of profit to cost:

$$\frac{\text{profit}}{\text{cost}} = \text{\% of profit based on cost.}$$

2. Percent of profit **based on selling price** is the ratio of profit to selling price:

$$\frac{\text{profit}}{\text{selling price}} = \text{\% of profit based on selling price.}$$

2. A company makes t-shirts for \$5 apiece and sells them for \$12. For each shirt, find:

a. the profit on each t-shirt.

b. the percent of profit based on cost.

c. the percent of profit based on selling price.

Example 2

A company manufactures and sells plastic boxes that cost \$21 each to produce, and that sell for \$28 each.

a. What is the profit on each box?

b. What is the percent of profit based on cost?

c. What is the percent of profit based on selling price?

Solution

a. To find the profit, subtract the cost from the selling price.

$$\begin{array}{rl} \$28 & \text{selling price} \\ -\ 21 & \text{cost} \\ \hline \$\ 7 & \text{profit} \end{array}$$

Objective 3

b. To find the percent of profit based on cost, divide the profit by the cost.

$$\frac{\$7 \text{ profit}}{\$21 \text{ cost}} = \frac{1}{3} = 0.33\frac{1}{3} = 33\frac{1}{3}\% \text{ profit based on cost}$$

c. To find the percent of profit based on selling price, divide the profit by the selling price.

$$\frac{\$7 \text{ profit}}{\$28 \text{ selling price}} = \frac{1}{4} = 0.25 = 25\% \text{ profit based on selling price}$$

Note that in parts **b.** and **c.** the profit is \$7 and this does not change. What changes, and gives different percents, is the denominator.

Now work exercise 2 in the margin.

Note: Percent of profit **based on cost** is higher than percent of profit **based on selling price**. The business community reports whichever percent serves its purposes better. Your responsibility as an investor or consumer is to know which percent is reported and what it means to you.

Completion Example 3

Women's coats were on sale for $250.

a. If the coats cost the store owner $200, what was his percent of profit based on cost?

b. What was his percent of profit based on selling price?

Solution

a. First find the profit.

$ 250	selling price
– ____	cost
$ ___	profit

b. Find the percent of profit based on cost.

$$\frac{\_\_\_\_\text{ profit}}{\_\_\_\_\text{ cost}} = \_\_\_\_\_\_\% \text{ profit based on cost}$$

c. Find the percent of profit based on selling price.

$$\frac{\_\_\_\_\text{ profit}}{\_\_\_\_\text{ selling price}} = \_\_\_\_\_\% \text{ profit based on selling price}$$

Now work exercise 3 in the margin.

3. A used car salesman buys a car for $2400 and sells it for $4000.

a. Find the percent of profit based on cost.

b. Find the percent of profit based on selling price.

Completion Example Answers

3. a. First find the profit.

$250	selling price
– **200**	cost
$ **50**	profit

b. Find the percent of profit based on cost.

$$\frac{\mathbf{\$50}\text{ profit}}{\mathbf{\$200}\text{ cost}} = \mathbf{25\%} \text{ profit based on cost}$$

c. Find the percent of profit based on selling price.

$$\frac{\mathbf{\$50}\text{ profit}}{\mathbf{\$250}\text{ selling price}} = \mathbf{20\%} \text{ profit based on selling price}$$

Practice Problems

1. A salesman earns \$1100 a month plus a 7% commission on any amount above \$5000 in sales. What does he earn in a month during which he sells \$9300 worth of merchandise?

*In each case, find **a.** the percent of profit based on cost and **b.** the percent of profit based on selling price.*

2. A baker sells loaves of bread for \$3 apiece; they cost him \$1.25 apiece to make.

3. A shoe store buys pairs of dress shoes wholesale for \$50 and sells them for \$80.

4. You find an old Beatles album at a yard sale for \$2 and sell it on eBay for \$50.

Answers to Practice Problems:

1. \$1401

2. a. 140% **b.** $58\frac{1}{3}\%$ or $58.\overline{3}\%$

3. a. 60% **b.** 37.5%

4. a. 2400% **b.** 96%

Name ______________________ Section ____________ Date __________

ANSWERS

1. ______________

2. ______________

3. ______________

4. ______________

5. ______________

6. a. ______________

b. ______________

Exercises 7.6

1. **Commissions.** A realtor works on a 6% commission. What is his commission on a house he sold for $195,000?

2. **Commissions.** A realtor selling commercial property works on a 4% commission. What is her commission on a building she sold for $875,000?

3. **Commissions.** A sales clerk receives a monthly salary of $1295 plus a commission of 7% on all sales over $2500. What did the clerk earn the month she sold $16,000 in merchandise?

4. **Commissions.** A shoe saleswoman works on fixed salary of $940 per month plus a 5% commission. How much did she make during the month in which she sold $7500 worth of shoes?

5. **Commissions.** If a salesman works on a 10% commission only (no monthly salary), how much merchandise will he have to sell to earn $2800 in one month?

6. **Field goal percentage.** A basketball player made 120 of 300 shots she attempted.
   a. What percent of her shots did she make?
   b. What percent did she miss?

7. ______

8. a. ______

b. ______

c. ______

9. a. ______

b. ______

c. ______

10. a. ______

b. ______

c. ______

11. a. ______

b. ______

12. ______

7. **Free throw percentage.** In one season, a basketball player missed 15% of his free throws. How many free throws did he make if he attempted 180?

8. **Sports equipment.** A set of golf clubs cost a golf pro $400, and he sold them in the pro shop for $550.
   - **a.** What was his profit?
   - **b.** What was his percent of profit based on cost?
   - **c.** What was his percent of profit based on selling price?

9. **Formal wear.** Men's suits were on sale for $300. Each one cost the store owner $250.
   - **a.** What was the profit for the store?
   - **b.** What was the store's percent of profit based on cost?
   - **c.** What was the store's percent of profit based on selling price?

10. **Electronics.** The cost of a flat-screen television set to a store owner was $3300, and he sold the set for $4500.
    - **a.** What was his profit?
    - **b.** What was his percent of profit based on cost?
    - **c.** What was his percent of profit based on selling price?

11. **Automobiles.** A car dealer bought a used car for $1500. He marked up the price so he would make a profit of 25% based on his cost.
    - **a.** What was the selling price?
    - **b.** If the customer paid 8% of the selling price in taxes and fees, what did the customer pay for his car?

12. **Property tax.** The property taxes on a house were $750. What was the tax rate if the house was valued at $25,000 for tax purposes?

Name ____________ Section ____________ Date ____________

ANSWERS

13. ____________

14. ____________

15. a. ____________

b. ____________

c. ____________

16. ____________

17. ____________

13. Bonuses. A computer programmer was told he would be given a bonus of 5% of any money his programs could save the company. How much would he have to save the company to earn a bonus of $6000?

14. Home loans. You want to purchase a new home for $98,000. The bank will loan you 80% of the purchase price, but you must pay a loan fee of 2% of the amount of the loan, plus other fees totaling $850. How much cash do you need in order to purchase the home?

15. Book publishing. The author of a book was told she would have to cut the number of pages by 12% in order for the book to sell at a competitive price and still show a profit for the publisher.

a. What percent of the pages were in the final form of the book?

b. If the book contained 220 pages in its final form, how many pages did the original form contain?

c. How many pages were cut?

16. Incubation. A pigeon egg incubates in 18 days, while a duck egg requires 30 days to hatch. Expressed as a percent, how much longer does it take for the duck egg to hatch?

17. Automobile production. In 2004, Ford Motor Company produced 141,907 Mustangs, which was about 16.89% of its total car production. About how many cars did Ford produce in 2004 (to the nearest thousand)?

18. ______

19. ______

20. ______

21. a. ______

b. ______

c. ______

22. a. ______

b. ______

18. **Reindeer.** Along with cattle, sheep, and poultry, Alaskan farmers also raise reindeer as livestock. In fact, in a recent year, 37,000 reindeer accounted for 71% of Alaska's livestock population. To the nearest thousand, what was the total livestock population in Alaska that year?

19. **Stock trading.** In 1990 there were 39.7 billion shares of stock traded on the New York Stock Exchange. That volume increased steadily through 2003, when 352.4 billion shares were traded. Write the number of shares traded in 2003 as a percent of the number traded in 1990 (to the nearest one percent).

20. **Public debt.** On July 1, 2000 the U.S. government's public debt per capita was \$20,149.08 and by 2005 that debt had increased to \$26,406.99 per capita. Write the per capita debt in 2005 as a percent of the per capita dept in 2000, to the nearest one percent.

Writing and Thinking about Mathematics

21. **Commissions.** Joel worked in a men's clothing store on a straight 8% commission. His friend, who worked at the same store, earned a monthly salary of \$500 plus a 4% commission.
    - **a.** How much did each make during the month in which each sold \$18,500 worth of clothing?
    - **b.** What percent more did Joel make than his friend?
    - **c.** Explain why there is more than one answer to part **b**.

22. **a.** What topic (or topics) discussed in Chapter 7 have you found to be the most interesting? Briefly, explain why.
    - **b.** What topic (or topics) have you found to be the most difficult to learn? Briefly, explain why.

Name ____________________ Section ____________ Date ____________

ANSWERS

23. a. ____________

b. ____________

24. ____________

23. a. What topic (or topics) discussed in the text to this point have you found to be most interesting? Briefly, explain why.

b. What topic (or topics) have you found to be the most difficult to learn? Briefly, explain why.

Collaborative Learning

24. Radioactive decay. With the class separated into teams of two to four students, each team is to analyze the following discussion of atomic half-life and answer the related questions as best they can. A general classroom discussion should follow with the class coming to an understanding of the concepts involved.

The half-life of a radioactive chemical element is the time that it takes for 50% of the atoms in the element sample to decay. Scientists can use methods such as carbon-14 dating to determine the ages of archaeological artifacts by comparing the amount of carbon-14 remaining in an artifact with the level of carbon-14 that would have been present originally. The half-life of carbon-14 is 5730 years. What percent of 10 grams would be left after 5730 years? How many grams? What percent would be left after 11,460 years? How many grams? What percent would be left after 17,190 years? How many grams? Describe a pattern in your calculations over these periods of time. What percent would be left after six such periods of time?

Chapter 7 Index of Key Terms and Ideas

Section 7.1 Decimals and Percents

Percent pages 566 – 567

Percent means hundredths, or the ratio of a number to 100. The symbol % is called the **percent sign**.

Relating percent to fractions with denominator 100 page 567

If a fraction has a denominator of 100, then the numerator can be read as a percent by dropping the denominator and adding the % sign.

Percent of Profit page 568

The ratio of money made to money invested is the **percent of profit**.

To Change a Decimal to a Percent page 569

Step 1: Move the decimal point two places to the right.

Step 2: Write the % sign.

(These two steps have the effect of multiplying by 100 then dividing by 100.)

To Change a Percent to a Decimal page 570

Step 1: Move the decimal point two places to the left.

Step 2: Delete the % sign.

Section 7.2 Fractions and Percents

To Change a Fraction to a Percent page 579

Step 1: Change the fraction to a decimal. (Divide the numerator by the denominator.)

Step 2: Change the decimal to a percent.

To Change a Percent to a Fraction or Mixed Number page 582

Step 1: Write the percent as a fraction with denominator 100 and drop the % sign.

Step 2: Reduce the fraction.

Section 7.3 Solving Percent Problems Using the Proportion $\frac{P}{100} = \frac{A}{B}$

The Basic Proportion $\frac{P}{100} = \frac{A}{B}$ page 591

$P\%$ = Percent

B = Base (number for which we are finding the percent)

A = Amount or percentage (a part of the base)

The relationship among P, B, and A is given in the basic proportion

$$\frac{P}{100} = \frac{A}{B}.$$

Continued on next page...

Section 7.3 Solving Percent Problems Using the Proportion $\frac{P}{100}=\frac{A}{B}$ (cont.)

The Three Basic Types of Percent Problems and the Proportion $\frac{P}{100}=\frac{A}{B}$ page 592

Type 1: Find the amount, given the base and the percent.

What is 65% of 500? $\frac{65}{100}=\frac{A}{500}$

P and B are known. The object is to find A.

Type 2: Find the base, given the percent and the amount.

57% of what number is 51.3? $\frac{57}{100}=\frac{51.3}{B}$

P and A are known. The object is to find B.

Type 3: Find the percent, given the base and the amount.

What percent of 170 is 204? $\frac{P}{100}=\frac{204}{170}$

A and B are known. The object is to find P.

Section 7.4 Solving Percent Problems Using the Equation $R\times B=A$

The Basic Equation $R\times B=A$ pages 601 – 602

R = Rate or percent (as a decimal or fraction)

B = Base (number for which we are finding the percent)

A = Amount or percentage (a part of the base)

The relationship among R, B, and A is given in the basic equation $R\times B=A$.

The Three Basic Types of Percent Problems and the Equation $R\times B=A$ page 602

Type 1: Find the amount, given the base and the percent.

What is 45% of 70?

$$R\times B=A$$
$$\swarrow \quad \downarrow \quad \downarrow$$
$$0.45\times 70=A$$

R and B are known. The object is to find A.

Type 2: Find the base, given the percent and the amount.

30% of what number is 18.36?

$$R\times B=A$$
$$\swarrow \quad \downarrow \quad \searrow$$
$$0.30\times B=18.36$$

R and A are known. The object is to find B.

Type 3: Find the percent, given the base and the amount.

What percent of 84 is 16.8?

$$R\times B=A$$
$$\downarrow \quad \downarrow \quad \searrow$$
$$R\times 84=16.8$$

A and B are known. The object is to find R.

Section 7.5 Applications with Percent: Discount, Sales Tax, and Tipping

Basic Steps for Problem Solving: page 615

1. Understand the problem. For example,
   a. Read the problem carefully and identify the key words.
   b. Understand what information is given and what is to be found.
2. Devise a plan using, for example, one or all of the following:
   a. Guess, estimate, or make a list of possibilities.
   b. Draw a picture or diagram.
   c. Use a variable and form an equation.
3. Carry out the plan. For example,
   a. Try all the possibilities you have listed.
   b. Solve any equations that you may have set up.
4. Look back over the results. For example,
   a. Can you see an easier way to solve the problem?
   b. Does your solution actually work? Does it make sense?
   c. If there is an equation, check your answer in the equation.

Discount page 616

A **discount** is a reduction in the **original price** (or **marked price**) of an item. The new reduced price is called the **sale price** and the discount is the difference between the original price and the sale price.

Sales Tax page 616

Sales tax is a tax charged on goods sold by retailers.

Tipping page 618

Tipping is the custom of leaving a percent of a bill (usually at a restaurant) as a payment to the waiter or waitress for providing good service.

Rule of Thumb for Calculating a 15% Tip: page 618

1. For ease of calculation, round the amount of the bill to the nearest whole dollar.
2. Find 10% of the rounded amount by moving the decimal point one place to the left.
3. Divide the answer in step 2 by 2. (This represents 5% of the rounded amount, or one-half of 10%.)
4. Add the two amounts found in steps 2 and 3. This sum is the amount of the tip.

Section 7.6 Applications with Percent: Commission, Profit, and Others

Commission page 625

A **commission** is a fee paid to an agent or salesperson for a service. Commissions are usually a percent of a negotiated contract or a percent of sales.

Continued on next page...

Terms Related to Profit page 626

Profit: The difference between selling price and cost
(profit = selling price – cost)

Percent of profit: There are two types; both are ratios with profit in the numerator.

1. Percent of profit **based on cost** is the ratio of profit to cost:
   $$\frac{\text{profit}}{\text{cost}} = \text{\% of profit based on cost.}$$
2. Percent of profit **based on selling price** is the ratio of profit to selling price:
   $$\frac{\text{profit}}{\text{selling price}} = \text{\% of profit based on selling price.}$$

Name ______________________ Section ____________ Date __________

ANSWERS

Chapter 7 Review Questions

1. Percent means __________.

Change the following fractions to percents.

2. $\frac{85}{100}$ **3.** $\frac{18}{100}$ **4.** $\frac{37}{100}$

5. $\frac{16\frac{1}{2}}{100}$ **6.** $\frac{15.2}{100}$ **7.** $\frac{115}{100}$

Change the following decimals to percents.

8. 0.06 **9.** 0.3 **10.** 0.67

11. 0.027 **12.** 3 **13.** 1.2

Change the following percents to decimals.

14. 35% **15.** 4% **16.** 0.25%

17. $\frac{1}{4}\%$ **18.** 7.1% **19.** 132%

Change the following numbers to percents.

20. $\frac{6}{10}$ **21.** $\frac{3}{20}$ **22.** $\frac{4}{25}$

23. $\frac{3}{8}$ **24.** $\frac{5}{12}$ **25.** $1\frac{4}{15}$

1. ______
2. ______
3. ______
4. ______
5. ______
6. ______
7. ______
8. ______
9. ______
10. ______
11. ______
12. ______
13. ______
14. ______
15. ______
16. ______
17. ______
18. ______
19. ______
20. ______
21. ______
22. ______
23. ______
24. ______
25. ______

26. ____________
27. ____________
28. ____________
29. ____________
30. ____________
31. ____________
32. ____________
33. ____________
34. ____________
35. ____________
36. ____________
37. ____________
38. ____________
39. ____________
40. ____________
41. ____________
42. ____________
43. ____________
44. ____________
45. ____________
46. ____________
47. ____________

Change the following percents to fractions or mixed numbers.

26. 14% **27.** 40% **28.** 66%

29. $12\frac{1}{2}\%$ **30.** 400% **31.** $33\frac{1}{2}\%$

Solve each problem for the unknown quantity.

32. 30% of 52 is ________. **33.** 15% of 17 is ________.

34. 3% of ________ is 7. **35.** 42% of ________ is 18.

36. ________ %of 36 is 7.2. **37.** ________ % of 48 is 16.

38. 75 is ________ % of 300. **39.** ________ is 6% of 18.25.

40. 5 is 10% of ________. **41.** 14 is $5\frac{1}{2}\%$ of ________.

42. ________ is $6\frac{1}{2}\%$ of 15. **43.** 62 is ________ % of 31.

Choose the correct answer by reasoning and calculating mentally.

44. Write $37\frac{1}{2}\%$ as a fraction, reduced.

(a) $\frac{1}{4}$ (b) $\frac{1}{3}$ (c) $\frac{3}{8}$ (d) $\frac{5}{8}$

45. Find 10% of 72.6.

(a) 0.726 (b) 7.26 (c) 72.6 (d) 726

46. Write $5\frac{1}{2}\%$ in decimal form.

(a) 0.055 (b) 0.55 (c) 5.5 (d) 550

47. What is 200% of 18.3?

(a) 3.66 (b) 36.6 (c) 366 (d) 3660

48. Write $\frac{3}{4}$ as a percent.

(a) 0.75% (b) 7.5% (c) 75% (d) 750%

48. ______________

49. **Discounts and taxes.** A shirt was marked 25% off. What would you pay for the shirt if the original price was $15 and you had to pay 6% sales tax?

49. ______________

50. **Retail profit.** Men's topcoats were on sale for $180. This was a discount of $30 from the original price.

a. If the store owner paid $120 for the coats, what was his percent of profit based on his cost?

b. What was his percent of profit based on the selling price?

50. a. ______________

b. ______________

51. **Test grades.** A student received a grade of 75% on a statistics test. If there were 32 problems on the test, all of equal value, how many problems did the student miss?

51. ______________

52. **Commissions.** A salesman works on a 9% commission on his sales over $10,000 each month, plus a base salary of $600 per month. How much did he make the month he sold $25,000 in merchandise?

52. ______________

53. **Property tax.** The property taxes on a house were $1800. What was the tax rate if the house was valued at $150,000?

53. ______________

54. **Saving and spending.** Mary's allowance each week was $15. She saved for 6 weeks, then she spent $5 on a movie, $35 on clothes, and $20 on a gift for her parents' anniversary. What percent of her savings did she spend on each item?

54. ______________

55. **Automobile sales.** The discount on a new car was $1500, including a rebate from the company.

a. What was the original price of the car if the discount was 15% of the original price?

b. What would be paid for the car if taxes and license fees totaled $650?

55. a. ______________

b. ______________

Name ______________________ Section ____________ Date __________

ANSWERS

Chapter 7 Test

Change each number to a percent.

1. $\frac{101}{100}$ **2.** 0.003 **3.** 0.173

4. $\frac{8}{25}$ **5.** $2\frac{3}{8}$

Change each percent to a decimal.

6. 70% **7.** 180% **8.** 9.3%

Change each percent to a fraction or mixed number with the fraction part reduced.

9. 130% **10.** $35\frac{1}{2}\%$ **11.** 8.6%

Find the unknown quantity.

12. 10% of 56 is__________. **13.** 12 is______% of 36.

14. 33 is 110% of ________. **15.** _______ is 62% of 475.

16. 6.765 is 33% of ________. **17.** 16.55 is ________ % of 50.

1. ______________
2. ______________
3. ______________
4. ______________
5. ______________
6. ______________
7. ______________
8. ______________
9. ______________
10. ______________
11. ______________
12. ______________
13. ______________
14. ______________
15. ______________
16. ______________
17. ______________

18. ______

19. ______

20. ______

21. ______

22. a. ______

b. ______

c. ______

23. a. ______

b. ______

c. ______

Choose the correct answer by reasoning and calculating mentally.

18. Write 11% as a fraction.

(a) $\frac{11}{1}$ (b) $\frac{11}{10}$ (c) $\frac{11}{100}$ (d) $\frac{11}{1000}$

19. Write $\frac{2}{3}$ as a percent.

(a) 66% (b) 67% (c) $33\frac{1}{3}\%$ (d) $66\frac{2}{3}\%$

20. Find 25% of 100.

(a) 2.5 (b) 25 (c) 250 (d) 2500

21. 10% of what number is 62?

(a) 6.2 (b) 62 (c) 620 (d) 6200

22. Using the rule of thumb stated in the text, calculate the 15% tip for each of the following amounts.

a. \$6.72 **b.** \$25.35 **c.** \$17.95

23. A hardware store has light fixtures on the sale for \$50.00. This represents a 20% discount.

a. What was the original selling price?

b. If the fixtures cost the store \$40, what was the store's percent of profit based on cost?

c. What was the store's percent of profit based on selling price?

Name ______________________ Section ______________ Date ____________

ANSWERS

24. A customer received a 2% discount on the purchase of a new dining room set because she paid in cash.

a. If she paid $1176, what was the original selling price?

b. What was the amount of the discount?

24. a. ______________

b. ______________

25. A salesman earns $900 each month plus a commission of 8% of his sales over $20,000. How much did he earn for the month in which his sales were $50,000?

25. ______________

26. On a 120 question test, a student answered 102 correctly. What percent of the problems did the student answer correctly?

26. ______________

Name ______________________ Section ____________ Date __________

ANSWERS

Cumulative Review: Chapters 1 – 7

1. The number 0 is called the additive ____________.

2. Name the property of addition that is illustrated.
   $17 + 15 = 15 + 17$

3. Round as indicated:
   a. 12,943 (nearest thousand) **b.** 3.09672 (nearest thousandth)

First, estimate the result, then perform the indicated operations to find the actual result.

4. $\begin{array}{r} 8695 \\ 457 \\ +1206 \\ \hline \end{array}$

5. $\begin{array}{r} 8500 \\ -4675 \\ \hline \end{array}$

6. $72\overline{)6696}$

7. $\begin{array}{r} 182 \\ \times\ 36 \\ \hline \end{array}$

8. Evaluate the expression $3\left(5+2^2\right)-6-2\cdot 3^2$.

9. List the squares of the prime numbers less than 30.

10. **a.** Find the LCM of 56, 60, and 75.
    b. State the number of times each number divides into the LCM.

Evaluate each of the following expressions. Reduce all fractions.

11. $\frac{15}{28}\cdot\frac{7}{25}$

12. $\frac{3}{4}+\frac{7}{8}\div\frac{1}{2}$

13. $5.6+7\frac{1}{2}-3\frac{5}{8}$

1. ________
2. ________
3. a. ________
   b. ________
4. ________
5. ________
6. ________
7. ________
8. ________
9. ________
10. a. ________
    b. ________
11. ________
12. ________
13. ________

14. ______

15. ______

16. ______

17. ______

18. a. ______

b. ______

19. a. ______

b. ______

c. ______

d. ______

20. a. ______

b. ______

21. ______

22. a. ______

b. ______

c. ______

14. $\left(\frac{1}{5}+\frac{7}{20}\right)\div\left(2\frac{1}{2}-1\frac{3}{5}\right)$

15. Solve the proportion: $\frac{7}{x}=\frac{3\frac{1}{2}}{11}$

16. Find 75% of 104.

17. 82% of ______ is 328.

18. Write each of the following as ratios reduced to lowest form. Use common units in the numerator and denominator whenever possible.

a. 14 inches to 1 foot **b.** 4 months to 1 year

19. Find the equivalent measures in the metric system.

a. 345 cm = ____________ mm **b.** 1.6 m = ____________ cm

c. 830 mm = ____________ cm **d.** 5200 m = ____________ km

20. Use a caculator to find the following square roots accurate to four decimal places.

a. $\sqrt{175}$ **b.** $\sqrt{108}$

21. Find the length of the hypotenuse of a right triangle that has legs of length 15 feet and 20 feet.

22. A clothing store has dresses on sale for $80.00. This represents a 20% discount.

a. What was the original selling price?

b. If the dresses cost the store $60, what was the store's percent of profit based on cost?

c. What was the store's percent of profit based on selling price?

Name ______________________ Section ____________ Date __________

ANSWERS

23. You ordered two pizzas to be delivered to your home. One was for \$11.75 and the other was for \$15.80.

a. If you wanted to give the driver a 15% tip, what would be the amount of the tip?

b. How much would you pay the driver?

24. You paid \$4500 as a down payment on a new car and made 48 monthly payments of \$350 each.

a. Approximately how much did you pay for the car?

b. Exactly how much did you pay for the car?

c. How much more did you pay for the car than you would have if you had bought the car originally for \$17,040 in cash?

d. What percent more did you pay by making monthly payments?

25. Name the type of each of the following triangles based on the measures and lengths shown.

a.

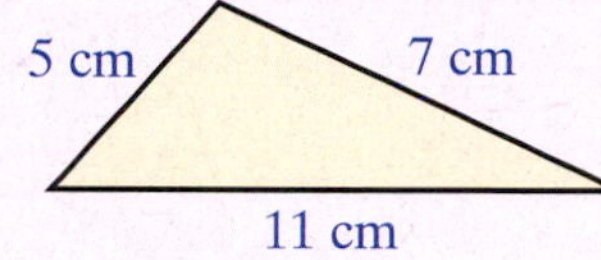

b.

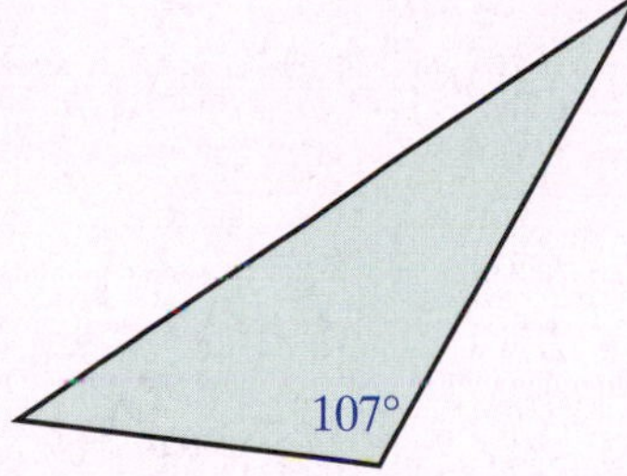

c.

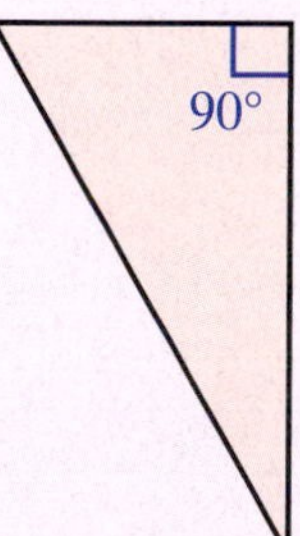

d.

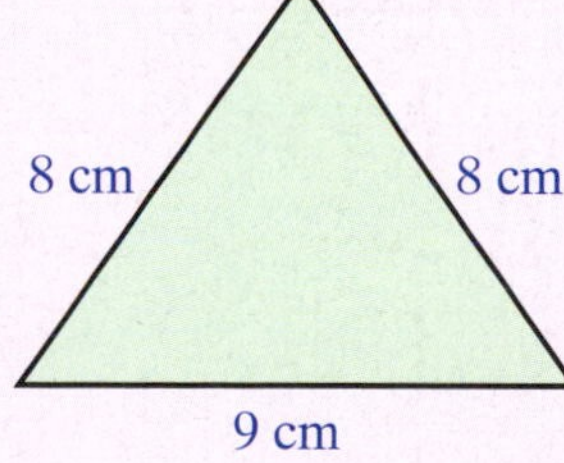

26. **a.** Find the measure of $\angle R$.

b. What kind of triangle is ΔRST?

c. Which side is opposite $\angle S$?

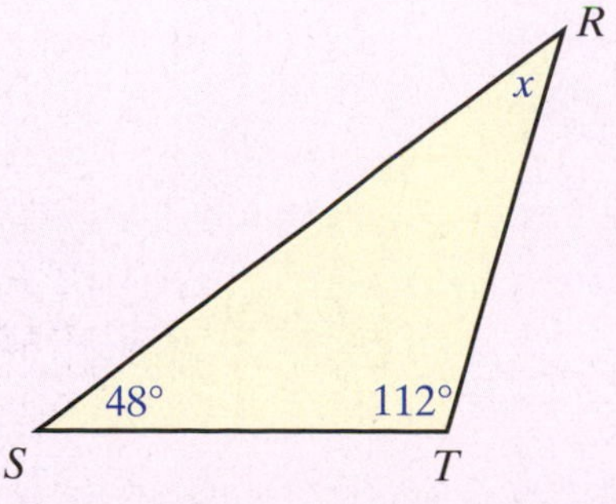

23. a. ____________

b. ____________

24. a. ____________

b. ____________

c. ____________

d. ____________

25. a. ____________

b. ____________

c. ____________

d. ____________

26. a. ____________

b. ____________

c. ____________

8

Consumer Applications

Chapter 8 Consumer Applications

WHAT TO EXPECT IN CHAPTER 8

Interest is money paid for the use of money. Every person should understand how interest is earned or paid because this is the basis for many daily life concepts – borrowing money, lending money, paying taxes, getting a pay raise, store discounts, and so on.

The related concepts of simple interest and compound interest are discussed in Sections 8.1 and 8.2. In Sections 8.3 – 8.5 we discuss everyday concerns related to balancing a checkbook, car expenses, and home expenses. A statistical topic involving reading graphs is developed in Section 8.6. As a consumer, you should find *all* the topics in this chapter useful for the rest of your life.

Chapter 8 Consumer Applications

Mathematics at Work!

Headlines in the L.A. Times on Saturday, April 16, 2005 read:

> **Fear of Slump Batters Wall St.**
> **The Dow sinks 3.6% for the week amid signs of a slowing economy**
> **Some analysts see respectable growth and little risk of recession**

If you had money in the stock market did you lose 3.6% on that day? Not necessarily. Headlines later in the same paper:

> **The firm's [Genetech] stock jumps 18% after a drug already shown to slow colon and lung tumors proves helpful against a third type of malignancy.**

The same day, in another article:

> **"The unemployment rate dropped to 5.4%, the lowest since 2001, but economists said that was probably because record numbers of people had been out of work so long that they no longer received unemployment insurance and were no longer counted."**

What do all of these percents and related terms mean? Obviously, in general, people lost money in the stock market that day. However, if you owned stock in Genetech you made money, lots! What is the meaning of a recession? What is the unemployment rate based on? You will need to take courses in business to be able to answer questions such as these. However, as you can see, understanding percent is fundamental to an informed citizen.

8.1 Simple Interest

Objective 1

Understanding Simple Interest

Interest is money paid for the use of money. The money that is invested or borrowed is called the **principal**. The **rate** is the **percent of interest** and is almost always stated as the **annual (yearly) rate**.

Interest is either paid or earned, depending on whether you are the borrower or the lender. In either case, the calculations involved are the same. Although interest rates can vary from year to year (or even daily, as in the case of home loans) and from one part of the world to another, the concept of interest is the same everywhere.

Some loans (called **notes**) are based on **simple interest** and involve only one payment (including interest and principal) at the end of the term of the loan. Such loans are usually made for a period of one year or less. **Compound interest**, which involves interest paid on interest, will be discussed in Section 8.2.

The following formula is used to calculate simple interest.

Formula for Calculating Simple Interest

$$I = P \times r \times t$$

where

I = Interest (earned or paid)

P = Principal (the amount invested or borrowed)

r = Rate of interest (stated as an annual, or yearly rate) and used in decimal or fraction form

t = Time (in years or fraction of a year)

Note: For calculation purposes, we will use 360 days in one year (30 days in a month). This is a common practice in business and banking; however, with the advent of computers, many lending institutions now base their calculations on 365 days per year and pay or charge interest on a daily basis.

Objective ②

Objectives

① Understand the concept of **simple interest**.

② Find simple interest using the formula $I = P \times r \times t$.

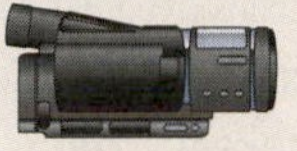

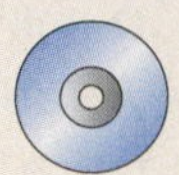

Although the rate of interest is generally given in the form of a percent, the calculations in the formula are made by changing the percent into decimal form or fraction form. For example, we can write 12% as the decimal 0.12 or the fraction $\frac{12}{100}$.

Example 1

If you were to borrow $2000 at 12% for one year, how much interest would you pay?

Solution

Use the formula for simple interest: $I = P \times r \times t$ with

$P = \$2000, \quad r = 12\% = 0.12, \quad \text{and} \quad t = 1 \text{ year}$

$I = \$2000 \times 0.12 \times 1 = \$240$

You would pay $240 interest.

1. If you were to borrow $1500 at 10% for one year, how much interest would you pay?

Now work exercise 1 in the margin.

Example 2

If you decided you would need the $2000 for only 90 days, how much interest would you pay? (The interest rate would still be stated at 12%, the annual rate.)

Continued on next page...

2. If you decided you would need the $1500 for only 30 days, how much interest would you pay? (The interest rate would still be stated at 10%, the annual rate.)

3. Ralph borrowed $3000 at 7% for sixty days. How much interest did he pay?

Solution

Use the formula for simple interest: $I = P \times r \times t$ with

$P = \$2000$, $r = 12\% = 0.12$, and

$t = 90$ days

$= \frac{90}{360}$ year $= \frac{1}{4}$ year

$I = \$2000 \times 0.12 \times \frac{1}{4} = \frac{240}{4} = \$60$

$I = \$2000 \times 0.12 \times 0.25 = \$60$

You would pay $60 interest if you borrowed the money for 90 days.

Now work exercise 2 in the margin.

Completion Example 3

Sylvia borrowed $2400 at 10% interest for 30 days. How much interest did she have to pay?

Solution

$P =$ ________ , $r =$ ________ % = ________ ,

$t = 30$ days = ________ year

$I =$ ________ × ________ × ________ = ________

Now work exercise 3 in the margin.

If you know the values of any three of the variables in the formula $I = P \times r \times t$, you can find the value of the fourth variable by substituting into the formula and solving the resulting equation for the unknown variable. This procedure is illustrated in Examples 4 and 5.

Example 4

How much (what principal) would you need to invest if your investment returned 9% interest and you wanted to make $100 in interest in 30 days?

Solution

Here the principal is unknown, while the interest ($I = \$100$), the rate of interest ($r = 9\% = 0.09$), and the time ($t = 30$ days) are all known. However, before substituting into the formula, we must change t to a fraction of a year: $t = \frac{30}{360} = \frac{1}{12}$ year.

$$I = P\times r\times t$$

$$100 = P\times 0.09\times\frac{1}{12}$$

$$100 = P\times\frac{9}{100}\times\frac{1}{12}$$

$$100 = P\times\frac{3}{400}$$

$$\frac{400}{3}\times\frac{100}{1} = P\times\frac{\cancel{3}}{\cancel{400}}\times\frac{\cancel{400}}{\cancel{3}}$$ Multiply both sides by $\frac{400}{3}$.

$$\frac{40,000}{3} = P$$

$$P = \$13,333.33,$$

or using $\frac{1}{12} = 0.083$ as a rounded decimal,

$$100 = P\times 0.09\times 0.083$$

$$100 = P\times 0.00747$$

$$\frac{100}{0.00747} = \frac{P\times \cancel{0.00747}}{\cancel{0.00747}}$$

$$P = \$13,386.88$$

Using the rounded decimal 0.083 leads to a "rounding error." You should understand that once a rounded value is used in any calculation there will be some error. The correct result can be found by using 0.083333333, which is the display on your calculator for $1\div 12$.

Now work exercise 4 in the margin.

4. How much (what principal) would you need to invest if your investment returned 8% interest and you wanted to make \$500 in interest in 90 days?

Example 5

Stuart wants to borrow \$1500 at 10% from his uncle and is willing to pay \$75 in simple interest. How long can he keep the money?

Solution

Here the time is unknown. The principal is \$1500; the interest is \$75; the rate is 10%. Substituting into the formula $I = P\times r\times t$ gives

$$75 = 1500\times 0.10\times t$$

$$75 = 150\times t$$

$$\frac{75}{150} = \frac{\cancel{150}\times t}{\cancel{150}}$$

$$\frac{1}{2} = t$$

or $t = \frac{1}{2}$ year [or $t = 0.5$ years or $t = (0.5)(360) = 180$ days]

Stuart can borrow the money for $t = \frac{1}{2}$ year (or 6 months).

Now work exercise 5 in the margin.

5. Rick wants to borrow \$1000 at 12% from his uncle and is willing to pay \$100 in simple interest. How long can he keep the money?

Completion Example Answers

3. $P = \mathbf{\$2400}$, $r = \mathbf{10\%} = \mathbf{0.10}$, $t = 30 \text{ days} = \frac{\mathbf{1}}{\mathbf{12}} \text{ year}$

$I = \mathbf{2400} \times \mathbf{0.10} \times \frac{\mathbf{1}}{\mathbf{12}} = \mathbf{\$20}$

Practice Problems

1. If you were to borrow \$1000 at 5% interest for nine months, how much interest would you pay?
2. What interest rate would you be paying if you borrowed \$1000 for 6 months and paid \$60 in interest?
3. How much (what principal) would you need to invest if your interest returned 10% interest and you wanted to make \$100 in one year?
4. You want to borrow \$2000 at 5% from a friend and you are willing to pay \$100 in simple interest. How long can you keep the money?

Answers to Practice Problems:

1. \$37.50 **2.** 12% **3.** \$1000 **4.** 1 year

Name ______________________ Section ____________ Date ____________

ANSWERS

Exercises 8.1

1. What is the simple interest paid on $500 at 6% for one year?

1. ____________

2. What is the simple interest paid on $2000 at 8% for one year?

2. ____________

3. How much interest would be paid on a loan of $1000 at 18% for 6 months? (**Note**: This interest rate may seem high, but the interest rates on some credit cards are even higher.)

3. ____________

4. How much interest would be paid on a loan of $3000 at 12% for 9 months?

4. ____________

5. You invested $2000 at 8% for 60 days. How much interest (to the nearest penny) did your money earn?

5. ____________

6. Stacey loaned her brother $1500 for 8 months at 10% interest. How much interest did she earn?

6. ____________

7. What principal will earn $50 in interest if it is invested at 8% for 90 days?

7. ____________

8. What principal will earn $75 in interest if it is invested for 60 days at 9%?

8. ____________

9. ______

9. How long will it take for $1000 invested at 5% to earn $50 in simple interest?

10. ______

10. What length of time will it take to earn $70 in simple interest if $2000 is invested at 7%?

11. ______

11. What will be the interest earned in one year on a savings account of $800 if the bank pays 4% interest?

12. ______

12. If interest is paid at 6% for one year, what will a principal of $1800 earn?

13. ______

13. If a principal of $900 is invested at a rate of 9% for 90 days, what will be the interest earned?

14. ______

14. A loan of $5000 is made at 8% for a period of 6 months. How much interest is paid?

15. ______

15. If you borrow $750 for 30 days at 18%, how much interest will you pay?

16. ______

16. How much interest is paid on 60-day loan of $500 at 12%?

Name ______________________ Section ______________ Date ________

ANSWERS

17. Find the simple interest paid on a savings account of $2800 for 120 days at 3.5%.

17. ______________

18. A savings account of $5300 is left for 90 days drawing interest at a rate of 5%.

a. How much interest is earned?

b. What is the amount in the account at the end of 90 days?

18. a. ______________

b. ______________

19. Every 6 months a stock pays 10% in dividends (interest on investment). What will be the earnings of $14,600 invested for 6 months? (Remember, the rates of interest are given as annual rates.)

19. ______________

20. If you charge $1000 worth of merchandise at a local department store at 18% interest, how much will you owe at the end of 60 days?

20. ______________

21. You buy an oven on sale from $500 to $450, but you don't make a payment for 60 days and are charged interest at a rate of 18%.

a. How much do you pay for the oven by waiting 60 days to pay?

b. How much do you save by buying the oven on sale? (Sales tax is not included here.)

21. a. ______________

b. ______________

22. A friend borrows $500 from you for a period of 8 months and pays you interest at 6%.

a. How much interest are you paid?

b. If you had asked 8%, how much more interest would you have earned?

22. a. ______________

b. ______________

23. What principal would have to be invested at 8% for 60 days to earn interest of $500?

23. ______________

24. ______

25. ______

26. a. ______

b. ______

c. ______

d. ______

27. a. ______

b. ______

c. ______

d. ______

28. a. ______

b. ______

24. What rate of interest is charged if a loan of $2500 for 90 days is paid off with $2562.50? (**Note**: The payoff is principal plus interest.)

25. How many days must you leave $1000 in a savings account at 5.5% to have a balance of $1011? (**Note**: The balance is the principal plus interest.)

26. Determine the missing item in each row.

Principal	Rate	Time	Interest
$400	16%	90 days	$ **a.** ___
$ **b.** ___	15%	120 days	$5.00
$560	12%	**c.** ___	$5.60
$2700	**d.** ___	40 days	$25.50

27. Determine the missing item in each row.

Principal	Rate	Time	Interest
$600	15%	30 days	$ **a.** ___
$500	18%	**b.** ___	$15.00
$450	**c.** ___	90 days	$22.50
$ **d.** ___	10%	30 days	$1.50

28. a. If Carlos has a savings account of $25,000 drawing interest at 8%, how much interest will he earn in 6 months?

b. How long must he leave the money in the account to earn $1500?

Name ______________________ Section ____________ Date __________

ANSWERS

29. Ms. Lee accumulated $240,000 and she wants to live on the interest each year. If she needs $2000 a month to live on, what interest rate must she earn on her money?

29. ____________

30. Mr. Smith has a savings account of $2500 that draws 4.5% interest. How many days will it take for him to earn $75?

30. ____________

31. A bank decides to loan $5 million to a contractor to build new homes. How much interest will the bank earn in one year if the interest rate is 9.2%?

31. ____________

32. A credit card company has $120 million loaned to its customers at 18.9%. How much interest will it earn in one month?

32. ____________

33. A small airline company borrowed $7.5 million to buy some new airplanes. The loan rate was 7.5%, and the airline paid $562,500 in interest. What was the length of time of the loan?

33. ____________

34. A department store keeps $15 million in merchandise in stock. If the store pays interest at 9% on a bank loan for this stock, how much interest will the store pay in 3 months' time?

34. ____________

35. a. ____________

b. ____________

c. ____________

d. ____________

35. Determine the missing item in each row.

Principal	Rate	Time	Interest
$1000	$10\frac{1}{2}\%$	60 days	$ **a.** ____
$800	$13\frac{1}{2}\%$	**b.** ____	$18.00
$2000	**c.** ____	9 months	$172.50
$ **d.** ____	$7\frac{1}{2}\%$	1 year	$85.00

8.2 Compound Interest

Understanding Compound Interest

Objective ①

Interest paid on interest earned is called **compound interest**. To calculate compound interest, we can calculate the simple interest for each period of time that interest is compounded, using a **new principal for each calculation.** This new principal is the **previous principal plus the earned interest**. The calculations can be performed in a step-by-step manner, as indicated in the following outline:

Objective ②

To Calculate Compound Interest:

Step 1: Using the formula for simple interest, $I = P \times r \times t$, calculate the simple interest where $t = \frac{1}{n}$ and n is the number of periods per year for compounding. For example,

for compounding **annually**, $n = 1$ and $t = \frac{1}{1} = 1$

for compounding **semiannually**, $n = 2$ and $t = \frac{1}{2}$

for compounding **quarterly**, $n = 4$ and $t = \frac{1}{4}$

for compounding **monthly**, $n = 12$ and $t = \frac{1}{12}$

for compounding **daily**, $n = 360$ and $t = \frac{1}{360}$

Step 2: Add this interest to the principal to create a new value for the principal.

Step 3: Repeat steps 1 and 2 however many times the interest is to be compounded.

Note: For the purpose of calculations, we will use 360 days in one year (30 days a month).

In Examples 1 and 2 we show how compound interest can be calculated in the step-by-step manner just outlined. This process serves to develop a basic understanding of the concept of compound interest. After Example 2 we will discuss another formula and show how to find compound interest with this formula and a calculator.

Objectives

① Understand the concept of **compound interest**.

② Find interest compounded periodically with repeated calculations.

③ Find compound interest using the formula $A = P\left(1 + \frac{r}{n}\right)^{nt}$.

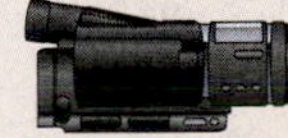

1. George deposits $500 in a savings account that pays 6% interest compounded quarterly. How much interest will he earn in 9 months?

Example 1

If an account is compounded annually (once a year) at 10%, how much interest will a principal of $1200 earn in three years?

Solution

Because the compounding is done annually, use $t = \frac{1}{n} = \frac{1}{1} = 1$ in the formula $I = P \times r \times t$. Also, since the compounding is for three years, calculate the interest three times with a new principal (the old principal plus interest) each time.

a. First year: the principal is $P = \$1200$.

$I = \mathbf{1200} \times 0.10 \times 1 = \$120.00$ interest for the first year

b. Second year: the principal is $P = \$1200 + \$120 = \$1320$.

$I = \mathbf{1320} \times 0.10 \times 1 = \$132.00$ interest for the second year

c. Third year: the principal is $P = \$1320 + \$132 = \$1452$.

$I = \mathbf{1452} \times 0.10 \times 1 = \$145.20$ interest for the third year

The total interest earned in three years will be

$$\begin{array}{r} \$\ 120.00 \\ 132.00 \\ +\ 145.20 \\ \hline \$\ 397.20 \end{array}$$

Now work exercise 1 in the margin.

Example 2

If a principal of $5000 is compounded monthly (12 times a year) at 6%, what will be the balance in the account at the end of four months?

Solution

Because the compounding is monthly, use $t = \frac{1}{n} = \frac{1}{12}$ in the formula $I = P \times r \times t$. Also, since the compounding is for four months, calculate the interest four times with a new principal (the old principal plus interest) each time.

a. First month: the principal is $P = \$5000$.

$I = \mathbf{5000} \times 0.06 \times \frac{1}{12} = \$25.00$ interest for the first month

b. Second month: the principal is $P = \$5000.00 + \$25.00 = \$5025.00$.

$I = \mathbf{5025} \times 0.06 \times \frac{1}{12} = \$25.13$ interest for the second month

c. Third month: the principal is $P = \$5025.00 + \$25.13 = \$5050.13$.

$$I = \mathbf{5050.13} \times 0.06 \times \frac{1}{12} = \$25.25 \text{ interest for the third month}$$

d. Fourth month: the principal is $P = \$5050.13 + \$25.25 = \$5075.38$.

$$I = \mathbf{5075.38} \times 0.06 \times \frac{1}{12} = \$25.38 \text{ interest for the fourth month}$$

The balance in the account at the end of four months will be

$$\$5075.38 + \$25.38 = \$5100.76.$$

And the total interest earned will be $\$5100.76 - \$5000 = \$100.76$.

Now work exercise 2 in the margin.

2. If a principal of $3000 is compounded bi-monthly (6 times a year) at 7%, what will be the balance in the account at the end of one year?

Example 3

Find the interest earned on $5000 if simple interest is calculated for a period of 4 months at 6%. What is the difference between this amount and the interest earned by compounding monthly (as shown in Example 2)?

Solution

For simple interest we use the formula $I = P \times r \times t$ just once with $t = \frac{4}{12} = \frac{1}{3}$.

This gives $I = 5000 \times 0.06 \times \frac{1}{3} = 100$

Thus the interest earned would be $100 and the difference in compound interest and simple interest for the four months would be

$$\$100.76 - \$100.00 = \$0.76.$$

Now work exercise 3 in the margin.

3. Find the interest earned on $1000 if simple interest is calculated for a period of 3 years at 5%. What is the difference between this amount and the interest earned by compounding monthly (as shown in Example 2)?

Finding Compound Interest by Using the Formula $A = P\left(1 + \frac{r}{n}\right)^{nt}$

Objective ③

The steps outlined in Examples 1 and 2 illustrate how the principal is adjusted for each time period of the compounding process and how the interest increases for each new time period. The interest is greater for each time period because the new adjusted principal is greater. In this process, we have calculated the actual interest for each time period.

The following compound interest formula does not calculate the actual interest for each time period. In fact, this formula does not calculate the interest directly. It does find the total accumulated **amount** (also called the **future value** of the principal). To find the total interest earned, subtract the initial principal from the accumulated amount.

Compound Interest Formula

When interest is compounded, the total **amount**, A, accumulated (including principal and interest) is given by the formula

$$A = P\left(1+\frac{r}{n}\right)^{nt}$$

where

P = the principal

r = the annual interest rate (in decimal or fraction form)

t = the length of time in years

n = the number of compounding periods in one year

In Examples 4 and 5 we will calculate the compound interest formula with a TI-84 graphing calculator. If you have a different calculator you may need to remember and apply the rules for order of operations:

First, operate inside the parentheses;
second, apply the exponent;
and third, multiply by the principal.

(Graphing calculators are programmed to automatically follow the rules for order of operations.)

Example 4

Christopher invests \$5000 at 6% compounded monthly. What will be the amount in his account in 10 years?

Solution

$P = \$5000, \quad r = 6\% = 0.06, \quad n = 12$ times per year, $\quad t = 10$ years

Substituting into the formula gives

$$A = 5000\left(1+\frac{0.06}{12}\right)^{12\cdot 10}$$

To use the TI-84 calculator, enter all of the numbers just as you see them in the formula. You can enter the exponent as 120 or enter it as the product of 12 times 10. If you enter the exponent as a product, it must be in parentheses as (12 * 10).

Step 1: Enter the numbers and parentheses so that the display appears as follows:

Step 2: Press .
The display should now read:

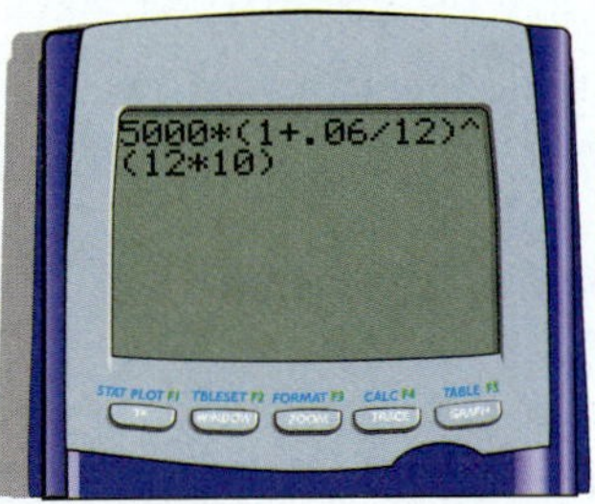

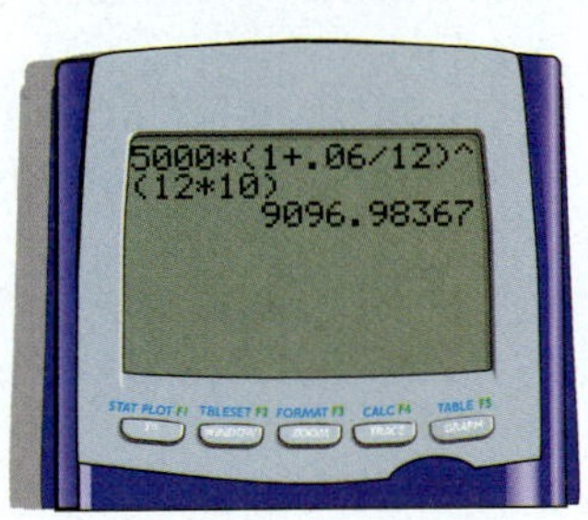

The amount in Christopher's account after 10 years is \$9096.98.

Now work exercise 4 in the margin.

4. Charley invests \$4000 at 4% compounded semi-annually (twice yearly). What will be the amount in his account in 8 years?

Example 5

Maria's grandmother was so pleased when she was born that she put \$10,000 in a savings account at 8% to be compounded daily for 20 years. Maria is to have the money for college when she is 20 years old. How much money would be in Maria's college fund when she is 20?

Solution

$P = \$10{,}000, \quad r = 8\% = 0.08, \quad n = 360$ times per year, $\quad t = 20$ years

Substituting into the formula gives

$$A = 10{,}000\left(1 + \frac{0.08}{360}\right)^{360 \cdot 20}$$

To use the TI-84 calculator, enter all of the numbers just as you see them in the formula. You can enter the exponent as 7200 or enter it as the product of 360 times 20. If you enter the exponent as a product, it must be in parentheses as (360 * 20).

Step 1: Enter the numbers and parentheses so that the display appears as follows:

Step 2: Press ENTER.
The display should now read:

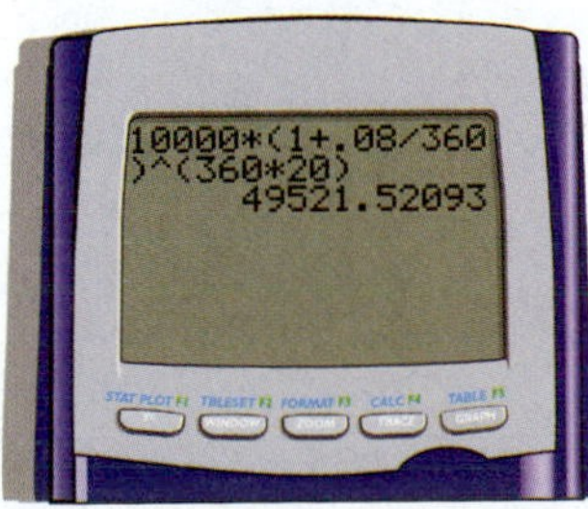

We find the amount in Maria's college fund in 20 years is \$49,521.52. (Thanks Grandma!)

Now work exercise 5 in the margin.

5. Carlos invests \$4000 at 5% compounded daily. What will be the amount in his account in 15 years? (Use the compound interest formula and a calculator.)

Practice Problems

1. You deposit \$800 at 7% to be compounded monthly and withdraw the money 2 months later. How much interest will you earn?

2. You deposit \$2500 at 6% to be compounded monthly. How much will the account be worth in one year?

3. You deposit \$1500 at 4% to be compounded semi-annually. How much interest will you earn in 3 years?

4. You deposit \$500 at 5% to be compounded daily. How much will the account be worth in 6 months?

Answers to Practice Problems:

1. \$9.36 **2.** \$2654.19 **3.** \$189.24 **4.** \$512.66

Name ______________________ Section ____________ Date ________

ANSWERS

Exercises 8.2

Use the formula for simple interest, $I = P \times r \times t$, repeatedly. Show all of the calculations for each period of compounding. See Examples 1 and 2.

1. You loan your cousin $2000 at 5% compounded annually for 3 years. How much interest will your cousin owe you?

   First year: $I = 2000 \times 0.05 \times 1 =$ _______

   Second year: $I =$ _______ $\times 0.05 \times 1 =$ _______

   Third year: $I =$ _______ $\times 0.05 \times 1 =$ _______

   The total interest is ________.

1. [Respond below exercise.] ____________

2. John borrowed $5000 from his uncle at 6% compounded annually for 4 years. How much interest will he owe his uncle at the end of 4 years?

   First year: $I = 5000 \times 0.06 \times 1 =$ _______

   Second year: $I =$ _______ $\times 0.06 \times 1 =$ _______

   Third year: $I =$ _______ $\times 0.06 \times 1 =$ _______

   Fourth year: $I =$ _______ $\times 0.06 \times 1 =$ _______

   The total interest is ________.

2. [Respond below exercise.] ____________

3. If $9000 is deposited in a savings account at 4% compounded monthly, what will be the balance in the account in 6 months?

   First month: $I = 9000 \times 0.04 \times \frac{1}{12} =$ _______

   Second month: $I =$ _______ $\times 0.04 \times \frac{1}{12} =$ _______

   Third month: $I =$ _______ $\times 0.04 \times \frac{1}{12} =$ _______

   Fourth month: $I =$ _______ $\times 0.04 \times \frac{1}{12} =$ _______

   Fifth month: $I =$ _______ $\times 0.04 \times \frac{1}{12} =$ _______

   Sixth month: $I =$ _______ $\times 0.04 \times \frac{1}{12} =$ _______

   The total interest earned is ________.

   The balance in the account is _______.

3. [Respond below exercise.] ____________

4. [Respond in exercise.] ______

5. a. ______

b. ______

6. a. ______

b. ______

7. ______

8. ______

9. a. ______

b. ______

c. [Respond below exercise.] ______

4. Jeremy put $3500 in a savings account at 5.5% compounded quarterly for 6 months. How much interest did he earn? Round to the nearest cent. (Remember that each quarter is 3 months.)

First quarter: $I = 3500 \times 0.055 \times \frac{1}{4} =$ ______

Second quarter: $I =$ ______ $\times 0.055 \times \frac{1}{4} =$ ______

The total interest earned is ______.

5. a. How much interest will be earned in 1 year on a loan of $4000 compounded quarterly at 4%?

b. How much will be owed by the borrower?

6. a. How much interest will be earned on a savings account of $3000 in two years if interest is compounded annually at 5.5%?

b. If interest is compounded semiannually (twice a year)?

7. Calculate the interest earned in six months on $20,000 compounded monthly at 8%.

8. If interest is calculated at 10% compounded quarterly, what will be the value of $15,000 in 9 months?

Use your calculator and the formula for compound interest, $A = P\left(1+\frac{r}{n}\right)^{n \cdot t}$, *to solve the following problems. Because this formula gives the total amount in an account, the interest earned can be found by subtracting the original principal from this amount:* $I = A - P$. *See Examples 4 and 5.*

9. a. Calculate the interest in one year on $5000 compounded monthly at 12%.

b. Suppose the interest is compounded semiannually. Is the accumulated value the same?

c. If not, explain why not in your own words.

Name ______________________ Section ______________ Date ______________

ANSWERS

10. a. What will be the interest on $10,000 compounded daily at 10% for one year?
    b. What is the difference between this and simple interest at 10% for one year?

11. a. Find the value of $5000 compounded quarterly at 8% for 4 years.
    b. What do you think the difference in interest would be if the money were compounded daily: about $5, $20, or over $100?
    c. Find the exact difference in interest.

12. a. What would be the value of a $20,000 savings account at the end of 5 years if interest were calculated at 7% compounded annually?
    b. How much more would be earned if the interest were compounded daily?

13. a. Suppose that $3000 is invested at 5% and compounded monthly for one year. Find the accumulated value.
    b. Is the accumulated amount the same if the original principal of $3000 is compounded annually for 12 years? If not, what is the difference?

14. a. Find the value of $25,000 compounded daily at 5% for 20 years.
    b. Do you think that the amount will be doubled or more than doubled if the rate is doubled to 10%?
    c. Find the amount if the rate is 10%.

Find the amount A and the interest earned I for the given information.

	Compounding Period	Principal	Annual Rate	Time	A	$I = A - P$
15.	Quarterly	$1000	10%	5 yr	a. ____	b. ____
16.	Monthly	$1000	10%	5 yr	a. ____	b. ____
17.	Daily	$2000	5%	10 yr	a. ____	b. ____
18.	Daily	$7500	8%	20 yr	a. ____	b. ____

10. a. ____________
    b. ____________
11. a. ____________
    b. ____________
    c. ____________
12. a. ____________
    b. ____________
13. a. ____________
    b. ____________
14. a. ____________
    b. ____________
    c. ____________
15. a. ____________
    b. ____________
16. a. ____________
    b. ____________
17. a. ____________
    b. ____________
18. a. ____________
    b. ____________

Writing and Thinking about Mathematics

19. [Respond below exercise.]

19. Use your calculator and choose the values of t (in years) to use in the formula for compound interest until you find how many years of daily compounding at 6% are needed for an investment of \$6000 to approximately double in value. Write down the values you chose for t, why you chose those particular values, and the corresponding accumulated values of money. Explain why you agree or disagree with the idea that \$10,000 would double in a shorter time period.

t	$\left(1+\frac{0.06}{360}\right)^{360\cdot t}$	A

20. [Respond below exercise.]

20. Use your calculator and choose the values of t (in years) to use in the formula for compound interest until you find how many years of daily compounding at 6% are needed for an investment of \$10,000 to approximately triple in value. Write down the values you chose for t, why you chose those particular values, and the corresponding accumulated values of money. Explain why you agree or disagree with the idea that \$30,000 would triple in a shorter time period.

t	$\left(1+\frac{0.06}{360}\right)^{360\cdot t}$	A

8.3 Balancing a Checking Account

Many people do not balance their checking accounts simply because they were never told how. Others trust the bank's calculations over their own. This is a poor practice because, for a variety of reasons, computers do make errors. Also, because the bank's statement comes only once a month, you should know your current balance so your account will not be overdrawn. Overdrawn accounts pay a penalty and can lead to a bad credit rating.

Like many applications with mathematics, balancing a checking account follows a pattern of steps. The procedure and related ideas are given in the following lists.

Objectives

① Become familiar with the information in a bank statement.

② Learn how to keep a checkbook register current.

③ Know how to balance a checking account.

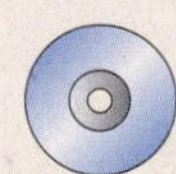

Objective ①

The bank or the savings and loan company sends you a statement of your checking account each month. This statement contains a record of:

1. The beginning balance;
2. A list of all checks paid by the bank;
3. A list of deposits you made;
4. Any interest paid to you (some checking accounts do pay interest);
5. Any service charge by the bank; and
6. The closing balance.

Your checkbook register contains a record of:

1. All the checks you have written;
2. All the deposits you have made; and
3. The current balance.

Objective ②

Your current balance may not agree with the closing balance on the bank statement because:

1. You have not recorded the interest.
2. You have not recorded the service charge.
3. The bank does not have a record of all the checks you have written. (Several checks will be *outstanding* because they have not been received yet for payment by the bank by the date given on the bank statement.)

Objective 3

To Balance your Checking Account

(sometimes called **reconciling** the bank statement with your checkbook register):

1. Go through your checkbook register and the bank statement and put a check mark (√) by each check paid and deposit recorded on the bank statement.

2. In your checkbook register:
 a. Add any interest paid to your current balance.
 b. Subtract any service charge from this balance.
 This number is your **true balance**.

3. Find the total of all the outstanding checks (no √ mark) in your checkbook register (checks not yet received by the bank).

4. On the reconciliation sheet:
 a. Enter the balance from the bank statement.
 b. Add any deposits you have made that are not recorded (no √ mark) on the bank statement.
 c. Subtract the total of your outstanding checks (found in Step 3). This number should agree with your **true balance**. Now your checking account is balanced.

If Your Checking Account Does Not Balance:

1. Go over your arithmetic in the balancing procedure.

2. Go over your arithmetic check by check in your checkbook register.

3. Make an appointment with bank personnel to find the reasons for any other errors.

Examples 1 and 2 show a checkbook register, a bank statement, and a reconciliation sheet. Steps A, B, and C have been followed and the true balance found on the reconciliation sheet matches the balance found on the checkbook register.

Example 1

YOUR CHECKBOOK REGISTER

Check No.	Date	Transaction Description	Payment (−)	(√)	Deposit (+)	Balance
			Balance brought forward			312.12
102	2-3	Painted Lady (cosmetics)	14.80	√		−14.80 297.32
103	2-4	Nu-Groceries (groceries)	76.51			−76.51 220.81
104	2-9	Wood Lumber (plywood)	53.70	√		−53.70 167.11
	2-16	Deposit		√	100.00	+100.00 267.11
105	2-21	The Metal Box (nails)	10.14	√		−10.14 256.97
	2-26	Deposit			250.00	+250.00 506.97
	2-28	Bank Interest		√		— 506.97
	2-28	Service Charge	3.00	√		−3.00 503.97
			True Balance			503.97

BANK STATEMENT
Checking Account Activity

Transaction Description	Amount	(√)	Running Balance	Date
Beginning Balance		√	312.12	2-1
Check #102	14.80	√	297.32	2-4
Check #104	53.70	√	243.62	2-9
Deposit	100.00	√	343.62	2-16
Check #105	10.14	√	333.48	2-24
Service Charge	3.00	√	330.48	2-28
Ending Balance			330.48	

RECONCILIATION SHEET

A. First, mark √ beside each check and deposit listed in both your checkbook register and on the bank statement.
B. Second, in your checkbook register, add any interest paid and subtract any service charge listed.
C. Third, find the total of all outstanding checks.

Outstanding Checks No.	Amount
103	76.51
Total	76.51

Statement Balance	330.48
Add deposits not credited	+250.00
Total	580.48
Subtract total amount of checks outstanding	−76.51
True Balance	503.97

After adding deposits not credited to the statement balance and subtracting the total amount of outstanding checks, the true balance is $503.97.

Example 2

YOUR CHECKBOOK REGISTER

Check No.	Date	Transaction Description	Payment (−)	(√)	Deposit (+)	Balance
			Balance brought forward			505.21
152	5-3	A1-Auto Lease (car lease)	198.60	√		−198.60 306.61
153	5-10	SCANG Power Co. (electric)	75.00			−75.00 231.61
154	5-10	Unite Airlines (plane ticket)	112.00	√		−112.00 119.61
	5-15	Deposit		√	500.00	+500.00 619.61
155	5-15	Safe Drugs (medicine)	5.60	√		−5.60 614.01
156	5-17	Juli's Boutique (dress)	49.80	√		49.80 564.21
157	6-03	A1-Auto Lease (car lease)	198.60			−198.60 365.61
	5-31	Bank Interest		√		— 503.97
	5-31	Service Charge	4.00	√		− 4.00 361.61
						—
			True Balance			361.61

BANK STATEMENT
Checking Account Activity

Transaction Description	Amount	(√)	Running Balance	Date
Beginning Balance		√	505.21	5-1
Check #152	198.60	√	306.61	5-6
Check #154	112.00	√	194.61	5-14
Deposit	500.00	√	694.61	5-15
Check #155	5.60	√	689.01	5-18
Check #156	49.80	√	639.21	5-20
Service Charge	4.00	√	635.21	5-31
Ending Balance			635.21	

RECONCILIATION SHEET

A. First, mark √ beside each check and deposit listed in both your checkbook register and on the bank statement.
B. Second, in your checkbook register, add any interest paid and subtract any service charge listed.
C. Third, find the total of all outstanding checks.

Outstanding Checks No.	Amount
153	75.00
157	198.60
Total	273.60

Statement Balance	635.21
Add deposits not credited	—
Total	635.21
Subtract total amount of checks outstanding	−273.60
True Balance	**361.61**

After adding deposits not credited to the statement balance and subtracting the total amount of outstanding checks, the true balance is \$361.61.

Practice Problems

1. You begin with a balance of \$575.14, write checks for \$100, \$75, and \$150.50; and deposit \$250. What is your balance?

2. You open a new account with a deposit of \$600. You write checks for \$120.50, \$320.05, and \$100.10; then you deposit \$250 and \$350. What is your balance?

3. You begin with a balance of \$250.34, deposit \$450, and write checks for \$200, \$35.87, and \$75.50. What is your balance?

4. Your beginning balance is \$4.14. You make two deposits of \$350 each; then you write checks for \$235.12, \$75.16, and \$129.98. What is your balance?

Answers to Practice Problems:

1. \$499.64 **2.** \$659.35 **3.** \$388.97 **4.** \$263.88

Name ______________________ Section ____________ Date __________

ANSWERS

Exercises 8.3

For each of the following problems, you are given a copy of a checkbook register, the corresponding bank statement, and a reconciliation sheet. You are to find the true balance of the account on both the checkbook register and on the reconciliation sheet as shown in Example 1 and 2. Follow the directions on the reconciliation sheet. For Exercise 1 and 2, step A has been done for you.

1.

YOUR CHECKBOOK REGISTER

Check No.	Date	Transaction Description	Payment (−)	(√)	Deposit (+)	Balance
			Balance brought forward			⊖
	7-15	Deposit		√	700.00	+700.00 700.00
1	7-15	Windy Acres Apt. (rent/deposit)	520.00	√		−520.00 180.00
2	7-15	Pa Bell Telephone (installation)	32.16	√		−32.16 147.84
3	2-15	A&E Power Co. (gas/electric)	46.49	√		−46.49 101.35
4	7-16	Foodway Stores (groceries)	51.90	√		−51.90 49.45
	7-20	Deposit		√	350.00	+350.00 399.45
5	7-23	Comfy Furniture (sofa, chair)	300.50			−300.50 98.95
	8-1	Deposit			350.00	+350.00 448.95
	7-31	Interest		√	2.50	+2.50
	7-31	Service	2.00	√		−2.00
			True Balance			

BANK STATEMENT
Checking Account Activity

Transaction Description	Amount	(√)	Running Balance	Date
Beginning Balance	0.00	√	0.00	7-01
Deposit	700.00	√	700.00	7-15
Check #1	520.00	√	180.00	7-16
Check #4	51.90	√	128.10	7-17
Check #2	32.16	√	95.94	7-18
Check #3	46.49	√	49.45	7-18
Deposit	350.00	√	399.45	7-20
Interest	2.50	√	401.95	7-31
Service Charge	2.00	√	399.95	7-31
Ending Balance			399.95	

RECONCILIATION SHEET

A. First, mark √ beside each check and deposit listed in both your checkbook register and on the bank statement.

B. Second, in your checkbook register, add any interest paid and subtract any service charge listed.

C. Third, find the total of all outstanding checks.

Outstanding Checks No.	Amount
Total	______

Statement Balance	______
Add deposits not credited	+ ______
Total	______
Subtract total amount of checks outstanding	− ______
True Balance	______

1. [Respond in exercise.]

2. [Respond in exercise.]

2.

YOUR CHECKBOOK REGISTER

Check No.	Date	Transaction Description	Payment (–)	(√)	Deposit (+)	Balance
			Balance brought forward			1610.39
1234	12-7	Pearl City (pearl ring)	524.00	√		–524.00 1086.39
1235	12-7	Compuway (home computer)	801.60	√		–801.60 284.79
1236	12-8	Sportz Hutz (skis)	206.25	√		–206.25 78.54
1237	12-8	Guild Card Shop (Christmas cards)	25.50	√		– 25.50 53.04
	12-10	Deposit			1000.00	+1000 1053.04
1238	12-14	Toy World (teddy bear)	13.41	√		–13.41
1239	12-24	Meat Markette (turkey)	35.50	√		–35.50
1240	12-24	Poodle Shop (puppy)	300.00	√		–300.00
1241	12-31	A.A. Sav. & Loan (mortgage)	600.00			–600.00
	12-31	Service Charge	⊖			
			True Balance			

BANK STATEMENT
Checking Account Activity

Transaction Description	Amount	(√)	Running Balance	Date
Beginning Balance	0.00	√	1610.39	12-01
Check #1234	524.00	√	1086.39	12-08
Check #1236	206.25	√	880.14	12-09
Check #1237	25.50	√	854.64	12-09
Deposit	1000.00	√	1854.64	12-10
Check #1235	801.60	√	1053.04	12-11
Check #1238	13.41	√	1039.63	12-15
Check #1239	35.50	√	1004.13	12-27
Check #1240	300.00	√	704.13	12-28
Ending Balance			704.13	

RECONCILIATION SHEET

A. First, mark √ beside each check and deposit listed in both your checkbook register and on the bank statement.
B. Second, in your checkbook register, add any interest paid and subtract any service charge listed.
C. Third, find the total of all outstanding checks.

Outstanding Checks No.	Amount
Total	______

Statement Balance	______
Add deposits not credited	+ ______
Total	______
Subtract total amount of checks outstanding	– ______
True Balance	______

3. [Respond in exercise.]

3.

YOUR CHECKBOOK REGISTER

Check No.	Date	Transaction Description	Payment (–)	(√)	Deposit (+)	Balance
			Balance brought forward			756.14
271	6-15	Parts, Parts, Parts (spark plugs)	12.72			–12.72 743.42
272	6-24	Firetread Tire Co. (2 tires)	121.40			
273	6-30	Dean's Gas (tune-up)	75.68			
	7-1	Deposit			250.00	
274	7-1	Prudent Ins Co. (car insurance)	300.00			
	6-30	Service Charge				
			True Balance			

BANK STATEMENT
Checking Account Activity

Transaction Description	Amount	(√)	Running Balance	Date
Beginning Balance	0.00		756.14	6-01
Check #271	12.72		743.42	6-16
Check #272	121.40		622.02	6-26
Service Charge	1.00		621.02	6-30
Ending Balance			621.02	

RECONCILIATION SHEET

A. First, mark √ beside each check and deposit listed in both your checkbook register and on the bank statement.
B. Second, in your checkbook register, add any interest paid and subtract any service charge listed.
C. Third, find the total of all outstanding checks.

Outstanding Checks No.	Amount
Total	______

Statement Balance	______
Add deposits not credited	+ ______
Total	______
Subtract total amount of checks outstanding	– ______
True Balance	______

Name ______________________ Section ____________ Date __________

ANSWERS

4. [Respond in exercise.]

5. [Respond in exercise.]

4.

YOUR CHECKBOOK REGISTER

Check No.	Date	Transaction Description	Payment (–)	(√)	Deposit (+)	Balance
			Balance brought forward			12.14
419	1-2	U.S. Post Office (stamps)	10.00			–10.00 2.14
	1-3	Deposit			525.50	
420	1-17	Gregg Smith, DDS (dentist)	63.50			
421	1-26	Cash	100.00			
422	1-31	High Rise Apts. (rent)	350.00			
	1-31	Service Charge				
		True Balance				

BANK STATEMENT
Checking Account Activity

Transaction Description	Amount	(√)	Running Balance	Date
Beginning Balance			12.14	1-01
Deposit	525.50		537.64	1-03
Check #419	10.00		527.64	1-03
Check #420	63.50		464.14	1-20
Check #421	100.00		364.14	1-26
Service Charge	2.00		362.14	1-31
Ending Balance			362.14	

RECONCILIATION SHEET

A. First, mark √ beside each check and deposit listed in both your checkbook register and on the bank statement.
B. Second, in your checkbook register, add any interest paid and subtract any service charge listed.
C. Third, find the total of all outstanding checks.

Outstanding Checks No.	Amount
Total	______

Statement Balance ______
Add deposits not credited + ______
Total ______
Subtract total amount of checks outstanding – ______
True Balance ______

5.

YOUR CHECKBOOK REGISTER

Check No.	Date	Transaction Description	Payment (–)	(√)	Deposit (+)	Balance
			Balance brought forward			967.22
772	4-13	Janet Poppy, CPA (accountant)	85.00			
	4-14	Deposit			1200.00	
773	4-14	Pharm X (aspirin)	4.71			
774	4-15	I.R.S. (income tax)	2000.00			
775	4-30	Well Finance Co. (loan payment)	52.50			
	5-1	Deposit			600.00	
	4-30	Interest				
	4-30	Service Charge				
		True Balance				

BANK STATEMENT
Checking Account Activity

Transaction Description	Amount	(√)	Running Balance	Date
Beginning Balance			967.22	4-01
Deposit	1200.00		2167.22	4-14
Check #772	85.00		2082.22	4-15
Check #773	4.71		2077.51	4-15
Interest	2.82		2080.33	4-30
Service Charge	4.00		2076.33	4-30
Ending Balance			2076.33	

RECONCILIATION SHEET

A. First, mark √ beside each check and deposit listed in both your checkbook register and on the bank statement.
B. Second, in your checkbook register, add any interest paid and subtract any service charge listed.
C. Third, find the total of all outstanding checks.

Outstanding Checks No.	Amount
Total	______

Statement Balance ______
Add deposits not credited + ______
Total ______
Subtract total amount of checks outstanding – ______
True Balance ______

6. [Respond in exercise.]

6.

YOUR CHECKBOOK REGISTER

Check No.	Date	Transaction Description	Payment (–)	(√)	Deposit (+)	Balance
		Balance brought forward				1403.49
86	9-1	School Works (school supplies)	17.12			
87	9-2	Gina's Closet (clothes)	192.50			
88	9-4	Campus S&B (books)	56.28			
89	9-7	Bursar's Office (tuition)	380.00			
90	9-7	University Apts. (rent)	380.00			
91	9-27	Veritell (phone bill)	49.99			
92	9-30	Food Town (groceries)	47.80			
	9-30	Service Charge				
		True Balance				

BANK STATEMENT
Checking Account Activity

Transaction Description	Amount	(√)	Running Balance	Date
Beginning Balance			1403.49	9-01
Check #86	17.12		1386.37	9-02
Check #87	192.50		1193.87	9-05
Check #88	56.28		1137.59	9-05
Check #90	380.00		757.59	9-10
Service Charge	4.00		753.59	9-30
Ending Balance			753.59	

RECONCILIATION SHEET

A. First, mark √ beside each check and deposit listed in both your checkbook register and on the bank statement.
B. Second, in your checkbook register, add any interest paid and subtract any service charge listed.
C. Third, find the total of all outstanding checks.

Outstanding Checks	
No.	Amount
Total	

Statement Balance	
Add deposits not credited	+
Total	
Subtract total amount of checks outstanding	–
True Balance	

7. [Respond in exercise.]

7.

YOUR CHECKBOOK REGISTER

Check No.	Date	Transaction Description	Payment (–)	(√)	Deposit (+)	Balance
		Balance brought forward				602.82
14	6-20	Jane Bridal (flowers)	402.40			
	6-22	Deposit			1000.00	
15	6-24	Tuxedo Junction (tuxedo)	155.65			
16	6-28	D. Lohengrin (organist)	55.00			
17	6-28	Lee's Limo (limo rental)	125.00			
18	6-30	C. C. Catering (food caterer)	700.00			
19	7-1	Halloway Gifts (cards)	35.20			
	6-30	Service Charge				
		True Balance				

BANK STATEMENT
Checking Account Activity

Transaction Description	Amount	(√)	Running Balance	Date
Beginning Balance			602.82	6-01
Deposit	1000.00		1602.82	6-22
Check #14	402.40		1200.42	6-22
Check #15	155.65		1044.77	6-26
Service Charge	1.00		1043.77	6-30
Ending Balance			1043.77	

RECONCILIATION SHEET

A. First, mark √ beside each check and deposit listed in both your checkbook register and on the bank statement.
B. Second, in your checkbook register, add any interest paid and subtract any service charge listed.
C. Third, find the total of all outstanding checks.

Outstanding Checks	
No.	Amount
Total	

Statement Balance	
Add deposits not credited	+
Total	
Subtract total amount of checks outstanding	–
True Balance	

Name ______________ Section ______________ Date ______________

ANSWERS

8. [Respond in exercise.]

9. [Respond in exercise.]

8.

YOUR CHECKBOOK REGISTER

Check No.	Date	Transaction Description	Payment (–)	(√)	Deposit (+)	Balance
			Balance brought forward			278.32
326	8-12	J. J. Miller (birthday check)	40.00			____
	8-15	Deposit			500.00	____
327	8-15	Jay Wright, MD (physical exam)	260.00			____
328	8-15	Local Waterworks (water/trash)	27.40			____
	8-22	Deposit			400.12	____
329	8-27	Time Digest (magazine subs.)	12.50			____
330	8-29	Lower Mortgage (house payment)	750.00			____
	8-31	Interest				____
	8-31	Service Charge				____
						____
			True Balance			

BANK STATEMENT
Checking Account Activity

Transaction Description	Amount	(√)	Running Balance	Date
Beginning Balance			278.32	8-01
Deposit	500.00		778.32	8-15
Check #326	40.00		738.32	8-15
Check #327	260.00		478.32	8-15
Deposit	400.12		878.44	8-22
Check #328	27.40		851.04	8-22
Interest	1.82		852.86	8-31
Service Charge	4.00		848.86	8-31
Ending Balance			848.86	

RECONCILIATION SHEET

A. First, mark √ beside each check and deposit listed in both your checkbook register and on the bank statement.
B. Second, in your checkbook register, add any interest paid and subtract any service charge listed.
C. Third, find the total of all outstanding checks.

Outstanding Checks No.	Amount
Total	____

Statement Balance	____
Add deposits not credited	+ ____
Total	____
Subtract total amount of checks outstanding	– ____
True Balance	____

9.

YOUR CHECKBOOK REGISTER

Check No.	Date	Transaction Description	Payment (–)	(√)	Deposit (+)	Balance
			Balance brought forward			147.02
203	2-3	Food Stoppe (groceries)	26.90			____
204	2-8	Ekkon Oil (gasoline bill)	71.45			____
205	2-14	Lily's Roses (flowers)	25.00			____
206	2-14	Alumni Assoc. (alumni dues)	20.00			____
	2-15	Deposit			600.00	____
207	2-26	SRO (theatre tickets)	52.50			____
208	2-28	MPG Mtg. (house payment)	500.00			____
	2-28	Service Charge				____
						____
						____
			True Balance			

BANK STATEMENT
Checking Account Activity

Transaction Description	Amount	(√)	Running Balance	Date
Beginning Balance			147.02	2-01
Check #203	26.90		120.12	2-04
Check #204	71.45		48.67	2-14
Deposit	600.00		648.67	2-15
Check #205	25.00		623.67	2-15
Service Charge	3.00		620.67	2-28
Ending Balance			620.67	

RECONCILIATION SHEET

A. First, mark √ beside each check and deposit listed in both your checkbook register and on the bank statement.
B. Second, in your checkbook register, add any interest paid and subtract any service charge listed.
C. Third, find the total of all outstanding checks.

Outstanding Checks No.	Amount
Total	____

Statement Balance	____
Add deposits not credited	+ ____
Total	____
Subtract total amount of checks outstanding	– ____
True Balance	____

10. [Respond in exercise.]

10.

Check No.	Date	Transaction Description	Payment (–)	(√)	Deposit (+)	Balance
			Balance brought forward			4071.82
996	10-1	Credit World (loan payment)	200.75			
997	10-10	United Ways (donation)	25.00			
998	10-21	McIntosh Farms (barrel of apples)	42.20			
999	10-26	Spook Me (costume rental)	35.00			
1000	10-28	Yum-Yum's (candy)	12.14			
1001	10-29	B-Sharp Inc. (piano tuner)	20.00			
1002	10-30	Food -2-Go (party platter)	78.50			
1003	10-31	Cash				
1004	10-31	Principal S&L (house payment)	1250.00			
	10-31	Interest				
			True Balance			

YOUR CHECKBOOK REGISTER

BANK STATEMENT
Checking Account Activity

Transaction Description	Amount	(√)	Running Balance	Date
Beginning Balance			4071.82	10-01
Check #996	200.75		3871.07	10-05
Check #998	42.20		3828.87	10-23
Check #999	35.00		3793.87	10-30
Check #1003	300.00		3493.87	10-31
Interest	16.29		3510.16	10-31
Ending Balance			3510.16	

RECONCILIATION SHEET

A. First, mark √ beside each check and deposit listed in both your checkbook register and on the bank statement.
B. Second, in your checkbook register, add any interest paid and subtract any service charge listed.
C. Third, find the total of all outstanding checks.

Outstanding Checks No.	Amount
Total	

Statement Balance	
Add deposits not credited	+
Total	
Subtract total amount of checks outstanding	–
True Balance	

8.4 Buying and Owning a Car

Buying a Car

Buying a car is not as expensive or complicated as buying a home. However, more people buy cars than homes, and in many cases, buying a car is the most expensive purchase of a person's life. As with finances in general, paying cash for a car is cheaper than financing the car with a bank or savings and loan. If you are going to finance the purchase of a car, at least be aware of the expenses involved and study all the papers so that you know the total amount you are paying for the car.

Objective ①

Expenses in Buying a Car:

Purchase price: The selling price agreed on by the seller and the buyer.

Sales tax: A fixed percent that varies from state to state.

License fee: Fixed by the state, often based on the type of car and its value.

Objectives

① Become aware of and learn how to calculate the expenses involved in buying a car.

② Know how to calculate the percent of your income that is spent on your car.

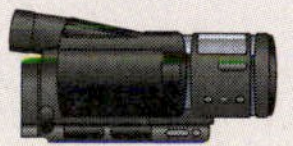
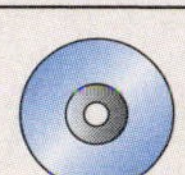

Example 1

You are going to buy a new car for \$18,500. The bank will loan you 70% of all the related expenses, including taxes and fees. If sales tax is figured at 8% and there is a license fee of \$250, how much cash do you need in order to buy the car?

Solution

a. First, find the total of all related expenses.

$$\begin{array}{rl} \$18,500 & \text{selling price} \\ \times\ \ 0.08 & \text{tax rate} \\ \hline \$1480 & \text{sales tax} \end{array} \qquad \begin{array}{rl} \$18,500 & \text{selling price} \\ 1\,480 & \text{sales tax} \\ +\ \ 250 & \text{license fee} \\ \hline \$20,230 & \text{total expenses} \end{array}$$

b. Find 30% of all expenses. (Since the bank will loan you 70%, you must provide 100% – 70% = 30% of the total expenses in the cash.)

$$\begin{array}{rl} \$20,230 & \text{total expenses} \\ \times\ \ 0.30 & \\ \hline \$6069.00 & \text{cash} \end{array}$$

Now work exercise 1 in the margin.

1. You are going to buy a new car for \$25,450. The bank will loan you 80% of all the related expenses, including taxes and fees. If sales tax is figured at 7% and there is a license fee of \$200, how much cash do you need in order to buy the car?

Owning a Car

Actually, the bank owns your car until all payments on the loan have been paid. The car is their security for the loan. However, in addition to the monthly payments, you must pay for insurance, any necessary repairs, and general maintenance costs.

Objective 2

Expenses in Owning a Car:

Monthly payments: Payments made if you borrowed money to buy the car.

Auto insurance: Covers a variety of situations (liability, collisions, towing, theft, and so on).

Operating costs: Basic items such as gasoline, oil, tires, tune-ups.

Repairs: Replacing worn or damaged parts.

2. Last November Becky's car expenses were as follows: loan payment, \$250; insurance, \$120; gasoline, \$155; battery, \$65.

a. What were her total car expenses for that month?

b. What percent of her car expenses was for the loan payment?

Example 2

In one month, Bonnie's car expenses were as follows: loan payment, \$350; insurance, \$80; gasoline, \$100; oil and filter, \$22; and a new headlight, \$32.

a. What were her total car expenses for that month?

b. What percent of her car expenses was for the loan payment?

Solution

a. Find her total expenses.

\$350	loan payment
80	insurance
100	gasoline
22	oil and filter
+ 32	headlight
\$584	total expenses

b. Find the percent of the total spent on the loan payment.

$$\frac{350}{584} \approx 0.5993 \approx 60\%$$

Bonnie spent \$584 on her car, and the loan payment was about 60% of her expenses.

Now work exercise 2 in the margin.

Practice Problems

1. You are going to buy a new truck for $31,450. The bank will loan you 90% of all the related expenses, including taxes and fees. If sales tax is figured at 5% and there is a license fee of $250, how much cash do you need in order to buy the truck?

2. A new car costs $24,500. The bank will loan you 75% of all the related expenses, including taxes and fees. If sales tax is figured at 10% and there is a $50 license fee, how much cash do you need in order to buy the car?

3. In September your car expenses included the following: loan payment, $575; insurance, $120; tires, $375; and gasoline $245. What were your car expenses for the month?

4. In October your loan payment is $1000 and your additional expenses consist of $350 for gasoline, $125 for insurance, and $12.95 for wiper blades. What percent (to the nearest hundredth) of your car expenses were for the loan payment?

Answers to Practice Problems:

1. $3327.25 **2.** $6750 **3.** $1315 **4.** 67.21%

Name ______________ Section ______________ Date ______________

ANSWERS

Exercises 8.4

1. To buy a used car for $6800, you must pay a 6% sales tax and a license fee of $120. If the bank will loan you 85% of your expenses, how much cash do you need to buy the car?

2. John wants to buy a new convertible for $25,000. His credit union will loan him 80% of his expenses. What amount of cash does he need to buy the car if the sales tax is 8.5% and the license fee is $250?

3. How much cash do you need to buy a car for $18,000 if the sales tax is calculated at 6%, the license fee is $200, and the loan company will let you borrow 75% of your expenses?

4. A used car is priced at $4500. Your old car is worth $800 on a trade-in. The sales tax is figured at 7.5% of the selling price, and the license fee is $80. If the savings and loan will lend you $3000, how much cash do you need to buy the car?

5. Your old car is worth $1500 if you trade it in for a new car priced at $11,200. Sales tax is 6% and the license fee is $225. If the bank will loan you 80% of your expenses, how much cash do you need to buy the new car?

6. **a.** If the car expenses for one month were $325 for the loan payment, $35 for insurance, $120 for gasoline, and $145 for two new tires, what were your total car expenses for the month?
   b. If your income was $2000, what percent of your income was used for car expenses?

1. ______________

2. ______________

3. ______________

4. ______________

5. ______________

6. a. ______________

b. ______________

7. ______________

8. a. ______________

b. ______________

9. ______________

10. a. ______________

b. ______________

c. ______________

7. Nancy decided her old car needed painting. If the paint job was priced at $1650, including repairing some dents, and she figured that driving cost an average of 24¢ per mile, including gas, oil and insurance, what were her car expenses that month if she drove 1200 miles?

8. Art owns his car, but it needs a new transmission for $1200 (installed).

a. What were his car expenses the month that he had the new transmission installed if he also spent $65 for insurance, $75 for gas, $15 for oil and filter, and $300 for a tune-up?

b. If he took $1000 from his savings account to help pay for the transmission, what percent of his income, $2600, was used for the remainder of his car expenses?

9. Danielle decided she would like to have a new car, but she could not afford one if the car expenses averaged more than 20% of her monthly income. If she figured the expenses would average $300 for a loan payment, $70 for insurance, $65 for gas, $8 for oil, $10 for tire wear, and $40 for general repairs, could she afford the car with a monthly income of $2100?

10. Suppose that you owned a car, your monthly income was $1800, and you figured you could spend 15% of this to operate a car.

a. What would you be able to spend on gas if insurance cost $85, oil and filter cost $25, and you estimated $40 per month for other expenses?

b. How many gallons of gas could you buy if gas cost $2.40 per gallon?

c. How many miles could you drive if your car averaged 19 miles per gallon?

8.5 Buying and Owning a Home

Objectives

① Become aware of and learn how to calculate the expenses involved in buying a home.

② Know how to calculate the percent of your income that is spent on your home.

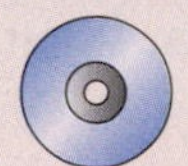

Buying a Home

Buying a home is probably the largest financial investment in one's life. If you are thinking of buying a home someday, this chapter will help you become an informed buyer. This section provides a base for understanding some of the terminology used by realtors and bankers in selling and financing homes and for developing a "feeling" for the amount of cash needed to actually buy a home. Some terms related to home purchasing are explained briefly in the following box.

Objective ①

Expenses in Buying a Home

Purchase price: The selling price (what you have agreed to pay).

Down payment: Cash you pay to the seller (usually 20% to 30% of the purchase price).

Mortgage loan (1st trust deed): Loan to you by bank or savings and loan (difference between purchase price and down payment).

Mortgage fee (or points): Loan fee charged by the lender (usually 1% to 3% of the mortgage loan).

Fire insurance: Insurance against the loss of your home by fire (required by almost all lenders).

Recording fees: Fees for recording you as the legal owner.

Property taxes: Taxes that must be prepaid before the lender will give you the loan (usually six months in advance).

Legal fees: Fees charged by a lawyer or escrow company for completing all forms in a legal manner.

Example 1

You buy a home for \$150,000. Your down payment is 20% of the selling price, and the mortgage fee is 2 points (2% of the new mortgage). You also have to pay \$500 for fire insurance, \$350 for taxes, \$50 for recording fees, and \$310 for legal fees.

a. What is the amount of your mortgage?

b. How much cash must you provide to complete the purchase?

Solution

a. To find the amount of the mortgage, find 80% of the selling price. (Since the down payment is 20%, the mortgage will be 100% – 20% = 80% of the selling price.)

Continued on next page...

1. You buy a home for $275,000. You pay 10% of the selling price as a down payment, and the mortgage fee is 3 points (3% of the new mortgage). You also have to pay $650 for fire insurance, $2200 for taxes, $50 for recording fees, and $600 for legal fees.

a. What is the amount of the mortgage?

b. How much cash must you provide to complete the purchase?

$$\begin{array}{rl} \$150{,}000 & \text{selling price} \\ \times \quad 0.80 & \\ \hline \$120{,}000.00 & \text{mortgage (or 1st trust deed)} \end{array}$$

b. Add the mortgage fee, the down payment, and all the other fees.

$$\begin{array}{rl} \$120{,}000 & \text{mortgage} \\ \times \quad 0.02 & \\ \hline \$2400.00 & \text{mortgage fee} \end{array}$$

$$\begin{array}{rl} \$150{,}000 & \text{selling price} \\ \times \quad 0.20 & \\ \hline \$30{,}000.00 & \text{down payment} \end{array}$$

$$\begin{array}{rl} \$30{,}000 & \text{down payment} \\ 2400 & \text{mortgage fee} \\ 500 & \text{fire insurance} \\ 350 & \text{property taxes} \\ 50 & \text{recording fees} \\ + \quad 310 & \text{legal fees} \\ \hline \$33{,}610 & \text{cash needed to complete purchase} \end{array}$$

The mortgage will be $120,000 and you will need $33,610 in cash.

Now work exercise 1 in the margin.

Owning a Home

After you have bought your home, you must make the payments on the mortgage, pay property taxes, pay for utilities (water, electricity, and gas), and pay for repairs. (Don't be too discouraged. One advantage of this is that many expenses are deductible from your income taxes.)

Expenses in Owning a Home

Monthly mortgage payment: Payment to mortgage holder includes both principal and interest.

Property taxes: May be paid monthly, semiannually, or annually.

Homeowner's insurance: Can be included with your fire insurance; includes insurance against theft and liability.

Utilities: Monthly payments for water, electricity, and gas.

Maintenance: Repairs, yard work, painting, and so on.

Objective ②

Example 2

a. What were the total expenses for your home the month you paid $1250 on your mortgage loan, $100 in taxes, $50 for fire insurance, $200 for all utilities, and $75 for maintenance?

b. If your salary was $3650, what percentage of your salary was spent on your home?

Solution

a. Find your total expenses.

$1250	loan payment
100	taxes
50	fire insurance
200	utilities
+ 75	maintenance
$1675	total spent on home

b. Calculate the percent of your salary spent on your home.

$$\frac{1675}{3650} \approx 0.4589 = 45.89\%$$

Now work exercise 2 in the margin.

Practice Problems

1. You buy a home for $225,000. Your down payment is 20% of the selling price, and the mortgage fee is 3 points (3% of the new mortgage). You also have to pay $750 for fire insurance, $1125 for taxes, and $325 for recording and legal fees. What is the amount of your mortgage?

2. Your new home is priced at $125,000. Your down payment is 10% of the selling price, and the mortgage fee is 2 points (2% of the new mortgage). You also have to pay $650 for fire insurance, $1000 for taxes, $250 for legal fees and $75 for recording fees. How much cash must you provide to purchase the home?

3. What were the total expenses for your home the month you paid $1650 on your mortgage loan, $80 in taxes, $75 for fire and flood insurance, $200 for all utilities, and $25 for maintenance?

4. If your salary was $4250.50 and your total home expenses were $2545, what was the percentage (to the nearest hundredth) of your salary spent on your home?

Answers to Practice Problems:

1. $180,000 **2.** $16,725 **3.** $2030 **4.** 59.88%

2. One month you paid $880 on your mortgage, $80 in taxes, $150 for fire and flood insurance, and $320 in utilities, and $125 for maintenance. Your monthly salary is $2250.

a. What were your total expenses for your home that month?

b. What percentage of your salary was spent on your home?

Name ______________ Section ______________ Date ______________

Exercises 8.5

1. A home is sold for $162,500. The buyer has to make a down payment of 25% of the selling price, pay a loan fee of 2% of the mortgage, $200 for fire insurance, $50 for recording fees, $580 for taxes, and $570 for legal fees.
   a. What is the amount of the down payment?
   b. What is the amount of the mortgage?
   c. How much cash does the buyer need to complete the purchase?

2. The purchase price on a home is $98,000 and the buyer makes a down payment of $9800 (10% of the selling price) and pays a loan fee of $2\frac{1}{2}\%$ of the new mortgage. He also pays $250 for legal fees, $320 for taxes, and $425 for fire insurance. (The seller agrees to pay all recording fees.)
   a. What is the amount of the loan fee?
   b. How much does the buyer owe in order to complete the purchase?

3. You bought a home for $85,000 and made a down payment of $17,000. If you paid a mortgage fee (loan fee) of $1360,
   a. What percent of the first trust deed was this fee?
   b. You also paid $250 for fire insurance, $35 for recording fees, $170 for taxes, and $195 for legal fees. How much cash did you need to complete the purchase?

4. A house is sold for $125,000 and the buyer makes a 30% down payment and is charged a 1% loan fee on the trust deed. The buyer is also charged $220 for recording fees, $345 for legal fees, $450 for taxes, and $520 for fire insurance.
   a. How much is the down payment?
   b. How much is the first trust deed?
   c. How much is the loan fee?
   d. How much cash does the buyer need?

ANSWERS

1. a. ______________

b. ______________

c. ______________

2. a. ______________

b. ______________

3. a. ______________

b. ______________

4. a. ______________

b. ______________

c. ______________

d. ______________

5. a. ______

b. ______

c. ______

d. ______

6. a. ______

b. ______

7. a. ______

b. ______

8. a. ______

b. ______

9. a. ______

b. ______

5. A condominium sold for $96,000. The buyer made a down payment of 25% of the selling price and paid a loan fee of $1\frac{1}{2}$% of the amount of the mortgage. She also paid $420 for fire insurance, $35 for recording fees, and $380 for taxes, (The seller paid all legal fees.)

a. What was the amount of the down payment?
b. What was the amount of the mortgage?
c. How much was the loan fee?
d. How much cash did she need to complete the purchase?

6. In March Ms. Smith made a mortgage payment of $625 and paid $25 for taxes, $10 for water, $35 for electricity, $40 for gas, and $55 for a plumber's bill.

a. What were her home expenses in March?
b. If her income was $1975, what percent of her income did she spend on her home?

7. During July, the Johnsons made a loan payment of $875 and paid $90 for taxes, $85 for utilities, $70 for the fire insurance, and $285 for a painter and other repairs.

a. How much did the Johnsons spend on their home in July?
b. If the combined income of Mr. and Mrs. Johnson was $4620, what percent of their income did they spend on their home?

8. In one month Sam paid $150 for home repairs, $390 for his home loan, $60 for utilities, $20 for fire insurance, and $40 for taxes.

a. What were his home expenses for that month?
b. What percent of his $2000 income did he spend on his home?

9. Your income for one month was $1800.

a. If you made a mortgage payment of $578 and paid $25 in taxes, a water bill of $15, an electric bill of $35, a gas bill of $45, and a fire insurance premium of $40, how much did you spend on your home?
b. What percent of your income was this?

Name ________________ Section ________ Date ________

ANSWERS

10. ____________

10. Your income was \$2800 per month and you figured you could afford to spend 30% of this each month on a home. What mortgage payment could you make if you estimated taxes at \$60 per month, utilities at \$80 per month, fire insurance at \$65 per month, and repairs at 7% of your income per month?

Upper class limit: the largest whole number that belongs to a class.

Class boundaries: numbers that are halfway between the upper limit of one class and the lower limit of the next class.

Class width: the difference between the class boundaries of a class (the width of each bar).

Frequency: the number of data items in a class.

Example 4

Histogram

Figure 8.4 shows a histogram that summarizes the scores of 50 students on an English placement test. Answer the following questions by referring to the graph.

a. How many classes are represented? 6

b. What are the class limits of the first class? 201 and 250

c. What are the class boundaries of the second class? 250.5 and 300.5

d. What is the width of each class? 50

e. Which class has the greatest frequency? second class

f. What is this frequency? 16

g. What percentage of the scores are between 200.5 and 250.5? $\frac{2}{50} = 4\%$

h. What percentage of the scores are above 400? $\frac{12}{50} = 24\%$

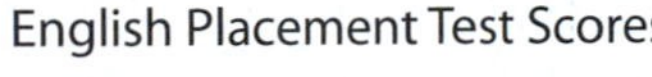

English Placement Test Scores

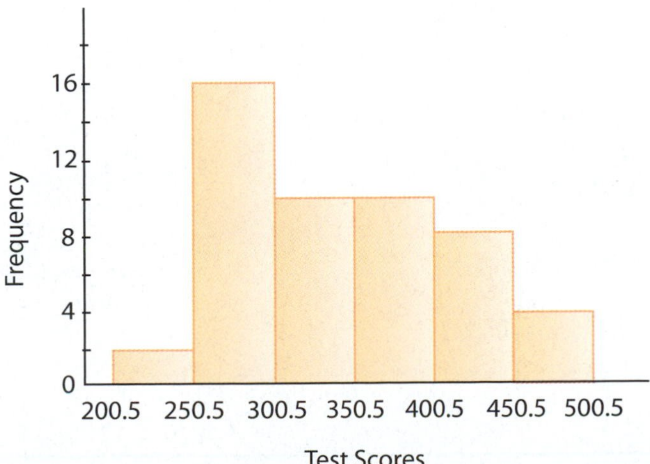

Figure 8.4: Histogram

Now work exercise 4 in the margin.

4. Use the histogram in Figure 8.4 to answer parts **a.**, **b.**, and **c.**

a. Which class has the least frequency?

b. What percentage of the scores are below 300.5?

c. Which two classes appear to be equal?

Practice Problems

1. Using the bar graph in Example 1, what was the amount of increase in sales between March and April?

2. Using the circle graph in Example 2, if the person's income increases to $30,000 and the percentage spent on each item does not change, what is the amount spent on entertainment over the year?

3. Using the line graph in Example 3, find the minimum difference between the high and low temperatures of a single day in the week shown.

4. Using the histogram in Example 4, what is the frequency of the class with the least frequency?

Answers to Practice Problems:

1. $25,000 **2.** $1500 **3.** 6° **4.** 2

Name ______________ Section ______________ Date ______________

Exercises 8.6

Answer the questions related to each of the graphs. Some questions can be answered directly from the graphs; others may require some calculations. See Examples 1 through 4.

1. The following bar graph shows the numbers of students in five fields of study at a university.

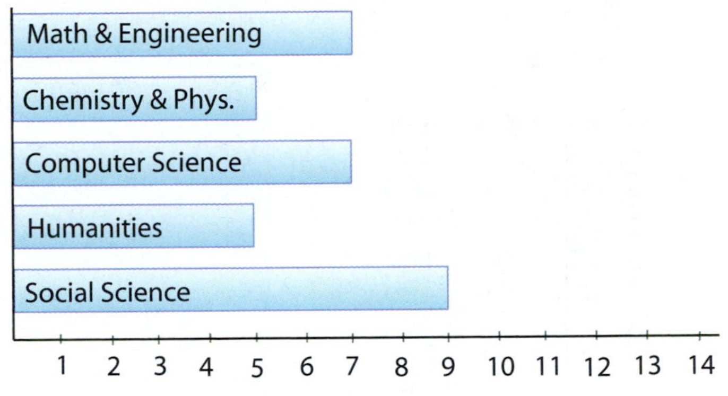

a. Which field of study has the largest number of declared majors?
b. Which field of study has the smallest number of declared majors?
c. How many declared majors are indicated in the entire graph?
d. What percent are computer science majors?

2. The bar graph shows the number of vehicles that crossed one intersection during a two-week period.

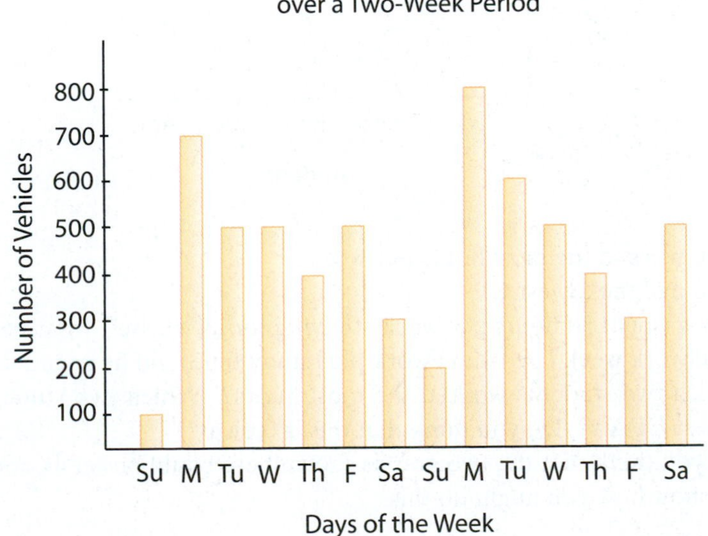

ANSWERS

1. a. ______________

b. ______________

c. ______________

d. ______________

2. a. ______

b. ______

c. ______

d. ______

3. a. ______

b. ______

c. ______

d. ______

e. ______

a. On which day did the highest number of vehicles cross the intersection? How many crossed that day?

b. What was the average number of vehicles that crossed the intersection on the two Sundays?

c. What was the total number of vehicles that crossed the intersection during the two weeks?

d. About what percent of the total traffic was counted on Saturdays?

3. In comparing the following two graphs, assume that all five students graduated with comparable grades from the same high school.

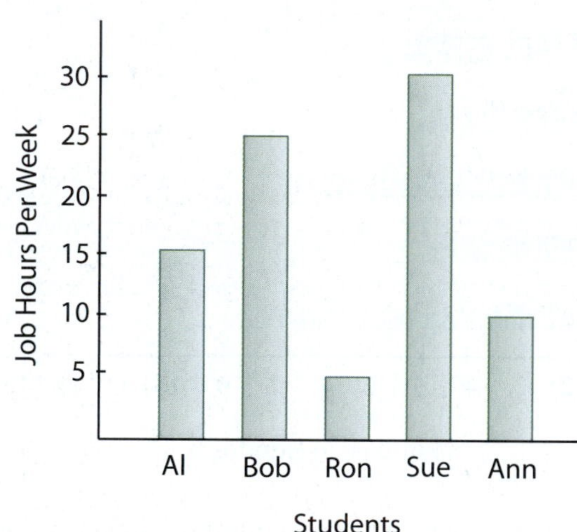

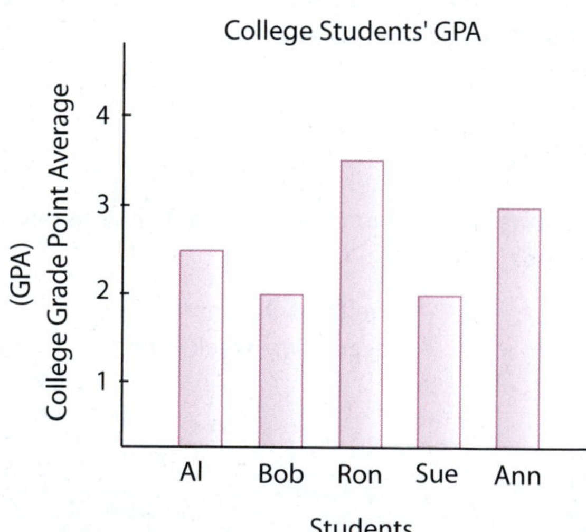

a. Who worked the most hours per week?

b. Who had the lowest GPA?

c. If Ron spent 30 hours per week studying for his classes, what percent of his total work week (part-time work plus study time) did he spend studying?

d. Which two students worked the most hours? Which two students had the lowest GPA's? Do you think that this is typical?

e. Do you think that the two graphs shown here could be set as one graph? If so, show how you might do this.

Name ______________________ Section ____________ Date __________

ANSWERS

8.

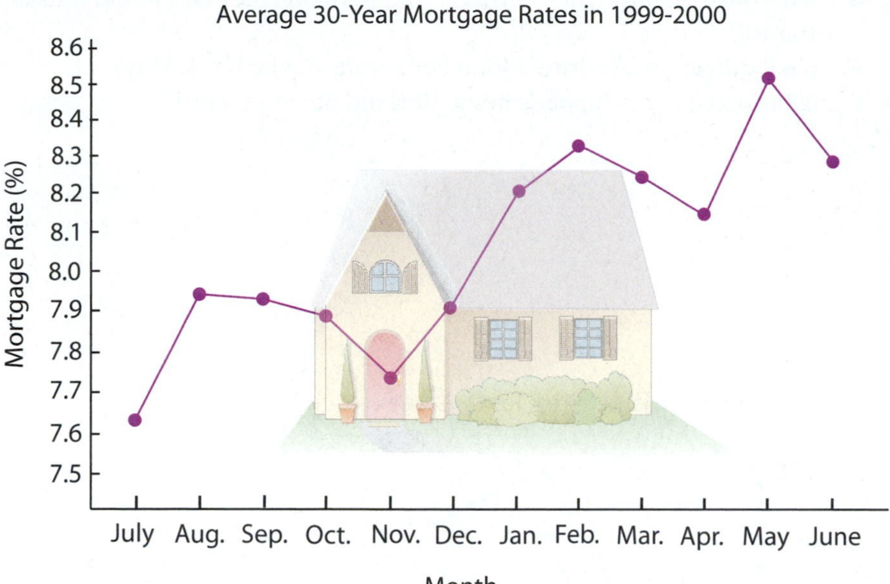

a. During what month or months in 1999-2000 were mortgage rates highest?

b. Lowest?

c. What was the average of the interest rates over the entire 12-month period? Round each value to the nearest tenth of a percent.

Source: http://www.huduser.org/

8. a. ______________

b. ______________

c. ______________

9.

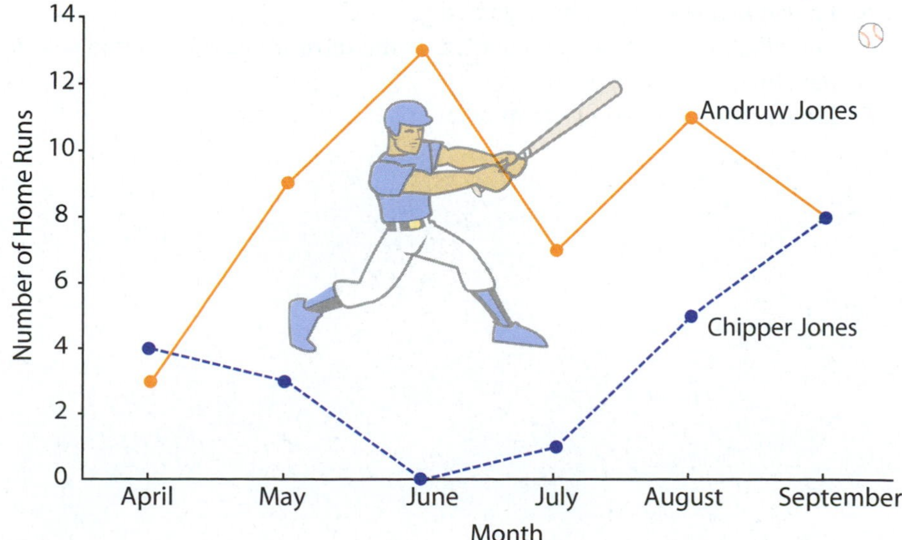

9. a. ____________

b. ____________

c. ____________

d. ____________

e. ____________

f. ____________

10. a. ____________

b. ____________

c. ____________

d. ____________

e. ____________

f. ____________

a. During which month did Andruw hit the most home runs?
b. How much higher was his total for that month than for the lowest month?
c. In what month did Chipper and Andruw hit the same amount of home runs?
d. What was the difference between Chipper and Andruw's home runs in August?
e. What percent of Andruw's total home runs did he hit in May?
f. What percent of Chipper's home runs did he hit in April?

10.

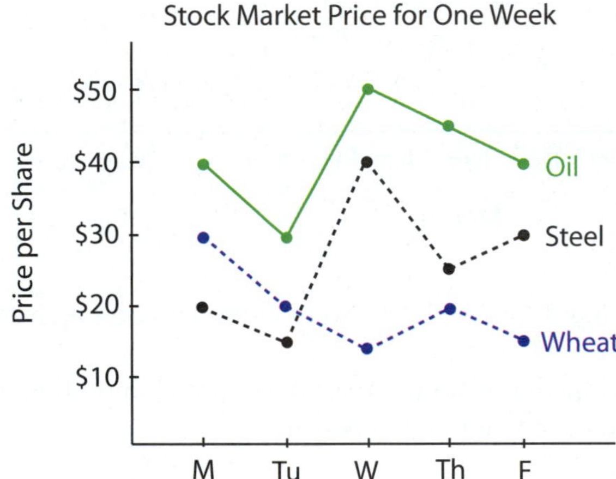

a. If on Monday morning you had 100 shares of each of the three stocks shown (oil, steel, wheat), and you held the stock all week, on which stock would you have lost money?
b. How much would you have lost?
c. On which stock would you have gained money?
d. How much would you have gained?
e. On which stock could you have made the most money if you had sold at the best time?
f. How much could you have made?

Name ______________________ Section ____________ Date __________

ANSWERS

11. a. ____________

b. ____________

c. ____________

d. ____________

e. ____________

f. ____________

g. ____________

h. ____________

11.

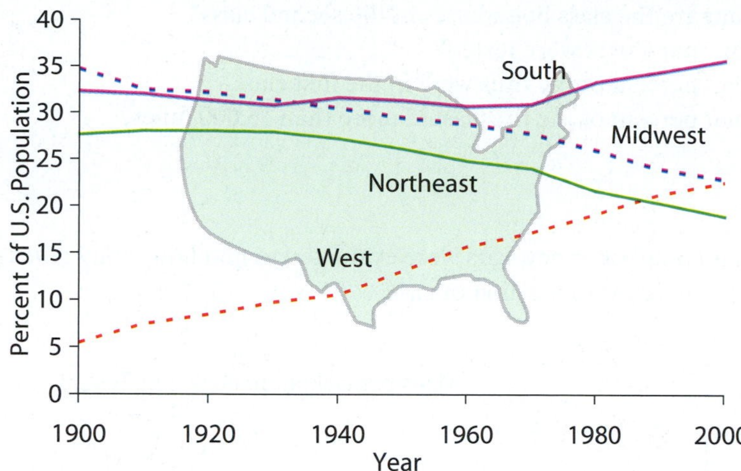

a. Approximately what percent of the population was in each of the four regions in 1900?

b. In 2000?

c. Which region seems to have had the most stable percent of the population between 1900 and 2000?

d. What is the difference between the highest and lowest percents for this region?

e. Which region has had the most growth?

f. What was its lowest percent and when?

g. What was its highest percent and when?

h. Which region has had the most decline?

Source: U.S. Census Bureau, decennial census of population, 1900 to 2000.

12.

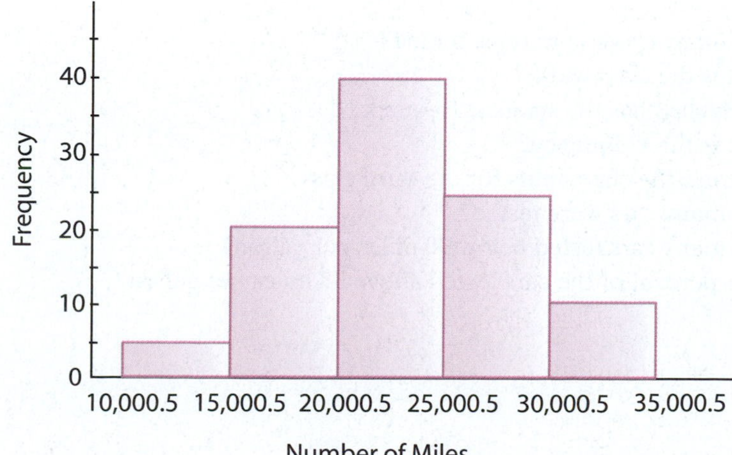

12. a. ____________

b. ____________

c. ____________

d. ____________

e. ____________

f. ____________

g. ____________

h. ____________

13. a. ____________

b. ____________

c. ____________

d. ____________

e. ____________

f. ____________

g. ____________

h. ____________

a. How many classes are represented?
b. What is the width of each class?
c. Which class has the highest frequency?
d. What is this frequency?
e. What are the class boundaries of the second class?
f. How many tires were tested?
g. What percent of the tires were in the first class?
h. What percent of the tires lasted more than 25,000 miles?

13. A certain number of new cars were evaluated to find how many miles per gallon could be driven with a gallon of gas.

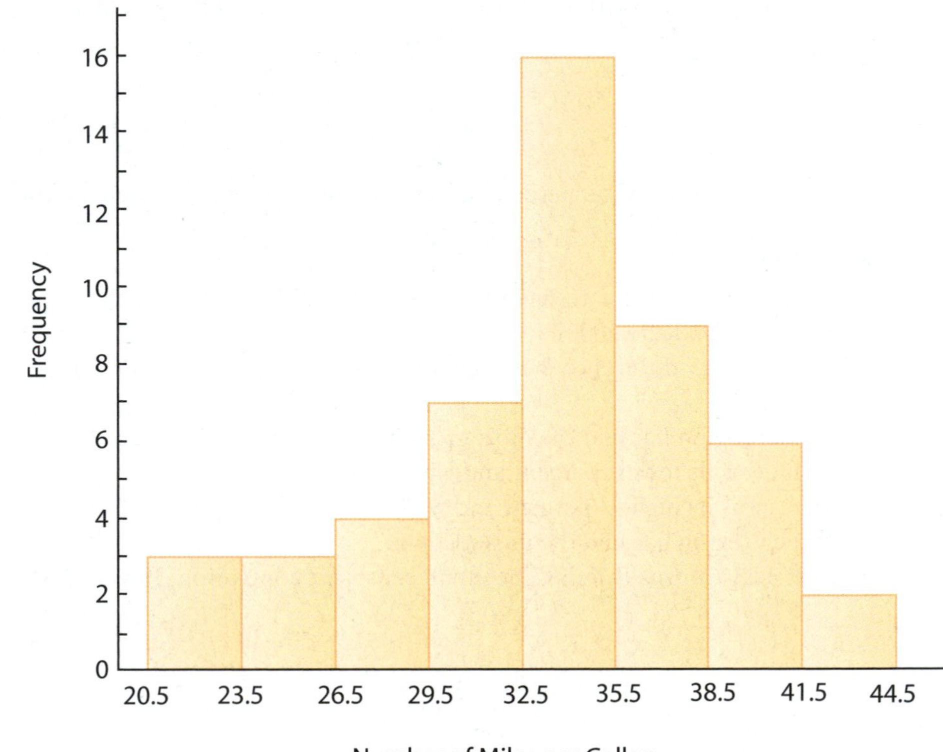

a. How many classes are represented?
b. What is the class width?
c. Which class has the smallest frequency?
d. What is this frequency?
e. What are the class limits for the third class?
f. How many cars were tested?
g. How many cars tested below 30 miles per gallon?
h. What percent of the cars tested above 38 miles per gallon?

Name ______________________ Section ____________ Date __________

ANSWERS

14. This circle graph represents the various sources of income for a city government with a total income of $100,000,000.

a. What is the city's largest source of income?
b. What percent of income comes from Goods and Services?
c. What is the ratio of income from Taxes to the total income?

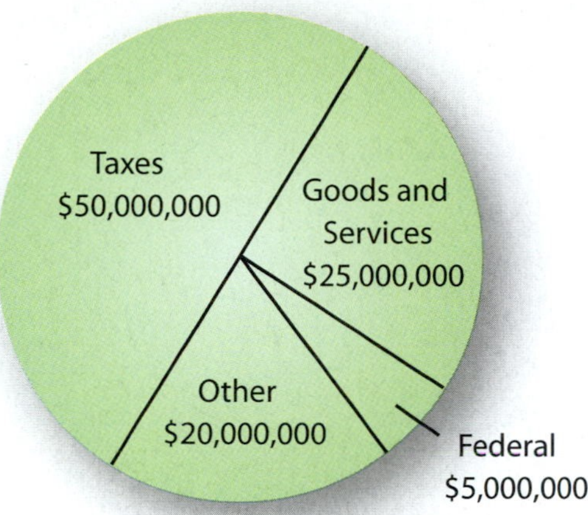

14. a. ______________

b. ______________

c. ______________

15. Sally's car expenses for the month of June are shown in the circle graph below.

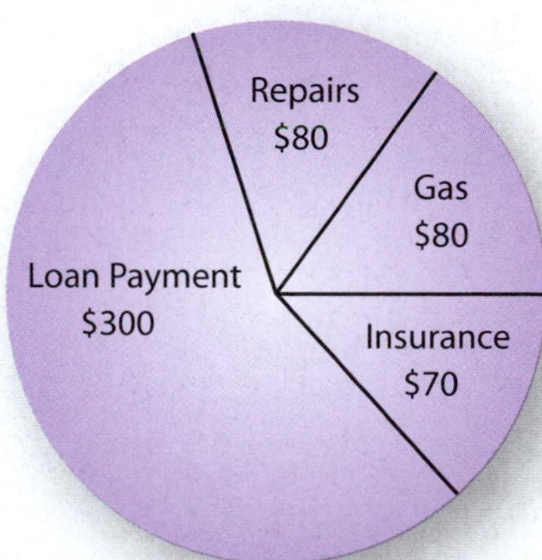

a. What were her total car expenses for the month?
b. What percent of her expenses did she spend on each category?
c. What was the ratio of her insurance expenses to her gas expenses?

15. a. ______________

b. ______________

c. ______________

Chapter 8 Index of Key Terms and Ideas

Section 8.1 Simple Interest

Interest page 652

Interest is money paid for the use of money. The money that is invested or borrowed is called the **principal**. The **rate** is the percent of interest and is almost always stated as an annual rate.

Simple Interest page 652

Loans that are based on **simple interest** involve only one payment (including interest and principal) at the end of the term of the loan.

Formula for Calculating Simple Interest page 653

$\boldsymbol{I = P \times r \times t}$

where I = Interest (earned or paid)

P = Principal (the amount invested or borrowed)

r = Rate of interest (stated as an annual, or yearly, rate and used in decimal or fraction form)

t = Time (in years or fractions of a year)

Section 8.2 Compound Interest

Compound Interest page 663

Interest paid on interest earned is **compound interest**.

To Calculate Compound Interest page 663

Step 1: Using the formula for simple interest, $I = P \times r \times t$, calculate the simple interest where $t = \frac{1}{n}$ and n is the number of times compounded per year.

Step 2: Add this interest to the principal to create a new value for the principal.

Step 3: Repeat Steps 1 and 2 however many times the interest is to be compounded.

Compound Interest Formula pages 665 – 666

When interest is compounded, the total **amount**, $\boldsymbol{A}$, accumulated (including principal and interest) is given by the formula

$$\boldsymbol{A = P\left(1 + \frac{r}{n}\right)^{nt}}$$

where P = the principal

r = the annual interest rate (in decimal or fraction form)

t = the length of time in years

n = the number of compounding periods in one year

Section 8.3 Balancing a Checking Account

Checking Account and Register page 673

To Balance your Checking Account page 674

1. Go through your checkbook register and the bank statement and put a check mark (√) by each check paid and deposit recorded on the bank statement.
2. In your checkbook register:
   a. Add any interest paid to your current balance.
   b. Subtract any service charge from this balance.
   This number is your **true balance**.
3. Find the total of all the outstanding checks in your checkbook register.
4. On the reconciliation sheet:
   a. Enter the balance from the bank statement.
   b. Add any deposits you have made that are not recorded on the bank statement.
   c. Subtract the total of your outstanding checks (found in Step 3). This number should agree with your true balance. Now your checking account is balanced.

If your Checking Account Does Not Balance page 674

1. Go over your arithmetic in the balancing procedure.
2. Go over your arithmetic check by check in your checkbook register.
3. Make an appointment with bank personnel to find the reasons for any other errors.

Section 8.4 Buying and Owning a Car

Expenses in Buying a Car page 683

Purchase price: The selling price agreed on by the seller and the buyer.
Sales tax: A fixed percent that varies from state to state.
License fee: Fixed by the state, often based on the type of car and its value.

Expenses in Owning a Car page 684

Monthly payments: Payments made if you borrowed money to buy the car.
Auto insurance: Covers a variety of situations (liability, collisions, towing, theft, and so on).
Operating costs: Basic items such as gasoline, oil, tires, tune-ups.
Repairs: Replacing worn or damaged parts.

Section 8.5 Buying and Owning a Home

Expenses in Buying a Home page 689

Purchase price: The selling price (what you have agreed to pay).
Down payment: Cash you pay to the seller (usually 20% to 30% of the purchase price).
Mortgage loan (1st trust deed): Loan to you by bank or savings and loan (difference between purchase price and down payment).

Continued on next page...

Section 8.5 Buying and Owning a Home (continued)

Expenses in Buying a Home (continued) page 689

Mortgage fee (or points): Loan fee charged by the lender (usually 1% to 3% of the mortgage loan).

Fire Insurance: Insurance against the loss of your home by fire (required by almost all lenders).

Recording fees: Fees for recording you as the legal owner.

Property taxes: Taxes that must be prepaid before the lender will give you the loan (usually six months in advance).

Legal fees: Fees charged by a lawyer or escrow company for completing all forms in a legal manner.

Expenses in Owning a Home page 690

Monthly mortgage payment: Payment to mortgage holder includes both principal and interest.

Property taxes: May be paid monthly, semiannually, or annually.

Homeowner's insurance: Can be included with your fire insurance; includes insurance against theft and liability.

Utilities: Monthly payments for water, electricity, and gas.

Maintenance: Repairs, yard work, painting, and so on.

Section 8.6 Reading Graphs

The Purpose of the Four Types of Graphs page 697

1. **Bar Graphs:** to emphasize comparative amounts.
2. **Circle Graphs:** to help in understanding percents or parts of a whole.
3. **Line Graphs:** to indicate tendencies or trends over a period of time.
4. **Histograms:** to indicate data in classes (a range or interval of numbers).

Terms Related to Histograms pages 700–701

Class: A range (or interval) of numbers that contain data items.

Lower class limit: The smallest number that belongs to a class.

Upper class limit: The largest number that belongs to a class.

Class boundaries: Numbers that are halfway between the upper limit of one class and the lower limit of the next class.

Class width: The difference between the class boundaries of a class (the width of each bar).

Frequency: The number of data items in a class.

Name ______________ Section ______________ Date ______________

Chapter 8 Review Questions

ANSWERS

1. ______________
2. ______________
3. ______________
4. ______________
5. a. ______________
   b. ______________
   c. ______________
   d. ______________

1. What will be the interest earned in one year on a savings account of \$1500 if the bank pays $6\frac{1}{2}\%$ simple interest?

2. If a stock pays 12% dividends every 6 months, what will be the dividend paid on an investment of \$13,450 in 6 months?

3. If a principal of \$1000 is invested at a rate of 9% for 30 days, what will be the interest earned?

4. You made \$50 on an investment at 10% simple interest for 3 months. What principal did you invest?

5. Determine the missing items in each row if the interest is simple interest.

Principal	Rate	Time	Interest
\$200	12%	180 days	**a.** ____
\$300	18%	**b.** ____	\$81
\$1000	**c.** ____	1 year	\$85
d. ____	9%	18 months	\$270

6. [Respond below exercise.] ______

7. [Respond below exercise.] ______

8. a. ______

b. ______

c. ______

d. ______

9. a. ______

b. ______

c. ______

d. ______

e. ______

f. ______

g. ______

h. ______

In Exercises 6 and 7, show the step-by-step calculations of compound interest.

6. You borrow $5000 from your aunt to buy a new motorcycle and agree to pay her interest at 5.5% compounded annually for 3 years. How much interest will your owe your aunt?

First year: $I = 5000 \times 0.055 \times 1 =$ ______

Second year: $I =$ ______ $\times 0.055 \times 1 =$ ______

Third year: $I =$ ______ $\times 0.055 \times 1 =$ ______

The total interest is ______.

7. Joan loaned her brother $10,000 for 4 years with interest to be compounded annually at 4%. How much will her brother pay her at the end of the 4 years?

First year: $I = 10{,}000 \times 0.04 \times 1 =$ ______

Second year: $I =$ ______ $\times 0.04 \times 1 =$ ______

Third year: $I =$ ______ $\times 0.04 \times 1 =$ ______

Fourth year: $I =$ ______ $\times 0.04 \times 1 =$ ______

The total interest is ______.

The total he will pay is ______.

8. How much interest will be earned on a savings account of $6500 in 5 years if interest is compounded at 6% **a.** annually? **b.** quarterly? **c.** monthly? **d.** daily?

9. Complete the following table of values if interest is compounded annually at 8%.

	Principal	Amount in Account at the End of the Year	Interest Earned
1st year	$10,000	a. ______	b. ______
2nd year	c. ______	d. ______	e. ______
3rd year	f. ______	g. ______	h. ______

Name ______________________ Section ____________ Date __________

ANSWERS

10. Find the true balance of the account on both the checkbook register and on the reconciliation sheet. Follow the directions on the reconciliation sheet.

10. [Respond in exercise.]

YOUR CHECKBOOK REGISTER

Check No.	Date	Transaction Description	Payment (−)	(√)	Deposit (+)	Balance
			Balance brought forward			2271.52
1001	8-3	Fred's Boats (loan payment)	500.00			
1002	8-5	Super Foods (groceries)	20.24			
1003	8-6	Super Drug (medication)	78.20			
1004	8-6	OK Tackle Shop (rod and reel)	65.05			
1005	8-18	Rapid Mart (gas for car)	14.34			
1006	8-29	A.J.'s Boat Repair (tune up for boat)	90.00			
1007	8-30	Eat em up (snacks for party)	93.50			
	8-31	Deposit				
1008	8-31	UABPC Banking (house payment)	1450.00			
	8-31	Interest				
		True Balance				

BANK STATEMENT
Checking Account Activity

Transaction Description	Amount	(√)	Running Balance	Date
Beginning Balance			2271.52	8-01
Check #1001	500.00		1771.52	8-05
Check #1003	78.20		1693.32	8-06
Check #1004	65.05		1628.27	8-12
Deposit	300.00		1928.27	8-31
Ending Balance			1928.27	

RECONCILIATION SHEET

A. First, mark √ beside each check and deposit listed in both your checkbook register and on the bank statement.
B. Second, in your checkbook register, add any interest paid and subtract any service charge listed.
C. Third, find the total of all outstanding checks.

Outstanding Checks No.	Amount
Total	______

Statement Balance	______
Add deposits not credited	+ ______
Total	______
Subtract total amount of checks outstanding	− ______
True Balance	______

11. The price of a new pickup truck that you wish to buy is \$24,200. Tax is 6.5% of the purchase price, and the license fee is \$350. If the bank will loan you 75% of the total expenses, how much cash will you need to buy the car?

11. ______________

12. You buy a new home for \$275,000. Your down payment is 35% of the selling price, and the mortgage fee is 4 points (4% of the new mortgage). You also have to pay \$1250 for fire insurance, \$1072.50 for taxes, and \$50 in recording and legal fees. What is the amount of your mortgage?

12. ______________

13. ____________

____

____

____

____

____

____

____

14. ____________

15. ____________

16. ____________

Refer to the circle graph for Exercises 13 – 16.

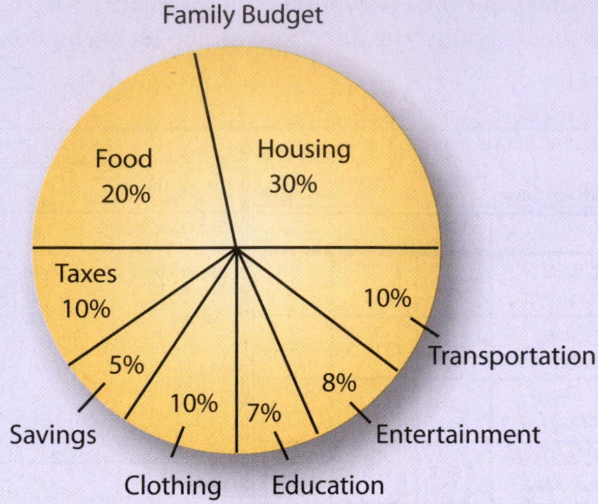

13. The circle graph shows a home budget. What amount will be spent on each category if the family income is $35,000?

14. How much more will the family spend for food than for clothing if their income is increased to $40,000 (to the nearest dollar)?

15. What fractional part of the family income is spent for food, housing, and taxes combined?

16. How much will the family spend for food, housing, and transportation combined if their income is reduced to $30,000 (to the nearest dollar)?

Name ______________________ Section ____________ Date __________

ANSWERS

Chapter 8 Test

1. ______________

2. ______________

3. ______________

4. ______________

5. a. [Respond below exercise.]

b. ______________

1. How much simple interest will be earned in one year on a savings account of \$1500 if the bank pays 4.5% interest?

2. You made \$42 in interest on an investment at 7% for 9 months. What principal did you invest?

3. If Ms. King had a savings account of \$3000 earning 5.5% simple interest, how long would it take for her to earn \$82.50 in interest?

4. If an investment of \$7500 earns simple interest of \$131.25 in 45 days, what is the rate of interest?

5. **a.** Find the total interest earned on a loan of \$6000 at 5% compounded quarterly for one year? Calculate the interest by using the step-by-step method.

First period: $I = 6000 \times 0.05 \times \frac{1}{4} =$ ________

Second period: $I =$ ________ $\times 0.05 \times \frac{1}{4} =$ ________

Third period: $I =$ ________ $\times 0.05 \times \frac{1}{4} =$ ________

Fourth period: $I =$ ________ $\times 0.05 \times \frac{1}{4} =$ ________

The total interest earned is ________.

b. What is the difference between the interest earned in part **a.** and simple interest on \$6000 for one year? Again, use 5% as the rate.

6. a. ____________

b. ____________

7. ____________

8. [Respond in exercise.] ____________

9. a. ____________

b. ____________

6. A certificate of deposit is earning 6% compounded quarterly.

a. If $1000 is deposited in the account, what will be the balance in the account at the end of one year?

b. At the end of five years?

7. How much more would be in the balance of the account in five years in Exercise 6 if the money was compounded daily?

8. Find the true balance of the account on both the checkbook register and on the reconciliation sheet. Follow the directions on the reconciliation sheet.

YOUR CHECKBOOK REGISTER

Check No.	Date	Transaction Description	Payment (–)	(√)	Deposit (+)	Balance
			Balance brought forward			1416.05
1011	7-3	Mike's Pet Shop (dog)	600.00			____
1012	7-5	Great Foods (groceries)	56.84			____
1013	7-6	Record Room (compact disc)	14.92			____
1014	7-6	Finch's (gym socks)	8.33			____
1015	7-18	Fast Fred's (dinner)	19.38			____
1016	7-29	Post Office (stamps)	10.00			____
1017	7-30	Ann Jones (yardwork)	35.00			____
1018	7-31	UPC Mgmt. (Rent)	600.00			____
	7-31	Service	1.25			____
	7-31	Interest				____
			True Balance			

BANK STATEMENT
Checking Account Activity

Transaction Description	Amount	(√)	Running Balance	Date
Beginning Balance			1416.05	7-01
Check #1011	600.00		816.05	7-04
Check #1013	14.92		801.13	7-09
Check #1014	8.33		792.80	7-12
Check #1018	600.00		192.80	7-31
Service Charge	1.25		191.55	7-31
Interest	.03		191.58	7-31
Ending Balance			191.58	

RECONCILIATION SHEET

A. First, mark √ beside each check and deposit listed in both your checkbook register and on the bank statement.

B. Second, in your checkbook register, add any interest paid and subtract any service charge listed.

C. Third, find the total of all outstanding checks.

Outstanding Checks No.	Amount		
		Statement Balance	____
		Add deposits not credited	+ ____
		Total	____
	____	Subtract total amount of checks outstanding	– ____
Total	____	True Balance	____

9. Fred is buying a car for $17,500. Sales tax is 6% and his license and transfer fees total $513.

a. If he receives a $4000 trade-in allowance and a $1200 factory rebate, how much cash will he need to buy the car?

b. If he finances 70% of his expenses before the trade-in and the rebate, how much cash will he need?

Name ____________ Section ____________ Date ____________

ANSWERS

10. You bought a home for $125,000 and made a down payment of $12,000. If you paid a mortgage fee (loan fee) of $1450:

a. What percent of the first trust deed was this fee?

b. You also paid $275 for fire insurance, $55 for recording fees, $1250 for taxes, and $55 for legal fees. How much cash did you need to complete the purchase?

10. a. ____________

b. ____________

Refer to the circle graph for Exercises 11 – 13.

Budget for Apartment Complex

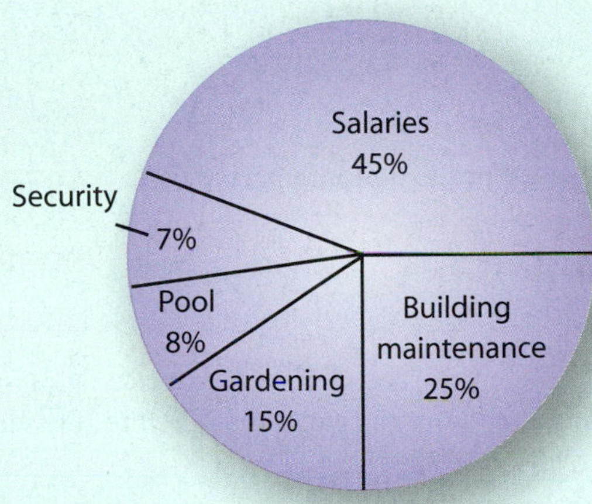

11. The budget for an apartment complex is shown in the graph. What fractional part of the budget is spent for salaries and security combined?

11. ____________

12. How much will be spent for building maintenance in one year if the budget is $250,000 per month?

12. ____________

13. How much will be spent for salaries and security combined in 6 months if the monthly budget is $150,000?

13. ____________

14. ____________

15. ____________

16. ____________

Refer to the bar graph for Exercises 14 – 16.

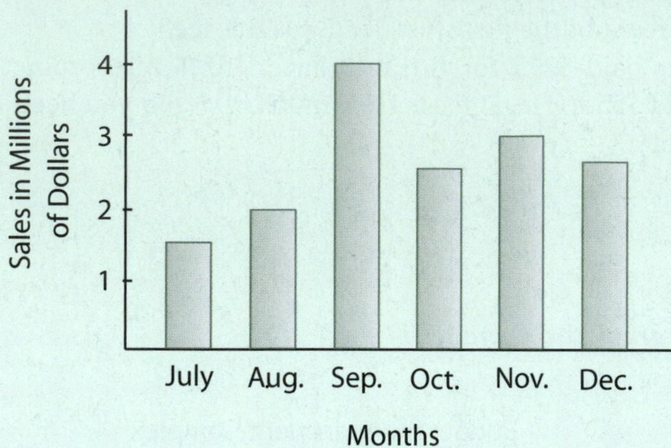

14. What were the total sales for the 6-month period?

15. What percent of the total sales for the 6 months were the July sales (to the nearest percent)?

16. What was the percent growth between August and September?

Name ____________ Section ________ Date ________

ANSWERS

Cumulative Review: Chapters 1 – 8

All fractions should be reduced to lowest terms.

1. Write the following in standard notation: two hundred thousand, sixteen.

2. Write the following in decimal notation: three hundred and four thousandths.

3. Round 16.996 to the nearest hundredth.

4. Find the decimal equivalent to $\frac{14}{35}$.

5. Find the decimal equivalent to $\frac{21}{40}$.

6. Write $\frac{9}{5}$ as a percent.

7. Write $1\frac{1}{2}\%$ as a decimal.

Perform the indicated operations for Exercises 8 – 19.

8. $\frac{2}{15}+\frac{11}{15}+\frac{7}{15}$

9. $4-\frac{3}{11}$

10. $\frac{2}{5}\cdot\frac{1}{3}\cdot\frac{4}{7}$

11. $2\frac{4}{15}+3\frac{1}{6}+4\frac{7}{10}$

1. ____________
2. ____________
3. ____________
4. ____________
5. ____________
6. ____________
7. ____________
8. ____________
9. ____________
10. ____________
11. ____________

12. ______

13. ______

14. ______

15. ______

16. ______

17. ______

18. ______

19. ______

20. ______

21. ______

22. ______

23. ______

24. ______

25. ______

12. $70\frac{1}{4} - 23\frac{5}{6}$

13. $4\frac{5}{7} \cdot 2\frac{6}{11}$

14. $6 \div 3\frac{1}{3}$

15. $(700)(8000)$

16. $403 - 4.012$

17. $71 + 0.354 + 4.39$

18. $(0.27)(0.043)$

19. $27.404 \div 0.34$

20. Evaluate $(36 \div 3^2 \cdot 2) + 12 \div 4 - 2^2$.

21. Use the tests for divisibility to determine if 732 can be divided exactly by 2, 3, 4, 5, 9, and 10.

22. Find the prime factorization of 396.

23. Find the LCM of 14, 21, and 30.

24. Division by ________ is undefined.

25. In the proportion $\frac{7}{8} = \frac{140}{160}$, the extremes are ________ and ________.

Name ____________ Section ____________ Date ____________

ANSWERS

26. 15% of ________ is 7.5.

27. $9\frac{1}{4}\%$ of 200 is ________.

28. 65 is ________% of 26.

29. Solve for x: $\dfrac{1\frac{2}{3}}{x} = \dfrac{10}{2\frac{1}{4}}$

30. In a certain company, three out of every five employees are male. How many female employees are there out of the 490 people working for this company?

31. The sum of two numbers is 521. If one of the numbers is 196, what is the other number?

32. James has the following scores on his first three math tests: 75, 87, and 79. To earn a B grade for the course, he must have at least an 80 average. What is the least score that he can get on his fourth test to maintain a B average?

33. An investment pays $6\frac{1}{4}\%$ simple interest. What is the interest on \$4800 invested for 8 months?

34. How much is put into a savings account that pays $5\frac{1}{2}\%$ simple interest if after 100 days, the interest earned is \$69.30?

35. Find the simple interest rate if an investment of \$6200 earns \$1395 in $2\frac{1}{2}$ years.

36. Five thousand dollars is invested in a savings account.

a. What is in the balance of the account after 9 years if interest is earned at 8% compounded quarterly?

b. If interest is earned at 8% compounded daily?

26. ________

27. ________

28. ________

29. ________

30. ________

31. ________

32. ________

33. ________

34. ________

35. ________

36. a. ________

b. ________

37. ____________

[Respond in
38. exercise.] ____________

39. ____________

40. a. ____________

b. ____________

37. Ten thousand dollars is placed in an account paying 4% interest, compounded monthly. How much is in the account at the end of 20 years (to the nearest cent)?

38. Find the true balance of the account on both the checkbook register and on the reconciliation sheet. Follow the directions on the reconciliation sheet.

YOUR CHECKBOOK REGISTER

Check No.	Date	Transaction Description	Payment (–)	(√)	Deposit (+)	Balance
			Balance brought forward			2032.28
885	9-5	Gary's Tires (new tires)	250.75			____
886	9-10	Mick's Music (compact disc)	14.29			____
887	9-21	Tim's Barbershop (haircut)	12.20			____
888	9-26	Custom Lawncare (yardwork)	35.00			____
889	9-28	Slack's Grocery (groceries)	32.14			____
890	9-29	Red Oak LLC. (stump removal)	75.00			____
891	9-30	Big Bill's Chains (jewelry)	150.50			____
892	9-30	Cash				____
893	9-30	Fred Clair (rent)	850.00			____
	9-30	Interest				____
		True Balance				

BANK STATEMENT
Checking Account Activity

Transaction Description	Amount	(√)	Running Balance	Date
Beginning Balance			2032.28	9-01
Check #885	250.75		1781.53	9-06
Check #886	14.29		1767.24	9-12
Check #888	35.00		1732.24	9-30
Check #892	100.00		1632.24	9-30
Interest	7.57		1639.81	9-30
Ending Balance			1639.81	

RECONCILIATION SHEET

A. First, mark √ beside each check and deposit listed in both your checkbook register and on the bank statement.
B. Second, in your checkbook register, add any interest paid and subtract any service charge listed.
C. Third, find the total of all outstanding checks.

Outstanding Checks No.	Amount
	____
Total	____

Statement Balance	____
Add deposits not credited	+ ____
Total	____
Subtract total amount of checks outstanding	– ____
True Balance	____

39. A scale on an architect's drawing indicates $2\frac{1}{2}$ inches represents 50 feet. How many feet apart are two points 6 inches apart on the drawing?

40. Find **a.** the perimeter and **b.** the area of the figure shown.

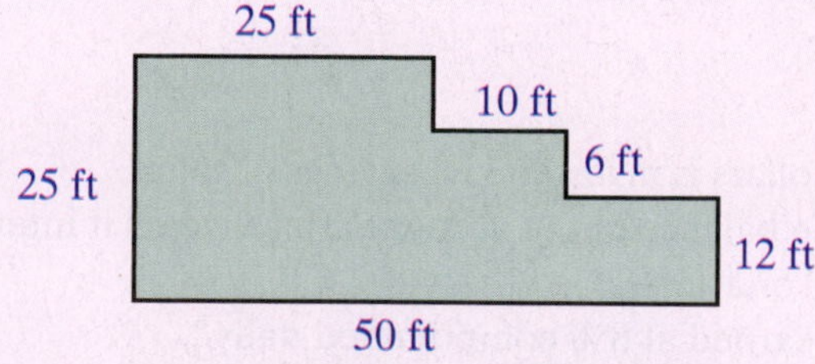

Name ______________________ Section ____________ Date __________

ANSWERS

41. Find **a.** the circumference of the inner circle, **b.** the circumference of the outer circle, and **c.** the area of the shaded figure. (Use $\pi = 3.14$.)

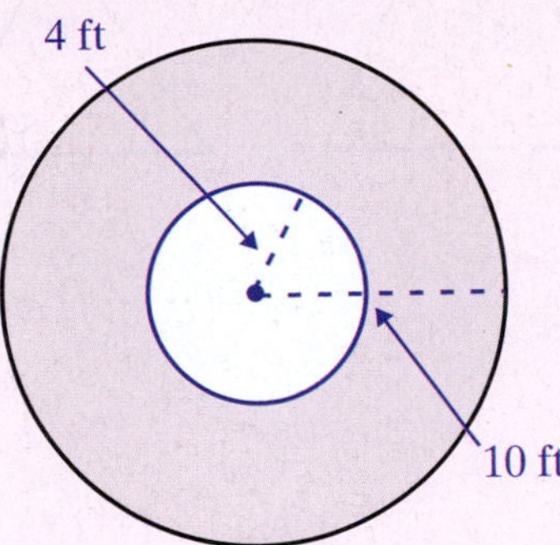

41. a. ______________

b. ______________

c. ______________

42. Find **a.** the perimeter, and **b.** the area of the semicircle shown.

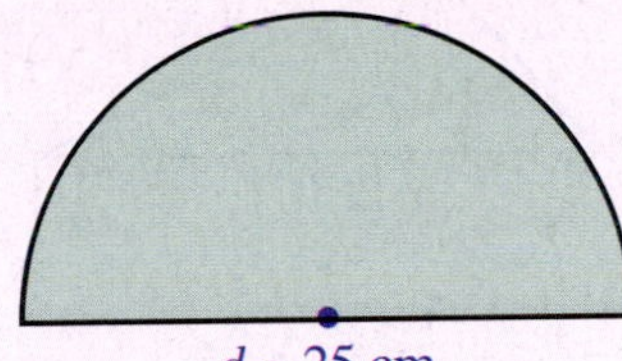

42. a. ______________

b. ______________

43. A tree has a shadow of 30 feet. At the same time of day a 6 foot tall man has a shadow of 2.5 feet. How tall is the tree?

43. ______________

44. A cereal box in the shape of a rectangular solid is 5 cm wide, 19 cm long, and 27 cm high. What is the volume of the box?

44. ______________

45. A can (in the shape of a cylinder) of orange juice has a circular base with a 2.5 inch diameter and a height of 4.5 inches. What is the volume of the can?

45. ______________

46. ______________

46. Two similar triangles are shown. Find the values of x and y.

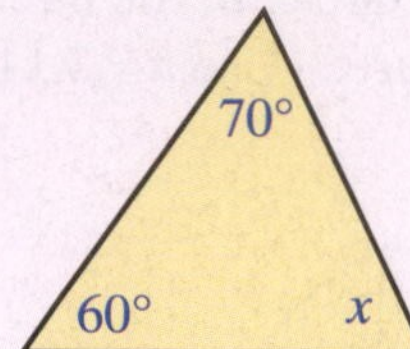

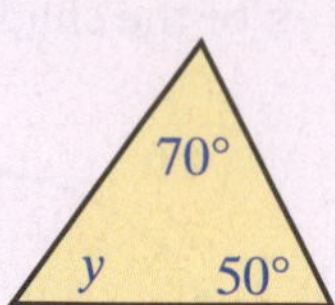

47. ______________

47. The Pythagorean Theorem states that in a right triangle, the square of the hypotenuse is equal to the sum of the squares of the two legs. Is the triangle shown a right triangle? Explain.

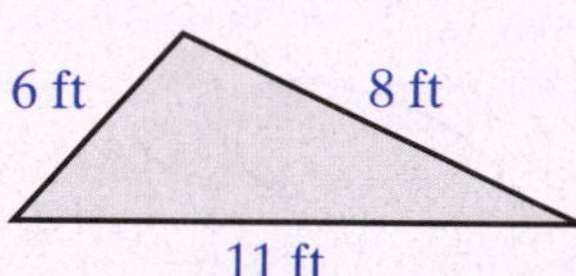

9

INTRODUCTION TO ALGEBRA

Chapter 9 Introduction to Algebra

WHAT TO EXPECT IN CHAPTER 9

Chapter 9 provides an introduction to several topics from algebra. The concept of negative numbers is introduced, along with the rules for operating with a new type of number (called an integer). In Section 9.1 number lines are used to provide a "picture" of integers and their relationships. Integers are graphed (plotted) on number lines, and number lines are used to give an intuitive understanding of the very important idea of the magnitude (absolute value) of an integer. This idea is the foundation for the rules of addition and subtraction with integers developed in Sections 9.2 and 9.3.

Multiplication and division with integers are discussed in Section 9.4, and the fact that division by 0 is undefined is reinforced–this time in terms of integers. Section 9.5 discusses methods for simplifying and evaluating algebraic expressions by using the distributive property to combine like terms. Section 9.6 provides a discussion of translating English phrases into algebraic expressions and presents basic techniques for solving equations with integers as solutions.

The chapter closes with a section on problem solving by using the translating skills developed in Section 9.6 and then setting up and solving the related equations.

Chapter 9 Introduction to Algebra

Mathematics at Work!

Suppose that you are a traveling salesperson and your company has allowed you \$65 per day for car rental. You rent a car for \$45 per day plus 5¢ per mile. How many miles can you drive for the \$65 allowed?

Plan: Set up an equation relating the amount of money spent and the money allowed in your budget and solve the equation.

Solution: Let x = number of miles driven.

Money spent = Money budgeted

$$\begin{aligned} 45 + 0.05x &= 65 \\ 45 + 0.05x - 45 &= 65 - 45 \\ 0.05x &= 20 \\ \frac{0.05x}{0.05} &= \frac{20}{0.05} \\ x &= 400 \text{ miles} \end{aligned}$$

You would like to get the most miles for your money. The car rental agency down the street charges only \$40 per day and 8¢ per mile. How many miles would you get for your \$65 with this price structure? Is this a better deal for you? (See Exercise 43, Section 9.7.)

9.1 Introduction to Integers

Number lines

The concepts of positive and negative numbers occur frequently in our daily lives:

	Negative	**Zero**	**Positive**
Temperatures are recorded as:	below zero	zero	above zero
Businesses report:	losses	no gain	profits
Golf scores are recorded as:	below par	par	above par

In this chapter we will develop techniques for understanding and operating with positive and negative numbers. We begin with the graphs of numbers on horizontal lines called **number lines**. For example, choose some point on a horizontal line and label it with the number 0 (Figure 9.1).

0

Figure 9.1

Now choose another point on the line to the right of 0 and label it with the number 1 (Figure 9.2).

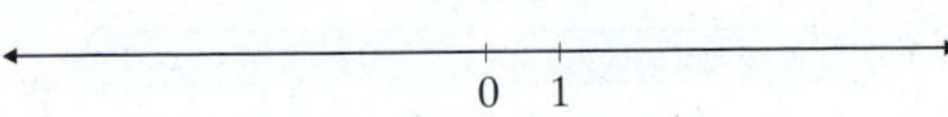

Figure 9.2

Points corresponding to all remaining whole numbers are now determined and are all to the right of 0. That is, the point for 2 is the same distance from 1 as 1 is from 0, and so on (Figure 9.3).

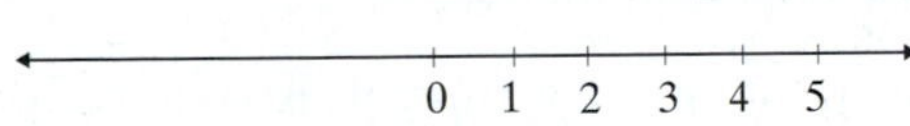

Figure 9.3

The **graph** of a number is the point on a number line that corresponds to that number, and the number is called the **coordinate** of the point. The terms **number** and **point** are used interchangeably. Thus we might refer to the **point** 6. The graphs of numbers are indicated by marking the corresponding points with large dots (Figure 9.4).

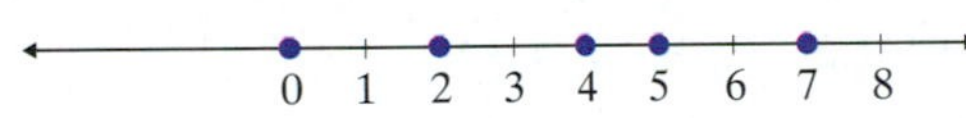

The graph of the set of numbers $A = \{0, 2, 4, 5, 7\}$

Figure 9.4

The point one unit to the left of 0 is the **opposite of 1**. It is also called **negative 1** and is symbolized as **–1**. Similarly, the **opposite of 2** is called **negative 2** and is symbolized **–2**, the **opposite of 3** is **–3**, and so on (Figure 9.5).

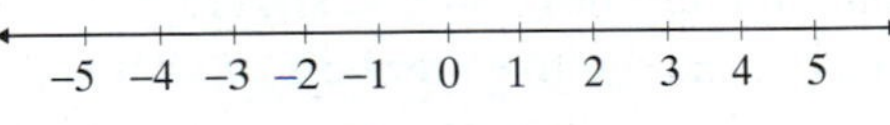

Figure 9.5

Objectives

① Understand what the integers are.

② Know how to find the opposite of an integer.

③ Be able to graph a set of integers on a number line.

④ Know how to use the inequality symbols $<, >, \leq,$ and $\geq$.

⑤ Learn the definition of absolute value.

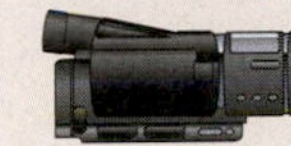

Objective ①

Integers

The set of **integers** is the set of whole numbers and their opposites:

Integers: $\ldots, -4, -3, -2, -1, 0, 1, 2, 3, 4, \ldots$

The counting numbers (all whole numbers except 0) are called **positive integers** and may be written as

+1, +2, +3, +4, and so on.

The **opposites** of the counting numbers are called **negative integers** and are written as

–1, –2, –3, –4, and so on.

The number 0 is neither positive nor negative (Figure 9.6).

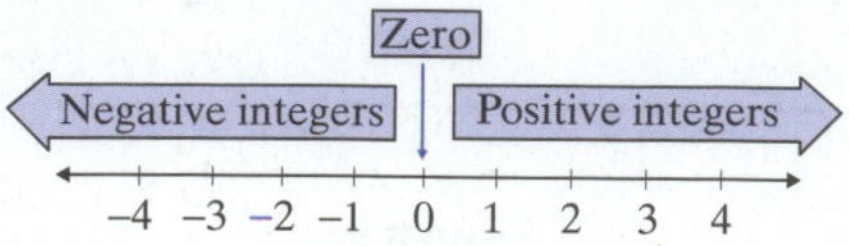

Figure 9.6

Objective 2

Opposites of Integers

Note the following facts about integers:

1. The opposite of a positive integer is a negative integer. For example,

   $-(+2) = -2$ and $-(+7) = -7.$ (opposite)

2. The opposite of a negative integer is a positive integer. For example,

   $-(-3) = +3$ and $-(-4) = +4.$ (opposite)

3. The opposite of 0 is 0. (That is, $-0 = 0$.)

Note: Integers are not the only type of number that can be graphed on number lines. We will see later in this chapter that fractions, mixed numbers, decimals, square roots, and the opposites of all these types of numbers can also be graphed on number lines. So you should be aware that the set of integers does not include all positive numbers or all negative numbers.

1. Find the opposite of each of the following numbers.

a. +12

b. –5

c. 0

Example 1

Find the opposite of each of the following numbers.

a. –6 **b.** –20 **c.** +8

Solution

a. $-(-6) = 6$ **b.** $-(-20) = 20$ **c.** $-(+8) = -8$

Now work exercise 1 in the margin.

Objective 3

2. Graph the set of integers:

$\{-3, 0, 1, 2\}$

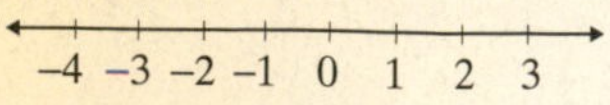

–4 –3 –2 –1 0 1 2 3

Example 2

Graph the set of integers $B = \{-4, -2, 0, 1, 2\}$.

Solution

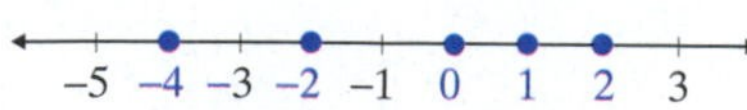

Now work exercise 2 in the margin.

Example 3

Graph the set of positive odd integers $C = \{1, 3, 5, 7, 9, \ldots\}$.

Solution

The three dots above the number line indicate that the pattern in the graph continues without end. The set of positive odd integers is an **infinite** set.

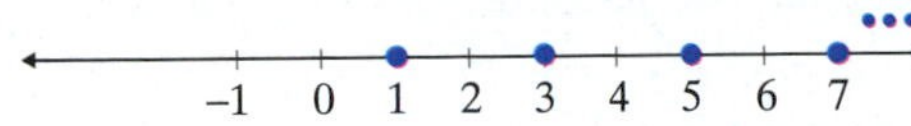

Now work exercise 3 in the margin.

3. Graph the set of odd integers:

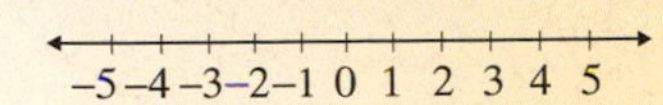

–5 –4 –3 –2 –1 0 1 2 3 4 5

Inequality Symbols

Objective 4

On a horizontal number line, **smaller numbers are always to the left of larger numbers**. Each number is smaller than any number to its right and larger than any number to its left. We use the following **inequality symbols** to indicate the order of numbers on the number line and their relative sizes.

Symbols for Order

$<$ is read "is less than"	$\leq$ is read "is less than or equal to"
$>$ is read "is greater than"	$\geq$ is read "is greater than or equal to"

The following relationships can be observed on the number line in Figure 9.7.

Using $<$ or $>$		Number Line
$2 < 5$ $5 > 2$	2 is less than 5, or 5 is greater than 2	–5 –4 –3 –2 –1 0 1 2 3 4 5
$-3 < 0$ $0 > -3$	–3 is less than 0, or 0 is greater than –3	–5 –4 –3 –2 –1 0 1 2 3 4 5
$-4 < -2$ $-2 > -4$	–4 is less than –2, or –2 is greater than –4	–5 –4 –3 –2 –1 0 1 2 3 4 5

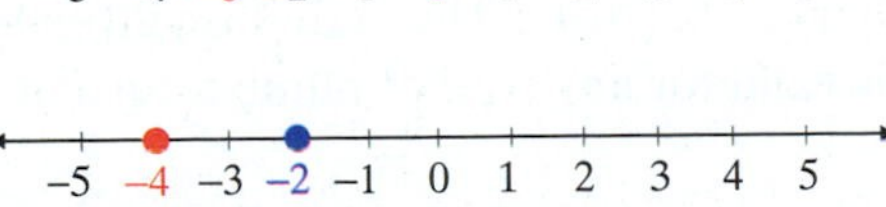

Figure 9.7

One useful idea implied by the previous discussion is that expressions containing the symbols < and > can be read either from right to left or from left to right. For example, we might read

$1 < 8$ as "1 is less than 8." (reading from left to right)

or

$1 < 8$ as "8 is greater than 1." (reading from right to left)

Also note that for the symbol $\geq$ (or the symbol $\leq$) we can write an expression such as

$5 \geq -10$ since 5 is greater than -10

and

$5 \geq 5$ since 5 is equal to 5.

The symbol $\geq$ is read **greater than or equal to**, and a true relationship is represented if either

the first number is **greater** than the second

or

the first number is **equal** to the second.

4. Determine whether each of the following statements is true or false. Rewrite any false statement so that it is true.

a. $-5 < 0$

b. $-4 \geq 4$

c. $-10 \leq -10$

Example 4

Determine whether the following statements are true or false. Rewrite any false statement so that it is true.

a. $3 \leq 11$ **b.** $6 > -1$ **c.** $-7 > 0$ **d.** $0 \geq 0$ **e.** $-6 \leq -6$

Solution

a. $3 \leq 11$ is true since 3 is less than 11.

b. $6 > -1$ is true since 6 is greater than -1.

c. $-7 > 0$ is false.
We can change the inequality to read $-7 < 0$ or $0 > -7$.

d. $0 \geq 0$ is true since $0 = 0$.

e. $-6 \leq -6$ is true since $-6 = -6$.

Now work exercise 4 in the margin.

Absolute Value

Another concept related to **signed numbers** (positive numbers, negative numbers, and zero) is that of **absolute value**, symbolized by two vertical bars, $|\ \ |$. (**Note**: The definition given here for the absolute value of an integer is valid for any type of number on a number line.)

Absolute value is used to indicate the distance a number is from 0. For example, we know that -4 and $+4$ are both 4 units from 0. Thus we can write $|-4| = |+4| = 4$. (See Figure 9.8.)

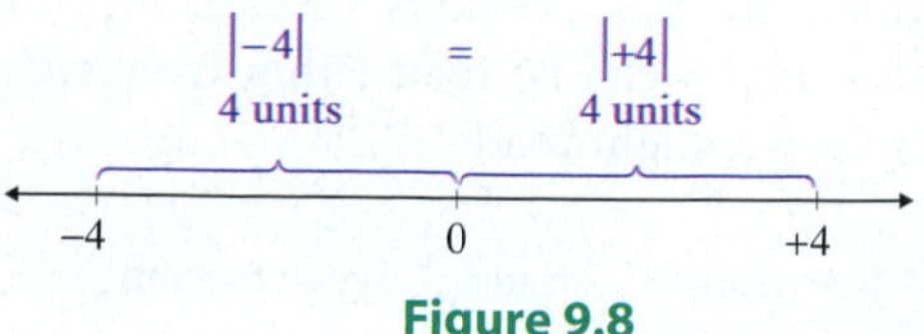

Figure 9.8

Objective 5

Absolute Value

The **absolute value** of an integer is its distance from 0.
The absolute value of an integer is never negative.
We can express the definition symbolically, for any integer a, as follows:

If a is a positive integer or 0, $|a| = a$.

If a is a negative integer, $|a| = -a$.

Note: When $\boldsymbol{a}$ represents a negative number, the symbol $-\boldsymbol{a}$ represents a positive number. **That is, the opposite of a negative number is a positive number**. For example,

If $a = -8$, then $-a = -(-8) = +8$.

And, for $a = -8$, we can write

$|a| = -a.$
$|-8| = -(-8) = +8$

The absolute value of –8 is the opposite of –8.

Example 5

$|-3| = |+3| = 3$

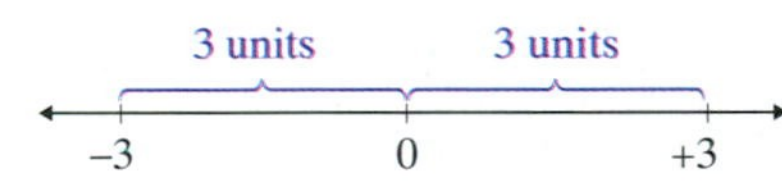

Now work exercise 5 in the margin.

Example 6

$|0| = 0$

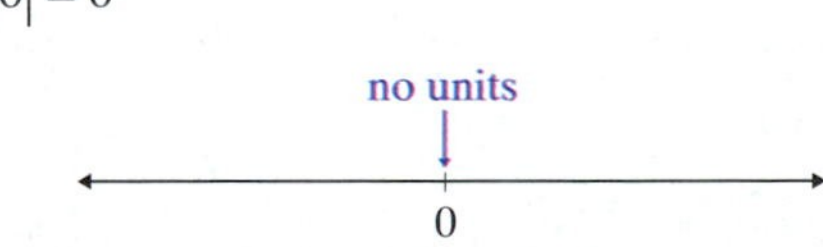

Now work exercise 6 in the margin.

Example 7

True or false: $|-12| \geq 12$.

Solution

True, since $|-12| = 12$ and $12 \geq 12$. (Remember that the symbol $\geq$ is read **greater than or equal to** so that "equal to" is valid with this symbol.)

Now work exercise 7 in the margin.

Find each absolute value.

5. $|-15|$

6. $|9|$

7. Determine whether each of the following statements is true or false. Rewrite any false statement so that it is true.

a. $|-5| \geq |+5|$

b. $|-38| > |-39|$

Practice Problems

1. Graph $\{-3, -1, 2, 4\}$ on the number line below.

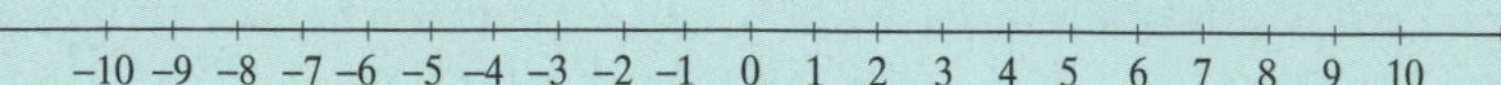

2. Find the opposite of 15.

3. Find the absolute value of $|-5|$.

4. Identify the appropriate symbol (<, >, or =) that will make
2 ________ −7 true.

Answers to Practice Problems:

1. −10 −9 −8 −7 −6 −5 −4 −3 −2 −1 0 1 2 3 4 5 6 7 8 9 10

2. −15 **3.** 5 **4.** $2 > -7$

Name ______________________ Section ____________ Date __________

Exercises 9.1

Graph each of the following sets on a number line. See Examples 2 and 3.

1. $\{0, 1, 2\}$

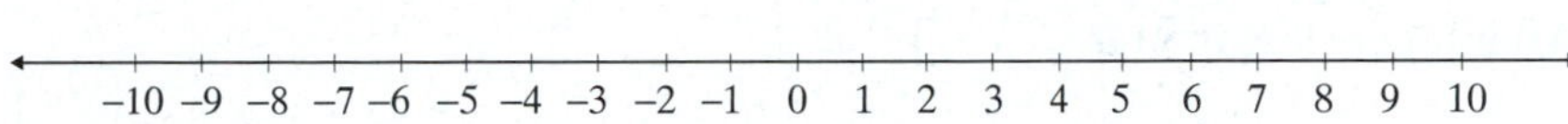

2. $\{0, 2, 4\}$

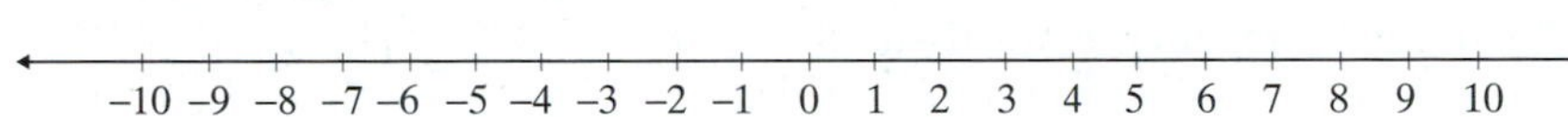

3. $\{-3, -1, 1\}$

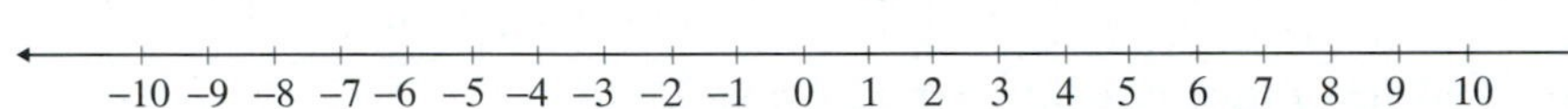

4. $\{-3, -2, 0\}$

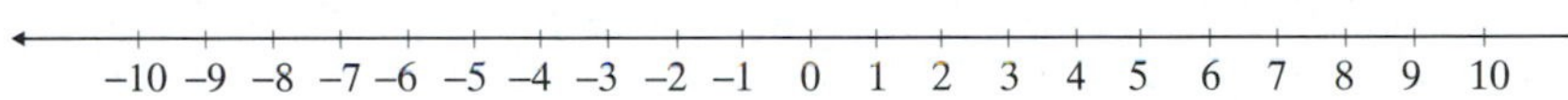

5. $\{-10, -9, -8, -7\}$

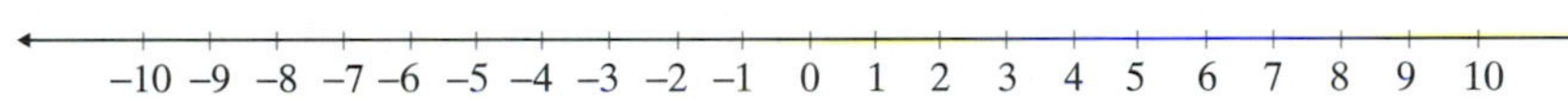

6. $\{-5, -4, -2, -1\}$

−10 −9 −8 −7 −6 −5 −4 −3 −2 −1 0 1 2 3 4 5 6 7 8 9 10

7. $\{-5, 0, 5\}$

−10 −9 −8 −7 −6 −5 −4 −3 −2 −1 0 1 2 3 4 5 6 7 8 9 10

8. $\{-3, -1, 0, 1, 3\}$

−10 −9 −8 −7 −6 −5 −4 −3 −2 −1 0 1 2 3 4 5 6 7 8 9 10

9. $\{1, 2, 3, \ldots, 10\}$

−10 −9 −8 −7 −6 −5 −4 −3 −2 −1 0 1 2 3 4 5 6 7 8 9 10

10. $\{-2, -1, 0, \ldots, 5\}$

−10 −9 −8 −7 −6 −5 −4 −3 −2 −1 0 1 2 3 4 5 6 7 8 9 10

11. $\{0, 2, 4, 6, 8, \ldots\}$

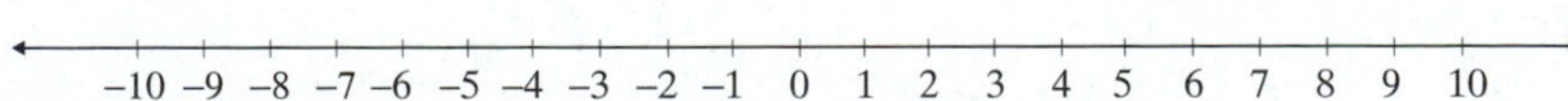

12. $\{\ldots, -7, -5, -3, -1\}$

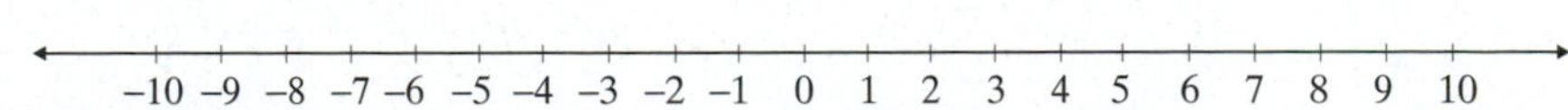

ANSWERS

1. [Respond in exercise.]

2. [Respond in exercise.]

3. [Respond in exercise.]

4. [Respond in exercise.]

5. [Respond in exercise.]

6. [Respond in exercise.]

7. [Respond in exercise.]

8. [Respond in exercise.]

9. [Respond in exercise.]

10. [Respond in exercise.]

11. [Respond in exercise.]

12. [Respond in exercise.]

13. [Respond in exercise.]
14. [Respond in exercise.]
15. [Respond in exercise.]
16. [Respond in exercise.]
17. ________
18. ________
19. ________
20. ________
21. ________
22. ________
23. ________
24. ________
25. ________
26. ________
27. ________
28. ________
29. ________
30. ________
31. [Respond in exercise.]
32. [Respond in exercise.]
33. [Respond in exercise.]
34. [Respond in exercise.]
35. [Respond in exercise.]
36. [Respond in exercise.]
37. [Respond in exercise.]
38. [Respond in exercise.]

13. All whole numbers less than 4

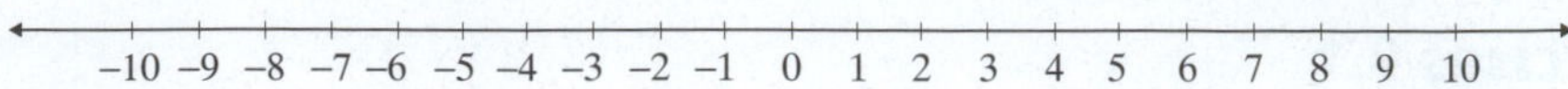

14. All integers less than 4

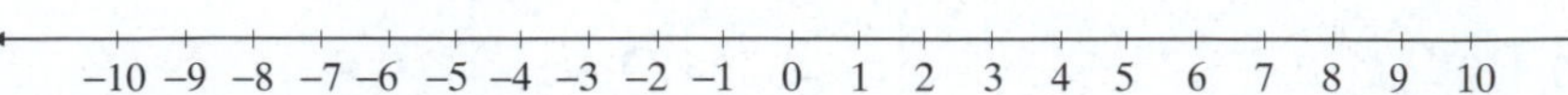

15. All integers less than 0

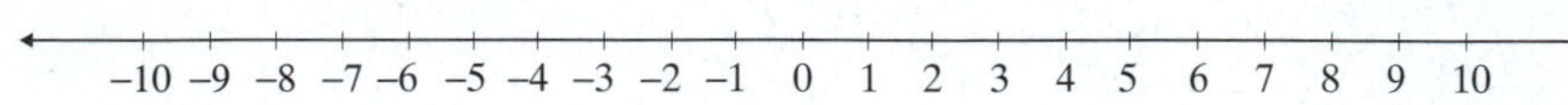

16. All negative integers greater than –5

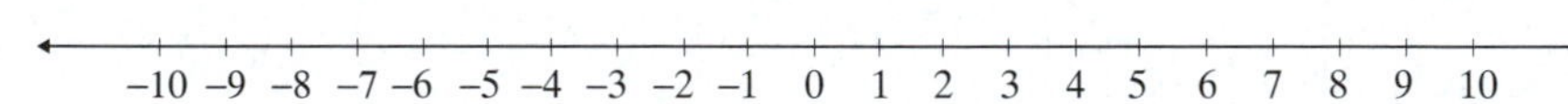

Find the opposite of each number. See Example 1.

17. –10 **18.** –9 **19.** 14

20. –6 **21.** 0 **22.** 40

Find each absolute value as indicated. See Examples 5 and 6.

23. $|-6|$ **24.** $|-10|$ **25.** $|+24|$ **26.** $|+16|$

27. $|-20|$ **28.** $|-50|$ **29.** $|0|$ **30.** $|27|$

Identify the appropriate symbol that will make each statement true: <, >, or =. See Examples 4 and 7.

31. 5 ________ 0 **32.** 7 ________ –2 **33.** –57 ________ –50

34. –30 ________ –29 **35.** $|19|$ ________ 19 **36.** $|-3|$ ________ 3

37. $|-21|$ ________ –21 **38.** –63 ________ $|-63|$

Name ______________________ Section ____________ Date __________

Determine whether each statement is true or false. If the statement is false, change the inequality or equality symbol so that the statement is true. (There may be more than one change that will make a false statement correct.) See Examples 4 and 7.

39. $0 = -0$

40. $-10 < -11$

41. $21 > |-21|$

42. $|-5| < 5$

43. $-6 = |-6|$

44. $19 = |-19|$

45. $-4 \geq 4$

46. $-12 \geq 12$

47. $|-3| \leq 3$

48. $-45 < |-45|$

49. $0 < |0|$

50. $|-4| > 0$

ANSWERS

39. ____________

40. ____________

41. ____________

42. ____________

43. ____________

44. ____________

45. ____________

46. ____________

47. ____________

48. ____________

49. ____________

50. ____________

9.2 Addition with Integers

An Intuitive Approach

We have learned how to add, subtract, multiply, and divide with whole numbers, fractions, mixed numbers, and decimal numbers. In the next three sections we will discuss the basic rules and techniques for these same four operations with integers.

As an intuitive approach to addition with integers, consider an open field with a straight line marked with integers (much like a football field is marked every 10 yards). Imagine that a ball player stands at the point marked 0 and throws the ball to the point marked +5 and then stands at +5 and throws the ball 3 more units in the positive direction (to the right). Where will the ball land? (See Figure 9.9.)

Objectives

1. Develop an intuitive understanding of adding integers along a number line.
2. Learn how to add with integers.

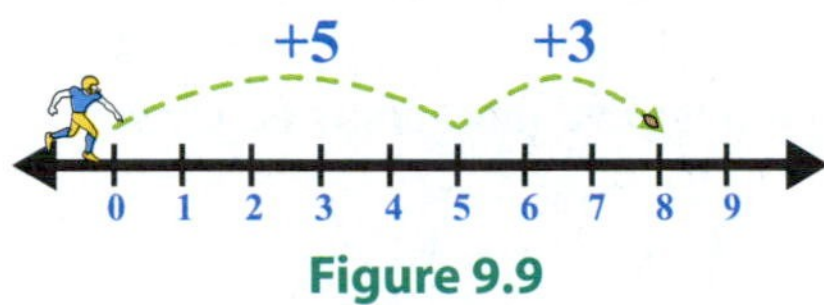

Figure 9.9

The ball will land at the point marked +8. We have essentially added the two positive integers +5 and +3.

$$(+5)+(+3)=+8 \qquad \text{or} \qquad (5)+(3)=8$$

Now, if the same player stands at 0 and throws the ball the same two distances in the opposite direction (to the left), where will the ball land on the second throw? The ball will land at –8. We have just added two negative integers, –5 and –3. (See Figure 9.10.)

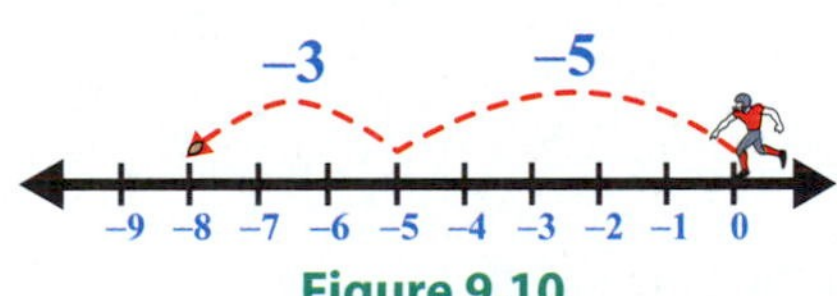

Figure 9.10

Sums involving both positive and negative integers are illustrated in Figures 9.11 and 9.12.

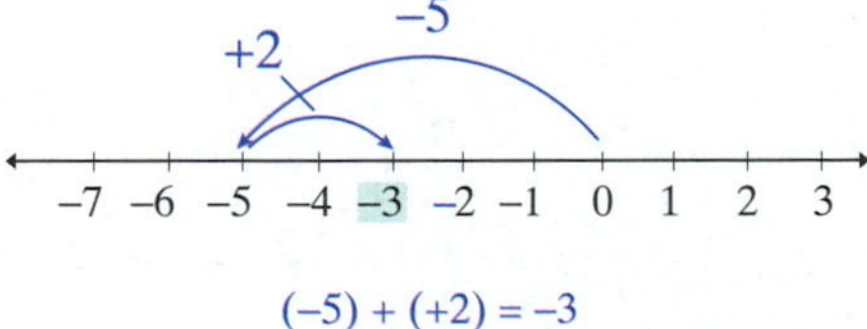

Figure 9.11

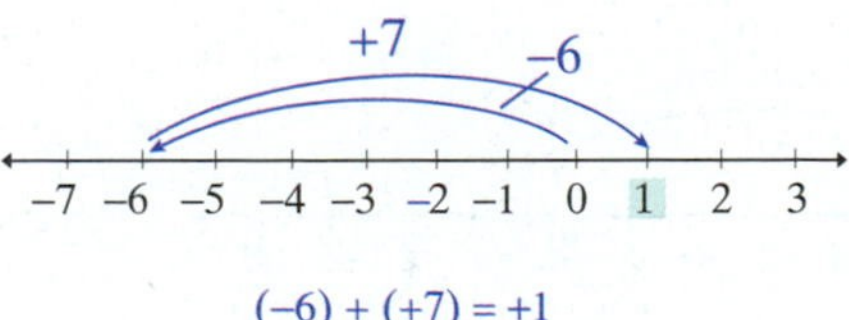

Figure 9.12

Although we are concerned here mainly with developing the techniques for adding integers, we note that addition with integers is **commutative** and **associative**.

To help you understand the rules for addition, we make the following suggestions for reading expressions with + and – signs:

+ used as the sign of a number is read **positive**.

+ used as an operation is read **plus**.

– used as the sign of a number is read **negative**.

– used as an operation is read **minus**. (See Section 9.3)

In summary:

1. The sum of two positive integers is positive:

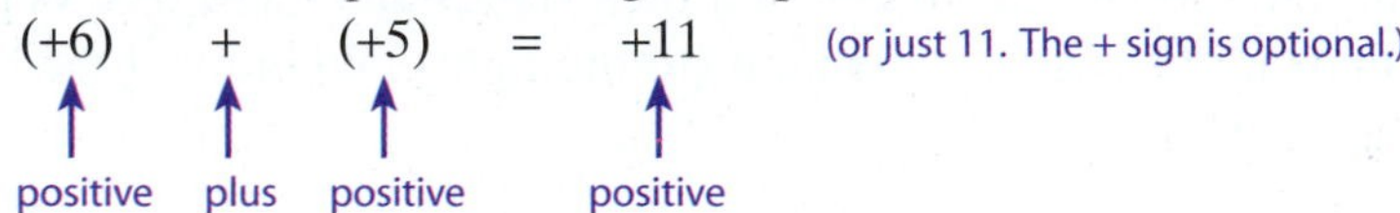

2. The sum of two negative integers is negative:

$(-4) \quad + \quad (-3) \quad = \quad -7$

negative plus negative negative

3. The sum of a positive and a negative integer may be negative or positive (or zero), depending on which number is farther from 0.

$(+5) \quad + \quad (-7) \quad = \quad -2$

positive plus negative negative

$(+9) \quad + \quad (-3) \quad = \quad +6$ (or just 6. The + sign is optional.)

positive plus negative positive

Example 1

Find each of the following sums.

a. $(-5) + (+4)$ **b.** $(-3) + (-8)$

c. $(+6) + (-10)$ **d.** $(-7) + (+7)$

Solution

a. 4 units right

–5 –4 –3 –2 –1 0

$(-5) + (+4) = -1$

b.

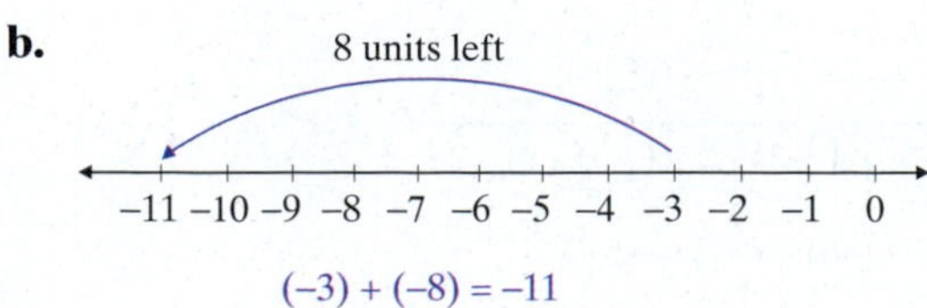

$(-3) + (-8) = -11$

c.

10 units left

−4 −3 −2 −1 0 1 2 3 4 5 6

$(+6) + (-10) = -4$

d.

7 units right

−7 −6 −5 −4 −3 −2 −1 0

$(-7) + (+7) = 0$

Now work exercise 1 in the margin.

1. Find each of the following sums.

a. $(-6) + (-10)$

b. $13 + (-3)$

c. $(-21) + (+8)$

d. $(-30) + (-5)$

Rules for Addition with Integers

Objective 2

Saying that one number is farther from 0 than another number is the same as saying that the first number has a larger absolute value. With this basic understanding, the rules for adding integers can be written out formally in terms of absolute value as follows.

Rules for Addition with Integers:

1. To add two integers with like signs, add their absolute values and use the common sign.

common sign

$$(+7)+(+2)=+(|+7|+|+2|)=+(7+2)=+9$$

common sign

$$(-7)+(-2)=-(|-7|+|-2|)=-(7+2)=-9$$

2. To add two integers with unlike signs, subtract their absolute values (the smaller from the larger) and use the sign of the integer with the larger absolute value.

$$(-16)+(+10)=-(|-16|-|+10|)=-(16-10)=-6$$

$$(+16)+(-10)=+(|+16|-|-10|)=+(16-10)=+6$$

$$(-30)+(+30)=+(|-30|-|+30|)=+(30-30)=0$$

Equations in algebra are almost always written horizontally, so addition and subtraction with integers are done in the horizontal format much of the time. However, there are situations (such as in long division) where sums and differences are written in a vertical format with one number directly under another, as illustrated in Example 2.

2. Find each of the following sums.

a. $\begin{array}{r} -10 \\ \underline{+\ 3} \end{array}$

b. $\begin{array}{r} -5 \\ \underline{+9} \end{array}$

c. $\begin{array}{r} -2 \\ -3 \\ \underline{-15} \end{array}$

d. $\begin{array}{r} -20 \\ +6 \\ \underline{-1} \end{array}$

Example 2

Find each of the following sums.

a. $\begin{array}{r} -30 \\ 8 \\ \underline{-12} \end{array}$ **b.** $\begin{array}{r} 52 \\ -20 \\ \underline{-15} \end{array}$ **c.** $\begin{array}{r} -10 \\ -11 \\ \underline{-9} \end{array}$

Solution

One technique for adding several integers is to mentally add the positive and negative integers separately and then add these results. (We are, in effect, using the commutative and associative properties of addition.)

a. $\begin{array}{r} -30 \\ 8 \\ \underline{-12} \\ -34 \end{array}$ or $\begin{array}{r} -42 \\ \underline{8} \\ -34 \end{array}$

b. $\begin{array}{r} 52 \\ -20 \\ \underline{-15} \\ 17 \end{array}$ or $\begin{array}{r} 52 \\ \underline{-35} \\ 17 \end{array}$

c. $\begin{array}{r} -10 \\ -11 \\ \underline{-9} \\ -30 \end{array}$

Now work exercise 2 in the margin.

Practice Problems

Find each of the following sums.

1. (+4) + (−12) **2.** (+10) + (−8) **3.** $\begin{array}{r} -12 \\ \underline{+\ 8} \end{array}$ **4.** $\begin{array}{r} -7 \\ -8 \\ \underline{-20} \end{array}$

Answers to Practice Problems:

1. −8 **2.** 2 **3.** −4 **4.** −35

Name ______________ Section ______________ Date ______________

ANSWERS

Exercises 9.2

Find each of the indicated sums. See Examples 1 and 2.

1. (+6) + (–4)

2. (+8) + (–7)

3. (+4) + (+6)

4. (+5) + (–8)

5. (+16) + (+3)

6. (–8) + (–2)

7. (–3) + (–6)

8. (–2) + (+2)

9. (+4) + (–4)

10. (+13) + (+12)

11. (+6) + (–10)

12. (+14) + (–17)

13. (+5) + (–3)

14. (+15) + (–18)

15. (–4) + (–12)

16. (–8) + (+8)

17. (+2) + (–6)

18. (–9) + (+5)

19. (–16) + (+3) + (+13)

20. (–5) + (+5) + (14)

21. (–1) + (–2) + (+7)

22. (+3) + (–4) + (–5)

23. (+6) + (+3) + (+5)

24. (–18) + (–5) + (–7)

25. (–1) + (+2) + (–4) + (+2)

1. ______________
2. ______________
3. ______________
4. ______________
5. ______________
6. ______________
7. ______________
8. ______________
9. ______________
10. ______________
11. ______________
12. ______________
13. ______________
14. ______________
15. ______________
16. ______________
17. ______________
18. ______________
19. ______________
20. ______________
21. ______________
22. ______________
23. ______________
24. ______________
25. ______________

26. ______

27. ______

28. ______

29. ______

30. ______

31. ______

32. ______

33. ______

34. ______

35. ______

36. ______

37. ______

38. ______

39. ______

40. ______

41. ______

42. ______

43. ______

44. ______

45. ______

26. $\begin{array}{r} -4 \\ \underline{+8} \end{array}$ **27.** $\begin{array}{r} -5 \\ \underline{-10} \end{array}$ **28.** $\begin{array}{r} -13 \\ \underline{-6} \end{array}$

29. $\begin{array}{r} +16 \\ \underline{+25} \end{array}$ **30.** $\begin{array}{r} +14 \\ \underline{-8} \end{array}$ **31.** $\begin{array}{r} +20 \\ \underline{-7} \end{array}$

32. $\begin{array}{r} +2 \\ -5 \\ \underline{-3} \end{array}$ **33.** $\begin{array}{r} +8 \\ +3 \\ \underline{-1} \end{array}$ **34.** $\begin{array}{r} +10 \\ -4 \\ \underline{+2} \end{array}$

35. $\begin{array}{r} -16 \\ -8 \\ \underline{+12} \end{array}$ **36.** $\begin{array}{r} -15 \\ -20 \\ \underline{-6} \end{array}$ **37.** $\begin{array}{r} -4 \\ -17 \\ \underline{+11} \end{array}$

38. $\begin{array}{r} +13 \\ -5 \\ +17 \\ \underline{-25} \end{array}$ **39.** $\begin{array}{r} +14 \\ -14 \\ +37 \\ \underline{-37} \end{array}$ **40.** $\begin{array}{r} -8 \\ -5 \\ -13 \\ \underline{-22} \end{array}$

Choose the response that correctly completes each statement.

41. If x and y are integers, then $x + y$ is (never, sometimes, always) equal to 0.

42. If x and y are integers, then $x + y$ is (never, sometimes, always) positive.

43. If x and y are integers, then $x + y$ is (never, sometimes, always) negative.

44. If x is a positive integer and y is a negative integer, then $x + y$ is (never, sometimes, always) equal to 0.

45. If x and y are both negative integers, then $x + y$ is (never, sometimes, always) equal to 0.

Name ______________________ Section ____________ Date __________

ANSWERS

46. ______________

47. ______________

48. ______________

49. ______________

50. ______________

51. [Respond below exercise.] ______________

52. [Respond below exercise.] ______________

53. [Respond below exercise.] ______________

Use a calculator to find the value of each expression. See Example 2. (**Note:** On the TI-84 Plus calculator, the key marked **(–)** is used to change the sign of a number from positive to negative. On other calculators, there may be a key marked **+/-** which can also be used to change the sign of a number from positive to negative or from negative to positive.)

46. 8305 + (–4783) + (–5400)

47. –4785 + (–7300) + (–2540)

48. –11,970 + (–4375) + (25,000) + (15,000)

49. 39,500 + (–34,100) + (–53,700) + (60,000)

50. –35,632 + (–10,658) + (17,344) + (–3479)

Writing and Thinking about Mathematics

51. Give three illustrations of the associative property of addition with integers.

52. Describe in your own words the conditions under which the sum of two integers will be 0.

53. Explain in your own words how the expression $-x$ can possibly represent a positive integer.

9.3 Subtraction with Integers

Additive Inverse

To define subtraction with integers, we need to emphasize and clarify the relationship between any integer and its opposite.

Additive Inverse

The **opposite** of an integer is called its **additive inverse**. The sum of an integer and its additive inverse is 0. Symbolically, for any integer a,

$$a + (-a) = 0$$

Objectives

1. Understand the concept of an additive inverse.
2. Learn how to subtract with integers.

 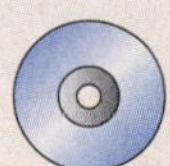

Note: The symbol $-a$ should be read as **the opposite of *a***. Since a is a variable, $-a$ might be positive, negative, or 0. For example,

if $a = 13$, then $-a = -13$ and $-a$ is **negative**.

However,

if $a = -4$, then $-a = -(-4) = +4$ and $-a$ is **positive**.

(opposite of) (negative 4)

Example 1

Find the additive inverse (opposite) of each integer.

a. 7 **b.** −3 **c.** −22

Solution

a. The additive inverse of 7 is −7, and

$7 + (-7) = 0$

b. The additive inverse of −3 is +3, and

$(-3) + (+3) = 0$

c. The additive inverse of −22 is +22, and

$(-22) + (+22) = 0$

Now work exercise 1 in the margin.

1. Find the additive inverse of each integer.

a. −8

b. 9

Objective 2

Subtraction with Integers

Just as subtraction with whole numbers is defined in terms of addition, subtraction with integers is also defined in terms of addition. The distinction is that now we can subtract larger numbers from smaller numbers, and we can subtract negative numbers and positive numbers from each other. In effect, we can now study differences that result in negative numbers. For example,

$10 - 15 = -5$	because	$15 + (-5) = 10$
$4 - 9 = -5$	because	$9 + (-5) = 4$
$10 - 13 = -3$	because	$13 + (-3) = 10$
$8 - (-6) = 14$	because	$(-6) + 14 = 8$
$-7 - (-5) = -2$	because	$(-5) + (-2) = -7.$

Difference

For any integers a and b, we define the **difference** between a and b as follows:

$$a - b = a + (-b)$$

This equation is read

"a minus b is equal to a plus the opposite of b."

Thus, as we have discussed and as the following examples illustrate, subtraction is accomplished by adding the opposite of the subtrahend.

Example 2

Find the following differences.

a. $(+2) - (-6)$ **b.** $(-3) - (-7)$

c. $(-5) - (-2)$ **d.** $(-9) - (-9)$

Solution

a. $(+2) - (-6) = (+2) + (+6) = +8$

minus changed to plus   opposite of –6

b. $(-3) - (-7) = (-3) + (+7) = +4$

minus changed to plus   opposite of –7

c. $(-5)-(-2)=(-5)+(+2)=-3$

minus changed to plus   opposite of −2

d. $(-9)-(-9)=(-9)+(+9)=0$

Now work exercise 2 in the margin.

The interrelationship between addition and subtraction with integers allows us to eliminate the use of so many parentheses. This simpler notation is illustrated as follows:

$$9-12 \;=\; 9 \;-\; (+12) \;=\; 9 \;+\; (-12) \;=\; -3$$

9 ↑ minus ↑ positive 12 ↑ 9 ↑ plus ↑ negative 12 ↑

Thus the expression $9-12$ can be thought of as subtraction or addition. The result is the same in either case. Understanding this notation takes some practice, but it is quite important since it is commonly used in all mathematics textbooks. Study the following examples carefully.

Example 3

a. $7-13=-6$

b. $10-8=2$

c. $-4-8=-12$

d. $6-9=-3$

e. $-20+15-10=-5-10=-15$

Now work exercise 3 in the margin.

The following examples show that subtraction is **not commutative** and **not associative**.

$12-5=7$ and $5-12=-7$

So $12-5 \neq 5-12$, and in general,

$$a-b \neq b-a.$$

Similarly,

$9-(4-1)=9-(3)=6$ and $(9-4)-1=5-1=4$.

So $9-(4-1) \neq (9-4)-1$, and in general,

$$a-(b-c) \neq (a-b)-c.$$

2. Find the following differences.

a. $+2-(-3)$

b. $(-8)-(-4)$

3. Evaluate each of the following expressions.

a. $-12-8$

b. $6-5$

c. $1-6$

d. $-2+1$

e. $7-3-10$

Find each of the following differences.

4. $\begin{array}{r} -19 \\ \underline{-(-13)} \end{array}$

5. $\begin{array}{r} -2 \\ \underline{-(+7)} \end{array}$

6. $\begin{array}{r} -10 \\ \underline{-(-21)} \end{array}$

As with addition, integers can be written vertically in subtraction. One number is written underneath the other, and the sign of the integer being subtracted (the subtrahend) is changed and the resulting numbers are added. That is, we add the opposite of the integer that is subtracted.

Example 4

To Subtract: Add the opposite of the bottom number.

$\begin{array}{r} -35 \\ \underline{-(-22)} \end{array}$ $\begin{array}{rl} -35 & \\ \underline{+22} & \text{sign is changed} \\ -13 & \text{difference} \end{array}$

Now work exercise 4 in the margin.

Example 5

To Subtract: Add the opposite of the bottom number.

$\begin{array}{r} -16 \\ \underline{-(+14)} \end{array}$ $\begin{array}{rl} -16 & \\ \underline{-14} & \text{sign is changed} \\ -30 & \text{difference} \end{array}$

Now work exercise 5 in the margin.

Example 6

To Subtract: Add the opposite of the bottom number.

$\begin{array}{r} -20 \\ \underline{-(-87)} \end{array}$ $\begin{array}{rl} -20 & \\ \underline{+87} & \text{sign is changed} \\ +67 & \text{difference} \end{array}$

Now work exercise 6 in the margin.

Practice Problems

Find the additive inverse of each number.

1. −10 **2.** 17

Subtract.

3. $(-1) - 17$ **4.** $\begin{array}{r} 22 \\ \underline{-(-7)} \end{array}$ **5.** $-20 - 2 + 6$

Answers to Practice Problems:

1. 10 **2.** −17 **3.** −18 **4.** 29 **5.** −16

Name ______________________ Section ____________ Date ____________

Exercises 9.3

Find the additive inverse of each number. See Example 1.

1. 16 **2.** 26 **3.** –45

4. –75 **5.** –8 **6.** 37

Perform the indicated operations. See Example 2.

7. (+5) – (+2) **8.** (+16) – (+3) **9.** (+8) – (–3)

10. (+12) – (–4) **11.** (–5) – (+2) **12.** (–10) – (+3)

13. (–10) – (–1) **14.** (–15) – (–1) **15.** (–3) – (–7)

16. (–2) – (–12) **17.** (–4) – (+6) **18.** (–9) – (+13)

19. (–13) – (–14) **20.** (–12) – (–15) **21.** (+9) – (–9)

22. (+11) – (–11) **23.** (+15) – (–2) **24.** (+20) – (–3)

25. (–7) – (–7) **26.** (–5) – (–5) **27.** (+8) – (+8)

28. (+5) – (+5) **29.** (–16) – (+10) **30.** (–17) – (+14)

ANSWERS

1. ____________
2. ____________
3. ____________
4. ____________
5. ____________
6. ____________
7. ____________
8. ____________
9. ____________
10. ____________
11. ____________
12. ____________
13. ____________
14. ____________
15. ____________
16. ____________
17. ____________
18. ____________
19. ____________
20. ____________
21. ____________
22. ____________
23. ____________
24. ____________
25. ____________
26. ____________
27. ____________
28. ____________
29. ____________
30. ____________

31. ______
32. ______
33. ______
34. ______
35. ______
36. ______
37. ______
38. ______
39. ______
40. ______
41. ______
42. ______
43. ______
44. ______
45. ______
46. ______
47. ______
48. ______
49. ______
50. ______
51. ______
52. ______
53. ______
54. ______
55. ______
56. ______
57. ______
58. ______

Subtract the bottom number from the top number. See Examples 4 through 6.

31. $\begin{array}{r} 18 \\ \underline{-(-12)} \end{array}$

32. $\begin{array}{r} 24 \\ \underline{-(16)} \end{array}$

33. $\begin{array}{r} -8 \\ \underline{-(-12)} \end{array}$

34. $\begin{array}{r} -13 \\ \underline{-(-18)} \end{array}$

35. $\begin{array}{r} -4 \\ \underline{-(+5)} \end{array}$

36. $\begin{array}{r} 32 \\ \underline{-(-48)} \end{array}$

37. $\begin{array}{r} -6 \\ \underline{-(-30)} \end{array}$

38. $\begin{array}{r} -25 \\ \underline{-(-13)} \end{array}$

39. $\begin{array}{r} -45 \\ \underline{-(-16)} \end{array}$

40. $\begin{array}{r} 28 \\ \underline{-(-15)} \end{array}$

Evaluate each of the following expressions. See Examples 2 and 3.

41. $6 + 2$

42. $4 + 8$

43. $7 - 1$

44. $9 - 4$

45. $4 + 6$

46. $8 + 9$

47. $-3 - 1$

48. $-2 - 6$

49. $12 - 6$

50. $-13 + 4$

51. $-20 + 14$

52. $-10 + 9$

53. $24 - 32$

54. $14 - 17$

55. $-12 - 6$

56. $-2 - 8$

57. $-6 + 6$

58. $-7 + 7$

Name ________________ Section ________ Date ________

ANSWERS

59. $-5 + 12 - 3$ **60.** $-20 - 2 + 6$ **61.** $-4 - 10 - 7$

62. $13 - 4 + 6 - 5$ **63.** $-4 + 10 - 12 + 1$ **64.** $19 - 5 - 8 - 6$

65. $-8 + 14 - 10 + 4$.

Identify the correct symbol that will make each statement true: <, >, or =.

66. $-8 + (-2)$ _____ $-5 - 6$ **67.** $-7 + (-4)$ _____ $-3 - 3$

68. $-10 - 10$ _____ $0 - 20$ **69.** $-9 - 9$ _____ $0 - 18$

70. $5 - (-1)$ _____ $5 - 1$

71. Temperature. Beginning with a temperature of 10° above zero, the temperature was measured hourly for five hours. It rose 4°, dropped 3°, dropped 2°, rose 1°, and dropped 5°. What was the final temperature recorded?

72. Investing. In a five-day week the stock market showed a gain of 28 points, a gain of 12 points, a loss of 19 points, a loss of 3 points, and a loss of 16 points. What was the net change in the stock market for the week?

73. Sports. In seven passing plays in a football game, the quarterback threw passes that gained 12 yards, lost 2 yards, lost 5 yards, gained 3 yards, gained 35 yards, lost 4 yards, and gained 15 yards. What was his net passing yardage for these plays?

59. ________
60. ________
61. ________
62. ________
63. ________
64. ________
65. ________
66. ________
67. ________
68. ________
69. ________
70. ________
71. ________
72. ________
73. ________

74.

75.

76.

77.

78.

79. [Respond below exercise.]

80. [Respond below exercise.]

81. [Respond below exercise.]

74. Sports. Suppose that par for a certain golf course is 72. In a four-day tournament, Nancy scored 3 under par, even par, 2 under par, and 1 over par. What was her total number of strokes for the tournament?

Use a calculator to find the value of each of the following expressions. See Examples 3 through 6.

75. $15{,}855 - 18{,}273$

76. $-302{,}500 - 257{,}600 + 207{,}300$

77. $400{,}000 - 1{,}780{,}350 + 542{,}000$

78. $354{,}750 - 425{,}792 - 356{,}425 + 321{,}273$

Writing and Thinking about Mathematics

79. Give two examples that illustrate why subtraction is not commutative.

80. Give two examples that illustrate why subtraction is not associative.

81. Explain, in your own words, how the difference of two negative integers might be a positive integer.

9.4 Multiplication, Division, and Order of Operations with Integers

Multiplication with Integers

Objective 1

Multiplication with whole numbers is a shorthand form of repeated addition. Multiplication with integers can be thought of in the same manner. For example,

$$4+4+4+4+4+4 = 6\cdot 4 = 24$$

and

$$(-4)+(-4)+(-4)+(-4)+(-4)+(-4) = 6(-4) = -24$$

and

$$(-2)+(-2)+(-2) = 3(-2) = -6$$

Repeated addition with a negative integer results in a product of a positive integer and a negative integer. Therefore, because the sum of negative integers is negative, we can reason that **the product of a positive integer and a negative integer is a negative integer**.

Objectives

1. Learn how to multiply with integers.
2. Learn how to divide with integers.
3. Correctly simplify expressions using order of operations.

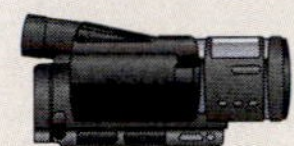
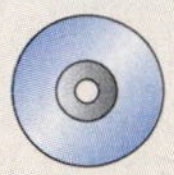

Example 1

a. $5(-5) = -25$ **b.** $3(-20) = -60$

c. $8(-11) = -88$ **d.** $-1(7) = -7$

Now work exercise 1 in the margin.

1. Find the following products.

   a. $6(-3)$

   b. $9(-4)$

   c. $-9(3)$

   d. $(7)(-10)$

The product of two negative integers can be explained in terms of opposites and the fact that, for any integer a,

$$\mathbf{-a = -1(a)}$$

That is, the opposite of a can be treated as -1 times a.
Now consider the product $(-3)(-5)$ and the following procedure:

$$\underset{\uparrow \text{ opposite of 3}}{(-3)}(-5) = \underbrace{-1(3)}_{-1\cdot 3}(-5) = -1\cdot[(3)(-5)]$$

$$= \underbrace{-1\cdot[-15]}_{-1\cdot[-15]} = \underset{\uparrow \text{ opposite of } -15}{-}[-15] = +15$$

Although one example does not prove a rule, this process is such that we can use it as a general procedure and come to the following correct conclusion: **The product of two negative integers is a positive integer.**

2. Find the following products.

a. $(-8)(-2)$

b. $(-14)(0)$

Example 2

a. $(-7)(-2) = +14$

b. $-8(-5) = +40$

c. $-9(-11) = +99$

d. $0(-12) = 0$

e. $0(+12) = 0$

f. $0(0) = 0$

Now work exercise 2 in the margin.

As Example **2d.** illustrates, the product of 0 and any integer is 0. The rules for multiplication with integers can be stated as follows.

Rules for Multiplication with Integers

1. The product of two positive integers (here, a and b) is **positive**:
   $a \cdot b = ab$
2. The product of two negative integers is **positive**:
   $(-a)(-b) = ab$
3. The product of a positive integer and a negative integer is **negative**:
   $a(-b) = -ab$
4. The product of 0 and any integer is **0**:
   $a \cdot 0 = 0$ and $(-a) \cdot 0 = 0$

The **commutative** and **associative** properties of multiplication are valid for multiplication with integers, just as they are with whole numbers. Thus, to find the product of more than two integers, we can multiply any two, then continue to multiply until all the integers have been multiplied.

3. Find the following products.

a. $(-2)(-2)(-2)$

b. $(-3)(-3)(0)$

c. $(-6)(-1)(4)(-3)$

Example 3

a. $(-4)(-2)(+5) = [(-4)(-2)](+5) = [+8](+5) = 40$

b. $(-2)(-2)(-3)(-6) = +4(-3)(-6) = -12(-6) = 72$

c. $(-1)(5)(-2)(-1)(7)(3) = -5(-2)(-1)(7)(3) = +10(-1)(7)(3)$
$= -10(7)(3) = -70(3) = -210$

Now work exercise 3 in the margin.

Division with Integers

Objective 2

Remember that division can be indicated in the form of a fraction with the numerator to be divided by the denominator. That is,

$$a \div b = \frac{a}{b}$$

Thus we can write $63 \div 7$ as $\frac{63}{7}$. Now, since the operation of division is defined in terms of multiplication, we know that

$$\frac{63}{7} = 9 \quad \text{because} \quad 63 = 7 \cdot 9$$

This relationship between division and multiplication is true for integers as well as whole numbers.

Relating Division and Multiplication:

For integers a, b, and x (where $b \neq 0$),

$$\frac{a}{b} = x \quad \text{means that} \quad a = b \cdot x.$$

Special Note:

In this text, we are emphasizing the rules for signs when operating (adding, subtracting, multiplying, and dividing) with integers. With this emphasis in mind, the problems are set up in such a way that the results are integers. However, you should be aware of the fact that the rules for operating with integers are valid for operating with any type of signed number, including positive and negative fractions, mixed numbers, and decimal numbers. Operations with these types of signed numbers are discussed in later courses.

Example 4

a. $\frac{+28}{+4} = +7$ because $+28 = (+4)(+7)$.

b. $\frac{-28}{-4} = +7$ because $-28 = (-4)(+7)$.

c. $\frac{+28}{-4} = -7$ because $+28 = (-4)(-7)$.

d. $\frac{-28}{+4} = -7$ because $-28 = (+4)(-7)$.

Now work exercise 4 in the margin.

4. Find the following quotients.

a. $\frac{+20}{+10}$

b. $\frac{-20}{-10}$

c. $\frac{+20}{-10}$

d. $\frac{-20}{+10}$

The rules for division with integers can be stated as follows.

Rules for Division with Integers

1. The quotient of two positive integers (here, a and b) is **positive**:

$$\frac{a}{b} = +\frac{a}{b}$$

2. The quotient of two negative integers is **positive**:

$$\frac{-a}{-b} = +\frac{a}{b}$$

3. The quotient of a positive integer and a negative integer is **negative**:

$$\frac{-a}{b} = -\frac{a}{b} \quad \text{and} \quad \frac{a}{-b} = -\frac{a}{b}$$

The rules for multiplication and division with nonzero integers can be summarized in the following two statements:

1. **If two nonzero integers have the same sign, then both their product and their quotient will be positive.**

2. **If two nonzero integers have unlike signs, then both their product and their quotient will be negative.**

Order of Operations with Integers

The rules for order of operations are the same for expressions with integers as they were stated in Sections 2.2, 3.6, and 4.5 for expressions with whole numbers, expressions with fractions, and expressions with mixed numbers, respectively. The rules are restated here for easy reference.

Rules for Order of Operations

1. First, simplify within grouping symbols, such as parentheses ( ), brackets [ ], or braces { }. Start with the innermost grouping.

2. Second, find any powers indicated by exponents.

3. Third, moving from **left to right**, perform any multiplications or divisions in the order in which they appear.

4. Fourth, moving from **left to right**, perform any additions or subtractions in the order in which they appear.

In Section 1.6 we stated that division by 0 is undefined. For completeness and understanding, we explain this fact again here.

Division by 0 is not Defined

Case 1: Suppose that $a \neq 0$ and $\frac{a}{0} = x$. Then, by the meaning of division, $a = 0 \cdot x$. But this is not possible since $0 \cdot x = 0$ for any value of x and we stated that $a \neq 0$.

Case 2: Suppose that $\frac{0}{0} = x$. Then $0 = 0 \cdot x$, which is true for all values of x. But this is not allowed since we must have a unique answer for division.

Therefore, we conclude that, in every case, **division by 0 is not defined**.

Find the value of each expression.

5. $\frac{-42}{0}$

Example 5

a. $\frac{8}{0}$ is undefined. If $\frac{8}{0} = x$, then $8 = 0 \cdot x$, which is not possible.

b. $\frac{-17}{0}$ is undefined. If $\frac{-17}{0} = x$, then $-17 = 0 \cdot x$, which is not possible.

Now work exercise 5 in the margin.

Objective 3

Example 6

Evaluate each of the following expressions by using the rules for order of operations.

a. $14 + 3(-5)$ **b.** $-4(5-6) - 5 \cdot 2$ **c.** $(-8)^2 + 6(-2) \div 4$

Solution

a. $14 + 3(-5) = 14 + (-15)$ Multiply first.

$= -1$ Add.

b. $-4(5-6) - 5 \cdot 2 = -4(-1) - 5 \cdot 2$ Simplify within parentheses.

$= +4 - 10$ Multiply from left to right.

$= -6$ Add (or subtract).

c. $(-8)^2 + 6(-2) \div 4 = +64 + 6(-2) \div 4$ Use the exponent to evaluate.

$= 64 + (-12) \div 4$ Multiply.

$= 64 + (-3)$ Divide.

$= 61$ Add.

Now work exercise 6 in the margin.

6. a. $4^3 + 5(2^3 - 10)$

b. $6 \div (-3) \cdot 2 + 7(-2)$

c. $[5 + 3(-2-4)] \div 13$

Practice Problems

Find the value of each expression.

1. $(-5)(5)$

2. $(-3)(-3)(0)$

3. $\frac{-4}{0}$

4. $\frac{-15}{-5}$

5. $\left(16-5^2\right)\div 3^2-1$

6. $\left[2(-3-5)+3\cdot 2\right]\div 5\cdot 3$

Answers to Practice Problems:

1. -25 **2.** 0 **3.** undefined

4. 3 **5.** -2 **6.** -6

Name ______________ Section ______________ Date ______________

Exercises 9.4

Find the following products. See Examples 1 through 3.

1. 5(–3)

2. 4(–6)

3. –6(–4)

4. –2(–7)

5. –5(4)

6. –8(3)

7. 14(2)

8. 13(3)

9. –10(5)

10. –11(3)

11. (–7)3

12. (–2)9

13. 6(–8)

14. 9(–4)

15. –7(–9)

16. –8(–9)

17. 0(–6)

18. 0(–4)

19. (–6)(–5)(3)

20. (–2)(–1)(7)

21. 4(–2)(–3)

22. 5(–6)(–1)

23. (–5)(3)(–4)

24. (–3)(7)(–5)

25. (–7)(–2)(–3)

26. (–4)(–4)(–4)

27. (–3)(–3)(–5)

28. (–2)(–2)(–8)

29. (–3)(–4)(–5)

30. (–2)(–5)(–7)

31. (–5)(0)(–6)

32. (–6)(0)(–2)

33. (–1)(–1)(–1)

ANSWERS

1. ______
2. ______
3. ______
4. ______
5. ______
6. ______
7. ______
8. ______
9. ______
10. ______
11. ______
12. ______
13. ______
14. ______
15. ______
16. ______
17. ______
18. ______
19. ______
20. ______
21. ______
22. ______
23. ______
24. ______
25. ______
26. ______
27. ______
28. ______
29. ______
30. ______
31. ______
32. ______
33. ______

34. ____
35. ____
36. ____
37. ____
38. ____
39. ____
40. ____
41. ____
42. ____
43. ____
44. ____
45. ____
46. ____
47. ____
48. ____
49. ____
50. ____
51. ____
52. ____
53. ____
54. ____
55. ____
56. ____
57. ____
58. ____
59. ____
60. ____
61. ____
62. ____
63. ____
64. ____
65. ____
66. ____
67. ____
68. ____
69. ____
70. ____

34. $(-3)(-3)(-3)$

35. $(-2)(-2)(-2)(-2)$

36. $(-2)(-4)(-4)$

37. $(-1)(-4)(-7)(+3)$

38. $(-5)(-2)(-1)(+5)$

39. $(-2)(-3)(-10)(-5)$

40. $(-11)(-2)(-4)(-1)$

Find the following quotients. See Example 4.

41. $\frac{-12}{4}$

42. $\frac{-18}{2}$

43. $\frac{-14}{7}$

44. $\frac{-28}{7}$

45. $\frac{-20}{-5}$

46. $\frac{-30}{-3}$

47. $\frac{-50}{-10}$

48. $\frac{-30}{-5}$

49. $\frac{30}{-6}$

50. $\frac{40}{-8}$

51. $\frac{75}{-25}$

52. $\frac{80}{-4}$

53. $\frac{12}{6}$

54. $\frac{24}{8}$

55. $\frac{36}{9}$

56. $\frac{22}{11}$

57. $\frac{-39}{13}$

58. $\frac{27}{-9}$

59. $\frac{32}{-4}$

60. $\frac{23}{-23}$

61. $\frac{-34}{-17}$

62. $\frac{-60}{-15}$

63. $\frac{-8}{-8}$

64. $\frac{26}{-13}$

65. $\frac{-31}{0}$

66. $\frac{17}{0}$

67. $\frac{0}{-20}$

68. $\frac{0}{-16}$

69. $\frac{35}{0}$

70. $\frac{0}{25}$

Name ______________________ Section ____________ Date __________

ANSWERS

Evaluate the expressions in Exercises 71 – 84 by using the rules for order of operations. See Example 6.

71. $8 \div 4 + 2 \cdot 7$

72. $2(-5) + 12 \div 3$

73. $3^2 - 14 \div 7 \cdot 2$

74. $5^2 \div 5 \cdot 3 + (-8)$

75. $6(2 - 7) \div 5$

76. $12 \cdot 2 - 3(5 - 7)$

77. $(-4)^2 + (-4)^3$

78. $(8 - 10)(15 - 12)$

79. $(8 - 4)(-6 - 5)$

80. $[4 + 3(-1 - 5)] \div 2$

81. $[10 + 2(-3 - 8)] \div 3$

82. $(4 \cdot 11 - 4) \div 10 \cdot 4 - 5^2$

83. $5^2 - 6 \cdot 3 + 2(5 + 2^3)$

84. $(-6)^2 - 4 \cdot 2 + 5(7 - 3^2)$

85. If the product of –12 and –3 is added to the product of –25 and 2, what is the sum?

86. If the quotient of –35 and 7 is subtracted from the product of –11 and –5, what is the difference?

87. What number should be added to –15 to get a sum of –29?

88. What number should be added to –40 to get a sum of –10?

89. What is the difference between the product of –16 and 2 and the quotient of –45 and 5?

71. ____________

72. ____________

73. ____________

74. ____________

75. ____________

76. ____________

77. ____________

78. ____________

79. ____________

80. ____________

81. ____________

82. ____________

83. ____________

84. ____________

85. ____________

86. ____________

87. ____________

88. ____________

89. ____________

[Respond below
90. exercise.]

[Respond below
91. exercise.]

[Respond below
92. exercise.]

Writing and Thinking about Mathematics

90. If you multiply an odd number of negative numbers together, do you think that the product will be positive or negative? Explain your reasoning.

91. If you multiply an even number of negative numbers together, do you think that the product will be positive or negative? Explain your reasoning.

92. In your own words, explain why division by 0 is not possible in arithmetic.

9.5 Combining Like Terms and Evaluating Algebraic Expressions

Combining Like Terms

Objective ①

A single number is called a **constant**, and a symbol (usually a letter) used to indicate an unknown is called a **variable**. Any constant or variable or the indicated product of constants and powers of variables is called a **term**. Examples of terms are

$$18, \quad \frac{3}{4}, \quad 5xy, \quad -4x^2, \quad -10x, \quad \text{and} \quad \frac{5}{9}x^3y$$

As we discussed in Section 6.3, a number written next to a variable (as in $-10x$) or a variable written next to another variable (as in xy) indicates multiplication. In the term $-4x^2$, the constant -4 is called the **numerical coefficient** of x^2 (or simply the **coefficient** of x^2). If a term has only one variable, then the exponent of the variable is called the **degree** of the term. Constants are said to be of zero degree.

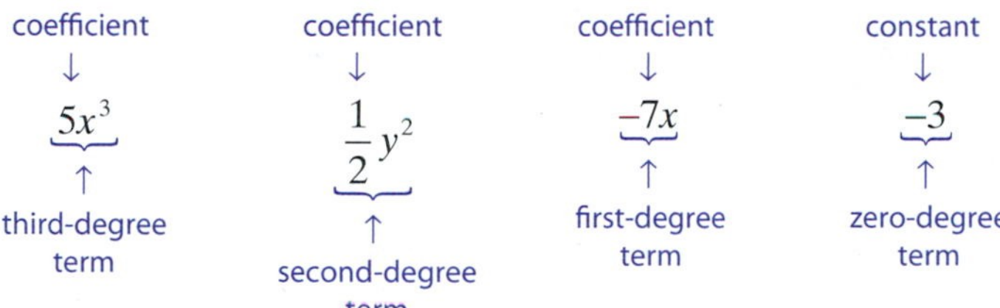

Objectives

① Know the meanings of the terms **constant**, **variable**, **term**, and **coefficient**.

② Recognize like terms and be able to combine like terms.

③ Understand the distributive property of multiplication over addition.

④ Be able to evaluate algebraic expressions for given values of the variables by using the rules for order of operations.

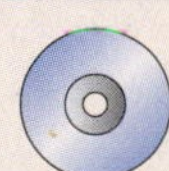

Like terms (or **similar terms**) are terms that contain the same variables (if any) raised to the same powers. Whatever power a variable is raised to in one term, it is raised to the same power in other like terms. Constants are like terms.

Objective ②

Like Terms

-7, 19, and $\frac{7}{8}$	are like terms because each term is a constant.
$-3a$, $14a$, and $5a$	are like terms because each term contains the same variable, a, raised to the same power, 1.
$4xy^2$ and $8xy^2$	are like terms because each term contains the same two variables, x and y, with x first-degree in both terms and y second-degree in both terms.

Unlike Terms

$9x$ and $3x^2$	are unlike terms (**not** like terms) because x is not of the same power in both terms.
$15ab^2$ and $-4a^2b$	are unlike terms since the variables are **not** of the same power in both terms.

If no number is written next to a variable, then the coefficient is understood to be 1. If a (–) sign is written next to a variable, then the coefficient is understood to be –1. For example,

$$x = 1 \cdot x, \quad a^3 = 1 \cdot a^3, \quad \text{and} \quad y = 1 \cdot y$$

$$-x = -1 \cdot x, \quad -a^3 = -1 \cdot a^3, \quad \text{and} \quad -y = -1 \cdot y$$

1. From the following list of terms, pick out the like terms.
$\{7, 5x, 9y^2z, -13y^2z, 13, 9x\}$

Example 1

From the following list of terms, pick out the like terms.

$$5, \quad 4y, \quad -12, \quad -y, \quad 3x^2z, \quad -\frac{1}{2}y, \quad -2x^2z, \quad 0$$

Solution

a. 5, –12, and 0 are like terms. All are constants.

b. $4y$, $-y$, and $-\frac{1}{2}y$ are like terms. All have the same variable factor, y.

c. $3x^2z$ and $-2x^2z$ are like terms. All have the same variable factor, x^2z.

Now work exercise 1 in the margin.

Algebraic expressions involving the sums or differences of terms, such as

$$8 + x, \quad 5y^2 - 11y, \quad \text{and} \quad 2x^2 - 5x - 2,$$

are called **polynomials**. To simplify this type of algebraic expression, we **combine like terms** whenever possible. For example,

$$7x + 2x = 9x \quad \text{and} \quad 8n + 3n + 2n = 13n$$

The technique that we use for combining like terms is based on the following **distributive property of multiplication over addition** (or simply the **distributive property**).

Objective 3

Distributive Property of Multiplication over Addition

For any numbers a, b, and c,

$$a(b + c) = ab + ac.$$

Another form of the distributive property is

$$ba + ca = (b + c)a$$

This form is particularly useful when b and c are numerical coefficients because it leads directly to the explanation of combining like terms. Thus

$$ba + ca = (b + c)a$$

$7x + 2x = (7 + 2)x = 9x$ by the distributive property

(Intuitively, we can think of x as a label or name. Thus, just as 7 oranges + 2 oranges is 9 oranges, $7x + 2x$ is $9x$.) Similarly,

$6n - n + 3n = (6 - 1 + 3)n = 8n$ Note that -1 is the coefficient in $-n$.

Simplify each expression by combining like terms.

2. a. $5x + 6x$

b. $-3y - 4y$

c. $6x + 6 + x - 8$

d. $2(n + 4) + 3(n - 3)$

Example 2

Simplify each expression by combining like terms whenever possible.

a. $6x + 10x + 3$ **b.** $4x^2 + x^2 - 8x^2$

c. $2(x + 3) + 4(x - 7)$ **d.** $3a - 52b + 9$

Solution

a. $6x + 10x + 3 = (6 + 10)x + 3$
$= 16x + 3$

Use the distributive property with $6x + 10x$. Note that the constant 3 is not combined with $16x$ because they are not like terms.

b. $4x^2 + x^2 - 8x^2 = (4 + 1 - 8)x^2$
$= -3x^2$

Note that $+1$ is the coefficient of x^2.

c. $2(x + 3) + 4(x - 7) = 2x + 6 + 4x - 28$ First, use the distributive property directly.
$= 2x + 4x + 6 - 28$
$= (2 + 4)x - 22$ Combine like terms.
$= 6x - 22$

d. $3a - 52b + 9$ This expression is already simplified since it has no like terms.

Now work exercise 2 in the margin.

Evaluating Algebraic Expressions

Objective 4

Algebraic expressions can be evaluated for given values of the variables by substituting one value for each variable into the expression and then following the rules for order of operations. When like terms are present, the process of evaluation can be made easier by first combining like terms.

To Evaluate an Algebraic Expression:

1. Combine like terms.
2. Substitute the values given for any variables. (**Note:** To indicate multiplication, enclose the numbers substituted in parentheses.)
3. Follow the rules for order of operations.

The rules for order of operations are restated here for easy reference.

Rules for Order of Operations

1. First, simplify within grouping symbols, such as parentheses ( ), brackets [ ], or braces { }. Start with the innermost grouping.
2. Second, find any powers indicated by exponents.
3. Third, moving from **left to right**, perform any multiplications or divisions in the order in which they appear.
4. Fourth, moving from **left to right**, perform any additions or subtractions in the order in which they appear.

Evaluate each expression for $x = 4$ and $a = -5$.

3. a. $3a + 14$

b. $7a + 3(x - 1)$

Example 3

Evaluate the expression $6ab + ab + 4a - a$ for $a = -2$ and $b = +3$.

Solution

Combining like terms gives

$$\begin{aligned} 6ab + ab + 4a - a &= (6 + 1)ab + (4 - 1)a \\ &= 7ab + 3a \end{aligned}$$

Substituting $a = -2$ and $b = +3$ into the simplified expression and following the rules for order of operations gives the value of the expression.

$$\begin{aligned} &7ab + 3a \\ &= 7(-2)(3) + 3(-2) \\ &= -42 - 6 \\ &= -48 \end{aligned}$$

Note that the substituted numbers must be in parentheses to indicate multiplication.

Now work exercise 3 in the margin.

Completion Example 4

Evaluate the expression $7x^2 - 2x^2 + 2x - x^2 - 14x$ for $x = 3$.

Solution

Combining like terms

$$7x^2 - 2x^2 + 2x - x^2 - 14x$$
$$= (\_\_\_\_\_\_\_\_\_\_)x^2 + (\_\_\_\_\_\_)x$$
$$= \_\_\_\_\_\_ x^2 - \_\_\_\_\_\_ x$$

Substituting $x = 3$ and evaluating:

$$4x^2 - 12x = 4(\_\_\_\_\_\_)^2 - 12(\_\_\_\_\_\_)$$
$$= \_\_\_\_\_\_ - \_\_\_\_\_\_ = \_\_\_\_\_\_$$

Now work exercise 4 in the margin.

Evaluate each expression for $x = 4$.

4. a. $2x - 8$

b. $6x^2 - 5x^2 + 3x - 2x + 8$

Completion Example Answers

4. $7x^2 - 2x^2 + 2x - x^2 - 14x = (\mathbf{7 - 2 - 1})x^2 + (\mathbf{2 - 14})x$
$= \mathbf{4}x^2 - \mathbf{12}x$

Substituting $x = 3$ and evaluating:

$4x^2 - 12x = 4(\mathbf{3})^2 - 12(\mathbf{3})$
$= \mathbf{36 - 36 = 0}$

Practice Problems

Simplify the following expressions by combining like terms whenever possible.

1. $-7x - 15x$ **2.** $-x + 4x + 3x + 1$

Evaluate each of the following expressions for $x = -2$, $y = 1$, and $a = -3$.

3. $x - 3$ **4.** $2a - a + 9$ **5.** $2y^2 - 3y + 8$

Answers to Practice Problems:

1. $-22x$ **2.** $6x + 1$ **3.** -5 **4.** 6 **5.** 7

Name ____________________ Section ____________ Date __________

ANSWERS

1. ______
2. ______
3. ______
4. ______
5. ______
6. ______
7. ______
8. ______
9. ______
10. ______
11. ______
12. ______
13. ______
14. ______
15. ______
16. ______
17. ______
18. ______
19. ______
20. ______
21. ______
22. ______
23. ______
24. ______
25. ______
26. ______
27. ______

Exercises 9.5

Simplify each of the following algebraic expressions by combining like terms whenever possible. See Example 2.

1. $6x + 2x$

2. $4x - 3x$

3. $5x + x$

4. $7x - 3x$

5. $-10a + 3a$

6. $-11y + 4y$

7. $-18y + 6y$

8. $-2x - 5x$

9. $-5x - 4x$

10. $-x - 2x$

11. $-7x - x$

12. $2x - 2x$

13. $3x - 5x + 12x$

14. $2a + 14a - 25a$

15. $6c - 13c + 5c$

16. $40x - 30x - 10x$

17. $16n - 15n - 3n$

18. $12y - 12y + 4y$

19. $5x^2 - 3x^2 + 2x$

20. $-2x^2 - x^2 - x$

21. $7x^2 - 4x^2 + 20$

22. $4x + 7 - 8 + 3x$

23. $-5x - 1 + 8 + 9x$

24. $10y + 3 - 4 + 6y$

25. $2(x + 1) + 3(x - 1)$

26. $4(x - 1) + 5(x - 2)$

27. $3ab + 6a + 2b$

28. ______

29. ______

30. ______

31. ______

32. ______

33. ______

34. ______

35. ______

36. ______

37. ______

38. ______

39. ______

40. ______

41. ______

42. ______

43. ______

44. ______

45. ______

46. ______

47. ______

48. ______

49. ______

50. ______

28. $9xy - 2x + 5y$ **29.** $x^2 - 5x + 6$ **30.** $x^2 - 7x + 12$

31. $-4n + n + 1 - 3$ **32.** $-2c + 5c + 6 - 5$ **33.** $3x^2 - 4x^2 + 2x - 1$

34. $-5x^2 + 4x^2 - 17x + 20x + 42 + 3$

35. $12x^2 - 2x^2 + 15x + 13x - 35 - 41$

Evaluate each of the following expressions for $x = -3$, $y = 2$, $z = 3$, $a = -1$, and $c = -2$. See Examples 3 and 4.

36. $x - 2$ **37.** $y - 2$ **38.** $z - 3$

39. $2x + 3x - 7$ **40.** $7a - a + 3$ **41.** $-3y - 4y + 6 - 2$

42. $-2c - 3c + 1 - 4$ **43.** $5y - 2y - 3y + 4$ **44.** $2x - 3x + x - 8$

45. $3a + 2x + 7x - a + x$ **46.** $6 + 2(a + 1) + 3z$

47. $5y + 14 + 3(c - 1) - 4$ **48.** $5x^2 - 8x + 2x + 7 - 9$

49. $4z^2 + 2z - 5z + 2 + 10$ **50.** $2c^2 + 8 + 2c + 3c + c^2$

9.6 Translating English Phrases and Solving Equations

Translating English Phrases

Objectives

① Be able to translate English phrases into algebraic expressions.

② Be able to translate algebraic expressions into English phrases.

③ Know how to solve linear equations that are of the form $ax + b = c$ where the constants a, b, and c are integers.

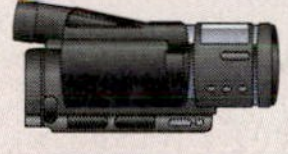

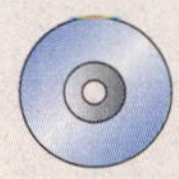

Many word problems can be solved by using the following three basic skills:

1. Know how to translate English phrases into algebraic expressions.
2. Be able to set up an equation using the translated expressions.
3. Solve the resulting equation.

In this section, we will learn how to translate English phrases by finding key words and how to develop techniques for solving equations. In Section 9.7, we will see how these skills can be used to set up equations using translated expressions to problem solve.

The following list contains some of the key words that appear in word problems. These words indicate what operations are to be performed either with known numbers (called **constants**) or with variables and constants.

Key Words (that indicate operations)

Addition	**Subtraction**	**Multiplication**	**Division**
add	subtract (from)	multiply	divide
sum	difference	product	quotient
plus	minus	times	
more than	less than	twice, double	
increased by	decreased by	of (with fractions)	

Objective ①

The following examples illustrate how English phrases can be translated into algebraic expressions. Different letters are used as variables to represent the unknown numbers, and the key words are in boldface print.

English Phrase | **Algebraic Expression**

1. A number **plus** 7
   The **sum** of a number and 7
   7 **more than** a number
   A number **increased by** 7
   7 **added to** a number

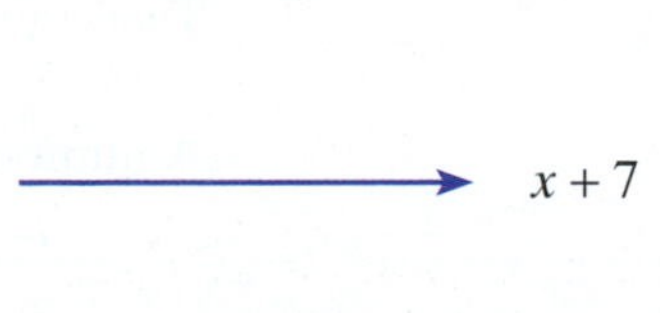

$x + 7$

2. 8 **times** a number
The **product** of a number and 8
A number **multiplied by** 8 } → $8n$

3. The **quotient** of a number and 6
A number **divided by** 6 } → $\frac{n}{6}$

4. **Twice** a number **decreased by** 51
51 **less than twice** a number
51 **subtracted from two times** a number }  $2y - 51$

The words **difference** and **quotient** deserve special mention because their use implies that the numbers are to be operated on in the order given. For example,

"the **difference** between x and 10" means $x - 10$, not $10 - x$;

"the **quotient** of y and 20" means $\frac{y}{20}$, not $\frac{20}{y}$.

Objective 2

A skill closely related to translating English phrases into algebraic expressions is the reverse process-that is, looking at an algebraic expression and creating an English phrase that has the same meaning. In such cases, there is often more than one correct phrase.

Example 1

For each expression, write an English phrase that has the same meaning.

a. $14x$ **b.** $n + 35$

c. $2x + 5x$ **d.** $\frac{y}{3}$

Solution

Algebraic Expression	Possible English phrase
a. $14x$	The product of 14 and a number
b. $n + 35$	35 more than a number
c. $2x + 5x$	Twice a number plus 5 times the number
d. $\frac{y}{3}$	A number divided by 3

Now work exercise 1 in the margin.

1. For parts **a.** and **b.**, write an English phrase that means the same as each expression. For parts **c.** and **d.**, write an algebraic expression that means the same as the English phrase.

a. $13x$

b. $n - 45$

c. Twice the sum of a number and 7.

d. 50 divided by a number

Solving Equations ($ax + b = c$)

Objective 3

If an equation contains a variable, such as the equation $5x + 1 = 16$, then any number may be substituted for x. For example,

Substituting $x = 8$ gives $5 \cdot 8 + 1 = 16$ — a false statement

Substituting $x = 3$ gives $5 \cdot 3 + 1 = 16$ — a true statement

We can substitute as many values for x as we choose. **However, the object is not just to substitute values for the variable x, but to find the value for x that results in a true statement when this value is substituted for x in the equation.** This procedure is called **solving the equation**. The value found is called the **solution** of the equation.

We just found that substituting $x = 3$ in the equation $5x + 1 = 16$ gives a true statement, and $x = 3$ is a solution of the equation. In fact, $x = 3$ is the *only* solution.

First-Degree Equation in x

A **first-degree equation in x** (or **linear equation in x**) is any equation that can be written in the form

$\boldsymbol{ax + b = c}$ where a, b, and c are constants and $a \neq 0$.

(**Note:** A variable other than x may be used.)

Example 2

Given the equation $3n + 15 = 27$, **a.** show that 4 is a solution, and **b.** show that 2 is not a solution.

Solution

a. Since $3 \cdot 4 + 15 = 27$ is a true statement, 4 is a solution.

b. Since $3 \cdot 2 + 15 = 27$ is a false statement, 2 is not a solution.

Now work exercise 2 in the margin.

2. Given the equation $-7n + 21 = 28$, show **a.** that -1 is a solution, and **b.** that 3 is not a solution.

a.

b.

A fundamental fact of algebra, stated here without proof, is that **every first-degree equation has exactly one solution**. Sometimes, with simple equations, this solution can be found by observation or by trial and error. However, by using the following principles, we can guarantee finding the solution in an efficient, organized manner that works, regardless of how complicated the equation is. The basic idea is that **the same operation is to be performed on both sides of the equal sign (that is, both sides of the equation)**.

For example, if you add 17 to one side of an equation, you must add 17 to the other side as well. You cannot add 17 to one side and subtract 17 from the other side. As illustrated in Figure 9.13, you can think of the equal sign as the fulcrum (or balance point) on a balance scale.

A = B $x - 17 = -10$

Each equation is balanced.

The equation will remain balanced if C is added to both sides.

(a)

The equation will remain balanced if 17 is added to both sides.

(b)

Figure 9.13

Principles Used in Solving a First-Degree Equation

In the four basic principles stated here, A and B represent algebraic expressions or constants.

1. **The Addition Principle:**
   If $A = B$ is true, then $A + C = B + C$ is also true for any number C. (The same number may be added to both sides of an equation.)

2. **The Subtraction Principle:**
   If $A = B$ is true, then $A - C = B - C$ is also true for any number C. (The same number may be subtracted from both sides of an equation.)

3. **The Multiplication Principle:**
   If $A = B$ is true, then $C \cdot A = C \cdot B$ is also true for any number C. (Both sides of an equation may be multiplied by the same number.)

4. **The Division Principle:**
   If $A = B$ is true, then $\frac{A}{C} = \frac{B}{C}$ is also true for any nonzero number C. (Both sides of an equation may be divided by the same nonzero number.)

Note that in the Division Principle, division by C is the same as multiplication by the reciprocal of C, namely $\frac{1}{C}$. Thus we could list the Division and Multiplication Principles as one principle. The use of any of these principles with an equation gives a new equation with the same solution. Such equations are said to be **equivalent**. The process of solving equations involves applying the principles just listed to find equivalent equations until the solution is obvious, as in $x = 3$, $y = 15$, or $700 = A$.

Example 3

Solve the equation $x + 13 = 25$.

Solution

$x + 13 = 25$	Write the equation.
$x + 13 - 13 = 25 - 13$	Use the Subtraction Principle and subtract 13 from both sides.
$x + 0 = 12$	
$x = 12$	Simplify both sides.

Now work exercise 3 in the margin.

Example 4

Solve the equation $5y = 180$.

Solution

$5y = 180$	Write the equation.
$\frac{\cancel{5}y}{\cancel{5}} = \frac{180}{5}$	Use the Division Principle and divide both sides by 5. (Or use the Multiplication Principle and multiply both sides by $\frac{1}{5}$.)
$1 \cdot y = 36$	Simplify both sides.
$y = 36$	

Now work exercise 4 in the margin.

Example 5

Solve the equation $\frac{n}{6} = 31$.

Solution

$\frac{n}{6} = 31$	Write the equation.
$\frac{\cancel{6}}{1} \cdot \frac{n}{\cancel{6}} = \frac{31}{1} \cdot \frac{6}{1}$	Use the Multiplication Principle and multiply both sides by $\frac{6}{1}$.
$n = 186$	Simplify both sides.

Now work exercise 5 in the margin.

More difficult problems may involve several steps. Keep in mind the following general guidelines.

General Guidelines for Solving Equations

1. The goal is to isolate the variable on one side of the equation (either the right side or the left side).
2. First use the Addition and Subtraction Principles whenever numbers are subtracted or added to the variable.
3. Use the Multiplication and Division Principles to make 1 (or +1) the coefficient of the variable. (The coefficient 1 is not usually written in the solution.)

Solve each of the following equations.

3. $y - 2 = -5.$

4. $3x = -12$

5. $\frac{y}{3} = 17.$

6. Solve the equation $3x - 35 = 1$.

Example 6

Solve the equation $4x - 16 = 64$.

Solution

$4x - 16 = 64$	Write the equation.
$4x - 16 + 16 = 64 + 16$	Use the Addition Principle and add 16 to both sides.
$4x + 0 = 80$	Simplify both sides.
$4x = 80$	
$\frac{1}{\not{4}} \cdot \not{4}x = \frac{1}{4} \cdot 80$	Use the Multiplication Principle and multiply both sides by $\frac{1}{4}$. (Or use the Division Principle and divide both sides by 4.)
$1 \cdot x = 20$	Simplify both sides.
$x = 20$	

Now work exercise 6 in the margin.

Checking Solutions to Equations

Checking can be done by substituting the solution found back into the original equation to see if the resulting statement is true.

Checking Example 6

$$4x - 16 = 64$$
$$4 \cdot 20 - 16 \stackrel{?}{=} 64$$
$$80 - 16 \stackrel{?}{=} 64$$
$$64 = 64 \quad \text{a true statement}$$

7. Solve the equation $5n + 22 = 32$.

Completion Example 7

Explain each step in the solution process shown here.

Equation	**Explanation**
$13 = 5n - 7$	Write the equation.
$13 + 7 = 5n - 7 + 7$	______
$20 = 5n$	______
$\frac{20}{5} = \frac{\not{5}n}{\not{5}}$	______
$4 = n$	______

Now work exercise 7 in the margin.

Solving Equations Involving Integers

The same principles used for solving first-degree equations with whole number constants and coefficients can be used to solve equations with integer (or decimal or fraction) constants and coefficients. The following examples illustrate solving equations with integer constants and coefficients and integer solutions.

Solve the following equations.

8. $13 + y = 8$

Example 8

Solve the equation $x + 17 = 14$.

Solution

$$x + 17 = 14$$ Write the equation.

$$x + 17 - 17 = 14 - 17$$ Use the Subtraction Principle and subtract 17 from both sides. (This is the same as adding –17 to both sides.)

$$x + 0 = -3$$

$$x = -3$$ Simplify both sides.

Check:

$$(-3) + 17 = 14$$

Now work exercise 8 in the margin.

9. $-4x = 72$

Example 9

Solve the equation $-5y = 80$.

Solution

$$-5y = 80$$ Write the equation.

$$\frac{\cancel{-5}y}{\cancel{-5}} = \frac{80}{-5}$$ Use the Division Principle and divide both sides by –5. (We want the coefficient of y to be +1.)

$$y = -16$$ Simplify both sides.

Check:

$$-5(-16) = 80$$

Now work exercise 9 in the margin.

Solve each of the following equations.

10. $53 = 17 - 6x$

11. $4x - 10 = -38$

Example 10

Solve the equation $12 = -4x + 16$.

Solution

$12 = -4x + 16$	Write the equation.
$12 - 16 = -4x + 16 - 16$	Use the Subtraction Principle and subtract 16 from both sides. (This is the same as adding −16 to both sides.)
$-4 = -4x$	Simplify both sides.
$\frac{-4}{-4} = \frac{\cancel{-4}x}{\cancel{-4}}$	Use the Division Principle and divide both sides by −4. (We want the coefficient of *x* to be +1.)
$1 = x$	Simplify both sides.

Check:

$12 = -4(1) + 16$

Now work exercise 10 in the margin.

Example 11

Solve the equation $5n + 2 = 3n - 8$.

Solution

$5n + 2 = 3n - 8$	Write the equation.
$5n + 2 - 2 = 3n - 8 - 2$	Add −2 to both sides.
$5n = 3n - 10$	Simplify both sides.
$5n - 3n = 3n - 10 - 3n$	Add −3*n* to both sides.
$2n = -10$	Simplify both sides.
$\frac{1}{\cancel{2}} \cdot \cancel{2}n = \frac{1}{2} \cdot (-10)$	Multiply both sides by $\frac{1}{2}$. (This is the same as dividing both sides by 2.)
$n = -5$	Simplify both sides.

Check:

$5(-5) + 2 = 3(-5) - 8 = -23$

Now work exercise 11 in the margin.

If an expression on one side or the other of an equation contains a number times a quantity in parentheses (or other grouping symbol), we use the distributive property to **clear** the parentheses, and then we proceed as before.

Example 12

Solve the equation $3(x + 2) = 18 - x$.

Solution

$3(x+2)=18-x$	Write the equation.
$3x+6=18-x$	Use the distributive property.
$3x+6-6=18-x-6$	Subtract 6 from both sides.
$3x=12-x$	Simplify both sides.
$3x+x=12-x+x$	Add x to both sides.
$4x=12$	Simplify both sides.
$\frac{1}{\cancel{4}}\cdot\cancel{4}x=\frac{1}{4}\cdot 12$	Multiply both sides by $\frac{1}{4}$.
$x=3$	Simplify both sides.

Check:

$3(3 + 2) = 18 - 3$

Now work exercise 12 in the margin.

12. $14 = -2n + 16$

Completion Example 13

Complete the solution process for the equation $y - 1 = 2(y - 5)$.

Solution

Equation	**Explanation**
$y - 1 = 2(y - 5)$	Write the equation.
______	Use the distributive property.
______	______
______	______
______	______
______	______

Now work exercise 13 in the margin.

13. Solve the equation $2(x + 1) = 17 - x$.

Completion Example Answers

7. Equation | **Explanation**

Equation	Explanation
$13 = 5n - 7$	Write the equation.
$13 + 7 = 5n - 7 + 7$	**Add 7 to both sides.**
$20 = 5n$	**Simplify.**
$\frac{20}{5} = \frac{\cancel{5}n}{\cancel{5}}$	**Divide both sides by 5.**
$4 = n$	**Simplify.**

13. Equation | **Explanation**

Equation	Explanation
$y - 1 = 2(y - 5)$	Write the equation.
$\mathbf{y - 1 = 2y - 10}$	Use the distributive property.
$\mathbf{y - 1 + 10 = 2y - 10 + 10}$	**Add 10 to both sides.**
$\mathbf{y + 9 = 2y}$	**Simplify.**
$\mathbf{y + 9 - y = 2y - y}$	**Subtract *y* from both sides. (Or add −*y* to both sides.)**
$\mathbf{9 = y}$	**Simplify.**

Practice Problems

1. Write an algebraic phrase describing three times a number plus four times the number.

2. Write an English phrase that means $5x + 7$.

Solve each of the following equations.

3. $x - 13 = 19$ **4.** $55 = 2x - 7$ **5.** $3(x - 7) = x + 11$

Answers to Practice Problems:

1. $3x + 4x$ **2.** five times a number plus 7

3. $x = 32$ **4.** $x = 31$ **5.** $x = 16$

Name ____________________ Section __________ Date __________

ANSWERS

1. __________

2. __________

3. __________

4. __________

5. __________

6. __________

7. __________

8. __________

9. __________

10. __________

11. __________

12. __________

13. __________

14. __________

15. __________

16. __________

17. __________

18. __________

Exercises 9.6

Write an algebraic expression described by each of the following English phrases. (Any variable may be used to represent the unknown number.)

1. a number plus 5

2. 8 more than a number

3. the sum of a number and 9

4. the product of a number and 9

5. the quotient of a number and 9

6. the difference of a number and 9

7. 13 less than a number

8. 55 decreased by a number

9. 13 decreased by a number

10. 16 plus a number

11. 4 more than twice a number

12. 15 increased by twice a number

13. 8 times a number plus 6

14. 6 more than 5 times a number

15. 18 less than the quotient of a number and 2

16. 25 more than the product of a number and 8

17. the difference of 3 and twice a number

18. the sum of 1 and three times a number

19. ______

20. ______

21. [Respond below exercise.]

22. [Respond below exercise.]

23. [Respond below exercise.]

24. [Respond below exercise.]

25. [Respond below exercise.]

26. [Respond below exercise.]

27. [Respond below exercise.]

28. [Respond below exercise.]

29. [Respond below exercise.]

30. [Respond below exercise.]

31. [Respond below exercise.]

32. [Respond below exercise.]

33. [Respond below exercise.]

34. [Respond below exercise.]

35. [Respond below exercise.]

36. ______

37. ______

38. ______

39. ______

40. ______

19. the product of 6 and a number increased by 4 times the number

20. 7 times a number decreased by the product of the number and 5

Write an English phrase that means the same as each expression. (There may be more than one correct phrase.) See Example 1.

21. $5x$

22. $n + 15$

23. $x - 6$

24. $y - 32$

25. $y + 491$

26. $7y$

27. $\frac{n}{17}$

28. $\frac{x}{10}$

29. $2x + 1$

30. $3x - 1$

31. $\frac{y}{3} + 20$

32. $\frac{a}{5} + 12$

33. $15x - 15$

34. $\frac{6}{x}$

35. $13 - x$

A set of numbers follows the equation in each problem below. Determine which number in the set is the solution of the equation by substituting the numbers into the equation. See Example 2.

36. $x - 17 = 10;\ \{27, 30, 33\}$

37. $2n + 3 = 13;\ \{0, 2, 5, 7\}$

38. $5n + 8 = 23;\ \{0, 1, 2, 3\}$

39. $\frac{n}{3} + 6 = 8;\ \{3, 6, 9, 12\}$

40. $\frac{x}{2} - 4 = 0;\ \{2, 4, 6, 8, 10\}$

Name ______________________ Section ____________ Date __________

Give an explanation (or a reason) for each step in the solution process. See Examples 6 through 13.

41. $3x + 10 = 22$ ______________________

$3x + 10 - 10 = 22 - 10$ ______________________

$3x = 12$ ______________________

$\frac{3x}{3} = \frac{12}{3}$ ______________________

$x = 4$ ______________________

42. $2x - 15 = -35$ ______________________

$2x - 15 + 15 = -35 + 15$ ______________________

$2x = -20$ ______________________

$\frac{2x}{2} = \frac{-20}{2}$ ______________________

$x = -10$ ______________________

Solve each of the following equations. See Examples 3 through 13.

43. $x - 12 = 6$

44. $x - 10 = 9$

45. $n + 3 = -13$

46. $n + 7 = -15$

47. $7y = 42$

48. $9y = 72$

49. $\frac{n}{7} = 32$

50. $\frac{x}{3} = 10$

51. $2x + 1 = -11$

52. $3x - 1 = -16$

ANSWERS

41. [Respond beside exercise.]

42. [Respond beside exercise.]

43. ______

44. ______

45. ______

46. ______

47. ______

48. ______

49. ______

50. ______

51. ______

52. ______

53. ______

54. ______

55. ______

56. ______

57. ______

58. ______

59. ______

60. ______

61. ______

62. ______

63. ______

64. ______

65. ______

66. ______

67. ______

68. ______

69. ______

70. ______

71. [Respond below exercise.] ______

53. $18 = -4y - 6$

54. $28 = -5n - 7$

55. $\frac{n}{2} + 3 = 15$

56. $\frac{x}{5} - 9 = 6$

57. $\frac{x}{3} + 12 = 20$

58. $\frac{x}{4} - 35 = 15$

59. $70 = 2x - 6$

60. $82 = 3x + 4$

61. $3x + 5 = 2x - 5$

62. $7x - 10 = 5x + 20$

63. $4n + 7 = 6n + 11$

64. $8n - 14 = 11n - 5$

65. $2(x + 3) = x - 1$

66. $3(x - 1) = x + 13$

67. $3(n - 5) = 2(n + 2)$

68. $2(n + 1) = 4(n - 1)$

69. $5x - 10 = 2(x - 20)$

70. $9x + 30 = 7(x - 2)$

Writing and Thinking about Mathematics

71. The solution of an equation is the same whether the variable appears on the right side of the equation or on the left side of the equation. For example, the solution of the equation $3x + 1 = 25$ is the same as the solution of the equation $25 = 3x + 1$. Does this seem reasonable to you? Explain briefly.

9.7 Problem Solving

Solving Number Problems

Objective 1

The word problems discussed here can be solved by translating English phrases into algebraic expressions–a skill that we developed in Section 9.6. The four-step problem-solving approach is outlined below:

Objectives

① Be able to solve number problems by using equations.

② Be able to solve problems involving geometric concepts by using equations.

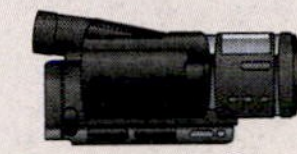
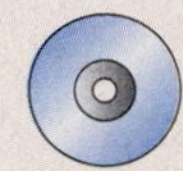

Basic Steps for Solving Number Problems:

1. Read the problem carefully at least twice. Look for key words and phrases that can be translated into algebraic expressions.
2. Assign a variable as the unknown quantity and form an equation using the expressions you translated.
3. Solve the equation.
4. Look back over the problem and check that the answer is reasonable.

Example 1

Four less than twice a number is equal to 20. Find the number.

Solution

Step 1: Let n = the unknown number.

Step 2: Translate "four less than twice a number" to $\underline{2n-4}$.

Step 3: Form the equation: $\underline{2n-4=20}$.

Step 4: Solve the equation:

$$
\begin{aligned}
2n-4&=20\\
2n-4+4&=20+4\\
2n&=24\\
\frac{1}{2}\cdot 2n&=\frac{1}{2}\cdot 24\\
1\cdot n&=12\\
n&=12
\end{aligned}
$$

Check:

$$
\begin{aligned}
2(12)-4&\stackrel{?}{=}20\\
24-4&\stackrel{?}{=}20\\
20&=20
\end{aligned}
$$

The number is 12.

Now work exercise 1 in the margin.

1. A number times 15 plus 3 is equal to 48. Find the number.

2. A number is decreased by 25 and the result is 37. What is the number?

Completion Example 2

A number is increased by 45 and the result is 30. What is the number?

Solution

Step 1: Let x = the unknown number.

Step 2: Translate "a number is increased by 45" to ______________.

Step 3: Form the equation: __________________.

Step 4: Solve the equation:

$$\_\_\_\_\_\_ = \_\_\_\_\_\_$$
$$x + 45 - \_\_\_ = 30 - \_\_\_$$
$$x + \_\_\_ = \_\_\_\_$$
$$x = \_\_\_\_$$

Check:

$$\_\_\_\_\_\_ + 45 \stackrel{?}{=} 30$$
$$\_\_\_\_\_\_ - 30$$

So _______ is the correct number.

Now work exercise 2 in the margin.

Example 3

A span of a suspension bridge is the distance between its supports. The longest span on the Tacoma Narrows bridge in Washington State is 2800 feet. This is 40 feet more than twice the longest span of the Triboro bridge in New York City. What is the longest span of the Triboro bridge?

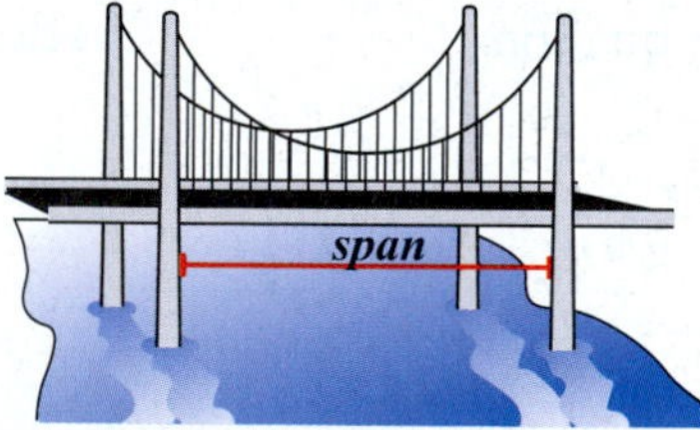

Solution

Step 1: Let x = the length of the longest span of the Triboro bridge.

Step 2: Translate "40 feet more than twice the longest span of the Triboro bridge" to $\underline{2x + 40}$.

Step 3: Form the equation: $\underline{2x + 40 = 2800}$.

Step 4: Solve the equation:

$$2x + 40 = 2800$$
$$2x + 40 - 40 = 2800 - 40$$
$$2x + 0 = 2760$$
$$2x = 2760$$
$$\frac{\not{2}x}{\not{2}} = \frac{2760}{2}$$
$$x = 1380$$

The longest span of the Triboro bridge is 1380 feet. (The student should check this result in the original equation.)

Now work exercise 3 in the margin.

3. A span of the Arthur Ravenel Jr. Bridge in Charleston, SC is 1546 feet. If this is 554 feet less than twice the longest span of the John C. Grace Bridge nearby, what is the length of the longest span of the John C. Grace Bridge?

Solving Geometry Problems

Objective 2

The concepts of perimeter, area, and volume of geometric figures are related to various formulas in the form of equations. In this section the corresponding formula will be provided in each problem and all the values except one will be known. By substituting the known values, the unknown value can be found by using the techniques we have learned for solving equations.

For easy reference, Table 9.1 contains the formulas for perimeter (P) and area (A) of six geometric figures.

Table 9.1 Formulas for Finding Perimeter and Area of Six Geometric Figures

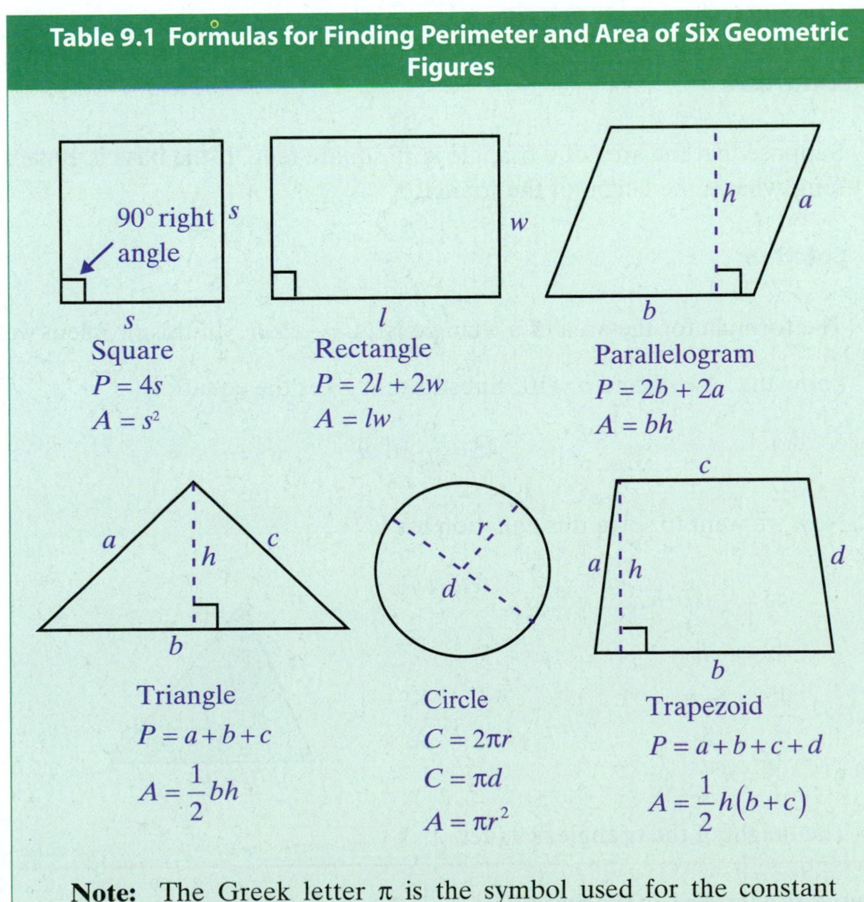

Note: The Greek letter π is the symbol used for the constant number 3.1415926535.... This constant is an infinite decimal with no pattern to its digits. For our purposes, we will use $\pi = 3.14$, but you should be aware that 3.14 is only an approximation for π.

4. A rectangular dog pen has a perimeter of 36 feet and a length that is 2 feet more than its width. Find its **a.** width and **b.** length.

a.

b.

Example 4

A rectangular swimming pool has a perimeter of 160 meters and a length that is 20 meters more than its width. Find its width and length.

Solution

$w =$ the width

$w + 20 =$ the length

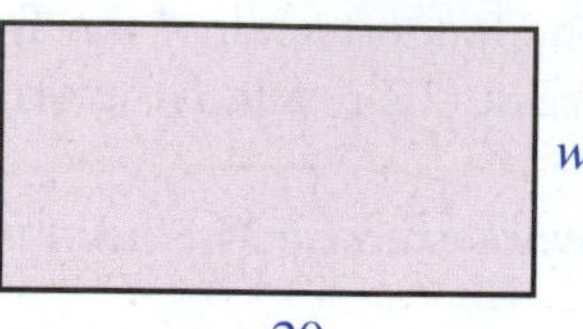

Substituting into the formula gives the equation to be solved:

$$160 = 2(w+20) + 2w$$
$$160 = 2w + 40 + 2w$$
$$160 = 4w + 40$$
$$160 - 40 = 4w + 40 - 40$$
$$120 = 4w$$
$$\frac{120}{4} = \frac{4w}{4}$$
$$30 = w$$

The width of the pool is 30 meters, and the length is $w + 20 = 50$ meters.

Now work exercise 4 in the margin.

5. Suppose the area of a triangle is 300 square feet. If the height of the triangle is 30 feet, how long is the base of the triangle?

Example 5

Suppose that the area of a triangle is 45 square feet. If the base is 10 feet long, what is the height of the triangle?

Solution

The formula for the area of a triangle is $A = \frac{1}{2} \cdot b \cdot h$. In this problem we know that $A = 45$ and $b = 10$. Substitution gives the equation

$$45 = \frac{1}{2} \cdot 10 \cdot h$$

Now we want to solve this equation for h.

$$45 = \frac{1}{2} \cdot 10 \cdot h$$
$$45 = 5 \cdot h$$
$$\frac{45}{5} = \frac{5 \cdot h}{5}$$
$$9 = h$$

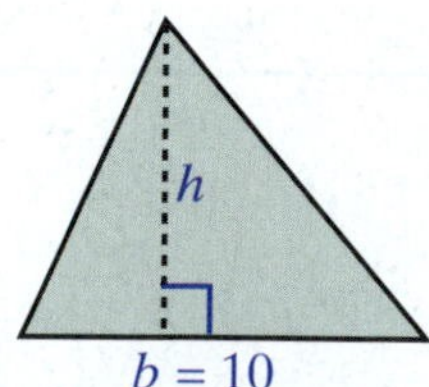

The height of the triangle is 9 feet.

Now work exercise 5 in the margin.

Completion Example Answers

2. Step 1: Let x = the unknown number.

Step 2: Translate "a number is increased by 45" to $\boldsymbol{x} + \mathbf{45}$.

Step 3: Form the equation: $\boldsymbol{x} + \mathbf{45} = \mathbf{30}$.

Step 4: Solve the equation:

$$\begin{aligned} \boldsymbol{x} + \mathbf{45} &= \mathbf{30} \\ x + 45 - \mathbf{45} &= 30 - \mathbf{45} \\ x + \mathbf{0} &= \mathbf{-15} \\ x &= \mathbf{-15} \end{aligned}$$

Check:

$$\begin{aligned} \mathbf{-15} + 45 &\stackrel{?}{=} 30 \\ \mathbf{30} &= 30 \end{aligned}$$

So $\mathbf{-15}$ is the correct number.

Practice Problems

1. Eight less than four times a number is equal to 20. Find the number.

2. Thirteen more than five times a number is equal to 28. Find the number.

3. If the product of a number and 8 is decreased by 16, the result is 32. Find the number.

4. A rectangle has a perimeter of 120 meters and a length that is 10 meters more than its width. Find its width and length.

Answers to Practice Problems:

1. 7 **2.** 3 **3.** 6

4. width is 25 m and length is 35 m

Name ____________________ Section __________ Date __________

ANSWERS

1. ____________

2. ____________

3. ____________

4. ____________

5. ____________

6. ____________

7. [Respond in exercise.] ____________

8. [Respond in exercise.] ____________

Exercises 9.7

(**Note:** Some of the exercises in this section involve the use of decimal numbers. You should understand that the rules and methods for solving equations with decimal numbers are the same as those for solving equations involving only integers.)

Translate each of the following equations into a sentence in English. Do not solve the equation.

1. $n + 5 = 16$ **2.** $x - 7 = 8$ **3.** $4x - 8 = 20$

4. $\frac{y}{2} = 61$ **5.** $\frac{n}{3} + 1 = 13$ **6.** $15 = 25 - 2x$

Follow the steps outlined in this section to solve Exercises 7 – 10. See Examples 1 and 2.

7. $\underbrace{\text{The difference between a number and twelve}}_{\text{(a)}}$ is equal to 16. What is the number?

Let x = the unknown number.

Translation of (a): ____________

Equation: ____________

Solve the equation.

8. $\underbrace{\text{The sum of a number and fifteen}}_{\text{(a)}}$ is equal to 35. What is the number?

Let n = the unknown number.

Translation of (a): ____________

Equation: ____________

Solve the equation.

9. [Respond in exercise.]

10. [Respond in exercise.]

11. ______

12. ______

13. ______

14. ______

9. Four less than three times a number is equal to 26. What is the number?

(a) = "Four less than three times a number"

Let y = the unknown number.

Translation of (a): ______________

Equation: ______________

Solve the equation.

10. Twenty is equal to one more than twice a number. What is the number?

(a) = "one more than twice a number"

Let x = the unknown number.

Translation of (a): ______________

Equation: ______________

Solve the equation.

First, make a guess at the answer. [For example, do you think the number is positive (greater than 0) or negative (less than 0)? Do you think the number is greater than 50 or less than 50?] Do not be afraid to make a "wrong" guess. Then translate the phrases, form an equation, and solve the equation.

11. The sum of a number and 45 is equal to 56. What is the number?

12. The difference between a number and 14 is 27. What is the number?

13. What is the number whose product with 3 is equal to 51?

14. If the quotient of a number and 6 is 24, what is the number?

Name ______________ Section ______________ Date ______________

ANSWERS

15. If the product of a number and four is increased by 3, the result is 23. Find the number.

15. ______________

16. If the product of a number and 7 is decreased by 5, the result is 44. Find the number.

16. ______________

17. Eleven is the result if 9 is subtracted from the quotient of a number and 5. What is the number?

17. ______________

18. Ten and eight tenths is equal to four less than twice a number. What is the number?

18. ______________

19. The product of a number and 8 increased by 24 is equal to twice the number. Find the number.

19. ______________

20. Find a number such that three times the sum of the number and 4 is equal to –60.

20. ______________

21. Five more than twice a number is equal to 20 more than the number. What is the number?

21. ______________

22. Twenty plus a number is equal to the sum of twice the number and three times the same number. Find the number?

22. ______________

23. Twice a number decreased by 6 is equal to 5 less than the number. What is the number?

23. ______________

24. ______

25. ______

26. ______

27. ______

28. ______

29. ______

Use your knowledge of geometric figures to set up an equation and solve each of the following problems.

24. The perimeter of a rectangle is 40 meters. If the length is 15 meters, what is the width of the rectangle? $(P = 2l + 2w)$

25. If the base of a triangle is 18 inches long and its area is 54 square inches, what is the height of the triangle? $\left(A = \frac{1}{2}bh\right)$

26. A parallelogram is a four-sided figure in which opposite sides are parallel. We can also show (in geometry) that these opposite sides have the same length. Suppose that the perimeter of a parallelogram is 180 yards. If one of two equal sides has a length of 50 yards, what is the length of each of the other two sides? $(P = 2a + 2b)$

27. A trapezoid is a four-sided figure with two parallel sides. If a trapezoid has an area of 195 square millimeters and parallel sides of length 20 millimeters and 10 millimeters, what is the height of the trapezoid? $\left[A = \frac{1}{2}h(b+c)\right]$

28. The perimeter of a circle is called its circumference. If the circumference of a circle is 18.84 inches, what is the radius of the circle? $(C = 2\pi r)$ (Use $\pi = 3.14$.)

29. In any triangle the sum of the angles equals 180°. If two of the angles have measures of 30° and 50°, what is the measure of the third angle?

Name ______________________ Section ____________ Date __________

ANSWERS

30. A rectangular solid is in the shape of a box. Consider a rectangular solid with a volume of 60 cubic inches. If the box is 3 inches wide and 4 inches long. What is its height? ($V = lwh$)

30. ______________

31. A rectangular solid is in the shape of a box. If the box has a volume of 360 cubic feet and is 4 feet high and 15 feet long, what is its width? ($V = lwh$)

31. ______________

32. ______________

For Exercises 32 – 35, refer to Section 5.6 for the discussion of square roots. Use a calculator to find the answers to the nearest tenth of a unit.

32. If the area of a square is 28 square meters, what is the length of one side of the square? $\left(A = s^2\right)$

33. ______________

33. Find the length of one side of a square with an area of 450 square inches. $\left(A = s^2\right)$

34. ______________

34. Find the radius of a circle that encloses an area of 28.26 square feet. $\left(A = \pi r^2\right)$ (Use $\pi = 3.14$.)

35. What is the radius of a circle that has an area of 37.68 square centimeters? $\left(A = \pi r^2\right)$ (Use $\pi = 3.14$.)

35. ______________

36. __________

36. The perimeter of a triangle is 36 centimeters. If two sides are equal in length and the third side is 8 centimeters long, what is the length of each of the other two sides? ($P = a + b + c$)

37. __________

37. Real Estate. A real estate agent says that the current value of a home is $40,000 more than twice its value when it was new. If the current value is $200,000, what was the value of the home when it was new?

38. __________

38. Transportation. The BART Trans-Bay Tubes, underwater rapid transit tunnels below San Francisco Bay, are 19,008 feet long and are the longest underwater tunnels in North America. They are 2058 feet longer than three times the length of the Summer Tunnel under Boston Harbor. How long is the Summer Tunnel?

39. __________

39. Cost of Living. According to the U.S. Department of Agriculture, the projected average annual expenditure by a middle-income family to raise a child born in 2003 to age 18 is $13,863. What is the projected total expenditure for raising a child to age 18?

40. __________

40. Real Estate. According to the U.S. Census Department, in 2004 the median price of an owner occupied housing unit in Honolulu, Hawaii was $408,859. This was $49,823 more than four times the median price of a home in Detroit at that time. What was the median price of a home in Detroit?

41. __________

41. Geography. In the National Park System, the smallest National Seashore is Fire Island in New York, which is 19,579 acres. That is 3024 acres less than one sixth of the number of acres in the Gulf Islands in Florida and Mississippi, which make up the largest National Seashore Park. How many acres are there in the Gulf Islands?

Name ______________________ Section ____________ Date __________

ANSWERS

42. Home design. A saltbox-style house (see diagram) is characterized by a long sloping roof in back. If such a roof measures 60 feet along its peak, 25 feet along one edge from top to front, and has a total area of 3900 square feet, what is the length along one edge from top to back?

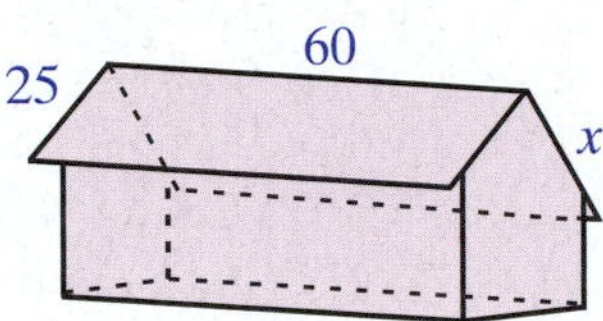

42. ____________

43. Auto Rental. Suppose that you are a traveling salesperson and your company has allowed you \$65 per day for car rental. You rent a car for \$45 per day plus 5¢ per mile. How many miles can you drive for the \$65 allowed?

Plan: Set up an equation relating the amount of money spent and the money allowed in your budget and solve the equation.

Solution: Let x = number of miles driven.

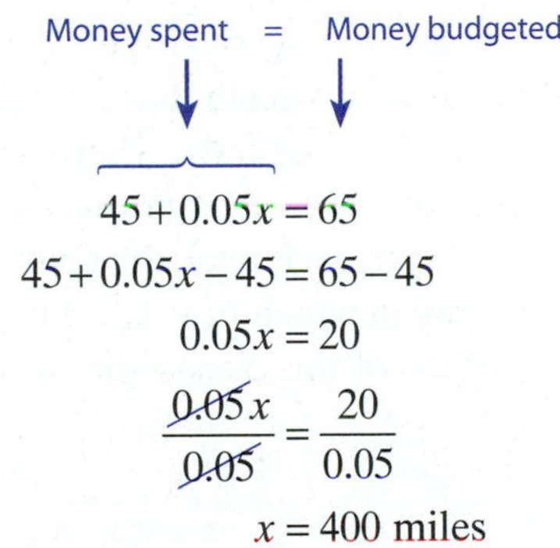

$$45 + 0.05x = 65$$
$$45 + 0.05x - 45 = 65 - 45$$
$$0.05x = 20$$
$$\frac{0.05x}{0.05} = \frac{20}{0.05}$$
$$x = 400 \text{ miles}$$

43. ____________

You would like to get the most miles for your money. The car rental agency down the street charges only \$40 per day and 8¢ per mile. How many miles would you get for your \$65 with this price structure? Is this a better deal for you?

44. Auto Rental. Under which of the following programs for a truck rental would you get the greatest number of miles for your \$200 budget if you want the truck for two days?

Plan 1: A fee of \$35 per day plus 50¢ per mile

Plan 2: A fee of \$55 per day plus 40¢ per mile

44. ____________

45. [Respond below exercise.]

Writing and Thinking about Mathematics

45. Auto Rental. If you planned to travel **a.** less than 400 miles or **b.** more than 400 miles with a truck under the rental conditions discussed in Exercise 44, which of the two programs would you use? Explain why.

Collaborative Learning

46. [Respond below exercise.]

46. As a result of his studies in discovery learning, George Pólya (1887–1985), a famous professor at Stanford University, developed the following four-step process as an approach to problem solving:

1. Understand the problem.
2. Devise a plan.
3. Carry out the plan.
4. Look back over the results.

With the class separated into teams of three to four students, each team is to discuss some nonmathematical problem that a member of the team solved today (for example: what to have for breakfast, where to study, what route to take to school). Can this person's thoughts and actions be described in terms of Pólya's four steps? As the team looks back over the solution, would there have been a better or more efficient way in which to solve this person's problem? The team leader is to report the results of this discussion to the class.

Chapter 9 Index of Key Terms and Ideas

Section 9.1 Introduction to Integers

Integers page 733

The set of **integers** is the set of whole numbers and their opposites:

Integers: $\ldots, -4, -3, -2, -1, 0, 1, 2, 3, 4, \ldots$

Opposites of Integers page 734

Note the following facts about integers:

1. The opposite of a positive integer is a negative integer. For example, $-(+2) = -2$ and $-(+7) = -7$.
2. The opposite of a negative integer is a positive integer. For example, $-(-3) = +3$ and $-(-4) = +4$.
3. The opposite of 0 is 0. (That is, $-0 = 0$.)

Symbols for Order page 735

$<$ is read "is less than"	$\leq$ is read "is less than or equal to"
$>$ is read "is greater than"	$\geq$ is read "is greater than or equal to"

Absolute Value page 737

The **absolute value** of an integer is its distance from 0.

The absolute value of an integer is never negative.

We can express the definition symbolically, for any integer a, as follows:

If a is a positive integer or 0, $|a| = a$.

If a is a negative integer, $|a| = -a$.

Section 9.2 Addition with Integers

Adding Integers Along a Number Line pages 743 – 744

Rules for Addition with Integers page 745

1. To add two integers with like signs, add their absolute values and use the common sign.
2. To add two integers with unlike signs, subtract their absolute values (the smaller from the larger) and use the sign of the integer with the larger absolute value.

Section 9.3 Subtraction with Integers

Additive Inverse page 751

The **opposite** of an integer is called its **additive inverse**. The sum of an integer and its additive inverse is 0. Symbolically, for any integer a, $a + (-a) = 0$.

Difference page 752

For any integers a and b, we define the **difference** between a and b as follows: $a - b = a + (-b)$.

This equation is read "a minus b is equal to a plus the opposite of b."

Section 9.4 Multiplication, Division, and Order of Operations with Integers

Rules for Multiplication with Integers page 760

1. The product of two positive integers (here, a and b) is **positive**:
   $a \cdot b = ab$
2. The product of two negative integers is **positive**:
   $(-a)(-b) = ab$
3. The product of a positive integer and a negative integer is **negative**.
   $a(-b) = -ab$
4. The product of 0 and any integer is **0**.
   $a \cdot 0 = 0$ and $(-a) \cdot 0 = 0$

Relating Division and Multiplication page 761

For integers a, b, and x (where $b \neq 0$),

$$\frac{a}{b} = x \quad \text{means that} \quad a = b \cdot x.$$

Rules for Division with Integers page 762

1. The quotient of two positive integers (here, a and b) is **positive**:
   $\frac{a}{b} = +\frac{a}{b}$
2. The quotient of two negative integers is **positive**: $\frac{-a}{-b} = +\frac{a}{b}$
3. The quotient of a positive integer and a negative integer is **negative**:
   $\frac{-a}{b} = -\frac{a}{b}$ and $\frac{a}{-b} = -\frac{a}{b}$

Rules for Order of Operations page 762

Division by 0 is Not Defined page 763

Case 1: Suppose that $a \neq 0$ and $\frac{a}{0} = x$. Then, by the meaning of division, $a = 0 \cdot x$. But this is not possible since $0 \cdot x = 0$ for any value of x and we stated that $a \neq 0$.

Case 2: Suppose that $\frac{0}{0} = x$. Then $0 = 0 \cdot x$, which is true for all values of x. But this is not allowed since we must have a unique answer for division.

Therefore, we conclude that, in every case, **division by 0 is not defined.**

Section 9.5 Combining Like Terms and Evaluating Algebraic Expressions

Constants, Variables, Terms page 769

A single number is called a **constant**, and a symbol (usually a letter) used to indicate an unknown number is called a **variable**. Any constant or varible or the indicated product of constants and powers of variables is called a **term**.

Coefficient page 769

In the term $-4x^2$, the constant -4 is called the **numerical coefficient of x^2**.

Continued on next page...

Section 9.5 Combining Like Terms and Evaluating Algebraic Expressions (cont.)

Like Terms page 769

Like terms (or **similar terms**) are terms that contain the same variables (if any) raised to the same powers.

Distributive Property of Multiplication over Addition page 770

For any numbers a, b, and c, $a(b + c) = ab + ac$.

To Evaluate an Algebraic Expression page 772

1. Combine like terms.
2. Substitute the values given for any variables. (**Note:** To indicate multiplication, enclose the numbers substituted in parentheses.)
3. Follow the rules for order of operations.

Section 9.6 Translating English Phrases and Solving Equations

Translating English Phrases into Algebraic Expressions pages 777 – 778

Translating Algebraic Expressions into English Phrases page 778

First-degree Equation in x page 779

A **first-degree equation in x** (or **linear equation in x**) is any equation that can be written in the form

$\boldsymbol{ax + b = c}$ where a, b, and c are constants and $a \neq 0$.

Principles Used in Solving a First-Degree Equation page 780

In the four basic principles stated here, A and B represent algebraic expressions or constants.

1. **The Addition Principle:**
   If $A = B$ is true, then $A + C = B + C$ is also true for any number C. (The same number may be added to both sides of an equation.)
2. **The Subtraction Principle:**
   If $A = B$ is true, then $A - C = B - C$ is also true for any number C. (The same number may be subtracted from both sides of an equation.)
3. **The Multiplication Principle:**
   If $A = B$ is true, then $C \cdot A = C \cdot B$ is also true for any number C. (Both sides of an equation may be multiplied by the same number.)
4. **The Division Principle:**
   If $A = B$ is true, then $\frac{A}{C} = \frac{B}{C}$ is also true for any nonzero number C.
   (Both sides of an equation may be divided by the same nonzero number.)

Continued on next page...

Section 9.6 Translating English Phrases and Solving Equations (cont.)

General Guidelines for Solving Equations page 781

1. The goal is to isolate the variable on one side of the equation (either the right side or the left side).
2. First use the Addition and Subtraction Principles whenever numbers are subtracted or added to the variable.
3. Use the Multiplication and Division Principles to make 1 (or +1) the coefficient of the variable.

Checking Solutions to Equations page 782

Checking can be done by substituting the solution found back into the original equation to see if the resulting statement is true.

Section 9.7 Problem Solving

Basic Steps for Solving Number Problems page 791

1. Read the problem carefully at least twice. Look for key words and phrases that can be translated into algebraic expresions.
2. Assign a variable as the unknown quantity and form an equation using the expressions you translated.
3. Solve the equation.
4. Look back over the problem and check that the answer is reasonable.

Formulas in Geometry page 793

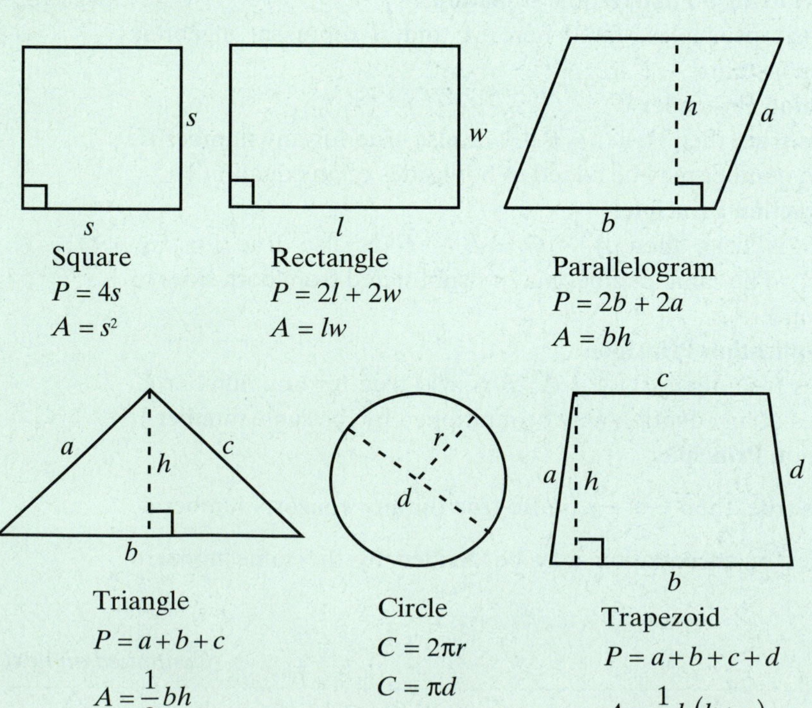

Square
$P = 4s$
$A = s^2$

Rectangle
$P = 2l + 2w$
$A = lw$

Parallelogram
$P = 2b + 2a$
$A = bh$

Triangle
$P = a + b + c$
$A = \frac{1}{2}bh$

Circle
$C = 2\pi r$
$C = \pi d$
$A = \pi r^2$

Trapezoid
$P = a + b + c + d$
$A = \frac{1}{2}h(b + c)$

Name ______________________ Section ____________ Date __________

ANSWERS

1. ____________
2. ____________
3. ____________
4. ____________
5. ____________
6. ____________
7. ____________
8. ____________
9. ____________
10. ____________
11. ____________
12. ____________
13. ____________
14. ____________
15. ____________
16. ____________
17. ____________
18. ____________

Chapter 9 Review Questions

Find the opposite of each integer.

1. -4 **2.** -8 **3.** $|-21|$

Graph each of the following sets of integers on a number line.

4. $\{-1, 0, 2\}$ **5.** $\{-2, -1, 0, |-1|\}$ **6.** $\{-5, -4, 1, |-2|, |3|\}$

Fill in the blank with the appropriate symbol: <, >, or =.

7. -4 ____ $|-4|$ **8.** -5 ____ -10 **9.** $-(-6)$ ____ $|-6|$

Determine whether each statement is true or false. If the statement is false, rewrite it in a true form. (There may be more than one change that will make a false statement true.)

10. $-(-1) \le 1$ **11.** $-18 < -12$ **12.** $|-5| \le -5$

Find each of the indicated sums.

13. $(-14) + (-18)$ **14.** $(-17) + (+12)$ **15.** $(-15) + (+4) + (-9)$

16. $(+14) + (-23) + (+9)$ **17.** $(-2) + (-1) + (-10) + (-3)$

18. $(-5) + (-8) + (+19) + (-6)$

19. ______

20. ______

21. ______

22. ______

23. ______

24. ______

25. ______

26. ______

27. ______

28. ______

29. ______

30. ______

31. ______

32. ______

33. ______

34. ______

35. ______

36. ______

37. ______

38. ______

39. ______

40. ______

41. ______

42. ______

19. $\begin{array}{r} -312 \\ -422 \\ \underline{+610} \end{array}$

20. $\begin{array}{r} -763 \\ +1101 \\ \underline{+224} \end{array}$

Find each of the indicated differences.

21. $(+12)-(+7)$ **22.** $(+8)-(-8)$ **23.** $(-16)-(+5)$ **24.** $(-9)-(-9)$

25. $\begin{array}{r} 16 \\ \underline{-(-10)} \end{array}$

26. $\begin{array}{r} 16 \\ \underline{-(+10)} \end{array}$

27. $\begin{array}{r} -62 \\ \underline{-(-12)} \end{array}$

28. $\begin{array}{r} -62 \\ \underline{-(+12)} \end{array}$

Perform the indicated operations.

29. $-12+5-3$ **30.** $5-2-1-6$ **31.** $-8-14+29-7$

32. $12(-6)$ **33.** $-4(-16)$ **34.** $(-2)(-6)(7)$

35. $(-4)(17)(0)(-3)$ **36.** $3(4)(-1)(-1)(-2)$ **37.** $(-5)^3$

38. $\frac{-51}{3}$ **39.** $\frac{-24}{-8}$ **40.** $\frac{0}{-73}$

41. $\frac{-52}{0}$ **42.** $\frac{450}{-15}$

Name ______________________________ Section ______________ Date ___________

ANSWERS

Evaluate each of the following expressions by using the rules for order of operations.

43. $15+2(-5)$ **44.** $-10+4(-3)$ **45.** $-3(5-7)-6\cdot 2$

46. $3(-4)+12\div 2$ **47.** $(-4)^2\div(8-2\cdot 3)$ **48.** $(-5)^2-16\div 2\cdot 4$

49. $\left[4+3(1-2^2)\right]\div 4$ **50.** $\left[6-3(2-3^2)\right]\div(-4)$

Fill in each blank with the correct term: (positive, negative, zero, undefined).

51. If any number is divided by zero, the quotient is ___________.

52. The product of a negative integer and zero is ___________.

53. The quotient of two negative integers is ___________.

54. The product of two negative integers is ___________.

Simplify each of the following algebraic expressions by combining like terms whenever possible.

55. $6y+y$ **56.** $-12x+x$ **57.** $5a^2-a^2+3a$

58. $2(x+4)+3(x-5)$ **59.** $y^2+5y-10$ **60.** $7c-12c+4c$

43. ______________
44. ______________
45. ______________
46. ______________
47. ______________
48. ______________
49. ______________
50. ______________
51. ______________
52. ______________
53. ______________
54. ______________
55. ______________
56. ______________
57. ______________
58. ______________
59. ______________
60. ______________

61. ______

62. ______

63. ______

64. ______

65. ______

66. ______

67. ______

68. ______

69. ______

70. ______

71. ______

72. ______

73. ______

74. ______

75. ______

Evaluate each of the following expressions for x = 3 and y = − 2.

61. $3x+5x-7$

62. $6y-2y+8$

63. $y^2+y^2-5y+16y$

64. $-3x^2-x^2-4x$

Write an algebraic expression described by each of the following English phrases.

65. 16 more than the product of a number and 9.

66. –25 increased by the quotient of a number and 2.

Solve each of the following equations.

67. $y+10=26$

68. $x-13=12$

69. $2n=-60$

70. $-35m=-70$

71. $3(x-1)=x+15$

72. $5n+20=2(n-5)$

73. Three times a number plus 15 is equal to –18. What is the number?

74. Twice a number decreased by 7 is equal to 4 less than the number. What is the number?

75. A cereal box is in the shape of a rectangular solid. If the volume of the box is 3600 cubic centimeters and the base of the box is 5 cm by 18 cm, what is the height of the box? ($V=lwh$)

Name ______________ Section ______________ Date ______________

ANSWERS

76. If the area of a square is 40 square feet, what is the length of one side of the square? (Use a calculator to find the answer to the nearest tenth of a foot.) $(A = s^2)$

76. ______________

77. The perimeter of a soccer field (rectangular in shape) is 320 yards. If the length is 110 yards, what is the width of the field? $(P = 2l + 2w)$

77. ______________

78. In any triangle, the sum of the measures of the angles equals $180°$. If two of the angles have measures of $20°$ and $70°$, what is the measure of the third angle?

78. ______________

Name ______________________ Section ____________ Date __________

ANSWERS

Chapter 9 Test

1. Graph the following set of integers on a number line.
$\{-3, -1, 0, |-2|, |-4|\}$

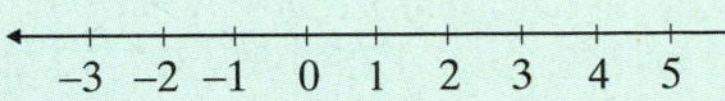

2. The opposite of –11 is ________.

3. State whether each of the following statements is true or false. If the statement is false, rewrite it in a true form.

a. $-12 > -15$ **b.** $|-15| \leq |-12|$

Perform the indicated operations.

4. $(+14) + (-6)$ **5.** $(-9) - (+3)$

6. $(-3) + (+7) + (-4) + (-9)$ **7.** $-10 - 5 - 1$

8. $17 - 4 - 12 + 15$ **9.** $(-4)(-7)$

10. $(-3)(-5)(2)(-3)$ **11.** $\dfrac{-56}{-7}$

12. $\dfrac{-3}{0}$ **13.** $\dfrac{65}{-13}$

1. [Respond in exercise.]

2. __________

3. a. __________

b. __________

4. __________

5. __________

6. __________

7. __________

8. __________

9. __________

10. __________

11. __________

12. __________

13. __________

14. __________

15. __________

16. __________

17. __________

18. a. __________

b. __________

19. a. __________

b. __________

20. __________

21. __________

22. __________

23. __________

24. __________

25. __________

Use the rules for order of operations to evaluate each of the following expressions.

14. $(14-20)(12-15)$

15. $(-6)^2 \div 12 \cdot 4$

16. $14+16\div 4\cdot 2-13\cdot 2$

17. $\left[5+3\left(1-2^2\right)\right]\div(-2)$

18. Simplify each of the following expressions by combining like terms.

a. $4x-5-x+10$

b. $5x^2+x-3+x^2-4x+8$

19. **a.** Simplify the following expression, and
b. Evaluate it for $x=2$ and $y=-1$.

$2x+2\,(y+3)-(3y+4)+3(x-3)$

Solve each of the following equations.

20. $5x+1=36$

21. $3x-4=-19$

22. $y-8=5y-32$

23. $3(n+3)=2(7-n)$

24. If the product of a number and 4 is decreased by 10, the result is 50. Find the number.

25. Four times the difference between a number and 7 is equal to 2 plus the number. Find the number.

Name ______________________ Section ____________ Date __________

ANSWERS

26. What is the number whose product with 6 is equal to twice the number minus 12?

26. __________

27. The perimeter of a rectangle is 60 feet. If the length is 20 feet, what is the width of the rectangle? $(P = 2l + 2w)$

27. __________

28. The base of a triangle is 30 centimeters long and its area is 210 square centimeters. What is the height of the triangle? $\left(A = \frac{1}{2}bh\right)$

28. __________

Name ____________ Section ____________ Date ____________

Cumulative Review: Chapters 1 – 9

Perform the indicated operations and simplify.

1. $1-\frac{1}{16}$

2. $\frac{3}{10}-\frac{2}{15}$

3. $25\frac{1}{6} \times 13\frac{3}{8}$

4. $340\frac{1}{10} - 150\frac{3}{5}$

5. $\left(4-\frac{1}{2}\right)\div\left(\frac{1}{24}+\frac{1}{6}\right)$

6. Multiply: (45.8)(16.73)

7. Divide (to the nearest hundredth): $176 \div 32.1$

Evaluate the following expressions. Reduce all fractions.

8. $\frac{16}{0}$

9. $\frac{1}{2}+\frac{3}{4}\cdot\left(\frac{1}{2}\right)^2-\frac{1}{16}$

10. $(-7)^2\cdot 5+2(-6)\div 4$

Choose the correct answer by reasoning and calculating mentally.

11. Write 7% as a fraction.

(a) $\frac{7}{1}$ (b) $\frac{7}{10}$ (c) $\frac{7}{100}$ (d) $\frac{7}{1000}$

12. Find 20% of 730.

(a) 7.3 (b) 73 (c) 14.6 (d) 146

ANSWERS

1. ____________
2. ____________
3. ____________
4. ____________
5. ____________
6. ____________
7. ____________
8. ____________
9. ____________
10. ____________
11. ____________
12. ____________

13. ______

14. ______

15. ______

16. ______

17. ______

18. ______

19 a. ______

b. ______

13. What percent of 100 is 25?
(a) 0.25% (b) 2.5% (c) 25% (d) 250%

14. 245 is approximately 50% of what number?
(a) 120 (b) 500 (c) 1000 (d) 2000

15. 135% of ______ is 270.

16. 25% of 200 is ______.

17. Find the average (mean) of the numbers 5000, 6250, 9475, and 8672.

18. A triangle has sides of 15 ft, 20 ft, and 25 ft. Is this a right triangle? Explain.

19. ΔABC and ΔDEF are similar right triangles with sides and angles as marked.
a. What are the measures of angles 1 and 2?
b. What are the values of x and y?

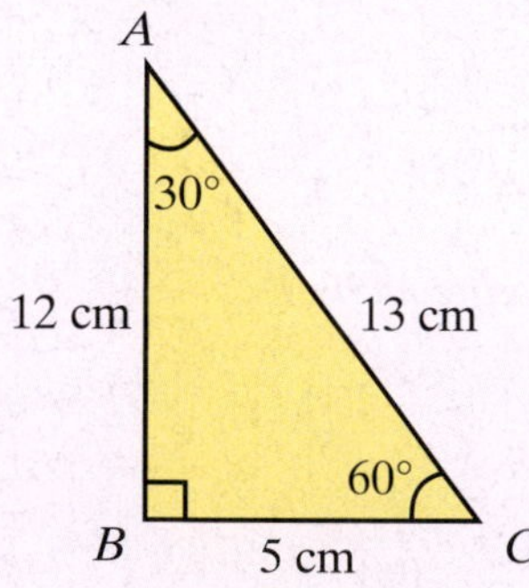

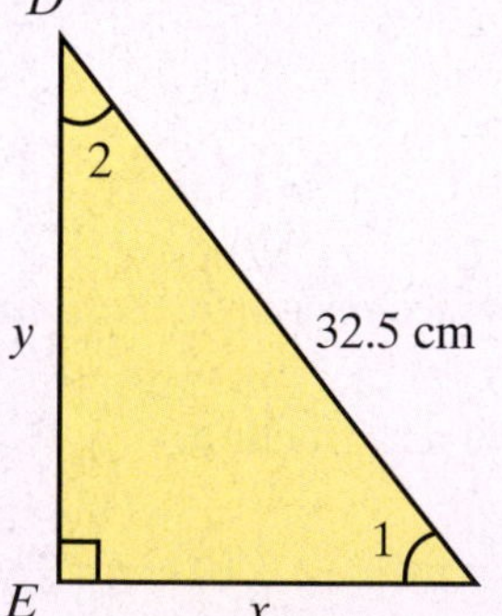

Name ______________________ Section ____________ Date __________

ANSWERS

20. Food. A pepperoni and cheese pizza has a diameter of 12 inches and a cheese only pizza has a diameter of 14 inches. What is the difference in the areas of the two pizzas? (Use $\pi = 3.14$.)

21. Use your calculator to find the value of the following radical expressions accurate to four decimal places.

a. $\sqrt{600}$ **b.** $2\sqrt{10}+\sqrt{90}$ **c.** $3\sqrt{2}-5\sqrt{3}$

22. Name each of the triangles shown given the indicated measures of angles and lengths of sides.

a.

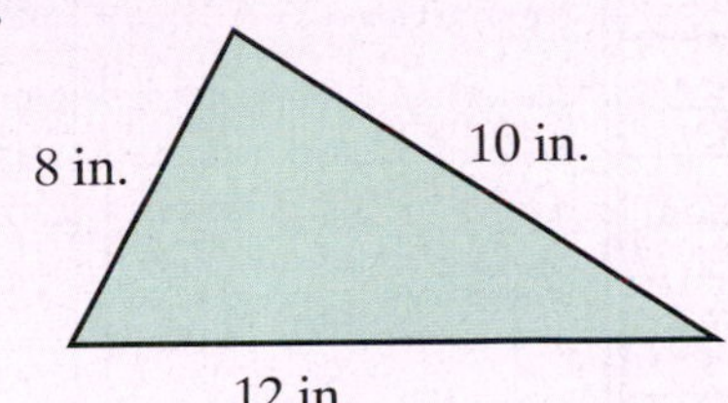

b.

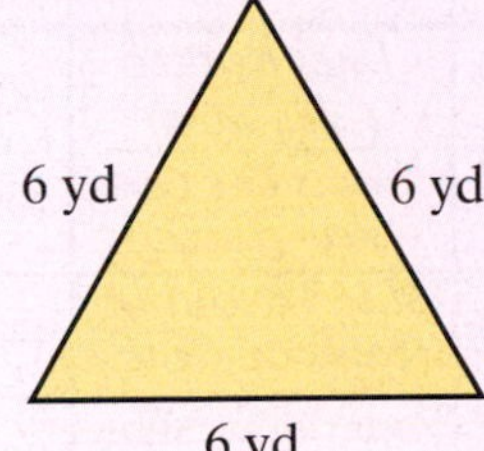

c.

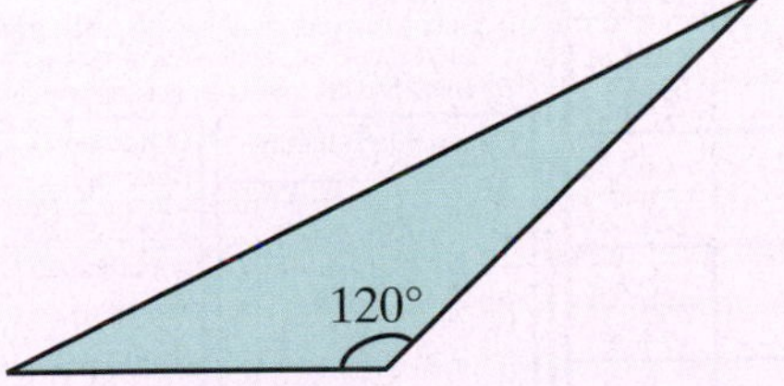

d.

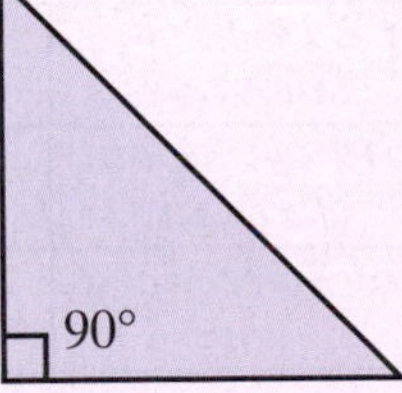

23. Find the volume of the cone with the indicated measures. (Use the formula $V = \frac{1}{3}\pi r^2 h$ and $\pi = 3.14$.)

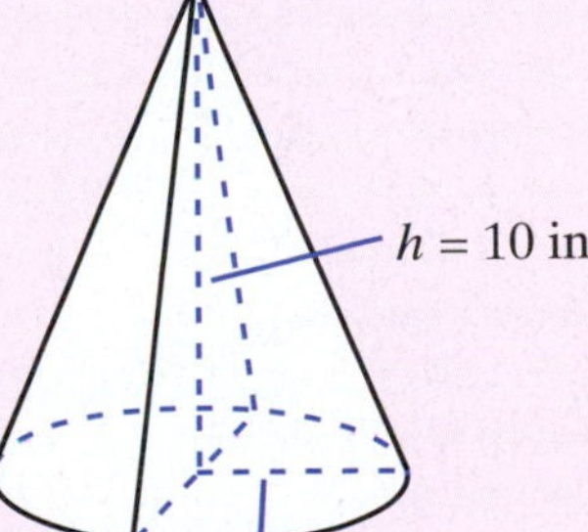

20. ________

21 a. ________

b. ________

c. ________

22 a. ________

b. ________

c. ________

d. ________

23. ________

24 a. ______

24. Find **a.** the perimeter and **b.** the area of the figure shown.

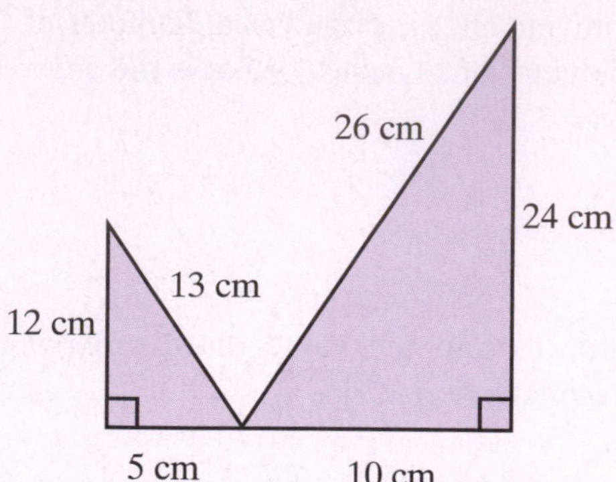

b. ______

25. Find the balance in the checking account with the given information:

YOUR CHECKBOOK REGISTER

Check No.	Date	Transaction Description	Payment (−)	(√)	Deposit (+)	Balance
			Balance brought forward			2615.59
1210	3-7	Jones Kennel (dog care)	124.00	√		− 124.00 2491.59
1211	3-7	Simms Car Care (new clutch)	601.40	√		− 601.40 1890.19
1212	3-8	Blue Mountain (camping tent)	102.25	√		− 102.25 1787.94
1213	3-8	Ann's Craft Shop (craft supplies)	15.50			− 15.50 1772.44
	3-9	Deposit		√	600.00	+600 2372.44
1214	3-14	Kitchen Help (Cateri	858.34	√		− 858.34 1514.10
1215	3-24	EZ Lawn Care (lawn care)	45.50	√		− 45.50 1468.60
1216	3-24	Henson Repair (fixed tv)	40.00	√		− 40.00 1428.60
1217	3-31	Red River Mgmt. (rent)	800.00			− 800.00 628.60
	3-31	Service Charge	⊖			
			True Balance			

BANK STATEMENT
Checking Account Activity

Transaction Description	Amount	(√)	Running Balance	Date
Beginning Balance	0.00	√	2615.59	3-01
Check #1210	124.00	√	2491.59	3-07
Check #1211	601.40	√	1890.19	3-09
Check #1212	102.25	√	1787.94	3-09
Check #1213	15.50	√	1772.44	3-11
Deposit	600.00	√	2372.44	3-11
Check #1214	858.34	√	1514.10	3-15
Check #1215	45.50	√	1468.60	3-27
Check #1216	40.00	√	1428.60	3-28
Ending Balance			1428.60	

RECONCILIATION SHEET

A. First, mark √ beside each check and deposit listed in both your checkbook register and on the bank statement.
B. Second, in your checkbook register, add any interest paid and subtract any service charge listed.
C. Third, find the total of all outstanding checks.

Outstanding Checks		Statement Balance	______
No.	Amount	Add deposits not credited	+ ______
		Total	______
	______	Subtract total amount of checks outstanding	− ______
Total	______	True Balance	______

25. [Respond in exercise.]

26. ______

26. Interest. If you borrow \$5000 from your aunt and agree to repay her the principal plus 5.5% simple interest in one year, how much will you pay her at the end of the year?

Name ______________________ Section ______________ Date __________

ANSWERS

27. ______________

28. ______________

29 a. ______________

b. ______________

c. ______________

30. ______________

31. ______________

32. ______________

33. ______________

34. ______________

27. Compound interest. Using the formula $A = P\left(1+\frac{r}{n}\right)^{nt}$, find the value of \$10,000 invested at 6% compounded daily for 5 years. (Use $n = 360$.)

28. Travel. If a car averages 23.6 miles per gallon, how many miles will it go on 15 gallons of gas?

29. Retail. The cost of a sofa to a furniture store was \$750. The sofa was sold for \$1250.

a. What was the store's profit?

b. What was the percent of profit based on cost?

c. What was the percent of profit based on selling price?

30. Geography. The scale on a map indicates that $1\frac{3}{4}$ inches represents 50 miles. How many miles apart are two cities that are 2 inches apart on the map?

Solve each of the following equations.

31. $2x+1=-11$

32. $5(n-1)=3n+15$

33. $33=-5y-7$

34. $9x-20=7(x-4)$

35. ______________

35. If one is added to a number, the result is equal to eleven more than three times the number. What is the number?

36. ______________

36. The perimeter of a rectangle is 100 yards. If the width is 15 yards, what is the length of the rectangle? $(P = 2l + 2w)$

APPENDIX I

THE METRIC SYSTEM

I.1 Metric System: Weight and Volume

Mass (Weight)

Mass is the amount of material in an object. Regardless of where the object is in space, its mass remains the same. (See Figure I.1.) **Weight** is the force of the Earth's gravitational pull on an object. The farther an object is from Earth, the less the gravitational pull of the Earth. Thus astronauts experience weightlessness in space, but their mass is unchanged.

The two objects have the same **mass** and balance on an equal arm balance, regardless of their location in space.

Figure I.1

Because most of us do not stray far from the Earth's surface, weight and mass will be used interchangeably in this text. Thus a **mass** of 20 kilograms will be said to **weigh** 20 kilograms.

The basic unit of mass in the metric system is the **kilogram**,* about 2.2 pounds. In some fields, such as medicine, the **gram** (about the mass of a paper clip) is more convenient as a basic unit than the kilogram.

Large masses, such as loaded trucks and railroad cars, are measured by the **metric ton** (1000 kilograms, or about 2200 pounds). (See Tables I.1 and I.2.)

* Technically, a kilogram is the mass of a certain cylinder of platinum-iridium alloy kept by the International Bureau of Weights and Measures in Paris.
Originally, the basic unit was a gram, defined to be the mass of 1 cm^3 of distilled water at 4° Celsius. This mass is still considered accurate for many purposes, so that
1 cm^3 of water has a mass of 1 g.
1 dm^3 of water has a mass of 1 kg.
1 m^3 of water has a mass of 1000 kg, or 1 metric ton.

Table I.1	Measures of Mass
1 **milli**gram (mg)	= 0.001 gram
1 **centi**gram (cg)	= 0.01 gram
1 **deci**gram (dg)	= 0.1 gram
1 gram (g)	= 1.0 gram
1 **deka**gram (dag)	= 10 grams
1 **hecto**gram (hg)	= 100 grams
1 **kilo**gram (kg)	= 1000 grams
1 metric ton (t)	= 1000 kilograms

Table I.2	Equivalent Measures of Mass
1000 mg = 1 g	0.001 g = 1 mg
1000 g = 1 kg	0.001 kg = 1 g
1000 kg = 1 t	0.001 t = 1 kg
1t = 1000 kg = 1,000,000 g = 1,000,000,000 mg	

The centigram, decigram, dekagram, and hectogram have little practical use and are not included in the exercises. For completeness, they are all included in the headings of the charts used to change units.

A chart can be used to change from one unit of mass to another.

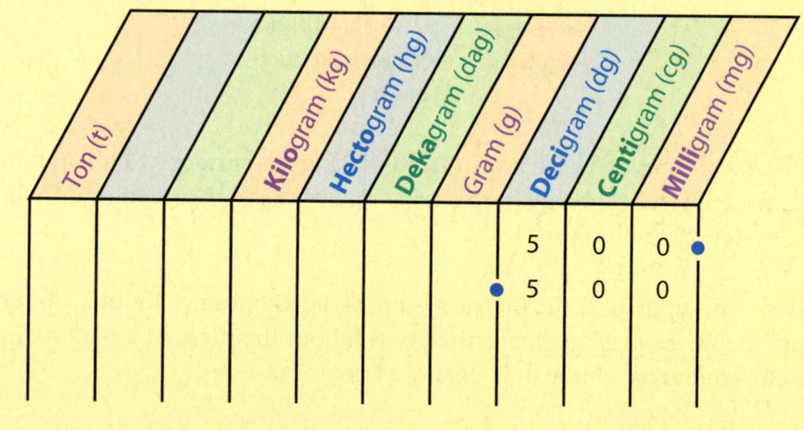

1. List each unit across the top.
2. Enter the given number so that there is **one** digit in each column with the decimal point on the given unit line.
3. Move the decimal point to the desired unit line.
4. Fill in the spaces with 0's using one digit per column.

The use of the chart below shows how the following equivalent measures can be found.

Example 1

a.	23 mg	=	0.023 g
b.	6 g	=	6000 mg
c.	49 kg	=	49,000 g
d.	5 t	=	5000 kg
e.	70 kg	=	0.07 t

Now work exercise 1 in the margin.

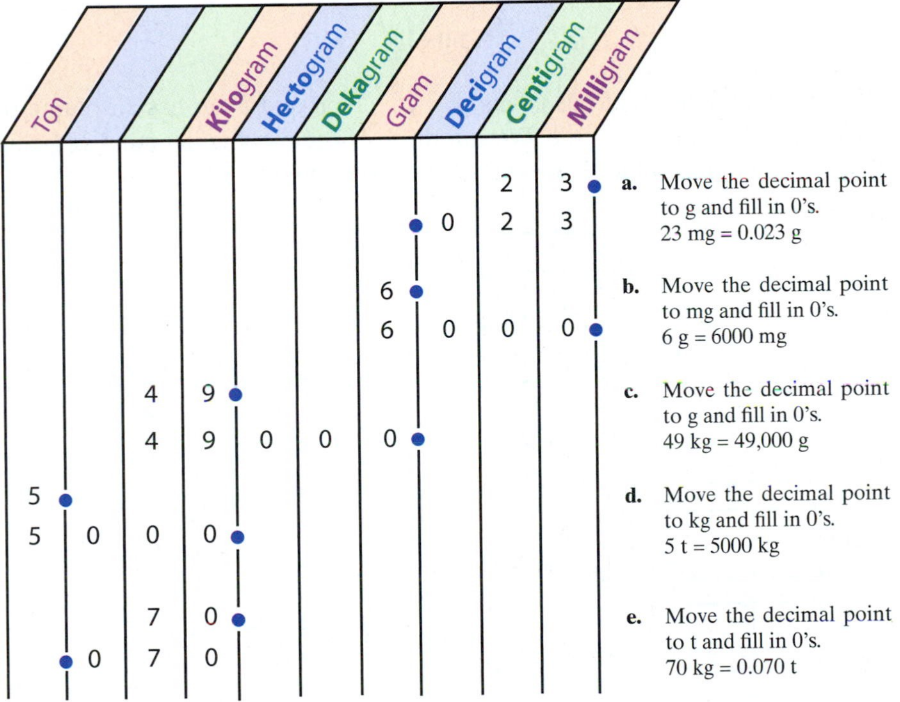

Ton
Kilogram
Hectogram
Dekagram
Gram
Decigram
Centigram
Milligram
a. Move the decimal point to g and fill in 0's. 23 mg = 0.023 g
b. Move the decimal point to mg and fill in 0's. 6 g = 6000 mg
c. Move the decimal point to g and fill in 0's. 49 kg = 49,000 g
d. Move the decimal point to kg and fill in 0's. 5 t = 5000 kg
e. Move the decimal point to t and fill in 0's. 70 kg = 0.070 t

Make your own chart on a piece of paper and see if you agree with the results in Example 2.

Example 2

a.	60 mg	=	0.06 g
b.	135 mg	=	0.135 g
c.	5700 kg	=	5.7 t
d.	100 g	=	0.1 kg
e.	78 kg	=	78,000 g

Now work exercise 2 in the margin.

1. a. Change 121 kilograms to grams

b. Change 3500 kilograms to tons.

c. Change 4,576,000 grams to tons.

d. Change 6700 milligrams to grams.

2. a. Change 43 kilograms to grams.

b. Change 250 kilograms to tons.

c. Change 23 grams to milligrams.

Volume

Volume is a measure of the space enclosed by a three-dimensional figure and is measured in **cubic units**. The volume or space contained within a cube that is 1 centimeter on each edge is **one cubic centimeter**, or 1 cm^3, as shown in Figure I.2. A cubic centimeter is about the size of a sugar cube.

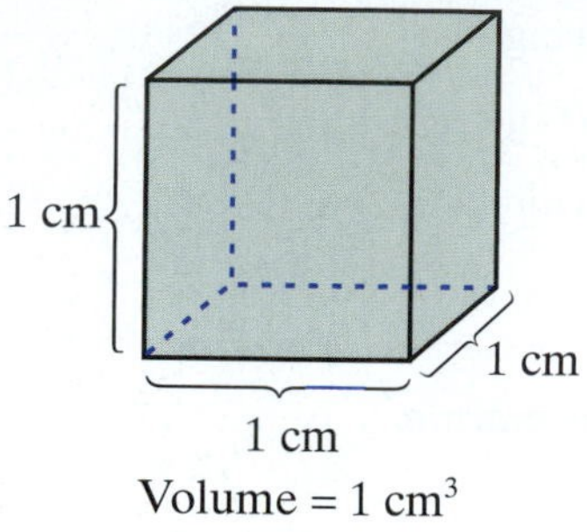

Volume = 1 cm^3

Figure I.2

A rectangular solid that has edges of 3 cm, 2 cm and 5 cm has a volume of 3 cm × 2 cm × 5 cm = 30 cm^3. We can think of the rectangular solid as being three layers of ten cubic centimeters, as shown in Figure I.3.

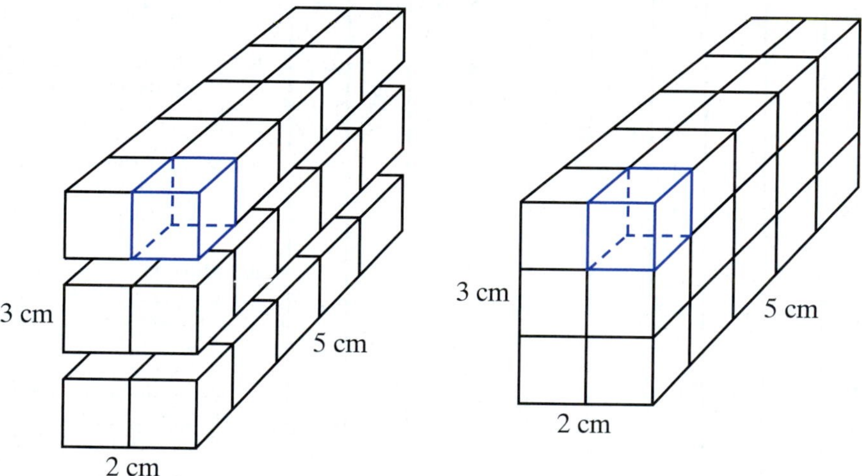

Figure I.3

If a cube is 1 decimeter along each edge, then the volume of the cube is 1 cubic decimeter (or 1 dm^3). In terms of centimeters, this same cube has volume

$$10 \text{ cm} \times 10 \text{ cm} \times 10 \text{ cm} = 1000 \text{ cm}^3.$$

That is, as shown in figure I. 4,

$$1 \text{ dm}^3 = 1000 \text{ cm}^3.$$

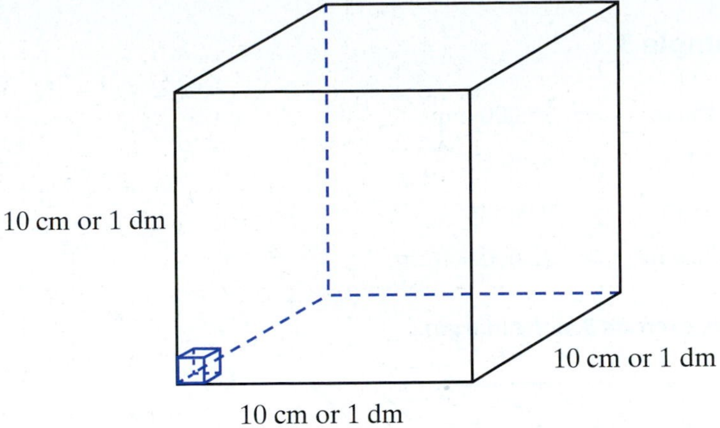

$1 \text{ dm}^3 = 1000 \text{ cm}^3$

Figure I.4

This relationship is true of cubic units in the metric system: equivalent cubic units can be found by multiplying the larger units by 1000 to get the smaller units. Again, we can use a chart; however, this time **there must be three digits in each column**.

A chart can be used to change from one unit of volume to another.

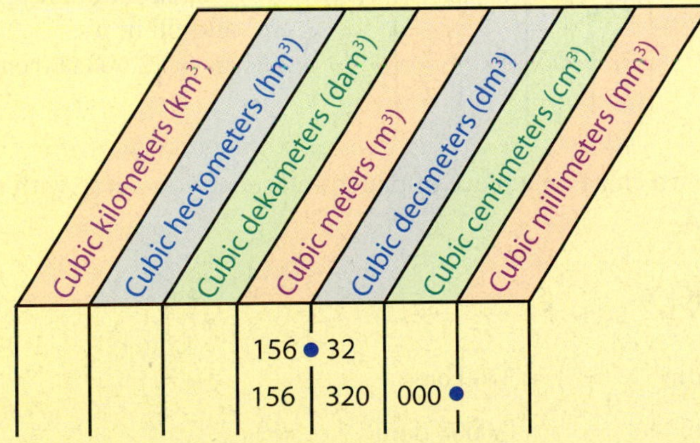

$156.32 \text{ m}^3 = 156\,320\,000 \text{ cm}^3$

1. List each volume unit across the top. (Abbreviations will do.)
2. Enter the given number so that there are **three** digits in each column with the decimal point on the given unit line.
3. Move the decimal point to the desired unit line.
4. Fill in the spaces with 0's using three digits per column.

The chart below shows how the following equivalent measures can be found.

3. a. Change 750 cubic millimeters to cubic centimeters.

b. Change 19 cubic decimeters to cubic centimeters.

c. Change 1.6 cubic meters to cubic centimeters.

d. Change 63.7 cubic meters to cubic centimeters.

Example 3

a. 15 cm³ = 15,000 mm³

b. 4.1 dm³ = 4100 cm³

c. 8 dm³ = 0.008 m³

d. 22.6 m³ = 22,600,000 cm³

Now work exercise 3 in the margin.

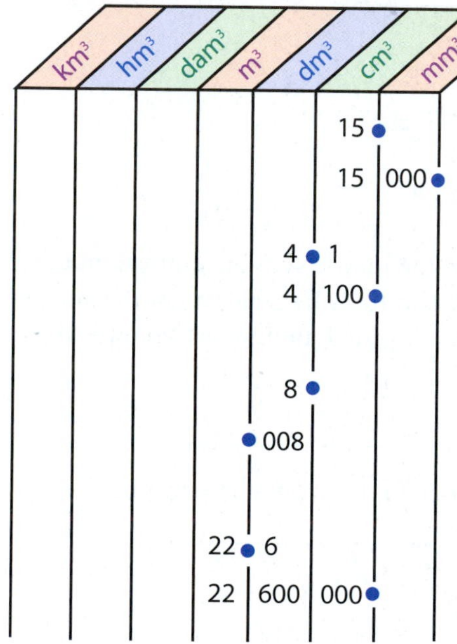

a. Move decimal point to mm³ and fill in 0's.
15 cm³ = 15,000 mm³

b. Move decimal point to cm³ and fill in 0's.
4.1 dm³ = 4100 cm³

c. Move decimal point to m³ and fill in 0's.
8 dm³ = 0.008 m³

d. Move decimal point to cm³ and fill in 0's.
22.6 m³ = 22,600,000 cm³

Make your own chart on a piece of paper and see if you agree with the results in Example 4.

4. a. Change 6.3 cubic centimeters to cubic meters.

b. Change 192 cubic decimeters to cubic centimeters.

Example 4

a. 3.7 dm³ = 3700 cm³

b. 0.8 m³ = 0.0008 dam³

c. 4 m³ = 4000 dm³ = 4,000,000 cm³ = 4,000,000,000 mm³

Now work exercise 4 in the margin.

Volumes measured in cubic kilometers are so large that they are not used in everyday situations. Many scientists work with these large volumes. More practically, we are only interested in m³, dm³, cm³, and mm³.

Liquid Volume

Liquid volume is measured in **liters** (abbreviated L). You are probably familiar with 1 L and 2 L bottles of soda on your grocer's shelf. A **liter** is the volume enclosed in a cube that is 10 cm on each edge. So 1 liter is equal to

$$10 \text{ cm} \times 10 \text{ cm} \times 10 \text{ cm} = 1000 \text{ cm}^3 \quad \text{or} \quad 1 \text{ liter} = 1000 \text{ cm}^3$$

That is, the cubic box shown in Figure I.4 would hold 1 liter of liquid.

The prefixes kilo-, hecto-, deka-, deci-, centi-, and milli- all indicate the same part of a liter as they do of the meter. **One digit per column** will be helpful for changing units. The centiliter (cL), deciliter(dL), and dekaliter (daL) are not commonly used and are not included in the tables or exercises.

Table I.3 Measures of Liquid Volume

1 **milli**liter (mL)	=	0.001 liter
1 liter (L)	=	1.0 liter
1 **hecto**liter (hL)	=	100 liters
1 **kilo**liter (kL)	=	1000 liters

Table I.4 Equivalent Measures of Volume

1000 mL	=	1 L	1 mL	=	1 cm^3
1000 L	=	1 kL	1 L	=	1 dm^3
10 hL	=	1 kL	1 kL	=	1 m^3

The use of the chart below shows how the following equivalent measures can be found.

Example 5

a. 6 L = 6000 mL = 0.06 hL

b. 500 mL = 0.5 L

c. 3 kL = 3000 L

d. 72 hL = 7.2 kL

Now work exercise 5 in the margin.

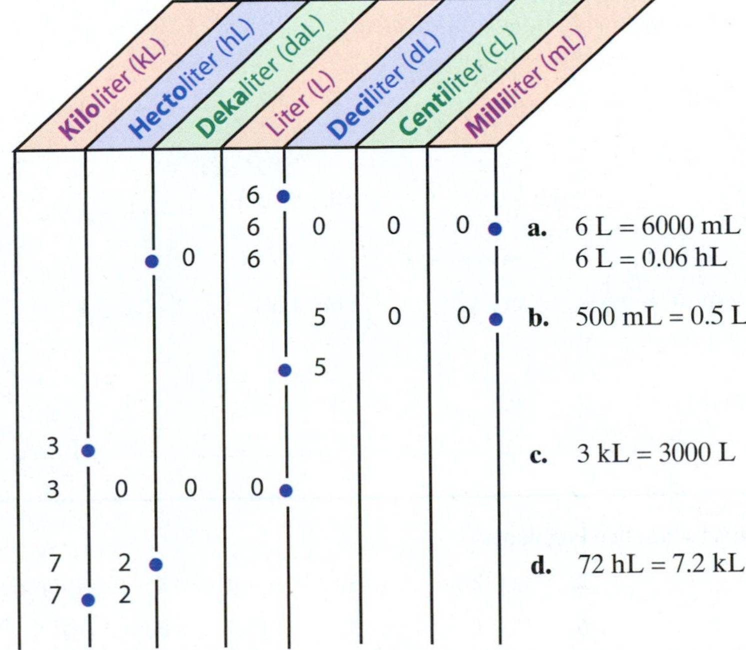

5. a. Change 9.37 liters to hectoliters.

b. Change 353 milliliters to liters.

c. Change 12 kiloliters to liters.

There is an interesting "crossover" relationship between liquid volume measures and cubic volume measures. Since

$$1 \text{ L} = 1000 \text{ mL} \quad \text{and} \quad 1\text{L} = 1000 \text{ cm}^3$$

we have

$$\mathbf{1 \text{ mL} = 1 \text{ cm}^3}$$

Also,

$$1 \text{ kL} = 1000 \text{ L} = 1{,}000{,}000 \text{ cm}^3 \quad \text{and} \quad 1{,}000{,}000 \text{ cm}^3 = 1 \text{ m}^3$$

This gives

$$\mathbf{1 \text{ kL} = 1000 \text{ L} = 1 \text{ m}^3}$$

Use Table I.4 on page A7 to confirm the following conversions.

6. a. Change 1952 milliliters to liters.

b. Change 124 milliliters to cubic centimeters.

c. Change 19.75 kiloliters to cubic meters.

Example 6

a. 6000 mL = 6 L

b. 3.2 L = 3200 mL

c. 60 hL = 6 kL

d. 637 mL = 0.637 L

e. 70 mL = 70 cm^3

f. 3.8 kL = 3.8 m^3

Now work exercise 6 in the margin.

Practice Problems

Change the following units as indicated.

1. 500 mg = ________ g

2. 43 g = ________ mg

3. 62 g = ________ kg

4. 18 cm^3 = ________ mm^3

5. 7.9 dm^3 = ________ m^3

6. 2 mL = ________ L

7. 500 mL = ________ L

8. 16 mL = ________ cm^3

Answers to Practice Problems:

1. 0.5 g **2.** 43,000 mg **3.** 0.062 kg **4.** 18,000 mm^3

5. 0.0079 m^3 **6.** 0.002 L **7.** 0.5 L **8.** 16 cm^3

Name ______________________ Section ____________ Date __________

ANSWERS

Exercises I.1

Change the following units of mass (weight) as indicated.

1. 2 g = ________ mg

2. 7 kg = ___________ g

3. 3700 kg = ________ t

4. 34.5 mg = _________ g

5. 5600 g = ________ kg

6. 4000 kg = _________ t

7. 91 kg = _________ t

8. 73 kg = ___________ mg

9. 0.7 g = _________ mg

10. 0.54 g = ___________ mg

11. How many kilograms are there in 5 tons?

12. How many kilograms are there in 17 tons?

13. Change 2 tons to kilograms.

14. Change 896 mg to grams.

15. Express 896 g in milligrams.

16. Express 342 kg in grams.

17. Convert 75,000 g to kilograms.

18. Convert 3000 mg to grams.

19. Convert 7 tons to grams.

20. Convert 0.4 t to grams.

21. Change 0.34 g to kilograms.

22. Change 0.78 g to milligrams.

1. [Respond in exercise.]

2. [Respond in exercise.]

3. [Respond in exercise.]

4. [Respond in exercise.]

5. [Respond in exercise.]

6. [Respond in exercise.]

7. [Respond in exercise.]

8. [Respond in exercise.]

9. [Respond in exercise.]

10. [Respond in exercise.]

11. ____________

12. ____________

13. ____________

14. ____________

15. ____________

16. ____________

17. ____________

18. ____________

19. ____________

20. ____________

21. ____________

22. ____________

23. ______

24. ______

25. ______

26. ______

27. ______

28. ______

29. ______

30. ______

31. ______

32. ______

33. ______

34. ______

35. ______

36. ______

37. ______

23. How many grams are in 16 mg?

24. How many milligrams are in 2.5 g?

25. 92.3 g = ______ kg

26. 3.94 g = ______ mg

27. 7.58 t = ______ kg

28. 5.6 t = ______ kg

29. 2963 kg = ______ t

30. 3547 kg = ______ t

Complete the following table.

31. 1 cm^3 = ______ mm^3

1 dm^3 = ______ cm^3

1 m^3 = ______ dm^3

1 km^3 = ______ m^3

Change the following units of volume as indicated.

32. 73 m^3 = ______ dm^3

33. 0.9 m^3 = ______ dm^3

34. 525 cm^3 = ______ m^3

35. 400 m^3 = ______ cm^3

36. 8.7 m^3 = ______ cm^3

37. 63 dm^3 = ______ m^3

Name ______________________________ Section ______________ Date ____________

ANSWERS

38. How many cm^3 are in 45 mm^3?

39. How many mm^3 are in 3.1 cm^3

40. Change 19 mm^3 to dm^3.

41. Change 5 cm^3 to mm^3.

42. Convert 2 dm^3 to cm^3.

43. Convert 76.4 mL to liters.

44. Change 5.3 L to milliliters.

45. Change 30 cm^3 to milliliters.

46. Change 30 cm^3 to liters.

47. Change 5.3 mL to liters.

48. 48 kL = __________ L

49. 72,000 L = __________ kL

50. 290 L = __________ kL

51. 569 mL = __________ L

52. 80 L = __________ mL = __________ cm^3

53. 7.3 L = __________ mL = __________ cm^3

38. ______________

39. ______________

40. ______________

41. ______________

42. ______________

43. ______________

44. ______________

45. ______________

46. ______________

47. ______________

48. [Respond in exercise.]

49. [Respond in exercise.]

50. [Respond in exercise.]

51. [Respond in exercise.]

52. [Respond in exercise.]

53. [Respond in exercise.]

I.2 U.S. Customary and Metric Equivalents

U.S. Customary and Metric Equivalents

We begin the following discussion of equivalent measures between the U.S. customary and metric systems with measures of temperature.

Temperature: U.S. customary measure is in **degrees Fahrenheit** $(°F)$.
Metric measure is in **degrees Celsius** $(°C)$.

The two scales are shown here on thermometers. Approximate conversions can be found by reading along a ruler or the edge of a piece of paper held horizontally across the page.

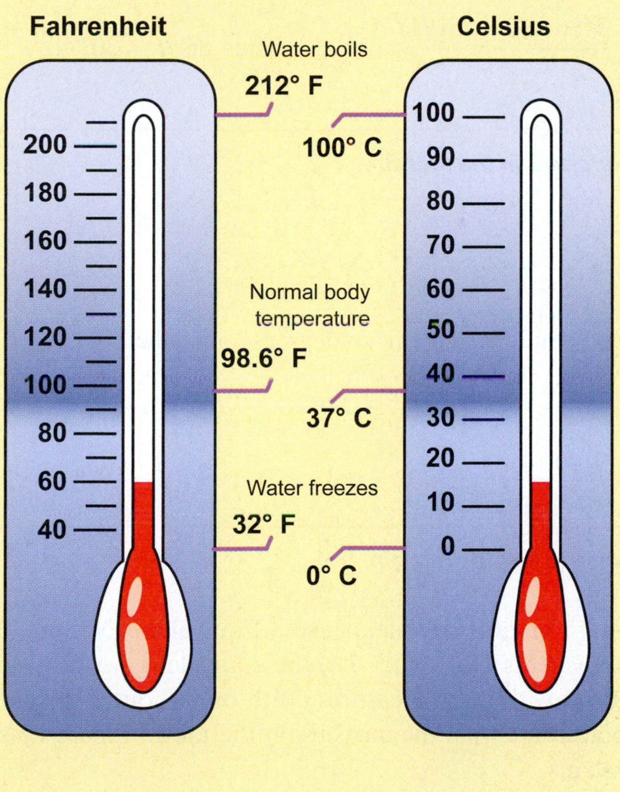

Example 1

Hold a straight edge horizontally across the two thermometers and you will read:

$100°\ C = 212°\ F$ Water boils at sea level.

$40°\ C = 104°\ F$ hot day in the desert

$20°\ C = 68°\ F$ comfortable room temperature

Continued on next page...

Use a straight edge across the two thermometers to convert each temperature.

1. a. $0° C$

b. $50° F$

Use the conversion formulas given to convert each temperature.

2. $77° F$

3. $45° C$

Two formulas that give exact conversions are given here.

F = Fahrenheit temperatures and C = Celsius temperature.

$$C = \frac{5(F-32)}{9} \qquad F = \frac{9 \cdot C}{5} + 32$$

A calculator will give answers accurate to eight digits. Answers that are not exact may be rounded to whatever place of accuracy you choose.

Now work exercise 1 in the margin.

Example 2

Let $F = 86°$ and find the equivalent measure in Celsius.

Solution

$$C = \frac{5(86-32)}{9} = \frac{5(54)}{9} = 30 \qquad \text{Thus } 86°\ F = 30°\ C.$$

Now work exercise 2 in the margin.

Example 3

Let $C = 40°$ and convert this to degrees Fahrenheit.

Solution

$$F = \frac{9 \cdot 40}{5} + 32 = 72 + 32 = 104 \qquad \text{Thus } 40°\ C = 104°\ F.$$

Now work exercise 3 in the margin.

In the tables of Length Equivalents, Area Equivalents, Volume Equivalents, and Mass Equivalents (Tables I.5–I.8) the equivalent measures are rounded. Any calculations with these measures (with or without a calculator) cannot be any more accurate than the measure in the table. Figure I.5 shows some length equivalents.

Table I.5 Length Equivalents

U.S. to Metric	Metric to U.S.
1 in. = 2.54 cm (exact)	1 cm = 0.394 in.
1 ft = 0.305 m	1 m = 3.28 ft
1 yd = 0.914 m	1 m = 1.09 yd
1 mi = 1.61 km	1 km = 0.62 mi

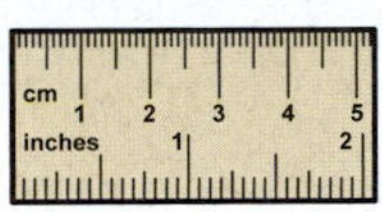

1 in. = 2.54 cm

miles
0 40 80
0 50 125
kilometers

80 mi = 128.8 km

5 ft 9 in. = 175 cm

Figure I.5

In Examples 4 – 7, use Table I.5 on page A14 to convert measurements as indicated. (Also see Figure I.5.)

Use Table I.5 to convert each of the following measurements as indicated.

Example 4

6 ft = ______ cm

6 ft = 72 in. = 72(2.54 cm) = 183 cm (rounded)

or

6 ft = 6(0.305 m) = 1.83 m = 183 cm

Now work exercise 4 in the margin.

4. 84 in. = ______ m

Example 5

25 mi = ______ km

25 mi = 25(1.61 km) = 40.25 km

Now work exercise 5 in the margin.

5. 100 yd = ______ m

Example 6

30 m = ______ ft

30 m = 30(3.28 ft) = 98.4 ft

Now work exercise 6 in the margin.

6. 75 cm = ______ in.

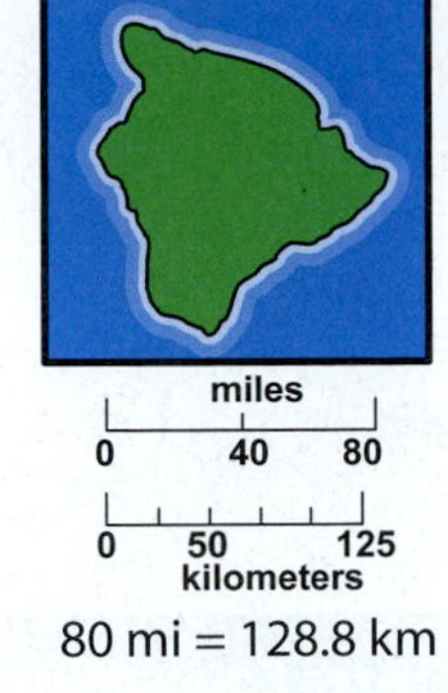

7. Use Table I.5 to convert the following measurement as indicated.

500 m = ______ ft

Example 7

10 km = ______ mi

10 km = 10(0.62 mi) = 6.2 mi

Now work exercise 7 in the margin.

Table I.6 Area Equivalents	
U.S. to Metric	**Metric to U.S.**
$1 \text{ in.}^2 = 6.45 \text{ cm}^2$	$1 \text{ cm}^2 = 0.155 \text{ in.}^2$
$1 \text{ ft}^2 = 0.093 \text{ m}^2$	$1 \text{ m}^2 = 10.764 \text{ ft}^2$
$1 \text{ yd}^2 = 0.836 \text{ m}^2$	$1 \text{ m}^2 = 1.196 \text{ yd}^2$
1 acre = 0.405 ha	1 ha = 2.47 acres

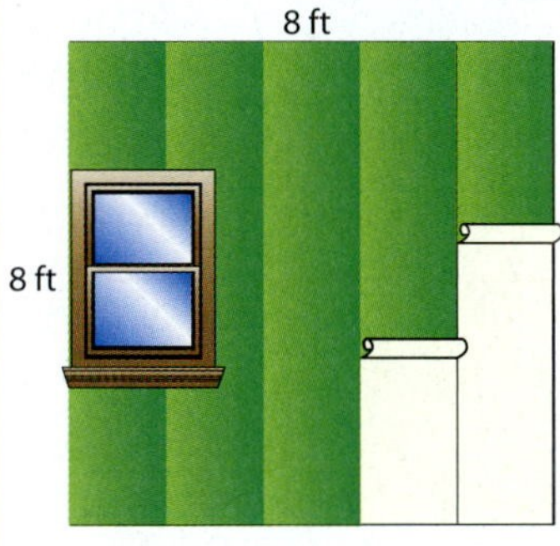

$64 \text{ ft}^2 = 5.952 \text{ m}^2$

$0.875 \text{ in.}^2 = 5.64 \text{ cm}^2$

1 ha = 2.47 acres

Figure I.6

In Examples 8 – 11, use Table I.6 to convert the measures as indicated. (Also see Figure I.6.)

8. Use Table I.6 to convert the following measurement as indicated.

$53 \text{ in.}^2 =$ ______ cm^2

Example 8

$40 \text{ yd}^2 =$ ______ m^2

$$40 \text{ yd}^2 = 40\left(0.836 \text{ m}^2\right) = 33.44 \text{ m}^2$$

Now work exercise 8 in the margin.

Example 9

5 acres = _____ ha

$$5 \text{ acres} = 5(0.405 \text{ ha}) = 2.025 \text{ ha}$$

Now work exercise 9 in the margin.

Example 10

5 ha = _____ acres

$$5 \text{ ha} = 5(2.47 \text{ acres}) = 12.35 \text{ acres}$$

Now work exercise 10 in the margin.

Example 11

$100 \text{ cm}^2 =$ _____ in.^2

$$100 \text{ cm}^2 = 100\left(0.155 \text{ in.}^2\right) = 15.5 \text{ in.}^2$$

Now work exercise 11 in the margin.

Table I.7 Volume Equivalents

U.S. to Metric	Metric to U.S.
$1 \text{ in.}^3 = 16.387 \text{ cm}^3$	$1 \text{ cm}^3 = 0.06 \text{ in.}^3$
$1 \text{ ft}^3 = 0.028 \text{ m}^3$	$1 \text{ m}^3 = 35.315 \text{ ft}^3$
$1 \text{ qt} = 0.946 \text{ L}$	$1 \text{ L} = 1.06 \text{ qt}$
$1 \text{ gal} = 3.785 \text{ L}$	$1 \text{ L} = 0.264 \text{ gal}$

5 gal = 18.925 L

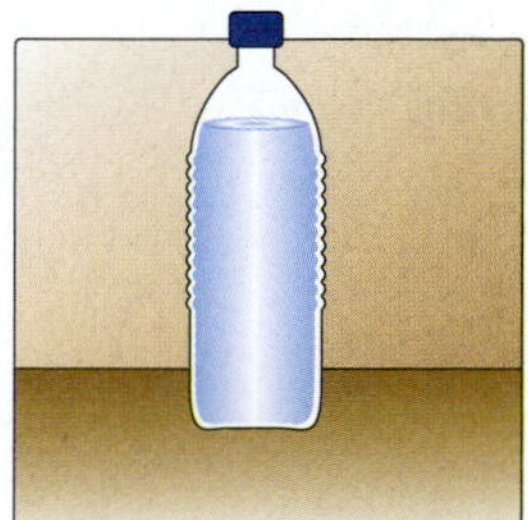

1 L = 1.06 qt

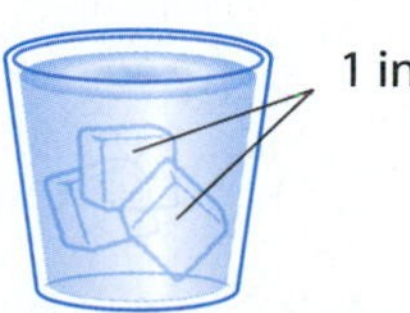

$3 \text{ in.}^3 = 49.161 \text{ cm}^3$

Figure I.7

Use table I.6 to convert each of the following measurements as indicated.

9. 16 acres = ______ ha

10. 3 ha = ______ acres

11. $38 \text{ m}^2 =$ _____ ft^2

Use Table I.7 to convert each of the following measurements as indicated.

12. 4 gal = ______ L

13. 16 L = ______ qt

14. 28 qt = ______ L

15. $12 \text{ m}^3 =$ _____ ft^3

In Examples 12 – 15, use Table I.7 to convert the measures as indicated. (Also see Figure I.7.)

Example 12

20 gal = _____ L

20 gal = 20(3.785 L) = 75.7 L

Now work exercise 12 in the margin.

Example 13

42 L = _____ gal

42 L = 42(0.264 gal) = 11.088 gal

or

42 L = 11.1 gal (rounded)

Now work exercise 13 in the margin.

Example 14

6 qt = _____ L

6 qt = 6(0.946 L) = 5.676 L

or

6 qt = 5.7 L (rounded)

Now work exercise 14 in the margin.

Example 15

$10 \text{ cm}^3 =$ _____ in.^3

$$10 \text{ cm}^3 = 10\left(0.06 \text{ in.}^3\right) = 0.6 \text{ in.}^3$$

Now work exercise 15 in the margin.

Table I.8	Mass Equivalents
U.S. to Metric	**Metric to U.S.**
1 oz = 28.35 g	1 g = 0.035 oz
1 lb = 0.454 kg	1 kg = 2.205 lb

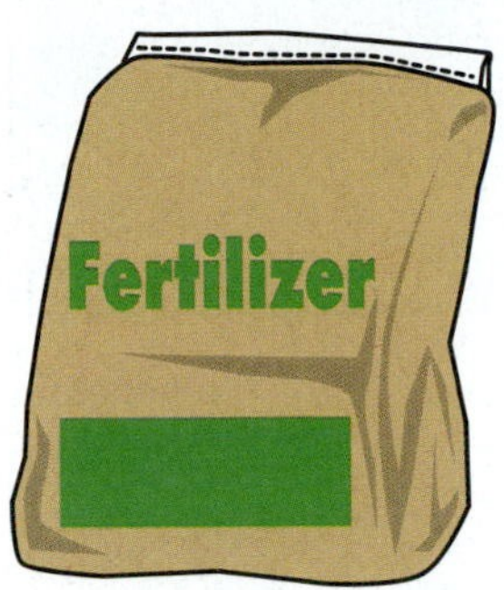

25 lb = 11.35 kg

9 kg = 19.85 lb

Figure I.8

In Examples 16 and 17, use Table I.8 to convert the measures as indicated. (Also see Figure I. 8.)

Example 16

5 lb = _____ kg

5 lb = 5(0.454 kg) = 2.27 kg

Now work exercise 16 in the margin.

Example 17

15 kg = _____ lb

15 kg = 15(2.205 lb) = 33.075 lb

or

15 kg = 33.1 lb (rounded)

Now work exercise 17 in the margin.

Use Table I.8 to convert each of the following measurements as indicated.

16. 3 oz = ______ g

17. 42 g = ______ oz

Practice Problems

Convert the following measures as indicated. You may need to refer to the tables throughout this lesson.

1. $23° F =$ _____ $° C$

2. 14 in. = ______ cm

3. $36 \text{ ft}^2 =$ _____ m^2

4. $24 \text{ cm}^3 =$ _____ in.^3

5. 33 kg = ______ lb

Answers to Practice Problems:

1. $-5° C$ **2.** 35.56 cm **3.** 3.348 m^2 **4.** 1.44 in.^3 **5.** 72.765 lb

Name ______________________ Section ______________ Date ___________

Exercises I.2

Convert the following measures as indicated, rounding to the nearest hundredth when necessary. Use the tables in Section I.2 as a reference.

1. $25^\circ\ C =$ _____ $^\circ F$

2. $80^\circ\ C =$ _____ $^\circ F$

3. $50^\circ\ C =$ _____ $^\circ F$

4. $35^\circ\ C =$ _____ $^\circ F$

5. $50^\circ\ F =$ _____ $^\circ C$

6. $100^\circ\ F =$ _____ $^\circ C$

7. Change $32^\circ\ F$ to degrees Celsius.

8. Change $41^\circ\ F$ to degrees Celsius.

9. How many meters are in 3 yds?

10. How many meters are in 5 yds?

11. Change 60 miles to kilometers.

12. Change 100 miles to kilometers.

13. Convert 200 kilometers to miles.

14. Convert 65 kilometers to miles.

15. How many inches are in 50 cm?

16. How many inches are in 100 cm?

17. $3\text{ in.}^2 =$ _____ cm^2

18. $16\text{ in.}^2 =$ _____ cm^2

19. $600\text{ ft}^2 =$ _____ m^2

20. $300\text{ ft}^2 =$ _____ m^2

21. $100\text{ yd}^2 =$ _____ m^2

22. $250\text{ yd}^2 =$ _____ m^2

23. 1000 acres = _____ ha

24. 250 acres = _____ ha

ANSWERS

1. [Respond in exercise.]

2. [Respond in exercise.]

3. [Respond in exercise.]

4. [Respond in exercise.]

5. [Respond in exercise.]

6. [Respond in exercise.]

7. ____________

8. ____________

9. ____________

10. ____________

11. ____________

12. ____________

13. ____________

14. ____________

15. ____________

16. ____________

17. [Respond in exercise.]

18. [Respond in exercise.]

19. [Respond in exercise.]

20. [Respond in exercise.]

21. [Respond in exercise.]

22. [Respond in exercise.]

23. [Respond in exercise.]

24. [Respond in exercise.]

25. ______

26. ______

27. ______

28. ______

29. ______

30. ______

31. [Respond in exercise.]

32. [Respond in exercise.]

33. [Respond in exercise.]

34. [Respond in exercise.]

35. [Respond in exercise.]

36. [Respond in exercise.]

37. [Respond in exercise.]

38. [Respond in exercise.]

39. [Respond in exercise.]

40. [Respond in exercise.]

41. ______

42. ______

25. How many acres are in 300 ha?

26. How many acres are in 400 ha?

27. Change 5 m^2 to square feet.

28. Change 10 m^2 to square feet.

29. Change 30 cm^2 to square inches.

30. Change 50 cm^2 to square inches.

31. 10 qt = _____ L

32. 20 qt = _____ L

33. 10 L = _____ qt

34. 25 L = _____ qt

35. 42 L = _____ gal

36. 50 L = _____ gal

37. 10 lb = _____ kg

38. 500 kg = _____ lb

39. 16 oz = _____ g

40. 100 g = _____ oz

41. Suppose that the home you are considering buying sits on a rectangular shaped lot that is 270 feet by 121 feet. Convert this area to square meters.

42. A new manufacturing building covers an area of 3 acres. How many hectares of ground does the new building cover?

Name ______________________ Section ____________ Date __________

43. A painting of a landscape is on a rectangular canvas that measures 3 feet by 4 feet.

a. How many square centimeters of wall space will the painting cover when it is hanging?

b. How many square meters?

ANSWERS

43. a. ______________

b. ______________

APPENDIX II

Greatest Common Divisor (GCD)

Consider the two numbers 12 and 18. Is there a number (or numbers) that will divide into **both** 12 and 18? To help answer this question, the divisors for 12 and 18 are listed below.

Set of divisors for 12: $\{1, 2, 3, 4, 6, 12\}$
Set of divisors for 18: $\{1, 2, 3, 6, 9, 18\}$

The **common divisors** for 12 and 18 are 1, 2, 3, and 6. The **greatest common divisor (GCD)** for 12 and 18 is 6: that is, of all the common divisors of 12 and 18, 6 is the largest divisor.

Example 1

List the divisors of each number in the set $\{36, 24, 48\}$ and find the greatest common divisor (GCD).

Set of divisors for 36: {**1, 2, 3, 4, 6,** 9, **12,** 18, 36}
Set of divisors for 24: {**1, 2, 3, 4, 6,** 8, **12,** 24}
Set of divisors for 48: {**1, 2, 3, 4, 6,** 8, **12,** 16, 24, 48}

The common divisors are **1, 2, 3, 4, 6,** and **12**. **GCD = 12.**

Now work exercise 1 in the margin.

1. List the divisors of each number in the set $\{45, 60, 90\}$ and find the greatest common divisor (GCD).

The Greatest Common Divisor

The **greatest common divisor** (GCD)* of a set of natural numbers is the largest natural number that will divide into all the numbers in the set.

As Example 1 illustrates, listing the divisors of each number before finding the GCD can be tedious and difficult. **The use of prime factorizations leads to a simple technique for finding the GCD.**

* The largest common divisor is, of course, the largest common factor, and the GCD could be called the **greatest common factor**, and be abbreviated GCF.

Technique for Finding the GCD of a Set of Counting Numbers

1. Find the prime factorization of each number.
2. Find the prime factors common to all factorizations.
3. Form the product of these primes, using each prime the number of times it is common to all factorizations.
4. This product is the GCD. If there are no primes common to all factorizations, the GCD is 1.

2. Find the GCD for {26, 52, 104}.

Example 2

Find the GCD for {36, 24, 48}.

$$\left.\begin{aligned} 36 &= \mathbf{2}\cdot\mathbf{2}\cdot 3\cdot\mathbf{3} \\ 24 &= \mathbf{2}\cdot\mathbf{2}\cdot 2\cdot\mathbf{3} \\ 48 &= \mathbf{2}\cdot\mathbf{2}\cdot 2\cdot 2\cdot\mathbf{3} \end{aligned}\right\} \text{GCD} = 2\cdot 2\cdot 3 = 12$$

In all the prime factorizations, the factor 2 appears twice and the factor 3 appears once. The GCD is 12.

Now work exercise 2 in the margin.

3. Find the GCD for {49, 343, 147}.

Example 3

Find the GCD for {360, 75, 30}.

$$\left.\begin{aligned} 360 &= 36\cdot 10 = 4\cdot 9\cdot 2\cdot 5 = 2\cdot 2\cdot 2\cdot 3\cdot\mathbf{3}\cdot\mathbf{5} \\ 75 &= 3\cdot 25 = \mathbf{3}\cdot\mathbf{5}\cdot 5 \\ 30 &= 6\cdot 5 = 2\cdot\mathbf{3}\cdot\mathbf{5} \end{aligned}\right\} \text{GCD} = 3\cdot 5 = 15$$

Each of the factors 3 and 5 appears only once in **all** the prime factorizations. The GCD is 15.

Now work exercise 3 in the margin.

4. Find the GCD for {60, 90, 210}.

Example 4

Find the GCD for {168, 420, 504}.

$$\left.\begin{aligned} 168 &= 8\cdot 21 = \mathbf{2}\cdot\mathbf{2}\cdot 2\cdot\mathbf{3}\cdot\mathbf{7} \\ 420 &= 10\cdot 42 = 2\cdot 5\cdot 6\cdot 7 \\ &= \mathbf{2}\cdot\mathbf{2}\cdot\mathbf{3}\cdot 5\cdot\mathbf{7} \\ 504 &= 4\cdot 126\cdot = 2\cdot 2\cdot 6\cdot 21 \\ &= \mathbf{2}\cdot\mathbf{2}\cdot 2\cdot 3\cdot\mathbf{3}\cdot\mathbf{7} \end{aligned}\right\} \text{GCD} = 2\cdot 2\cdot 3\cdot 7 = 84$$

In **all** prime factorizations, 2 appears twice, 3 once, and 7 once. The GCD is 84.

Now work exercise 4 in the margin.

If the GCD of two numbers is 1 (that is, they have no common prime factors), then the two numbers are said to be **relatively prime**. The numbers themselves may be prime or they may be composite.

Example 5

Find the GCD for $\{15, 8\}$.

$$\left.\begin{aligned} 15 &= 3 \cdot 5 \\ 8 &= 2 \cdot 2 \cdot 2 \end{aligned}\right\} \text{GCD} = 1 \quad \text{8 and 15 are relatively prime.}$$

Now work exercise 5 in the margin.

Example 6

Find the GCD for $\{20, 21\}$.

$$\left.\begin{aligned} 20 &= 2 \cdot 2 \cdot 5 \\ 21 &= 3 \cdot 7 \end{aligned}\right\} \text{GCD} = 1 \quad \text{20 and 21 are relatively prime.}$$

Now work exercise 6 in the margin.

5. Find the GCD for $\{26, 19\}$.

6. Find the GCD for $\{33, 343\}$.

Practice Problems

Find the GCD of the following sets of numbers.

1. $\{22, 33, 77\}$

2. $\{70, 35, 210\}$

3. $\{27, 54, 81\}$

4. $\{18, 35\}$

Answers to Practice Problems:

1. 11 **2.** 35 **3.** 27 **4.** 1

Name ______________________ Section ____________ Date __________

ANSWERS

Exercises II

Find the GCD for each of the following sets of numbers.

1. $\{12, 8\}$ **2.** $\{16, 28\}$ **3.** $\{85, 51\}$

4. $\{20, 75\}$ **5.** $\{20, 30\}$ **6.** $\{42, 48\}$

7. $\{15, 21\}$ **8.** $\{27, 18\}$ **9.** $\{18, 24\}$

10. $\{77, 66\}$ **11.** $\{182, 184\}$ **12.** $\{110, 66\}$

13. $\{8, 16, 64\}$ **14.** $\{121, 44\}$ **15.** $\{28, 52, 56\}$

16. $\{98, 147\}$ **17.** $\{60, 24, 96\}$ **18.** $\{33, 55, 77\}$

19. $\{25, 50, 75\}$ **20.** $\{30, 78, 60\}$ **21.** $\{17, 15, 21\}$

22. $\{520, 220\}$ **23.** $\{14, 55\}$ **24.** $\{210, 231, 84\}$

25. $\{140, 245, 420\}$

1. ________
2. ________
3. ________
4. ________
5. ________
6. ________
7. ________
8. ________
9. ________
10. ________
11. ________
12. ________
13. ________
14. ________
15. ________
16. ________
17. ________
18. ________
19. ________
20. ________
21. ________
22. ________
23. ________
24. ________
25. ________

26. ____________

27. ____________

28. ____________

29. ____________

30. ____________

31. ____________

32. ____________

33. ____________

34. ____________

35. ____________

Decide whether each set of numbers is relatively prime using the GCD. If it is not relatively prime, state the GCD.

26. $\{35, 24\}$ **27.** $\{11, 23\}$ **28.** $\{14, 36\}$ **29.** $\{72, 35\}$

30. $\{42, 77\}$ **31.** $\{8,15\}$ **32.** $\{20, 21\}$ **33.** $\{16, 22\}$

34. $\{16, 51\}$ **35.** $\{10, 27\}$

APPENDIX III

Statistics (Mean, Median, Mode, and Range)

Statistics is the study of how to gather, organize, analyze, and interpret numerical information. In this section we will study four measures (or four statistics) that are easily found or calculated: **mean**, **median**, **mode**, and **range**. Other measures that you might read about or study in a course in statistics are:

standard deviation and **variance** (both are measures of how "spread out" data are)

***z*-score** (a measure that compares numbers that are expressed in different units–for example, your score on a biology exam and your score on a mathematics exam)

correlation coefficient (a measure of how two different types of data might be related–for example, to determine if there is a relationship between the amount of schooling a person has and the amount of that person's lifetime earnings)

You will need at least one semester of algebra to be able to study and understand these topics, so keep working hard.

The following terms and their definitions are necessary for understanding the topics and related problems in this section. They are listed here in one place for easy reference.

Terms Used in the Study of Statistics

Data: Value(s) measuring some characteristic of interest. (We will consider only numerical data.)

Mean: The arithmetic average of the data. (Add all the data and divide by the number of data items.)

Median: The middle data item. (Arrange the data in order and pick out the middle item.)

Continued on next page...

Mode: The single data item that appears the most number of times. (Some data may have more than one mode. We will leave the discussion of such a situation to a course in statistics. In this text, if the data have a mode, there will be only one mode.)

Range: The difference between the largest and smallest data items.

In Examples 1–3 answer the questions about statistics using the data from Group A and Group B.

Group A: Annual Income for 8 Families

\$18,000; \$12,000; \$15,000; \$17,000; \$35,000; \$70,000; \$15,000; \$20,000

Group B: Grade Point Averages (GPA) for 11 students

2.0; 2.0; 1.9; 3.1; 3.5; 2.9; 2.5; 3.6; 2.0; 2.4; 3.4

The top ten home run totals in the National League for the 2005 season were 51, 46, 41, 40, 37, 36, 35, 34, 33, and 33. Call this Group C. (Use this group for Margin Examples 1, 2, and 3.)

1. Find the mean home run total in Group C.

Example 1

Find the mean income for the families in Group A.

Solution

Find the sum of the 8 incomes and divide by 8.

$$\begin{array}{r} \$\ 18,000 \\ 12,000 \\ 15,000 \\ 17,000 \\ 35,000 \\ 70,000 \\ 15,000 \\ +\ \ 20,000 \\ \hline \$202,000 \end{array} \qquad \begin{array}{r} \$25,250 \\ 8\overline{)202,000} \\ \underline{16} \\ 42 \\ \underline{40} \\ 2\,0 \\ \underline{1\,6} \\ 40 \\ \underline{40} \\ 00 \\ \underline{0} \\ 0 \end{array}$$

You may want to use a calculator to do this arithmetic. The mean annual income is \$25,250.

Now work exercise 1 in the margin.

To Find the Median:

1. Arrange the data in order.
2. If there is an **odd** number of items, the median is the middle item.
3. If there is an **even** number of items, the median is the average of the two middle items.

Example 2

Find the median income for Group A and the median GPA for Group B.

Solution

Arrange both sets of data in order.

Group A

$12,000; $15,000; $15,000; $17,000; $18,000; $20,000; $35,000; $70,000

Group B

1.9; 2.0; 2.0; 2.0; 2.4; 2.5; 2.9; 3.1; 3.4; 3.5; 3.6

For Group A, the median is the average of the 4th and 5th items because there is an even number of items (8 items).

$$\text{Median} = \frac{\$17,000 + \$18,000}{2} = \frac{35,000}{2} = \$17,500$$

For Group B, the median is the 6th item because, with an **odd** number of 11 items, the 6th item is the middle item.

$$\text{Median} = 2.5$$

Now work exercise 2 in the margin.

2. Find the median of Group C.

Example 3

Find the mode and the range for both Group A and Group B.

Solution

The mode is the most frequent item. From the arranged data in Example 2, we can see that:

for Group A, the mode is 15,000.

for Group B, the mode is 2.0.

The range is the difference between the largest and smallest items:

Group A range = $70,000 – $12,000 = $58,000

Group B range = 3.6 – 1.9 = 1.7

Now work exercise 3 in the margin.

3. Find the mode and range of Group C.

Commentary: Of the four statistics mentioned in this section, the mean and median are most commonly used. Many people feel that the mean (or arithmetic average) is relied on too much in reporting central tendencies for data such as income, housing costs, and taxes where a few very high items can *distort* the picture of a central tendency. As you can see in Group A data, the median of $17,500 is probably more representative of the data than the mean of $25,250 because the one high income of $70,000 raises the mean considerably.

When you read an article in a magazine or newspaper that reports means or medians, you should now have a better understanding of the implications.

Practice Problems

Ben Wallace of the Detroit Pistons had the following rebounding totals in the first 10 games of the 2006 NBA season: 8, 17, 15, 7, 10, 13, 11, 12, 9, and 17.

Find the following measures.

1. the mean

2. the median

3. the mode

4. the range

Answers to Practice Problems:

1. 11.9 **2.** 11.5 **3.** 17 **4.** 10

Name ______________________ Section ____________ Date __________

Exercises III

For each of the following problems, find **a.** *the mean,* **b.** *the median,* **c.** *the mode, and* **d.** *the range of the given data.*

1. Ten math students had the following scores on a final exam:

75, 83, 93, 65, 85,
85, 88, 90, 55, 71

2. Joe did the following number of sit-ups each morning for a week:

25, 52, 48, 42, 38, 58, 52

3. Fifteen college students reported the following hours of sleep the night before an exam:

4, 6, 6, 7, 6.5, 6.5, 7.5, 8.5
5, 6, 4.5, 5.5, 9, 3, 8

4. The local high school basketball team scored the following points per game during their 20-game season:

85, 60, 62, 70, 75, 52, 88, 50, 80, 72,
90, 85, 85, 93, 70, 75, 68, 73, 65, 82

5. Stacey went to six different repair shops to get the following estimates to repair her car. (The accident was not her fault; her car was parked at the time.)

$425, $525, $325, $300, $500, $325

6. Mike kept track of his golf scores for twelve rounds of eighteen holes each. His scores were:

85, 90, 82, 85, 87, 80,
78, 82, 88, 82, 86, 81

ANSWERS

1. a. __________
b. __________
c. __________
d. __________
2. a. __________
b. __________
c. __________
d. __________
3. a. __________
b. __________
c. __________
d. __________
4. a. __________
b. __________
c. __________
d. __________
5. a. __________
b. __________
c. __________
d. __________
6. a. __________
b. __________
c. __________
d. __________

7. a. ______

b. ______

c. ______

d. ______

8. a. ______

b. ______

c. ______

d. ______

9. a. ______

b. ______

c. ______

d. ______

10. a. ______

b. ______

c. ______

d. ______

11. a. ______

b. ______

c. ______

d. ______

12. a. ______

b. ______

c. ______

d. ______

7. The local weather station recorded the following daily high temperatures for one month:

75, 76, 76, 78, 85, 82, 85, 88, 90, 90,
88, 95, 96, 92, 88, 88, 80, 80, 78, 80,
78, 76, 77, 75, 75, 74, 70, 70, 72, 73

8. The Big City fire department reported the following mileage for tires used on their nine fire trucks:

14,000; 14,000; 11,000; 15,000;
9,000; 14,000; 12,000; 10,000; 9,000

9. The city planning department issued the following numbers of building permits over a three-week period:

17, 19, 18, 35, 30, 29, 23, 14,
18, 16, 20, 18, 18, 25, 30

10. Police radar measured the following speeds of 35 cars on one street:

28, 24, 22, 38, 40, 25, 24, 35, 25,
23, 22, 50, 31, 37, 45, 28, 30, 30,
30, 25, 35, 32, 45, 52, 24, 26, 18,
20, 30, 32, 33, 48, 58, 30, 25

11. On a one-day fishing trip, Mr. and Mrs. Johnson recorded the following lengths of fish they caught (measured in inches):

14.3; 13.6; 10.5; 15.5; 20.1;
10.9; 12.4; 25.0; 30.2; 32.5

12. A machine puts out parts measured in thickness to the nearest hundredth of an inch. One hundred parts were measured and the results are tallied in the following chart. Find **a.** the mean, **b.** the median, **c.** the mode, and **d.** the range of thickness for the 100 parts.

Thickness Measured	0.80	0.83	0.84	0.85	0.87
Number of Parts	22	41	14	20	3

APPENDIX IV

Base Two and Base Five

IV.1 The Binary System (Base Two)

In the decimal system, ten is the base. You might ask if another number could be chosen as the base in a place value system. And, if so, would the system be any better or more useful than the decimal system? The fact is that computers do operate under a place value system with base two. In the **binary system** (or base two system), only two digits are needed, 0 and 1. These two digits correspond to the two possible conditions of an electric current, either on or off.

Any number can be represented in base two or in base ten. However, base ten has a definite advantage when large numbers are involved, as you will see. The advantage of base two over base ten is that for base two only two digits are needed, while ten digits are needed for base ten.

If the base of a place value system were not ten but two, then the beginning point would not be a decimal point but a **binary point**. The value of each place would be a power of two, as shown in Figure IV.1.

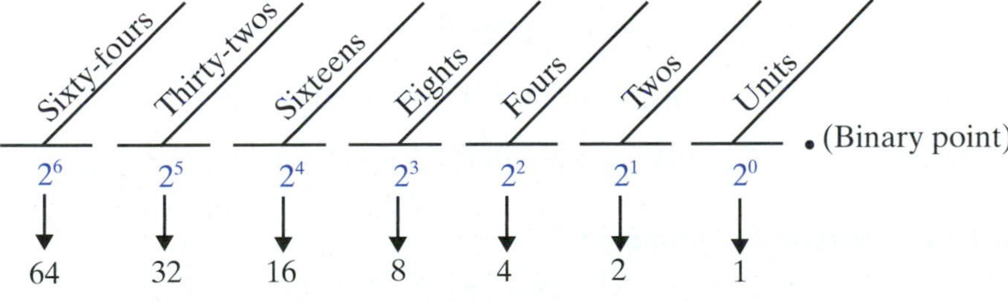

Figure IV.1

To Write Numbers in the Base Two System, Remember Three Things:

1. {0, 1} is the set of digits that can be used.
2. The value of each place from the binary point is in powers of two.
3. The symbol 2 does not exist in the binary system, just as there is no digit for ten in the decimal system.

To avoid confusion with the base ten numerals, we will write $_{(2)}$ to the lower right of each base two numeral. We could write $_{(10)}$ to the lower right of each base ten numeral, but this would not be practical since most of the numerals we work with are in base ten. Therefore, **if no base is indicated, the numeral will be understood to be in base ten**.

1. Find the value of the numeral $1110_{(2)}$.

Example 1

Find the value of the numeral $1101_{(2)}$.

Solution

Writing the value of each place under the digit gives

$$\frac{1}{2^3}\ \frac{1}{2^2}\ \frac{0}{2^1}\ \frac{1}{2^0}\ {}_{(2)}$$

In expanded notation,

$$\begin{aligned}1101_{(2)} &= 1(2^3)+1(2^2)+0(2^1)+1(2^0)\\ &= 1(8)+1(4)+0(2)+1(1)\\ &= 8+4+0+1\\ &= 13\end{aligned}$$

Thus, to a computer, the symbol $1101_{(2)}$ means "thirteen."

Now work exercise 1 in the margin.

2. Find the value of the numeral $11010_{(2)}$.

Example 2

a. $1_{(2)} = 1$

b. $10_{(2)} = 1(2)+0 = 2$

c. $100_{(2)} = 1(2^2)+0(2)+0(1) = 4+0+0 = 4$

d. $110_{(2)} = 1(2^2)+1(2)+0(1) = 4+2+0 = 6$

e. $1001_{(2)} = 1(2^3)+0(2^2)+0(2)+1(1) = 8+0+0+1 = 9$

f. $1010_{(2)} = 1(2^3)+0(2^2)+1(2)+0(1) = 8+0+2+0 = 10$

Now work exercise 2 in the margin.

Do **not** read $100_{(2)}$ as "one hundred" because the 1 is not in the hundreds place. The 1 is in the fours place. So $100_{(2)}$ is read "four" or "one, zero, zero-base two." Similarly, $111_{(2)}$ is read "seven" or "one, one, one–base two."

Practice Problems

Find the value for each of the following numerals.

1. $11011_{(2)}$

2. $100011_{(2)}$

3. $1010100_{(2)}$

4. $1110111_{(2)}$

Answers to Practice Problems:

1. 27 **2.** 35 **3.** 84 **4.** 119

Name ______________________ Section ____________ Date __________

ANSWERS

Exercises IV.1

Write the following base ten numerals in expanded form using exponents.

Example: $273 = 2(10^2)+7(10^1)+3(10^0)$

1. 35

2. 761

3. 8469

4. 500

5. 62,322

Write the following base two numerals in expanded form and find the value of each numeral.

Example: $$\begin{aligned}110_{(2)} &= 1(2^2)+1(2^1)+0(2^0)\\ &= 1(4)+1(2)+0(1)\\ &= 4+2+0\\ &= 6\end{aligned}$$

6. $11_{(2)}$

7. $101_{(2)}$

8. $111_{(2)}$

9. $1011_{(2)}$

10. $1101_{(2)}$

11. $110111_{(2)}$

1. [Respond below exercise.] ____

2. [Respond below exercise.] ____

3. [Respond below exercise.] ____

4. [Respond below exercise.] ____

5. [Respond below exercise.] ____

6. [Respond below exercise.] ____

7. [Respond below exercise.] ____

8. [Respond below exercise.] ____

9. [Respond below exercise.] ____

10. [Respond below exercise.] ____

11. [Respond below exercise.] ____

12. [Respond below exercise.]

13. [Respond below exercise.]

14. [Respond below exercise.]

15. [Respond below exercise.]

16. [Respond below exercise.]

17. [Respond below exercise.]

18. [Respond below exercise.]

19. [Respond below exercise.]

20. [Respond below exercise.]

21. ______

12. $11110_{(2)}$

13. $101011_{(2)}$

14. $11010_{(2)}$

15. $1000_{(2)}$

16. $1000010_{(2)}$

17. $11101_{(2)}$

18. $10110_{(2)}$

19. $111111_{(2)}$

20. $1111_{(2)}$

21. A computer is instructed to place some information in memory space number $1101111_{(2)}$. What is the number of this memory space in base ten?

IV.2 The Quinary System (Base Five)

Many numbers may be used as bases for place value systems. To illustrate this point and to emphasize the concept of place value, we will discuss one more base system, base five. Interested students may want to try writing numerals in base three or base eight or base eleven.

Again, the system relies on powers of the base and a set of digits. In the **quinary system** (base five system), the powers of 5 are $5^0, 5^1, 5^2, 5^3, 5^4$, and so on, and the digits to be used are {0, 1, 2, 3, 4}. The **quinary point** is the beginning point, as shown in Figure IV.2.

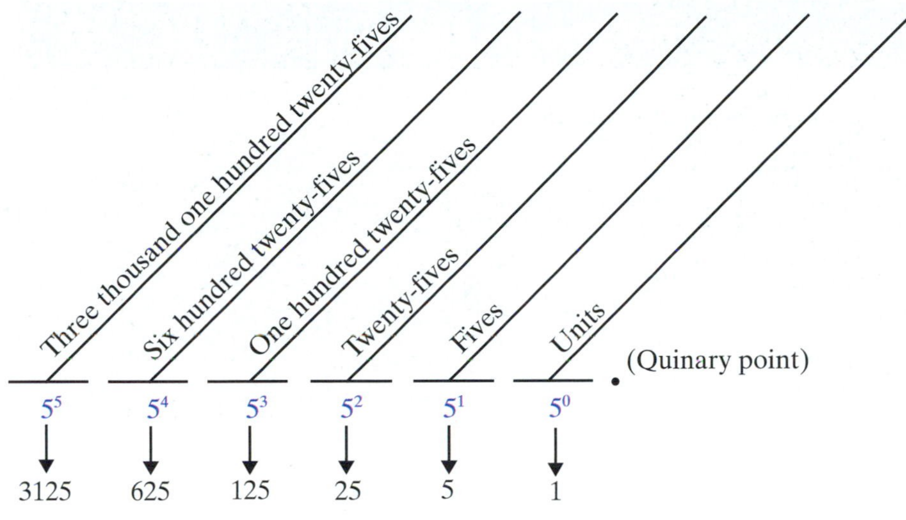

Figure IV.2

Example 1

a. $1_{(5)} = 1$

b. $2_{(5)} = 2$

c. $3_{(5)} = 3$

Now work exercise 1 in the margin.

1. Find the value of the numeral $4_{(5)}$.

Example 2

a. $11_{(5)} = 1(5)+1(1) = 5+1=6$

b. $12_{(5)} = 1(5)+2(1) = 5+2=7$

c. $14_{(5)} = 1(5)+4(1) = 5+4=9$

d. $20_{(5)} = 2(5)+0(1) = 10+0=10$

e. $21_{(5)} = 2(5)+1(1) = 10+1=11$

f. $22_{(5)} = 2(5)+2(1) = 10+2=12$

Now work exercise 2 in the margin.

2. Find the value of the following numerals.

a. $23_{(5)}$

b. $31_{(5)}$

c. $44_{(5)}$

3. Find the value of the numeral $1003_{(5)}$.

Example 3

$$\begin{aligned}324_{(5)} &= 3(5^2)+2(5^1)+4(5^0)\\ &= 3(25)+2(5)+4(1)\\ &= 75+10+4\\ &= 89\end{aligned}$$

Now work exercise 3 in the margin.

Practice Problems

Find the following numerals.

1. $210_{(5)}$

2. $412_{(5)}$

3. $330_{(5)}$

4. $123_{(5)}$

Answers to Practice Problems:

1. 55 **2.** 107 **3.** 90 **4.** 38

Name ______________________ Section ____________ Date __________

ANSWERS

1. [Respond below exercise.]
2. [Respond below exercise.]
3. [Respond below exercise.]
4. [Respond below exercise.]
5. [Respond below exercise.]
6. [Respond below exercise.]
7. [Respond below exercise.]
8. [Respond below exercise.]
9. [Respond below exercise.]
10. [Respond below exercise.]
11. [Respond below exercise.]
12. [Respond below exercise.]

Exercises IV.2

Write the following base five numerals in expanded form and find the value of each.

1. $24_{(5)}$

2. $13_{(5)}$

3. $10_{(5)}$

4. $43_{(5)}$

5. $104_{(5)}$

6. $312_{(5)}$

7. $32_{(5)}$

8. $230_{(5)}$

9. $423_{(5)}$

10. $444_{(5)}$

11. $1034_{(5)}$

12. $4124_{(5)}$

13. [Respond below exercise.]

14. [Respond below exercise.]

15. [Respond below exercise.]

16. ______

17. ______

18. ______

19. ______

20. ______

13. $244_{(5)}$

14. $3204_{(5)}$

15. $13042_{(5)}$

16. Do the numerals $101_{(2)}$ and $10_{(5)}$ represent the same number? If so, what is the number?

17. Do the numerals $1101_{(2)}$ and $23_{(5)}$ represent the same number? If so, what is the number?

18. Do the numerals $11100_{(2)}$ and $103_{(5)}$ represent the same number? If so, what is the number?

19. What set of digits do you think would be used in a base eight system?

20. What set of digits do you think would be used in a base twelve system? [HINT: New symbols for some new digits must be introduced.]

IV.3 Addition and Multiplication in Base Two and Base Five

Now that we have two new numeration systems, base two and base five, a natural question to ask is, how are addition and multiplication* performed in these systems? The basic techniques are the same as for base ten, since place value is involved. However, because different bases are involved, the numerals will be different. For example, to add five plus seven in base ten, we write $5+7=12$. In base two, this same sum is written $101_{(2)}+111_{(2)}=1100_{(2)}$.

Writing the numerals vertically (one under the other) gives

$$\begin{array}{r} 5 \\ \underline{7} \\ 12 \end{array} \qquad \begin{array}{r} 101_{(2)} \\ \underline{111_{(2)}} \\ 1100_{(2)} \end{array}$$

Now a step-by-step analysis of the sum in base two will be provided.

Example 1

Step1: $\begin{array}{r} 101_{(2)} \\ \underline{111_{(2)}} \end{array}$ The numerals are written so that the digits of the same place value line up.

Step2: $\begin{array}{r} \overset{1}{1}01_{(2)} \\ \underline{111_{(2)}} \\ 0_{(2)} \end{array}$ Adding 1 + 1 in the units column gives "two," which is written $10_{(2)}$. 0 is written in the units column, and 1 is "carried" to the "twos" column.

Step3: $\begin{array}{r} \overset{1\,1}{10}1_{(2)} \\ \underline{111_{(2)}} \\ 00_{(2)} \end{array}$ Now, in the twos column, 1 + 0 + 1 is again "two," or $10_{(2)}$. Again 0 is written, and 1 is "carried" to the next column, the "fours" column.

Step4: $\begin{array}{r} \overset{1\,1}{10}1_{(2)} \\ \underline{111_{(2)}} \\ 1100_{(2)} \end{array}$ In the fours column (or third column), 1 + 1 + 1 is "three," or $11_{(2)}$. Since there are no digits in the "eights" column (or fourth column), 11 is written, and the sum is $1100_{(2)}$.

Check:

$$\begin{array}{rclclr} 101_{(2)} &=& 1(2^2)+0(2^1)+1(2^0) &=& 4+0+1=5 & 5 \\ \underline{111_{(2)}} &=& 1(2^2)+1(2^1)+1(2^0) &=& 4+2+1=7 & \underline{7} \\ 1100_{(2)} &=& 1(2^3)+1(2^2)+0(2^1)+0(2^0) &=& 8+4+0+0=12 & 12 \end{array}$$

Now work exercise 1 in the margin.

1. Find the sum $\begin{array}{r} 1001_{(2)} \\ \underline{1110_{(2)}} \end{array}$

* Subtraction and division may also be performed in base two and base five, but will not be discussed here for reasons of time. Some students may want to investigate these operations on their own.

Addition in base five is similar. Although the thinking is done in base ten, which is familiar to us, the numerals written must be in base five.

2. Find the sum $\begin{array}{r} 121_{(5)} \\ \underline{231_{(5)}} \end{array}$

Example 2

Step1: $\begin{array}{r} 143_{(5)} \\ \underline{34_{(5)}} \end{array}$

The numerals are written so that the digits of the same place value line up.

Step2: $\begin{array}{r} \overset{1}{1}43_{(5)} \\ \underline{34_{(5)}} \\ 2_{(5)} \end{array}$

Adding 3 + 4 in the units column gives "seven," which is written $12_{(5)}$. 2 is written in the units column, and 1 is carried to the fives column.

Step3: $\begin{array}{r} \overset{1}{1}\overset{1}{4}3_{(5)} \\ \underline{34_{(5)}} \\ 32_{(5)} \end{array}$

In the fives column, 1 + 4 + 3 gives "eight," which is $13_{(5)}$. The 3 is written, and 1 is carried to the next column (the twenty-fives column).

Step4: $\begin{array}{r} \overset{1}{1}\overset{1}{4}3_{(5)} \\ \underline{34_{(5)}} \\ 232_{(5)} \end{array}$

In the third column, 1 + 1 gives 2. The sum is $232_{(5)}$.

Check:

$$\begin{array}{rcrcrcrr} 143_{(5)} & = & 1(5^2)+4(5^1)+3(5^0) & = & 25+20+3 & = & 48 & 48 \\ \underline{34_{(5)}} & = & 3(5^1)+4(5^0) & = & 15+4 & = & 19 & \underline{19} \\ 232_{(5)} & = & 2(5^2)+3(5^1)+2(5^0) & = & 50+15+2 & = & 67 & 67 \end{array}$$

Now work exercise 2 in the margin.

Multiplication in each base is performed and checked in the same manner as addition. Of course, the difference is that you multiply instead of add. When you multiply, be sure to write the correct symbol for the number in the base being used. Also, remember to add in the correct base.

Example 3

$$\begin{array}{r} 101_{(2)} \\ \times\ 111_{(2)} \\ \hline 101 \\ \overset{1}{1}01 \\ \overset{1}{1}01 \\ \hline 100011_{(2)} \end{array}$$

Multiplication in base two is easy, since we are multiplying by only 1's or 0's. The adding must be done in base two.

Check:

$$\begin{array}{rcr} 101_{(2)} & = & 5 \\ \times\ \ 111_{(2)} & = & \times\ 7 \\ \hline 101 & = & 35 \\ 101\ \ & & \\ 101\ \ \ \ & & \\ \hline 100011_{(2)} & & \end{array}$$

Remember to **multiply** the checking numbers.

$100{,}011_{(2)} = 1(2^5)+0(2^4)+0(2^3)+0(2^2)+1(2)+1(1)$
$= 32+2+1 = 35$

Now work exercise 3 in the margin.

3. Find the product $\begin{array}{r} 110_{(2)} \\ \underline{100_{(2)}} \end{array}$

Example 4

$$\begin{array}{r} {}^{1}\overset{2}{3}4_{(5)} \\ \times\ 23_{(5)} \\ \hline 212 \\ 123\ \ \\ \hline 1442_{(5)} \end{array}$$

Multiplying 3×4 gives "twelve," which is $22_{(5)}$. Write 2 and carry 2 just as in regular multiplication. Then, 3×3 is "nine," and "nine" plus 2 is "eleven"; but in base five, "eleven" is $21_{(5)}$. Similarly, 2×4 is "eight," or $13_{(5)}$. Write the 3, carry the 1. 2×3 is "six," and "six" plus 1 is "seven," or $12_{(5)}$.

Check:

$$\begin{array}{rcr} 34_{(5)} & = & 19 \\ \times\ \ 23_{(5)} & = & \times 13 \\ \hline 212 & & 57 \\ 123\ \ & & 19\ \ \\ \hline 1442_{(5)} & & 247 \end{array}$$

Remember to **multiply** the checking numbers.

$1442_{(5)} = 1(5^3)+4(5^2)+4(5)+2(1)$
$= 125+100+20+2 = 247$

Now work exercise 4 in the margin.

4. Find the product $\begin{array}{r} 12_{(5)} \\ \underline{21_{(5)}} \end{array}$

Practice Problems

Find the sums.

1. $\begin{array}{r} 1111_{(2)} \\ 1011_{(2)} \\ \hline \end{array}$

2. $\begin{array}{r} 3240_{(5)} \\ 1244_{(5)} \\ \hline \end{array}$

Find the products.

3. $\begin{array}{r} 1001_{(2)} \\ \times\, 1100_{(2)} \\ \hline \end{array}$

4. $\begin{array}{r} 131_{(5)} \\ \times\, 120_{(5)} \\ \hline \end{array}$

Answers to Practice Problems:

1. $11010_{(2)}$ **2.** $10034_{(5)}$ **3.** $1101100_{(2)}$ **4.** $21220_{(5)}$

Check:

$$\begin{array}{rcr} 101_{(2)} & = & 5 \\ \times\ 111_{(2)} & = & \times\ 7 \\ \hline 101 & = & 35 \\ 101 & & \\ 101 & & \\ \hline 100011_{(2)} & & \end{array}$$

Remember to **multiply** the checking numbers.

$100{,}011_{(2)} = 1(2^5)+0(2^4)+0(2^3)+0(2^2)+1(2)+1(1)$
$= 32+2+1 = 35$

Now work exercise 3 in the margin.

3. Find the product $\begin{array}{r} 110_{(2)} \\ 100_{(2)} \\ \hline \end{array}$

Example 4

$$\begin{array}{r} {}^{1}\overset{2}{3}4_{(5)} \\ \times\ 23_{(5)} \\ \hline 212 \\ 123 \\ \hline 1442_{(5)} \end{array}$$

Multiplying 3×4 gives "twelve," which is $22_{(5)}$. Write 2 and carry 2 just as in regular multiplication. Then, 3×3 is "nine," and "nine" plus 2 is "eleven"; but in base five, "eleven" is $21_{(5)}$. Similarly, 2×4 is "eight," or $13_{(5)}$. Write the 3, carry the 1. 2×3 is "six," and "six" plus 1 is "seven," or $12_{(5)}$.

Check:

$$\begin{array}{rcr} 34_{(5)} & = & 19 \\ \times\ 23_{(5)} & = & \times 13 \\ \hline 212 & & 57 \\ 123 & & 19 \\ \hline 1442_{(5)} & & 247 \end{array}$$

Remember to **multiply** the checking numbers.

$1442_{(5)} = 1(5^3)+4(5^2)+4(5)+2(1)$
$= 125+100+20+2 = 247$

Now work exercise 4 in the margin.

4. Find the product $\begin{array}{r} 12_{(5)} \\ 21_{(5)} \\ \hline \end{array}$

Practice Problems

Find the sums.

1. $\begin{array}{r} 1111_{(2)} \\ \underline{1011_{(2)}} \end{array}$

2. $\begin{array}{r} 3240_{(5)} \\ \underline{1244_{(5)}} \end{array}$

Find the products.

3. $\begin{array}{r} 1001_{(2)} \\ \underline{\times\, 1100_{(2)}} \end{array}$

4. $\begin{array}{r} 131_{(5)} \\ \underline{\times\, 120_{(5)}} \end{array}$

Answers to Practice Problems:

1. $11010_{(2)}$ 2. $10034_{(5)}$ 3. $1101100_{(2)}$ 4. $21220_{(5)}$

Name ____________________ Section ____________ Date ____________

Exercises IV.3

Add in the base indicated and check your work in base ten.

1. $101_{(2)}$ + $11_{(2)}$
2. $43_{(5)}$ + $213_{(5)}$
3. $1101_{(2)}$ + $1011_{(2)}$
4. $111_{(2)}$ + $1010_{(2)}$
5. $134_{(5)}$ + $243_{(5)}$
6. $11_{(2)}$ + $10_{(2)}$ + $11_{(2)}$
7. $11_{(2)}$ + $11_{(2)}$ + $101_{(2)}$
8. $214_{(5)}$ + $343_{(5)}$
9. $14_{(5)}$ + $321_{(5)}$ + $43_{(5)}$
10. $431_{(5)}$ + $314_{(5)}$ + $102_{(5)}$
11. $11_{(2)}$ + $101_{(2)}$ + $111_{(2)}$ + $101_{(2)}$
12. $111_{(2)}$ + $11_{(2)}$ + $110_{(2)}$ + $111_{(2)}$
13. $101_{(2)}$ + $101_{(2)}$ + $101_{(2)}$ + $101_{(2)}$
14. $23_{(5)}$ + $103_{(5)}$ + $214_{(5)}$ + $322_{(5)}$
15. $414_{(5)}$ + $211_{(5)}$ + $334_{(5)}$ + $222_{(5)}$

Multiply in the base indicated and check your work in base ten.

16. $1101_{(2)} \times 111_{(2)}$
17. $1011_{(2)} \times 101_{(2)}$
18. $423_{(5)} \times 30_{(5)}$
19. $104_{(5)} \times 23_{(5)}$
20. $223_{(5)} \times 44_{(5)}$
21. $423_{(5)} \times 32_{(5)}$
22. $1111_{(2)} \times 111_{(2)}$
23. $111_{(2)} \times 111_{(2)}$
24. $2212_{(5)} \times 43_{(5)}$
25. $10111_{(2)} \times 110_{(2)}$

ANSWERS

1. ____________
2. ____________
3. ____________
4. ____________
5. ____________
6. ____________
7. ____________
8. ____________
9. ____________
10. ____________
11. ____________
12. ____________
13. ____________
14. ____________
15. ____________
16. ____________
17. ____________
18. ____________
19. ____________
20. ____________
21. ____________
22. ____________
23. ____________
24. ____________
25. ____________

APPENDIX V

Ancient Numeration Systems

V.1 Egyptian, Mayan, Attic Greek, and Roman Systems

The number systems used by ancient peoples are interesting from a historical point of view, but from a mathematical point of view, they are difficult to work with. One of the many things that determine the progress of any civilization is its system of numeration. Humankind has made its most rapid progress since the invention of the zero and the place value system (which we will discuss in the next section) by the Hindu-Arabic peoples around A.D. 800.

Egyptian Numerals (Hieroglyphics)

The ancient Egyptians used a set of symbols called **hieroglyphics** as early as 3500 B.C. (See Table V.1.) To write the numeral for a number, the Egyptians wrote the symbols next to each other from left to right, and the number represented was the sum of the values of the symbols. The most times any symbol was used was nine. Instead of using a symbol ten times, they used the symbol for the next higher number. They also grouped the symbols in threes or fours.

Table V.1 Egyptian Hieroglyphic Numerals

Symbol	Name	Value	
|	Staff (vertical stroke)	1	one
∩	Heel bone (arch)	10	ten
𓍢	Coil of rope (scroll)	100	one hundred
𓆼	Lotus flower	1000	one thousand
𓂭	Pointing finger	10,000	ten thousand
𓆐	Bourbot (tadpole)	100,000	one hundred thousand
𓁨	Astonished man	1,000,000	one million

1. Find the number represented by the following hieroglyphic symbols:

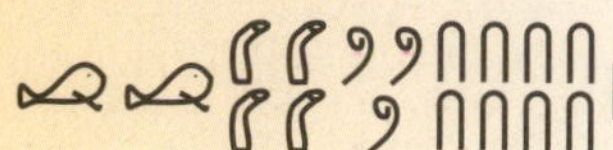

Example 1

represents the number one thousand six hundred twenty-seven, or $1000 + 600 + 20 + 7 = 1627$

Now work exercise 1 in the margin.

The Mayan System

The Mayans used a system of dots and bars (for numbers from 1 to 19) combined with a place value system. A dot represented one and a bar represented five. They had a symbol, , for zero and based their system, with one exception, on twenty. (See Table V.2.) The symbols were arranged vertically, smaller values starting at the bottom. The value of the third place up was 360 (18 times the value of the second place), but all other places were 20 times the value of the previous place.

Table V.2	Mayan Numerals
Symbol	**Value**
•	1 one
—	5 five
	0 zero

Find the numbers represented by the following Mayan symbols:

2.

3.

4.

Example 2

$(3 + 5 = 8)$

Now work exercise 2 in the margin.

Example 3

$(3 \times 5 + 4 = 19)$

Now work exercise 3 in the margin.

Example 4

three 20's

0 units

$(3 \times 20 + 0 = 60)$

Now work exercise 4 in the margin.

Example 5

two 7200's
zero 360's
six 20's
seven units

(2 × 7200 + 0 × 360 + 6 × 20 + 7 = 14,527)

(**Note:** [shell symbol] is used as a place holder.)

Now work exercise 5 in the margin.

5.

Attic Greek System

The Greeks used two numeration systems, the Attic (see Table V.3) and the Alexandrian. (See Section V.2 for information on the Alexandrian system.) In the Attic system, no numeral was used more than four times. When a symbol was needed five or more times, the symbol for five was used, as shown in the examples.

Table V.3 Attic Greek Numerals

Symbol		Value
I	1	one
Γ	5	five
Δ	10	ten
H	100	one hundred
X	[illegible]	one thousand
M	10,000	ten thousand

Find the numbers represented by the following Attic Greek symbols:

Example 6

X X Γ^{H} H H Γ^{Δ} I I I I (2 × 1000 + 7 × 100 + 5 × 10 + 4 = 2754)

Now work exercise 6 in the margin.

6. Γ^{H} H Δ I I I

Example 7

Γ^{X} H H H H Δ Δ Γ (5 × 1000 + 4 × 100 + 2 × 10 + 5 = 5425)

Now work exercise 7 in the margin.

7. Γ^{X} X X X H H Γ^{Δ} Δ Δ Δ I

Roman System

The Romans used a system (Table V.4) that we still see today in places such as hour symbols on clocks and dates on buildings.

Table V.4		Roman Numerals
Symbol		**Value**
I	1	one
V	5	five
X	10	ten
L	50	fifty
C	100	one hundred
D	500	five hundred
M	1000	one thousand

The symbols were written largest to smallest, from left to right. The value of the numeral was the sum of the values of the individual symbols. Each symbol was used as many times as necessary, with the following exceptions: When the Romans got to 4, 9, 40, 90, 400, or 900, they used a system of subtraction.

I V = 5 − 1 = 4
I X = 10 − 1 = 9
X L = 50 − 10 = 40

X C = 100 − 10 = 90
C D = 500 − 100 = 400
C M = 1000 − 100 = 900

Find the numbers represented by the following Roman symbols:

8. X X I

Example 8

V I I represents 7

Now work exercise 8 in the margin.

9. C C C I X

Example 9

D X L I V represents 544

Now work exercise 9 in the margin.

10. M M C M X L I V

Example 10

M C C C X X V I I I represents 1328

Now work exercise 10 in the margin.

Example 5

two 7200's
zero 360's
six 20's
seven units

$(2 \times 7200 + 0 \times 360 + 6 \times 20 + 7 = 14{,}527)$

(**Note:**  is used as a place holder.)

Now work exercise 5 in the margin.

5. 

Attic Greek System

The Greeks used two numeration systems, the Attic (see Table V.3) and the Alexandrian. (See Section V.2 for information on the Alexandrian system.) In the Attic system, no numeral was used more than four times. When a symbol was needed five or more times, the symbol for five was used, as shown in the examples.

Table V.3 Attic Greek Numerals

Symbol		Value
I	1	one
Γ	5	five
Δ	10	ten
H	100	one hundred
X	1000	one thousand
M	10,000	ten thousand

Find the numbers represented by the following Attic Greek symbols:

Example 6

X X Γ^{H} H H Γ^{Δ} I I I I $(2 \times 1000 + 7 \times 100 + 5 \times 10 + 4 = 2754)$

Now work exercise 6 in the margin.

6. Γ^{H} H Δ I I I

Example 7

Γ^{X} H H H H Δ Δ Γ $(5 \times 1000 + 4 \times 100 + 2 \times 10 + 5 = 5425)$

Now work exercise 7 in the margin.

7. Γ^{X} X X X H H Γ^{Δ} Δ Δ Δ I

Roman System

The Romans used a system (Table V.4) that we still see today in places such as hour symbols on clocks and dates on buildings.

Table V.4	Roman Numerals	
Symbol		**Value**
I	1	one
V	5	five
X	10	ten
L	50	fifty
C	100	one hundred
D	500	five hundred
M	1000	one thousand

The symbols were written largest to smallest, from left to right. The value of the numeral was the sum of the values of the individual symbols. Each symbol was used as many times as necessary, with the following exceptions: When the Romans got to 4, 9, 40, 90, 400, or 900, they used a system of subtraction.

I V = 5 − 1 = 4
I X = 10 − 1 = 9
X L = 50 − 10 = 40

X C = 100 − 10 = 90
C D = 500 − 100 = 400
C M = 1000 − 100 = 900

Find the numbers represented by the following Roman symbols:

8. X X I

9. C C C I X

10. M M C M X L I V

Example 8

V I I represents 7

Now work exercise 8 in the margin.

Example 9

D X L I V represents 544

Now work exercise 9 in the margin.

Example 10

M C C C X X V I I I represents 1328

Now work exercise 10 in the margin.

Practice Problems

Find the values of the following ancient numbers.

1. 𓁨 𓆐 𓂭𓂭 𓍢𓍢 𓎆𓎆𓎆𓎆𓎆𓎆 𓏺𓏺𓏺𓏺𓏺

2. 𝋭
   𝋱

3. M X X 𐅅 H H H Δ Δ Γ I I

4. C M L X I X

Answers to Practice Problems:

1. 1,120,265 **2.** 277 **3.** 12,827 **4.** 969

Name ______________ Section ______________ Date ______________

Exercises V.1

Find the values of the following ancient numbers.

1. 𓍢𓍢𓎆𓎆𓎆𓎆𓎆𓏺𓏺𓏺

2. 𓆐𓂭𓂭𓂭𓂭𓂭𓂭𓆼𓆼𓆼𓎆𓎆𓎆𓎆𓏺

3. 𓂭𓂭𓆼𓍢𓍢𓍢𓎆𓎆𓏺𓏺𓏺𓏺𓏺𓏺𓏺

4. 𝋭

5. 𝋣
𝋠

6. 𝋬
𝋫

7. 𐅄III

8. 𐅅HH𐅄ΔΔΔI

9. XXHHH𐅄ΔII

10. XCVII

11. DCCXLIV

12. MMMCDLXV

13. CMLXXVIII

14. Write 64
    a. as an Egyptian numeral
    b. as a Mayan numeral
    c. as an Attic Greek numeral
    d. as a Roman numeral

ANSWERS

1. ______
2. ______
3. ______
4. ______
5. ______
6. ______
7. ______
8. ______
9. ______
10. ______
11. ______
12. ______
13. ______
14. [Respond below exercise.] ______

15. [Respond below exercise.]

15. Follow the instructions for exercise 14, using 532 in place of 64.

16. [Respond below exercise.]

16. Follow the same instructions, using 1969.

17. [Respond below exercise.]

17. Follow the same instructions, using 846.

V.2 Babylonian, Alexandrian Greek, and Chinese-Japanese Systems

Babylonian System (Cuneiform Numerials)

The Babylonians (about 3500 B.C.) used a place value system based on the number sixty, called a **sexagesimal system**. They had only two symbols, v and <. (See Table V.5.) These wedge shapes are called **Cuneiform numerals**, since **cuneus** means "wedge" in Latin.

Table V.5		Cuneiform
Symbol		**Value**
v	1	one
<	10	ten

The symbol for one was used as many as nine times, and the symbol for ten as many as five times; however, since there was no symbol for zero, many Babylonian numbers could be read several ways. For our purposes, we will group the symbols to avoid some of the ambiguities inherent in the system.

Find the numbers represented by the following Cuneiform symbols:

1.
```
         vvvv
vv    <<  vvv
```

Example 1

```
            vvv
vvv    <<   vv     <<< vv
```

$(3 \times 60^2) + (25 \times 60) + (32 \times 1)$
$= (3 \times 3600) + (25 \times 60) + 32$
$= 10{,}800 + 1500 + 32$
$= 12{,}332$

Now work exercise 1 in the margin.

2.
```
v   < vvv   << vv
```

Example 2

```
         vvv        vvvv
v   <<<<  vvv   <    vvv
```

$(1 \times 60^2) + (46 \times 60) + (17 \times 1) = 3600 + 2760 + 17 = 6377$

Now work exercise 2 in the margin.

Alexandrian Greek System

The Greeks used two numeration systems, the Attic and the Alexandrian. We discussed the Attic Greek system in Section V.1.

In the Alexandrian system (Table V.6), the letters were written next to each other, largest to smallest, from left to right. Since the numerals were also part of the Greek alphabet, an accent mark or bar was sometimes used above a letter to indicate that it represented a number. Multiples of 1000 were indicated by strikes (/) in front of the unit symbols, and multiples of 10,000 were indicated by placing the unit symbols above the symbol M.

Table V.6				Alexandrian Greek Symbols			
Symbol	**Name**	**Value**		**Symbol**	**Name**	**Value**	
Α	Alpha	1	one	Ν	Nu	50	fifty
Β	Beta	2	two	Ξ	Xi	60	sixty
Γ	Gamma	3	three	Ο	Omicron	70	seventy
Δ	Delta	4	four	Π	Pi	80	eighty
Ε	Epsilon	5	five	C	Koppa	90	ninety
F	Digamma	6	six	Ρ	Rho	100	one hundred
	(or Vau)			Σ	Sigma	200	two hundred
Ζ	Zeta	7	seven	Τ	Tau	300	three hundred
Η	Eta	8	eight	Υ	Upsilon	400	four hundred
Θ	Theta	9	nine	Φ	Phi	500	five hundred
Ι	Iota	10	ten	Χ	Chi	600	six hundred
Κ	Kappa	20	twenty	Ψ	Psi	700	seven hundred
Λ	Lambda	30	thirty	Ω	Omega	800	eight hundred
Μ	Mu	40	forty	ϡ	Sampi	900	nine hundred

Find the numbers represented by the following Alexandrian Greek symbols:

3. $\overline{/\Delta \mathrm{X} \Theta}$

Example 3

$\overline{\Phi\, \Xi\, \mathrm{Z}}$ (500 + 60 + 7 = 567)

Now work exercise 3 in the margin.

4. $\overline{\Gamma}$
Μ / Α Ρ Π Β

Example 4

$\overline{\mathrm{B}}$
Μ Τ Ν Δ (20,000 + 300 + 50 + 4 = 20,354)

Now work exercise 4 in the margin.

Chinese - Japanese System

The Chinese-Japanese system (Table V.7) uses a different numeral for each of the digits up to ten, then a symbol for each power of ten. A digit written above a power of ten is to be multiplied by that power, and all such results are to be added to find the value of the numeral.

Table V.7		Chinese-Japanese Numerals	
Symbol	**Value**	**Symbol**	**Value**
一	1 one	七	7 seven
二	2 two	八	8 eight
三	3 three	九	9 nine
四	4 four	十	10 ten
五	5 five	百	100 one hundred
六	6 six	千	1000 one thousand

Find the numbers represented by the following Chinese-Japanese symbols:

Example 5

三
十 } 30

九 9

(30 + 9 = 39)

Now work exercise 5 in the margin.

5. 五
百
七

Example 6

五
千 } 5000

四
百 } 400

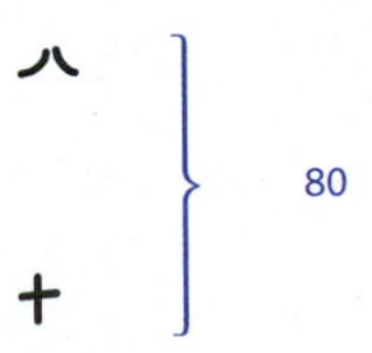
八
十 } 80

二 2

(5000 + 400 + 80 + 2 = 5482)

Now work exercise 6 in the margin.

6. 九
百
四
十
一

Practice Problems

Find the values of the following ancient numbers.

1. << vv < vvvv/vvvv

2. $\overline{\mathrm{X}\,\Lambda\,\Delta}$

3. 六
十
四

Answers to Practice Problems:

1. 1338 **2.** 634 **3.** 64

Name ______________________ Section ____________ Date __________

Exercises V.2

Find the value of each of the following ancient numerals.

1. ∨ <<∨∨∨ ∨∨∨ **2.** <<<∨∨ **3.** <<< ∨∨

4. $\overline{\text{Y N E}}$ **5.** $\overline{\Delta}$
M /Zπ **6.** $\overline{\Sigma\text{ K B}}$

7. 四
十
六

8. 五
千
一
百
八

9. 九
百
九
十
九

Write the following numbers as ***a.*** *Babylonian numerals,* ***b.*** *Alexandrian Greek numerals, and* ***c.*** *Chinese-Japanese numerals.*

10. 472 **11.** 596

ANSWERS

1. ______________

2. ______________

3. ______________

4. ______________

5. ______________

6. ______________

7. ______________

8. ______________

9. ______________

10. [Respond below exercise.]

11. [Respond below exercise.]

12. [Respond below exercise.]

13. [Respond below exercise.]

14. [Respond below exercise.]

15. [Respond below exercise.]

12. 5047

13. 3665

14. 7293

15. 10,852

Answer Key

CHAPTER 1

Margin Exercises 1.1

1. 400 + 60 + 3 **2.** 7000 + 300 + 0 + 2 **3.** 20,000 + 9000 + 500 + 20 + 4 **4.** 800,000 + 0 + 8000 + 400 + 90 + 1 **5.** 500 + 60 + 7; five hundred sixty-seven **6.** 20,000 + 5000 + 400 + 0 + 0; twenty-five thousand, four hundred **7.** 6000 + 700 + 90 + 1; six thousand seven hundred ninety-one **8.** 20,000 + 3000 + 500 + 0 + 7; twenty-three thousand, five hundred seven **9.** 28,642 **10.** 363,075 **11.** 6,300,500 **12.** 4,875,000

Exercises 1.1, page 9

1. 76 **3.** 2005 **5.** 33,333 **7.** 281,300,501 **9.** 82,700,000 **11.** 757 **13.** left; decimal point **15.** and **17.** 21 to 99; 0 **19.** 80 + 4 **21.** 800 + 20 + 1 **23.** 2000 + 0 + 50 + 9 **25.** 30,000 + 2000 + 300 + 40 + 1 **27.** one hundred twenty-two **29.** six hundred eighty-three thousand, one hundred **31.** sixteen million, three hundred two thousand, five hundred ninety **33.** 75 **35.** 142 **37.** 3834 **39.** 400,736 **41.** 63,251,065 **43.** three million, eight hundred forty-five thousand, five hundred forty-one **45.** one hundred ten; two thousand six hundred fifty
47. one hundred one; one thousand seven hundred sixty-one **49. a.** 6 **b.** 8 **c.** 0

Margin Exercises 1.2

1. 110 **2.** 13; commutative property **3.** 20; associative property **4.** 6889 **5.** 7797 **6.** 1217 **7.** 1614 **8.** $1377

Exercises 1.2, page 21

1. 19 **3.** 13 **5.** 21 **7.** 10 **9.** 12 **11.** 18 **13.** 12 **15.** 17 **17.** 27 **19.** 24 **21.** commutative **23.** associative **25.** associative **27.** additive identity **29.** additive identity **31.** 108 **33.** 239 **35.** 1298 **37.** 1127 **39.** 6869 **41.** 329,134 **43.** 1,238,914 **45.** 2762 miles **47.** $55,085 **49.** 1518 students

51.

+	5	8	7	9
3	8	11	10	12
6	11	14	13	15
5	10	13	12	14
2	7	10	9	11

Margin Exercises 1.3

1. 11 **2.** 233 **3.** 48 **4.** 363 **5.** 539 **6.** 38 **7.** 6577 **8.** 386 **9.** $67 **10.** 300 + 40 + 1 = 341 **11.** 2556 people **12.** $18,103

Exercises 1.3, page 33

1. 3 **3.** 0 **5.** 9 **7.** 9 **9.** 5 **11.** 0 **13.** 13 **15.** 20 **17.** 5 **19.** 13 **21.** 94 **23.** 218 **25.** 475 **27.** 593 **29.** 188 **31.** 1569 **33.** 1568 **35.** 0 **37.** 694 **39.** 2517 **41.** 2,806,644 **43.** 1,006,958 **45.** 5,671,011 **47.** 140 **49.** 16 years **51.** $250,404

Exercises 1.3 (cont.)

53. \$976 **55.** \$39,100 **57.** \$3700

Margin Exercises 1.4

1. 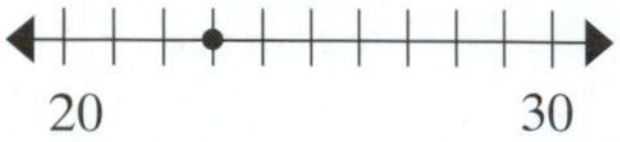

23 is closer to 20 than it is to 30, so 23 rounded to the nearest 10 is 20.

2. 

782 is closer to 800 than it is to 700, so 782 rounds to 800. **3.** 2000 **4.** 5000 **5.** 300 **6.** 2000 (estimate); 1976 (sum) **7.** 200 (estimate); 205 (difference) **8.** 9000 (estimate); 8861 (sum)

Exercises 1.4, page 45

1. 760 **3.** 80 **5.** 300 **7.** 990 **9.** 4200 **11.** 500 **13.** 600 **15.** 76,500 **17.** 7000 **19.** 8000 **21.** 13,000 **23.** 62,000 **25.** 80,000 **27.** 260,000 **29.** 120,000 **31.** 180,000 **33.** 100 **35.** 1000 **37.** 180 (estimate); 167 (sum) **39.** 1200 (estimate); 1173 (sum) **41.** 1900 (estimate); 1881 (sum) **43.** 9500 (estimate); 9224 (sum) **45.** 6000 (estimate); 5467 (difference) **47.** 5000 (estimate); 5931 (difference) **49.** 20,000 (estimate); 20,804 (difference) **51.** 700,000,000 people **53.** \$250

Margin Exercises 1.5

1. 56; commutative property **2.** 5; commutative property **3.** 0; zero factor law **4.** 56; associative property **5.** 18; associative property **6.** 8000 **7.** 40,000 **8.** 15,000,000 **9.** 5 and 10 or 50 and 1 **10.** 800 and 1, or 80 and 10, or 8 and 100 **11.** 68 and 1000, or 680 and 100, or 6800 and 10, or 68,000 and 1 **12.** 1792 **13.** 4067 **14.** 23,808 **15.** 252 **16.** 3145 **17.** 1752 **18.** 5022 **19.** 1,416,000 **20.** 8,767,000 **21.** 1400 (estimate); 1332 (product) **22.** 24,000 (estimate); 19,285 (product)

Exercises 1.5, page 61

1. 72 **3.** 24 **5.** 7 **7.** 0 **9.** 28; commutative **11.** 5; multiplicative identity **13.** 250 **15.** 4000 **17.** 16,000 **19.** 900 **21.** 16,000,000 **23.** 5000 **25.** 36,000 **27.** 150,000 **29.** 240 (estimate); 224 (product) **31.** 450 (estimate); 432 (product) **33.** 240 (estimate); 252 (product) **35.** 2400 (estimate); 2352 (product) **37.** 1000 (estimate); 960 (product) **39.** 7200 (estimate); 7055 (product) **41.** 600 (estimate); 544 (product) **43.** 800 (estimate); 880 (product) **45.** 600 (estimate); 375 (product) **47.** 1800 (estimate); 2064 (product) **49.** 100 (estimate); 156 (product) **51.** 4000 (estimate); 5166 (product) **53.** 3000 (estimate); 2850 (product) **55.** 20,000 (estimate); 29,601 (product) **57.** 10,000 (estimate); 9800 (product) **59.** 140,000 (estimate); 125,178 (product) **61.** 231 (sum); 58 (difference); 13,398 (product) **63.** \$2760 per year; \$165,600 over five years **65.** \$284,400 **67.** \$23 **69.** 240 min; 1680 min **71.** 3,844,757,974,500 miles **73.** (a) C; (b) D; (c) B; (d) E; (e) A

Margin Exercises 1.6

1. 16 with a remainder of 1 **2.** 94 with a remainder of 3 **3.** 46 R 5 **4.** 20 R 4 **5.** 308 R 23 **6.** 155 R 9 **7.** 214 R 24 **8.** 14 divides into 294 with a quotient of 21 and a remainder of 0; thus both 21 and 14 are divisors of 294. **9.** \$115 **10.** 11 (estimate); 11 R 46 **11.** 666 (estimate); 682 R 46 **12.** 37 R 8; 25 (estimate); 12 (difference)

Exercises 1.6, page 79

1. 40 **3.** 30 **5.** 21 **7.** 12 **9.** 5 **11.** 6 R 4 **13.** 24 **15.** 10 R 11 **17.** 11 R 7 **19.** 41 **21.** 5 R 2 **23.** 6 R 1 **25.** 6 **27.** 9 **29.** 32 R 2 **31.** 9 **33.** 8 **35.** 11 R 8 **37.** 42 R 3 **39.** 20 **41.** 30 **43.** 300 R 13 **45.** 301 R 4 **47.** 3 R 3 **49.** 61 R 15 **51.** 3 R 6 **53.** 22 R 74 **55.** 7 R 358 **57.** 196 R 370 **59.** 107 R 215 **61.** \$3000; \$2548 **63.** $1008 \div 28 = 36$; $1008 \div 36 = 28$ **65.** 75 people/mi^2 **67.** (a) B; (b) D; (c) A; (d) C

Margin Exercises 1.7

1. \$21 **2.** \$414 **3.** 36 home runs **4.** 109 **5.** 200,000 people **6.** 200,000 people **7.** 12 points

Exercises 1.7, page 89

1. \$160 **3.** \$786 **5.** \$874 **7.** \$485 **9.** \$316 **11. a.** 2000 **b.** 40,030 cars **13.** \$125 **15.** 103 **17.** 6 **19.** 485 **21.** 85 **23.** \$26/share; \$500 profit **25. a.** \$665 **b.** \$2495 **27.** 62,291 people **29. a.** 24,383 students **b.** \$15,350 **c.** \$22,175 **d.** \$8008

Margin Exercises 1.8

1. 32 ft **2.** 58 cm **3.** 130 in. **4.** 28 mm^2 **5.** 100 cm^2 **6.** 80 cm^2

Exercises 1.8, page 103

1. point, line, and plane **3. a.** True **b.** False **5.** 12 in. **7.** 15 ft **9.** 70 in. **11.** 108 ft **13.** 48 in. **15.** 56 ft **17.** 360 cm^2 **19.** 176 yd^2 **21.** 1200 ft^2 **23.** 25 yd^2 **25.** 1 ft^2 **27.** 120 ft^2 **29.** 500 yd^2 **31.** 120 in. **33.** 96 in.

Chapter 1 Review Questions, page 113

1. 400 + 90 + 5; four hundred ninety-five **2.** 1000 + 900 + 70 + 5; one thousand nine hundred seventy-five **3.** 60,000 + 0 + 300 + 0 + 8; sixty thousand, three hundred eight **4.** 4856 **5.** 15,032,197 **6.** 672,340,083 **7.** 630 **8.** 15,000 **9.** 700 **10.** 2600 **11.** commutative property of addition **12.** associative property of multiplication **13.** associative property of addition **14.** commutative property of multiplication **15.** 9800 (estimate); 10,541 (sum) **16.** 1740 (estimate); 1674 (sum) **17.** 508 **18.** 2384 **19.** 2102 **20.** 0 **21.** 5600 **22.** 360,000 **23.** 5000 (estimate); 5096 (product) **24.** 3,600,000 (estimate); 3,913,100 (product) **25.** 25,000,000 (estimate); 24,185,000 (product) **26.** 285 (estimate); 292 (quotient) **27.** 500 (estimate); 606 (quotient) **28.** 140 (estimate); 135 R 81 **29.** 1059 **30.** 35 **31.** 9 **32.** \$1485; \$99 **33.** 83 **34.** 70 **35.** \$7700

36.

Given no.	Add 100	Double	Subtract 200
3	103	206	6
20	120	240	40
15	115	230	30
8	108	216	16

37. 82 m **38.** 104 in. **39. a.** 96 cm **b.** 336 cm^2 **40.** 6600 yd^2

Chapter 1 Test, page 119

1. 8000 + 900 + 50 + 2; eight thousand nine hundred fifty-two **2.** identity **3.** $7 \cdot 9 = 9 \cdot 7$ (Answers may vary) **4.** 1000 **5.** 140,000 **6.** 12,300 (estimate); 12,009 (sum) **7.** 1840 (estimate); 1735 (sum) **8.** 4,000,000 (estimate); 4,057,750 (sum) **9.** 488 **10.** 1229 **11.** 5707 **12.** 2400 (estimate); 2584 (product) **13.** 270,000 (estimate); 220,405 (product) **14.** 240,000 (estimate); 210,938 (product) **15.** 403 **16.** 172 R 388 **17.** 2005 **18.** 74 **19.** 54 **20. a.** \$306 **b.** \$51 **c.** \$224 **21.** \$2025 **22.** 80 miles **23. a.** 152 in. **b.** 1428 $in.^2$ **24. a.** 60 cm **b.** 120 cm^2

CHAPTER 2

Margin Exercises 2.1

1. a. $5^2 = 25$ **b.** $2^4 = 16$ **c.** $3^3 = 27$ **d.** $4^3 = 64$ **2. a.** 2; 8; 64 **b.** 3; 6; 216 **c.** 2; 5; 25 **d.** 6; 2; 64 **e.** 0; 12; 1 **f.** 4; 10; 10,000

Exercises 2.1, page 129

1. a. 3 **b.** 2 **c.** 8 **3. a.** 2 **b.** 5 **c.** 25 **5. a.** 0 **b.** 7 **c.** 1 **7. a.** 4 **b.** 1 **c.** 1 **9. a.** 0 **b.** 4 **c.** 1 **11. a.** 2 **b.** 3 **c.** 9 **13. a.** 0 **b.** 5 **c.** 1 **15. a.** 1 **b.** 62 **c.** 62 **17. a.** 2 **b.** 10 **c.** 100 **19. a.** 2 **b.** 4 **c.** 16 **21. a.** 4 **b.** 10 **c.** 10,000 **23. a.** 3 **b.** 6 **c.** 216 **25. a.** 0 **b.** 19 **c.** 1 **27.** 5^2 **29.** 3^3 **31.** 11^2 **33.** 2^3 **35.** 6^2 **37.** 9^2 or 3^4 **39.** 12^2 **41.** 10^2 **43.** 100^2 or 10^4 **45.** 3^5 **47.** 15^2 **49.** 7^3 **51.** 6^5 **53.** $2^2 \cdot 7^2$ **55.** $2^2 \cdot 3^3$ **57.** $7^2 \cdot 13$ **59.** $2 \cdot 3^2 \cdot 11^2$ **61.** 64 **63.** 49 **65.** 225 **67.** 324 **69.** 144 **71.** 100

Exercise 2.1 (cont.)

73. 900 **75.** 2500

Margin Exercises 2.2

1. 8 **2.** 13 **3.** 11 **4.** 5 **5.** 54 **6.** 21 **7.** 36 **8.** 19

Exercises 2.2, page 139

1. muliplication; division; addition **3.** division (within parentheses); addition (within parentheses); division **5.** 3 **7.** 7 **9.** 22 **11.** 3 **13.** 5 **15.** 5 **17.** 5 **19.** 3 **21.** 7 **23.** 0 **25.** 6 **27.** 3 **29.** 27 **31.** 26 **33.** 1 **35.** 69 **37.** 68 **39.** 140 **41.** 5 **43.** 0 **45.** 9 **47.** 12 **49.** 24 **51.** 70 **53.** 230 **55.** 110 **57.** 110 **59.** 2980 **61.** 17 **63.** 2072 **65.** 198

Margin Exercises 2.3

1. a. Yes, the last digit is 8, an even digit. **b.** No, $7 + 9 + 1 + 2 = 19$ and 19 is not divisible by 3. **c.** No, 1576 is not divisible by 3. **d.** No, the last digit is not 0 or 5. **e.** Yes, $4 + 6 + 5 + 3 = 18$ and 18 is divisible by 9. **f.** Yes, the last digit is 0. **2.** 2, 3, 5, 6, and 10 **3.** 2, 3, 5, 6, 9, and 10 **4.** 3289 is not divisible by 2, 3, 5, 6, 9, or 10
5. a. Yes, $2 + 5 + 5 = 12$ and 12 is divisible by 3. **b.** Yes, the last digit is 5. **c.** No, 255 is not divisible by 2.
6. a. Yes, the last digit is 8, an even digit. **b.** No, the last digit is not 0 or 5. **c.** Yes, $3 + 7 + 8 = 18$ and 18 is divisible by 9.
7. 24 does divide the product, and it divides the product 315 time. **8.** 21 does not divide the given product.
9. 30 does divide the product, and it divides the product 21 times.
10. 55 does divide the product, and it divides the product 96 times.

Exercises 2.3, page 153

1. 2, 3, 6, 9 **3.** 3, 5 **5.** 2, 3, 5, 6, 10 **7.** 2 **9.** 2, 3, 6 **11.** none **13.** 3, 9 **15.** none **17.** none **19.** 3 **21.** 3, 5 **23.** 2, 3, 5, 6, 9, 10 **25.** 2, 3, 6 **27.** none **29.** 2 **31.** 3, 5 **33.** 3, 9 **35.** 2, 3, 6, 9 **37.** 2, 5, 10 **39.** 3, 9 **41.** 2, 3, 5, 6, 10 **43.** 2 **45.** 2, 3, 5, 6, 10 **47.** 3, 5 **49.** none **51.** 2, 3, 6, 9 **53.** 2, 3, 6 **55.** 2, 3, 6, 9 **57.** 3, 9 **59.** 2, 3, 6
61. Yes; $2 \cdot 3 \cdot 3 \cdot 5 = 6 \cdot 15$. The product is 90 and 6 divides the product 15 times. **63.** Yes; $2 \cdot 3 \cdot 5 \cdot 7 = 14 \cdot 15$. The product is 210 and 14 divides the product 15 times. **65.** No; $3 \cdot 3 \cdot 5 \cdot 7 = 315$ and 10 is not a factor of 315.
67. Yes; $2 \cdot 2 \cdot 3 \cdot 5 \cdot 5 = 25 \cdot 12$. The product is 300 and 25 divides the product 12 times.
69. Yes; $3 \cdot 3 \cdot 5 \cdot 7 \cdot 11 = 21 \cdot 165$. The product is 3465 and 21 divides the product 165 times.

Review Problems (from Section 1.4), page 156

1. 850 **2.** 1930 **3.** 400 **4.** 2600 **5.** 14,000 **6.** 21,000 **7.** 690 (estimate); 693 (sum) **8.** 4700 (estimate); 4450 (sum) **9.** 5000 (estimate); 5225 (difference) **10.** 20,000 (estimate); 28,522 (difference)

Margin Exercises 2.4

1. a. Prime; 13 has only two factors, 1 and 13. **b.** Composite; 1, 5, and 25 are all factors of 25. **c.** Composite; 1, 2, 4, 8, 16, and 32 are all factors of 32. **2. a.** Prime; 31 has only two factors, 1 and 31. **b.** Composite; 1, 3, 9, and 27 are all factors of 27. **c.** Prime; 19 has only two factors, 1 and 19.
3. No, since the digit in the units place is 4, 504 is divisible by 2. Therefore, 504 has more than two factors.
4. Yes, 113 is a prime number. Test the prime numbers less than 11 to come to this conclusion.
5. 247 is not a prime number because 1, 13, 19, and 247 are factors of 247.
6. 299 is not a prime number because 1, 13, 23, and 299 are factors of 299. **7.** $21 \cdot 4 = 84$ and $21 + 4 = 25$.

Exercises 2.4, page 165

1. 5, 10, 15, 20 **3.** 11, 22, 33, 44 **5.** 12, 24, 36, 48 **7.** 20, 40, 60, 80 **9.** 16, 32, 48, 64

11.

1	(2)	(3)	4	(5)	6	(7)	8	9	10
(11)	12	(13)	14	15	16	(17)	18	(19)	20
21	22	(23)	24	25	26	27	28	(29)	30
(31)	32	33	34	35	36	(37)	38	39	40
(41)	42	(43)	44	45	46	(47)	48	49	50
51	52	(53)	54	55	56	57	58	(59)	60
(61)	62	63	64	65	66	(67)	68	69	70
(71)	72	(73)	74	75	76	77	78	(79)	80
81	82	(83)	84	85	86	87	88	(89)	90
91	92	93	94	95	96	(97)	98	99	100

13. prime **15.** composite; $1 \cdot 32, 2 \cdot 16, 4 \cdot 8$ **17.** prime **19.** composite; $1 \cdot 63, 3 \cdot 21, 7 \cdot 9$ **21.** composite; $1 \cdot 51, 3 \cdot 17$ **23.** prime **25.** prime **27.** composite; $1 \cdot 57, 3 \cdot 19$ **29.** composite; $1 \cdot 86, 2 \cdot 43$ **31.** prime **33.** 3, 4 **35.** 12, 1 **37.** 25, 2 **39.** 8, 3 **41.** 12, 3 **43.** 21, 3 **45.** 5, 5 **47.** 12, 5 **49.** 9, 3

Review Problems (from Section 1.5), page 169

1. commutative property of multiplication **2.** associative property of multiplication **3.** 16,000 **4.** 150,000 **5.** 840,000 **6.** 22,922 **7.** 84,812 **8.** 178,500

Margin Exercises 2.5

1. $84 = 2 \cdot 2 \cdot 3 \cdot 7$ **2.** $110 = 2 \cdot 5 \cdot 11$ **3. a.** $13 \cdot 5$ **b.** $2^4 \cdot 5$ **c.** $2^6 \cdot 3$ **d.** $2^2 \cdot 3^2 \cdot 5$ **4.** $250 = 2 \cdot 5^3$ **5.** $775 = 5^2 \cdot 31$ **6.** $165 = 3 \cdot 5 \cdot 11$ **7.** 1, 2, 3, 6, 7, 14, 24, and 42 **8.** 1, 2, 4, 5, 8, 10, 16, 20, 32, 40, 80, 160

Exercises 2.5, page 177

1. $2^3 \cdot 3$ **3.** 3^3 **5.** $2^2 \cdot 3^2$ **7.** $2^3 \cdot 3^2$ **9.** 3^4 **11.** 5^3 **13.** $3 \cdot 5^2$ **15.** $2 \cdot 3 \cdot 5 \cdot 7$ **17.** $2 \cdot 5^3$ **19.** $2^3 \cdot 3 \cdot 7$ **21.** $2 \cdot 3^2 \cdot 7$ **23.** prime **25.** $3 \cdot 17$ **27.** 11^2 **29.** $3^2 \cdot 5^2$ **31.** 2^5 **33.** $2^2 \cdot 3^3$ **35.** prime **37.** $2 \cdot 5 \cdot 17$ **39.** $2^4 \cdot 5^4$ **41.** 1, 2, 3, 4, 6, 12 **43.** 1, 2, 4, 7, 14, 28 **45.** 1, 11, 121 **47.** 1, 3, 5, 7, 15, 21, 35, 105 **49.** 1, 97

Review Problems (from Sections 1.7 and 1.8), page 180

1. 1575 **2.** $12,942 **3.** 104 **4. a.** 30 meters **b.** 30 square meters

Margin Exercises 2.6

1. 300 **2.** 126 **3.** 180 **4.** 120 **5.** 480 **6.** 630 **7.** 210; $210 = 15 \cdot 14$; $210 = 35 \cdot 6$; $210 = 42 \cdot 5$ **8.** 450; $450 = 10 \cdot 45$; $450 = 18 \cdot 25$; $450 = 75 \cdot 6$ **9. a.** 900 **b.** $900 = 20 \cdot \mathbf{45}$; $900 = 25 \cdot \mathbf{36}$; $900 = 36 \cdot \mathbf{25}$ **10.** 3 hours (i.e., they met again at 11 am) **11.** The first intern will have made 6 rounds, the second intern will have made 5 rounds and the third intern will have made 10 rounds.

Exercises 2.6, page 189

1. 24 **3.** 36 **5.** 110 **7.** 60 **9.** 200 **11.** 196 **13.** 240 **15.** 100 **17.** 700 **19.** 252 **21.** 8 **23.** 1560 **25.** 60 **27.** 240 **29.** 2250 **31.** 726 **33.** 2610 **35.** 675 **37.** 120 **39.** 1680 **41. a.** 120 **b.** 15, 12, 8 **43. a.** 120 **b.** 12, 8, 5 **45. a.** 270 **b.** 45, 15, 10, 6 **47. a.** 4410 **b.** 98, 70, 45 **49. a.** 14,157 **b.** 143, 99, 39 **51. a.** 70 minutes **b.** 7 laps and 5 laps **53. a.** 48 hours **b.** 4 orbits and 3 orbits **55. a.** 180 days **b.** 18, 15, 12, and 10 trips

Chapter 2 Review Questions, page 197

1. base, exponent, 81 **2.** different factors **3.** prime **4.** composite **5.** 2^4 or 4^2 **6.** 3^3 **7.** 13^2 **8.** 20^2 **9.** 15 **10.** 46 **11.** 35 **12.** 13 **13.** 2 **14.** 2 **15.** 3, 5, 9 **16.** 2, 3, 6, 9 **17.** none **18.** 2, 3, 5, 6, 9, 10 **19.** 2 **20.** 3 **21.** 3, 6, 9, 12, 15, 18, 21, 24, 27, 30; yes, 3 is prime **22.** yes **23.** 2, 3, 5, 7, 11, 13, 17, 19, 23, 29, 31, 37, 41, 43, 47, 53, 59 **24.** 6 and 4 **25.** 5 and 12 **26.** $2 \cdot 3 \cdot 5^2$ **27.** $5 \cdot 13$ **28.** $2^2 \cdot 3 \cdot 7$ **29.** $2^2 \cdot 23$ **30.** 168 **31.** 1800 **32.** 270 **33.** 1638; 91, 42, 26 **34.** 210 seconds; 14 laps, 12 laps

Chapter 2 Test, page 199

1. base, exponent, 343 **2. a.** 7, 11, 13, 17, 19 **b.** 49, 121, 169, 289, 361 **3.** 13 **4.** 48 **5.** 0 **6.** 72 **7.** 2, 3, 5, 6, 9, 10 **8.** none **9.** 2, 3, 6, 9 **10.** 2, 5, 10 **11.** 13, 26, 39, 52, 65, 78, 91 **12.** 6, 9, 15, 27, 39, 51 **13.** $7452 \div 81 = 92$, so 81 is a factor of 7452 and $81 \cdot 92 = 7452$. **14.** Yes, $2 \cdot 5 \cdot 6 \cdot 7 \cdot 9 = 42 \cdot 90$. So 42 divides the product 90 times. **15.** 1, 2, 3, 4, 5, 6, 10, 12, 15, 20, 30, 60 **16.** $124 = 2 \cdot 2 \cdot 31 = 2^2 \cdot 31$ **17.** $165 = 3 \cdot 5 \cdot 11$ **18.** 107 is prime. **19.** 84; 21, 6, 4 **20.** 60; 10, 4, 1 **21.** 840; 105, 84, 56, 30 **22. a.** every 36 days **b.** every 180 days

Cumulative Review: Chapters 1 - 2, page 201

1. a. $50{,}000 + 0 + 700 + 30 + 2$ **b.** fifty thousand, seven hundred thirty-two **2.** commutative property of mulitplication **3.** associative property of multiplication **4.** additive identity **5.** commutative property of addition **6.** 42,000 **7. a.** 1980 **b.** 2065 **8. a.** 5200 **b.** 5244 **9. a.** 6300 **b.** 6364 **10. a.** 100 **b.** 130 **11.** 27,000 **12.** 3 **13.** 57 **14.** 2, 5, 10 (even and ends in zero) **15.** 307 is prime. **16.** 2 is the only even prime number. **17.** 1, 5, 13, 65 **18.** $475 = 5 \cdot 5 \cdot 19 = 5^2 \cdot 19$ **19.** Yes, 56 divides the product 162 times. **20. a.** 945 **b.** 63, 35, 27 **21. a.** 7260 **b.** 220, 165, 132, 60 **22.** 2333 km **23.** 1728 in.^{2} **24.** 85 **25. a.** 378 seconds **b.** 7 laps and 6 laps

CHAPTER 3

Margin Exercises 3.1

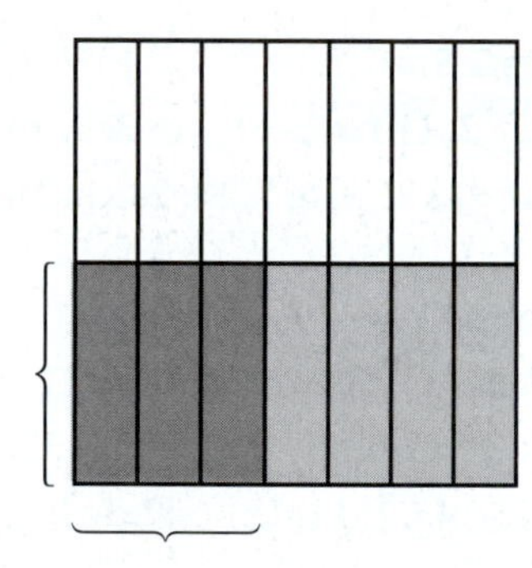

1. a. $\frac{1}{6}$ **b.** $\frac{2}{5}$ **2.** $\frac{3}{8}$ **3.** $\frac{3}{4}$ **4.** $\frac{3}{2}$ **5.** 0 **6.** undefined **7.** $\frac{5}{24}$ **8.** $\frac{3}{14}$; **9. a.** $\frac{3}{35}$ **b.** 0 **c.** $\frac{15}{14}$ **d.** $\frac{21}{40}$ **e.** $\frac{21}{32}$ **10. a.** commutative property of multiplication **b.** commutative property of multiplication **11.** associative property of multiplication **12.** 8 **13.** 15 **14.** 12 **15.** 12 **16.** $\frac{36}{63}$ **17.** $\frac{48}{60}$ **18.** $\frac{1}{160}$

$\frac{1}{2}$ $\frac{3}{7}$

Exercises 3.1, page 217

1. 0 **3.** 0 **5.** undefined **7.** undefined **9.** Please refer to the discussion on page 208. **11.** $\frac{1}{3}$ **13.** $\frac{1}{4}$ **15.** $\frac{1}{4}$

17. $\frac{3}{20}$

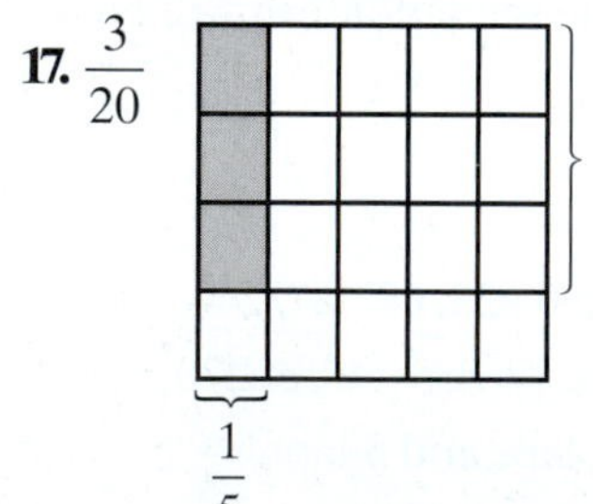

19. $\frac{8}{21}$

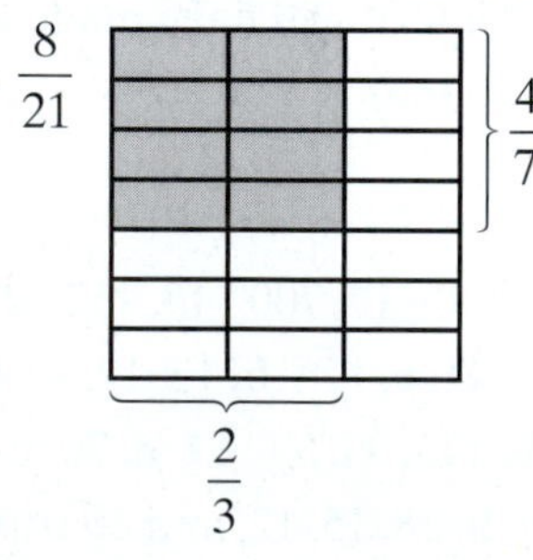

21. $\frac{3}{32}$ **23.** $\frac{9}{49}$ **25.** $\frac{15}{32}$ **27.** 0 **29.** $\frac{35}{12}$ **31.** $\frac{10}{1}$ or 10 **33.** $\frac{45}{2}$ **35.** $\frac{32}{15}$ **37.** $\frac{99}{20}$ **39.** $\frac{48}{455}$ **41.** $\frac{1}{10}$ **43.** $\frac{1}{8}$ **45.** $\frac{12}{35}$ **47.** commutative property of multiplication **49.** associative property of multiplication **51.** $\frac{7}{20}; \frac{13}{20}$ **53.** $\frac{3}{8}$ **55.** $\frac{5}{8} = \frac{5}{8} \cdot \frac{3}{3} = \frac{15}{24}$ **57.** $\frac{2}{5} = \frac{2}{5} \cdot \frac{5}{5} = \frac{10}{25}$ **59.** $\frac{1}{9} = \frac{1}{9} \cdot \frac{5}{5} = \frac{5}{45}$ **61.** $\frac{5}{8} = \frac{5}{8} \cdot \frac{2}{2} = \frac{10}{16}$ **63.** $\frac{14}{3} = \frac{14}{3} \cdot \frac{3}{3} = \frac{42}{9}$ **65.** $\frac{9}{16} = \frac{9}{16} \cdot \frac{6}{6} = \frac{54}{96}$ **67.** $\frac{10}{11} = \frac{10}{11} \cdot \frac{4}{4} = \frac{40}{44}$ **69.** $\frac{11}{12} = \frac{11}{12} \cdot \frac{4}{4} = \frac{44}{48}$ **71.** $\frac{2}{3} = \frac{2}{3} \cdot \frac{16}{16} = \frac{32}{48}$ **73.** $\frac{9}{10} = \frac{9}{10} \cdot \frac{10}{10} = \frac{90}{100}$

Exercises 3.1 (cont.)

75. $\frac{7}{10}=\frac{7}{10}\cdot\frac{7}{7}=\frac{49}{70}$ **77.** $\frac{5}{12}=\frac{5}{12}\cdot\frac{9}{9}=\frac{45}{108}$ **79.** $\frac{7}{6}=\frac{7}{6}\cdot\frac{6}{6}=\frac{42}{36}$

Margin Exercises 3.2

1. a. $\frac{2}{9}$ **b.** $\frac{6}{11}$ **2. a.** $\frac{40}{35}=\frac{\not{5}\cdot 8}{\not{5}\cdot 7}=\frac{8}{7}$ **b.** $\frac{40}{35}=\frac{\not{5}\cdot 2\cdot 2\cdot 2}{\not{5}\cdot 7}=\frac{8}{7}$ **3.** $\frac{12}{25}$ **4.** $\frac{3}{2}$ **5.** $\frac{1}{6}$ **6.** $\frac{3}{4}$ **7.** $\frac{2}{3}$ **8.** $\frac{4}{3}$ **9.** $\frac{8}{25}$ **10.** 2 **11.** $\frac{2}{3}$ **12.** $\frac{1}{3}$ **13.** $\frac{16}{9}$ **14.** $\frac{3}{8}$ **15.** 375

Exercises 3.2, page 229

1. $\frac{1}{3}$ **3.** $\frac{3}{4}$ **5.** $\frac{2}{5}$ **7.** $\frac{7}{18}$ **9.** 0 **11.** $\frac{2}{5}$ **13.** $\frac{5}{6}$ **15.** $\frac{2}{3}$ **17.** $\frac{2}{3}$ **19.** $\frac{6}{25}$ **21.** $\frac{2}{3}$ **23.** $\frac{12}{35}$ **25.** $\frac{2}{9}$ **27.** $\frac{25}{76}$ **29.** $\frac{1}{2}$ **31.** $\frac{3}{8}$ **33.** $\frac{12}{13}$ **35.** $\frac{11}{15}$ **37.** $\frac{8}{9}$ **39.** 2 **41.** $\frac{8}{9}$ **43.** $\frac{5}{7}$ **45.** $\frac{1}{3}$ **47.** 1 **49.** $\frac{1}{6}$ **51.** $\frac{1}{12}$ **53.** $\frac{2}{25}$ **55.** $\frac{10}{3}$ **57.** $\frac{4}{15}$ **59.** $\frac{5}{18}$ **61.** $\frac{1}{4}$ **63.** $\frac{21}{4}$ **65.** $\frac{77}{4}$ **67.** $\frac{8}{5}$ **69.** $\frac{2}{7}$ **71.** $\frac{625}{128}$ feet **73.** 10 feet, 5 feet, $\frac{5}{2}$ feet, and $\frac{5}{2}$ feet **75.** $\frac{1}{12}$ **77.** 225 "earth-days"

Review Problems (from Section 2.5), page 233

1. $3^2\cdot 5$ **2.** $2\cdot 3\cdot 13$ **3.** $2\cdot 5\cdot 13$ **4.** $2^2\cdot 5\cdot 23$ **5.** $2^6\cdot 5^6$ **6.** 1, 2, 3, 5, 6, 10, 15, 30 **7.** 1, 2, 3, 4, 6, 8, 12, 16, 24, 48 **8.** 1, 3, 7, 9, 21, 63

Margin Exercises 3.3

1. $\frac{8}{7}$ **2.** 10 **3.** $\frac{1}{16}$ **4.** 3 **5.** $\frac{21}{10}$ **6.** $\frac{1}{4}$ **7.** $\frac{5}{8}$ **8.** $\frac{6}{5}$ **9.** $\frac{7}{36}$ **10.** 1 **11.** $\frac{5}{21}$ **12.** $\frac{20}{63}$

Exercises 3.3, page 241

1. $\frac{13}{12}$ **3.** 0 **5.** $\frac{8}{9}$ **7.** $\frac{5}{7}$ **9.** $\frac{7}{5}$ **11.** $\frac{1}{3}$ **13.** $\frac{3}{4}$ **15.** $\frac{4}{9}$ **17.** $\frac{5}{8}$ **19.** $\frac{39}{32}$ **21.** $\frac{9}{10}$ **23.** 1 **25.** $\frac{32}{21}$ **27.** $\frac{10}{3}$ **29.** $\frac{16}{21}$ **31.** $\frac{7}{60}$ **33.** $\frac{63}{50}$ **35.** $\frac{7}{5}$ **37.** $\frac{5}{7}$ **39.** $\frac{8}{3}$ **41.** $\frac{3}{8}$ **43.** 448 **45.** $\frac{3}{8}$ **47. a.** $\frac{3}{8}$ **b.** $\frac{1}{4}$ **49. a.** $\frac{5}{16}$ **b.** $\frac{5}{4}$ **51.** $\frac{50}{27}$ **53.** 1250 freshmen **55.** More than 45; 60 passengers **57.** More; it is \$1800. Your new salary will be more than \$1320 because you will be getting $\frac{1}{10}$ of a number larger than \$1200 as your raise. This number will be more than \$120 and it will be added to a number larger than \$1200.

Review Problems (from Section 2.6), page 245

1. 390 **2.** 252 **3.** 100 **4.** 663 **5.** 300 **6.** 968 **7. a.** 250 **b.** 5 times, 2 times **8. a.** 120 **b.** 6 times, 5 times, 4 times **9. a.** 420 **b.** 35 times, 12 times, 6 times **10. a.** 3465 **b.** 77 times, 55 times, 35 times **11.** 90 days

Margin Exercises 3.4

1. $\frac{6}{7}$ **2.** $\frac{10}{13}$ **3.** $\frac{2}{3}$ **4.** $\frac{9}{5}$ **5.** 18 **6.** 252 **7.** $\frac{37}{63}$ **8.** $\frac{105}{88}$ **9.** $\frac{7}{18}$ **10.** $\frac{53}{40}$ **11.** $\frac{3309}{1000}$ **12.** $\frac{25}{12}$

Exercises 3.4, page 255

1. 16 **3.** 54 **5.** 1000 **7.** $\frac{5}{14}$ **9.** $\frac{3}{2}$ **11.** $\frac{10}{5}=2$ **13.** $\frac{5}{3}$ **15.** $\frac{13}{18}$ **17.** $\frac{11}{16}$ **19.** $\frac{3}{5}$ **21.** $\frac{12}{12}=1$ **23.** $\frac{17}{20}$ **25.** $\frac{17}{21}$ **27.** $\frac{9}{13}$ **29.** $\frac{23}{54}$ **31.** $\frac{23}{72}$ **33.** $\frac{173}{84}$ **35.** $\frac{1}{2}$ **37.** $\frac{5}{8}$ **39.** $\frac{5}{6}$ **41.** $\frac{22}{27}$ **43.** $\frac{13}{9}$ **45.** $\frac{1}{5}$ **47.** $\frac{317}{1000}$ **49.** $\frac{271}{10,000}$

Exercises 3.4 (cont.)

51. $\frac{8191}{1000}$ **53.** $\frac{753}{1000}$ **55.** $\frac{89}{200}$ **57.** $\frac{613}{1000}$ **59.** $\frac{5134}{1000}=\frac{2567}{500}$ **61.** $\frac{5}{3}$ **63.** $\frac{119}{72}$ **65.** $\frac{19}{100}$ **67.** $\frac{49}{36}$ **69.** $\frac{73}{96}$ **71.** 1 oz **73.** $\frac{23}{20}$ in. **75.** $\frac{5}{8}$ in. **77.** $6000

Margin Exercises 3.5

1. $\frac{3}{10}$ **2.** $\frac{11}{15}$ **3.** $\frac{1}{2}$ **4.** $\frac{2}{3}$ **5.** $\frac{2}{55}$ **6.** $\frac{59}{70}$ **7.** $\frac{17}{30}$ **8.** $\frac{31}{12}$ **9.** $\frac{1}{38}$ **10.** $\frac{11}{75}$ **11.** $\frac{37}{8}$ **12.** The Cavaliers lost $\frac{1}{15}$ of their games by exactly 5 runs.

Exercises 3.5, page 267

1. $\frac{3}{7}$ **3.** $\frac{3}{5}$ **5.** $\frac{1}{2}$ **7.** $\frac{1}{3}$ **9.** $\frac{3}{5}$ **11.** $\frac{1}{2}$ **13.** $\frac{13}{30}$ **15.** $\frac{1}{12}$ **17.** $\frac{9}{32}$ **19.** $\frac{13}{20}$ **21.** $\frac{7}{54}$ **23.** $\frac{1}{40}$ **25.** $\frac{7}{60}$ **27.** $\frac{27}{8}$ **29.** $\frac{3}{16}$ **31.** $\frac{87}{100}$ **33.** $\frac{3}{50}$ **35.** $\frac{1}{25}$ **37.** $\frac{3}{20}$ **39.** 0 **41.** $\frac{1}{10}$ **43.** $\frac{1}{3}$ **45.** $\frac{17}{20}$ **47.** $\frac{5}{24}$ **49.** $\frac{1}{50}$ **51.** $\frac{5}{16}$ **53.** $\frac{11}{12}$ **55.** $\frac{11}{10}$

Review Problems (from Section 2.1 and 2.2), page 270

1. 81 **2.** 169 **3.** 225 **4.** 324 **5.** 0 **6.** 29 **7.** 74 **8.** 0 **9.** 70 **10.** 4

Margin Exercises 3.6

1. $\frac{9}{22}$ is larger by $\frac{7}{66}$ **2.** $\frac{19}{24}$ is larger by $\frac{1}{72}$. **3.** $\frac{5}{9};\frac{7}{12};\frac{2}{3}$ **4.** $\frac{17}{45}$ **5.** $\frac{15}{8}$ **6.** $\frac{25}{84}$ **7.** $\frac{3}{38}$ **8.** $\frac{8}{7}$ **9.** $\frac{35}{12}$

Exercises 3.6, page 277

1. $\frac{3}{4}$ by $\frac{1}{12}$ **3.** $\frac{17}{20}$ by $\frac{1}{20}$ **5.** $\frac{13}{20}$ by $\frac{1}{40}$ **7.** equal **9.** $\frac{11}{48}$ by $\frac{1}{60}$ **11.** $\frac{3}{5},\frac{2}{3},\frac{7}{10};\frac{1}{10}$ **13.** $\frac{11}{12},\frac{19}{20},\frac{7}{6};\frac{1}{4}$ **15.** $\frac{1}{4},\frac{1}{3},\frac{1}{2};\frac{1}{4}$ **17.** $\frac{13}{18},\frac{7}{9},\frac{31}{36};\frac{5}{36}$ **19.** $\frac{20}{10{,}000},\frac{3}{1000},\frac{1}{100};\frac{1}{125}$ **21.** $\frac{2}{3}$ **23.** $\frac{13}{9}$ **25.** $\frac{187}{32}$ **27.** $\frac{1}{4}$ **29.** $\frac{8}{39}$ **31.** $\frac{15}{64}$ **33.** $\frac{29}{36}$ **35.** $\frac{3}{16}$ **37.** $\frac{11}{70}$ **39.** $\frac{25}{21}$ **41.** $\frac{33}{4}$ **43.** 2 **45.** 0 **47.** $\frac{13}{80}$ **49. a.** Yes. If both of the fractions are greater than one half, then their sum will be greater than 1. **b.** No. The two denominators being multiplied will only give a larger denominator and thus a number less than 1. **51.** Yes. For 0, the quotient will remain 0.

Chapter 3 Review Questions, page 285

1. 0 **2.** undefined **3.** $\frac{3}{2},\frac{2}{3}$ **4.** associative property of addition **5.** $\frac{4}{15}$ **6.** $\frac{1}{30}$ **7.** $\frac{3}{49}$ **8.** $\frac{1}{12}$ **9.** 2 **10.** 54 **11.** 75 **12.** $\frac{1}{2}$ **13.** $\frac{9}{8}$. **14.** 0 **15.** $\frac{5}{4}$ **16.** $\frac{5}{7}$ **17.** $\frac{2}{3}$ **18.** $\frac{1}{4}$ **19.** $\frac{49}{72}$ **20.** $\frac{7}{22}$ **21.** $\frac{25}{54}$ **22.** $\frac{7}{20}$ **23.** $\frac{1}{3}$ **24.** $\frac{7}{8}$ **25.** $\frac{1}{9}$ **26.** $\frac{5}{3}$ **27.** 1 **28.** $\frac{5}{4}$ **29.** $\frac{4}{5}$ **30.** $\frac{4}{5}$ by $\frac{2}{15}$ **31.** $\frac{11}{20},\frac{5}{9},\frac{7}{12}$ **32.** $\frac{25}{112}$ **33.** $\frac{7}{5}$ **34.** $\frac{19}{30}$ **35.** $\frac{7}{18}$ **36.** $\frac{35}{16}$ **37.** $\frac{1}{2}$ **38.** 14 students

Chapter 3 Test, page 289

1. $\frac{8}{5}$ **2.** $\frac{35}{80}$ **3.** commutative; multiplication **4.** $\frac{5}{6}$ **5.** $\frac{7}{5}$ **6.** $\frac{3}{4}$ **7.** $\frac{4}{15}$ **8.** $\frac{5}{72}$ **9.** 9 **10.** $\frac{2}{9}$ **11.** 25 **12.** $\frac{9}{80}$ **13.** $\frac{25}{44}$ **14.** $\frac{8}{11}$ **15.** $\frac{1}{2}$ **16.** $\frac{3}{8}$ **17.** $\frac{1}{2}$ **18.** $\frac{12}{5}$ **19.** 2 **20.** $\frac{1}{6}$ **21.** $\frac{1}{16}$ **22.** $\frac{13}{20}$ **23.** $\frac{11}{30}$ **24.** $\frac{1}{3}$ **25.** $\frac{2}{3}, \frac{3}{4}, \frac{7}{8}$ **26.** $\frac{2}{3}$ **27.** $\frac{9}{20}$

Cumulative Review: Chapters 1 - 3, page 291

1. 2,500,000 **2. a.** $5 + (6 + 7) = (5 + 6) + 7$ and $13 + (12 + 11) = (13 + 12) + 11$ **b.** $7 \cdot 1 = 7$ and $59 \cdot 1 = 59$
3. a. 3690 (estimate) **b.** 3413 (sum) **4. a.** 20,000 (estimate) **b.** 24,192 (product) **5. a.** 1000 (estimate)
b. 996 (difference) **6. a.** 250 (estimate) **b.** 316 R 10 (quotient and remainder) **7.** 350,000
8. All. 2: because the number is even. 3: because the sum of the digits is 18 and 18 is divisible by 3. 5: because the units digit is 0. 6: because the number is divisible by 2 and 3. 9: because the sum of the digits is 18 and 18 is divisible by 9. 10: because the units digit is 0.
9. 431 is prime. **10.** $780 = 2 \cdot 2 \cdot 3 \cdot 5 \cdot 13 = 2^2 \cdot 3 \cdot 5 \cdot 13$ **11.** $\text{LCM} = 2 \cdot 3 \cdot 3 \cdot 5 \cdot 7 = 630.$
$630 = 18 \cdot 35$; $630 = 42 \cdot 15$; $630 = 90 \cdot 7$ **12.** $\text{LCM} = 2 \cdot 2 \cdot 2 \cdot 2 \cdot 2 \cdot 3 \cdot 3 \cdot 5 \cdot 7 = 2^5 \cdot 3^2 \cdot 5 \cdot 7 = 10{,}080$;
$10{,}080 = 36 \cdot 280$; $10{,}080 = 60 \cdot 168$; $10{,}080 = 84 \cdot 120$; $10{,}080 = 96 \cdot 105$ **13.** 32 **14.** $\frac{1}{24}$ **15.** $\frac{139}{120}$ **16.** $\frac{79}{102}$ **17.** $\frac{9}{8}$
18. $\frac{9}{7}$ **19.** 95 **20.** \$3680 **21.** 775 **22.** 800 sq cm **23.** about 4000 miles **24. a.** $\frac{8}{105}$ **b.** $\frac{4}{35}$ **c.** $\frac{2}{3}$

25. a. 310 yd **b.** 5500 yd^2 **26. a.** 16 in. **b.** 12 in.^{2} **27. a.** 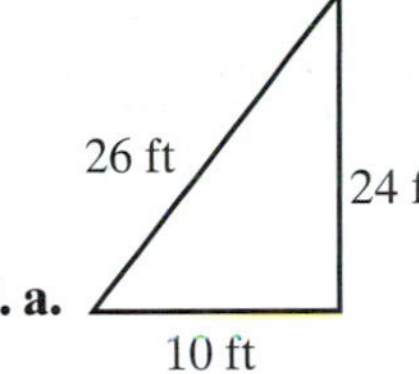

b. 60 ft **c.** 120 ft^2

CHAPTER 4

Margin Exercises 4.1

1. a. $\frac{57}{8}$ **b.** $\frac{47}{4}$ **2.** $\frac{86}{11}$ **3.** $\frac{73}{9}$ **4.** $\frac{11}{2}$ **5. a.** $2\frac{4}{7}$ **b.** $1\frac{1}{4}$ **6. a.** $12\frac{1}{2}$ **b.** $2\frac{1}{3}$ **7.** $2\frac{13}{24}$ pounds of seafood

Exercises 4.1, page 301

1. $\frac{4}{3}$ **3.** $\frac{4}{3}$ **5.** $\frac{3}{2}$ **7.** $\frac{7}{5}$ **9.** $\frac{5}{4}$ **11.** $4\frac{1}{6}$ **13.** $1\frac{1}{3}$ **15.** $1\frac{1}{2}$ **17.** $6\frac{1}{7}$ **19.** $7\frac{1}{2}$ **21.** $3\frac{1}{9}$ **23.** 3 **25.** $4\frac{1}{2}$ **27.** 3
29. $1\frac{3}{4}$ **31.** $\frac{37}{8}$ **33.** $\frac{76}{15}$ **35.** $\frac{46}{11}$ **37.** $\frac{7}{3}$ **39.** $\frac{32}{3}$ **41.** $\frac{34}{5}$ **43.** $\frac{50}{3}$ **45.** $\frac{101}{5}$ **47.** $\frac{92}{7}$ **49.** 17 **51.** $2\frac{19}{24}$ feet
53. net gain of \$$1\frac{3}{4}$

Review Problems (from Section 3.2 and 3.3), page 304

1. $\frac{7}{9}$ **2.** $\frac{2}{5}$ **3.** 3 **4.** $\frac{6}{13}$ **5.** $\frac{1}{5}$ **6.** $\frac{3}{16}$ **7.** $\frac{5}{9}$ **8.** $\frac{15}{14}$

Margin Exercises 4.2

1. $\frac{25}{16}$ **2.** $\frac{114}{7}$ **3.** 226 ft^2 **4.** $\frac{7}{2}$ or $3\frac{1}{2}$ **5.** 14 **6.** 18 **7.** $\frac{8}{5}$ or $1\frac{3}{5}$ **8.** $1\frac{9}{16}$ **9.** 1 **10.** $\frac{392}{51}$ or $7\frac{35}{51}$

Margin Exercises 4.2 (cont.)

11. $\frac{133}{46}$ or $2\frac{41}{46}$

Exercises 4.2, page 311

1. $\frac{91}{12}$ or $7\frac{7}{12}$ **3.** $\frac{21}{2}$ or $10\frac{1}{2}$ **5.** $\frac{45}{2}$ or $22\frac{1}{2}$ **7.** $\frac{187}{6}$ or $31\frac{1}{6}$ **9.** 8 **11.** 30 **13.** $\frac{87}{4}$ or $21\frac{3}{4}$ **15.** 1 **17.** $\frac{55}{2}$ or $27\frac{1}{2}$ **19.** $\frac{77}{4}$ or $19\frac{1}{4}$ **21.** $\frac{1}{7}$ **23.** $\frac{1}{10}$ **25.** 17 **27.** $\frac{21}{10}$ or $2\frac{1}{10}$ **29.** $\frac{1539}{40}$ or $38\frac{19}{40}$ **31.** $\frac{3465}{169}$ or $20\frac{85}{169}$ **33.** 40 **35.** 20 **37.** $\frac{21}{16}$ or $1\frac{5}{16}$ **39.** $\frac{117}{35}$ or $3\frac{12}{35}$ **41.** $\frac{1}{3}$ **43.** $\frac{5}{9}$ **45.** $\frac{10}{39}$ **47.** $\frac{29}{155}$ **49.** $\frac{28}{17}$ or $1\frac{11}{17}$ **51.** $\frac{200}{301}$ **53.** $\frac{7}{5}$ or $1\frac{2}{5}$ **55.** $\frac{41}{3}$ or $13\frac{2}{3}$ **57.** $\frac{63}{5}$ or $12\frac{3}{5}$ **59.** $\frac{20}{11}$ or $1\frac{9}{11}$ **61.** 177 miles **63.** 90 pages; 18 hours **65.** 10 feet; 22 feet **67.** $\frac{50}{27}$ or $1\frac{23}{27}$ **69.** 175 passengers **71.** \$3200 **73. a.** 455 miles **b.** 3551 cents or \$35.51 **75.** $\frac{325}{2}$ miles or $162\frac{1}{2}$ miles **77. a.** More since $10\frac{1}{2} > 5\frac{7}{10}$ **b.** less; $\frac{35}{19}$

Review Problems (from Section 3.4 and 3.5), page 316

1. $\frac{17}{21}$ **2.** $\frac{61}{40}$ or $1\frac{21}{40}$ **3.** $\frac{31}{30}$ or $1\frac{1}{30}$ **4.** $\frac{59}{45}$ or $1\frac{14}{45}$ **5.** $\frac{17}{40}$ **6.** $\frac{67}{84}$ **7.** $\frac{17}{64}$ **8.** $\frac{1}{6}$

Margin Exercises 4.3

1. $5\frac{8}{9}$ **2.** $36\frac{9}{10}$ **3.** $15\frac{13}{40}$ **4. a.** 17 m **b.** $12\frac{3}{4}$ sq m

Exercises 4.3, page 321

1. $14\frac{5}{7}$ **3.** 8 **5.** $12\frac{3}{4}$ **7.** $8\frac{2}{3}$ **9.** $34\frac{7}{18}$ **11.** $7\frac{2}{3}$ **13.** $42\frac{7}{20}$ **15.** $10\frac{3}{4}$ **17.** $12\frac{7}{27}$ **19.** $18\frac{23}{45}$ **21.** $28\frac{1}{2}$ **23.** $9\frac{29}{40}$ **25.** $12\frac{47}{60}$ **27.** $63\frac{11}{24}$ **29.** $53\frac{83}{192}$ **31.** $18\frac{7}{8}$ **33.** $36\frac{23}{24}$ **35.** $90\frac{1}{8}$ **37.** $35\frac{7}{10}$ km **39.** $12\frac{1}{8}$ in. **41.** \$$174\frac{43}{50}$ million or \$174,860,000 **43. a.** 6 **b.** 600 **c.** 60,000

Margin Exercises 4.4

1. $6\frac{2}{7}$ **2.** $6\frac{1}{2}$ **3.** $7\frac{9}{10}$ **4.** $3\frac{13}{28}$ **5.** $7\frac{5}{11}$ **6.** $4\frac{1}{6}$ **7.** $5\frac{23}{40}$ **8.** $1\frac{13}{20}$ minutes

Exercises 4.4, page 331

1. $4\frac{1}{2}$ **3.** $1\frac{5}{12}$ **5.** 4 **7.** $3\frac{1}{2}$ **9.** $3\frac{1}{4}$ **11.** $3\frac{5}{8}$ **13.** $4\frac{2}{3}$ **15.** $6\frac{7}{12}$ **17.** $10\frac{4}{5}$ **19.** $3\frac{9}{10}$ **21.** $7\frac{4}{5}$ **23.** $1\frac{11}{16}$ **25.** 17 **27.** $13\frac{20}{21}$ **29.** $16\frac{5}{24}$ **31. a.** $\frac{3}{5}$ hour **b.** 36 minutes **33.** $\frac{5}{6}$ hour **35.** \$$4\frac{5}{8}$ **37.** $219\frac{13}{16}$ pounds **39.** $4\frac{13}{20}$ parts

Review Problems (from Section 2.2 and 3.6), page 334

1. 18 **2.** 101 **3.** $\frac{47}{18}$ or $2\frac{11}{18}$ **4.** 0

Margin Exercises 4.5

1. 2 **2.** $\frac{7}{3}$ or $2\frac{1}{3}$ **3.** $\frac{83}{10}$ or $8\frac{3}{10}$ **4.** $\frac{13}{7}$ or $1\frac{6}{7}$ **5.** $\frac{61}{18}$ or $3\frac{7}{18}$

Exercises 4.5, page 339

1. 4 **3.** $\frac{7}{45}$ **5.** 1 **7.** $\frac{52}{75}$ **9.** $\frac{1}{5}$ **11.** $\frac{138}{65}$ or $2\frac{8}{65}$ **13.** $\frac{23}{30}$ **15.** $\frac{1}{5}$ **17.** $\frac{172}{9}$ or $19\frac{1}{9}$ **19.** $\frac{791}{120}$ or $6\frac{71}{120}$ **21.** $\frac{3}{5}$
23. $\frac{15}{4}$ or $3\frac{3}{4}$ **25.** $\frac{543}{40}$ or $13\frac{23}{40}$ **27.** $\frac{42}{5}$ or $8\frac{2}{5}$ **29.** $\frac{47}{40}$ or $1\frac{7}{40}$ **31.** $\frac{223}{32}$ or $6\frac{31}{32}$ **33.** $3\frac{1}{2}$ hours

Margin Exercises 4.6

1. 11 ft 1 in. **2.** 3 lb 6 oz **3.** 4 hr 15 min **4.** 18 ft 1 in. **5.** 24 min 10 sec **6.** 5 gal 2 qt **7.** 2 lb 6 oz **8.** 8 ft 5 in. **9.** 30 min

Exercises 4.6, page 349

1. 4 ft 8 in. **3.** 7 lb 4 oz **5.** 6 min 20 sec **7.** 3 days 6 hr **9.** 9 gal 1 qt **11.** 5 pt 4 fl oz **13.** 9 ft 7 in. **15.** 18 lb 2 oz
17. 18 min 10 sec **19.** 4 hr 11 min 15 sec **21.** 6 days 2 hr 35 min **23.** 13 gal 3 qt
25. 7 gal 2 qt 20 fl oz (or 7 gal 2 qt 1 pt 4 fl oz) **27.** 12 yd 2 ft 6 in. **29.** 2 yd 2 ft 2 in.
31. 3 gal 2 qt 30 fl oz (or 3 gal 2 qt 1 pt 14 fl oz) **33.** 2 hr 45 min **35.** 4 min 50 sec **37.** 4 lb 8 oz
39. 3 ft 8 in. (or 1 yd 8 in.) **41.** 7 ft 1 in. **43.** 3 hr 40 min **45.** 7 gal 1 qt **47.** 15 ft

Chapter 4 Review Questions, page 355

1. $6\frac{5}{8}$ **2.** 7 **3.** $3\frac{21}{50}$ **4.** $\frac{51}{10}$ **5.** $\frac{35}{12}$ **6.** $\frac{67}{5}$ **7.** $\frac{123}{4}$ or $30\frac{3}{4}$ **8.** $\frac{11}{15}$ **9.** 12 **10.** $\frac{2}{9}$ **11.** $\frac{59}{8}$ or $7\frac{3}{8}$ **12.** $\frac{56}{3}$ or $18\frac{2}{3}$
13. 104 **14.** $\frac{7}{10}$ **15.** $\frac{79}{12}$ or $6\frac{7}{12}$ **16.** $\frac{13}{8}$ or $1\frac{5}{8}$ **17.** $10\frac{2}{3}$ or $\frac{32}{3}$ **18.** $120\frac{17}{18}$ or $\frac{2177}{18}$ **19.** $\frac{90}{7}$ or $12\frac{6}{7}$
20. $\frac{49}{15}$ or $3\frac{4}{15}$ **21.** $\frac{11}{70}$ **22.** $\frac{33}{4}$ **23.** $\frac{91}{120}$ **24.** $\frac{7}{4}$ or $1\frac{3}{4}$ **25.** 64 **26.** $\frac{5}{2}$ or $2\frac{1}{2}$ **27.** $\frac{39}{124}$ **28.** $\frac{11}{4}$ or $2\frac{3}{4}$ lb
29. $625 **30.** $\frac{59}{8}$ or $7\frac{3}{8}$ **31.** $\frac{61}{8}$ or $7\frac{5}{8}$ yd **32.** $6000 **33.** 6 ft **34.** 6 students **35.** $\frac{379}{2}$ or $189\frac{1}{2}$ miles
36. 46 min **37.** 16 ft 4 in. **38.** 5 days 11 hours 20 minutes

Chapter 4 Test, page 359

1. 0 **2. a.** $3\frac{2}{5}$ **b.** $3\frac{1}{33}$ **3. a.** $\frac{53}{8}$ **b.** $\frac{43}{10}$ **4.** $6\frac{7}{12}$ **5.** $2\frac{5}{6}$ **6.** $24\frac{1}{6}$ **7.** $10\frac{5}{24}$ **8.** $1\frac{2}{5}$ **9.** $7\frac{1}{20}$ **10.** $1\frac{37}{42}$ **11.** $4\frac{17}{40}$
12. 20 **13.** $\frac{15}{4}$ or $3\frac{3}{4}$ **14.** $\frac{1}{3}$ **15.** $\frac{3}{2}$ or $1\frac{1}{2}$ **16.** $\frac{35}{3}$ or $11\frac{2}{3}$ **17.** $\frac{83}{105}$ **18.** $\frac{495}{32}$ or $15\frac{15}{32}$ **19.** 38 min 10 sec
20. 19 gal 2 qt 12 oz **21.** 5 hr 40 min **22.** 19 ft 3 in. (or 6 yd 1 ft 3 in.) **23. a.** $16\frac{3}{5}$ cm **b.** $9\frac{9}{25}$ cm
24. a. $28\frac{3}{4}$ miles **b.** $9\frac{7}{12}$ miles **25. a.** 39 in. **b.** 92 sq in.

Cumulative Review: Chapters 1 - 4, page 363

1. 50,000 + 3000 + 400 + 60; fifty-three thousand, four hundred sixty **2.** 270,000 **3.** commutative property of addition **4.** (a) B; (b) A; (c) D; (d) C **5.** 5,600,000 **6.** 158 **7.** 2, 3, 5, 7, 11, 13, 17, 19, 23, 29 **8. a.** $170 = 2 \cdot 5 \cdot 17$ **b.** $305 = 5 \cdot 61$ **9.** 0; undefined **10.** 43 **11.** 13,250 **12.** 2527 **13.** 26,352 **14.** 203 **15.** $\frac{7}{48}$ **16.** $\frac{1}{20}$ **17.** $\frac{7}{24}$ **18.** $\frac{28}{15}$ or $1\frac{13}{15}$ **19.** $\frac{183}{35}$ or $5\frac{8}{35}$ **20.** $\frac{7}{4}$ or $1\frac{3}{4}$ **21.** $\frac{315}{2}$ or $157\frac{1}{2}$ **22.** 3 **23.** $\frac{241}{90}$ or $2\frac{61}{90}$ **24.** $\frac{7}{10}, \frac{3}{4}, \frac{5}{6}, \frac{8}{9}$
25. a. 5 hr 11 min 15 sec **b.** 28 lb 6 oz

Cumulative Review: Chapters 1 - 4 (cont.)

26. \$1513 **27.** $16\frac{1}{2}$ gallons **28.** \$57.86 **29.** \$225 **30.** 20 min; 10 hours

CHAPTER 5

Margin Exercises 5.1

1. 19.3, nineteen and three tenths **2.** 8.743, eight and seven hundred forty-three thousandths **3.** 39.0184, thirty-nine and one hundred eighty-four ten-thousandths **4.** 0.91 **5.** 200.021 **6.** 0.221 **7.** \$200 **8.** 13.25 **9.** 58.945 **10.** 21.9236 **11.** 40,000

Exercises 5.1, page 377

1. 37.498 **3.** 4.11 **5.** 95.2 **7.** 62.7 **9.** $82\frac{56}{100}=82\frac{14}{25}$ **11.** $10\frac{576}{1000}=10\frac{72}{125}$ **13.** $65\frac{3}{1000}$ **15.** 0.3 **17.** 0.17 **19.** 60.028 **21.** 850.0036 **23.** five tenths **25.** five and six hundredths **27.** seven and three thousandths **29.** ten and four thousand six hundred thirty-eight ten-thousandths **31. a.** 7 **b.** 3 **c.** 3, 7, 3, 2 **d.** 8.47 **33.** 4.8 **35.** 76.3 **37.** 89.0 **39.** 18.0 **41.** 0.39 **43.** 5.72 **45.** 7.00 **47.** 0.08 **49.** 0.067 **51.** 0.634 **53.** 32.479 **55.** 0.002 **57.** 479 **59.** 20 **61.** 650 **63.** 6333 **65.** 5160 **67.** 500 **69.** 1000 **71.** 92,540 **73.** 7000 **75.** 48,000 **77.** 217,000 **79.** 4,501,000 **81.** 0.00058 **83.** 90 **85.** 500 **87.** 3.230 **89.** 80,000 **91.** two and fifty-four hundredths **93.** one hundred fifty-seven and six hundred three thousandths **95.** three and fourteen thousand, one hundred fifty-nine hundred-thousandths **97.** four hundred five thousandths **99.** 356.45; three hundred fifty-six and $\frac{45}{100}$ **101.** 2506.64; two thousand five hundred six and $\frac{64}{100}$ **103.** Answers will vary.

Review Problems (from Sections 1.2, 1.3, and 1.4), page 384

1. a. 9000 (estimate) **b.** 9107 (sum) **2. a.** 1600 (estimate) **b.** 1555 (sum) **3. a.** 550,000 (estimate) **b.** 562,712 (sum) **4. a.** 200 (estimate) **b.** 188 (difference) **5. a.** 4000 (estimate) **b.** 4001 (difference) **6. a.** 85,000 (estimate) **b.** 84,891 (difference) **7.** 350

Margin Exercises 5.2

1. 28.96 **2.** 42.0667 **3.** 62.673 **4.** \$156.58 **5.** 6.192 **6.** 141.467 **7.** 127 (estimated); 132.12 (sum)

Exercises 5.2, page 389

1. 2.3 **3.** 7.55 **5.** 72.31 **7.** 276.096 **9.** 44.6516 **11.** 118.333 **13.** 7.148 **15.** 93.877 **17.** 103.429 **19.** 137.150 **21.** 1.44 **23.** 15.89 **25.** 64.947 **27.** 4.7974 **29.** 2.9434 **31.** \$4.50 **33.** 12.28 in. **35. a.** \$110 **b.** \$80.65 **37. a.** \$4200 **b.** \$3930 **c.** \$270 off (This is a good estimate.) **39.** \$2.193 billion **41.** 2.26 in.; 37.9 in. **43. a.** No; the difference cannot be more than 1 becuase each number is less than 1. **b.** Yes; the difference can be less than 1 because each number is less than 1. **c.** No; the difference cannot be equal to 1 because both numbers are less than 1.

Review Problems (from Section 1.5), page 394

1. 350,000 **2.** 525 **3.** 7020 **4.** 2,000,000 (estimate); 1,714,125 (product) **5.** 168; 79; 13,272

Margin Exercises 5.3

1. 0.16 **2.** 0.224 **3.** 11.38081 **4.** 3.306 **5.** 0.05814 **6. a.** 1.3 **b.** 8780 **7. a.** 560 cm **b.** 352.5 mm **c.** 16,430 m **8.** 30 (estimate); 38.3758 (product) **9.** 20.16; 18 **10.** \$600

Exercises 5.3, page 405

1. (a) B; (b) C; (c) E; (d) A; (e) D **3.** 0.42 **5.** 0.04 **7.** 21.6 **9.** 0.42 **11.** 0.004 **13.** 0.108 **15.** 0.0276 **17.** 0.0486 **19.** 0.0006 **21.** 0.375 **23.** 1.4 **25.** 1.725 **27.** 5.063 **29.** 0.08 **31.** 346 **33.** 782 **35.** 1610 **37.** 4.35 **39.** 18.6 **41.** 380 **43.** 50 **45.** 74,000 **47.** 130 mm **49.** 1500 cm **51.** 6170 mm **53.** 16,000 m **55.** 600 m = 60,000 cm = 600,000 mm **57.** 20 m = 2000 cm = 20,000 mm **59.** 0.009 (estimate), 0.00954 (product) **61.** 2 (estimate), 2.032 (product) **63.** 0.0014 (estimate), 0.0013845 (product) **65.** 2 (estimate), 1.717202 (product) **67.** 1.6 (estimate), 1.4094 (product) **69.** 3.6 (estimate), 3.4314 (product) **71.** \$240.90 **73. a.** \$636 **b.** \$617.98 **75.** 12.8 ft (perimeter); 10.24 sq ft (area) **77.** 48.58 m (perimeter); 112.75 sq m (area) **79.** \$936,945 **81.** The estimate says you pay \$1000 more with cash (\$15,000 instead of \$14,000). This is not reasonable. **83.** (a) 689.7 in.; (b) 689.5 in.; The results are different because different rounding techniques were used to find the answers.

Review Problems (from Section 1.6), page 410

1. 8 R 0 **2.** 15 R 15 **3.** 81 R 7 **4.** 203 R 100 **5. a.** 1 square mile per person **b.** 1 square kilometer per person

Margin Exercises 5.4

1. 6.1 **2.** 13.3 **3.** 11.56 **4.** 5.2 **5.** 4.7 **6.** 1.55 **7.** 728.683 **8.** 0.00576 **9. a.** 3.50 **b.** 6.8 **c.** 3.952 **10. a.** 5 **b.** 5.0 **11.** 10 meters per second **12.** 1080 words

Exercises 5.4, page 421

1. (a) B; (b) E; (c) D; (d) C; (e) A **3.** 2.34 **5.** 0.99 **7.** 0.08 **9.** 2056 **11.** 20 **13.** 56.9 **15.** 0.7 **17.** 0.1 **19.** 21.0 **21.** 0.01 **23.** 5.70 **25.** 2.74 **27.** 5.04 **29.** 0.784 **31.** 0.5036 **33.** 0.07385 **35.** 16.7 **37.** 0.785 **39.** 0.01699 **41.** 0.5 cm **43.** 0.83 m **45.** 0.344 m **47.** 1.5 km **49.** 9.72 mm **51.** 3.2 cm = 0.32 dm = 0.032 m **53.** 3.5 (estimate), 3.087 (quotient) **55.** 0.2 (estimate), 0.285 (quotient) **57.** 0.005 (estimate), 0.007 (quotient) **59. a.** 400 miles **b.** 442.8 miles **61. a.** about \$1 per pound **b.** \$1.25 per pound **63.** \$239.56 **65.** \$295 **67.** 5670 at bats **69.** 268 free throws **71.** 42.2 kilometers **73. a.** \$5500 **b.** \$6500

Margin Exercises 5.5

1. $\frac{12}{25}$ **2.** $\frac{19}{20}$ **3.** $\frac{381}{500}$ **4.** $\frac{3}{100}$ **5.** $\frac{79}{5}=15\frac{4}{5}$ **6.** $\frac{31}{4}=7\frac{3}{4}$ **7.** 0.4 **8.** 0.65 **9.** 0.5 **10.** 0.2727… **11.** 0.13333… **12.** 0.444… **13.** $0.\overline{037}$ **14.** $0.08\overline{3}$ **15.** 17.03 **16.** $\frac{9}{16}$ is larger than 0.52. The difference is 0.0425.

Exercises 5.5, page 437

1. $\frac{9}{10}$ **3.** $\frac{5}{10}$ **5.** $\frac{62}{100}$ **7.** $\frac{57}{100}$ **9.** $\frac{526}{1000}$ **11.** $\frac{16}{1000}$ **13.** $\frac{51}{10}$ **15.** $\frac{815}{100}$ **17.** $\frac{1}{8}$ **19.** $\frac{9}{50}$ **21.** $\frac{9}{40}$ **23.** $\frac{17}{100}$ **25.** $\frac{16}{5}$ or $3\frac{1}{5}$ **27.** $\frac{25}{4}$ or $6\frac{1}{4}$ **29.** $0.\overline{6}$ **31.** $0.\overline{63}$ **33.** 0.6875 **35.** $0.\overline{428571}$ **37.** $0.1\overline{6}$ **39.** $0.\overline{5}$ **41.** 0.292 **43.** 0.417 **45.** 0.031 **47.** 1.231 **49.** 1.429 **51.** 0.7 **53.** 1.635 **55.** 14.98 **57.** 1.125 **59.** 13.51 **61.** 0.089 **63.** 11.083 **65.** 2.638 **67.** 27.3 **69.** 2.3 is larger; 0.05 **71.** 0.28 is larger; $0.0\overline{72}$ **73.** 3.3 is larger; $0.1\overline{571428}$ **75.** $3\frac{2}{3}$ is larger; $0.1\overline{6}$ **77.** $\frac{7}{10}, \frac{3}{4}, 0.76$ **79.** $\frac{5}{16}, 0.3126, 0.314$ **81.** $83\frac{4}{5}$ **83.** $26\frac{3}{10}$; $24\frac{1}{10}$ **85.** $21\frac{1}{2}$ **87.** 0.14 **89.** 6.86

Margin Exercises 5.6

1. a. 324 **b.** 13 **2.** 2.2361 **3.** 2.7386 **4.** 33.6006 **5.** no, $8^2 \neq 7^2 + 4^2$ **6.** 17 cm **7.** $\sqrt{208}$ or 14.42 yd. **8.** $\sqrt{2125}$ or 46.10 ft

Exercises 5.6, page 451

1. 144 **3.** 15 **5.** 6 **7.** 400 **9.** 13 **11.** $(1.732)^2 = 2.999824$ and $(1.733)^2 = 3.003289$ **13.** 3.4641 **15.** 4.8990 **17.** 6.9282 **19.** 19.0526 **21.** 0.5 **23.** 0.9 **25.** 1.2 **27.** 1.5 **29.** 90 **31.** 30 **33.** 0.05 **35.** 2.8284 **37.** 3.1623 **39.** 6.7082 **41.** 1.23 **43.** 3.01 **45.** 28.2843 **47.** 0.0548 **49.** 0.03 **51.** yes; $6^2 + 8^2 = 10^2$ **53.** no; $6^2 \neq 4^2 + 3^2$ **55.** $c = 2.2361$ **57.** $x = 7.0711$ **59.** $c = 10.7703$ **61.** 63.2 feet **63.** 522.0 feet **65. a.** 127.3 feet **b.** closer to home plate **67.** 44.9 inches **69.** 7.1 cm

Margin Exercises 5.7

1. 69.08 m; 379.94 m^2 **2.** 66.82 in. **3.** 47.1 ft; 176.625 ft^2 **4.** 140.67 yd^2 **5.** 810 in.^3 **6.** 678.24 cm^3

Exercises 5.7, page 465

1. a. 31.4 ft **b.** 78.5 ft^2 **3. a.** 19.468 yd **b.** 30.175 yd^2 **5. a.** 6.28 ft **b.** 3.14 ft^2 **7. a.** 10.71 cm **b.** 7.065 cm^2 **9. a.** 27.42 m **b.** 50.13 m^2 **11. a.** 19.42 ft **b.** 26.13 ft^2 **13. a.** 28.56 in. **b.** 57.12 in.^2 **15.** 7536 m^2 **17.** 70 in.^3 **19.** 381.51 cm^3 **21.** 12.56 dm^3 **23.** 224 cm^3 **25.** 282.6 cm^3 **27.** 7.33 ft^3

Chapter 5 Review Questions, page 473

1. decimal **2.** four tenths **3.** seven and eight hundredths **4.** ninety-two and one hundred thirty-seven thousandths **5.** eighteen and five thousand five hundred twenty-six ten-thousandths **6.** $81\frac{47}{100}$ **7.** $100\frac{3}{100}$ **8.** $9\frac{74}{125}$ **9.** $200\frac{1}{2}$ **10.** 2.17 **11.** 84.075 **12.** 3003.003 **13.** 5900 **14.** 7.6 **15.** 0.039 **16.** 2.06988 **17.** 26.82 **18.** 93.418 **19.** 9.02 **20.** 3.9623 **21.** 104.272 **22.** 22.708 **23.** 0.72 **24.** 0.0064 **25.** 235 **26.** 1.7632 **27.** 5964.1 **28.** 1.728 **29.** 171.55 **30.** 0.71 **31.** 880.53 **32.** 2.961 **33.** 0.00567 **34.** 23.5 **35.** 1820 mm **36.** 135 cm **37.** 35 cm **38.** 1.8 km **39.** $\frac{7}{100}$ **40.** $2\frac{1}{40}$ **41.** $\frac{3}{200}$ **42.** $0.\overline{3}$ **43.** 0.625 **44.** $2.\overline{4}$ **45.** 0.882 **46.** 0.980 **47.** $568,650 **48.** $900 **49.** 7.7460 **50.** 3.2 **51.** 9.5917 **52.** 0.63 **53.** yes, because $26^2 = 24^2 + 10^2$ **54.** 5.6569 **55. a.** 37.68 cm **b.** 113.04 cm^2 **56.** 87.92 in.^2 **57.** 89.25 mm **58.** 30.56 mm **59.** 1436.03 in.^3 **60.** 66.67 mm^3 **61.** 14.13 ft^3 **62.** 806.67 m^3

Chapter 5 Test, page 479

1. thirty and six hundred fifty-seven thousandths **2.** $\frac{3}{8}$ **3.** 0.3125 **4.** 203.02 **5.** 1972.805 **6.** 0.1 **7.** 103.758 **8.** 1.888 **9.** 122.11 **10.** 18.145 **11.** 37.313 **12.** 0.90395 **13.** 0.294 **14.** 12.80335 **15.** 1920 **16.** 0.03614 **17.** 17.83 **18.** 66.28 **19.** 681 m = 68,100 cm = 681,000 mm **20.** 35.5 cm = 0.355 m **21.** 392 miles **22. a.** 2.4495 **b.** 28.2843 **c.** 1.325 **23.** 107.7 yards **24. a.** 18.84 in. **b.** 28.26 in.^2 **25.** 5572 cm^3

Cumulative Review: Chapters 1 - 5, page 483

1. a. commutative property of addition **b.** associative property of multiplication **c.** additive identity property **2.** 13 **3.** 28 **4.** 52 **5.** 45 **6.** $2^2 \cdot 3 \cdot 5 \cdot 7$ **7. a.** 140 **b.** 64 **8.** $\frac{1}{30}$ **9.** $8\frac{21}{40}$ **10.** $55\frac{29}{40}$ **11.** $32\frac{7}{20}$ **12.** 2 **13.** $\frac{11}{30}$ **14.** 6 **15.** $\frac{32}{9} = 3\frac{5}{9}$ **16.** 72 **17.** 18.9 **18.** 2.185 **19. a.** $200,000 **b.** $207,700 **20.** $940 **21. a.** 3.3166 **b.** 21 **c.** 44.7214 **d.** 1.8868 **22.** $20^2 = 12^2 + 16^2$ **23.** 61.6 in. **24. a.** 65.7 ft **b.** 239.25 ft^2 **25. a.** 60 m **b.** 177.75 m^2 **26. a.** yes, because $74^2 = 70^2 + 24^2$; 168 m **b.** 840 m^2 **27.** 1256 ft^3 **28.** 234.45 cm^3

CHAPTER 6

Margin Exercises 6.1

1. 4 apples**:** 5 oranges or 4 apples **to** 5 oranges or $\frac{4 \text{ apples}}{5 \text{ oranges}}$ **2.** $\frac{3 \text{ quarters}}{1 \text{ dollar}} = \frac{3 \text{ quarters}}{4 \text{ quarters}} = \frac{3}{4}$

Margin Exercises 6.1 (cont.)

3. $\frac{500 \text{ washers}}{4000 \text{ bolts}} = \frac{1 \text{ washer}}{8 \text{ bolts}}$ **4.** $\frac{36 \text{ inches}}{5 \text{ feet}} = \frac{3 \text{ feet}}{5 \text{ feet}} = \frac{3}{5}$ **5.** 3-liter bottle for \$1.49
6. 16-ounce box for \$3.00; 20¢/oz (12.5-ounce box); 18.8¢/oz (16-ounce box)

Exercises 6.1, page 495

1. $\frac{1}{2}$ **3.** $\frac{4}{1}$ or 4 **5.** $\frac{50 \text{ miles}}{1 \text{ hour}}$ **7.** $\frac{25 \text{ miles}}{1 \text{ gallon}}$ **9.** $\frac{6 \text{ chairs}}{5 \text{ people}}$ **11.** $\frac{3}{4}$ **13.** $\frac{8}{7}$ **15.** $\frac{\$2 \text{ profit}}{\$5 \text{ invested}}$ **17.** $\frac{1}{1}$ or 1
19. $\frac{1 \text{ hit}}{4 \text{ times at bat}}$ **21.** $\frac{7}{25}$ **23.** $\frac{9}{41}$ **25.** 69.8¢/lb; 58.9¢/lb; 10 lb at \$5.89 **27.** 12.5¢/oz; 15¢/oz; 48 oz at \$5.99
29. 34.7¢/oz; 23.1¢/oz; 39 oz at \$8.99 **31.** 18.7¢/fl oz; 18.3¢/fl oz; 12 fl oz at \$2.19 **33.** 4.3¢/oz; 6.7¢/oz; 155 oz at \$6.59
35. 32.2¢/oz; 25.6¢/oz; 39 oz at \$9.99 **37.** 3.7¢/sq ft.; 4.9¢/sq ft.; 6.4¢/sq ft.; 6¢/sq ft.; 200 sq ft. at \$7.39
39. 4.1¢/fl oz; 2.4¢/fl oz; 2.3¢/fl oz; 174 fl oz at \$3.99 **41.** 16.8¢/oz; 15.3¢/oz; 32 oz at \$4.89 **43.** 25¢; 16.5¢; 40 at \$6.59
45. 3.8¢/oz; 4¢/oz; 29 oz at \$1.09 **47.** 29.1¢/oz; 33.3¢/oz; 27.5¢/oz; 40 oz at \$10.99 **49.** 11.7¢/oz; 9.3¢/oz; 15 oz at \$1.39
51. 12.8¢/oz; 10.7¢/oz; 26 oz at \$2.79 **53.** 10.3¢/oz; 9.3¢/oz; 175 oz at \$16.19 **55.** 5.9¢/oz; 5.5¢/oz; 31 oz at \$1.69
57. 8.9¢/oz; 6.8¢/oz; 28 oz at \$1.89 **59.** 13.2¢/oz; 14.2¢/oz; 9 oz at \$1.19

Review Problems (from Sections 3.2, 4.2, 5.3, and 5.4), page 498

1. 11 **2.** $\frac{185}{304}$ **3.** $\frac{1}{20}$ **4.** 80 **5.** 149.76 or $\frac{3744}{25}$ **6.** 0.02118 or $\frac{1059}{50{,}000}$ **7.** 5.5 or $\frac{11}{2}$ **8.** 400

Margin Exercises 6.2

1. 1.3 and 2.1 are the extremes; 1.5 and 1.82 are the means **2.** $5\frac{1}{2}$ and 3 are the extremes; 2 and $8\frac{1}{4}$ are the means
3. true **4.** true **5.** false **6.** true

Exercises 6.2, page 503

1. a. 7 and 544 **b.** 8 and 476 **3.** true **5.** true **7.** false **9.** true **11.** true **13.** true **15.** true **17.** true **19.** false
21. false **23.** true **25.** false **27.** true **29.** true **31.** true **33.** true **35.** true **37.** true **39.** false **41.** true **43.** true

Margin Exercises 6.3

1. $x = 20$ **2.** $R = 30$ **3.** $x = 60$ **4.** $x = 0.8$ **5.** $z = 3\frac{3}{5}$ **6.** $z = \frac{3}{8}$

Exercises 6.3, page 513

1. $x = 12$ **3.** $x = 20$ **5.** $B = 40$ **7.** $x = 50$ **9.** $A = \frac{21}{2}$ (or 10.5) **11.** $D = 100$ **13.** $x = 1$ **15.** $x = \frac{3}{5}$ (or 0.6) **17.** $w = 24$
19. $y = 6$ **21.** $x = \frac{3}{2}\left(\text{or } 1\frac{1}{2}\right)$ **23.** $R = 60$ **25.** $A = 2$ **27.** $B = 120$ **29.** $R = \frac{100}{3}\left(\text{or } 33\frac{1}{3}\right)$ **31.** $x = 22$ **33.** $x = 1$
35. $x = 15$ **37.** $B = 7.8$ **39.** $x = 1.56$ **41.** $R = 50$ **43.** $B = 48$ **45.** $A = 27.3\left(\text{or } 27\frac{3}{10}\right)$ **47.** $A = 255$ **49.** 5800 **51.** (b)
53. (c) **55.** (c)

Review Problems (from Section 5.2 and 5.5), page 516

1. 8.51 **2.** 42.6 **3.** 13.7 **4.** 18.2925 **5.** 21.95 **6.** 2.98

Margin Exercises 6.4

1. 6 pounds **2.** 126 voted in Precinct 2 **3.** 7 hours **4.** 14 students **5.** $x = \frac{5}{3}$

Exercises 6.4, page 523

1. 45 miles **3.** 7 gallons **5.** \$180 **7.** 20 yards **9.** 437.5 grams **11.** \$12 **13.** \$1275 **15.** \$34 **17.** 42.86 ft or $42\frac{6}{7}$ ft
19. 7.5 mph; 52.5 mph **21.** 3360 miles **23.** 9 hours **25.** 10.81 gallons **27.** \$648 **29.** $1\frac{7}{8}$ in. **31.** 90 lb **33.** $1\frac{1}{2}$ cups
35. 4700 grams **37. a.** 118.8 pounds **b.** \$84 **39. a.** 96 lb **b.** 5 bags **41.** (c) **43.** (a)

Review Problems (from Sections 4.2 to 4.5), page 528

1. close to 6; $\frac{63}{10}\left(\text{or } 6\frac{3}{10}\right)$ **2.** close to 20; $19\frac{7}{10}$ **3.** less than 6; 5 **4.** $\frac{4}{3}\left(\text{or } 1\frac{1}{3}\right)$

Margin Exercises 6.5

1. $m\angle 1 = 120°$; $m\angle 2 = 60°$ **2. a.** right **b.** obtuse **c.** straight **3.** neither; because $m\angle 2 + m\angle 3 = 110°$ **4. a.** 110° **b.** no **5.** scalene **6.** yes **7. a.** 90° **b.** right, scalene **c.** $\overline{BO}$ **d.** $\overline{RB}$ and $\overline{RO}$ **e.** yes; because $m\angle R = 90°$ **8.** Statement (b) is correct **9.** 2.5 cm

Exercises 6.5, page 541

1. a. 75° **b.** 87° **c.** 45° **d.** 15° **3. a.** a right angle **b.** an acute angle **c.** an obtuse angle **5.** 25° **7.** 15° **9.** 40° **11.** 110° **13.** $m\angle A = 55°$; $m\angle B = 75°$; $m\angle C = 50°$ **15.** $m\angle L = 122°$; $m\angle M = 93°$; $m\angle N = 105°$; $m\angle O = 109°$; $m\angle P = 111°$ **17. a.** straight angle **b.** right angle **c.** obtuse angle **19.** scalene (and obtuse) **21.** right (and scalene) **23.** isoceles (and acute) **25.** isoceles (and right) **27.** acute (and scalene) **29.** acute (and equilateral) **31.** $x = 50°, y = 60°$ **33.** $x = 20°, y = 100°$ **35.** The triangles are not similar. The corresponding sides are not proportional. **37.** $\Delta ABC \sim \Delta CDE$. Corresponding angles have the same measure. **39. a.** Each of the other four angles measures 75°. **b.** The triangles are similar since the three pairs of corresponding angles are equal.

Chapter 6 Review Questions, page 553

1. $\frac{4 \text{ nickels}}{5 \text{ nickels}} = \frac{4}{5}$ **2.** $\frac{5 \text{ in.}}{9 \text{ in.}} = \frac{5}{9}$ **3.** $\frac{10 \text{ cm}}{200 \text{ cm}} = \frac{1}{20}$ **4.** $\frac{17 \text{ miles}}{1 \text{ gallon}}$ **5.** $\frac{3 \text{ hours}}{8 \text{ hours}} = \frac{3}{8}$ **6.** $\frac{2 \text{ girls}}{3 \text{ boys}}$ **7.** 21.5¢/oz; 23.1¢/oz; packaged cheese **8.** 6.2¢/fl oz; 2.6¢/fl oz; 1 gallon of milk **9.** \$1.20/lb; \$1.10/lb; $1\frac{1}{2}$ lb **10.** 33.2¢/oz; 31.1¢/oz; 8 oz bologna **11.** means: 5, 16; extremes: 4, 20 **12.** means: 3, $\frac{1}{9}$; extremes: 2, $\frac{1}{6}$ **13.** means: 7, 0.75; extremes: 3, 1.75 **14.** true **15.** true **16.** false **17.** $x = 5$ **18.** $y = 300$ **19.** $w = \frac{81}{16}$ **20.** $a = \frac{7}{15}$ **21.** 149.8 miles **22.** $2.\overline{6}$ **23.** 166.85 miles **24.** \$700 **25.** 40,000 safety pins **26.** 45 mph **27.** $26\frac{2}{3}$ ft **28.** $266\frac{2}{3}$ miles **29.** 64 km per hour **30.** 468 girls **31. a.** isosceles **b.** obtuse **c.** right **d.** scalene **32.** $x = 3, y = 4$

Chapter 6 Test, page 557

1. $\frac{7}{6}$ **2.** extremes **3.** false; $4 \cdot 14 \neq 6 \cdot 9$ **4.** $\frac{3}{5}$ **5.** $\frac{2}{5}$ **6.** $\frac{55 \text{ miles}}{1 \text{ hour}}$ **7.** 28.3¢/oz; 24.9¢/oz; 16 oz at \$3.99 **8.** $x = 27$ **9.** $y = \frac{50}{3}$ **10.** $x = 75$ **11.** $y = 19.5$ **12.** $x = \frac{5}{6}$ **13.** $x = \frac{1}{90}$ **14.** $y = 4.5$ **15.** $x = 36$ **16.** 171.4 miles **17.** 24 miles **18.** \$295 **19.** 45 miles per hour **20.** 25 weeks **21.** 312 pounds **22.** \$7.50 **23. a.** right angle **b.** obtuse angle **c.** $\angle AOB$ and $\angle AOE$; $\angle AOB$ and $\angle BOD$; (or $\angle AOE$ and $\angle DOE$; $\angle DOE$ and $\angle DOB$) **24 a.** 55° **b.** 165°

25. a. $x = 30°$ **b.** obtuse (and scalene) **c.** $\overline{RT}$ **26.** $x = 2.5$

Cumulative Review: Chapters 1 - 6, page 561

1. 5740 **2.** three and seventy-five thousandths **3.** $2 \cdot 5 \cdot 11 \cdot 41$ **4.** 79 **5.** 26 **6.** 210 **7.** 60 **8.** $\frac{1}{16}$ **9.** $\frac{37}{16}$ or $2\frac{5}{16}$ **10.** $\frac{44}{3}$ or $14\frac{2}{3}$ **11.** $\frac{18}{5}$ or $3\frac{3}{5}$ **12.** $\frac{13}{4}$ or $3\frac{1}{4}$ **13.** $\frac{153}{4}$ or $38\frac{1}{4}$ **14.** 57 **15.** 0.036 **16. a.** six hundred and six thousandths **b.** six hundred six thousandths **17.** 1639.438 **18.** 16.172 **19.** 4077.36 **20.** 0.79 **21. a.** 0.8485 **b.** 9.7468 **22.** $x = \frac{5}{2}$ **23.** $A = 0.32$ **24. a.** 45.5 miles per hour **b.** 318.5 miles

Cumulative Review: Chapters 1–6, page 561 (cont.)

25. \$2600 **26.** $57\frac{1}{7}$ miles **27.** 21.2132 **28. a.** $\frac{121}{3}$ in. **b.** $\frac{595}{6}$ in.^{2} **29.** 153.86 in.^{2} **30.** 10 cm^2 **31.** 785 cm^3 **32.** 16 dm^3

CHAPTER 7

Margin Exercises 7.1

1. a. 9% **b.** 1.25% **c.** 125% **d.** $6\frac{1}{4}$% **2.** (b) is the better investment because $13.\overline{3}$% is more than 12%. **3.** (a) is the better investment because 31.25% is more than 30%. **4. a.** 34% **b.** 0.82% **c.** 100% **d.** 579.9% **5. a.** 0.40 **b.** 2.11 **c.** 0.006 **d.** 0.2937

Exercises 7.1, page 573

1. 60% **3.** 65% **5.** 100% **7.** 20% **9.** 15% **11.** 53% **13.** 125% **15.** 336% **17.** 0.48% **19.** 2.14% **21. a.** $\frac{18}{100} = 18\%$ **b.** $\frac{17}{100} = 17\%$; **a.** is better **23. a.** $\frac{8}{100} = 8\%$ **b.** $\frac{8}{100} = 8\%$; both are equal **25.** 2% **27.** 10% **29.** 36% **31.** 40% **33.** 2.5% **35.** 5.5% **37.** 110% **39.** 200% **41.** 0.02 **43.** 0.18 **45.** 0.3 **47.** 0.0026 **49.** 1.25 **51.** 2.32 **53.** 0.173 **55.** 0.132 **57.** 0.0725 **59.** 0.085 **61.** 28% **63.** 0.45 **65.** 6.5% **67.** 4.76

Review Problems (from Section 6.3), page 577

1. $R = 80$ **2.** $R = 66\frac{2}{3}$ **3.** $A = 115$ **4.** $A = 45$ **5.** $B = 40$ **6.** $B = 74$

Margin Exercises 7.2

1. 65% **2.** 25% **3.** 150% **4.** 27.3%; $27\frac{3}{11}$% or $27.\overline{27}$% **5.** 93.3% or $93.\overline{3}$% **6.** 77.8%; $77\frac{7}{9}$% or $77.\overline{7}$% **7. a.** $\frac{4}{5}$ **b.** $\frac{4}{25}$ **c.** $\frac{11}{200}$ **d.** $2\frac{7}{20}$

Exercises 7.2, page 585

1. 3% **3.** 7% **5.** 50% **7.** 25% **9.** 55% **11.** 30% **13.** 20% **15.** 80% **17.** 26% **19.** 48% **21.** 12.5% **23.** 87.5% **25.** $55.\overline{5}$% **27.** 85% **29.** $63.\overline{63}$% **31.** 131.25% **33.** 105% **35.** 175% **37.** 137.5% **39.** 210% **41.** $\frac{1}{10}$ **43.** $\frac{3}{20}$ **45.** $\frac{1}{4}$ **47.** $\frac{1}{2}$ **49.** $\frac{3}{8}$ **51.** $\frac{1}{3}$ **53.** $\frac{33}{100}$ **55.** $\frac{1}{400}$ **57.** 1 **59.** $1\frac{1}{5}$ **61.** $\frac{3}{1000}$ **63.** $\frac{5}{8}$ **65.** $\frac{3}{400}$ **67. a.** 0.55 **b.** 55% **69. a.** $\frac{7}{4}\left(\text{or } 1\frac{3}{4}\right)$ **b.** 175% **71. a.** $\frac{21}{200}$ **b.** 0.105 **73.** 4% **75. a.** $8\frac{1}{3}$% (or $8.\overline{3}$%) **b.** 25% **c.** $6\frac{1}{4}$% (or 6.25%) **77.** $19\frac{2}{3}$% (or $19.\overline{6}$%)

Review Problems (from Section 6.4), page 589

1. \$7200 **2.** 216 feet **3.** 8.08 gallons **4. a.** 24 pounds **b.** 3 bags

Margin Exercises 7.3

1. 12 **2.** 500 **3.** 22 **4.** 40% **5.** 25%

Exercises 7.3, page 597

1. 9 **3.** 7.5 **5.** 24 **7.** 47 **9.** 250 **11.** 900 **13.** 62 **15.** 46 **17.** 150% **19.** 30% **21.** 150% **23.** $33\frac{1}{3}$% (or $33.\overline{3}$%) **25.** 17.5 **27.** 160 **29.** 230% **31.** 120 **33.** 620 **35.** 200% **37.** 33.6 **39.** 76.5 **41.** 62.1 **43.** 105 **45.** 160 **47.** 66.5%

Exercises 7.3 (cont.)

49. $16\frac{2}{3}\%\left(\text{or } 16.\overline{6}\%\right)$ **51.** 852,854 people **53.** 57,370 fans

Margin Exercises 7.4

1. 13.7 **2.** 66.6 **3.** 25 **4.** 146 **5.** 120 **6.** 76.1 **7.** 59.4 **8.** 156 **9.** 210 **10.** 90 **11.** 126 **12.** 16

Exercises 7.4, page 609

1. 7 **3.** 9 **5.** 9 **7.** 36 **9.** 150 **11.** 700 **13.** 75 **15.** 42 **17.** 150% **19.** 20% **21.** 50% **23.** $33\frac{1}{3}\%\left(\text{or } 33.\overline{3}\%\right)$ **25.** 12.5 **27.** 110 **29.** 180% **31.** 72 **33.** 520 **35.** 200% **37.** 16.32 **39.** 58.5 **41.** 43.92 **43.** 72 **45.** 80 **47.** 40 **49.** 10 **51.** 37.5 **53.** 70 **55.** 76.3 **57.** 6800 **59.** 44% **61.** (c) **63.** (d) **65.** (c) **67.** (c) **69.** (c)

Margin Exercises 7.5

1. a. \$13 **b.** \$39 **2.** \$41.73 **3.** \$150 **4.** \$5.70

Exercises 7.5, page 621

1. a. \$465 **b.** \$15,035 **3. a.** \$6.93 **b.** \$122.38 **5. a.** \$750 **b.** \$600 **c.** \$636 **7. a.** \$10.53 **b.** 60% **9.** \$4.30 **11.** \$1.95 **13.** \$5.70; \$44.10 **15.** \$29.90 **17. a.** the paperback **b.** \$4.12 **19.** \$32,800 **21. a.** \$161 **b.** \$5474

Margin Exercises 7.6

1. \$192 **2. a.** \$7 **b.** 140% **c.** $58\frac{1}{3}\%$ or $58.\overline{3}\%$ **3. a.** $66\frac{2}{3}\%$ or $66.\overline{6}\%$ **b.** 40%

Exercises 7.6, page 629

1. \$11,700 **3.** \$2240 **5.** \$28,000 **7.** 153 **9. a.** \$50 **b.** 20% **c.** $16\frac{2}{3}\%\ \left(\text{or } 16.\overline{6}\%\right)$ **11. a.** \$1875 **b.** \$2025 **13.** \$120,000 **15. a.** 88% **b.** 250 pages **c.** 30 pages **17.** 840,000 **19.** 888% **21. a.** Joel: \$1480; Friend: \$1240 **b.** 19.4% more, 16.2% more **c.** There is more than one base to calculate percentage. **23. a.** Answers may vary. **b.** Answers may vary.

Chapter 7 Review Questions, page 639

1. hundredths **2.** 85% **3.** 18% **4.** 37% **5.** $16\frac{1}{2}\%$ **6.** 15.2% **7.** 115% **8.** 6% **9.** 30% **10.** 67% **11.** 2.7% **12.** 300% **13.** 120% **14.** 0.35 **15.** 0.04 **16.** 0.0025 **17.** 0.0025 **18.** 0.071 **19.** 1.32 **20.** 60% **21.** 15% **22.** 16% **23.** 37.5% **24.** $41\frac{2}{3}\%\left(\text{or } 41.\overline{6}\%\right)$ **25.** $126\frac{2}{3}\%\left(\text{or } 126.\overline{6}\%\right)$ **26.** $\frac{7}{50}$ **27.** $\frac{2}{5}$ **28.** $\frac{33}{50}$ **29.** $\frac{1}{8}$ **30.** 4 **31.** $\frac{67}{200}$ **32.** 15.6 **33.** 2.55 **34.** $233\frac{1}{3}\left(\text{or } 233.\overline{3}\%\right)$ **35.** $42\frac{6}{7}\left(\text{or } 42.\overline{857142}\right)$ **36.** 20% **37.** $33\frac{1}{3}\%\left(\text{or } 33.\overline{3}\%\right)$ **38.** 25 **39.** 1.095 **40.** 50 **41.** $254\frac{6}{11}$ **42.** $\frac{39}{40}\left(\text{or } 0.975\right)$ **43.** 200% **44.** (c) **45.** (b) **46.** (a) **47.** (b) **48.** (c) **49.** \$11.93 **50. a.** 50% **b.** $33\frac{1}{3}\%\left(\text{or } 33.\overline{3}\%\right)$ **51.** 8 problems **52.** \$1950 **53.** 1.2% **54.** $5.\overline{5}\%$ or $5\frac{5}{9}\%$ – movie; $38.\overline{8}\%$ or $38\frac{8}{9}\%$ – clothes; $22.\overline{2}\%$ or $22\frac{2}{9}\%$ – anniv. **55. a.** \$10,000 **b.** \$9150

Chapter 7 Test, page 643

1. 101% **2.** 0.3% **3.** 17.3% **4.** 32% **5.** 237.5% **6.** 0.7 **7.** 1.8 **8.** 0.093 **9.** $\frac{13}{10}$ **10.** $\frac{71}{200}$ **11.** $\frac{43}{500}$ **12.** 5.6 **13.** $33\frac{1}{3}\%\left(\text{or } 33.\overline{3}\%\right)$ **14.** 30 **15.** 294.5 **16.** 20.5 **17.** 33.1% **18.** (c) **19.** (d) **20.** (b) **21.** (c) **22. a.** \$1.05 **b.** \$3.75 **c.** \$2.70 **23. a.** \$62.50 **b.** 25% **c.** 20% **24. a.** \$1200 **b.** \$24 **25.** \$3300 **26.** 85%

Cumulative Review: Chapters 1 - 7, page 647

1. identity **2.** Commutative property of addition **3. a.** 13,000 **b.** 3.097 **4.** 10,500 (estimate); 10,358 (sum)

Cumulative Review: Chapters 1–7, page 647 (cont.)

5. 4000 (estimate); 3825 (difference) **6.** 100 (estimate); 93 (quotient) **7.** 8000 (estimate); 6552 (product) **8.** 3 **9.** 4, 9, 25, 49, 121, 169, 289, 361, 529, 841 **10. a.** 4200 **b.** 75, 70, 56 **11.** $\frac{3}{20}$ **12.** $\frac{5}{2}$ **13.** $\frac{379}{40}\left(\text{or } 9\frac{19}{40}\right)$ **14.** $\frac{11}{18}$ **15.** $x = 22$ **16.** 78 **17.** 400 **18. a.** $\frac{7}{6}$ **b.** $\frac{1}{3}$ **19. a.** 3450 mm **b.** 160 cm **c.** 83 cm **d.** 5.2 km **20. a.** 13.2288 **b.** 10.3923 **21.** 25 feet **22. a.** \$100 **b.** $33.\overline{3}\%$ **c.** 25% **23. a.** \$4.20 **b.** \$31.75 **24. a.** \$25,000 **b.** \$21,300 **c.** \$4260 **d.** 25% **25. a.** scalene **b.** obtuse **c.** right **d.** isosceles **26. a.** 20° **b.** obtuse **c.** $\overline{RT}$

CHAPTER 8

Margin Exercises 8.1

1. \$150 **2.** \$12.50 interest for 30 days **3.** \$35 interest **4.** The principal is \$25,000 **5.** 300 days or $\frac{5}{6}$ year

Exercises 8.1, page 657

1. \$30 **3.** \$90 **5.** \$26.67 **7.** \$2500 **9.** 1 year **11.** \$32 **13.** \$20.25 **15.** \$11.25 **17.** \$32.67 **19.** \$730 **21. a.** \$463.50 **b.** \$36.50 **23.** \$37,500 **25.** 72 days or $\frac{1}{5}$ year **27. a.** \$7.50 **b.** 60 days or $\frac{1}{6}$ year **c.** 20% **d.** \$180 **29.** 10% **31.** \$460,000 **33.** 1 year **35. a.** \$17.50 **b.** 60 days or $\frac{1}{6}$ year **c.** 11.5% **d.** \$1133.33

Margin Exercises 8.2

1. \$22.84 **2.** \$3216.22 **3.** \$150; \$11.47 **4.** \$5491.14 **5.** \$8467.56

Exercises 8.2, page 669

1. \$100; \$2100, \$105; \$2205, \$110.25; \$315.25 **3.** \$30; \$9030, \$30.10; \$9060.10, \$30.20; \$9090.30, \$30.30; \$9120.60, \$30.40; \$9151.00, \$30.50; \$181.50; \$9181.50 **5. a.** \$162.42 **b.** \$4162.42 **7.** \$813.45 **9. a.** \$634.13 **b.** no **c.** Monthly compounding allows interest earned the previous month to gain interest the following month, but semiannual compounding waits six months for this process to begin. **11. a.** \$6863.93 **b.** about \$20 **c.** \$21.47 **13. a.** \$3153.49 **b.** no; \$2234.08 **15. a.** \$1638.62 **b.** \$638.62 **17. a.** \$3297.33 **b.** \$1297.33

Exercises 8.3, page 677

1.

YOUR CHECKBOOK REGISTER

Check No.	Date	Transaction Description	Payment (−)	(√)	Deposit (+)	Balance
			Balance brought forward			⊖
	7-15	Deposit		√	700.00	+700.00 700.00
1	7-15	Windy Acres Apt. (rent/deposit)	520.00	√		−520.00 180.00
2	7-15	Pa Bell Telephone (installation)	32.16	√		−32.16 147.84
3	2-15	A&E Power Co. (gas/electric)	46.49	√		−46.49 101.35
4	7-16	Foodway Stores (groceries)	51.90	√		−51.90 49.45
	7-20	Deposit		√	350.00	+350.00 399.45
5	7-23	Comfy Furniture (sofa, chair)	300.50			−300.50 98.95
	8-1	Deposit			350.00	+350.00 448.95
	7-31	Interest		√	2.50	+2.50 451.45
	7-31	Service	2.00	√		−2.00 449.45
					True Balance	449.45

BANK STATEMENT
Checking Account Activity

Transaction Description	Amount	(√)	Running Balance	Date
Beginning Balance	0.00	√	0.00	7-01
Deposit	700.00	√	700.00	7-15
Check #1	520.00	√	180.00	7-16
Check #4	51.90	√	128.10	7-17
Check #2	32.16	√	95.94	7-18
Check #3	46.49	√	49.45	7-18
Deposit	350.00	√	399.45	7-20
Interest	2.50	√	401.95	7-31
Service Charge	2.00	√	399.95	7-31
Ending Balance			399.95	

RECONCILIATION SHEET

A. First, mark √ beside each check and deposit listed in both your checkbook register and on the bank statement.
B. Second, in your checkbook register, add any interest paid and subtract any service charge listed.
C. Third, find the total of all outstanding checks.

Outstanding Checks No.	Amount
5	300.50
Total	300.50

Statement Balance	399.95
Add deposits not credited	+350.00
Total	749.95
Subtract total amount of checks outstanding	−300.50
True Balance	449.45

3.

YOUR CHECKBOOK REGISTER

Check No.	Date	Transaction Description	Payment (−)	(√)	Deposit (+)	Balance
			Balance brought forward			756.14
271	6-15	Parts, Parts, Parts (spark plugs)	12.72	√		−12.72 743.42
272	6-24	Firetread Tire Co. (2 tires)	121.40	√		−121.40 622.02
273	6-30	Dean's Gas (tune-up)	75.68			−75.68 546.34
	7-1	Deposit			250.00	+250.00 796.34
274	7-1	Prudent Ins Co. (car insurance)	300.00			−300.00 496.34
	6-30	Service Charge	1.00	√		−1.00 495.34
					True Balance	495.34

BANK STATEMENT
Checking Account Activity

Transaction Description	Amount	(√)	Running Balance	Date
Beginning Balance	0.00		756.14	6-01
Check #271	12.72	√	743.42	6-16
Check #272	121.40	√	622.02	6-26
Service Charge	1.00	√	621.02	6-30
Ending Balance			621.02	

RECONCILIATION SHEET

A. First, mark √ beside each check and deposit listed in both your checkbook register and on the bank statement.
B. Second, in your checkbook register, add any interest paid and subtract any service charge listed.
C. Third, find the total of all outstanding checks.

Outstanding Checks No.	Amount
273	75.68
274	300.00
Total	375.68

Statement Balance	621.02
Add deposits not credited	+250.00
Total	871.02
Subtract total amount of checks outstanding	−375.68
True Balance	495.34

5.

YOUR CHECKBOOK REGISTER

Check No.	Date	Transaction Description	Payment (–)	(√)	Deposit (+)	Balance
			Balance brought forward			967.22
772	4-13	Janet Poppy, CPA (accountant)	85.00	√		– 85.00 882.22
	4-14	Deposit		√	1200.00	+1200.00 2082.22
773	4-14	Pharm X (aspirin)	4.71	√		– 4.71 2077.51
774	4-15	I.R.S. (income tax)	2000.00			–2000.00 77.51
775	4-30	Well Finance Co. (loan payment)	52.50			–52.50 25.01
	5-1	Deposit			600.00	+600.00 625.01
	4-30	Interest		√	2.82	+2.82 627.83
	4-30	Service Charge	4.00	√		– 4.00 623.83
			True Balance			623.83

BANK STATEMENT
Checking Account Activity

Transaction Description	Amount	(√)	Running Balance	Date
Beginning Balance			967.22	4-01
Deposit	1200.00	√	2167.22	4-14
Check #772	85.00	√	2082.22	4-15
Check #773	4.71	√	2077.51	4-15
Interest	2.82	√	2080.33	4-30
Service Charge	4.00	√	2076.33	4-30
Ending Balance			2076.33	

RECONCILIATION SHEET

A. First, mark √ beside each check and deposit listed in both your checkbook register and on the bank statement.
B. Second, in your checkbook register, add any interest paid and subtract any service charge listed.
C. Third, find the total of all outstanding checks.

Outstanding Checks

No.	Amount
774	2000.00
775	52.50
Total	2052.50

Statement Balance	2076.33
Add deposits not credited	+ 600.00
Total	2676.33
Subtract total amount of checks outstanding	–2052.50
True Balance	623.83

7.

YOUR CHECKBOOK REGISTER

Check No.	Date	Transaction Description	Payment (–)	(√)	Deposit (+)	Balance
			Balance brought forward			602.82
14	6-20	Jane Bridal (flowers)	402.40	√		402.40 200.42
	6-22	Deposit		√	1000.00	+1000.00 1200.42
15	6-24	Tuxedo Junction (tuxedo)	155.65	√		155.65 1044.77
16	6-28	D. Lohengrin (organist)	55.00			55.00 989.77
17	6-28	Lee's Limo (limo rental)	125.00			125.00 864.77
18	6-30	C. C. Catering (food caterer)	700.00			700.00 164.77
19	7-1	Halloway Gifts (cards)	35.20			35.20 129.57
	6-30	Service Charge	1.00	√		1.00 128.57
			True Balance			128.57

BANK STATEMENT
Checking Account Activity

Transaction Description	Amount	(√)	Running Balance	Date
Beginning Balance			602.82	6-01
Deposit	1000.00	√	1602.82	6-22
Check #14	402.40	√	1200.42	6-22
Check #15	155.65	√	1044.77	6-26
Service Charge	1.00	√	1043.77	6-30
Ending Balance			1043.77	

RECONCILIATION SHEET

A. First, mark √ beside each check and deposit listed in both your checkbook register and on the bank statement.
B. Second, in your checkbook register, add any interest paid and subtract any service charge listed.
C. Third, find the total of all outstanding checks.

Outstanding Checks

No.	Amount
16	55.00
17	125.00
18	700.00
19	35.20
Total	915.20

Statement Balance	1043.77
Add deposits not credited	+
Total	1043.77
Subtract total amount of checks outstanding	– 915.20
True Balance	128.57

9.

YOUR CHECKBOOK REGISTER

Check No.	Date	Transaction Description	Payment (−)	(√)	Deposit (+)	Balance
			Balance brought forward			147.02
203	2-3	Food Stoppe (groceries)	26.90	√		− 26.90 120.12
204	2-8	Ekkon Oil (gasoline bill)	71.45	√		− 71.45 48.67
205	2-14	Lily's Roses (flowers)	25.00	√		−25.00 23.67
206	2-14	Alumni Assoc. (alumni dues)	20.00			− 20.00 3.67
	2-15	Deposit		√	600.00	+600.00 603.67
207	2-26	SRO (theatre tickets)	52.50			− 52.50 551.17
208	2-28	MPG Mtg. (house payment)	500.00			−500.00 51.17
	2-28	Service Charge	3.00	√		− 3.00 48.17
			True Balance			48.17

BANK STATEMENT
Checking Account Activity

Transaction Description	Amount	(√)	Running Balance	Date
Beginning Balance			147.02	2-01
Check #203	26.90	√	120.12	2-04
Check #204	71.45	√	48.67	2-14
Deposit	600.00	√	648.67	2-15
Check #205	25.00	√	623.67	2-15
Service Charge	3.00	√	620.67	2-28
Ending Balance			620.67	

RECONCILIATION SHEET

A. First, mark √ beside each check and deposit listed in both your checkbook register and on the bank statement.

B. Second, in your checkbook register, add any interest paid and subtract any service charge listed.

C. Third, find the total of all outstanding checks.

Outstanding Checks

No.	Amount
206	20.00
207	52.50
208	500.00
Total	572.50

Statement Balance	620.67
Add deposits not credited	+
Total	620.67
Subtract total amount of checks outstanding	−572.50
True Balance	48.17

Margin Exercises 8.4

1. \$5486.30 **2. a.** \$590 **b.** 42.37%

Exercises 8.4, page 687

1. \$1099.20 **3.** \$4820 **5.** \$2119.40 **7.** \$1938 **9.** No, her total expenses would be \$493 which would be about 23% of her income.

Margin Exercises 8.5

1. a. \$247,500 **b.** \$38,425 **2. a.** \$1555 **b.** 69.11%

Exercises 8.5, page 693

1. a. \$40,625 **b.** \$121,875 **c.** \$44,462.50 **3. a.** 2% **b.** \$19,010 **5. a.** \$24,000 **b.** \$72,000 **c.** \$1080 **d.** \$25,915 **7. a.** \$1405 **b.** about 30.41% **9. a.** \$738 **b.** 41%

Margin Exercises 8.6

1. a. \$50,000 **b.** \$50,000 **c.** 50% **2.** \$2000 **3. a.** 80º **b.** 60º **c.** 8º **4. a.** first class **b.** $\frac{18}{50} = 36\%$ **c.** third and fourth classes

Exercises 8.6, page 703

1. a. Social Science **b.** Chemistry & Physics, and Humanities **c.** about 3300 **d.** about 21.21% **3. a.** Sue **b.** Bob and Sue **c.** 85.71% **d.** Bob and Sue; Bob and Sue; Yes in most cases **e.** No, the vertical scales represent two different types of quantities. **5.** News: 300 min; Movies: 120 min; Sitcoms: 156 min; Soaps: 180 min; Drama: 144 min; Childrens' shows: 120 min; Commercials: 180 min **7. a.** January and May **b.** 8 inches **c.** June **d.** 3.83 inches **9. a.** June **b.** 10 home runs **c.** September **d.** 6 home runs **e.** about 26% **f.** about 20% **11. a.** West-5%; Northeast-28%; Midwest-35%; South-32% **b.** West-22%; Northeast-19%; Midwest-23%; South-36% **c.** South **d.** 5% **e.** West **f.** 5%, 1900 **g.** 22%, 2000 **h.** Midwest **13. a.** 8 **b.** 3 **c.** 8th class **d.** 2 **e.** 27, 29 **f.** 50 **g.** 10 **h.** 16% **15. a.** \$530 **b.** Repairs: 15.09%; Gas: 15.09%; Insurance: 13.21%; Loan Payment: 56.60% **c.** 7 : 8

Chapter 8 Review Questions, page 717

1. $97.50 **2.** $807 **3.** $7.50 **4.** $2000 **5. a.** $12 **b.** 1.5 years or 18 months **c.** 8.5% **d.** $2000 **6.** $275; $5275, $290.13; $5565.13, $306.08; $871.21 **7.** $400; $10,400, $416; $10,816, $432.64; $11,248.64, $449.95; $1698.59; $11,698.59 **8. a.** $2198.47 **b.** $2254.56 **c.** $2267.53 **d.** $2273.86 **9. a.** $10,800 **b.** $800 **c.** $10,800 **d.** $11,664 **e.** $864 **f.** $11,664 **g.** $12,597.12 **h.** $933.12

10.

YOUR CHECKBOOK REGISTER

Check No.	Date	Transaction Description	Payment (−)	(√)	Deposit (+)	Balance
			Balance brought forward			2271.52
1001	8-3	Fred's Boats (loan payment)	500.00	√		500.00 1771.52
1002	8-5	Super Foods (groceries)	20.24			20.24 1751.28
1003	8-6	Super Drug (medication)	78.20	√		− 78.20 1673.08
1004	8-6	OK Tackle Shop (rod and reel)	65.05	√		65.05 1608.03
1005	8-18	Rapid Mart (gas for car)	14.34			14.34 1593.69
1006	8-29	A.J.'s Boat Repair (tune up for boat)	90.00			− 90.00 1503.69
1007	8-30	Eat em up (snacks for party)	93.50			− 93.50 1410.19
	8-31	Deposit		√	300.00	+300.00 1710.19
1008	8-31	UABPC Banking (house payment)	1450.00			−1450.00 260.19
	8-31	Interest		√		260.19
			True Balance			260.19

BANK STATEMENT
Checking Account Activity

Transaction Description	Amount	(√)	Running Balance	Date
Beginning Balance			2271.52	8-01
Check #1001	500.00	√	1771.52	8-05
Check #1003	78.20	√	1693.32	8-06
Check #1004	65.05	√	1628.27	8-12
Deposit	300.00	√	1928.27	8-31
Ending Balance			1928.27	

RECONCILIATION SHEET

A. First, mark √ beside each check and deposit listed in both your checkbook register and on the bank statement.
B. Second, in your checkbook register, add any interest paid and subtract any service charge listed.
C. Third, find the total of all outstanding checks.

Outstanding Checks

No.	Amount
1002	20.24
1005	14.34
1006	90.00
1007	93.50
1008	1450.00
Total	1668.08

Statement Balance	1928.27
Add deposits not credited	+
Total	1928.27
Subtract total amount of checks outstanding	−1668.08
True Balance	260.19

11. $6530.75 **12.** $178,750 **13.** Housing: $10,500; Food: $7000; Taxes: $3500; Transportation: $3500; Clothing: $3500; Savings: $1750; Entertainment: $2800; Education: $2450 **14.** $4000 **15.** $\frac{3}{5}$ **16.** $18,000

Chapter 8 Test, page721

1. $67.50 **2.** $800 **3.** 6 months **4.** 14% **5. a.** $75; $6075, $75.94; $6150.94, $76.89; $6227.83, $77.85; $305.68 **b.** $5.68 **6. a.** $1061.36 **b.** $1346.86 **7.** $2.97

8.

YOUR CHECKBOOK REGISTER

Check No.	Date	Transaction Description	Payment (−)	(√)	Deposit (+)	Balance
		Balance brought forward				1416.05
1011	7-3	Mike's Pet Shop (dog)	600.00	√		600.00 816.05
1012	7-5	Great Foods (groceries)	56.84			56.84 759.21
1013	7-6	Record Room (compact disc)	14.92	√		− 14.92 744.29
1014	7-6	Finch's (gym socks)	8.33	√		8.33 735.96
1015	7-18	Fast Fred's (dinner)	19.38			19.38 716.58
1016	7-29	Post Office (stamps)	10.00			− 10.00 706.58
1017	7-30	Ann Jones (yardwork)	35.00			−35.00 671.58
1018	7-31	UPC Mgmt. (Rent)	600.00	√		+600.00 71.58
	7-31	Service	1.25	√		− 1.25 70.33
	7-31	Interest		√	.03	.03 70.36
					True Balance	70.36

BANK STATEMENT
Checking Account Activity

Transaction Description	Amount	(√)	Running Balance	Date
Beginning Balance			1416.05	7-01
Check #1011	600.00	√	816.05	7-04
Check #1013	14.92	√	801.13	7-09
Check #1014	8.33	√	792.80	7-12
Check #1018	600.00	√	192.80	7-31
Service Charge	1.25	√	191.55	7-31
Interest	.03	√	191.58	7-31
Ending Balance			191.58	

RECONCILIATION SHEET

A. First, mark √ beside each check and deposit listed in both your checkbook register and on the bank statement.
B. Second, in your checkbook register, add any interest paid and subtract any service charge listed.
C. Third, find the total of all outstanding checks.

Outstanding Checks

No.	Amount
1012	56.84
1015	19.38
1016	10.00
1017	35.00
Total	121.22

Statement Balance	191.58
Add deposits not credited	+
Total	191.58
Subtract total amount of checks outstanding	− 121.22
True Balance	70.36

9. a. \$13,863 **b.** \$518.90 **10. a.** 1.28% **b.** \$15,085 **11.** $\frac{13}{25}$ **12.** \$750,000 **13.** \$468,000 **14.** \$15.5 million **15.** 10% **16.** 100%

Cumulative Review: Chapters 1 - 8, page 725

1. 200,016 **2.** 300.004 **3.** 17.00 **4.** 0.4 **5.** 0.525 **6.** 180% **7.** 0.015 **8.** $\frac{4}{3}=1\frac{1}{3}$ **9.** $\frac{41}{11}=3\frac{8}{11}$ **10.** $\frac{8}{105}$ **11.** $\frac{152}{15}=10\frac{2}{15}$ **12.** $46\frac{5}{12}$ **13.** 12 **14.** $\frac{9}{5}=1\frac{4}{5}$ **15.** 5,600,000 **16.** 398.988 **17.** 75.744 **18.** 0.01161 **19.** 80.6 **20.** 7 **21.** 2, 3, and 4 **22.** $2^2\cdot 3^2\cdot 11$ **23.** 210 **24.** 0 **25.** 7 and 160 **26.** 50 **27.** 18.5 **28.** 250% **29.** $x=\frac{3}{8}$ **30.** 196 **31.** 325 **32.** 79 **33.** \$200 **34.** \$4536 **35.** 9% **36. a.** \$10,199.44 **b.** \$10,271.34 **37.** \$22,225.82

38.

YOUR CHECKBOOK REGISTER

Check No.	Date	Transaction Description	Payment (–)	(√)	Deposit (+)	Balance
			Balance brought forward			2032.28
885	9-5	Gary's Tires (new tires)	250.75	√		250.75 1781.53
886	9-10	Mick's Music (compact disc)	14.29	√		14.29 1767.24
887	9-21	Tim's Barbershop (haircut)	12.20			12.20 1755.04
888	9-26	Custom Lawncare (yardwork)	35.00	√		35.00 1720.04
889	9-28	Slack's Grocery (groceries)	32.14			32.14 1687.90
890	9-29	Red Oak LLC. (stump removal)	75.00			75.00 1612.90
891	9-30	Big Bill's Chains (jewelry)	150.50			150.50 1462.40
892	9-30	Cash	100.00	√		100.00 1362.40
893	9-30	Fred Clair (rent)	850.00			850.00 512.40
	9-30	Interest		√	7.57	7.57 519.97
			True Balance			519.97

BANK STATEMENT
Checking Account Activity

Transaction Description	Amount	(√)	Running Balance	Date
Beginning Balance			2032.28	9-01
Check #885	250.75	√	1781.53	9-06
Check #886	14.29	√	1767.24	9-12
Check #888	35.00	√	1732.24	9-30
Check #892	100.00	√	1632.24	9-30
Interest	7.57	√	1639.81	9-30
Ending Balance			1639.81	

RECONCILIATION SHEET

A. First, mark √ beside each check and deposit listed in both your checkbook register and on the bank statement.
B. Second, in your checkbook register, add any interest paid and subtract any service charge listed.
C. Third, find the total of all outstanding checks.

Outstanding Checks

No.	Amount
887	12.20
889	32.14
890	75.00
891	150.50
893	850.00
Total	1119.84

Statement Balance	1639.81
Add deposits not credited	+
Total	1639.81
Subtract total amount of checks outstanding	–1119.84
True Balance	519.97

39. 120 ft **40. a.** 150 ft **b.** 985 sq. ft **41. a.** 25.12 ft **b.** 62.80 ft **c.** 263.76 ft^2 **42. a.** 64.25 cm **b.** 245.31 cm^2 **43.** 72 ft **44.** 2565 cm^3 **45.** 22.08 cm^3 **46.** $x = 50°, y = 60°$ **47.** no, because $6^2 + 8^2 \neq 11^2$

CHAPTER 9

Margin Exercises 9.1

1. a. –12 **b.** 5 **c.** 0 **2.** 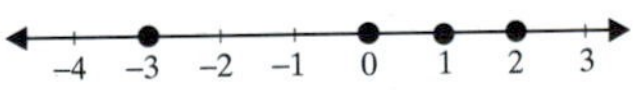 **3.** 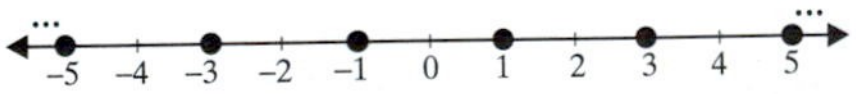 **4. a.** true **b.** false; we can change the inequality to read $-4 < 4$ **c.** true **5.** 15 **6.** 9 **7. a.** true **b.** false; $|-38| < |-39|$

Exercises 9.1, page 739

1.  **3.**  **5.** 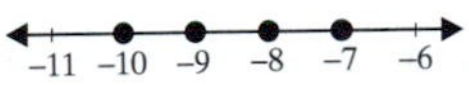
7.  **9.** 
11. 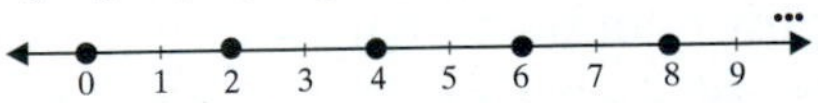 **13.** 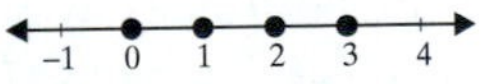 **15.** 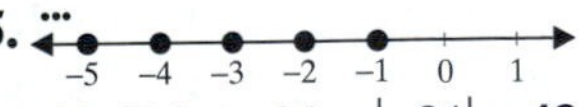 **17.** 10 **19.** –14
21. 0 **23.** 6 **25.** 24 **27.** 20 **29.** 0 **31.** > **33.** < **35.** = **37.** > **39.** True **41.** False: $21 = |-21|$ **43.** False: $-6 < |-6|$ **45.** False: $-4 \leq 4$ **47.** True **49.** False: $0 = |0|$

Margin Exercises 9.2

1. a. –16 **b.** 10 **c.** –13 **d.** –35 **2. a.** –7 **b.** 4 **c.** –20 **d.** –15

Exercises 9.2, page 747

1. 2 **3.** 10 **5.** 19 **7.** –9 **9.** 0 **11.** –4 **13.** 2 **15.** –16 **17.** –4 **19.** 0 **21.** 4 **23.** 14 **25.** –1 **27.** –15 **29.** 41 **31.** 13 **33.** 10 **35.** –12 **37.** –10 **39.** 0 **41.** sometimes **43.** sometimes **45.** never **47.** –14,625 **49.** 11,700

Margin Exercises 9.3

1. a. +8 **b.** –9 **2. a.** 5 **b.** –4 **3. a.** –20 **b.** 1 **c.** –5 **d.** –1 **e.** –6 **4.** –6 **5.** –9 **6.** 11

Exercises 9.3, page 755

1. -16 **3.** 45 **5.** 8 **7.** 3 **9.** 11 **11.** -7 **13.** -9 **15.** 4 **17.** -10 **19.** 1 **21.** 18 **23.** 17 **25.** 0 **27.** 0 **29.** -26 **31.** 30 **33.** 4 **35.** -9 **37.** 24 **39.** -29 **41.** 8 **43.** 6 **45.** 10 **47.** -4 **49.** 6 **51.** -6 **53.** -8 **55.** -18 **57.** 0 **59.** 4 **61.** -21 **63.** -5 **65.** 0 **67.** $<$ **69.** $=$ **71.** 5° **73.** 54 yards **75.** -2418 **77.** $-838{,}350$

Margin Exercises 9.4

1. a. -18 **b.** -36 **c.** -27 **d.** -70 **2. a.** 16 **b.** 0 **3. a.** -8 **b.** 0 **c.** -72 **4. a.** 2 **b.** 2 **c.** -2 **d.** -2 **5.** undefined **6. a.** 54 **b.** -18 **c.** -1

Exercises 9.4, page 765

1. -15 **3.** 24 **5.** -20 **7.** 28 **9.** -50 **11.** -21 **13.** -48 **15.** 63 **17.** 0 **19.** 90 **21.** 24 **23.** 60 **25.** -42 **27.** -45 **29.** -60 **31.** 0 **33.** -1 **35.** 16 **37.** -84 **39.** 300 **41.** -3 **43.** -2 **45.** 4 **47.** 5 **49.** -5 **51.** -3 **53.** 2 **55.** 4 **57.** -3 **59.** -8 **61.** 2 **63.** 1 **65.** undefined **67.** 0 **69.** undefined **71.** 16 **73.** 5 **75.** -6 **77.** -48 **79.** -44 **81.** -4 **83.** 33 **85.** -14 **87.** -14 **89.** -23

Margin Exercises 9.5

1. 7, 13; $5x, 9x$; $-13y^2z, 9y^2z$ **2. a.** $11x$ **b.** $-7y$ **c.** $7x-2$ **d.** $5n-1$ **3. a.** -1 **b.** -26 **4. a.** 0 **b.** 28

Exercises 9.5, page 775

1. $8x$ **3.** $6x$ **5.** $-7a$ **7.** $-12y$ **9.** $-9x$ **11.** $-8x$ **13.** $10x$ **15.** $-2c$ **17.** $-2n$ **19.** $2x^2+2x$ **21.** $3x^2+20$ **23.** $4x+7$ **25.** $5x-1$ **27.** $3ab+6a+2b$ **29.** x^2-5x+6 **31.** $-3n-2$ **33.** $-x^2+2x-1$ **35.** $10x^2+28x-76$ **37.** 0 **39.** -22 **41.** -10 **43.** 4 **45.** -32 **47.** 11 **49.** 39

Margin Exercises 9.6

1. a. The product of 13 and a number **b.** 45 less than a number **c.** $2(y+7)$ **d.** $\frac{50}{x}$ **2. a.** $-7(-1)+21=28, 28=28$ **b.** $-7(3)+21=0, 0\neq 28$ **3.** $y=-3$ **4.** $x=-4$ **5.** $y=51$ **6.** $x=12$ **7.** $n=2$ **8.** $y=-5$ **9.** $x=-18$ **10.** $x=-6$ **11.** $x=-7$ **12.** $n=1$ **13.** $x=5$

Exercises 9.6, page 787

1. $x+5$ **3.** $x+9$ **5.** $\frac{x}{9}$ **7.** $x-13$ **9.** $13-x$ **11.** $2y+4$ **13.** $8y+6$ **15.** $\frac{n}{2}-18$ **17.** $3-2n$ **19.** $6x+4x$
21. 5 times a number **23.** The difference between a number and 6 **25.** A number plus 491
27. A number divided by 17 **29.** 1 more than twice a number **31.** The quotient of a number and 3 increased by 20
33. The product of a number and 15 decreased by 15 **35.** 13 minus a number **37.** 5 **39.** 6 **41.** Write the equation; Subtract 10 from both sides; Simplify both sides; Divide both sides by 3; Simplify both sides. **43.** $x=18$ **45.** $n=-16$ **47.** $y=6$ **49.** $n=224$ **51.** $x=-6$ **53.** $y=-6$ **55.** $n=24$ **57.** $x=24$ **59.** $x=38$ **61.** $x=-10$ **63.** $n=-2$ **65.** $x=-7$ **67.** $n=19$ **69.** $x=-10$

Margin Exercises 9.7

1. 3 **2.** 62 **3.** 1050 ft **4. a.** 8 ft **b.** 10 ft **5.** 20 ft

Exercises 9.7, page 797

1. The sum of a number and 5 is equal to 16. **3.** Four times a number decreased by 8 is equal to 20. **5.** One more than the quotiient of a number and 3 is equal to 13. **7.** $x-12$; $x-12=16$; $x=28$ **9.** $3y-4$; $3y-4=26$; $y=10$ **11.** $x+45=56$; $x=11$ **13.** $3x=51$; $x=17$ **15.** $4y+3=23$; $y=5$ **17.** $\frac{y}{5}-9=11$; $y=100$ **19.** $8n+24=2n$; $n=-4$ **21.** $2n+5=n+20$; $n=15$ **23.** $2x-6=x-5$; $x=1$ **25.** $54=\frac{1}{2}\cdot 18\cdot h$; $h=6$ in. **27.** $195=\frac{1}{2}h(10+20)$; $h=13$ mm

Exercises 9.7, (cont.)

29. $30° + 50° + \gamma = 180°; \gamma = 100°$ **31.** $360 = 15 \cdot w \cdot 4; w = 6$ ft **33.** 21.2 in. **35.** 3.5 cm **37.** $80,000 **39.** $249,534 **41.** 135,618 acres **43.** 312.50 mi; no

Chapter 9 Review Questions, page 809

1. 4 **2.** 8 **3.** –21 **4.**  **5.**  **6.** 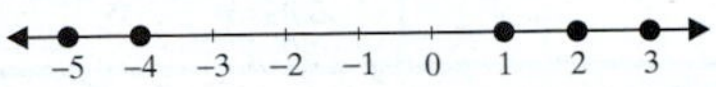 **7.** < **8.** > **9.** = **10.** true **11.** true **12.** false, $|-5| > -5$ **13.** –32 **14.** –5 **15.** –20 **16.** 0 **17.** –16 **18.** 0 **19.** –124 **20.** 562 **21.** 5 **22.** 16 **23.** –21 **24.** 0 **25.** 26 **26.** 6 **27.** –50 **28.** –74 **29.** –10 **30.** –4 **31.** 0 **32.** –72 **33.** 64 **34.** 84 **35.** 0 **36.** –24 **37.** –125 **38.** –17 **39.** 3 **40.** 0 **41.** undefined **42.** –30 **43.** 5 **44.** –22 **45.** –6 **46.** –6 **47.** 8 **48.** –7 **49.** $-\frac{5}{4}$ **50.** $-\frac{27}{4}$ **51.** undefined **52.** zero **53.** positive **54.** positive **55.** $7y$ **56.** $-11x$ **57.** $4a^2 + 3a$ **58.** $5x - 7$ **59.** $y^2 + 5y - 10$ **60.** $-c$ **61.** 17 **62.** 0 **63.** –14 **64.** –48 **65.** $9x + 16$ **66.** $-25 + \frac{x}{2}$ **67.** $y = 16$ **68.** $x = 25$ **69.** $n = -30$ **70.** $m = 2$ **71.** $x = 9$ **72.** $n = -10$ **73.** $x = -11$ **74.** $x = 3$ **75.** $h = 40$ cm **76.** 6.3 ft **77.** 50 yd **78.** 90°

Chapter 9 Test, page 815

1.  **2.** 11 **3. a.** True **b.** False: $|-15| > |-12|$ **4.** 8 **5.** –12 **6.** –9 **7.** –16 **8.** 16 **9.** 28 **10.** –90 **11.** 8 **12.** undefined **13.** –5 **14.** 18 **15.** 12 **16.** –4 **17.** 2 **18. a.** $3x + 5$ **b.** $6x^2 - 3x + 5$ **19. a.** $5x - y - 7$ **b.** 4 **20.** $x = 7$ **21.** $x = -5$ **22.** $y = 6$ **23.** $n = 1$ **24.** $x = 15$ **25.** $x = 10$ **26.** $x = -3$ **27.** 10 ft **28.** 14 cm

Cumulative Review: Chapters 1 - 9, page 819

1. $\frac{15}{16}$ **2.** $\frac{1}{6}$ **3.** $336\frac{29}{48}$ **4.** $189\frac{1}{2}$ **5.** $16\frac{4}{5}$ **6.** 766.234 **7.** 5.48 **8.** Undefined **9.** $\frac{5}{8}$ **10.** 242 **11.** (c) **12.** (d) **13.** (c) **14.** (b) **15.** 200 **16.** 50 **17.** 7349.25 **18.** Yes, because $25^2 = 15^2 + 20^2$ **19. a.** 60° and 30° **b.** $x = 12.5$ cm and $y = 30$ cm **20.** 13π in.^{2} or 40.82 in.^{2} **21. a.** 24.4949 **b.** 15.8114 **c.** –4.4176 **22. a.** scalene (and acute) **b.** equilateral (and acute) **c.** scalene (and obtuse) **d.** right triangle **23.** $\frac{250\pi}{3}$ in.^{3} (or $261.\overline{6}$ in.^{3}) **24. a.** 90 cm **b.** 15cm^2

25.

YOUR CHECKBOOK REGISTER

Check No.	Date	Transaction Description	Payment (−)	(√)	Deposit (+)	Balance
			Balance brought forward			2615.59
1210	3-7	Jones Kennel (dog care)	124.00	√		− 124.00 2491.59
1211	3-7	Simms Car Care (new clutch)	601.40	√		− 601.40 1890.19
1212	3-8	Blue Mountain (camping tent)	102.25	√		− 102.25 1787.94
1213	3-8	Ann's Craft Shop (craft supplies)	15.50	√		− 15.50 1772.44
	3-9	Deposit		√	600.00	+600 2372.44
1214	3-14	Kitchen Help (Catering)	858.34	√		− 858.34 1514.10
1215	3-24	EZ Lawn Care (lawn care)	45.50	√		− 45.50 1468.60
1216	3-24	Henson Repair (fixed tv)	40.00	√		− 40.00 1428.60
1217	3-31	Red River Mgmt. (rent)	800.00			− 800.00 628.60
	3-31	Service Charge	⊖			——
			True Balance			628.60

BANK STATEMENT
Checking Account Activity

Transaction Description	Amount	(√)	Running Balance	Date
Beginning Balance	0.00	√	2615.59	3-01
Check #1210	124.00	√	2491.59	3-07
Check #1211	601.40	√	1890.19	3-09
Check #1212	102.25	√	1787.94	3-09
Check #1213	15.50	√	1772.44	3-11
Deposit	600.00	√	2372.44	3-11
Check #1214	858.34	√	1514.10	3-15
Check #1215	45.50	√	1468.60	3-27
Check #1216	40.00	√	1428.60	3-28
Ending Balance			1428.60	

RECONCILIATION SHEET

A. First, mark √ beside each check and deposit listed in both your checkbook register and on the bank statement.
B. Second, in your checkbook register, add any interest paid and subtract any service charge listed.
C. Third, find the total of all outstanding checks.

Outstanding Checks No.	Amount
1217	800.00
Total	800.00

Statement Balance	1428.60
Add deposits not credited	+ 0
Total	1428.60
Subtract total amount of checks outstanding	− 800.00
True Balance	628.60

26. \$5275 **27.** \$13,498.25 **28.** 354 miles **29. a.** \$500 **b.** $66.\overline{6}\%$ **c.** 40% **30.** $\frac{400}{7}$ miles **31.** $x = -6$ **32.** $n = 10$ **33.** $y = -8$ **34.** $x = -4$ **35.** −5 **36.** 35 yd

APPENDIX I

Margin Exercises I.1

1. a. 121,000 g **b.** 3.5 t **c.** 4.576 t **d.** 6.7 g **2. a.** 43,000 g **b.** 0.250 t **c.** 23,000 mg **3. a.** 0.750 cubic centimeters **b.** 19,000 cubic centimeters **c.** 1,600,000 cubic centimeters **d.** 63,700 cubic decimeters **4. a.** 0.0000063 cubic meters **b.** 192,000 cubic centimeters **5. a.** 0.0937 hectoliters **b.** 0.353 liters **c.** 12,000 liters **6. a.** 1.952 liters **b.** 124 cubic centimeters **c.** 19.75 cubic meters

Exercises I.1, page A9

1. 2000 **3.** 3.7 **5.** 5.6 **7.** 0.091 **9.** 700 **11.** 5000 kg **13.** 2000 kg **15.** 896,000 mg **17.** 75 kg **19.** 7,000,000 g **21.** 0.00034 kg **23.** 0.016 g **25.** 0.0923 **27.** 7580 **29.** 2.963 t **31.** 1000; 1000; 1000; 1,000,000,000 **33.** 900 **35.** 400,000,000 **37.** 0.063 **39.** 3100 mm^3 **41.** 5000 mm^3 **43.** 0.0764 L **45.** 30 mL **47.** 0.0053 L **49.** 72 **51.** 0.569 **53.** 7300; 7300

Margin Exercises I.2

1. a. 32° F **b.** 10° C **2.** 25° C **3.** 113° F **4.** 2.135 m **5.** 91.4 m **6.** 29.55 in. **7.** 1640 ft **8.** 341.85 cm^2 **9.** 6.48 ha **10.** 7.41 acres **11.** 409.032 ft^2 **12.** 15.14 L **13.** 16.96 qt **14.** 26.488 L **15.** 423.78 ft^3 **16.** 85.05 g **17.** 1.47 oz

Exercises I.2, page A21

1. 77° F **3.** 122° F **5.** 10° C **7.** 0° C **9.** 2.74 m **11.** 96.6 km **13.** 124 mi **15.** 19.7 in. **17.** 19.35 cm^2 **19.** 55.8 m^2 **21.** 83.6 m^2 **23.** 405 ha **25.** 741 acres **27.** 53.82 ft^2 **29.** 4.65 in^2 **31.** 9.46 L **33.** 10.6 qt **35.** 11.08 gal **37.** 4.54 kg **39.** 453.6 g **41.** 3038.31 m^2 **43. a.** 11,145.6 cm^2 **b.** 1.115 m^2

APPENDIX II

Margin Exercises II

1. Set of divisors for 45: {1, 3, 5, 9, 15, 45}; Set of divisors for 60: {1, 2, 3, 4, 5, 6, 10, 12, 15, 20, 30, 60}; Set of divisors for 90: {1, 2, 3, 5, 6, 9, 10, 15, 18, 30, 45, 90}; GCD = 15 **2.** GCD = 26 **3.** GCD = 49 **4.** GCD = 30 **5.** GCD = 1 **6.** GCD = 1

Exercises II, page A29

1. 4 **3.** 17 **5.** 10 **7.** 3 **9.** 6 **11.** 2 **13.** 8 **15.** 4 **17.** 12 **19.** 25 **21.** 1 **23.** 1 **25.** 35 **27.** relatively prime **29.** relatively prime **31.** relatively prime **33.** not relatively prime; GCD is 2 **35.** relatively prime

APPENDIX III

Margin Exercises III

1. The mean home run total was 38.6. **2.** Median = 36.5 **3.** Mode = 33; Range = 18

Exercises III, page A35

1. a. 79 **b.** 84 **c.** 85 **d.** 38 **3. a.** 6.2 **b.** 6 **c.** 6 **d.** 6 **5. a.** $400 **b.** $375 **c.** $325 **d.** $225 **7. a.** 81 **b.** 79 **c.** 88 **d.** 26 **9. a.** 22 **b.** 19 **c.** 18 **d.** 21 **11. a.** 18.5 in. **b.** 14.9 in. **c.** none **d.** 22.0 in.

APPENDIX IV

Margin Exercises IV.1

1. 14 **2.** 26

Exercises IV.1, page A41

1. $3(10^1)+5(10^0)$ **3.** $8(10^3)+4(10^2)+6(10^1)+9(10^0)$ **5.** $6(10^4)+2(10^3)+3(10^2)+2(10^1)+2(10^0)$

7. $1(2^2)+0(2^1)+1(2^0)=1(4)+0(2)+1(1)=4+0+1=5$

9. $1(2^3)+0(2^2)+1(2^1)+1(2^0)=1(8)+0(4)+1(2)+1(1)=8+0+2+1=11$

11. $1(2^5)+1(2^4)+0(2^3)+1(2^2)+1(2^1)+1(2^0)=1(32)+1(16)+0(8)+1(4)+1(2)+1(1)$

$=32+16+0+4+2+1=55$

13. $1(2^5)+0(2^4)+1(2^3)+0(2^2)+1(2^1)+1(2^0)=1(32)+0(16)+1(8)+0(4)+1(2)+1(1)$

$=32+0+8+0+2+1=43$

15. $1(2^3)+0(2^2)+0(2^1)+0(2^0)=1(8)+0(4)+0(2)+0(1)=8+0+0+0=8$

17. $1(2^4)+1(2^3)+1(2^2)+0(2^1)+1(2^0)=1(16)+1(8)+1(4)+0(2)+1(1)=16+8+4+0+1=29$

19. $1(2^5)+1(2^4)+1(2^3)+1(2^2)+1(2^1)+1(2^0)=1(32)+1(16)+1(8)+1(4)+1(2)+1(1)$

$=32+16+8+4+2+1=63$

21. 111

Margin Exercises IV.2

1. 4 **2. a.** 13 **b.** 16 **c.** 24 **3.** 128

Exercises IV.2, page A45

1. $2(5^1)+4(5^0)=2(5)+4(1)=10+4=14$ **3.** $1(5^1)+0(5^0)=1(5)+0(1)=5+0=5$

5. $1(5^2)+0(5^1)+4(5^0)=1(25)+0(5)+4(1)=25+0+4=29$

7. $3(5^1)+2(5^0)=3(5)+2(1)=15+2=17$ **9.** $4(5^2)+2(5^1)+3(5^0)=4(25)+2(5)+3(1)=100+10+3=113$

11. $1(5^3)+0(5^2)+3(5^1)+4(5^0)=1(125)+0(25)+3(5)+4(1)=125+0+15+4=144$

13. $2(5^2)+4(5^1)+4(5^0)=2(25)+4(5)+4(1)=50+20+4=74$

15. $1(5^4)+3(5^3)+0(5^2)+4(5^1)+2(5^0)=1(625)+3(125)+0(25)+4(5)+2(1)=625+375+0+20+2=1022$

17. Yes; 13 **19.** $\{0,1,2,3,4,5,6,7\}$

Margin Exercises IV.3

1. $10111_{(2)}$ **2.** $402_{(5)}$ **3.** $11000_{(2)}$ **4.** $302_{(5)}$

Exercises IV.3, page A51

1. $1000_{(2)}=8$ **3.** $11000_{(2)}=24$ **5.** $432_{(5)}=117$ **7.** $1011_{(2)}=11$ **9.** $433_{(5)}=118$ **11.** $10100_{(2)}=20$ **13.** $10100_{(2)}=20$
15. $2241_{(5)}=321$ **17.** $110111_{(2)}=55$ **19.** $3002_{(5)}=377$ **21.** $30141_{(5)}=1921$ **23.** $110001_{(2)}=49$
25. $10001010_{(2)}=138$

APPENDIX V

Margin Exercises V.1

1. 240,381 **2.** 11 **3.** 43 **4.** 636 **5.** 37,448 **6.** 613 **7.** 8281 **8.** 21 **9.** 309 **10.** 2944

Exercises V.1, page A59

1. 253 **3.** 21,327 **5.** 160 **7.** 53 **9.** 2362 **11.** 744 **13.** 978

15. a. 9 9 9 9 9 ∩ ∩ ∩ I I **b.** · / ·· / ·· = **c.** Γ^H Δ Δ Δ I I **d.** D X X X I I

17. a. 9 9 9 9 9 9 9 9 ∩ ∩ ∩ ∩ III III **b.** ·· / · — · **c.** Γ^H H H H Δ Δ Δ Δ Γ I **d.** D C C C X L V I

Margin Exercises V.2

1. 147 **2.** 4402 **3.** 4609 **4.** 31,182 **5.** 507 **6.** 941

Exercises V.2, page A65

1. 4983 **3.** 1802 **5.** 47,900 **7.** 46 **9.** 999 **11. a.** ∨∨∨∨∨ ∨∨∨∨ <<< << ∨∨∨ ∨∨∨ **b.** $\overline{\Phi\ C\ F}$ **c.** 五百九十六

13. a. ∨ ∨ ∨∨∨ ∨∨ **b.** $\overline{/\ \Gamma\ X\ \Xi\ E}$ **c.** 三千六百十五 **15. a.** ∨ ∨ ∨ < < < < < ∨ ∨ **b.** $\overline{A}$ MΩNB

c. 十千八百五十二

Index

D

E

N

O

P

Q

R

S

T

Powers, Roots, and Prime Factorizations

Number	Square	Square Root	Cube	Cube Root	Prime Factorization
1	1	1.0000	1	1.0000	—
2	4	1.4142	8	1.2599	prime
3	9	1.7321	27	1.4423	prime
4	16	2.0000	64	1.5874	2 · 2
5	25	2.2361	125	1.7100	prime
6	36	2.4495	216	1.8171	2 · 3
7	49	2.6458	343	1.9129	prime
8	64	2.8284	512	2.0000	2 · 2 · 2
9	81	3.0000	729	2.0801	3 · 3
10	100	3.1623	1000	2.1544	2 · 5
11	121	3.3166	1331	2.2240	prime
12	144	3.4641	1728	2.2894	2 · 2 · 3
13	169	3.6056	2197	2.3513	prime
14	196	3.7417	2744	2.4101	2 · 7
15	225	3.8730	3375	2.4662	3 · 5
16	256	4.0000	4096	2.5198	2 · 2 · 2 · 2
17	289	4.1231	4913	2.5713	prime
18	324	4.2426	5832	2.6207	2 · 3 · 3
19	361	4.3589	6859	2.6684	prime
20	400	4.4721	8000	2.7144	2 · 2 · 5
21	441	4.5826	9261	2.7589	3 · 7
22	484	4.6904	10,648	2.8020	2 · 11
23	529	4.7958	12,167	2.8439	prime
24	576	4.8990	13,824	2.8845	2 · 2 · 2 · 3
25	625	5.0000	15,625	2.9240	5 · 5
26	676	5.0990	17,576	2.9625	2 · 13
27	729	5.1962	19,683	3.0000	3 · 3 · 3
28	784	5.2915	21,952	3.0366	2 · 2 · 7
29	841	5.3852	24,389	3.0723	prime
30	900	5.4772	27,000	3.1072	2 · 3 · 5
31	961	5.5678	29,791	3.1414	prime
32	1024	5.6569	32,768	3.1748	2 · 2 · 2 · 2 · 2
33	1089	5.7446	35,937	3.2075	3 · 11
34	1156	5.8310	39,304	3.2396	2 · 17
35	1225	5.9161	42,875	3.2711	5 · 7
36	1296	6.0000	46,656	3.3019	2 · 2 · 3 · 3
37	1369	6.0828	50,653	3.3322	prime
38	1444	6.1644	54,872	3.3620	2 · 19
39	1521	6.2450	59,319	3.3912	3 · 13
40	1600	6.3246	64,000	3.4200	2 · 2 · 2 · 5
41	1681	6.4031	68,921	3.4482	prime
42	1764	6.4807	74,088	3.4760	2 · 3 · 7
43	1849	6.5574	79,507	3.5034	prime
44	1936	6.6332	85,184	3.5303	2 · 2 · 11
45	2025	6.7082	91,125	3.5569	3 · 3 · 5
46	2116	6.7823	97,336	3.5830	2 · 23
47	2209	6.8557	103,823	3.6088	prime
48	2304	6.9282	110,592	3.6342	2 · 2 · 2 · 2 · 3
49	2401	7.0000	117,649	3.6593	7 · 7
50	2500	7.0711	125,000	3.6840	2 · 5 · 5

U.S. Customary and Metric Equivalents

In the following tables, the equivalents are rounded.

U.S. to Metric		Metric to U.S.	
Length Equivalents			
1 in.	= 2.54 cm (exact)	1 cm	= 0.394 in.
1 ft	= 0.305 m	1 m	= 3.28 ft
1 yd	= 0.914 m	1 m	= 1.09 yd
1 mi	= 1.61 km	1 km	= 0.62 mi
Area Equivalents			
1 $in.^2$	= 6.45 cm^2	1 cm^2	= 0.155 $in.^2$
1 ft^2	= 0.093 m^2	1 m^2	= 10.764 ft^2
1 yd^2	= 0.836 m^2	1 m^2	= 1.196 yd^2
1 acre	= 0.405 ha	1 ha	= 2.47 acres
Volume Equivalents			
1 $in.^3$	= 16.387 cm^3	1 cm^3	= 0.06 $in.^3$
1 ft^3	= 0.028 m^3	1 m^3	= 35.315 ft^3
1 qt	= 0.946 L	1 L	= 1.06 qt
1 gal	= 3.785 L	1 L	= 0.264 gal
Mass Equivalents			
1 oz	= 28.35 g	1 g	= 0.035 oz
1 lb	= 0.454 kg	1 kg	= 2.205 lb

Celsius and Fahrenheit Equivalents

$$C = \frac{5(F-32)}{9}$$

$$F = \frac{9 \cdot C}{5} + 32$$

Celsius		Fahrenheit
100°	← water boils at sea level →	212°
95°		203°
90°		194°
85°		185°
80°		176°
75°		167°
70°		158°
65°		149°
60°		140°
55°		131°
50°		122°
45°		113°
40°		104°
35°		95°
30°	comfort range	86°
25°	comfort range	77°
20°	comfort range	68°
15°		59°
10°		50°
5°		41°
0°	← water freezes at sea level →	32°